CELLULAR *and* MOLECULAR PHYSIOLOGY *of* CELL VOLUME REGULATION

CELLULAR *and* MOLECULAR PHYSIOLOGY *of* CELL VOLUME REGULATION

Edited by

Kevin Strange, Ph.D.

Division of Nephrology
Children's Hospital
Harvard Medical School
Boston, Massachusetts

CRC

CRC Press
Boca Raton Ann Arbor London Tokyo

Library of Congress Cataloging-in-Publication Data

Cellular and molecular physiology of cell volume regulation / edited
by Kevin Strange.
p. cm.
Includes bibliographical references and index.
ISBN 0-8493-4448-4
1. Osmoregulation. I. Strange, Kevin, 1955– .
QP90.6.C44 1993
574.87—dc20 93-23981
CIP

Direct all inquiries to CRC Press, Inc., 2000 Corporate Blvd., N.W., Boca Raton, Florida 33431.

International Standard Book Number 0-8493-4448-4

Library of Congress Card Number 93-23981
Printed in the United States of America 1 2 3 4 5 6 7 8 9 0
Printed on acid-free paper

To John Parker.
His science
and contagious enthusiasm for it
will always serve as a stimulus
and a reminder
of why we pursue answers
to these questions.

CONTENTS

PREFACE

The ability to regulate cell volume in the face of osmotic challenge is one of the most fundamental of cellular homeostatic mechanisms. It is probably one of the evolutionarily oldest homeostatic mechanisms as well. When it became advantageous for primitive replicating molecules to surround themselves with a membrane and to control their immediate environment, it also became necessary to regulate "cell" volume.

This book is an integrated collection of articles describing the cellular and molecular physiology of cell volume control. It is designed to be accessible to students, researchers, and teachers who are interested in cellular osmoregulation and membrane transport physiology. Rather than focus on mechanisms of volume regulation in specific cell types, the book instead has been organized around concepts and cellular/molecular processes. An important goal of this organizational scheme was to bring diverse viewpoints and perspectives together in a single text and to take a comparative approach to the problem of cell volume regulation.

The book is divided into four parts, each of which focuses on a major topic: (1) fundamental principles and physiological roles of cell volume regulation, (2) organic osmolytes, (3) volume sensing and signal transduction, and (4) the role of gene expression in cell volume control. Each section begins with a detailed introductory chapter that covers basic concepts and principles (e.g., biophysics of membrane transport, physical chemistry of organic osmolytes, cellular signal transduction mechanisms, and control of gene expression in vertebrates). The introductory chapters are followed by shorter chapters that review a single, related topic in detail (e.g., transcellular crosstalk and volume regulation in epithelia, comparative mechanisms of organic osmolyte regulation, stretch-activated ion channels, macromolecular crowding, coordinated regulation of volume-activated transporters, genetic control of volume regulation in plants and bacteria, etc.).

Research on cellular osmoregulation is entering an extremely exciting phase. As Hallows and Knauf (Chapter 1) note, the study of cell volume control provides an intersection point for many different disciplines. That intersection is yielding unique and powerful insights, not only into volume regulation, but also into many other basic biological questions. For example, recent electrophysiological and molecular biological studies have generated novel information on the function and regulation of volume-sensitive ion channels (e.g., Gründer et al., *Nature,* 360:759–762, 1992) and ATP-driven pumps (e.g., Gill et al., *Cell,* 71:23–32, 1992). Enormous progress has been made in the last decade on understanding the thermodynamic characteristics, cellular function and control, and the molecular biology of membrane exchangers (e.g., Na^+/H^+ exchange) and cotransporters (e.g., $Na^+/K^+/2Cl^-$, K^+/Cl^-, Na^+/organic and NaCl/organic cotransport). The current understanding of these key membrane proteins had many of its beginnings in the study of cell volume control. Studies of cell volume regulation are continuing to provide important information on the control of gene expression. It is now clear that chronic adaptation to osmotic stress in cells ranging from bacteria to mammalian kidney and brain cells involves highly coordinated changes in the expression of osmoregulatory transporters and metabolic enzymes (Part IV). The cloning of the genes encoding these proteins will ultimately reveal how changes in a simple environmental parameter, osmotic pressure, induce complex molecular responses.

An area where the study of cell volume regulatory physiology will undoubtedly provide novel insights is in the field of cellular signal transduction. How does a cell sense its size and convert that "size signal" into an appropriate regulatory response? Can cells sense rates and magnitudes of volume change? What are the relationships and interactions between ligand-activated and volume-sensitive signaling pathways? Does cell size in and of itself function as a primary signal that controls cell metabolism, cell growth and proliferation, and salt and water transport? These fundamental questions remain to be answered, providing one of the major challenges to workers in the field.

The study of cellular osmoregulation is yielding many practical benefits as well. For example, in his chapter on organic osmolytes, Yancey notes how this fascinating group of solutes is now being used to improve reconstitution of biochemical preparations and for the stabilization of pharmaceuticals. The implications of organic osmolytes for agriculture are substantial as demonstrated by a recent paper in *Science.* Tarczynski and coworkers (*Science,* 259:508–510, 1993) have successfully introduced a bacterial gene that encodes for mannitol 1-phosphate dehydrogenase into tobacco plants. These transformed plants accumulate the organic osmolyte mannitol and withstand salinity stress much better than the untransformed variety. Genetic engineering of plants with the molecular machinery necessary for synthesizing or transporting organic osmolytes will likely produce crops better adapted to withstand not only high salinity soils and drought, but also perhaps temperature stress.

A fundamental understanding of cellular osmoregulation is of substantial clinical importance. McManus and Churchwell (Chapter 4) discuss how knowledge of cell volume control provides insight into patient care on a variety of levels, particularly when clinicians treat osmotic disorders or manipulate total body osmotic homeostasis. These authors also discuss how derangements and disruptions in cellular osmoregulation may give rise to a diverse group of disease states and their complications.

From the above discussion it should be obvious that the study of cell volume control has broad basic and practical implications. As a worker in the field, it is exciting to look back and see what has already been accomplished. It is more exciting to look to the future and see what needs to be done. I hope that this book serves as a stimulus for further research.

I would like to thank Dr. John Parker for helpful discussions on contributors and topics and Melissa Vicencio and Merry Lindsey for expert secretarial assistance. My laboratory was supported during the preparation of this book by grants from the NIH, American Diabetes Association, and the American Heart Association.

THE EDITOR

Dr. Kevin Strange, Ph.D., is Assistant Professor of Pediatrics in the Division of Nephrology at Children's Hospital, Harvard Medical School, Boston, Massachusetts, and Associate Director of the Renal Research Laboratories.

Dr. Strange graduated from the University of California at Davis with a B.S. degree in zoology (summa cum laude) in 1977. He obtained his M.A. and Ph.D. degrees in zoology in 1978 and 1983 from the University of California at Davis and the University of British Columbia, Vancouver, British Columbia, respectively. From 1983 to 1986, he was a research fellow in the Laboratory of Kidney and Electrolyte Metabolism at the National Institutes of Health and became Assistant Professor of Physiology and Biophysics at Wright State University School of Medicine in 1986. He assumed his present position at Children's Hospital and Harvard Medical School in 1990.

Dr. Strange is a member of the Society of General Physiologists, the American Physiological Society, the American Society of Nephrology, and the Society for Neuroscience. He has presented a number of invited lectures at international and national scientific meetings and has presented invited talks to the Foster Club at Cambridge University, Cambridge, England, and the Danish Biological Society in Copenhagen, Denmark. In 1993, he was an invited lecturer at the St. Göran's Children's Hospital and the Karolinska Institute in Stockholm, Sweden.

Dr. Strange has published more than 30 first or senior authored and numerous coauthored papers. The major research focus in his laboratory is on elucidating the cellular and molecular mechanisms of membrane solute transport and cell volume control in the central nervous system with an emphasis on understanding how these processes relate to various disease states. His research is funded by the National Institutes of Health, the American Heart Association, the National Science Foundation, the American Diabetes Association, and private corporations. He is currently an Established Investigator of the American Heart Association.

CONTRIBUTORS

Dr. James S. Ballantyne
Department of Zoology
University of Guelph
Guelph, Ontario
Canada

Dr. Joseph V. Bonventre
Medical Services
Massachusetts General Hospital East
Boston, Massachusetts

Dr. Walter F. Boron
Department of Cellular and Molecular Physiology
Yale University School of Medicine
New Haven, Connecticut

Dr. Mary E. Chamberlin
Department of Biological Sciences
Ohio University
Athens, Ohio

Dr. Kevin B. Churchwell
Children's Hospital
The Multidisciplinary Intensive Care Unit and Department of Anesthesia
Boston, Massachusetts

Dr. David M. Cohen
Renal Division
Brigham and Women's Hospital
Boston, Massachusetts

Dr. Calvin Cotton
Department of Pediatrics
Pulmonary Division
Rainbow Babies' and Children's Hospital
Case Western Reserve University
Cleveland, Ohio

Dr. Bruce A. Davis
Department of Cellular and Molecular Physiology
Yale University School of Medicine
New Haven, Connecticut

Dr. Julia Dragolovich
National Institutes of Health
Laboratory of Kidney and Electrolyte Metabolism
Bethesda, Maryland

Dr. Joan D. Ferraris
National Institutes of Health
Laboratory of Kidney and Electrolyte Metabolism
National Heart, Lung, and Blood Institute
Bethesda, Maryland

Dr. Thomas Force
Cardiac Unit of the Medical Services
Massachusetts General Hospital
Boston, Massachusetts

Dr. J. Kevin Foskett
Division of Cell Biology
Hospital for Sick Children
Toronto, Ontario
Canada

Dr. Arlyn García-Pérez
National Institutes of Health
Laboratory of Kidney and Electrolyte Metabolism
National Heart, Lung, and Blood Institute
Bethesda, Maryland

Dr. Steven R. Gullans
Renal Division
Brigham and Women's Hospital
Boston, Massachusetts

Dr. Kenneth R. Hallows
Department of Biophysics
University of Rochester Medical Center
Rochester, New York

Dr. Steven C. Hebert
Laboratory of Molecular Physiology and Biophysics
Renal Division
Department of Medicine
Brigham and Women's Hospital
Boston, Massachusetts

Emilia Hogan
Department of Cellular and Molecular Physiology
Yale University School of Medicine
New Haven, Connecticut

Dr. Philip A. Knauf
Department of Biophysics
University of Rochester Medical Center
Rochester, New York

Dr. H. Moo Kwon
Renal Division
Department of Medicine
Johns Hopkins University
Baltimore, Maryland

Dr. Ian Henry Lambert
Laboratory for Biological Chemistry A
University of Copenhagen
Copenhagen, Denmark

R. John MacLeod
Department of Pediatrics
McGill University
Montreal Children's Hospital Research Institute
Montreal, Quebec
Canada

Dr. Michael L. McManus
Children's Hospital
The Multidisciplinary Intensive Care Unit
and Department of Anesthesia
Boston, Massachusetts

Dr. John W. Mills
Department of Biology
Clarkson University
Potsdam, New York

Dr. Allen P. Minton
Laboratory of Biochemical Pharmacology
National Institutes of Health, NIDDKD
Bethesda, Maryland

Dr. H. Clive Palfrey
Department of Pharmacological
and Physiological Sciences
University of Chicago
Chicago, Illinois

Dr. John C. Parker
Department of Medicine
The University of North Carolina at Chapel Hill
Chapel Hill, North Carolina

Dr. Sidney K. Pierce
Department of Zoology
University of Maryland
College Park, Maryland

Dr. Luis Reuss
Department of Physiology and Biophysics
University of Texas Medical Branch
Galveston, Texas

Dr. David Rhodes
Horticulture Department
Purdue University
West Lafayette, Indiana

Dr. Henry Sackin
Department of Physiology and Biophysics
Cornell University Medical College
New York, New York

Dr. Yiannis Samaras
Horticulture Department
Purdue University
West Lafayette, Indiana

Dr. Jeff M. Sands
Renal Division, Department of Medicine
Emory University School of Medicine
Atlanta, Georgia

Dr. Erik M. Schwiebert
Department of Physiology
Johns Hopkins University School of Medicine
Baltimore, Maryland

Dr. Bruce A. Stanton
Department of Physiology
Dartmouth Medical School
Hanover, New Hampshire

Dr. Adam M. Sun
Laboratory of Molecular Physiology and
Biophysics
Renal Division, Department of Medicine
Brigham and Women's Hospital
Boston, Massachusetts

Dr. Merna R. Villarejo
Section of Microbiology
University of California
Davis, California

Dr. Paul H. Yancey
Biology Department
Whitman College
Walla Walla, Washington

Dr. Harry H. Yim
Department of Microbiology and Molecular
Genetics
University of California
Los Angeles, California

Part I
Fundamental Principles and Physiological Roles of Cell Volume Regulation

Chapter 1

Principles of Cell Volume Regulation

Kenneth R. Hallows and Philip A. Knauf

TABLE OF CONTENTS

I. INTRODUCTION

Early studies of osmotic phenomena investigated nonidealities in the volume changes of cells exposed to anisotonic conditions. In the 1930s and 1940s, Ponder and others observed that hypotonically swollen red cells do not always behave as "perfect osmometers".[1,2] Rather, they often swell less in hypotonic media than expected if

0-8493-4448-4/94/$0.00+$.50

water exchange alone were involved. Among other possible explanations, it was postulated that these cells could be losing osmotically active solutes to the surrounding media under hypotonic conditions. Although this hypothesis has since been borne out, additional factors also contribute to the nonideal osmotic behavior of red cells, such as concentration-dependent changes in the osmotic coefficient of hemoglobin.[3–5]

It has long been recognized that in animal cells, which lack rigid cell walls and thus cannot support substantial pressure gradients across the cell membrane, the control of cell volume is achieved through the regulation of intracellular solute content. The pump-and-leak concept for the long-term or chronic maintenance of cell volume was first formulated by Davson and Dean (1940)[6] and later developed quantitatively by Leaf (1959),[7] Ussing (1960),[8] and Tosteson and Hoffmann (1960).[9] In this model, cell volume control represents a balance between active ion pumps (e.g., the Na^+,K^+-ATPase) and passive "leak" fluxes to maintain a steady state. In recent years, it has been discovered that these leaks or downhill flows of ions often represent very sophisticated and highly regulated transport systems, such as coupled cotransporters and exchangers and regulated ion channels.[10] Kregenow was among the first to investigate the ionic bases of short-term volume-regulatory responses in animal cells in his evaluation of the responses of duck erythrocytes to hypotonic and hypertonic media.[11,12] Since then, the capacity for anisotonic volume regulation has been observed in many cell types (for recent reviews see References 10 and 13 to 16). Today, volume regulation research has branched out to involve many disciplines, from ion transport and electrophysiological studies to cellular signaling and molecular genetics.

The rest of this chapter will consider in detail: (1) the behavior of water and solutes in cells and the basis for their equilibrium and transport across cell membranes, and (2) the general principles, processes, and phenomena involved in the control of animal cell volume under both steady-state and anisotonic conditions.

II. EQUILIBRIUM AND TRANSPORT OF WATER ACROSS CELL MEMBRANES

A. STRUCTURE AND BEHAVIOR OF INTRACELLULAR WATER

Before embarking on a quantitative analysis of the equilibrium and transport properties of water across cell membranes, one should consider the extent to which it is expected to behave in an "ideal" manner from a fundamental thermodynamic standpoint. Specifically, it is known that sufficiently near to a surface, water behaves as if it is immobilized or bound to the surface, with diminished activity and solvent properties.[17] The proportion of such "nonideal" intracellular water to total cell water has been the subject of some debate.

Shporer and Civan's nuclear magnetic resonance data provided evidence that only a small portion (~5%) of cellular water behaves as if it were bound.[18] This would correspond only to water in the first one or two layers of molecules next to cell surfaces.

Clegg,[19] however, argues that a large fraction of total cell water exhibits properties markedly different from that of bulk water. To support this hypothesis, he cites the findings of Kempner and Miller that the composition and metabolic activities of the cytosol (obtained by methods of cell disruption and fractionation) bear almost no resemblance to those of the aqueous cytoplasm of intact cells.[20,21] Furthermore, measurements of long-range forces between surfaces suggest that the influence of surfaces (e.g., phospholipid bilayers and the cytoskeleton) on water may extend over distances as large as 50 Å (17 molecular layers of water).[22,23] This could have profound effects on the kinetics of enzymes in the cytoplasm. For example, the viscosity of aqueous microenvironments adjacent to enzymes may play a key role in rate enhancement. Indeed, certain microenvironments can cause the association or crowding of proteins. The effect of macromolecular crowding on intracellular solute association may play a key role in the signaling pathway for the activation of volume-regulatory responses (discussed below; see also Chapter 10).

Although a significant fraction of cell water may indeed have altered properties, osmotic experiments provide little evidence to suggest that the structure of intracellular water significantly affects the equilibrium thermodynamics or the kinetics of water flow in cells (i.e., most cell water appears to be available for dissolving solute). Therefore, in the following analysis we use the approximation that the macroscopic behavior of cellular water is ideal from a physical and thermodynamic standpoint.

B. FUNDAMENTAL EQUILIBRIUM RELATIONS

Although living cells as a whole exist in a dynamic steady state and are not in equilibrium with their environment, water and certain ions often behave as if they are at thermodynamic equilibrium across the cell membrane. For a particular species to be at equilibrium, its electrochemical potential inside the cell must equal its electrochemical potential outside the cell:

$$\tilde{\mu}_j^i = \tilde{\mu}_j^o \tag{1}$$

where $\tilde{\mu}_j$ is the electrochemical potential of species j, and i and o signify the inside and outside of the cell, respectively. This can be expanded as:

$$\mu^{oi}_j + RT\ \ln\ X_j^i + p^i v_j + z_j F\psi^i = \mu^{oo}_j + RT \ln X_j^o + p^o v_j + z_j F\psi^o \tag{2}$$

where μ°_j is the standard chemical potential, X_j is the mole fraction, v_j is the partial molar volume, and z_j is the valence of species j. ψ is the electrical potential, p is the pressure, R is the gas constant, and T is the absolute temperature (K).

C. WATER EQUILIBRIUM AND OSMOTIC PHENOMENA

The above analysis holds generally for the equilibrium distribution of any species, both charged and uncharged, across the cell membrane. In the case of water, z_j in Equation 2 is zero. If the temperature is the same at the two sides of the membrane, so that the standard chemical potential for water is the same, the terms in μ° cancel out. The equation thus simplifies to

$$RT\ \ln(X_w^i/X_w^o) = (p^o - p^i)v_w = \Delta p v_w \tag{3}$$

where X_w is the mole fraction of water and v_w is the partial molar volume of water; that is, the change in volume per mole of water added to the solution. Equation 3 states the conditions required for water to be at equilibrium across the membrane; that is, for there to be no driving force for net water flow. Thus, Δp is the hydrostatic pressure that is required to prevent net water flow if there is a difference in the mole fraction of water at the two sides of the membrane. Note that p must be higher on the side of the membrane where the mole fraction of water is lower, to prevent water flow down its concentration gradient. This pressure is often referred to as the osmotic pressure difference, $\Delta\pi$.

1. Relationship of Osmotic Pressure to Solute Concentration

The osmotic pressure of an individual solution, π, can be defined as the pressure required to prevent flow of the solvent, in this case water, into the solution if it is separated from pure solvent by an ideal semipermeable membrane, that is, one which is permeable to the solvent but not to the solute. From Equation 3, if the pure solvent is considered as the inside compartment and the solution as the outside compartment, $X_w^i = 1$, so $\ln X_w^i = 0$ and therefore:

$$RT(\ln X_w^i - \ln X_w^o) = -RT\ \ln X_w^o = \Delta p v_w = \pi v_w \tag{4}$$

Because Equation 4 applies to any solution, we will drop the superscript o in the following discussion. For any solution,

$$X_w = 1 - \Sigma X_j \tag{5}$$

where X_j is the mole fraction of each solute, j, in the solution. For dilute solutions, ΣX_j is sufficiently small so that we can use the Maclaurin series approximation that $\ln(1 - x) = -x$. Thus, from Equations 4 and 5 we obtain:

$$\pi v_w = RT\Sigma X_j \tag{6}$$

For dilute solutions,

$$v_w = V_w/N_w \tag{7}$$

where V_w is the total volume of water and N_w is the number of moles of water. The sum of the mole fractions of solutes, ΣX_j, can be expressed as:

$$\Sigma X_j = \Sigma N_j/(N_w + \Sigma N_j) \approx \Sigma N_j/N_w \tag{8}$$

Substituting Equations 7 and 8 into Equation 6:

$$\pi = RT\Sigma N_j/V_w \tag{9}$$

Because the density of water ≈ 1.0 kg/l, $N_j/V_w \approx C_j$, the *molal* concentration of the jth solute, so:

$$\pi = RT\Sigma C_j \tag{10}$$

For *very* dilute solutions, $V_w \approx V_T$, the total volume of the solution, so $N_j/V_w \approx N_j/V_T = c_j$, the *molar* concentration, so:

$$\pi = RT\Sigma c_j \tag{11}$$

which is the classical Boyle-van't Hoff formula for osmotic pressure. Note that in Equations 10 and 11, the summation includes all *species* present in the solution, that is, all dissociated particles are considered individually. In fact, the observation that for salts such as NaCl, the osmotic pressure behaves as if the concentration is doubled provided some of the evidence upon which Arrhenius first proposed the theory of electrolytic dissociation.

Some confusion can be caused by the fact that the osmotic pressure, $\Delta\pi$, is actually the *hydrostatic* pressure that is required to *prevent* net water flow. Thus, in the absence of any hydrostatic pressure, water would flow toward the side of the membrane with the *higher* π. If one thinks of π as an ordinary "pressure", this flies in the face of one's physical intuition that flow should be from the region with higher pressure to that with lower pressure. Realization that $\Delta\pi$ is the pressure required to *prevent* flow, however, resolves this apparent paradox.

The proper units of π are pressure units, that is, atmospheres or Pascals. In most experimental circumstances, one deals with ratios of π for different solutions at the same temperature, so R and T cancel out. For this reason, it is common to express osmotic pressures in units of concentration, as osmoles per kilogram. We will refer to osmotic pressures in concentration units, which really equal π/RT (Equation 10), as π'.

2. Corrections for Nonideal Behavior

Because the osmotic behavior of solutions is not generally ideal, ϕ, the molal osmotic coefficient[24] for each solute, is used as a correction factor for Equation 10:

$$\pi = RT\Sigma\phi_j C_j \tag{12}$$

Under physiological conditions, ϕ for small uncharged species (e.g., sucrose) is usually slightly greater than 1; for electrolytes ϕ is slightly less than 1; and for macromolecules ϕ is greater than 1 due to various factors,[5] including the large entropy of mixing which results when the volume is very large relative to the volume of water.[25] Furthermore, ϕ may not be constant; e.g., the osmotic coefficient of hemoglobin increases with increasing hemoglobin concentration in red blood cells.[3–5] Further discussion of the consequences of such behavior for water equilibrium and water flow may be found in References 5 and 26.

The osmotic coefficient, ϕ_j, is different from the activity coefficient, γ_j. γ_j corrects for the effect of nonideality of the solute on the activity (effective concentration) of *solute;* ϕ_j, on the other hand, corrects for the effect of nonideality of the solute on the activity of the *solvent.* The two are related through the Gibbs-Duhem equation and one can be calculated from the other.[24,27] Values of ϕ_j for electrolytes are closer to 1 (i.e., reflect more ideal behavior) than are values of γ_j. For example, the mean ionic activity coefficient, $\gamma_\pm$, for 0.1 molal KCl is 0.770, but the osmotic coefficient, ϕ, is 0.926.[27]

3. Implications of Osmotic Equilibrium

In contrast to the other colligative properties, such as freezing point depression or boiling point elevation, which are small in magnitude, the osmotic pressure is quite large. For example, RT = 25.4 atm · l/mol at 37°C, which means that a pressure difference of 25 atm is required to prevent water flow from pure water into a 1 Osmol solution. Even a seemingly small $\Delta\pi'$ of 10 mOsmol across a cell membrane would require a Δp of approximately 0.25 atm or 260 cm H_2O for water to be at equilibrium!

Plant cells enclosed in rigid cell walls routinely withstand much greater pressure gradients than this, a phenomenon known as turgor maintenance (see Chapter 21). Although animal cell membranes do have some mechanical strength, it is generally believed that they cannot support the kinds of hydrostatic pressures required to balance even relatively small differences in osmotic pressure (cf. Reference 28). However, recent experiments by Ito et al.[29] suggest that tension in the cortical actin cytoskeleton of cells may be sufficient under certain conditions to constrain water flow and prevent the cell from swelling or shrinking to the extent expected from osmotic considerations. They measured water flow as a function of Δp through actin filament networks *in vitro.* Actin filament networks cross-linked with actin-binding protein were able to prevent water flow driven by pressures of up to 64 cm H_2O, the maximum value tested. This result suggests that cortical actin networks may at least constitute a buffer against cell volume change in the face of small osmotic pressure fluctuations that constantly affect mammalian cells. If cells are presented with a sufficiently large $\Delta\pi$, however, it is clear that water will tend to flow across the membrane, with resulting cell volume changes.

Assuming that $\Delta p = 0$ in animal cells, according to Equation 3 water will be at equilibrium if, and only if, $\Delta\pi = 0$, that is:

$$\pi^i = \pi^o = \pi \tag{13}$$

According to Equation 9, with the addition of ϕ_j to correct for nonideality as in Equation 12, the intracellular osmotic pressure π^i is

$$\pi^i = RT\Sigma\phi_j C_j = RT\Sigma\phi_j N_j/V_w \quad \textbf{(14)}$$

where N_j and V_w in this case specifically refer to the amount of dissolved solute and the volume of water associated with a given volume of cells. If the cells are at osmotic equilibrium with an initial solution whose osmotic pressure is π_0, then:

$$\pi_0 = RT\Sigma\phi_j N_j/V_{w0} \quad \textbf{(15)}$$

where V_{w0} is the initial cell water volume. If the cells are now transferred to a solution with a different osmotic pressure, π, and if $\Sigma\phi_j N_j$ is constant (that is, the solutes do not leave the cell and the osmotic coefficients are constant), then:

$$\pi V_w = \pi_0 V_{w0} \quad \textbf{(16)}$$

where V_w represents the cell water volume in the new solution. If we define b as the volume of the cell that is not water, we can express this in terms of the total cell volume, V, after the change of solution and the original total cell volume, V_0:

$$\pi(V - b) = \pi_0(V_0 - b) \quad \textbf{(17)}$$

This modification of the Boyle-van't Hoff equation has proven to be very useful in describing osmotic phenomena and can be modified to give the volume V at any osmotic pressure π:

$$V = (\pi_0/\pi)(V_0 - b) + b \quad \textbf{(18)}$$

According to Equation 18, a plot of V against $1/\pi$ for different tonicity changes should be linear, and the intercept when $1/\pi$ equals zero should give b. The value of b indicates the nonsolvent or osmotically inactive volume of the cell (i.e., that part of the cellular volume that does not participate in osmotic swelling or shrinkage). A theoretical Boyle-van't Hoff plot is shown in Figure 1–1.

One might expect that b should always equal ($V - V_w$) and that the fit to the Boyle-van't Hoff plot should always be linear for cells, but these expectations are often not met. The classical example of experiments investigating osmotic phenomena in red blood cells illustrates this point. For red cells, it was postulated early on that nonideal osmotic behavior could be attributed to the properties of hemoglobin at high concentration.[3] Specifically, as mentioned above, the osmotic coefficient, ϕ, for hemoglobin increases with increasing concentration, and therefore, the $\Sigma\phi_j N_j$ term in Equation 14 increases as the cells shrink. A model that includes the concentration-dependent osmotic coefficient of hemoglobin accurately predicts the osmotic behavior of red blood cells whose ion permeabilities have been increased by the antibiotic nystatin.[5] That the anomalous behavior of hemoglobin is responsible for the nonideal osmotic behavior of in-

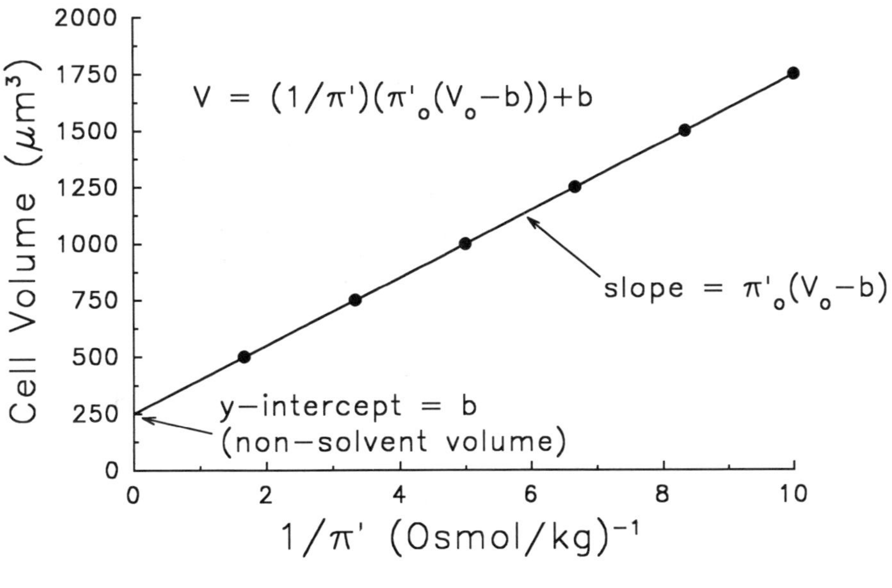

Figure 1–1 A theoretical Boyle-van't Hoff plot is shown for osmometric behavior of a cell with an apparent nonsolvent volume (b) of 250 μm³ and a resting volume (V_0) of 750 μm³ in isotonic ($\pi'_0 = \pi_0/RT = 0.3$ Osmol/kg) medium.

tact red cells was confirmed by osmotic studies conducted on resealed ghosts.[30] Ghosts that had lost most of their hemoglobin prior to osmotic challenge were found to obey the Boyle-van't Hoff law.

Although probably not applicable to osmotic behavior in red cells, other possible factors may contribute to nonidealities in the osmotic behavior of cells. First, as discussed above, the assumption that under equilibrium conditions $\Delta\pi = 0$ may not necessarily be true. In the preceding analysis, it was also assumed that there were no solute fluxes across the membrane during osmotic flow (e.g., volume-regulatory ion fluxes), and as discussed in depth later, these fluxes can be quite significant. Finally, as discussed above, in some cells the difference between b and $(V - V_w)$ could be due in part to the presence of nonsolvent (e.g., bound or immobilized) water within the cell.

D. WATER TRANSPORT: DEFINITIONS AND EQUATIONS

It is clear from the general equation for electrochemical potential (Equation 2) that, if the standard chemical potential for water is the same at each side of a membrane, the major factors affecting the flow of water will be the transmembrane hydrostatic pressure gradient, Δp, and the mole fraction of water at each side of the membrane. The latter can be expressed conveniently in terms of the osmotic pressure, π, which equals $RT\Sigma\phi_jC_j$ (Equation 12). The driving force for water flow across the membrane is usually expressed in terms of $\Delta\mu_w/v_w$, which leads to the expression:

$$J_v^w = L_p[\Delta p - RT(\Sigma\phi_jC_j^i - \Sigma\phi_jC_j^o)] \qquad \textbf{(19)}$$

where J_v^w is the water flow and L_p is the hydraulic conductivity of the membrane. This can be simplified to:

$$J_v^w = L_p(\Delta p - \Delta\pi) \qquad \textbf{(20)}$$

In fact, the parameter usually measured is not water flow, but *volume* flow across the membrane, J_v. For impermeant solutes, water flow is exactly equal to volume flow, and Equation 20 holds for J_v as well as for J_v^w. For solutes to which the membrane is highly permeable, however, a correction factor, σ, must be applied:

$$J_v = L_p(\Delta p - \sigma\Delta\pi) = L_p[\Delta p - RT(\Sigma\sigma_j\Delta\phi_jC_j)] \qquad \textbf{(21)}$$

The reflection coefficient, σ, represents the degree to which the membrane "reflects", that is, prevents passage of the solute, and is a property of the *membrane,* not of the solute (for example, different membranes can have very different σ values for the same solute). If the solute does not pass through at all, it is completely reflected and $\sigma = 1$, so Equation 21 becomes identical to Equation 20. If the solute passes through the membrane as easily as water, $\sigma = 0$, and a concentration difference of the solute produces no osmotic effect. For a more complete discussion of the factors that affect σ and water flow, see References 31 and 32.

E. MECHANISMS OF WATER TRANSPORT ACROSS CELL MEMBRANES

Two principal mechanisms are involved in water flow, diffusion through the lipid bilayer and permeation through specialized pores. The first mechanism is important in all cells. Although the partition coefficient (concentration in bilayer divided by concentration in aqueous phase) for water is only about 10^{-6}, the concentration of water in water is so large (55.5 mol/l) and the surface area of the lipid bilayer is so large that diffusion through the lipid represents a significant flow of water. Some appreciation of the magnitude of water flow through the lipid pathway compared to specialized channels can be gained from the fact that in the red blood cell, blockage of water channels with the sulfhydryl reagent PCMBS (*p*-chloromercuribenzene sulfonate), reduces water permeability to only one tenth of its normal value.[33] The half-time for osmotic equilibration of water at room temperature in the absence of the drug is about 0.3 s,[34] so even without the channels operating water equilibrates rapidly with a half-time of only 3 s! Of course, in other cells with smaller surface-to-volume ratios or different membrane lipid compositions, the half-time may be substantially longer.

As mentioned above, in some cell types, such as red blood cells and the epithelial cells of certain portions of the kidney, water permeability is greatly increased by the presence of channels for water flow. Such channels were first proposed on the basis of the biophysical studies of Solomon and coworkers.[34a] They proposed that if water channels exist in membranes, flow driven by a pressure gradient should have characteristics of Poiseuille flow and should depend on the radius of the tube to the fourth power. Diffusion of water, as measured by tritiated water, should depend on the area of the pore, and hence on the radius to the second power. Measurements indicated that the apparent water permeability driven by a pressure (filtration permeability, P_f) was greater than that measured by isotope diffusion (P_d), providing evidence for channels as well as a way for determining an effective pore radius for these channels.

Increasingly more precise measurements of P_f/P_d, however, predicted increasingly greater values for the effective pore radius, up to 12.5 Å for the PCMBS-sensitive component of water flux.[35] If the pore radius were this large, molecules such as sucrose, which does not enter red cells, would be expected to diffuse readily through the pores. An alternative interpretation of the P_f/P_d ratio is suggested by experiments with gramicidin, which forms a 4 Å diameter channel across the membrane. This channel is much too small to allow Poiseuille flow of water, but the observed P_f/P_d ratio is about 5. In such a channel, where water molecules flow in single file without passing each other, the P_f/P_d ratio is determined by the number of water molecules lined up in the channel. Such single-file flow of water seems to be a more plausible explanation for the high P_f/P_d ratio observed in red blood cells.[35]

The membrane protein that forms water channels in red blood cells and some kidney cells was first identified as a contaminant of about 28 kDa molecular weight in preparations of an Rh antigen-bearing polypeptide from red blood cells.[36] Cloning and sequencing showed that the 269-amino acid polypeptide was homologous to a family of proteins thought to function as channels in plants, insects, and animals.[37,38] When expressed in *Xenopus* oocytes, the protein, now known as CHIP28 (for channel-like integral protein), greatly increased osmotic water permeability, and the increase could be inhibited by PCMBS.[39] Only one of the four cysteine residues in CHIP28, Cys-189, is required for inhibition by PCMBS.[40] The purified CHIP28 dramatically increased water permeability in lipid vesicles, but the permeability to urea and protons was not increased,[41] suggesting that the channels are highly specific for water and that water molecules may not form a continuous chain within the channel. Each CHIP28 molecule has six putative transmembrane segments, and it is possible that four molecules form a noncovalent tetramer,[37] but each molecule seems to function as an independent water channel.[40] CHIP28 is also found in those regions of the kidney with a constitutively high water permeability, but not in those regions with regulated water permeability.[42]

The search for the molecular basis of specialized water channels is thus entering a new and interesting phase. From the standpoint of volume regulation, however, the most important point is that, even in the absence of specialized mechanisms, most membranes are sufficiently permeable to water so that water may be assumed to equilibrate rapidly down osmotic gradients.

III. EQUILIBRIUM AND TRANSPORT OF SOLUTES ACROSS CELL MEMBRANES

A. NERNST POTENTIAL AND DONNAN EQUILIBRIUM

For charged solutes, the general equation for thermodynamic equilibrium (Equation 2) is rather complex, but it can be simplified if certain reasonable assumptions are made. If we assume that in animal cells the transmembrane hydrostatic pressure difference is negligible and if there is no temperature difference across the cell membrane, the terms containing μ°_j and p drop out. Rearranging terms, we obtain:

$$z_j F E_m = RT \ln (X_j^o / X_j^i) \qquad \textbf{(22)}$$

where the membrane potential, $E_m \equiv \psi^i - \psi^o$.

From the definition of the mole fraction (Equation 8) as N_j/N_w, together with the fact that the mass of water in a solution is proportional to the number of moles of water, N_w, we can substitute the molal concentrations of j, C_j, for the mole fractions in Equation 22. If the activity coefficients for j are the same at both sides of the membrane, we obtain the familiar Nernst equation:

$$E_m = (RT/z_j F) \ln(C_j^o / C_j^i) \qquad \textbf{(23)}$$

which gives the distribution ratio for a substance that is at equilibrium across a cell membrane at a given E_m. Alternatively, knowing the transmembrane concentrations of an ion, one can calculate the Nernst equilibrium potential for the ion (E_j) from Equation 23.

The situation becomes more complex when one tries to account for the distribution of all charged solutes and solvent water in a living cell. In attempts to model this distribution, the classical Donnan equilibrium formalism is often used. As the following discussion will show, however, a true Donnan *equilibrium* does not exist in cells.

To see why animal cells are not at true Donnan equilibrium, consider a Donnan equilibrium situation in which a semipermeable membrane, permeable to ions but impermeable to protein, separates an intracellular compartment which contains protein at a concentration P_i, together with Na^+, K^+, and Cl^-, from an extracellular medium with ions but no protein. If there are no active pumps, the ions will diffuse down their electrochemical potential gradients and will eventually come into equilibrium with the membrane potential, E_m. From Equation 23,

$$Na_o/Na_i = K_o/K_i = Cl_i/Cl_o = r \qquad \textbf{(24)}$$

where r is the Donnan ratio. In addition to this, there is a second constraint on the system, macroscopic electroneutrality. This means that the sum of positive charge is equal to the sum of negative charge, that is

$$Na_i + K_i = Cl_i - zP_i \qquad \textbf{(25)}$$

for the intracellular compartment, where z is the number of *positive* charges per protein molecule. For the extracellular compartment:

$$Na_o + K_o = Cl_o \qquad \textbf{(26)}$$

Note that Equations 25 and 26 are properly written in terms of concentrations, because the total sum of positive and negative charges must be equal, regardless of activity coefficients. Equation 24, on the other hand, should properly be written in terms of ion activities, and only takes the form above if the activity coefficients at the two sides of the membrane are the same.

Next, we ask whether or not the system is at osmotic equilibrium. If so, then the sum of the solute concentrations at one side will equal that at the other side (if the osmotic coefficients are equal to 1). The question can be stated as:

$$Na_i + K_i + Cl_i + P_i \stackrel{?}{=} Na_o + K_o + Cl_o \qquad \textbf{(27)}$$

or, using Equations 24 and 26:

$$Cl_o + r^2Cl_o + rP_i \stackrel{?}{=} 2rCl_o \qquad \textbf{(28)}$$

Subtracting $2rCl_o$ from both sides, this becomes:

$$(1 - 2r + r^2)Cl_o + rP_i \stackrel{?}{=} 0 \qquad \textbf{(29)}$$

or:

$$(1 - r)^2 Cl_o + rP_i \stackrel{?}{=} 0 \qquad \textbf{(30)}$$

But r is >0 because it is the quotient of positive numbers and $(1 - r)^2$ is ≥0 because the square of any real number is ≥0. Cl_o and P_i are both >0, so the left-hand side of Equation 30 must be >0. This means that the total solute concentration inside the cell is greater than the total solute concentration outside, that is, $\pi_i > \pi_o$. Thus, water should flow into the cell (because water flows toward the side with the greater π). Note that the imbalance in osmotic pressure is the result of two terms. If the protein has zero net charge, then r = 1 and only the rP_i term contributes. If the protein has nonzero charge, then the first term in Equation 30 provides an additional contribution to the osmotic pressure imbalance. Thus, regardless of the sign of the charge, a charge on the protein increases its osmotic effect, due to the counterions which must be present to neutralize the charge.

One way of preventing net water influx would be to apply a hydrostatic pressure across the membrane, with high pressure inside, similar to the strategy of plant cells. This is the situation for true Donnan equilibrium. Animal cells, with a more flexible plasma membrane incapable of maintaining large hydrostatic pressure differences, should swell and eventually burst. The only other way of preventing this "Donnan explosion" would be to have nonpenetrating solute present in the outside medium. To prevent volume increase by this method, the concentration of external nonpenetrating solute must be effectively equal to that inside, when charge distribution effects are taken into account. Although cells normally live in a medium, such as plasma or interstitial fluid, with some external nonpenetrating proteins, their concentration is not sufficient to prevent volume expansion. Thus, this strategy could only work to prevent volume expansion if additional extracellular nonpenetrating solutes were present.

In fact, cells do use this strategy, but with a special twist (Figure 1–2). The membrane contains a Na^+,K^+-pump which pumps Na^+ out and K^+ in. This pump will increase the Na^+ and K^+ gradients across the membrane, until the net influx of Na^+ by the leak pathways (J_L^{Na}) exactly balances the net efflux of Na^+ by the pump (J_P^{Na}). If influx is defined as positive and efflux as negative,

$$J_L^{Na} + J_P^{Na} = 0 \qquad \textbf{(31)}$$

That is, the sum of the leak and pump fluxes is zero, despite the fact that there is a $\Delta\tilde{\mu}_{Na}$ across the membrane. This is the same phenomenological behavior that would be expected if the membrane were totally impermeable to Na^+! Because the

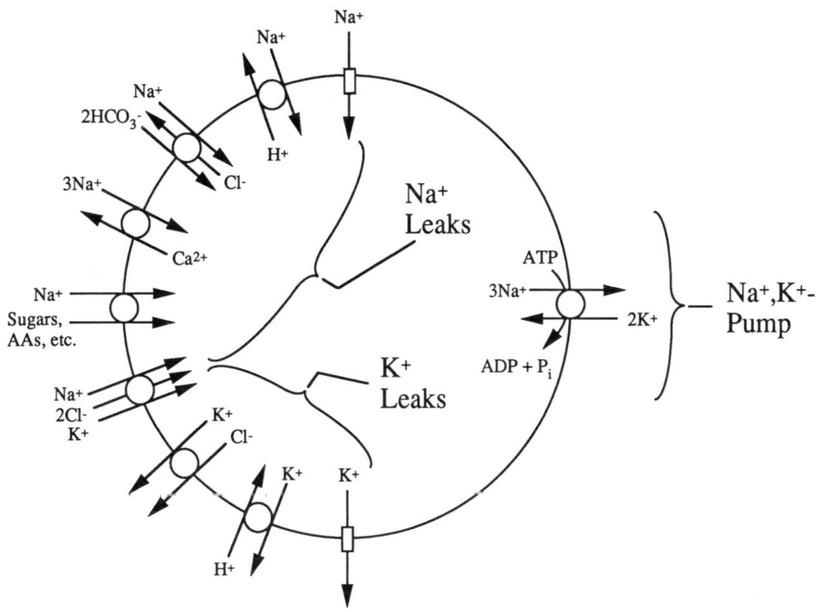

Figure 1–2 Pump-and-leak concept: schematic illustration of the Na^+,K^+-ATPase and examples of Na^+ and K^+ leak pathways in a typical cell.

membrane is *effectively impermeable* to Na^+, Na^+ behaves as a nonpenetrating solute in the extracellular solution, and osmotic balance can be achieved without a hydrostatic pressure gradient across the membrane.

At steady state, the same holds for K^+, that is, the sum of pump and leak fluxes with appropriate signs is zero, so the membrane is *effectively impermeable* to K^+ also. Because Na^+ and K^+ do not come to electrochemical equilibrium, the Donnan equilibrium is not achieved and the cells do not swell.

Although this system involves a more precarious lifestyle, with survival dependent on keeping a balance between pump and leak fluxes, it also provides a very flexible system for altering volume or ion content for the purposes of growth or other functions, such as transport across epithelia. Two features (see below) make the balance less precarious than it would otherwise be: the efflux of Na^+ by the pump is strongly increased in response to an increase in intracellular Na^+ concentration, and various transport systems exist which are activated upon changes in cell volume in such a way as to restore the volume to its original value.

B. NATURE OF THE TRANSPORT PATHWAYS (PUMPS AND LEAKS)

1. Cellular Pumps

The distinguishing feature of cellular pumps, as opposed to cellular leak pathways, is their requirement for the coupling of metabolic energy to translocate ions "uphill" against their electrochemical gradients. Examples of pumps critical for cellular function include H^+-ATPases found in vacuolar and plasma membranes,[43] the Ca^{2+}-ATPase found in mitochondrial and plasma membranes,[44] and the K^+/H^+ exchange pump found in parietal cells of the gastric mucosa.[45] However, the most important pump for cell volume regulation is the Na^+,K^+-ATPase, found in the plasma membrane of nearly all animal cells (for a recent review, see Reference 46).

It is widely believed that the Na^+,K^+-ATPase (or sodium pump) normally exchanges three intracellular Na^+ ions for two extracellular K^+ ions with the hydrolysis of one molecule of ATP for each full pumping cycle. This 3:2:1 stoichiometry has been suggested by kinetic studies in many cell types (but rigorously proven in only a few) through measurements of the dependence of rates of ouabain-sensitive Na^+ efflux and K^+ influx on intracellular Na^+ (Na_i) and extracellular K^+ (K_o) concentrations, respectively.[46] Specifically, it has been shown that the shape of the activation curve for the pump is sigmoidal, with a strong dependence on Na_i (Hill coefficient of ~3) and K_o (Hill coefficient of ~2). That this stoichiometry appears to remain constant under different physiological conditions has been suggested from studies in which simultaneous measurements of electrogenic pump current and Na^+ efflux over a wide range of Na^+ and K^+ concentrations were made.[46]

Although there may exist other naturally occurring substance(s) (e.g., endogenous ouabain analogues) that bind to and modulate pump activity,[47] regulation of the pump is primarily accomplished directly by the concentrations of ion substrates (particularly Na_i), provided that ATP and Mg^{2+} are also available. This system is essentially self-stabilizing; for example, if Na^+ entry into the cell should increase, then the intracellular Na^+ concentration will rise, which will activate the pump and thereby increase the rate of Na^+ extrusion until Na^+ efflux once again balances Na^+ influx, resulting in a new steady state. This means of pump regulation is elegant, involving a simple feedback system that is regulated by precisely that which it is trying to control. Because the efflux of Na^+ from the cell via the sodium pump is exquisitely sensitive to changes in Na_i, a balance between the sodium pump and leak fluxes is rigidly maintained. Although the ratio of the leak fluxes for Na^+ and K^+ at steady state must be 3:2, the overall permeability of the cell to K^+ is much greater than to Na^+ so the resting membrane potential (E_m) is generally much closer to the Nernst equilibrium potential for K^+ (E_K) than to the equilibrium potential for Na^+ (E_{Na}).

Perhaps the most crucial role of the sodium pump in cells is to maintain transmembrane ion gradients, which serve as energy sources that drive many other transport processes. For example, Na^+ and K^+ gradients maintained directly by the pump are the basis of electrical excitability in neurons. Also, these gradients, through their coupling with secondary transport processes, work to establish gradients for other ions and solutes (e.g., Cl^- [and HCO_3^-], Ca^{2+}, sugars, amino acids [AAs], metabolic intermediates, cofactors, and vitamins) that are crucial in maintaining an internal milieu compatible with normal cellular function. Many of these gradient-driven secondary transport processes (Figure 1–3) are Na^+-coupled. These include many ion fluxes that regulate cellular pH and volume (discussed below) and fluxes involved in transepithelial salt and water transport in the kidney, gastrointestinal tract, and other tissues. Furthermore, since the

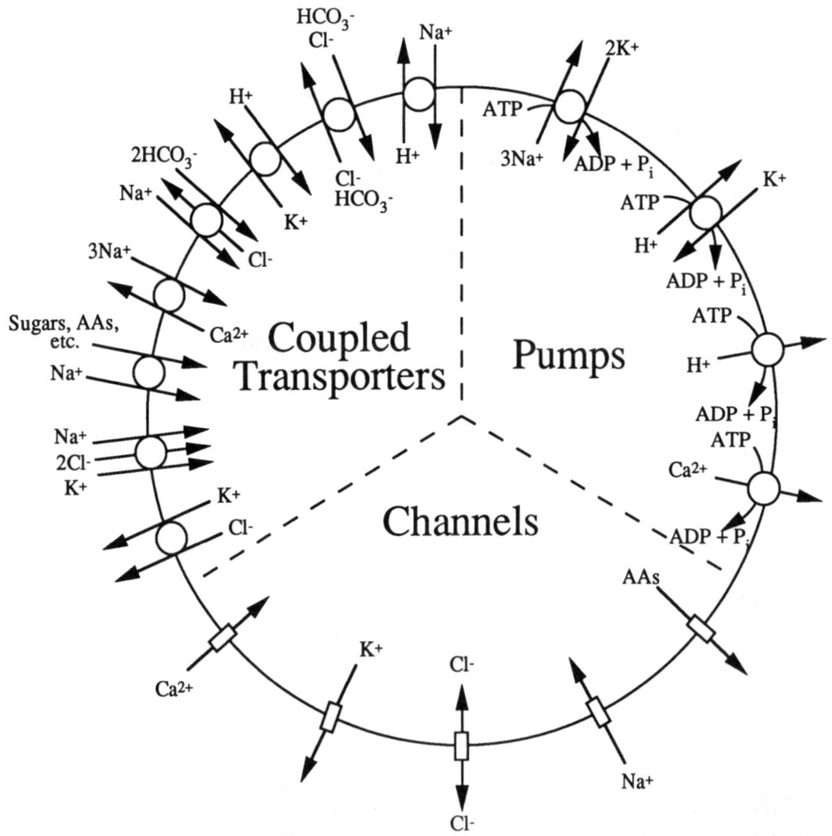

Figure 1–3 Cellular membrane transport systems: schematic illustration of many of the important fluxes mediated by pumps, channels, and coupled transporters in cells. Arrows show the usual directions of net flow for the various species.

pump is electrogenic, it plays an important role in the maintenance and modulation of the resting membrane potential in all cells.[46]

2. Cellular Leaks

By definition, cellular leak flux pathways are passive in the sense that the overall process involved is thermodynamically spontaneous; i.e., the overall change in Gibbs free energy (ΔG) is <0, so that direct coupling to an external energy source is not required. There are two main classes of leak flux pathways: channels and coupled transport pathways.

a. Channels

Membrane channels are transmembrane proteins, consisting functionally of a pore and some selectivity mechanism, that allow the chemically uncoupled passage of small ions, usually in response to various stimuli. Their selectivity is based primarily on the size and charge of the ion and is rather low relative to carriers or coupled transporters. For instance, most K^+ channels that have been characterized can effectively select against anions or Na^+, but will transport either K^+ or Rb^+ with similar kinetics.[48] As may be inferred from their relatively low selectivity, channels do not bind the transported ion very tightly, and so the maximum turnover rate is high relative to coupled transporters (often $>10^6$/s). Furthermore, in contrast to coupled transporters, major conformational changes in the channel are presumably not required for ion translocation to occur. An important practical characteristic pertaining to flow through a channel is the dependence of the rate of transport on substrate concentration, which appears linear under physiological conditions because the concentration required to saturate the transporter is usually relatively high. Coupled transporters, on the other hand, usually display saturation (e.g., Michaelis-Menten) kinetics.

Hodgkin and Huxley[49] were the first to identify voltage-gated Na^+ channels in squid giant axons. Since then, other voltage (or E_m)-sensitive channels have been characterized, including K^+, Cl^-, and Ca^{2+} channels.[50] A host of other factors have been shown to regulate the activity of membrane channels, including external ligands (e.g., acetylcholine, γ-aminobutyric acid [GABA], glycine, and L-glutamate in neurons),[51–54] various intracellular signals and second messengers (e.g., cyclic nucleotide-dependent phosphorylation and Ca^{2+}),[55–57] and mechanical stretch/deformation of the membrane (see Chapter 13).[58]

The best-characterized ion channels to date include Na^+, K^+, Ca^{2+}, and Cl^- channels (see Reference 48). Other simple leak pathways have been found that mediate the uncoupled transport of amino acids and other nonionic solutes.[59] Both ionic and nonionic leak pathways (Figure 1–3) have been implicated in the mechanisms of volume recovery in many cell types, as discussed below and in later chapters of this volume.

The following section provides a brief overview of some of the important equations often used to characterize ion fluxes through conductive channels, which are useful in describing the behavior of volume-regulatory transport systems and in modeling expected changes in various cellular parameters. For a more thorough treatment, see Reference 48.

b. Diffusive ion flux equations and their uses

Under nonsaturating conditions, ion fluxes through membrane channels can be modeled in terms of diffusion kinetics.[48] Thus, the unidirectional flux of an ion, J_j, through a channel depends on the product of its mobility, U_j, its concentration, C_j, and the driving force, which is just the gradient of the electrochemical potential, $d\tilde{\mu}_j/dx$, where x represents the distance through the membrane or channel:

$$J_j = -U_jC_j(d\tilde{\mu}_j/dx) \qquad \textbf{(32)}$$

By employing the expanded form for the electrochemical potential (Equation 2), one can derive the Nernst-Planck equation governing the flux of an ion under diffusive flow conditions:

$$J_j = -RTU_j[(dC_j/dx) + (z_jFC_j/RT)(d\psi/dx)] \qquad \textbf{(33)}$$

The Nernst-Planck equation has limited utility in describing diffusive ion fluxes because it cannot generally be solved with precision for physiological systems. Goldman made a simplifying assumption to derive a more generally solvable and therefore useful expression for diffusive ion fluxes across a membrane.[60] He assumed that both J_j and the electric field, $d\psi/dx$, are constant as one moves through the membrane (or channel). Using this constant field assumption, Equation 33 can be integrated with respect to x from one side of the membrane to the other. When the boundary conditions at the two sides of the membrane are inserted, the following expression, known as the Goldman flux equation, is obtained:

$$J_j = P_j(z_jFE_m/RT)\{[C_j^o - (C_j^i \exp(z_jFE_m/RT))]/[1-\exp(z_jFE_m/RT)]\} \qquad \textbf{(34)}$$

where P_j, the permeability of ion j, is defined as (RTU_j/h), and h is the thickness of the membrane. Thus, the permeability (P_j) can be calculated if the membrane potential (E_m), the flux (J_j), and transmembrane concentrations of j (C_j^o and C_j^i) are known.

If under steady-state conditions there is no net current flow across the membrane and the only permeant ions are Na^+, K^+, and Cl^-, the following equation, originally formulated by Hodgkin and Katz,[61] can be derived from Equation 34, provided that diffusive fluxes are the predominant factor in determining E_m:

$$E_m = (RT/F)\ \ln[(P_K K_o + P_{Na} Na_o + P_{Cl} Cl_i)/ (P_K K_i + P_{Na} Na_i + P_{Cl} Cl_o)] \quad \textbf{(35)}$$

From this expression, known as the Goldman-Hodgkin-Katz (GHK) equation, it is apparent that the permeabilities and transmembrane concentrations of the various ions work in concert to determine the steady-state membrane potential.

If the concentrations of an ion at each side of the membrane are known, together with the flux of the ion, J_j, and the membrane potential (measured by microelectrode or a membrane potential-sensing dye), Equation 34 can be used to compute the apparent permeability (P_j) for that ion. Even if techniques are not available for measuring E_m, P_j can be measured by equilibrium isotope exchange, provided that conditions can be set up where E_m is known. This can be done by setting the ion ratios so that all ions that contribute significantly to the GHK equation have the same Nernst potential, which must therefore be equal to E_m. For example, if P_{Na} and/or Na_i and Na_o are very low, this can be achieved by making $K_i/K_o = Cl_o/Cl_i$. In this case, Equation 34 cannot be used to determine P_j, because the net flux under equilibrium exchange conditions is zero. Instead, the permeability is calculated from a modified form of Equation 34 for unidirectional flux, which is obtained by setting the C^o_j or C^i_j term to zero.

Once permeabilities have been determined in this fashion, Equations 34 and 35 enable one to model expected flux, permeability, and E_m changes under conditions where large diffusive fluxes are expected to dominate, such as during volume regulation under hypotonic conditions in many cell types (discussed below). Predictions made using these equations can be compared to actual measurements under various conditions to test whether or not a model based on diffusive (i.e., mediated by channels) fluxes is adequate to explain the volume regulatory behavior.

c. Coupled transport

Coupled (or carrier-mediated) transport systems comprise the other major class of leak pathways found in cells. In contrast to channels, coupled transporters are more highly selective among substrates, have a lower maximal turnover rate ($<5 \times 10^5$/s), and are usually saturable at concentrations that are not much above the physiological range. The most powerful test for the presence of a coupled transport mechanism is the demonstration that an ion is transported against its electrochemical gradient through coupling with the flow of another ion down its gradient. The overall ΔG for the entire process, of course, must be <0. Many of the activation mechanisms mentioned above for channels are also shared by coupled transport systems.[10] In addition, the activities of some (most notably, the Na^+/H^+ and Cl^-/HCO_3^- exchangers) are also modulated by intra- and extracellular pH.[62–64]

The two main classes of coupled transporters are cotransporters (or symporters), which mediate the fluxes of two or more species across the membrane in the same direction, and exchangers (or antiporters), which mediate the fluxes of two or more species across the membrane in opposite directions. As with channels, coupled transporters play a prominent role in effecting cell volume regulation, as described below. Some examples of important cotransport and exchange systems are shown in Figure 1–3.

d. Equilibrium vs. kinetics of coupled transport

The most basic question concerning possible involvement of a coupled transport system in volume regulation is whether or not it is at equilibrium and, if not, in which direction spontaneous flow will occur. This can readily be calculated from equilibrium thermodynamics, using the principle that at equilibrium $\Delta G = 0$ and for spontaneous flow $\Delta G < 0$.

The total ΔG for a coupled system can be expressed as the product of the influx of each transported species (with appropriate sign, e.g., efflux is negative) times the gradient of electrochemical potential across the membrane for that species, $\Delta\tilde{\mu}_j$, where $\Delta\tilde{\mu}_j$ is defined as $\tilde{\mu}_j^i - \tilde{\mu}_j^o$. If temperature and pressure are equal at the two sides of the membrane, so the standard chemical potential terms in $\tilde{\mu}_j$ cancel out, and if the activity coefficients are the same at both sides of the membrane so that concentrations can be used instead of activities, from Equation 2:

$$\Delta\tilde{\mu}_j = RT\ \ln(C_j^i/C_j^o) + z_j F E_m \quad \textbf{(36)}$$

Consider the simple case of a Na^+/H^+ exchanger, which mediates the influx of one Na^+ ion in exchange for efflux of one H^+. Note that this is the only *allowed* process for this system; that is, it is the only way in which G can change. (Independent flow of Na^+ or H^+, for example, is not allowed.) If an infinitesimal amount of Na^+, δN_{Na}, flows in, this will be accompanied by an equal but opposite flow of H^+, so $\delta N_H = -\delta N_{Na}$. Thus, the infinitesimal change in free energy, δG, will be:

$$\begin{aligned}\delta G = \delta N_{Na}\Delta\tilde{\mu}_{Na} + \delta N_H \Delta\tilde{\mu}_H = \\ \delta N_{Na}(\Delta\tilde{\mu}_{Na} - \Delta\tilde{\mu}_H)\end{aligned} \quad \textbf{(37)}$$

Because δN_{Na} was assumed to be finite and positive (>0), if we consider equilibrium, where $\delta G = 0$, Equation 37 implies that:

$$(\Delta\tilde{\mu}_{Na} - \Delta\tilde{\mu}_H) = 0 \quad \textbf{(38)}$$

Substituting the expressions for $\Delta\tilde{\mu}_{Na}$ and $\Delta\tilde{\mu}_H$ from Equation 36 into Equation 38, the terms in E_m cancel out and we obtain:

$$Na_i/Na_o = H_i/H_o \quad \textbf{(39)}$$

Similar equilibrium expressions can be derived for any coupled transport system. If the transport process involves net charge flow, of course, such expressions will include the membrane potential, E_m.

To determine the direction of spontaneous flow, we simply set δG in Equation 37 to be <0. The corresponding expression in Equation 38 is <0, and so the condition for spontaneous flow of Na^+ in (and of H^+ out) is

$$Na_i/Na_o < H_i/H_o \quad \textbf{(40)}$$

That is, for spontaneous Na^+ influx, the tendency of Na^+ to go out, represented by the left-hand side of the inequality, is less than the tendency of H^+ to go out, represented by the right-hand side. It is important to emphasize that the contribution to δG from each ion's flow is determined by the *electrochemical* potential for that ion ($\Delta\tilde{\mu}_j$), which includes electrical contributions to the free energy (the z_jFE_m term in Equation 36). The electrical contributions cancel out in this case, but they must be included to obtain an accurate estimate of δG. Thus, coupled transport systems follow the usual thermodynamic laws, even though it may seem that they represent a special case because of cancellation of the electrical component of $\Delta\tilde{\mu}_j$ in the case of electroneutral transport.

Equation 40 only tells us whether or not Na^+ influx *can* occur, not how fast it takes place or even whether or not it takes place. Because Equation 40 was derived from principles of equilibrium thermodynamics, it tells us nothing about the rate of a process in a nonequilibrium situation. This is a much more difficult problem.

Calculation of flow rates requires detailed information about the kinetics of the particular transport system involved. To appreciate how difficult it is to make such calculations, consider the case of a simple ping-pong mechanism for exchange of two solutes, X and Y (such as Na^+ and H^+ in the example above, or Cl^- and HCO_3^-). The rate of exchange of X for Y, J_{ex}, is

$$\begin{aligned}J_{ex} = (A_1X_iY_o - A_2X_oY_i)/(A_3X_i + \\ A_4X_o + A_5Y_i + A_6Y_o + A_7X_iX_o + \\ A_8Y_iY_o + A_9X_iY_o + A_{10}X_oY_i)\end{aligned} \quad \textbf{(41)}$$

where $A_1 \ldots A_{10}$ are combinations of the individual association-dissociation constants for substrates, together with the rate constants for inward and outward transport.[65] Evidence or assumptions concerning the rate-limiting steps can simplify this problem somewhat, so that for thoroughly studied transport systems sufficient information may exist for calculation of the flow rate (see, e.g., Reference 66). In most cases where coupled transport systems have been included in calculations of volume regulatory behavior, however, much simpler approximations have been used,[67] such as that $J_{ex} = k_{ex}(X_iY_o - X_oY_i)$. This predicts zero flux at equilibrium, as it should (compare Equation 39), but is likely to be quite unrealistic under conditions far from equilibrium.

The situation is made even more complex when coupled transport systems have regulatory sites, whose influence must also be included in the flux equation. For example, in many cells the Na^+/H^+ exchange rate is nearly zero under resting conditions, despite a considerable gradient for Na^+ influx according to Equation 40. This is believed to be due to the existence of a regulatory site which is controlled by the internal H^+ concentration.[64] Because of such complexities, one cannot even conclude from equilibrium thermodynamics that flow via a particular coupled transport system actually takes place, only that it *could* take place in the presence of a transport system with suitable kinetic parameters.

IV. ANIMAL CELL VOLUME REGULATION

Having considered some of the important concepts involved in the equilibrium and transport of water and solutes across cell membranes, we now turn to the question of how a cell actually controls its own volume, which involves many of the principles discussed above. In the following sections, we first set out some general principles that apply to volume regulation, then discuss the long-term or steady-state maintenance of cell volume (chronic volume regulation) and then the more commonly studied short-term or transient regulation of cell volume that occurs, for example, under anisotonic conditions (acute volume regulation).

A. GENERAL PRINCIPLES — CELLULAR ACCOUNTING

To some extent, the problem of volume regulation in a cell is analogous to the regulation of cash flow in a business, and one can use accounting principles to do a kind of "cellular accounting" of the amount of various solutes in a cell and their possible involvement in changing cell volume. In particular, four criteria can be used to see whether or not a particular transport process may play a role in volume regulation and whether or not a volume regulatory model is adequate:

1. *Materiality:* This means that a process must move sufficient solute molecules in order to significantly affect cell volume, just as a particular economic transaction must involve a large enough amount of money to be worth considering in terms of the overall profit-and-loss status of a business. Minor components, like small change, may safely be ignored. In particular, this means that the transport system must transport solutes that are present in sufficiently large amounts to materially affect the total number of solute molecules in the cell compartment. One caution is required here: species such as H^+ or HCO_3^-, which can be formed or consumed by chemical reactions involving the intracellular buffer stores, may contribute importantly to volume regulation by altering the charge (e.g., of proteins) and the number of counterions that neutralize that charge, even though they themselves are present at low free concentrations.
2. *Direction of flow:* This means that the process must spontaneously transport solutes in the appropriate direction for volume regulation, just as in a business it is important whether a particular type of transaction represents income or disbursement. Flow direction can be calculated on the basis of equilibrium thermodynamic principles, as described above, but the amount of flow requires either experimental measurements or more detailed knowledge of the kinetics of the transport systems involved.
3. *Appropriate regulation:* This means that the proposed transport system must respond with increased or decreased activity depending on the volume status of the cell. Transport systems that are not so regulated may affect the volume-regulatory response and may be necessary in order that it can be fully expressed, but cannot initiate it. In business terms, this would represent transactions such as borrowing or lending, whose activity can be regulated to maintain overall cash balance.
4. *Overall balance:* If particular transport systems are proposed to effect volume regulation, the total net solute flow resulting from their operation must be equivalent to the total transport of solute molecules required for the observed change in volume. In business terms, the sum of debits and credits must equal the overall net profit or loss. If this is not true, something has been left out, and the "missing" osmolytes must be identified. (In a business, this search might lead to the missing embezzler who recently left for Argentina.)

B. CHRONIC VOLUME REGULATION

The steady-state maintenance of volume is critical for survival and normal functioning in all cells (for review, see Reference 68). In an organism with cells that cannot support substantial pressure gradients across their membranes, cell volume is determined by the combination of extracellular fluid osmolality and intracellular solute content. Since, for cells in most areas of the body, the surrounding plasma and interstitial fluid osmolality is tightly regulated by renal function under normal conditions, the question of how an individual cell regulates its volume must focus on how it regulates its solute content.

The two main classes of solutes found in cells include: (1) organic compounds (e.g., soluble proteins) generally synthesized within the cell and which for the most part are impermeant; and (2) small solutes, both organic and inorganic, which can potentially permeate across the cell membrane. Long-term maintenance of cell volume is achieved both when there is no net synthesis or consumption of the impermeant solutes and when the net active transport of the permeant solutes, primarily ions, balances the net leak in the opposite direction. Similarly, it follows that for cells to undergo long-term volume changes, a change in the balance of one or both of these processes must occur.

A more quantitative understanding of the mechanism of steady-state cell volume maintenance requires a detailed description of the rates of active (pump) and passive (leak) ion fluxes and the transmembrane permeabilities and electrochemical gradients of the principal ions. Several decades ago, Tosteson and Hoffmann quantitatively described volume maintenance under steady-state conditions as a balance between active transport by the sodium pump and passive leaks through unknown pathways in sheep red blood cells.[9] Since this pioneering work, further attempts have been made to refine and apply models to other simple cell systems. Using the constraints of macroscopic electroneutrality within the cell and osmotic balance across the membrane, along with the various flux equations for the principal ions (Na^+, K^+, and Cl^-), Jakobsson presented an integrated model relating cell volume, membrane potential, membrane transport, and the distribution of permeant solutes for a simplified, generic animal cell.[69] Jakobsson's model made several interesting predictions, including the idea that cell volume regulation by a sodium pump obligatorily induces a negative resting E_m and that volume regulation is dependent on both active pumping of Na^+ and passive permeability to K^+. More recently, Lew and coworkers have designed integrated models for cell volume, pH, and ion content regulation in both mature red blood cells[70] and reticulocytes.[67] Their models integrate the charge and nonideal osmotic behavior of hemoglobin and other impermeant species with the pump and leak processes of these cells. Moreover, some important predictions of these models have been tested and substantiated experimentally in these systems,[71,72] whereas for other predictions experimental agreement is only approximate at best.[73]

Let us now consider qualitatively what might be responsible for gradual, long-term changes in cell volume, as seen, for example, with cell growth between successive mitotic divisions. Clearly, sustained cell growth requires controlled adjustments in the volume set-point of the cell, resulting in a controlled net synthesis and/or accumulation of intracellular solutes over time. Since changes in the levels of cellular ions, especially cations, are tightly controlled by pump and leak transport processes, these processes could be centrally important in mediating cell growth. Because the number of sodium pumps must double during the cell cycle, it seems unlikely that the increase in cell solute content during growth within a cell cycle could result from a decrease in the number of sodium pumps present in the membrane. Although the concentrations of available ionic substrates (Na^+ and K^+) seem to be the principal regulators of pump activity (see above), it is conceivable that the pump activity during the cell cycle could be fine-tuned by other as yet unrecognized influences (e.g., cellular signaling mediators or changes in the ATP/ADP ratio). Changes in the set-points for leak flux pathways are perhaps more likely to be important in mediating cell growth. Many cellular signaling events are known to modulate the activities of various leak pathways, including cell volume-sensitive transport systems (see chapters in Section III), and the control of these signals could be cell cycle-dependent. Finally, if net protein synthesis occurs during progression through the cell cycle, the size of the cellular pool of impermeant macromolecules should tend to increase, thereby contributing to cell growth.

There is some evidence that volume-regulatory transport activities change during cell maturation or differentiation. For instance, it is known that the number of Na^+/H^+ exchangers expressed[74] and their kinetic parameters[75,76] change significantly during induced differentiation of human promyelocytic leukemic HL-60 cells. In addition, sodium pump expression in erythroid cell membranes decreases during maturation, as assessed by comparison of the number of [^{3}H]-ouabain binding sites in reticulocytes vs. mature red blood cells.[77] However, details of the underlying control mechanisms for these changes and any connections between them and concomitant changes in steady-state cell volume are not yet clear.

Ultimately, all putative mechanisms for adjustments in chronic cell volume regulation must be under genetic control. Interestingly, specific genes have recently been identified whose expression is thought to control progression through the different phases of the cell cycle.[78] It is possible that expression of these cyclins or other similar cell cycle-specific genes may serve as "master switches" that induce a generalized increase in protein synthesis and/or influence signaling systems that affect the activities of various pump and leak pathways.

In summary, although some progress has been made in understanding the various transport systems and the macromolecular synthesis and degradation processes that may be involved in chronic volume regulation, the relative importance of each factor under different conditions is not yet clear. Moreover, the underlying cellular and molecular mechanisms that are responsible for the control and integration of these processes remain obscure.

C. ACUTE VOLUME REGULATION

1. Background and Description

For more depth and background on the details of cell volume-regulatory phenomena presented in this section, one may refer to other chapters in this volume and several extensive reviews that have appeared in the recent literature.[10,13–16]

a. Significance

Whereas chronic volume regulation may involve synthesis and degradation of macromolecules and the steady-state interplay between pump and leak pathways, acute volume regulatory phenomena are non-steady-state processes, generally involving an activation of dormant leak flux pathways, which occur, for example, when cells are exposed to anisotonic media. Thus, acute volume regulation is critical for cells exposed to wide ranges of osmolality in their local environment (e.g., gastrointestinal and renal medullary epithelial cells [see Chapter 3] and presumably the blood cells traversing their underlying capillary beds).[15] In addition, the ion transport systems involved in epithelial cell volume regulation may provide a means of coupling between apical and basolateral membrane transport systems.[79] This communication between opposite surfaces of the cell (transcellular cross-talk) permits the coordinated transfer of specific solutes and water across epithelial cell barriers (see Chapter 2). However, since body fluid homeostatic mechanisms normally regulate the osmotic strength within narrow limits in most mammalian cellular environments, acute cell volume regulatory responses observed *in vitro* for such cells would be of questionable physiological significance. In fact, one might speculate that for many cells the capacity to regulate cell volume under anisotonic conditions may be either an atavism retained from an earlier ontogenic or phylogenic period, when osmoregulation was more critical, or the result of fortuitous activation of transport pathways normally involved in other cell functions, such as acid-base balance or control of the membrane potential.[80] Another possibility is that some of the mechanisms that are studied experimentally as acute volume regulatory responses may actually play a role in chronic volume regulation. For example, some of the leak flux pathways involved in acute volume regulation (e.g., K^+-Cl^- cotransport) may have tonic activity under steady-state conditions,[81] serving to buffer a cell's volume against the normal slight shifts in ambient osmolality. Such mechanisms could also compensate for occasional imbalances between pump and leak fluxes due, for example, to changes in the rate of Na^+-dependent nutrient uptake. Thus, the difference between what a cell may encounter *in vivo* and the acute and dramatic anisotonic stimuli often used experimentally to induce and characterize volume-regulatory responses may be one of degree only.

b. Classifications

When animal cells are exposed acutely to anisotonic media, they initially behave as osmometers, undergoing volume changes dictated by passive water flow across the cell membrane. In a hypotonic environment many types of cells, after initially swelling, then undergo a slower, compensatory shrinkage through the loss of intracellular solutes along with osmotically obliged water, a process known as regulatory volume decrease (RVD). Conversely, after first shrinking in a hypertonic environment, many cells then swell by taking up solutes from the medium, a process known as regulatory volume increase (RVI). Two different types of RVI can be distinguished, which we have termed primary (1°) and secondary (2°) RVI. 1° RVI (also referred to as "hypertonic RVI") occurs after cells are transferred from an isotonic to a hypertonic medium. 2° RVI (also known as "post-RVD RVI") occurs when cells that have undergone a previous RVD and have thus become acclimated to a hypotonic medium are then transferred back to "isotonic" medium, which is now hypertonic relative to these cells.

A large perturbation of cell volume is not necessarily required for cell volume regulation to occur. A process termed "isovolumetric regulation" by Lohr and Grantham has been shown to occur in rabbit kidney proximal tubules.[82] Employing sufficiently gradual changes in extracellular medium osmolality, they found that through the loss or uptake of largely uncharacterized solutes, tubular cell volume remained constant over a wide range of osmolalities. This phenomenon has clear physiological relevance to various pathological states where gradual alterations in plasma and interstitial fluid osmolality may occur (see Chapter 4).

c. Volume-regulatory transport systems

A wide variety of cells and tissues have been shown to regulate their volume, and there is great diversity in the transport systems and osmolytes involved. In cells of higher vertebrates, inorganic ions such as K^+, Na^+, Cl^- and HCO_3^- comprise the bulk of osmotically active solute, with K^+ and Cl^- concentrations generally the highest. Thus, it is not surprising that for the RVD response in hypotonic medium a net loss of K^+ and Cl^- ions generally occurs to lower cell volume. However, several different types of ion transporters have been shown to mediate this process (see Figure 1–4A). In *Amphiuma* red blood cells hypotonic swelling results in activation of an electroneutral K^+/H^+ exchanger which, coupled with Cl^-/HCO_3^- exchange, leads to the net loss of KCl along with cell water.[83] In avian and

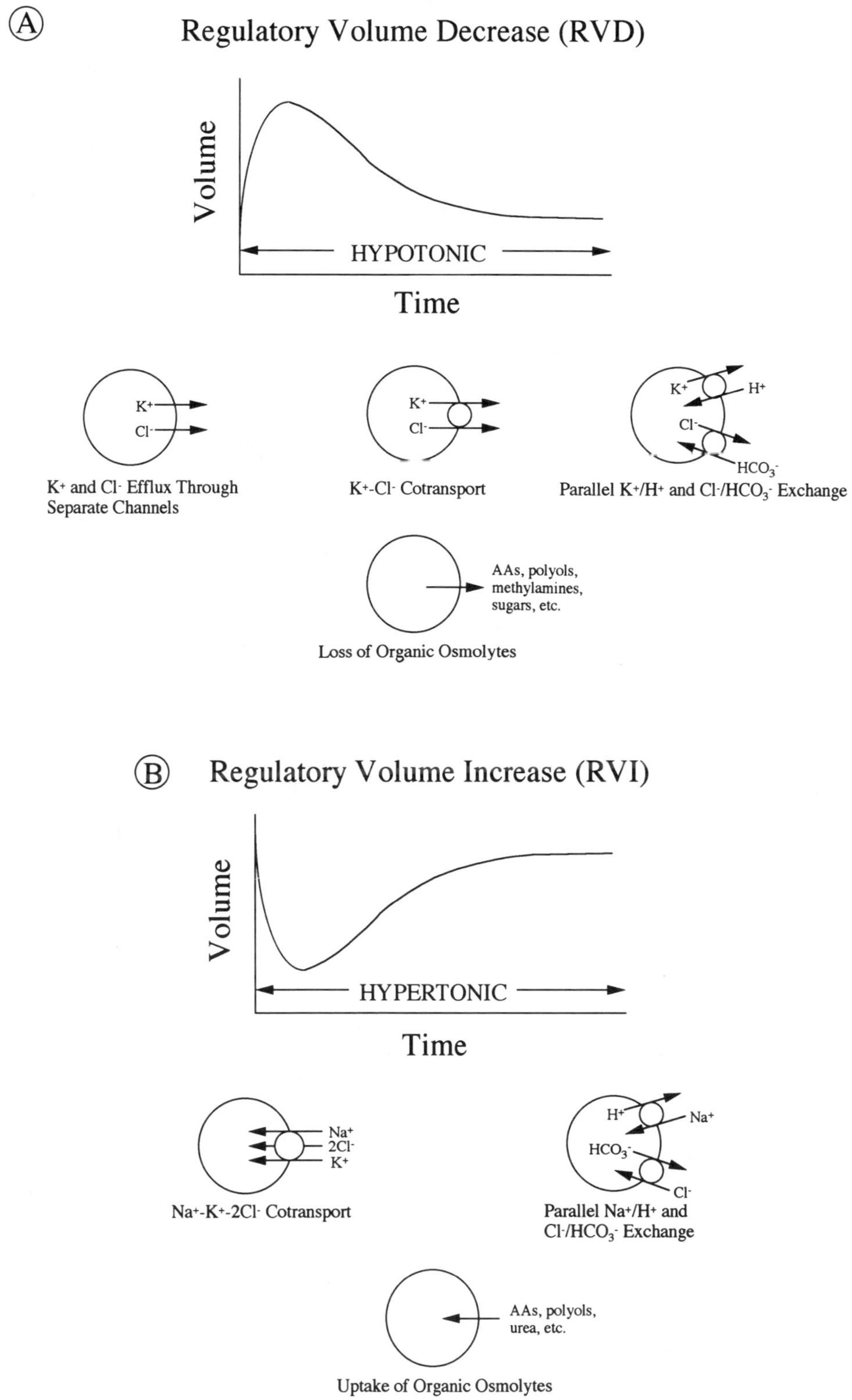

Figure 1–4 Illustrations of regulatory volume decrease (A) and regulatory volume increase (B) and the various classes of transport systems shown to be involved for each in different cell types.

most mammalian (including human) red blood cells, RVD occurs through the activation of K^+-Cl^- cotransport.[84] The activation of separate K^+ and Cl^- channels in the membrane has been shown to mediate RVD in most other cell types, including epithelial cells,[85–87] lymphocytes,[80] and Ehrlich ascites tumor cells.[88] An exception to the rule of KCl loss mediating RVD occurs with dog and other carnivore red blood cells that have high Na^+ and low K^+ content, where RVD occurs by Na^+ extrusion, probably through Na^+/Ca^{2+} exchange.[10]

The RVI response in higher vertebrates is generally characterized by an initial uptake of Na^+ and Cl^- in cells, which, followed by a secondary exchange of Na^+ for K^+ via the Na^+,K^+-ATPase, eventually results in a net gain of K^+ and Cl^-. Two different ionic mechanisms for RVI have been observed in different cell types (see Figure 1–4B). In avian red blood cells,[89] certain epithelial cell types such as Madin-Darby canine kidney (MDCK) cells,[90] and Ehrlich ascites tumor cells,[91] an Na^+-K^+-$2Cl^-$ cotransport mechanism has been implicated for RVI. On the other hand, amiloride-sensitive Na^+/H^+ exchange functionally coupled to Cl^-/HCO_3^- exchange has been shown to mediate RVI in *Amphiuma* red blood cells,[83] some epithelial cell types such as *Necturus* gallbladder cells,[92] and lymphocytes.[80]

Whereas cells in higher organisms generally follow these volume-regulatory patterns, fundamentally different mechanisms for anisotonic volume regulation have been observed in certain lower marine animal, plant, bacterial, fungal, and protist cells, whose complements of intracellular solutes are markedly different. Unlike the above systems, which primarily involve inorganic ions, these cell volume-regulatory responses make use of organic osmolytes, including sugars, polyols, amino acids (e.g., proline, glutamate, and taurine), methylamines, and urea (see chapters in Section II).[14] However, it should be noted that amino acid (especially taurine) transport has also been implicated for RVD in several vertebrate cell types, including MDCK cells[93] and Ehrlich ascites tumor cells.[94] Moreover, organic osmolyte synthesis or accumulation may supplement RVI in some renal cells,[16] and inositol accumulation is responsible for long-term volume adaptation during hypertonic stress in rat brain glioma cells.[95]

2. Analysis of Volume-Regulatory Processes

Now that some of the important background and descriptive concepts pertaining to acute and chronic cell volume regulation have been introduced, we focus on some of the interesting themes and questions faced by current researchers in the field.

a. Difference between primary and secondary RVI

The functional distinction made above between 1° and 2° RVI is important because many cell types show little or no 1° RVI response, whereas almost all cells show a significant volume recovery under conditions of 2° RVI.[15] Specifically, although a significant 1° RVI does occur in certain cell types, such as *Necturus* gallbladder epithelial cells[92] and *Amphiuma* red cells,[83] a clear deficiency in the strength of 1° RVI relative to 2° RVI exists in other cell types, including lymphocytes[96] and Ehrlich ascites tumor cells.[97]

The underlying mechanism for the difference in the strength between the two types of RVI is not known. It is clear that, to get a given degree of volume recovery under conditions of 1° RVI, more solute accumulation by the cell is required than for 2° RVI because of the greater ambient osmotic strength in the former case. Thus, one would expect the *rate* of volume recovery to be slower under 1° RVI conditions for a given degree of transporter flux activity. However, one would not expect this consideration to affect the *extent* of volume recovery attained at steady state.

To explain the decreased extent of 1° RVI, two general possibilities are conceivable: either there is an intrinsically increased activation of the ion transport system(s) for RVI after RVD has previously occurred, or the extrinsic driving forces for transport (e.g., due to different electrochemical gradients) are increased under conditions of 2° RVI relative to 1° RVI. Regarding the first possibility, it has been postulated that the lowered intracellular Cl^- content after a prior RVD may play a permissive role in the activation of RVI-associated fluxes.[98] A recent preliminary report suggests that lowered Cl_i is indeed required to activate the Na^+/H^+ exchanger and Na^+-K^+-$2Cl^-$ cotransporter in salivary gland epithelial cells.[99] It is also conceivable that changes in the post-RVD intracellular environment (e.g., related to a decrease in the osmotic pressure and/or the ionic strength) directly or indirectly enhance shrinkage-activated ion transport system(s). The second main possibility is that differences in electrochemical driving forces for the RVI-associated transport systems (e.g., an enhanced inward Cl^- gradient for $Cl^-_o/HCO^-_{3\,i}$ exchange) may play a role in producing the more robust 2° RVI response. A combination of several factors could be involved.

b. Sensor-transducer-effector systems in volume regulation

The questions of how a cell perceives its volume and how the volume signal is then amplified and transduced to the effector volume-regulatory transporters are crucial to the understanding of volume-regulatory phenomena. The sensor, transducer, and effector elements in this feedback system are not necessarily mutually exclusive; e.g., some entities, such as stretch-activated channels, have features of all three components. Furthermore, several of these elements often interact to produce an integrated volume-regulatory response. Clearly, different systems may be operative for RVD vs. RVI and in different cell types.

Regarding the question of how a cell perceives its volume, one can envision two broad classes of potential mechanisms. One class involves the role of mechanical deformation or stretch of the cell membrane or changes in the underlying cytoskeleton in the sensing of cell volume perturbation. For example, stretch-activated channels are of potential importance in mediating RVD responses in several cell types (for review, see Reference 58; see also Chapter 13). The cytoskeleton could serve as a component of both the volume-sensing and transducing apparatus, as its architecture is often sensitive to changes in cell volume,[100–102] and it may be involved in mediating the movement of cytoplasmic vesicles containing volume-regulatory transport proteins to and from the cell membrane.[103] Indeed, through the use of cytochalasins to inhibit actin assembly, a role for the cytoskeleton in volume regulation has been implicated in several cell types (see also Chapter 14).[100,102,104]

The other general possibility vis-à-vis the sensing of cell volume is that concentration or dilution of the cytoplasmic milieu through cell shrinkage or swelling leads to the activation of appropriate volume-regulation responses. In particular, changes in the concentration of a specific impermeant solute inside the cell could serve as the initial volume sensor. There is some evidence that intracellular Mg^{2+} and the activity of kinase/phosphatase systems may play a role in this regard.[105–107] Also, changes in the concentration of the soluble cytosolic proteins in a cell may nonspecifically affect the activity of certain enzymes through a phenomenon known as "macromolecular crowding", which could play a role in the sensing of cell volume if such affected enzymes (e.g., kinases and phosphatases) in turn modulate the activity of effector transport systems (see Chapter 10).

Since in many cases, relatively minor changes in cell volume can cause severalfold changes in ion transport,[108] an additional amplification process must be postulated. Some examples of known cellular signaling entities that have been linked to the transduction and amplification of the primary volume signal include Ca^{2+} transients (for review, see Reference 109), phosphoinositide turnover,[97,110] eicosanoid metabolism,[111,112] kinase/phosphatase systems,[108] cAMP,[113] and G-proteins.[114]

The chapters in Section III provide more details on cellular signaling events that may be important in the activation of volume regulatory responses. In addition, Sarkadi and Parker have recently provided a quite extensive review on various triggering events that may be important for volume regulation and their relationship to known volume-regulatory transport systems.[10]

c. Coordination of volume-regulatory transport processes

Many volume-regulatory processes involve parallel fluxes of two or more solutes; for example, an anion and a cation. For volume regulation to be effective, these fluxes must be properly coordinated and must both be terminated when the desired (set-point) volume is achieved.

The most obvious way of accomplishing this is to activate a process that directly couples the solute flows, such as the K^+-Cl^- and Na^+-K^+-$2Cl^-$ cotransport systems. In many cases, however, volume regulation involves flows of two or more solutes through independent pathways.

One way of achieving parallel flow in such systems is to have the flow of one solute establish a gradient for flow of the other solute. In the case of parallel ion channels, opening of one channel, e.g., for Cl^-, leads to a change in membrane potential, which then creates a driving force for the other solute, such as K^+. Another example is the case of parallel exchangers, where, for example, activation of Na^+/H^+ exchange transports Na^+ inward and H^+ outward, increasing cell pH. In the presence of CO_2, this increases cell HCO_3^- and creates a driving force for Cl^- influx via Cl^-/HCO_3^- exchange, with the net result of coordinated influx of Na^+ and Cl^-. Regulation in such systems is most easily achieved if one transport process is rate-limiting, while the other is so rapid as to be near equilibrium. Thus, only one (rate-limiting) process needs to be sensitive to volume and only one process must be shut off when the desired volume is reached.

An interesting variation on this theme is exemplified by lymphocytes,[115,116] in which swelling dramatically increases Cl^- permeability in an all-or-none fashion, but for a limited time. The K^+ permeability, on the other hand, steadily increases in a graded fashion as cell volume increases, but is lower than the Cl^- permeability. Thus, after cell swelling, the K^+ permeability becomes rate-limiting, and the regulatory volume decrease stops when the K^+ permeability decreases to a low value, which happens when the cell returns to its normal volume. This system performs well so long as normal volume is achieved within the limited time that the Cl^- permeability is elevated.

In some cases, activation of one transport process by the other may involve additional mechanisms, besides the establishment of gradients for net flow. For example, changes in E_m caused by opening of one channel may increase the open probability of channels for other solutes. A rise in intracellular Ca^{2+} due to influx through Ca^{2+} channels can increase K^+ flow through Ca^{2+}-activated K^+ channels.[117] Increases in pH caused by one exchanger, e.g., the Na^+/H^+ exchanger, may increase the transport rate of a parallel exchanger, e.g., Cl^-/HCO_3^- exchange, through the action of a modifier site, as is seen in lymphocytes.[118] Such relatively simple regulatory effects can greatly increase the speed of the volume-regulatory response.

Other more complex regulatory processes may also be involved. For example, the same signaling pathway may activate more than one transport system. Such processes might underlie the coordinate increase in permeability to several solutes sometimes seen during RVD.

Another important problem in the coordination of solute fluxes involves the push-pull interaction of volume-increasing and volume-decreasing mechanisms. It is important for the cell not only that the volume-regulatory systems be sensitive to small changes in volume, but also that the opposite systems not be activated at the same time. Because all volume-regulatory fluxes ultimately depend on solute gradients established at the cost of cellular metabolic activity, simultaneous activation of volume-increasing and volume-decreasing systems would be energetically wasteful and deleterious to the cell. Some of the mechanisms that prevent this, and which act to coordinate RVD and RVI through a central signaling system, are discussed in Chapter 18.

d. Identification of rate-limiting processes

Coordinated volume-regulatory processes can only proceed as fast as the slowest, rate-limiting transport system allows. Therefore, in determining the significance of a given transport system in the control of volume, identification of the rate-limiting flux pathway is a critical step.

One technique used to test whether a transporter is rate-limiting is the comparison of the dose-response effects of an inhibitor on the transport system alone with the dose-response effects on the overall observed volume-regulation response. The expected results of such a comparison are shown in Figure 1–5 with a theoretical example involving separate K^+ and Cl^- conductive pathways activated during RVD (see legend for details). The dose-response effects of a specific Cl^- channel inhibitor on both Cl^- permeability and net KCl efflux are compared under a variety of conditions, ranging from the case where K^+ permeability is largely rate-limiting ($P_{Cl} = 100\ P_K$) to the case where Cl^- permeability is largely rate-limiting ($P_{Cl} = 0.1\ P_K$). Under K^+ permeability-limiting conditions, there is little effect on the overall volume-regulatory flux, even at relatively high inhibitor doses (although the shape of the curve is still the same), whereas under Cl^- permeability-limiting conditions, the dose-response curve for the observed volume-regulatory flux approaches that for the isolated Cl^- permeability.

The other main strategy used to determine the rate-limiting volume-regulatory transport process involves the use of ionophores or penetrating ions to bypass the rate-limiting step, thus accelerating both the non-rate-limiting flux and the overall rate of volume recovery. This technique is particularly useful in comparing parallel conductive ionic fluxes (e.g., K^+ and Cl^- channels for RVD), where the constraint of macroscopic electroneutrality prevents an imbalance in the rates of the two net fluxes. For example, gramicidin, a Na^+ and K^+ ionophore, has been used successfully in lymphocytes to identify the K^+ conductance as rate-limiting for RVD in this particular system.[80]

e. Phenomenological effects of volume-regulatory fluxes

One of the more interesting and potentially confusing aspects of volume regulation is that the activation of one transporter in the membrane may lead to changes in the cellular concentration of different ions or solutes from those which are transported by the rate-limiting process.[83] An excellent example of this is seen in the case of RVI by means of activation of Na^+/H^+ exchange. This transporter catalyzes an inflow of Na^+ and outflow of H^+. The latter leads to cellular alkalinization, and therefore an increase of OH^- and, in the presence of CO_2, an increase in HCO_3^-. If the membrane contains an active Cl^-/HCO_3^- exchange system that is near equilibrium, even a slight increase in $HCO_3^-{}_i$ will cause bicarbonate to flow out in exchange for Cl^-. Thus, the Na^+/H^+ exchange will be transformed into an apparent Na^+-Cl^- cotransport.

Further, with a sodium pump in the membrane, the rise in Na_i will cause an increase in the pump flux, such that internal Na^+ will be exchanged for external K^+. The net result of this process will be the uptake of K^+ and Cl^- as a result of the initial activation of a Na^+/H^+ exchanger. The apparent

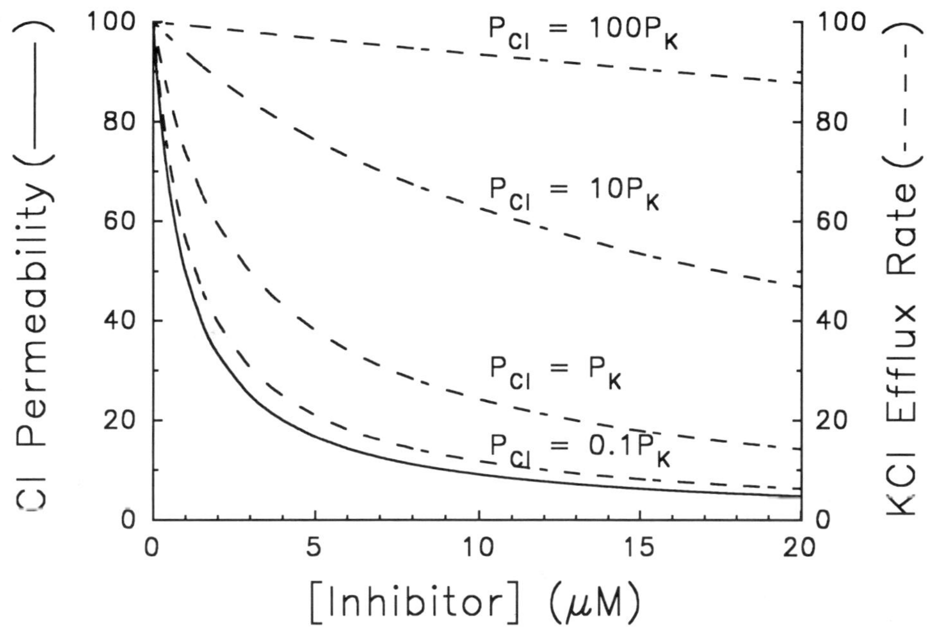

Figure 1–5 Comparison of the dose-response effects of a specific Cl^- channel inhibitor on the Cl^- permeability (P_{Cl}') (expressed as a percentage of the uninhibited permeability, P_{Cl}) and observed volume-regulatory KCl loss (expressed as a percentage of the uninhibited net KCl efflux), assuming diffusive kinetics. The effects of various P_{Cl}/P_K ratios on the net KCl efflux curves are shown. Cl^- permeability as a function of inhibitor dose was calculated from the equation for single-site hyperbolic inhibition:

$$P_{Cl}' = P_{Cl}/[1 + (I/K_I)]$$

where I is the inhibitor concentration and K_I is the half-inhibition constant for the inhibitor (= 1 μ*M* for this figure). Net KCl fluxes were modeled by the Goldman flux and GHK equations (Equations 34 and 35). As determined explicitly by the equation above, the Cl^- permeability inhibition curve is a rectangular hyperbola. Surprisingly, the shapes of the net KCl flux curves (determined from the more complex set of relationships among P_{Cl}, P_K, and E_m as defined in the Goldman flux and GHK equations) are also rectangular hyperbolae which, depending on the P_{Cl}/P_K ratio, have slightly greater or much greater apparent K_I values, as compared to the intrinsic K_I for inhibition of Cl^- permeability.

changes in ion content can therefore have little or nothing to do with the actual volume regulatory flux. In the example given, one might think that cell shrinkage had activated a K^+-Cl^- cotransport system, rather than the actual Na^+/H^+ exchange process!

Such effects are not really as misleading as they might at first seem. The possibility of net K^+-Cl^- inward cotransport could easily be ruled out by equilibrium thermodynamic considerations, which would show that the gradient for such flow is outward, not inward. Still, this and other similar examples provide an indication of the complexity of cellular homeostatic mechanisms, in which several transport systems may cooperate to produce a given result. It is perhaps also worth noting, in the example discussed above, that the result of all of this maneuvering is to keep cell pH and ion content nearly constant, so that the cellular interior environment is disturbed as little as possible during the volume-regulatory process.

f. Complicating effects of volume-regulatory fluxes

The ability of a cell to respond to acute osmotic shocks could be critical to its immediate survival. The cell may need to lose or gain a very large quantity of solutes within a relatively short time period to prevent lysis or irreversible damage. Therefore, the strategy that it uses to regulate its volume may involve making short-term sacrifices which need to be corrected in the long term for its continued prosperity. Specifically, animal cells of higher vertebrates generally respond to osmotic challenge via the loss or gain of primarily inorganic ions, conveniently using preexisting gradients created by the activity of cellular pumps. However, significant changes in the intracellular ionic environment that may result from acute volume-regulatory responses can adversely affect enzyme structure and function (see Reference 14). In contrast, many organic solutes appear to have stabilizing influences on proteins and are referred to as "compatible solutes" (see Chapter 5). The uptake of such organic osmolytes by certain cells is thus a means of long-term adaptation to osmotic stress. For instance, accumulated inorganic ions are partly replaced by inositol in brain glial cells as a means of adapting to the chronic hypertonic conditions that may prevail in certain pathological states.[95] Intracellular pH homeostasis is another cellular function that may be temporarily compromised in the face of osmotic shock, as significant excursions in intracellular pH may be the by-product of volume-regulatory processes.[119,120]

V. RELATIONSHIP OF VOLUME REGULATION TO CELLULAR PHYSIOLOGY

The process of volume regulation, involving as it does a number of transport systems and signaling mechanisms which must respond to and be compatible with various cell requirements, such as metabolism and growth, provides an excellent example of the importance of taking an integrated approach to the study of cellular regulatory phenomena. No matter how well each element of the process is understood in itself, a complete understanding of the system demands that the interrelationships of various systems and stimuli be taken into account. A better understanding of volume-regulatory phenomena requires much additional knowledge about the complex interactions of cell volume, pH, and ion regulation, which can only be achieved by constructing improved integrated models of cell function.

On the other hand, before such models can be sufficiently realistic, we need to know much more about the basic elements involved. The fundamental signaling mechanisms that sense volume and produce a response are still obscure, although current evidence points to the involvement of a bewildering variety of processes in various cell types. Put succinctly, we still don't know how a cell knows how big it is and how it manages to do something about it, although the other chapters in this volume provide some intriguing clues. Even the data on most of the effector transport systems are inadequate to permit calculation of fluxes under various experimental conditions for modeling purposes. In this connection, however, it should be noted that a great deal of important and pioneering work on various channels and coupled transport systems, which seem to play important physiological roles in a variety of cells, has come as a by-product of research on volume regulation. Volume regulation provides an intersection point, where information from fundamental thermodynamics, protein biochemistry, membrane biophysics, transport kinetics, molecular biology, and computer simulation converges on a problem of central importance to the cell. For this reason, it is a very complex field, but one with extraordinary potential for future progress.

REFERENCES

1. **MacLeod, J. and Ponder, E.,** The measurement of red cell volume. IV. Alterations in cell volume in hypotonic plasma, *J. Physiol.,* 77, 181, 1933.
2. **Guest, G. M.,** Osmometric behavior of normal and abnormal human erythrocytes, *Blood,* 3, 541, 1948.
3. **Adair, G. S.,** A theory of partial osmotic pressures and membrane equilibria, with special reference to the application of Dalton's Law to haemoglobin solutions in the presence of salts, *Proc. R. Soc. A (London),* 120, 573, 1928.
4. **McConaghey, P. D. and Maizels, M.,** The osmotic coefficients of haemoglobin in red cells under varying conditions, *J. Physiol.,* 155, 28, 1961.
5. **Freedman, J. C. and Hoffman, J. F.,** Ionic and osmotic equilibria of human red blood cells treated with nystatin, *J. Gen. Physiol.,* 74, 157, 1979.
6. **Davson, H.,** The permeability of the erythrocyte to cations, *Cold Spring Harbor Symp. Quant. Biol.,* 8, 255, 1940.
7. **Leaf, A.,** Maintenance of concentration gradients and regulation of cell volume, *Ann. N.Y. Acad. Sci.,* 72, 396, 1959.
8. **Ussing, H. H.,** Active and passive transport of the alkali metal ions, in *The Alkali Metal Ions in Biology,* Ussing, H. H., Kruhøffer, P., Hess Thaysen, J., and Thorn, N. A., Eds., Springer-Verlag, Berlin, 1960, 45.
9. **Tosteson, D. C. and Hoffman, J. F.,** Regulation of cell volume by active cation transport in high and low potassium sheep red cells, *J. Gen. Physiol.,* 44, 169, 1960.

10. **Sarkadi, B. and Parker, J. C.,** Activation of ion transport pathways by changes in cell volume, *Biochim. Biophys. Acta,* 1071, 407, 1991.
11. **Kregenow, F. M.,** The response of duck erythrocytes to nonhemolytic hypotonic media. Evidence for a volume-controlling mechanism, *J. Gen. Physiol.,* 58, 372, 1971.
12. **Kregenow, F. M.,** The response of duck erythrocytes to hypertonic media. Further evidence for a volume-controlling mechanism, *J. Gen. Physiol.,* 58, 396, 1971.
13. **Eveloff, J. L. and Warnock, D. G.,** Activation of ion transport systems during cell volume regulation, *Am. J. Physiol.,* 252, F1, 1987.
14. **Chamberlin, M. E. and Strange, K.,** Anisosmotic cell volume regulation: a comparative view, *Am. J. Physiol.,* 257, C159, 1989.
15. **Hoffmann, E. K. and Simonsen, L. O.,** Membrane mechanisms in volume and pH regulation in vertebrate cells, *Physiol. Rev.,* 69, 315, 1989.
16. **Spring, K. R. and Hoffmann, E. K.,** Cellular volume control, in *The Kidney: Physiology and Pathophysiology,* 2nd ed., Seldin, D. W. and Giebisch, G., Eds., Raven Press, New York, 1992, 147.
17. **House, C. R.,** *Water Transport in Cells and Tissues,* Edward Arnold (Publishers), London, 1974, 14.
18. **Shporer, M. and Civan, M. M.,** The state of water and alkali cations within intracellular fluids: the contribution of NMR spectroscopy, in *Current Topics in Membranes and Transport,* Bronner, F. and Kleinzeller, A., Eds., Academic Press, New York, 1977, 1.
19. **Clegg, J. S.,** Properties and metabolism of the aqueous cytoplasm and its boundaries, *Am. J. Physiol.,* 246, R133, 1984.
20. **Kempner, E. S. and Miller, J. H.,** The molecular biology of *Euglena gracilis.* IV. Cellular stratification by centrifuging, *Exp. Cell Res.,* 51, 141, 1968.
21. **Kempner, E. S. and Miller, J. H.,** The molecular biology of *Euglena gracilis.* V. Enzyme localization, *Exp. Cell Res.,* 51, 150, 1968.
22. **LeNeveu, D. M., Rand, R. P., and Parsegian, V. A.,** Measurements of forces between lecithin bilayers, *Nature,* 259, 601, 1976.
23. **Israelachvili, J. N. and Adams, G. E.,** Direct measurements of long range forces between two mica surfaces in aqueous KNO_3 solutions, *Nature,* 262, 774, 1976.
24. **Robinson, R. A. and Stokes, R. H.,** *Electrolyte Solutions,* 2nd ed., Butterworths, London, 1959.
25. **Dick, D. A. T.,** *Cell Water,* Butterworths, London, 1965.
26. **Siebens, A. W.,** Cellular volume control, in *The Kidney: Physiology and Pathophysiology,* Seldin, D. W. and Giebisch, G., Eds., Raven Press, New York, 1985, 91.
27. **Harned, H. S. and Owen, B. B.,** *The Physical Chemistry of Electrolytic Solutions,* 2nd ed., Reinhold, New York, 1950.
28. **Rand, R. P. and Burton, A. C.,** Mechanical properties of the red cell membrane. I. Membrane stiffness and intracellular pressure, *Biophys. J.,* 4, 115, 1964.
29. **Ito, T., Suzuki, A., and Stossel, T. P.,** Regulation of water flow by actin-binding protein-induced actin gelation, *Biophys. J.,* 61, 1301, 1992.
30. **Kwant, W. O. and Seeman, P.,** The erythrocyte ghost is a perfect osmometer, *J. Gen. Physiol.,* 55, 208, 1970.
31. **Dawson, D. C.,** Water transport: principles and perspectives, in *The Kidney: Physiology and Pathophysiology,* 2nd ed., Seldin, D. W. and Giebisch, G., Eds., Raven Press, New York, 1992, 301.
32. **Finkelstein, A.,** *Water Movement Through Lipid Bilayers, Pores, and Plasma Membranes: Theory and Reality,* Wiley-Interscience, New York, 1987.
33. **Macey, R. I. and Farmer, R. E. L.,** Inhibition of water and solute permeability in human red cells, *Biochim. Biophys. Acta,* 211, 104, 1970.
34. **Colombe, B. W. and Macey, R. I.,** Effects of calcium on potassium and water transport in human erythrocyte ghosts, *Biochim. Biophys. Acta,* 363, 226, 1974.
34a. **Paganelli, C. V. and Solomon, A. K.,** The rate of exchange of tritiated water across the human red cell membrane, *J. Gen. Physiol.,* 41, 259, 1957.
35. **Galey, W. R. and Brahm, J.,** The failure of hydrodynamic analysis to define pore size in cell membranes, *Biochim. Biophys. Acta,* 818, 425, 1985.
36. **Denker, B. M., Smith, B. L., Kuhajda, F. P., and Agre, P.,** Identification, purification, and partial characterization of a novel M_r 28,000 integral membrane protein from erythrocytes and renal tubules, *J. Biol. Chem.,* 263, 15634, 1988.
37. **Smith, B. L. and Agre, P.,** Erythrocyte M_r 28,000 transmembrane protein exists as a multisubunit oligomer similar to channel proteins, *J. Biol. Chem.,* 266, 6407, 1991.
38. **Preston, G. M. and Agre, P.,** Isolation of the cDNA for erythrocyte integral membrane protein of 28 kilodaltons: member of an ancient

channel family, *Proc. Natl. Acad. Sci. U.S.A.*, 88, 11110, 1991.

39. **Preston, G. M., Carroll, T. P., Guggino, W. B., and Agre, P.,** Appearance of water channels in *Xenopus* oocytes expressing red cell CHIP28 protein, *Science,* 256, 385, 1992.
40. **Preston, G. M., Jung, J. S., Guggino, W. B., and Agre, P.,** The Hg^{2+} sensitive residue at Cys-189 in the CHIP28 water channel, *Biophys. J.,* 64, A185, 1993.
41. **Zeidel, M. L., Ambudkar, S. V., Smith, B. L., and Agre, P.,** Reconstitution of functional water channels in liposomes containing purified red cell CHIP28 protein, *Biochemistry,* 31, 7436, 1992.
42. **Nielsen, S., Smith, B. L., Christensen, E. I., Knepper, M. A., and Agre, P.,** CHIP28 water channels are localized in constitutively water-permeable segments of the nephron, *J. Cell Biol.,* 120, 371, 1993.
43. **Swallow, C. J., Grinstein, S., and Rotstein, O. D.,** A vacuolar type H^+-ATPase regulates cytoplasmic pH in murine macrophages, *J. Biol. Chem.,* 265, 7645, 1990.
44. **Carafoli, E.,** The calcium pumping ATPase of the plasma membrane, *Annu. Rev. Physiol.,* 53, 531, 1991.
45. **Rabon, E. C. and Reuben, M. A.,** The mechanism and structure of the gastric H,K-ATPase, *Annu. Rev. Physiol.,* 52, 321, 1990.
46. **De Weer, P.,** Cellular sodium-potassium transport, in *The Kidney: Physiology and Pathophysiology,* 2nd ed., Seldin, D. W. and Giebisch, G., Eds., Raven Press, New York, 1992, 93.
47. **Hamlyn, J. M., Blaustein, M. P., Bova, S., DuCharme, D. W., Harris, D. W., Mandel, F., Mathews, W. R., and Ludens, J. H.,** Identification and characterization of a ouabain-like compound from human plasma, *Proc. Natl. Acad. Sci. U.S.A.,* 88, 6259, 1991.
48. **Hille, B.,** *Ionic Channels of Excitable Membranes,* 2nd ed., Sinauer Associates, Sunderland, MA, 1992.
49. **Hodgkin, A. L. and Huxley, A. F.,** Currents carried by sodium and potassium ions through the membrane of the giant axon of *Loligo, J. Physiol.,* 116, 449, 1952.
50. **Jan, L. Y. and Jan, Y. N.,** Voltage-sensitive ion channels, *Cell,* 56, 13, 1989.
51. **Raftery, M. A., Hunkapiller, M. W., Strader, C. D., and Hood, L. E.,** Acetylcholine receptor: complex of homologous subunits, *Science,* 208, 1454, 1980.
52. **Schofield, P. R., Darlison, M. G., Fujita, N., Burt, D. R., Stephenson, F. A., Rodriguez, H., Rhee, L. M., Ramachandran, J., Reale, V., Glencorse, T. A., Seeburg, P. H., and Barnard, E. A.,** Sequence and functional expression of the $GABA_A$ receptor shows a ligand-gated receptor super-family, *Nature,* 328, 221, 1987.
53. **Grenningloh, G., Rienitz, A., Schmitt, B., Methfessel, C., Zensen, M., Beyreuther, K., Gundelfinger, E. D., and Betz, H.,** The strychnine-binding subunit of the glycine receptor shows homology with nicotinic acetylcholine receptors, *Nature,* 328, 215, 1987.
54. **Westbrook, G. L. and Jahr, C. E.,** Glutamate receptors in excitatory neurotransmission, *Sem. Neurosci.,* 1, 103, 1989.
55. **Hartzell, H. C.,** Regulation of cardiac ion channels by catecholamines, acetylcholine, and second messenger systems, *Prog. Biophys. Mol. Biol.,* 52, 165, 1988.
56. **Trautwein, W. and Hescheler, J.,** Regulation of cardiac L-type calcium current by phosphorylation and G proteins, *Annu. Rev. Physiol.,* 52, 257, 1990.
57. **von Tcharner, V., Prod'hom, B., Baggiolini, M., and Reuter, H.,** Ion channels in human neutrophils activated by a rise in free cytosolic calcium concentration, *Nature,* 324, 369, 1986.
58. **Morris, C. E.,** Mechanosensitive ion channels, *J. Membr. Biol.,* 113, 93, 1990.
59. **Lambert, I. H.,** Taurine transport in Ehrlich ascites tumour cells. Specificity and chloride dependence, *Mol. Physiol.,* 7, 323, 1985.
60. **Goldman, D. E.,** Potential, impedance and rectification in membranes, *J. Gen. Physiol.,* 27, 37, 1943.
61. **Hodgkin, A. L. and Katz, B.,** The effect of sodium ions on the electrical activity of the giant axon of the squid, *J. Physiol.,* 108, 37, 1949.
62. **Restrepo, D., Cho, D. S., and Kron, M. J.,** Essential activation of Na^+-H^+ exchange by $[H^+]_i$ in HL-60 cells, *Am. J. Physiol.,* 259, C490, 1990.
63. **Olsnes, S., Tonnessen, T. I., and Sandvig, K.,** pH-regulated anion antiport in nucleated mammalian cells, *J. Cell Biol.,* 102, 967, 1986.
64. **Boron, W. F.,** Control of intracellular pH, in *The Kidney: Physiology and Pathophysiology,* 2nd ed., Seldin, D. W. and Giebisch, G., Eds., Raven Press, New York, 1992, 219.
65. **Frohlich, O. and Gunn, R. B.,** Erythrocyte anion transport: the kinetics of a single-site obligatory exchange system, *Biochim. Biophys. Acta,* 864, 169, 1986.
66. **Wieth, J. O. and Brahm, J.,** Cellular anion transport, in *The Kidney: Physiology and*

Pathophysiology, Seldin, D. W. and Giebisch, G., Eds., Raven Press, New York, 1985, 49.

67. **Lew, V. L., Freeman, C. J., Ortiz, O. E., and Bookchin, R. M.,** A mathematical model of the volume, pH, and ion content regulation in reticulocytes. Application to the pathophysiology of the sickle cell dehydration, *J. Clin. Invest.,* 87, 100, 1991.
68. **Macknight, A. D. C.,** Volume maintenance in isosmotic conditions, in *Current Topics in Membranes and Transport. Cell Volume Control: Fundamental and Comparative Aspects in Animal Cells,* Kleinzeller, A., Ed., Academic Press, New York, 1987, 3.
69. **Jakobsson, E.,** Interactions of cell volume, membrane potential, and membrane transport parameters, *Am. J. Physiol.,* 238, C196, 1980.
70. **Lew, V. L. and Bookchin, R. M.,** Volume, pH, and ion-content regulation in human red cells: analysis of transient behavior with an integrated model, *J. Membr. Biol.,* 92, 57, 1986.
71. **Freeman, C. J., Bookchin, R. M., Ortiz, O. E., and Lew, V. L.,** K-permeabilized human red cells lose an alkaline, hypertonic fluid containing excess K over diffusible anions, *J. Membr. Biol.,* 96, 235, 1987.
72. **Bookchin, R. M., Ortiz, O. E., Freeman, C. J., and Lew, V. L.,** Predictions and tests of a new integrated reticulocyte model: implications for dehydration of sickle cells, in *The Red Cell: Seventh Ann Arbor Conference,* Alan R. Liss, New York, 1989, 615.
73. **Bookchin, R. M., Ortiz, O. E., and Lew, V. L.,** Evidence for a direct reticulocyte origin of dense red cells in sickle cell anemia, *J. Clin. Invest.,* 87, 113, 1991.
74. **Rao, G. N., DeRoux, N., Sardet, C., Pouyssegur, J., and Berk, B.,** Na^+/H^+ antiporter gene expression precedes monocytic differentiation in HL60 cells, *FASEB J.,* 5, A671, 1991.
75. **Costa-Casnellie, M. R., Segel, G. B., Cragoe, E. J., Jr., and Lichtman, M. A.,** Characterization of the Na^+/H^+ exchanger during maturation of HL-60 cells induced by dimethyl sulfoxide, *J. Biol. Chem.,* 262, 9093, 1987.
76. **Costa-Casnellie, M. R., Segel, G. B., and Lichtman, M. A.,** The Na^+/H^+ exchanger in immature and mature granulocytic HL-60 cells, *J. Biol. Chem.,* 263, 11851, 1988.
77. **Wiley, J. S. and Shaller, C. C.,** Selective loss of calcium permeability on maturation of reticulocytes, *J. Clin. Invest.,* 59, 1113, 1977.
78. **Murray, A. W. and Kirschner, M. W.,** What controls the cell cycle?, *Sci. Am.,* 264, 56, 1991.
79. **Schultz, S. G.,** Membrane cross-talk in sodium-absorbing epithelial cells, in *The Kidney: Physiology and Pathophysiology,* 2nd ed., Seldin, D. W. and Giebisch, G., Eds., Raven Press, New York, 1992, 287.
80. **Grinstein, S., Rothstein, A., Sarkadi, B., and Gelfand, E. W.,** Responses of lymphocytes to anisotonic media: volume-regulating behavior, *Am. J. Physiol.,* 246, C204, 1984.
81. **Kaji, D.,** Volume-sensitive K transport in human erythrocytes, *J. Gen. Physiol.,* 88, 719, 1986.
82. **Lohr, J. W. and Grantham, J. J.,** Isovolumetric regulation of isolated S_2 proximal tubules in anisotonic media, *J. Clin. Invest.,* 78, 1165, 1986.
83. **Cala, P. M.,** Volume regulation by red blood cells. Mechanisms of ion transport, *Mol. Physiol.,* 4, 33, 1983.
84. **Knauf, P. A.,** Anion transport in erythrocytes, in *Physiology of Membrane Disorders,* 2nd ed., Andreoli, T. E., Hoffman, J. F., Fanestil, D. D., and Schultz, S. G., Eds., Plenum Press, New York, 1986, 191.
85. **Knoblauch, C., Montrose, M. H., and Murer, H.,** Regulatory volume decrease in cultured renal cells, *Am. J. Physiol.,* 256, C252, 1989.
86. **Welling, P. A. and O'Neil, R. G.,** Cell swelling activates basolateral membrane Cl and K conductances in rabbit proximal tubule, *Am. J. Physiol.,* 258, F951, 1990.
87. **Banderali, U. and Roy, G.,** Activation of K^+ and Cl^- channels in MDCK cells during volume regulation in hypotonic media, *J. Membr. Biol.,* 126, 219, 1992.
88. **Hoffmann, E. K., Lambert, I. H., and Simonsen, L. O.,** Separate, Ca^{2+}-activated K^+ and Cl^- transport pathways in Ehrlich ascites tumor cells, *J. Membr. Biol.,* 91, 227, 1986.
89. **Haas, M., Schmidt, W. F., III, and McManus, T. J.,** Catecholamine-stimulated ion transport in duck red cells. Gradient effects in electrically neutral (Na + K + 2Cl) cotransport, *J. Gen. Physiol.,* 80, 125, 1982.
90. **McRoberts, J. A., Erlinger, S., Rindler, M. J., and Saier, M. H., Jr.,** Furosemide-sensitive salt transport in the Madin-Darby canine kidney cell line: evidence for the cotransport of Na^+, K^+, and Cl^-, *J. Biol. Chem.,* 257, 2260, 1982.
91. **Levinson, C.,** Regulatory volume increase in Ehrlich ascites tumor cells is mediated by the 1Na: 1K: 2Cl cotransport system, *J. Membr. Biol.,* 126, 227, 1992.
92. **Ericson, A. C. and Spring, K. R.,** Volume regulation by *Necturus* gallbladder: apical Na^+-

H^+ and Cl^--HCO_3^- exchange, *Am. J. Physiol.*, 243, C146, 1982.

93. **Roy, G. and Sauvé, R.,** Effect of anisotonic media on volume, ion and amino acid content and membrane potential of kidney cells (MDCK) in culture, *J. Membr. Biol.*, 100, 83, 1987.
94. **Hoffmann, E. K. and Lambert, I. H.,** Amino acid transport and cell volume regulation in Ehrlich ascites tumour cells, *J. Physiol.*, 338, 613, 1983.
95. **Strange, K., Morrison, R., Heilig, C. W., DiPietro, S., and Gullans, S. R.,** Upregulation of inositol transport mediates inositol accumulation in hyperosmolar brain cells, *Am. J. Physiol.*, 260, C784, 1991.
96. **Grinstein, S. and Foskett, J. K.,** Ionic mechanisms of cell volume regulation in leukocytes, *Annu. Rev. Physiol.*, 52, 399, 1990.
97. **Hoffmann, E. K., Lambert, I. H., and Simonsen, L. O.,** Mechanisms in volume regulation in Ehrlich ascites tumor cells, *Renal Physiol. Biochem.*, 3–5, 221, 1988.
98. **Hoffmann, E. K., Sjoholm, C., and Simonsen, L. O.,** Na^+, Cl^- cotransport in Ehrlich ascites tumor cells activated during volume regulation (regulatory volume increase), *J. Membr. Biol.*, 76, 269, 1983.
99. **Robertson, M. A. and Foskett, J. K.,** Activation of Na^+/H^+ exchanger and NaK2Cl cotransporter in single acinar cells requires agonist-induced fall of $[Cl^-]_i$, *Biophys. J.*, 64, A81, 1993.
100. **Cornet, M., Delpire, E., and Gilles, R.,** Study of microfilaments network during volume regulation process of cultured PC12 cells, *Pflügers Arch.*, 410, 223, 1987.
101. **Hallows, K. R., Packman, C. H., and Knauf, P. A.,** Acute cell volume changes in anisotonic media affect F-actin content of HL-60 cells, *Am. J. Physiol.*, 261, C1154, 1991.
102. **Ziyadeh, F. N., Mills, J. W., and Kleinzeller, A.,** Hypotonicity and cell volume regulation in shark rectal gland: role of organic osmolytes and F-actin, *Am. J. Physiol.*, 262, F468, 1992.
103. **Cantiello, H. F., Stow, J. L., Prat, A. G., and Ausiello, D. A.,** Actin filaments regulate epithelial Na^+ channel activity, *Am. J. Physiol.*, 261, C882, 1991.
104. **Foskett, J. K. and Spring, K. R.,** Involvement of calcium and cytoskeleton in gallbladder epithelial cell volume regulation, *Am. J. Physiol.*, 248, C27, 1985.
105. **Parker, J. C., McManus, T. J., Starke, L. C., and Gitelman, H. J.,** Coordinated regulation of Na/H exchange and [K-Cl] cotransport in dog red cells, *J. Gen. Physiol.*, 96, 1141, 1990.
106. **Starke, L. C. and McManus, T. J.,** Intracellular free magnesium determines the volume regulatory set point in duck red cells, *FASEB J.*, 4, A818, 1990.
107. **Jennings, M. L. and Schulz, R. K.,** Okadaic acid inhibition of KCl cotransport. Evidence that protein dephosphorylation is necessary for activation of transport by either cell swelling or N-ethylmaleimide, *J. Gen. Physiol.*, 97, 799, 1991.
108. **Jennings, M. L. and Schulz, R. K.,** Swelling-activated KCl cotransport in rabbit red cells: flux is determined mainly by cell volume rather than shape, *Am. J. Physiol.*, 259, C960, 1990.
109. **Pierce, S. K. and Politis, A. D.,** Ca^{2+}-activated cell volume recovery mechanisms, *Annu. Rev. Physiol.*, 52, 27, 1990.
110. **McConnell, F. M. and Goldstein, L.,** Intracellular signals and volume regulatory response in skate erythrocytes, *Am. J. Physiol.*, 255, R982, 1988.
111. **Lambert, I. H., Hoffmann, E. K., and Christensen, O.,** Role of prostaglandins and leukotrienes in volume regulation by Ehrlich ascites tumor cells, *J. Membr. Biol.*, 98, 247, 1987.
112. **Lambert, I. H.,** Leukotriene-D_4 induced cell shrinkage in Ehrlich ascites tumor cells, *J. Membr. Biol.*, 108, 165, 1989.
113. **Mills, J. W. and Lubin, M.,** Effect of adenosine 3′,5′-cyclic monophosphate on volume and cytoskeleton of MDCK cells, *Am. J. Physiol.*, 250, C319, 1986.
114. **Davis, B. A., Hogan, E. M., and Boron, W. F.,** Role of G proteins in stimulation of Na-H exchange by cell shrinkage, *Am. J. Physiol.*, 262, C533, 1992.
115. **Sarkadi, B., Mack, E., and Rothstein, A.,** Ionic events during volume responses of human peripheral blood lymphocytes to hypotonic media. I. Distinctions between volume-activated Cl^- and K^+ conductance pathways, *J. Gen. Physiol.*, 83, 497, 1984.
116. **Sarkadi, B., Mack, E., and Rothstein, A.,** Ionic events during volume response of human peripheral blood lymphocytes to hypotonic media. II. Volume and time-dependent activation and inactivation of ion transport pathways, *J. Gen. Physiol.*, 83, 513, 1984.
117. **Christensen, O.,** Mediation of cell volume regulation by Ca^{2+} influx through stretch-activated channels, *Nature*, 330, 66, 1987.
118. **Mason, M. J., Smith, J. D., Garcia-Soto, J. D., and Grinstein, S.,** Internal pH-sensitive

site couples Cl^--HCO_3^- exchange to Na^+-H^+ antiport in lymphocytes, *Am. J. Physiol.*, 256, C428, 1989.

119. **Livne, A. and Hoffmann, E. K.,** Cytoplasmic acidification and activation of Na^+/H^+ exchange during regulatory volume decrease in Ehrlich ascites tumor cells, *J. Membr. Biol.*, 114, 153, 1990.

120. **Star, R. A., Zhang, B., Loessberg, P. A., and Muallem, S.,** Regulatory volume decrease in the presence of HCO_3^- by single osteosarcoma cell UMR-106-01, *J. Biol. Chem.*, 267, 17665, 1992.

Chapter 2

Volume Regulation in Epithelia: Transcellular Transport and Cross-Talk

Luis Reuss and Calvin U. Cotton

TABLE OF CONTENTS

I. INTRODUCTION

In all animal cells, the presence of nondiffusible anions in the intracellular compartment creates a Gibbs-Donnan equilibrium which tends to produce colloid-osmotic swelling. Such a swelling is prevented by so-called mechanisms of cell volume maintenance. In addition, epithelial cells face unique and potentially severe challenges for cell volume regulation during changes in transport rate. Primary volume changes in epithelial cells are ultimately caused by either a change in extracellular or in intracellular effective osmolality; *in vivo* changes in epithelial cell volume are largely the result of changes in the rate of solute entry or exit.

The intracellular and extracellular fluid osmolalities of higher pluricellular organisms are closely regulated. Thus, rapid changes in either epithelial (polar) or nonepithelial (nonpolar) cell volume are assumed to be rare *in vivo*. An exception is the epithelium of the small intestine, in which the luminal concentrations of organic solutes (glucose and amino acids) fluctuate with the digestive process. Since these solutes are cotransported with Na^+ across the apical membrane, net solute entry and, hence, epithelial cell volume will tend to vary with time. There are other clear instances of physiologic changes in cell volume. For example, circu-

0-8493-4448-4/94/$0.00+$.50

lating blood cells are challenged with large osmotic differences during passage through the renal medulla in which the interstitial osmolality may reach about 1200 mOsmol/kg. *In vitro* studies of epithelial cells have shown that large, rapid changes in cell volume can occur as a result of salt and water efflux following stimulation with cholinergic agonists or cyclic AMP.[1–4] Changes in cell volume of cultured neurons have also been reported to occur upon exposure to neurotransmitters.[5] The latter two examples, which represent changes in cell volume induced by alterations in membrane ion permeabilities rather than overt perturbations in extracellular or intracellular osmolality, must be considered a part of normal cell function. Changes in cell volume in response to pathological conditions[6] will not be discussed in this chapter.

Vectorial transport by epithelia requires cell polarity, i.e., the existence of two structurally and functionally distinct cell membrane domains. For instance, in Na^+-absorbing epithelial cells the apical (lumen-facing) membrane contains passive transporters permitting Na^+ transport into the cell, whereas the basolateral membrane expresses Na^+,K^+ pumps, which extrude Na^+ to the extracellular fluid of the animal. At the steady state, epithelial cell volume and composition are maintained constant, although there is continuous entry (of solute or solute and water) at one membrane and exit at the other one. Evidently, "cell homeostasis" under these conditions requires equal rates of entry and exit. This equality assures near-constancy of the intracellular contents, as illustrated in the central diagram of Figure 2–1.[7]

The steady state described above is perturbed by changes in solute flux at one of the two membranes. This alters both solute content and volume of the cell (Figure 2–1), unless there is a rapid and efficient adjustment of the transport rates at the two membranes. For instance, a primary increase in solute entry across the apical membrane of an absorptive epithelial cell will cause an increase in the intracellular solute concentration (and in the intracellular osmolality), followed by cell swelling because at least one cell membrane domain has a high osmotic water permeability.[8] In order for solute and water content of the cell to be regulated, the changes (in cell volume and/or composition) must be sensed. Finally, the sensing mechanism(s), with or without amplification, must result in one or more "corrective" effects on solute transport and osmotically coupled water transport. Inasmuch as the initial perturbation in this example is an increase in

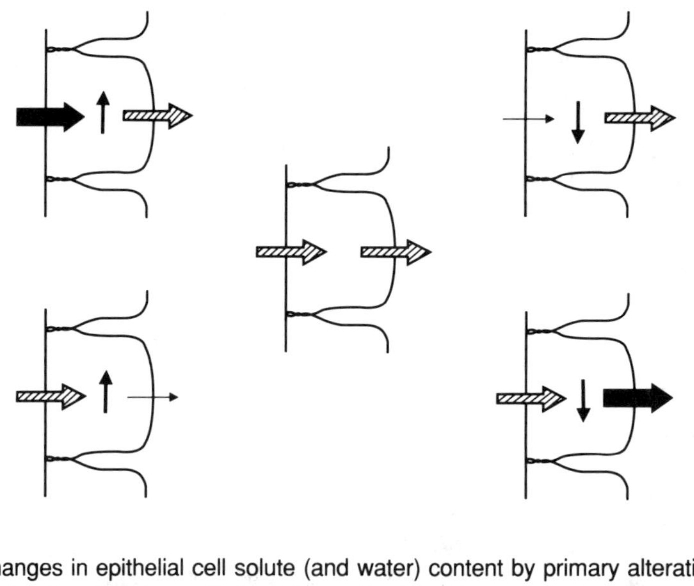

Figure 2–1 Changes in epithelial cell solute (and water) content by primary alterations of transport rate. The center drawing depicts a salt-absorbing epithelial cell at steady state, i.e., apical solute entry equals basolateral solute exit and cell solute content remains constant. The diagrams on the left-hand side denote the elementary mechanisms of increase in cell solute content, by a primary increase in entry (top) or a primary decrease in exit (bottom). The diagrams on the right-hand side depict the mechanisms of decrease in cell solute content, by a primary decrease in entry (top) or a primary increase in exit (bottom). In all four cases, the change in solute content will result in a change in cell water volume in the same direction, because of the high osmotic water permeability of one or both membranes. (From Reuss, L., *Renal Physiol. Biochem.*, 11, 187, 1988. © S. Karger AG, Basel. With permission.)

solute influx across the apical membrane, the appropriate response is an increase in solute efflux across the basolateral membrane. This sequence of events allows for an increase in the rate of transepithelial transport while "cell homeostasis" is preserved.

The example outlined above illustrates the theme of this article, namely the relationships between cell volume regulation and transepithelial transport regulation. The importance and uniqueness of this mechanism were first recognized by Schultz.[9] The expression *cross-talk* was coined to describe pari passu changes in transport rates at the two membrane domains of the transporting epithelial cell.[10] This concept has been the subject of a recent detailed and scholarly review of the literature for Na^+-absorbing epithelia by Schultz and Hudson,[11] which covers the field until 1988. Our aims are to provide an overview of the problem, to review salient experimental results since 1988, and to include secretory epithelia in our discussion.

II. EPITHELIAL-CELL "HOMEOSTASIS" IS CHALLENGED BY CHANGES IN TRANSEPITHELIAL TRANSPORT RATE

Consider the simplest absorbing epithelium, namely that described by the two-membrane hypothesis of Koefoed-Johnsen and Ussing,[12] depicted in Figure 2–2. The epithelium absorbs only Na^+, which enters by electrodiffusion (via channels) across the apical membrane and is transported solely by the Na^+,K^+ pump across the basolateral membrane. The basolateral membrane also contains K^+ channels. Na^+ entry can vary with physiologic stimuli (for instance, hormones that elevate apical membrane Na^+ permeability) and the amount transported per unit time can be quite large relative to the amount contained in the cell. The elevation in the rate of Na^+ entry will cause an increase in intracellular Na^+ concentration ($[Na^+]_i$) and hence will stimulate the Na^+,K^+ pump, and the increase in pump rate will elevate the rate of K^+ uptake, creating a need for faster K^+ recycling via the basolateral K^+ channels. Perfect cross-talk would require stimulation of both the Na^+,K^+ pump and the K^+ channels. As explained below, the increase in basolateral Na^+ pumping would require a stimulation of the pump by a mechanism other than an increase in $[Na^+]_i$, since the latter appears to remain constant. This scheme is complicated in "real" epithelia by the presence of apical membrane K^+ channels, apical or basolateral membrane Cl^- channels, basolateral KCl or $NaKCl_2$ cotransporters, and apical or basolateral membrane ion exchangers.

"Homocellular" regulatory mechanisms are intrinsic mechanisms set to maintain cell volume and composition in the face of changes in transepithelial transport rate.[9,11] Adjustments of the transport rates at the two membrane domains of the epithelial cells can result from one or more of the following mechanisms: (1) mass-action effects, which do not involve changes in the number of transporters or their kinetic properties; (2) changes in the number of active transporters in the membrane, by insertion of new copies, activation of quiescent copies present in the membrane, or a combination of both; and (3) changes in the turnover rate or active state of preexisting, operative transporters, i.e., carriers, pumps, or channels. A complete understanding of the process by which the transport rates at the two membranes are adjusted involves identification of the signal, the sensor, the transduction mechanism(s), and the nature and function of the effectors, i.e., the transporters.

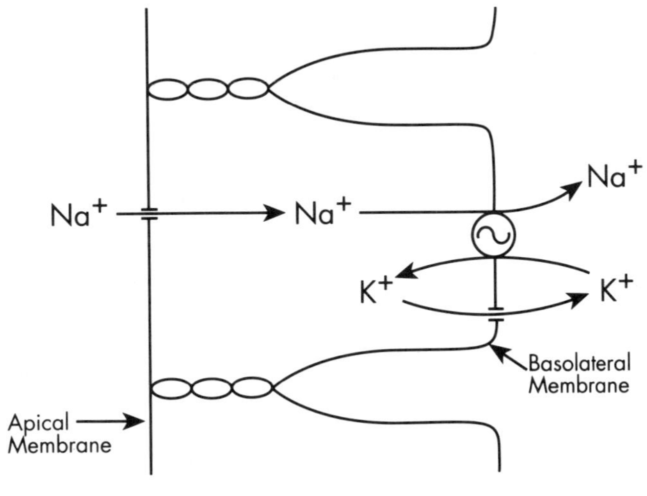

Figure 2–2 Koefoed-Johnsen and Ussing model for a Na^+-absorbing "tight" epithelium. Na^+ entry across the apical cell membrane is via an amiloride-sensitive Na^+-channel. Extrusion of Na^+ across the basolateral cell membrane is mediated by a ouabain-sensitive Na^+,K^+-ATPase. K^+ recycles across the basolateral cell membrane via a K^+-channel.

III. EXAMPLES OF CROSS-TALK AND REGULATION

Changes in transepithelial transport rate may be slow (minutes to hours), e.g., stimulation of Na^+ absorption in the renal collecting duct by aldosterone, or rapid (seconds), e.g., addition of transported sugars to the lumen of the intestine. The rate of transepithelial transport can also be altered by nonphysiological means such as exposure of epithelial cells to transport inhibitors or changes in osmolality or ionic composition of one or both bathing solutions. Experimental maneuvers of this type are particularly useful in studies of epithelial cross-talk and cell volume regulation.

The first experimental observation supporting the idea of cross-talk in epithelial cells was made by MacRobbie and Ussing.[13] They observed that inhibition of the basolateral Na^+ pump in isolated frog skin caused reductions in ion permeabilities of both the basolateral and apical cell membranes. Since these pioneering observations, many studies have confirmed the existence of regulatory mechanisms that respond to changes in transport rate to maintain relatively constant cell volume and ion concentrations. In the sections that follow, we will discuss examples in which primary changes in salt influx or efflux across either the apical or the basolateral cell membrane of transporting epithelial cells result in compensatory changes in the fluxes at the opposite membrane. The results from experiments with three fundamentally different types of epithelia will be presented: (1) "tight" Na^+-absorbing epithelia, such as renal collecting tubule and amphibian urinary bladder; (2) "leaky" Na^+-absorbing epithelia, such as small intestine and renal proximal tubule; and (3) Cl^--secreting exocrine acinar cells.

A. CROSS-TALK IN TIGHT EPITHELIA

The two-membrane hypothesis,[12] first proposed to account for Na^+ absorption across the isolated frog skin epithelium, remains as the model of choice to explain Na^+ absorption by so-called tight epithelia (Figure 2–2). Abundant experimental evidence suggests that the activities of the three major transport pathways (apical Na^+ channel, basolateral K^+ channel, and basolateral Na^+,K^+ pump) are coordinated. Inhibition of the basolateral Na^+,K^+-ATPase with G-strophanthidin or by lowering the pH of the inner bathing solution causes a decrease in the passive permeability of the apical cell membrane to Na^+ and of the basolateral cell membrane to K^+.[13] The idea that the properties of the apical and basolateral cell membranes are somehow linked was given additional support by electrophysiological studies of Finn and Reuss.[14] They found that in the toad urinary bladder changes in the osmolality of the serosal bathing solution cause parallel changes in transport properties of both apical and basolateral cell membranes.

The effect of Na^+ pump inhibition on apical cell membrane Na^+ permeability in tight epithelia was subsequently confirmed by several investigators. In toad urinary bladder, pump inhibition with ouabain did not cause cell swelling,[15] although swelling would be expected if apical entry of Na^+ (and net influx of accompanying anion) continued unabated. Exposure of tissues to metabolic inhibitors or a K^+-free serosal bathing solution, maneuvers expected to lead to pump inhibition, also resulted in a decrease in apical membrane Na^+ permeability, P_{Na}.[16,17] In contrast, rapid cell swelling was observed when isolated, perfused rabbit renal cortical collecting ducts were exposed to ouabain.[18] The cells swelled within 2 min to about 230% of their resting volume and thereafter their volume returned slowly toward control levels. Several important experimental details should be noted. First, the studies with collecting ducts were done at 36°C, whereas the studies with amphibian skin and bladder were done at room temperature in tissues with a much lower transport rate. Second, the rate and extent of cell swelling was directly related to the pretreatment of the rabbits with the mineralocorticoid deoxycorticosterone acetate (DOCA) to increase Na^+ absorption. In tubules from animals not treated with DOCA, ouabain induced slow cell swelling.[18a] Regardless of the quantitative differences, Strange concluded that pump inhibition reduced apical membrane P_{Na} and that the volume regulatory decrease (RVD) that followed cell swelling resulted from K^+ loss involving activation of an apical membrane K^+ channel silent under control conditions,[19] and Cl^- loss via a basolateral Cl^- channel.[20]

The fact that increases in $[Na^+]_i$ cause reductions in apical membrane P_{Na} suggests that falls in $[Na^+]_i$ increase apical membrane P_{Na}. In agreement with this notion, removal followed by readdition of Na^+ to the solution bathing the apical surface of colonic epithelia from rabbits causes a large Na^+ current that relaxes over several minutes to a lower steady-state level.[21,22] The increase in current and its subsequent decline were attributed to changes, in the same direction as the current, in apical membrane amiloride-sensitive Na^+ permeability. Similar conclusions were reached in studies on frog skin.[23] These results suggest that the rate of transepithelial Na^+ absorption by "tight" epithelia is dependent upon apical cell membrane P_{Na}, which can be affected by maneuvers that would be expected to produce changes in $[Na^+]_i$. In some circumstances (see below) changes in $[Ca^{2+}]_i$ and/or pH_i may mediate these effects.

In addition to conductive Na^+ entry across the apical membrane, transepithelial Na^+ absorption in tight epithelia involves Na^+ extrusion across the basolateral membrane, mediated by the Na^+,K^+-ATPase. The simplest signal coupling Na^+ transport and pump activity would be the change in $[Na^+]_i$. However, studies of the relationship between $[Na^+]_i$ and transepithelial net Na^+ flux (J_{Na}) have yielded unclear results. In *Necturus* urinary bladder epithelium the $[Na^+]_i$ was nearly constant (about 10 m*M*), despite large differences in Na^+ absorption produced by changing luminal solution $[Na^+]$.[24] In contrast, a positive correlation exists between J_{Na} and $[Na^+]_i$ in frog skin under certain conditions.[11] Studies in rabbit colon, rabbit urinary bladder, and toad urinary bladder have yielded mixed results.[11]

Amiloride addition to the solution bathing the apical surface of toad or frog urinary bladder epithelium induces a rapid (seconds) fall in basolateral membrane conductance with no change in cell volume.[25] Furthermore, the rate of RVD after transient exposure to hyposmotic solutions was slowed by amiloride pretreatment. It was suggested[25] that inhibition of apical membrane P_{Na} resulted in inhibition of basolateral membrane K^+ permeability (P_K) and thereby slowed KCl efflux during RVD. Conversely, exposure of *Necturus* urinary bladder to Ba^{2+} on the basolateral surface to reduce K^+ conductance (G_K) caused a near-simultaneous fall in apical membrane P_{Na}.[26]

In conclusion, the results discussed in this section provide convincing demonstration of several instances of cross-talk in tight epithelia. In a later section we will discuss the sensing and transduction systems involved in these responses.

B. CROSS-TALK IN LEAKY EPITHELIA

In this section we will focus primarily on the epithelia of small intestine and renal proximal tubule. Unlike tight epithelia, in which apical membrane Na^+ influx is via an amiloride-sensitive Na^+ channel, uptake of Na^+ in leaky epithelia in most cases is by coupled transport. The apical membrane entry step may be electroneutral (Na^+/H^+ and Cl^-/HCO_3^- exchanges, NaCl cotransport, $NaKCl_2$ cotransport), see Figure 2–3, or rheogenic (Na^+-sugar, Na^+-amino acid cotransport), as illustrated in Figure 2–4. As is the case with tight epithelia, the rate of apical membrane solute entry and basolateral membrane solute efflux are under rapid regulation in the normal physiologic state, and there is good evidence for cross-talk.

The addition of certain organic solutes, such as transported sugars and amino acids, to the luminal solution of small intestine or renal proximal tubule causes a rapid, sustained influx of organic solute coupled to Na^+ influx. The initial electrophysiological responses (apical membrane depolarization and fall in resistance) are consistent with activation of a rheogenic solute entry.[11] The initial depolarization of the apical membrane voltage and the fall in the ratio of apical to basolateral membrane resistance (R_a/R_b) are followed by a slower repolarization, concomitant with an increase in R_a/R_b. Addition of phloridzin (to block Na^+-sugar cotransport) raises R_a/R_b to a value greater than that observed prior to exposure to the sugar. Furthermore, addition of Ba^{2+} to the serosal bathing solution prevents

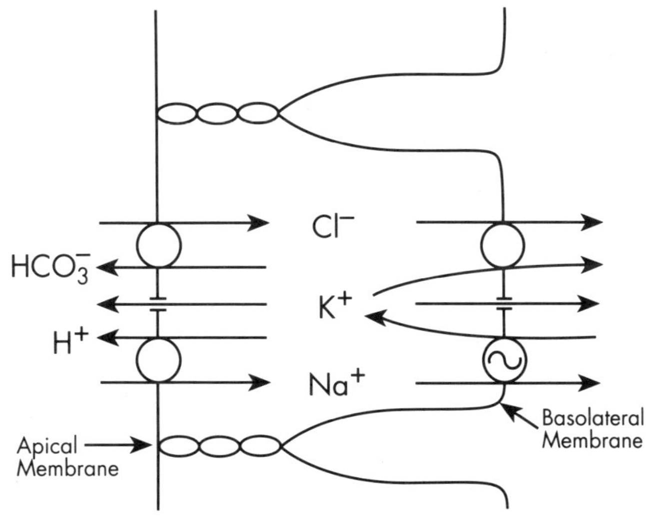

Figure 2–3 Essential transport mechanisms in a "leaky" electroneutral Na^+-absorbing epithelium. Entry of Na^+ and Cl^- across the apical cell membrane is coupled to efflux of H^+ and HCO_3^-, respectively. Cl^- exits across the basolateral cell membrane coupled to K^+ efflux. Na^+ extrusion is mediated by the Na^+,K^+-ATPase and K^+ recycles via a K^+ channel.

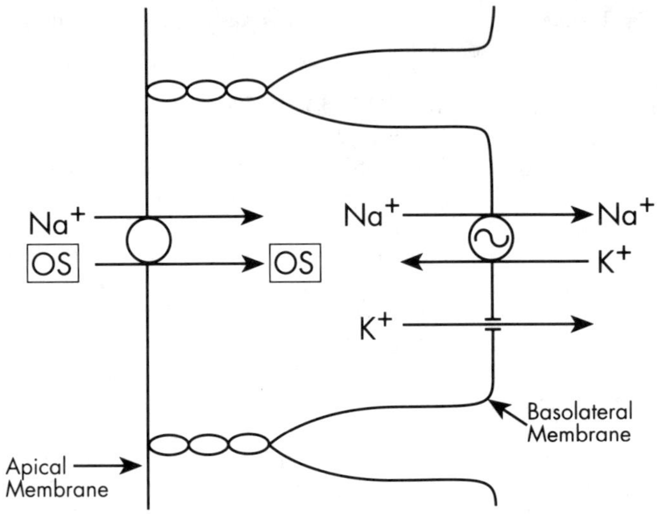

Figure 2–4 Essential transport mechanisms in a "leaky" rheogenic Na^+-absorbing epithelium. Na^+ influx is coupled to the movement of an organic solute (e.g., sugar or amino acid), denoted as OS. The transporters at the basolateral cell membrane are as shown in Figure 2–3.

the secondary rise in R_a/R_b. The explanation for the secondary changes in voltage and membrane-resistance ratio is an increase in basolateral G_K, presumably in response to the increased solute influx across the apical cell membrane.[9,11,27] An increase in apical membrane Na^+ and organic-solute influx, plus a likely gain of anion because of the membrane depolarization, would be expected to cause an increase in cell volume. Compensatory changes in basolateral membrane efflux pathways may serve to limit the cell swelling. Consistent with this interpretation, exposure of small intestine epithelium to a hyposmotic bathing solution increased the Ba^{2+}-sensitive basolateral membrane K^+ conductance.[28] The change in basolateral membrane permeability was similar to that observed when apical solute influx was acutely increased by exposure to galactose or alanine. Although the changes in basolateral membrane permeability can be prevented by Ba^{2+} addition to the serosal solution, it is not clear whether the rate of Na^+ and coupled solute influx is diminished beyond what would be expected from the fall in driving force (i.e., the Ba^{2+}-induced fall in apical membrane potential difference due to current flow across the low-resistance paracellular pathway). Recently, it was shown that isolated guinea pig intestinal epithelial cells swell and subsequently volume regulate upon exposure to alanine and glucose.[29] The RVD appeared to involve activation of K^+ and Cl^- conductive pathways. Finally, cultured hepatocytes respond to alanine with increases in Na^+ entry, cellular Na^+ content, Na^+,K^+-ATPase activity, and K^+ efflux.[30] Similar changes were observed upon addition of alanine to cells shrunken with extracellular sucrose. Thus, it was concluded that the stimulation of K^+ efflux was not a response to cell swelling *per se*.

The link between $[Na^+]_i$ and transepithelial Na^+ transport is clearer in leaky than in tight epithelia. Studies with intracellular Na^+-selective microelectrodes demonstrate that in several leaky epithelia maneuvers that produce a sustained increase in Na^+ absorption by 2- to 4-fold do not cause a significant steady-state increase in $[Na^+]_i$.[11] Thus, a kinetic effect on the Na^+ pump is needed to account for the elevation in transepithelial Na^+ transport. This effect could result from an increase in the number of operative pumps expressed in the membrane (insertion of new pumps or activation of quiescent ones) or from an increase in turnover rate of operative pumps. The compensatory mechanisms do not require a sustained increase in $[Na^+]_i$, but a transient rise could play a role.

The pumping rate depends on the $[Na^+]_i$, the extracellular $[K^+]$, the intracellular concentrations of ATP, ADP, and organic phosphate, the number of active pumps in the membrane, and several modulatory factors. The best-characterized ones are $[Na^+]_i$ and pH_i. It is likely that a relatively high intracellular $[Na^+]_i$ is necessary for surface expression of pumps synthesized upon stimulation with aldosterone.[31] Intracellular acidification inhibits the pump[32,33] near the physiologic pH_i range. Intracellular $[Ca^{2+}]$ levels are unlikely to play a role in rapid regulation of the pump, although increases in $[Ca^{2+}]_i$ stimulate transcription of both α- and β-subunits of the Na^+,K^+-ATPase.[34] Phosphorylation has been recognized recently as a pump modulator (see below). Changes in pump stoichiometry, claimed to occur during stimulation of transepithelial Na^+ transport (see Schultz and Hudson[11] for review), remain a possibility difficult to prove unambiguously.

In sum, the mechanism of stimulation of the Na^+,K^+ pump in response to elevated Na^+ entry is unknown. The number of pump sites in the membrane can be assessed by ouabain-specific binding,[35,36] but few relevant studies have been published. The number of operational pumps appears to be regulated, since in frog skin at low transport rates ouabain binding causes recruitment of other pumps.[37] In avian salt gland cells, methacholine stimulation caused an increase in the rate of ouabain binding (i.e., pump turnover), but no change in the number of pumps assessed by steady-state binding.[35] In lacrimal gland, most of the Na^+,K^+-ATPase activity is not in the surface membrane, but in an intracellular pool; stimulation of secretion with a cholinergic agonist increases the enzymatic activity in the basolateral membrane and decreases that in the intracellular pool.[38]

In the renal proximal tubule, a sizable fraction of solute entry is by Na^+-solute cotransport pathways. Several investigators[27,39] have shown that the basolateral membrane G_K changes as a function of the rate of transepithelial Na^+ absorption. Direct measurements of cell $[Ca^{2+}]$ demonstrated that the transport-induced increase in G_K was not coupled to an increase in cell calcium.[39] It was suggested, instead, that an "unidentified volume-sensitive mechanism is responsible for the regulation of G_K." Consistent with this conclusion, exposure of proximal tubules to a hyposmotic bathing solution causes cell swelling followed by partial RVD. In contrast, exposure to a hyperosmotic bathing solution causes cell shrinkage with no recovery of cell volume. The RVD induced by exposure to a hyposmotic bathing solution appears to result from activation of Cl^- and K^+ conductances and is absolutely dependent upon extracellular Ca^{2+}.[40] In perfused rabbit proximal convoluted tubule, cell swelling activates KCl efflux via K^+ and Cl^- channels in the basolateral cell membrane.[41] It was found that isosmotic changes in bathing solution $[K^+]$ and $[Cl^-]$ caused changes in cell volume, the rate of which depended upon whether the cells were shrunken or swollen with anisosmotic solutions. In mouse proximal tubule cells, RVD appears to depend on the presence of K^+ and HCO_3^-, but not Cl^-.[42] Recently, Beck and Potts[43] demonstrated that cell swelling in perfused rabbit proximal tubules caused by either exposure to hyposmotic solutions or by addition of glucose and alanine to the luminal perfusate resulted in a sustained increase in basolateral G_K. Later studies[44] revealed that cytosolic $[Ca^{2+}]$ increased upon hyposmotic swelling, but not in response to swelling induced by enhanced Na^+-organic solute influx. Interestingly, the cells exhibited RVD only with the first of these perturbations. The RVD responses elicited by hyposmotic solution or by increased solute entry appear also to be different in *Necturus* gallbladder epithelium[45] and in isolated guinea pig small intestine cells.[46] The signals involved in volume regulation/maintenance will be discussed below.

Another form of cross-talk well studied in leaky epithelia is the one resulting from pump inhibition. Inhibition of the Na^+,K^+-ATPase tends to cause an increase in $[Na^+]_i$, a decrease in cell K^+, depolarization, and accumulation of Cl^- in the cell. However, in proximal renal tubules the rate of swelling can be very slow, and depends on the presence of organic cotransported solutes in the lumen. In their absence, V_m[47,48] and cell volume[49] remain stable for hours. In the presence of organic cotransported solutes, K^+ loss reduces basolateral G_K, and the ensuing depolarization reduces Na^+ entry and impairs basolateral extrusion of HCO_3^-. This alkalinizes the cell, inhibiting Na^+/H^+ exchange and reducing Na^+ entry.[50] In *Necturus* gallbladder, the cells swell upon pump inhibition, but then NaCl entry undergoes feedback inhibition.[51]

C. CROSS-TALK IN EXOCRINE Cl^--SECRETORY ACINAR EPITHELIA

Cholinergic stimulation of isolated salivary gland acinar cells causes pronounced cell shrinkage.[1,52] A similar response has been observed in isolated sweat gland secretory coil cells.[4] Foskett and co-workers[1,53,54] used simultaneous differential interference contrast and fluorescence imaging to study the relationship between cell volume and intracellular $[Ca^{2+}]$, $[Cl^-]$, and $[Na^+]$. A strong inverse correlation was found between changes in cell volume and intracellular Ca^{2+} activity, i.e., as intracellular Ca^{2+} increased within the physiological range, cell volume decreased. The relationship was retained during spontaneous oscillations in intracellular Ca^{2+}. Cell volume and intracellular Cl^- activity were tightly coupled, consistent with the notion that cell shrinkage was a consequence of loss of Cl^- (and K^+).

These observations are consistent with the currently accepted model, based primarily on electrophysiological studies, for cholinergic stimulation of salt secretion by acinar cells (Figure 2–5). It is thought that cholinergic stimulation elevates intracellular $[Ca^{2+}]$, which either directly or indirectly increases the apical membrane Cl^- conductance (G_{Cl}) and the basolateral membrane G_K. It is uncertain at present whether there is any "cross-talk" between apical and basolateral membranes in these cells. It appears that cell shrinkage is limited by the availability of intracellular Cl^-. Presumably, for sustained Cl^- secretion to occur, basolateral Cl^- entry must be accelerated. The most likely mecha-

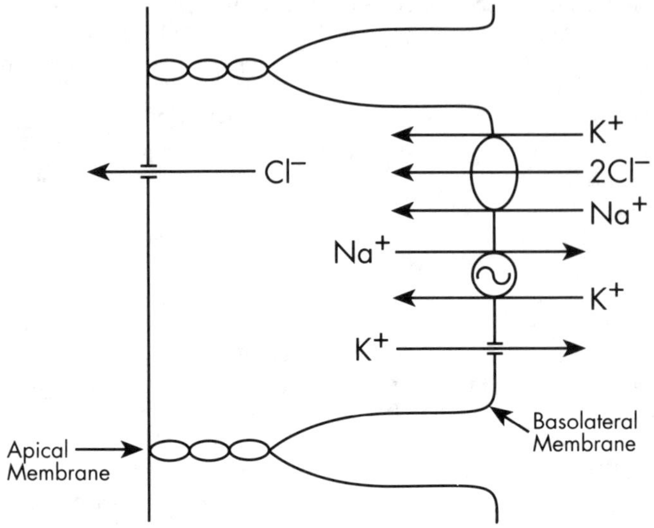

Figure 2–5 Transport model for an exocrine-gland Cl^--secreting epithelium. Intracellular $[Cl^-]$ is maintained above electrochemical equilibrium by electroneutral influx with Na^+ and K^+ across the basolateral cell membrane. Na^+ and K^+ are recycled across the same membrane by the Na^+,K^+-ATPase and a K^+ channel, respectively. Net transepithelial secretion is induced by activation of an apical cell membrane Cl^- channel.

nism for basolateral Cl^- entry is via $NaKCl_2$ cotransport. Activation of this transporter would result in net Na^+ influx and thereby in an increase in $[Na^+]_i$. An increase in $[Na^+]_i$ would in turn stimulate the Na^+,K^+-ATPase to extrude Na^+ and accumulate K^+. A mechanism for the reaccumulation of K^+ is essential in order to maintain E_K and hence the driving force (V_m) for Cl^- efflux. Direct tests of these proposals are, however, lacking. We are unaware of published studies of volume regulation of acinar cells following exposure to anisosmotic solutions. Foskett and co-workers[54a] have found that cell swelling induced by exposure of resting cells to hyposmotic solutions is followed by RVD that is essentially complete within 5 min. They also found that the volume of unstimulated cells was unchanged after a 10-min exposure to ouabain.

Exocrine Cl^- secretory acinar epithelial cells may be fundamentally different in their regulation of cell volume in response to agonist stimulation. Unlike the Na^+-absorbing epithelia described above, exocrine epithelia are, in general, at rest until stimulated by hormones or neurotransmitters. Their requirement for intermittent secretion, under extrinsic control, may obviate the need for elaborate mechanisms of homocellular regulation. Additional studies specifically designed to address this issue will be required before firm conclusions can be drawn.

IV. SIGNALLING MECHANISMS IN CROSS-TALK

The mechanisms of cross-talk in epithelial cells are not fully understood. However, studies in a variety of epithelia appear to implicate changes in cell volume as the initial signal that is sensed by the cell and transduced to physical or chemical events that ultimately alter one or more membrane transport processes. In this section we will first discuss this possibility and then consider other mechanisms of functional coupling between apical and basolateral membranes of transporting epithelial cells.

A. SENSING MECHANISMS DEPENDENT ON CELL VOLUME CHANGES

The first mechanistic question is in what way cell volume is sensed. The two major theories propose a *mechanical sensing mechanism* and a *chemical sensing mechanism,* respectively.

1. Mechanical Sensor

It is thought that a mechanical-strain sensor in the membrane or the cytoskeleton detects changes in cell volume and modulates the activity of the membrane transporters responsible for the regulatory response. Stretch-activated channels have been identified in a number of cell types. Clearly, their open probability (P_o) increases not only in response to membrane stretch brought about by changes in pipette hydrostatic pressure, but also by cell swelling.[55–60] We are not aware of observations of other changes in channel properties other than P_o (i.e., the channel conductance, selectivity, and pharmacological characteristics appear to be unchanged). The effect of membrane stretch on P_o is reversible.[59] There is no evidence supporting a causal relationship between membrane stretch and transport rate via either membrane carriers or membrane pumps. In fact, the observations of Jennings and

Schulz[61] in red blood cells indicate that changes in cell shape do not result in alterations in the sensor, transduction mechanism, or KCl efflux elicited by cell swelling. In skeletal muscle, the Na^+,K^+-ATPase is stimulated by cell swelling[62] but there is no evidence for or against a role of membrane stretch in this response.

From the point of view of their ionic selectivity, stretch-activated channels in animal cells can be classified in three groups, namely nonselective cation channels, K^+ channels, and Cl^- channels. Interestingly, no Na^+- or Ca^{2+}-selective stretch-activated channels have been identified.[59] Two possible roles for these channels in RVD are evident:

1. Direct activation of the channel(s) mediating the ion effluxes responsible for RVD: for this mechanism to operate, it is only necessary that one channel be activated if the other one (permeable to the counterion) is already expressed in the membrane. For example, in the case of a preexisting high Cl^- conductance, membrane activation of K^+-selective channels will cause a net K^+ efflux if the electrochemical potential difference is of the right orientation, but will also cause Cl^- efflux via the preexisting channels because of membrane hyperpolarization (V_m becomes closer to E_K), which increases the electrochemical gradient favoring electrodiffusive Cl^- loss. By a similar mechanism, activation of Cl^- channels in a K^+-conductive membrane will tend to cause KCl loss. Channel activation can alter the driving force for the counterion because of the change in membrane voltage, but can also change the activity of voltage-dependent channels.[3]
2. Activation of nonselective cation channels, which in turn activate K^+ (and perhaps Cl^-) channels: these channels conduct monovalent (K^+,Na^+) and divalent (Ca^{2+}) cations. Their role in cell volume regulation is well exemplified by work on the choroid plexus of *Necturus maculosus*. Using this preparation, Christensen[63] demonstrated that membrane stretch activates a nonselective cation channel, and postulated that the resulting Ca^{2+} influx activates maxi-K^+ channels, thus contributing to the loss of salt needed for RVD.

Less direct, but compelling, indications for a role of cytosolic Ca^{2+} in RVD have been obtained in a variety of tight and leaky epithelia.[40,64,65]

2. Chemical Sensor

It is clear that RVD takes place upon very small changes in cell volume. For instance, a 6% decrease in osmolality can elicit a large increase in G_K.[28] Further, if the bathing solution osmolality is changed very slowly, volume regulation can occur at virtually the same rate and hence there will be no measurable change in cell volume.[66] From these observations, it was suggested that dilution of an intracellular solute could not be the signal for RVD, because such a signal is too small.[11] However, the alternative possibility is that the signal is amplified, hence causing a fractional change in solute transport much greater than the fractional change in cell volume and in concentration of the putative signal solute.[67]

The central tenet of this theory is that the signal denoting changes in cell volume is the concentration of one or more cytoplasmic molecules. This process can be specific (e.g., change in the concentration of a second messenger) or nonspecific (e.g., change in the concentration of macromolecules). In any event, the change in concentration is amplified and results in activation or inactivation of one or more membrane transporters responsible for the compensatory response. One possibility is that the initial signal is amplified by an enzymatic cascade whose end result is the chemical modification of the transporter(s), for example by phosphorylation or dephosphorylation.[67–69]

a. Specific chemical sensor hypothesis

Jennings and Schulz[67] have discussed the possibility[70] that the intracellular [Mg^{2+}] may be the signal responsible for the initiation of cell volume regulatory responses. Although this is an interesting possibility, demonstration that changes in cell volume alter intracellular free Mg^{2+} levels is lacking and the role of Mg^{2+} in cell volume regulation is currently controversial.[71] A crucial missing experiment is to assess whether "clamping" $[Mg^{2+}]_i$ prevents cell volume regulation.

b. Macromolecular crowding hypothesis

An alternative view is that the mediator is the protein concentration in the cytoplasm.[72,73] The two main points of these studies were first, that changes in the activity of transporters involved in cell volume regulation in dog red blood cell ghosts did not depend on the cell water volume, but on the intracellular protein concentration; and second, that the effect of protein concentration on transporter activity was nonspecific, since it could be obtained with albumin, hemoglobin, or a mixture of both. From these results, Colclasure and Parker[72,73] proposed that the intracellular signal initiating cell volume regulation is the concentration of macromolecules, acting on transporter activity via the process known as "macromolecular crowding", namely the change in reactivity (or activity coefficient) of macromolecules produced by changes in the molar fraction of solvent. In other words, the space occupancy by large solute particles can exert large effects (one to three orders of magnitude) on the activity coefficients of other large solute molecules ("probe" molecules). This interesting hypothesis has been discussed in detail by Minton et al.[74] and Minton.[75] The activation or inactivation of membrane transporters would result from changes in the reactivity of one or more soluble components in the cytoplasm, perhaps enzymes determining the level of phosphorylation of regulatory proteins or of the transporters themselves. The amplification of the signal is provided by the steep relationship between macromolecule concentration and probe activity coefficient.[75] The experimental data presented so far do not prove, however, that the effect is exclusively via a nonspecific role of macromolecules, since all experiments were carried out with proteins.

B. SENSING MECHANISMS INDEPENDENT OF CELL VOLUME CHANGES

Several mechanisms other than changes in cell volume can, in principle, be part of the sensing or the transduction mechanism in cross-talk. Here we will consider possible roles of intracellular $[Na^+]$ and [ATP].

Changes in $[Na^+]_i$ alter apical membrane P_{Na} in tight epithelia in a direction consistent with negative feedback, i.e., when $[Na^+]_i$ rises apical membrane P_{Na} falls.[76] This effect accounts for cross-talk from basolateral to apical membrane under some circumstances. The most likely explanation is that the change in P_{Na} is mediated by changes in pH_i or $[Ca^{2+}]_i$. When the Na^+ pump is inhibited, $[Na^+]_i$ rises; via alterations in the rate of Na^+/H^+ and/or Na^+/Ca^{2+} exchange, the intracellular $[H^+]$ and $[Ca^{2+}]$ levels are also elevated, reducing the open probability of apical Na^+ channels.[77]

The demonstration of Na^+-sensitive K^+ channels in neurons[78] suggests the possibility that changes in $[Na^+]_i$ may modulate the activity of basolateral membrane K^+ channels. The direction of the effects is such that increases in $[Na^+]_i$ would reduce P_K, i.e., this process could account for the pump-leak parallelism observed in tight and leaky epithelia upon pump inhibition,[11] but would be in the wrong direction to account for the increase in basolateral membrane P_K elicited by an elevation of Na^+ entry across the apical membrane. We are not aware of reports supporting the existence of Na^+-sensitive K^+ channels in epithelia.

Observations in pancreatic beta cells, heart muscle, neurons, and other cells indicate that changes in $[ATP]_i$ regulate K^+ channel P_o by a purinergic-receptor mechanism, with no ATP hydrolysis.[79] Hence, it was speculated that this mechanism could play a role in coupling the rate of the pump to the basolateral membrane K^+ conductance in salt-transporting epithelia.[11] This possibility has been elegantly confirmed by Tsuchiya et al.[80] in studies in rabbit proximal tubule. These investigators demonstrated, first, the existence of a basolateral K^+ channel sensitive to $[ATP]_i$, and showed that this channel is the main determinant of the basolateral membrane G_K. Second, they found that $[ATP]_i$ does in fact decrease when transepithelial Na^+ transport (and pump activity) are stimulated. Third, they showed that intracellular ATP loading uncoupled the above response. Inasmuch as ATP-sensitive K^+ channels have been found in a number of epithelia, this mechanism may account for the parallelism between pump rate and K^+ leak.[80] ATP-sensitive K^+ channels are also expressed in the apical membranes of Na^+-absorbing epithelia.[81] Hence, $[ATP]_i$ might also couple basolateral pumping rate to apical K^+ secretion.

C. TRANSDUCTION MECHANISMS IN CROSS-TALK

Our knowledge of intracellular signaling mechanisms has expanded considerably in recent years. Several pathways have been proposed to play some role in cross-talk as such, stimulation of transport leading to changes in cell volume, or cell volume regulation in transporting epithelial cells.[82] Here we will discuss the roles of intracellular Ca^{2+}, intracellular pH, protein phosphorylation, and membrane voltage.

1. Intracellular Ca^{2+}

The two best-studied roles of cytosolic $[Ca^{2+}]$ in cross-talk are in the activation of ion channels upon cell swelling, which accounts for, or contributes to, RVD, and in the control of the apical membrane P_{Na} in tight epithelia, which could explain the inhibition of Na^+ entry upon inhibition of the Na^+,K^+-ATPase.

a. Role of Ca^{2+} in RVD

In many epithelial cells $[Ca^{2+}]_i$ rises upon osmotic swelling (toad urinary bladder,[83] human intestinal cells in culture,[84] rabbit proximal tubule,[85,86] and cultured medullary thick ascending loop of Henle cells[87]). In addition, the presence of extracellular Ca^{2+} is necessary for RVD upon hyposmotic cell swelling (*Necturus* gallbladder,[88] frog urinary bladder,[89] human intestinal cells in culture,[84] mammalian proximal renal tubule,[91] cultured medullary thick ascending loop of Henle cells[87]). Taken together, these observations suggest that cell swelling causes Ca^{2+} entry, via plasma membrane channels, and that the increase in cytosolic [Ca^{2+}], perhaps by a mechanism involving calmodulin,[88] activates volume-regulatory mechanisms. In choroid plexus epithelium, as explained above, cell swelling does activate nonselective cation channels which allow for Ca^{2+} entry.[63] A similar mechanism operates in opossum kidney cells.[55,90,91] Initial membrane depolarization, due to the Na^+ and Ca^{2+} conductances of the nonselective cation channels, is a fingerprint for their activation.[91] In contrast, activation of stretch-sensitive K^+ channels displaces the membrane voltage toward the K^+ equilibrium potential, and hence causes cell membrane hyperpolarization.[43] The properties of the Ca^{2+} channels involved in RVD have not yet been studied in detail, although in proximal straight tubule it appears that the pharmacological properties of swelling-activated and native Ca^{2+} channels differ.[92]

Release of Ca^{2+} from intracellular stores appears to at least contribute to RVD in nonpolar cells and a possible role for the phosphatidyl inositol system in the elevation of cytosolic Ca^{2+} has been proposed for such cells.[93] Changes in inositol phosphate turnover have also been observed in renal proximal tubule cells exposed to hyposmotic solutions,[40,86,94] but Ca^{2+} release from intracellular stores is not sufficient to sustain RVD.[86]

Eventually, the elevated cytosolic [Ca^{2+}] causes activation of plasma membrane K^+ and/or Cl^- channels.[95] This results in loss of KCl provided that both conductive pathways coexist and that the electrochemical driving forces promote efflux of both ions (see section of voltage coupling, below). The activation of voltage-dependent maxi-K^+ channels by Ca^{2+} appears to be direct,[96] but the activation of other Ca^{2+}-sensitive K^+ channels or of Cl^- channels could be indirect. In addition, in some epithelia it is clear that the main anion efflux accompanying net K^+ loss is HCO_3^- and not Cl^- (reviewed by Montrose-Rafizadeh and Guggino[97]).

As discussed above, it is unclear whether changes in cytosolic [Ca^{2+}] mediate the activation of basolateral K^+ conductance elicited by luminal addition of glucose and alanine in rabbit proximal tubule,[39,44] or of glucose in *Necturus* gallbladder.[45] It is possible that the mechanisms of activation of basolateral G_K upon cell swelling depend on how the cells are swollen, i.e., by primary water entry or by water entry secondary to an increase in solute content. Therefore, the sensing mechanism and/or the transduction mechanism may differ in these two experimental conditions.

b. Role of Ca^{2+} in the control of apical membrane P_{Na}

The demonstration of Na^+/Ca^{2+} exchange in basolateral membranes of epithelial cells[76] suggests a Ca^{2+}-mediated mechanism for cross-talk between the two cell membranes. If the basolateral Na^+ efflux is reduced, e.g., by inhibition of the Na^+,K^+-pump, then the $[Na^+]_i$ will rise, and this will reduce or even reverse Ca^{2+} transport via the exchanger.[76] The elevation in intracellular [Ca^{2+}] in turn inhibits Na^+ entry by reducing apical membrane P_{Na} and the cross-talk mechanism is completed. Direct studies addressing this issue became possible with the application of the patch-clamp technique to epithelial cells. The results support the notion that increases in both intracellular [H^+] and [Ca^{2+}] inhibit amiloride-sensitive Na^+ channels by reducing P_o, but the effect of Ca^{2+} appears to be indirect, i.e., it was not present in excised membrane patches.[77]

2. Intracellular pH

Transepithelial transport of H^+ equivalents, as well as intracellular pH regulation, are Na^+-dependent in several epithelia. Hence, pH_i can undergo changes in response to alterations in $[Na^+]_i$ and in the rate of transepithelial Na^+ transport.

a. Effects of transepithelial transport rate on pH_i

Two mechanisms can be envisioned for this relationship if one considers the role of Na^+/H^+ exchangers as: (1) pathways for apical Na^+ entry and transepithelial Na^+ transport (Figure 2–3), and (2) transporters involved in pH_i regulation ("housekeeping" Na^+/H^+ exchangers in basolateral membranes). In the first instance, changes in the Na^+ transport rate are directly coupled to changes in the rate of H^+ transport in the opposite direction and the changes in H^+ flux alter pH_i. The pH_i change, in turn, can have effects on other transporters, in the same or in the opposite membrane. In the second case, the change in pH_i is secondary to an alteration in $[Na^+]_i$ brought about by a primary change in either entry or exit. An elevation of $[Na^+]_i$ tends to cause intracellular acidification (because the driving force for H^+ extrusion is reduced), whereas a decrease in $[Na^+]_i$ tends to cause cell alkalinization because the driving force for exchange is increased.

b. Effects of pH_i on membrane transporters

Changes in pH_i can alter transport by two mechanisms, namely by altering the driving forces, or by exerting allosteric effects on the transporters. When the effects are on the same membrane, the result is intramembrane coordination of transport. A good example is the adjustment of the rates of apical membrane Na^+/H^+ and Cl^-/HCO_3^- exchange in *Necturus* gallbladder epithelium.[98] The two exchangers operate independently and therefore a change in the flux via one of the exchangers has no immediate effect on the flux via the other one. However, changes in either Na^+/H^+ or Cl^-/HCO_3^- exchange alter pH_i, and this in turn will exert an effect on the other exchanger, by a change in driving force, an allosteric mechanism, or both. The end result is an adjustment of the rates of Na^+ and Cl^- transport across the apical membrane.

When the effect of pH_i is exerted on the opposite membrane, the result is a form of cross-talk. Changes in pH_i have been shown to modulate all three kinds of transporters, namely channels, carriers, and pumps. Some of these effects are large, occur in the physiologic range of pH_i, and appear to play major roles in cross-talk in epithelia. Three examples follow:

1. The amiloride-sensitive Na^+ channels present in the apical membrane of tight epithelia are highly pH_i-sensitive. Intracellular acidification decreases their P_o and thus inhibits the apical membrane P_{Na}.[77]
2. Many K^+ channels are also pH_i-sensitive. For instance, intracellular acidification inhibits Ca^{2+}-sensitive maxi-K^+ channels in renal and in other epithelial cells (e.g., Oberleithner et al.[99]). Hence, cytoplasmic acidification in response to a reduction in Na^+ transport will reduce electrodiffusive K^+ transport, and alkalinization will have the opposite effect. Interestingly, stimulation of Na^+ transport with phenylalanine in frog proximal tubule causes cell alkalinization.[100] Further, in rabbit proximal tubule cell swelling brought about by exposure to a hyposmotic solution in the presence of bicarbonate elevates pH_i and increases basolateral membrane G_K.[101]
3. The Na^+ pump is also highly pH_i-sensitive, both in epithelial[32] and in nonepithelial cells.[33] These effects of pH_i could contribute to some forms of cross-talk in epithelial cells, namely the coupling of Na^+ entry to pump rate (via the modulatory effect of pH_i on the pump), the parallelism between Na^+ entry and K^+ recycling across the basolateral membrane (via the pH_i-dependence of basolateral membrane K^+ channels), and the reduction in apical conductive Na^+ entry effected by pump inhibition.

The role of pH_i in modulating transport in epithelial cells can be complicated, because in some cases pH_i changes will not promote cross-talk between the membranes. For instance, in a leaky epithelium transporting Na^+ via a Na^+/H^+ exchanger in series with a Na^+,K^+-ATPase (Figure 2–3), a primary intracellular acidification would activate the Na^+/H^+ exchanger by the two mechanisms explained above, and inhibit the Na^+,K^+-pump; the overall effect could be a gain in cell Na^+, which, if accompanied by net anion entry, would cause cell swelling.

3. Protein Phosphorylation

Jennings and Schulz[67] have shown that a phosphorylation-dephosphorylation cycle is an integral part of RVD in rabbit red blood cells; most likely, cell swelling inhibits a protein kinase (neither A nor C), causing phosphatase-dependent dephosphorylation of a key substrate and eventually activation of the KCl cotransporter. Regulatory volume increase (RVI) in avian red blood cells, which occurs by activation of $NaKCl_2$ cotransport, appears to be due to phosphorylation of the transporter or an activating protein.[102] These observations suggest the possibility of a single mechanism of control of cell volume, mediated by the effects of protein kinases and phosphatases on ion transporters or regulators.[69] In addition to its role in regulating carrier-mediated transport, phosphorylation has been shown to modulate transport via ion channels[103] and pumps.[104]

4. Membrane Voltage

Changes in membrane voltage can couple the function of membrane transporters by two mechanisms, namely the change in driving force for rheogenic (i.e., current-carrying) transport processes and the change in activity of voltage-sensitive transporters. The second mechanism appears to be significant only in the case of ion channels. In this section we will discuss two examples illustrating the role of changes in membrane voltage in the coordination of ion transport in epithelia.

In *Necturus* gallbladder epithelium, elevation of intracellular cAMP levels activates an apical membrane Cl^- conductance not present under control conditions. This depolarizes the membrane by about 20 mV, which increases the driving force for K^+ exit and also increases 5-fold the P_o of the maxi-K^+ channels. The end result is conductive loss of K^+ and Cl^- and cell shrinkage.[3]

Voltage-mediated coordination of the function of two channels could be a form of cross-talk in leaky epithelia, since the two cell membranes are electrically connected via a low-resistance paracellular pathway. Hence, activation of one channel, in one membrane, can change the voltage of both membranes.

In several epithelia, the basolateral membrane expresses so-called inwardly rectifying K^+ channels.[11,105] In contrast to the maxi-K^+ channel referred to above, the open probability of this channel decreases with depolarization. This channel is thought to participate in the coupling of basolateral membrane pump rate and K^+ conductance.[105] Upon stimulation of electrogenic cotransport of Na^+ and organic solutes across the apical membrane, the channel closes because both the apical and the basolateral membrane depolarize, the latter due to current flow via the paracellular pathway. This increases the electrical resistance of the basolateral membrane. Shortly thereafter, activation of the pump results in current flow, which in turn hyperpolarizes the membrane, increasing the P_o of the inwardly rectifying channel, and hence increasing basolateral membrane P_K.[105]

In leaky epithelia, the increase in basolateral G_K upon entry of Na^+ and cotransported organic solutes across the apical membrane is very important in maintaining the driving force for continued transcellular solute transport. The hyperpolarization of the basolateral membrane hyperpolarizes the apical membrane as well, because of intraepithelial current flow via the paracellular pathway. The end result is the preservation of an electrical driving force favoring the entry of Na^+ and cotransported solute.[11]

V. CONCLUSIONS

The notion of cross-talk in epithelia[9,10] has stimulated much experimental work designed to identify and characterize these processes, and to unravel the mechanisms involved. Recent efforts have addressed mechanistic questions. There are many forms of cross-talk and other modes of coordinated operation of transporters in epithelial cells. It seems also clear that there is no single, unifying mechanism accounting for all of these. The information available suggests major roles of cell volume and cytoplasmic ion concentrations as primary signals.

The precise mechanism by which changes in cell volume alter membrane transport processes is beginning to be understood. A reasonable hypothesis is a dual control system involving a physical sensing mechanism, i.e., membrane/cytoskeleton tension and a chemical sensing mechanism, either specific or nonspecific (e.g., macromolecular crowding). Cytoplasmic $[Ca^{2+}]$ and protein phosphorylation certainly play important roles in the transduction stages.

In epithelial cells, it is likely that mechanisms other than changes in cell volume could account for some modes of cross-talk and intramembrane coordination of ion transport. Intracellular pH, intracellular [ATP], membrane voltage, and other factors can alter the rate of transport via plasma membrane channels, carriers, and pumps. As suggested before,[11] the signals and the transducing events could be multiple and interactive.

Recent results summarized above suggest that simple experimental perturbations such as cell swelling can have different effects, depending on the mechanism of the volume change. This observation underscores the need for detailed cell physiology studies of epithelial cross-talk. New experimental systems will have to be examined, in which, ideally, all relevant experimental parameters can be quantified under the same experimental condition. Implicit or explicit simplifying assumptions, e.g., that the consequences of cell swelling are independent of the swelling mechanism, may be incorrect, at least in some cases. Finally, progress in the molecular biology and biophysics of ion channels and other transporters underscores the insufficiency of assessments of "macroscopic" membrane properties. In other words, single-channel studies with detailed fingerprinting must replace conductance measurements.

ACKNOWLEDGMENTS

We thank Drs. E. Bello-Reuss, J. Copello, M. L. Jennings, and S. A. Lewis for comments on the manuscript. This work was supported in part by National Institutes of Health Grants DK-38734 (L.R.) and DK-43000 (C.U.C.) and by Grant Z-435 from the Cystic Fibrosis Foundation (C.U.C.)

REFERENCES

1. **Foskett, J. K. and Melvin, J. E.,** Activation of salivary secretion: coupling of cell volume and $[Ca^{2+}]_i$ in single cells, *Science,* 244, 1582, 1989.
2. **Cotton, C. U. and Reuss, L.,** Electrophysiological effects of extracellular ATP on *Necturus* gallbladder epithelium, *J. Gen. Physiol.,* 97, 949, 1991.
3. **Cotton, C. U. and Reuss, L.,** Effects of changes in mucosal solution Cl^- or K^+ concentration on cell water volume of *Necturus* gallbladder epithelium, *J. Gen. Physiol.,* 97, 667, 1991.
4. **Takemura, T., Sato, F., Saga, K., Suzuki, Y., and Sato, K.,** Intracellular ion concentrations and cell volume during cholinergic stimulation of eccrine secretory coil cells, *J. Membr. Biol.,* 119, 211, 1991.
5. **Ballanyi, K. and Grafe, P.,** Cell volume regulation in the nervous system, *Renal Physiol. Biochem.,* 11, 142, 1988.

6. **McManus, M. L. and Churchwell, K. B.,** Chapter 4, this volume.
7. **Reuss, L.,** Cell volume regulation in nonrenal epithelia, *Renal Physiol. Biochem.,* 11, 187, 1988.
8. **Spring, K. R.,** Mechanism of fluid transport by epithelia, in *Handbook of Physiology, The Gastrointestinal System. IV. Intestinal Absorption and Secretion,* 1991, 195.
9. **Schultz, S. G.,** Homocellular regulatory mechanisms in sodium-transporting epithelia: avoidance of extinction by "flush-through", *Am. J. Physiol.,* 241, F579, 1981.
10. **Diamond, J. M.,** Transcellular cross-talk between epithelial cell membranes, *Nature (London),* 300, 683, 1982.
11. **Schultz, S. G. and Hudson, R. L.,** Biology of sodium-absorbing epithelial cells: dawning of a new era, in *Handbook of Physiology, The Gastrointestinal System. IV. Intestinal Absorption and Secretion,* 1991, 45.
12. **Koefoed-Johnsen, V. and Ussing, H. H.,** The nature of the frog skin potential, *Acta Physiol. Scand.,* 42, 298, 1958.
13. **MacRobbie, E. A. C. and Ussing, H. H.,** Osmotic behavior of the epithelial cells of frog skin, *Acta Physiol. Scand.,* 53, 348, 1961.
14. **Finn, A. L. and Reuss, L.,** Effects of changes in the composition of the serosal solution on the electrical properties of the toad urinary bladder epithelium, *J. Physiol. London,* 250, 541, 1975.
15. **Davis, C. W. and Finn, A. L.,** Sodium transport effects on the basolateral membrane in toad urinary bladder, *J. Gen. Physiol.,* 80, 733, 1982.
16. **Lewis, S. A., Eaton, D. C., and Diamond, J. M.,** The mechanism of Na^+ transport by rabbit urinary bladder, *J. Membr. Biol.,* 28, 41, 1976.
17. **Chase, H. S., Jr. and Al-Awqati, Q.,** Regulation of the sodium permeability of the luminal border of toad bladder by intracellular sodium and calcium, *J. Gen. Physiol.,* 77, 693, 1981.
18. **Strange, K.,** Ouabain-induced cell swelling in rabbit cortical collecting tubule: NaCl transport by principal cells, *J. Membr. Biol.,* 107, 249, 1989.
18a. **Strange, K.,** unpublished observations.
19. **Strange, K.,** Volume regulation following Na^+ pump inhibition in CCT principal cells: apical K^+ loss, *Am. J. Physiol.,* 258, F732, 1990.
20. **Strange, K.,** Volume regulatory Cl^- loss after Na^+ pump inhibition in CCT principal cells, *Am. J. Physiol.,* 260, F225, 1991.
21. **Schultz, S. G., Frizzell, R. A., and Nellans, H. N.,** Active sodium transport and the electrophysiology of rabbit colon, *J. Membr. Biol.,* 33, 351, 1977.
22. **Turnheim, K., Frizzell, R. A., and Schultz, S. G.,** Interaction between cell sodium and the amiloride-sensitive sodium entry step in rabbit colon, *J. Membr. Biol.,* 39, 233, 1978.
23. **Leblanc, G. and Morel, F.,** Na and K movements across the membranes of frog skin epithelia associated with transient current changes, *Pfluegers Arch. Eur. J. Physiol.,* 358, 159, 1975.
24. **Thomas, S. R., Suzuki, Y., Thompson, S. M., and Schultz, S. G.,** The electrophysiology of *Necturus* urinary bladder. I. "Instantaneous" current-voltage relations in the presence of varying mucosal sodium concentrations, *J. Membr. Biol.,* 73, 157, 1983.
25. **Davis, C. W. and Finn, A. L.,** Sodium transport inhibition by amiloride reduces basolateral membrane potassium conductance in tight epithelia, *Science Wash. DC,* 216, 525, 1982.
26. **Demarest, J. R. and Finn, A. L.,** Interaction between the basolateral K^+ and apical Na^+ conductances in *Necturus* urinary bladder, *J. Gen. Physiol.,* 89, 563, 1987.
27. **Lang, F., Messner, G., and Rehwald, W.,** Electrophysiology of sodium-coupled transport in proximal renal tubules, *Am. J. Physiol.,* 250, F953, 1986.
28. **Lau, K. R., Hudson, R. L., and Schultz, S. G.,** Cell swelling induces a barium-inhibitable potassium conductance in the basolateral membrane of *Necturus* small intestine, *Proc. Natl. Acad. Sci. U.S.A.,* 81, 3591, 1984.
29. **MacLeod, R. J. and Hamilton, J. R.,** Separate K^+ and Cl^- transport pathways are activated for regulatory volume decrease in jejunal villus cells, *Am. J. Physiol.,* 260, G405, 1991.
30. **Cohen, B. J. and Lechene, C.,** Alanine stimulation of passive potassium efflux in hepatocytes is independent of Na^+-K^+ pump activity, *Am. J. Physiol.,* 258, C24, 1990.
31. **O'Neil, R. G. and Hayhurst, R. A.,** Sodium-dependent modulation of the renal Na-K-ATPase: influence of mineralocorticoids on the cortical collecting duct, *J. Membr. Biol.,* 85, 169, 1985.
32. **Eaton, D. C., Hamilton, K. L., and Johnson, K. E.,** Intracellular acidosis blocks the basolateral Na-K pump in rabbit urinary bladder, *Am. J. Physiol.,* 247, F946, 1984.
33. **Breitwieser, G. E., Altamirano, A. A., and Russell, J. M.,** Effects of pH changes on sodium pump fluxes in squid giant axon, *Am. J. Physiol.,* 253, C547, 1987.

34. **Rayson, B. M.,** $[Ca^{2+}]_i$ regulates transcription rate of the Na^+K^+-ATPase α_1 subunit, *J. Biol. Chem.,* 266, 21335, 1991.
35. **Hootman, S. R. and Ernst, S. A.,** Effect of methacholine on Na^+ pump activity and ion content of dispersed avian salt gland cells, *Am. J. Physiol.,* 241, R77, 1981.
36. **Widdicombe, J. H., Basbaum, C. B., and Highland, E.,** Sodium-pump density of cells from dog tracheal mucosa, *Am. J. Physiol.,* 248, C389, 1985.
37. **Cala, P. M., Cogswell, N., and Mandel, L. J.,** Binding of [^{3}H]-ouabain to split frog skin: the role of the Na,K-ATPase in the generation of the short circuit current, *J. Gen. Physiol.,* 71, 347, 1978.
38. **Yiu, S. C., Lambert, R. W., Bradley, M. E., Ingham, C. E., Hales, K. L., Wood, R. L., and Mircheff, A. K.,** Simulation-associated redistribution of Na,K-ATPase in rat lacrimal gland, *J. Membr. Biol.,* 102, 185, 1988.
39. **Lapointe, J., Garneau, L., Bell, P. D., and Cardinal, J.,** Membrane crosstalk in the mammalian proximal tubule during alterations in transepithelial sodium transport, *Am. J. Physiol.,* 258, F339, 1990.
40. **McCarty, N. A. and O'Neil, R. G.,** Calcium signalling in cell volume regulation, *Physiol. Rev.,* 72, 1037, 1992.
41. **Schild, L., Aronson, P. S., and Giebisch, G.,** Basolateral transport pathways for K^+ and Cl^- in rabbit proximal tubule: effects on cell volume, *Am. J. Physiol.,* 260, F101, 1991.
42. **Völkl, H. and Lang, F.,** Ionic requirement for regulatory cell volume decrease in renal straight proximal tubules, *Pfluegers Arch. Eur. J. Physiol.,* 412, 1, 1988.
43. **Beck, J. S. and Potts, D. J.,** Cell swelling, co-transport activation and potassium conductance in isolated perfused rabbit kidney proximal tubules, *J. Physiol. London,* 425, 369, 1990.
44. **Beck, J. S., Breton, S., Laprade, R., and Giebisch, G.,** Volume regulation and intracellular calcium in the rabbit proximal convoluted tubule, *Am. J. Physiol.,* 260, F861, 1991.
45. **Furlong, T. J. and Spring, K. R.,** Mechanisms underlying volume regulatory decrease by *Necturus* gallbladder epithelium, *Am. J. Physiol.,* 258, C1016, 1990.
46. **MacLeod, R. J., Lembessis, P., and Hamilton, J. R.,** Differences in Ca^{2+}-mediation of hypotonic and Na^+-nutrient regulatory volume decrease in suspensions of jejunal enterocytes, *J. Membr. Biol.,* 130, 23, 1992.
47. **Lang, F., Messner, G., Wang, W., Paulmichl, M., Oberleithner, H., and Deetjen, P.,** The influence of intracellular sodium activity on the transport of glucose in proximal tubule of frog kidney, *Pfluegers Arch. Eur. J. Physiol.,* 401, 14, 1984.
48. **Völkl, H., Geibel, J., Greger, R., and Lang, L.,** Effects of ouabain and temperature on cell membrane potentials in isolated perfused straight proximal tubules of the mouse kidney, *Pfluegers Arch. Eur. J. Physiol.,* 407, 252, 1986.
49. **MacKnight, A. D. C. and Leaf, A.,** Regulation of cellular volume, *Physiol. Rev.,* 57, 510, 1977.
50. **Lang, F., Friedrich, F., Paulmichl, M., Schobersberger, W., Jungwirth, A., Ritter, M., Steidl, M., Weiss, H., Wöll, E., Tschernko, E., Paulmichl, R., and Hallbrucker, C.,** Ion channels in Madin-Darby canine kidney cells, *Renal Physiol. Biochem.,* 13, 82, 1990.
51. **Koch-Jensen, P., Fisher, R. S., and Spring, K. R.,** Feedback inhibition of NaCl entry in *Necturus* gallbladder epithelial cells, *J. Membr. Biol.,* 82, 95, 1984.
52. **Nakahari, T., Murakami, M., Saski, Y., Kataoka, T., Imai, Y., Shiba, Y., and Kanno, Y.,** Dose effects of acetylcholine on the cell volume of rat mandibular salivary acini, *Jpn. J. Physiol.,* 41, 153, 1991.
53. **Foskett, J. K.,** $[Ca^{2+}]_i$ modulation of Cl^- content controls cell volume in single salivary acinar cells during fluid secretion, *Am. J. Physiol.,* 259, C998, 1990.
54. **Wong, M. M. Y. and Foskett, J. K.,** Oscillations of cytosolic sodium during calcium oscillations in exocrine acinar cells, *Science Wash. DC,* 254, 1014, 1991.
54a. **Foskett, J. K.,** personal communication.
55. **Ubl, J., Murer, H., and Kolb, H.-A.,** Ion channels activated by osmotic and mechanical stress in membranes of opossum kidney cells, *J. Membr. Biol.,* 104, 223, 1988.
56. **Sackin, H.,** A stretch-activated K^+ channel sensitive to cell volume, *Proc. Natl. Acad. Sci. U.S.A.,* 86, 1731, 1989.
57. **Filipovic, D. and Sackin, H.,** Stretch- and volume-activated channels in isolated proximal tubule cells, *Am. J. Physiol.,* 262, F857, 1992.
58. **Lewis, S. A. and Donaldson, P.,** Ion channels and cell volume regulation: chaos in an organized system, *News Physiol. Sci.,* 5, 112, 1990.
59. **Sachs, F.,** Stretch-sensitive ion channels. An update, in *Sensory Transduction,* 45th Annu. Symp. of the Society of General Physiologists, Rockefeller Press, New York, 1992, 241.

60. **Sackin, H.**, Chapter 13, this volume.
61. **Jennings, M. L. and Schulz, R. K.**, Swelling-activated KCl cotransport in rabbit red cells: flux is determined mainly by cell volume rather than shape, *Am. J. Physiol.*, 259, C960, 1990.
62. **Venosa, R. A.**, Hypo-osmotic stimulation of active Na^+ transport in frog muscle: apparent upregulation of Na^+ pumps, *J. Membr. Biol.*, 120, 97, 1991.
63. **Christensen, O.**, Mediation of cell volume regulation by Ca^{2+} influx through stretch-activated channels, *Nature (London)*, 330, 66, 1987.
64. **Pierce, S. K. and Politis, A. D.**, Ca^{2+}-activated cell volume recovery mechanisms, *Annu. Rev. Physiol.*, 52, 27, 1990.
65. **Chamberlin, M. E. and Strange, K.**, Anisosmotic cell volume regulation: a comparative review, *Am. J. Physiol.*, 257, C159, 1989.
66. **Lohr, J. W. and Grantham, J. J.**, Isovolumetric regulation of isolated S_2 proximal tubules in anisotonic media, *J. Clin. Invest.*, 78, 1165, 1986.
67. **Jennings, M. L. and Schulz, R. K.**, Okadaic acid inhibition of KCl cotransport. Evidence that protein dephosphorylation is necessary for activation of transport by either cell swelling or N-ethylmaleimide, *J. Gen. Physiol.*, 97, 799, 1991.
68. **Parker, J. C., Colclasure, G. C., and McManus, T. J.**, Coordinated regulation of shrinkage-induced Na/H exchange and swelling-induced [K-Cl] cotransport in dog red cells. Further evidence from activation kinetics and phosphatase inhibition, *J. Gen. Physiol.*, 98, 869, 1991.
69. **Grinstein, S., Furuya, W., and Bianchini, L.**, Protein kinases, phosphatases, and the control of cell volume, *News Physiol. Sci.*, 7, 232, 1992.
70. **Starke, L. C.**, Regulation of Na/K/2Cl and KCL Cotransport in Duck Red Cells, Ph.D. thesis, Duke University, Durham, NC, 1989.
71. **Sarkadi, B. and Parker, J. C.**, Activation of ion transport pathways by changes in cell volume. *Biochim. Biophys. Acta*, 1071, 407, 1991.
72. **Colclasure, G. C. and Parker, J. C.**, Cytosolic protein concentration is the primary volume signal in dog red cells, *J. Gen. Physiol.*, 98, 881, 1991.
73. **Colclasure, G. C. and Parker, J. C.**, Cytosolic protein concentration is the primary volume signal for swelling-induced [K-Cl] cotransport in dog red cells, *J. Gen. Physiol.*, 100, 1, 1992.
74. **Minton, A. P., Colclasure, G. C., and Parker, J. C.**, Model for the role of macromolecular crowding in regulation of cellular volume, *Proc. Natl. Acad. Sci. U.S.A.*, 89, 10504, 1992.
75. **Minton, A. P.**, Chapter 10, this volume.
76. **Windhager, E. E. and Taylor, A.**, Regulatory role of intracellular calcium ions in epithelial Na transport, *Annu. Rev. Physiol.*, 45, 519, 1983.
77. **Palmer, L. G. and Frindt, G.**, Effects of cell Ca and pH on Na channels from rat cortical collecting tubule, *Am. J. Physiol.*, 253, F333, 1987.
78. **Martin, A. R. and Dryer, S. E.**, Potassium channels activated by sodium, *Q. J. Exp. Physiol.*, 74, 1033, 1989.
79. **Bean, B. P.**, Pharmacology and electrophysiology of ATP-activated ion channels, *Trends Pharmacol. Sci.*, 13, 87, 1992.
80. **Tsuchiya, K., Wang, W., Giebisch, G., and Welling, P. A.**, ATP is a coupling modulator of parallel Na,K-ATPase-K-channel activity in the renal proximal tubule, *Proc. Natl. Acad. Sci. U.S.A.*, 89, 6418, 1992.
81. **Wang, W., Sackin, H., and Giebisch, G.**, Renal potassium channels and their regulation, *Annu. Rev. Physiol.*, 54, 81, 1992.
82. **Foskett, J. K.**, Chapter 15, this volume.
83. **Wong, S. M. E. and Chase, H. S., Jr.**, Role of intracellular calcium in cellular volume regulation, *Am. J. Physiol.*, 250, C841, 1986.
84. **Hazama, A. and Okada, Y.**, Ca^{2+} sensitivity of volume-regulatory K^+ and Cl^- channels in cultured human epithelial cells, *J. Physiol. London*, 402, 687, 1988.
85. **McCarty, N. A. and O'Neil, R. G.**, Dihydropyridine-sensitive cell volume regulation in proximal tubule: the calcium window, *Am. J. Physiol.*, 259, F950, 1990.
86. **McCarty, N. A. and O'Neil, R. G.**, Calcium-dependent control of volume regulation in renal proximal tubule cells. I. Swelling-activated Ca^{2+} entry and release, *J. Membr. Biol.*, 123, 149, 1991.
87. **Montrose-Rafizadeh, C. and Guggino, W. B.**, Role of intracellular calcium in volume regulation by rabbit medullary thick ascending limb cells, *Am. J. Physiol.*, 260, F402, 1991.
88. **Foskett, J. K. and Spring, K. R.**, Involvement of calcium and cytoskeleton in gallbladder epithelial cell volume regulation, *Am. J. Physiol.*, 248, C27, 1985.
89. **Davis, C. W. and Finn, A. L.**, Effects of mucosal sodium removal on cell volume in *Necturus* gallbladder epithelium, *Am. J. Physiol.*, 249, C304, 1985.

90. **Ubl, J., Murer, H., and Kolb, H.-A.,** Hypotonic shock evokes opening of Ca^{2+}-activated K channels in opossum kidney cells, *Pfluegers Arch. Eur. J. Physiol.,* 412, 551, 1988.
91. **Ubl, J., Murer, H., and Kolb, H.-A.,** Simultaneous recording of cell volume, membrane current and membrane potential: effect of hypotonic shock, *Pfluegers Arch. Eur. J. Physiol.,* 415, 381, 1989.
92. **McCarty, N. A. and O'Neil, R. G.,** Calcium-dependent control of volume regulation in renal proximal tubule cells. II. Roles of dihydropyridine-sensitive and -insensitive Ca^{2+} entry pathways, *J. Membr. Biol.,* 123, 161, 1991.
93. **Hoffmann, E. K. and Kolb, H.-A.,** Mechanisms of activation of regulatory volume responses after cell swelling, *Adv. Comp. Environ. Physiol.,* 9, 140, 1991.
94. **Suzuki, M., Kawahara, K., Ogawa, A., Morita, T., Kawaguchi, Y., Kurihara, S., and Sakai, O.,** $[Ca^{2+}]_i$ rises via G protein during regulatory volume decrease in rabbit proximal tubule cells, *Am. J. Physiol.,* 258, F690, 1990.
95. **Iwatsuki, N., Maruyama, Y., Matsumoto, O., and Nishiyama, A.,** Activation of Ca^{2+}-dependent Cl^- and K^+ conductances in rat and mouse parotid acinar cells, *Jpn. J. Physiol.,* 35, 933, 1985.
96. **Latorre, R. and Alvarez, O.,** Voltage-dependent channels in planar lipid bilayer membranes, *Physiol. Rev.,* 61, 77, 1981.
97. **Montrose-Rafizadeh, C. and Guggino, W. B.,** Cell volume regulation in the nephron, *Annu. Rev. Physiol.,* 52, 761, 1990.
98. **Reuss, L.,** Ion transport across gallbladder epithelium, *Physiol. Rev.,* 69, 503, 1989.
99. **Oberleithner, H., Kersting, U., and Hunter, M.,** Cytoplasmic pH determines K^+ conductance in fused renal epithelial cells, *Proc. Natl. Acad. Sci. U.S.A.,* 85, 8345, 1988.
100. **Messner, G., Koller, A., and Lang, F.,** The effect of phenylalanine on intracellular pH and sodium activity in proximal convoluted tubule cells of the frog kidney, *Pfluegers Arch. Eur. J. Physiol.,* 404, 145, 1985.
101. **Beck, J. S., Breton, S., Giebisch, G., and Laprade, R.,** Potassium conductance regulation by pH during volume regulation in rabbit proximal convoluted tubules, *Am. J. Physiol.,* 263, F453, 1992.
102. **Pewitt, E. B., Hegde, R. S., Haas, M., and Palfrey, H. C.,** The regulation of Na/K/2Cl cotransport and bumetanide binding in avian erythrocytes by protein phosphorylation and dephosphorylation. Effects of kinase inhibitors and okadaic acid, *J. Biol. Chem.,* 265, 20747, 1990.
103. **Reuter, H.,** Modulation of ion channels by phosphorylation and second messengers, *News Physiol. Sci.,* 2, 168, 1987.
104. **Bertorello, A., Aperia, A., Walaas, S. I., Nairn, A. C., and Greengard, P.,** Phosphorylation of the catalytic subunit of Na^+,K^+-ATPase inhibits the activity of the enzyme, *Proc. Natl. Acad. Sci. U.S.A.,* 88, 11359, 1991.
105. **Costantin, J., Alcalen, S., Otero, A. S., Dubinsky, W. P., and Schultz, S. G.,** Reconstitution of an inwardly rectifying potassium channel from the basolateral membranes of *Necturus* enterocytes into planar lipid bilayers, *Proc. Natl. Acad. Sci. U.S.A.,* 86, 5212, 1989.

Chapter 3

Volume Regulation in Renal Medullary Nephron Segments

Adam M. Sun and Steven C. Hebert

TABLE OF CONTENTS

I. INTRODUCTION

The maintenance of cell volume is one of the fundamental properties of all animal cells, including renal epithelial cells. Over the last few years, much has been learned about how renal epithelial cells (both cortical and medullary) maintain a constant, or nearly constant, cell volume under certain well-defined conditions.

The events of returning cell volume toward its original value after cell swelling or shrinkage are termed cell volume regulatory increase (RVI) and decrease (RVD), respectively. Both inorganic ions and certain organic solutes (termed osmolytes) such as sugars, amino acids, polyols, and methylamines have been implicated in cell volume regulation in a variety of cell types, including renal epithelial cells.[1–9] Furthermore, the cellular accumulation of inorganic and organic solutes during osmotic stress has been shown to represent a highly coordinated and ordered event in some cell types.[10] Figure 3–1 shows a proposed sequence of events resulting in hypertonic RVI in medullary thick ascending limb of Henle. Environmental hypertonic stress induces cell shrinkage, which in concert with vasopressin activates Na^+/H antiporters located in the basolateral membrane. This antiporter, functionally coupled with a Cl^-/HCO^-_3 antiporter, leads to net NaCl uptake and return of cell volume to normal (RVI response). Under certain circumstances a $Na^+/K^+/2Cl^-$ cotransporter also contributes to ion accumulation during recovery from cell shrinkage. This is followed by synthesis and accumulation of organic osmolytes with parallel decreases in intracellular NaCl. On returning to the isotonic condition, cells swell initially due to an increase in intracellular osmolality from solute accumulation during the

0-8493-4448-4/94/$0.00+$.50

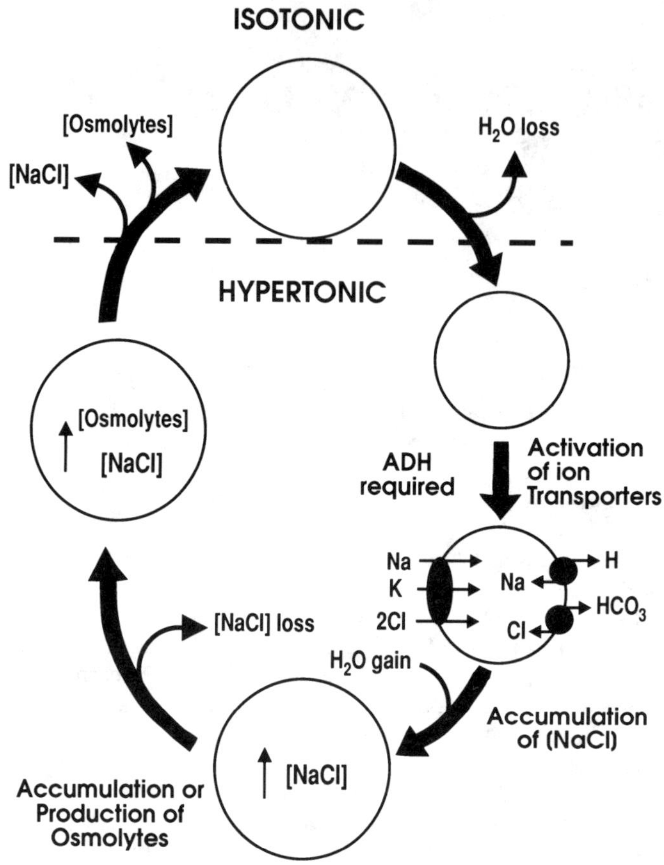

Figure 3–1 Sequence of events involved in adaptation of renal medullary epithelial cells to their hypertonic environment.

RVI response. Then organic osmolytes and NaCl are extruded from the cells, returning cell volume to the normal state.

This chapter will focus exclusively on the rapid cell volume regulatory responses that are mediated by uptake or loss of inorganic ions in two renal medullary nephron segments, i.e., mouse medullary thick ascending limb of Henle (MTAL) and rat inner medullary collecting duct (IMCD). In particular, the results of the studies on cell volume regulation in mouse MTAL will be emphasized, because we believe that this nephron segment provides a good model for study of cell volume regulation by medullary kidney cells. Specifically, we will describe (1) the ion transport systems that are activated to maintain volume constant following perturbations in extracellular osmolality; (2) the interrelationships between cell volume regulation and transepithelial salt absorption, and (3) the role of vasopressin in cell volume regulation.

II. TRANSPORT PROPERTIES OF THE MEDULLARY THICK ASCENDING LIMB OF HENLE

Since the mouse MTAL will be used as a model epithelium for discussion of volume-regulatory responses in the kidney, it is important to review our current understanding of the membrane transport properties of this nephron segment.

A. POLARITY OF WATER PERMEABILITY IN THE MTAL

The ability of kidney epithelia to transport solutes and water requires that apical and basolateral membranes of these cells form distinct structural and functional domains. This cell membrane polarity is not only accomplished by targeting transporters to either apical or basolateral membranes, but also by establishing differences in the basic permeability properties of these lipid membrane domains. Most renal medullary epithelial cell types such as the MTAL and medullary collecting duct are "water tight". Their apical cell membranes are the primary barrier that limits transcellular water movement and accounts for the negligibly low hydraulic conductance of these nephron segments.[15] In contrast, the basolateral cell membrane of these water-tight renal medullary cells are highly permeable to water and differences in the activity of water between cell cytoplasm and the basolateral medium (interstitial fluid *in vivo*) provides the driving force for water movement into or out of these cells (i.e., changes cell volume).

This disparity in apical-basolateral water permeability for the mouse MTAL is shown in Figure 3–2. In this figure it is apparent that reduction in osmolality of the apical (luminal) fluid had no effect on cell volume while a similar change in the osmolality of the medium bathing the basolateral membrane leads to dramatic cell swelling. This polarity in water permeability is required for the functional separation of salt and water absorption, a process required for both dilution and concentration of urine. As a consequence, MTAL cells undergo shrinkage or swelling when exposed to hyper- or hypo-osmotic medium, respectively, on the basolateral side.

B. MODEL FOR NaCl ABSORPTION IN THE MTAL

The thick ascending limb of Henle (TAL) absorbs 30% of the NaCl filtered at the glomerulus. The TAL is divided into two distinct segments: the cortical TAL (CTAL) and the medullary TAL (MTAL). They are different morphologically and physiologically. For example, NaCl absorption mechanisms in the mouse CTAL and MTAL segments are similar but not identical. Apical Na^+/H^+ and Cl^-/HCO_3^- antiporters mediate half of the apical NaCl entry in the CTAL but not in the MTAL.[11]

Over the last decade, NaCl transport in both CTAL and MTAL in different mammalian species have been studied extensively. Figure 3–3 shows a model that summarizes our current understanding of NaCl transport in the MTAL. According to this model, NaCl absorption is a secondary active transport process. Basolateral Na^+/K^+-ATPase maintains an inwardly directed Na^+ gradient to drive NaCl and K^+ entry into cells from the lumen via the apical $Na^+/K^+/2Cl^-$ cotransporter. Na^+ and Cl^- exit

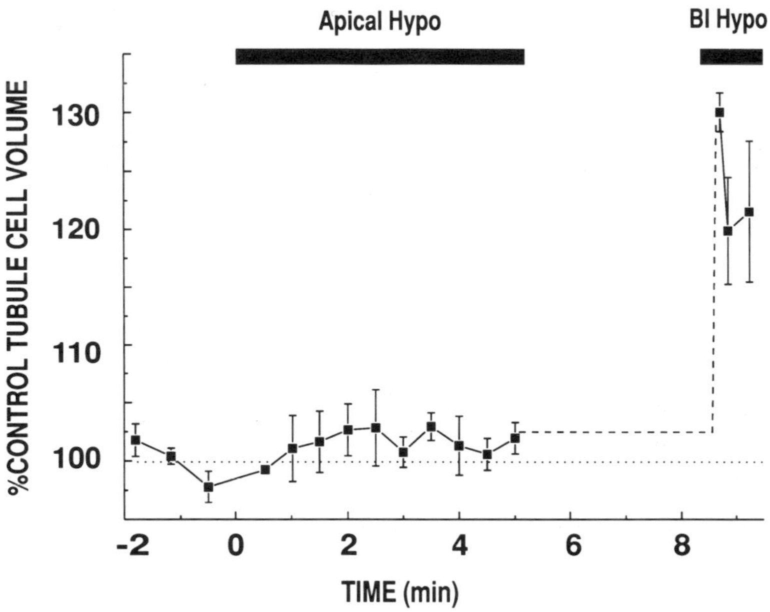

Figure 3–2 Disparity in apical-basolateral membrane water permeability in the mouse medullary thick ascending limb of Henle (MTAL). MTAL tubules were perfused *in vitro* at 37°C with isotonic (290 mOsmol/kg H_2O) HEPES-buffered media. Where indicated by the solid bars, the apical or basolateral perfusion medium osmolality was reduced by 50 mOsmol/kg H_2O. The dotted line indicates the isotonic control cell volume. Relative cell volume (percent; [cell volume/control cell volume] × 100) is plotted over time in minutes; zero time is when the apical (luminal) perfusion medium was changed to the hypotonic medium.

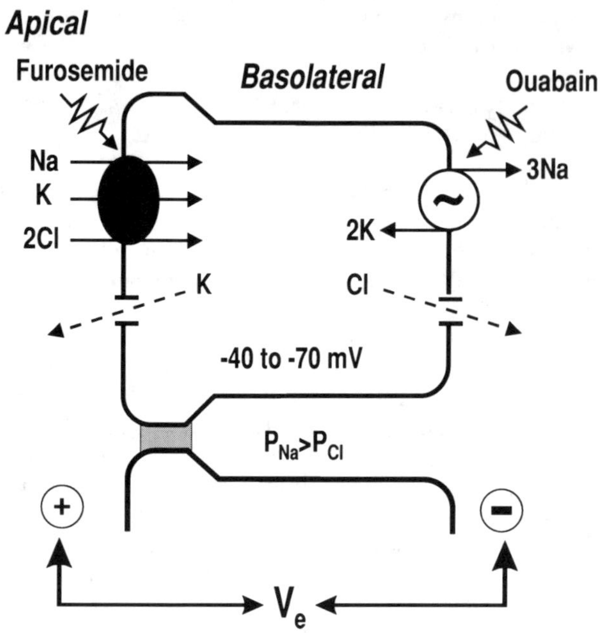

ADH	Ve (mV)	J_{Cl} (nEq/s cm^2)	P_f (μM/s)	G_e (mS/cm^2)
–	5	5	10-15	80
+	10	10	10-15	100

Figure 3–3 Model for NaCl absorption by the mouse medullary thick ascending limb of Henle. V_e, Transepithelial voltage referenced to basolateral ground; J_{Cl}, rate of transepithelial NaCl absorption; P_f, epithelial water permeability; G_e, transepithelial electrical conductance.

to the interstitium via a basolateral Na^+,K^+-ATPase and Cl^- channel, respectively. However, the majority of K^+ exits back to the lumen via an apical K^+ channel and is then recycled into cells. This conductive K^+ recycling, coupled with conductive Cl^- exit to the interstitium via a basolateral Cl^- channel, creates a lumen-positive transepithelial voltage which is sufficient to drive 50% of the absorbed Na^+ via the cation-selective paracellular pathway ($P_{Na} > P_{Cl}$; Figure 3–4). Since the transparacellular Na^+ absorption process is driven by the lumen-positive transepithelial voltage, it does not result in oxygen consumption to form ATP. Therefore, this mechanism of NaCl absorption enhances the metabolic efficiency of NaCl transport, which is of particular importance for the MTAL since this epithelium is highly vulnerable to ischemic damage and is located in a region of the kidney with a low oxygen tension.[12] In the mouse and rat MTAL, vasopressin, via a V_2 receptor effect, stimulates NaCl absorption by increasing the functional expression of an apical $Na^+/K^+/2Cl^-$ cotransporter and K^+ channel.[13] Vasopressin does not, however, alter the low water permeability of the apical membrane (P_f unchanged; Figure 3–4). In addition, Sun et al.[14] have recently demonstrated in the mouse MTAL that vasopressin alters the mode of NaCl transport from Na^+/Cl^- to $Na^+/K^+/2Cl^-$ cotransport, an effect that results in a doubling in the rate of NaCl absorption without a parallel increase in the rate of transcellular NaCl absorption or transport-related oxygen consumption.

III. CHALLENGES TO MAINTENANCE OF CELL VOLUME IN THE RENAL MEDULLARY CELLS

Renal medullary tubular cells may be subjected to at least three basic challenges to the maintenance of their volume [see **(i)** in Figure 3–4]. Charged and uncharged poorly permeable solutes in these cells tend to swell cells by creating not only a colloid osmotic pressure, but in addition, the fixed anion species tend to increase the intracellular concentration of permeant cations across cell membranes, according to the principles of Gibbs-Donnan equilibrium. This propensity toward cell swelling is counterbalanced by the operation of ion pumps that extrude ions at a rate equal to their entry via passive "leak" pathways (the pump-leak model), a theory which developed from the fundamental observa-

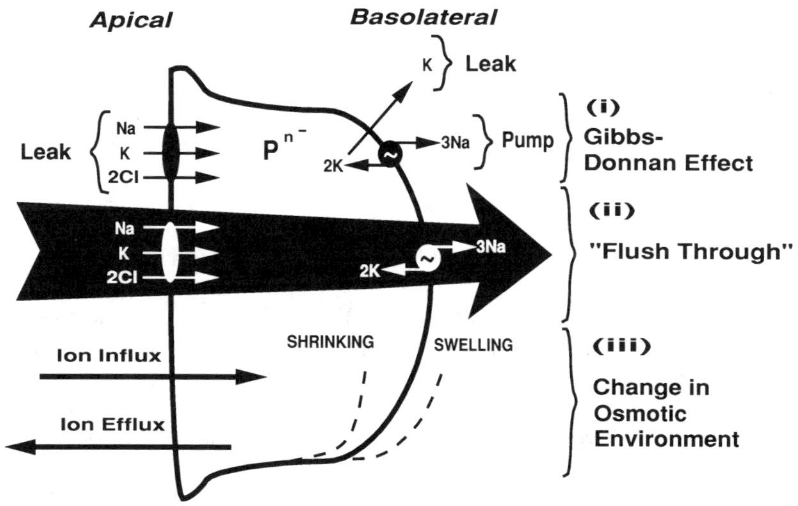

Figure 3–4 Challenges to cell volume in renal medullary epithelial cells. See text for discussion.

tions of Leaf,[16] Ussing,[17] and Tosteson and Hoffman.[18] In recent years it has become clear that adjustments of both pump and leak rates are required to maintain a constant cell volume under isotonic conditions.[4]

The pump-leak concept can be illustrated in the mouse MTAL.[19] As shown in Figure 3–6, under isotonic conditions, blockade of the basolateral Na^+,K^+-ATPase (pump) with ouabain results in a 50 to 60% increase in cell volume, which is reversible upon removal of ouabain. This effect of pump blockade on cell volume is abolished by inhibition of the apical $Na^+/K^+/2Cl^-$ cotransporter leak pathway with "loop" diuretics like furosemide. Solute-transporting cells like the mouse MTAL are faced with the additional problem of coping with large changes in the solute movement or traffic through cells (called by Schultz[20] the "flush through effect") as a consequence of increases or decreases in solute absorptive rates [(**ii**) in Figure 3–4]. Schultz used the term "homocellular" volume regulation to refer to the intrinsic cellular events that maintain a constant cell volume during changes in the rate of transcellular solute transport. The homocellular volume regulatory capability of the MTAL can be exemplified by examining the effects of vasopressin on NaCl absorption and cell volume. Addition of vasopressin to the basal side of the mouse MTAL leads to a doubling of the rate of NaCl absorption in 10 to 15 min by enhancing the rate of ion entry across the apical membrane via the $Na^+/K^+/2Cl^-$ cotransporter.[13,21] In this state of enhanced NaCl absorption, the intracellular pool of Na^+ has been estimated to turn over once every 26 s.[19] Yet this dramatic increase in transcellular ion transport occurs without changing the steady-state cell volume.[19]

Finally, variations in the tonicity of the fluid surrounding cells provide a driving force for cell water gain or loss [(**iii**) in Figure 3–4]. Although most mammalian cells (e.g., renal cortical cells like the cortical thick ascending limb [CTAL]) are not exposed to changes in extracellular tonicity *in vivo,* except under certain pathophysiological conditions,[22] they demonstrate volume-regulatory responses with certain anisotonic challenges.[6,7,9] The mechanisms involved in the volume-regulatory responses to anisotonic media may play a role in coping with Gibbs-Donnan effects and the "flush-through" effects of changes in the rates of transcellular salt transport. In addition, volume-regulatory responses may also play essential roles in metabolic events (e.g., osmotic effects related to uptake of organic metabolic fuels) and in cell volume expansion during cell growth and proliferation.[23] In contrast, renal medullary cells (interstitial, vascular, and epithelial) are subjected to major alterations in their osmotic environment (i.e., urine and interstitial osmolality) *in vivo* under normal physiological conditions, i.e., when shifting between water diuretic and antidiuretic states. As we will discuss below, renal medullary cells have evolved special mechanisms to cope with this latter challenge.

IV. CELL VOLUME REGULATION: REGULATORY VOLUME INCREASE (RVI)

A. HYPERTONIC RVI AND POST-RVD RVI

Cell shrinkage may occur with a sudden decrease in cell net solute entry or with exposure of cells to an absolute or relative increase in extracellular osmolality. Cell shrinkage activates specific solute entry mechanisms that result in osmotic water entry, and consequently, cell volume increases toward the baseline value observed under isotonic conditions. This type of volume recovery is called a regulatory volume increase (RVI) response. In studying RVI, cell shrinkage can be elicited by an absolute increase in external osmolality. This type of RVI is termed hypertonic RVI. Alternatively, cell shrinkage can be evoked by re-exposure of the cells to isotonic media after a regulatory volume decrease, RVD, response. Cell shrinkage occurring under this condition is due to a net solute loss during the RVD process, which renders the intracellular milieu hypotonic relative to the external medium on re-exposure to the isotonic medium. This type of RVI is termed post-RVD RVI.

Most of the mammalian renal (both cortical and medullary) epithelial cells studied to date do not show hypertonic RVI response. However, some renal tubular cells are capable of hypertonic RVI under certain conditions. For example, Lohr and Grantham[7] have shown in the *in vitro* nonperfused rabbit proximal convoluted tubule that these cells are capable of hypertonic RVI response provided that the osmolality of the medium is increased slowly at a rate less than 1.5 mOsmol/min. Interestingly, they also showed that the capability of hypertonic RVI in this nephron segment was enhanced with addition of certain short-chain fatty acids, such as butyrate, to the medium.[24] The role of these fatty acids in the RVI response for renal tubular cells *in vivo* is not clear. In contrast to hypertonic RVI, post-RVD RVI has been demonstrated in a number of renal cell types,[9,25,26] including those cells that are incapable of a rapid hypertonic RVI response, such as rabbit proximal convoluted tubules.[26] It is also not clear why certain cells are capable of post-RVD RVI but not hypertonic RVI.

Another example of hypertonic RVI occurring under certain conditions was found in the mouse MTAL[19] and in IMCD cells.[5] Both cell types are capable of hypertonic RVI only in the presence, but not absence, of vasopressin.[5,19] Figure 3–5 illustrates a hypertonic RVI response in mouse MTAL cells induced by a 50 mOsmol increase in the osmolality of basolateral medium with mannitol or NaCl. Furthermore, the post-RVD RVI response in mouse MTAL cells is also vasopressin-dependent (see Figure 3–10).[25]

B. A UNIQUE RVI PROCESS IN RENAL MEDULLARY CELLS: VASOPRESSIN DEPENDENCY

In mouse MTAL tubules, basolateral hypertonicity induces cell shrinkage as a result of osmotic water loss across the basolateral membrane. In addition, a decrease in the rate of transepithelial NaCl absorption by an increase in peritubular osmolality is

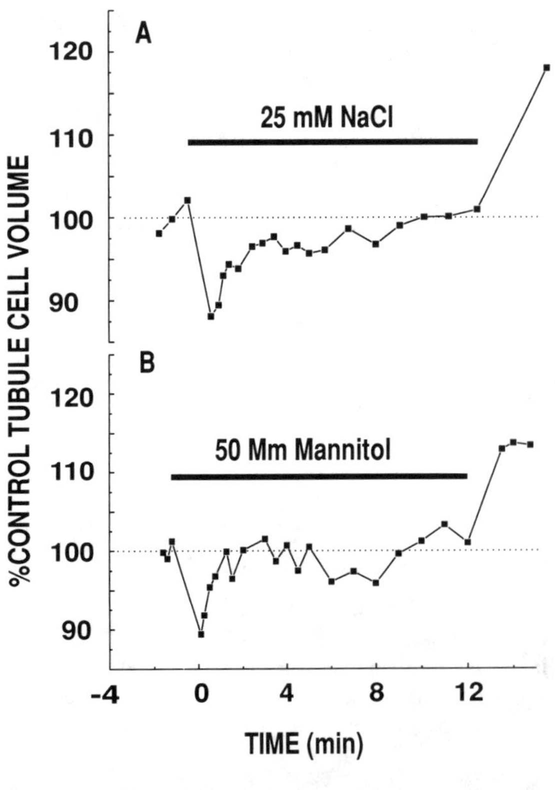

Figure 3–5 Hypertonic cell volume regulation in the mouse medullary thick ascending limb of Henle MTAL. MTAL tubules were perfused *in vitro* at 37°C in symmetrical (CO_2-HCO_3^-)-buffered isotonic media. The MTAL tubule was exposed to a basolateral perfusion medium made hypertonic either by addition of 25 m*M* NaCl (A) or 50 m*M* mannitol (B) over the period indicated by the solid bar. The dotted line indicated the isotonic control volume. Note the rapid regulatory volume increase, RVI, response and the cell volume "overshoot" after returning to the isotonic medium in each case.

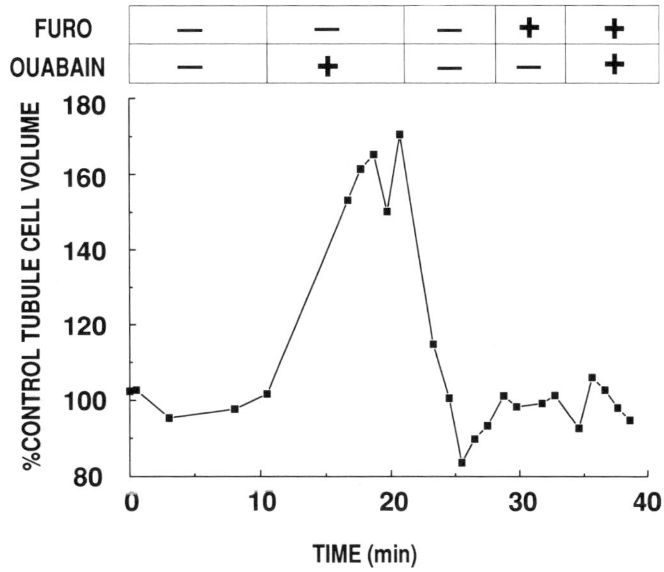

Figure 3–6 Interactions of luminal furosemide and basolateral ouabain in the MTAL illustrating the Gibbs-Donnan and "flush through" challenges to maintenance of cell volume [see Figure 3–4 **(i)** and **(ii)**]. MTAL tubule was perfused *in vitro* at 37°C in HEPES-buffered media. Where indicated the luminal or basolateral solutions contained 10^{-4} *M* furosemide or 10^{-3} *M* ouabain, respectively. See text for discussion.

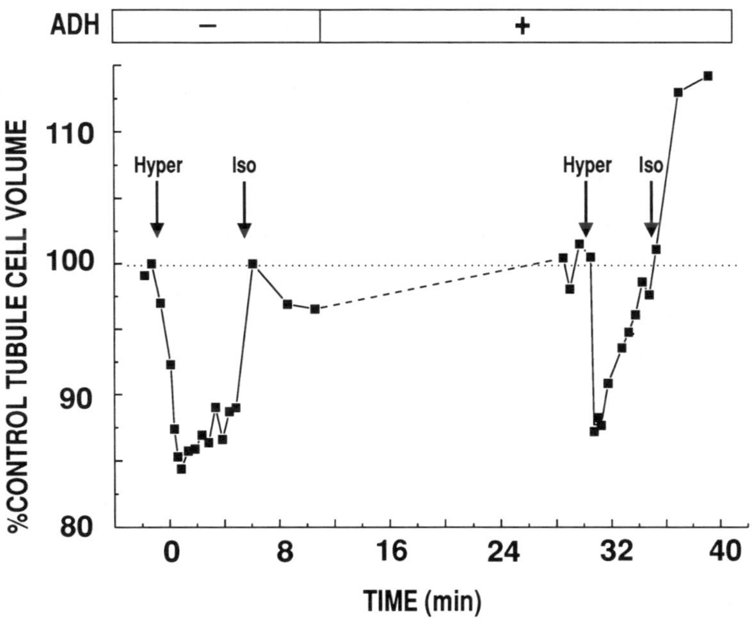

Figure 3–7 Cell hypertonic cell volume regulation requires vasopressin in the mouse MTAL. Experimental conditions as in Figure 3–6. Iso, Isotonic 290 mOsmol/kg H_2O; Hyper, isotonic solution plus 50 m*M* mannitol. Arginine vasopressin added to the basolateral medium at a concentration of 25 μU/ml where indicated.

also contributory. As shown in Figure 3–7, mouse MTAL tubules behave as simple osmometers in the absence of vasopressin, showing no evidence of hypertonic RVI response after cell shrinkage induced by a 50 mOsmol increase in the osmolality of basolateral medium with mannitol. In contrast, following exposure of MTAL cells to physiological concentrations of vasopressin (25 to 100 μU/ml), a rapid hypertonic RVI response is observed with nearly complete recovery of cell volume in 2 to 5 min (Figure 3–7). The effect of vasopressin on the ability of mouse MTAL cells to express a hypertonic RVI response is mimicked by the addition of dibutyryl cyclic AMP, and by direct activation of the catalytic subunit of adenylate cyclase with forskolin.[19] These results indicate that the hypertonic RVI response of this medullary epithelial cell type requires, at a minimum, an effect of vasopressin working through V_2 receptors. The increase in cell volume above baseline ("overshoot"; Figure 3–7; compare +AVP with –AVP responses) upon returning to the isotonic basolateral medium is consistent with net solute uptake into these cells during the hypertonic RVI response.

In addition to the mouse MTAL, another renal medullary cell type, the rat IMCD, was found to be capable of a hypertonic RVI response only in the presence of vasopressin.[5] Therefore, it is likely that vasopressin (via cyclic AMP) plays a general role in the RVI responses of renal medullary cells. In general, nonrenal cells that are capable of hypertonic RVI or post-RVD RVI response do not require exposure to vasopressin or other hormones/factors.[9,26] However, Hebert and Sun[25] have shown in the mouse MTAL that the post-RVD RVI response, like hypertonic RVI, also requires vasopressin (see Figure 3–10). Therefore, the requirement for vasopressin in expressing a RVI response appears to be a unique feature of renal medullary cells.

C. SOLUTE TRANSPORT MECHANISMS OF RVI

In animal cells, two types of electroneutral NaCl transport mechanisms have been proposed to mediate rapid RVI responses.[27,28] One process requires CO_2/HCO_3^- and involves parallel Na^+/H^+ and Cl^-/HCO_3^- exchangers; the other process is independent of CO_2/HCO_3^- and involves a loop diuretic-sensitive $Na^+/K^+/2Cl^-$ (or Na^+/Cl^-) cotransporter. It has been demonstrated that both hypertonic RVI and post-RVD RVI responses in mouse MTAL cells[3,25] and hypertonic RVI response in rat IMCD[5] cells are mediated by NaCl uptake via coupled Na^+/H^+ and Cl^-/HCO_3^- exchangers located in basolateral membranes. In addition, the apical Na^+ entry pathway in both cell types, i.e., Na^+ channel in the rat IMCD[5] and $Na^+/K^+/2Cl^-$ cotransporter in the mouse MTAL,[25] appears to play an important role in the post-RVD RVI response of mouse MTAL cells and in the hypertonic RVI response of the rat IMCD. This section will present the evidence for the apical and basolateral Na^+-coupled ion transporters mediating RVI processes in the mouse MTAL and rat IMCD.

1. Evidence for Coupled Basolateral Na^+/H^+ and Cl^-/HCO_3^- Exchangers Mediating RVI

In the hypertonic RVI response of the mouse MTAL, NaCl uptake occurs via coupled Na^+/H^+ and Cl^-/HCO_3^- exchangers located in basolateral membranes (Figure 3–8). The evidence for this is as follows. Both the RVI response and the post-RVI isotonic "overshoot" are abolished either by symmetrical Na^+ deletion from extracellular media or by basolateral, but not luminal, Cl^- removal. Since either addition of ouabain to the basolateral solution or cooling the tissue to 15°C reduces or abolishes the hypertonic RVI response, NaCl uptake during RVI is an active (i.e., secondary active) process ultimately dependent on basolateral Na^+,K^+-ATPase. RVI can also be abolished by basolateral, but not luminal, amiloride (an inhibitor of Na^+/H^+ exchange), by basolateral addition of 4-acetamido-4′-isothiocyanostilbene-2,2′-disulfonic acid (SITS; an inhibitor of Cl^-/HCO_3^- exchange) or by the isohydric replacement of CO_2/HCO_3^- with HEPES. Carbonic anhydrase activity also appears to be re-

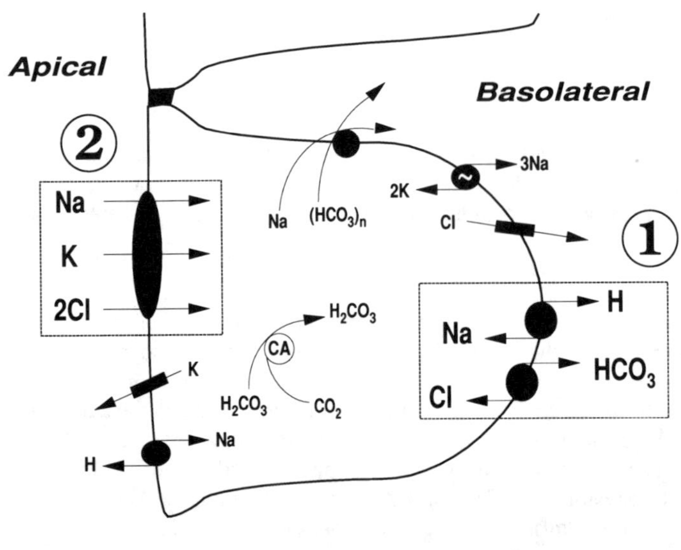

Figure 3–8 Ion transport mechanisms mediating the regulatory volume increase, RVI, response in the mouse medullary thick ascending limb of Henle, MTAL. See text for explanation.

quired in order to maintain an adequate supply of H^+ and HCO_3^- to the basolateral exchangers, since the carbonic anhydrase inhibitor, ethoxzolamide, blocks the hypertonic RVI response of the mouse MTAL.

Using similar ion substitution and transport inhibition techniques performed in the presence of vasopressin, Hebert and Sun have also shown that post-RVD RVI in the mouse[25] and hypertonic RVI in the rat IMCD[5] involve basolateral Na^+/H^+ and Cl^-/HCO_3^- exchangers. However, there is a difference between the basolateral Na^+/H^+ exchanger involved in hypertonic RVI in mouse MTAL and rat IMCD. In the mouse MTAL, the basolateral Na^+/H^+ exchanger is not functional or as active during steady-state isotonic condition.[32,34] Sun et al.[32] have demonstrated recently in the mouse MTAL that the basolateral Na^+/H^+ exchanger has a very low set point of pH 6.7 under isotonic conditions. In other words, the basolateral Na^+/H^+ exchanger is inactive if cell pH is above 6.7. And since the steady-state cell pH is 7.2 under isotonic conditions, the basolateral Na^+/H^+ exchanger is normally inactive. However, both external hypertonicity and vasopressin are capable of activating this exchanger by shifting the set point toward a more alkaline pH. In contrast, our preliminary data indicate that basolateral Na^+/H^+ exchanger in the rat IMCD is active under the isotonic condition even in the absence of vasopressin. The reasons for these differences are not clear at the present time. In addition, whether the Na^+/H^+ exchangers in these two cell types are different isoforms of the same exchanger is not known.

2. The Role of Apical Na^+ Entry Pathways in RVI

In the mouse MTAL, apical Na^+ entry is mediated by a $Na^+/K^+/2Cl^-$ cotransporter, which is enhanced by vasopressin.[13] The contribution of this cotransporter to RVI depends on the method used to elicit the RVI response. Inhibition of the apical $Na^+/K^+/2Cl^-$ cotransporter with furosemide (or the substitution of luminal Cl^- with isethionate) has no effect on the rate of RVI (Table 1). In contrast, the rate of post-RVD RVI is significantly reduced by this loop diuretic (Table 1). It is also apparent from Table 1 that the rate of post-RVD RVI in the absence of luminal furosemide is about 3-fold higher than the rates of hypertonic RVI, with or without luminal furosemide, or the rate of post-RVD RVI in the presence of furosemide. Thus, the apical $Na^+/K/2Cl^-$ cotransporter provides a mechanism for augmenting ion uptake only during the post-RVD RVI response. On the other hand, the contribution of the vasopressin-dependent basolateral Na^+/H^+ and

Table 1 **Role of the Apical $Na^+/K^+/2Cl^-$ Cotransporter in the RVI Responses of the Mouse MTAL**

	ADH[a]	Furo	Rate of RVI (nl/min/cm)
Hypertonic			
	+	–	0.105±0.032
	+	+	0.167±0.032
Post-RVD RVI			
	+	–	0.310±0.050[b]
	+	+	0.138±0.096

[a] ADH, Antidiuretic hormone; [b] Indicates significantly different from the rate of post-RVD RVI in the presence of furosemide.

From Sun, A. M. et. al., *Kidney Int*, 38, 1019, 1990. With permission.

Cl^-/HCO_3^- exchangers to RVI does not appear to be altered by the method used to shrink MTAL cells.[3,25] Possible reasons for the different roles of the luminal $Na^+/K^+/2Cl^-$ cotransporter in these two types of RVI in the mouse MTAL are as follows.

At least three factors may account for the functional impairment of the apical $Na^+/K^+/2Cl^-$ cotransporter during hypertonic, but not isotonic, cell shrinkage. First, the electrochemical driving force for ion uptake by the cotransporter may be reduced to a greater extent in the hypertonic than in the post-RVD RVI condition. This is because in the latter condition total intracellular ionic content of MTAL cells is reduced after the RVD response (probably due to loss of cell K^+ and Cl^- during RVD). In addition, hypertonicity may increase cell Cl^- concentration by a direct reduction in basolateral Cl^- conductance.[33] Finally, since the activity of water is lower in the hypertonic state than in the isotonic state, the absolute quantity of Na^+, K^+, and Cl^- that must enter the cell to give the same amount of water entry (and cell volume increase) would be greater in the hypertonic condition. Thus, the combined effects of a lower driving force for ion uptake via the apical $Na^+/K^+/2Cl^-$ cotransporter and the necessity for a greater absolute quantity of solute uptake to drive water entry are likely sufficient to reduce the functional role of the apical cotransporter in hypertonic RVI.

In contrast to the hypertonic RVI in the mouse MTAL, apical Na^+ entry appears to play an important role in the hypertonic RVI response in the rat IMCD. In this renal segment, apical Na^+ entry has been shown to be mediated by an amiloride-sensitive channel.[29] In addition, this apical Na^+ entry process is enhanced by vasopressin.[30,31] Sun and Hebert[5] have shown in *in vitro* perfused rat IMCD tubules that apical addition of amiloride abolishes vasopressin-dependent hypertonic RVI. It is not clear at present how apical Na^+ entry affects hypertonic RVI mediated by the basolateral Na^+/H^+ and Cl^-/HCO_3^- exchangers in this nephron segment.

D. ACTIVATING MECHANISMS

Transport processes during RVI have been studied extensively in a number of cell types. However, it is not clear how cells sense volume changes and transduce this signal to ion transporters. In the mouse MTAL, both hypertonicity and vasopressin are crucial in activating solute transport during RVI, since vasopressin by itself does not change cell volume and hypertonicity induces cell shrinkage without a subsequent RVI in the absence of vasopressin.[19] In several nonrenal cell types, it has been demonstrated that the Na^+/H^+ exchanger can be activated by external hypertonicity.[36] In the mouse MTAL, recent studies by Sun and Hebert[35] have demonstrated that peritubular hypertonicity increases the activity of basolateral Na^+/H^+ exchangers in the absence of vasopressin. The cellular mechanism of hypertonicity-induced activation of the Na^+/H^+ exchanger is not known. However, a few recent studies point out the possibility that hypertonicity may stimulate a cell signaling pathway which leads to phosphorylation and, subsequently, activation of Na^+/H^+ exchangers. Bianchini et al.[37] have shown in the rat lymphocyte that external hypertonicity activates Na^+/H^+ exchangers via phosphorylation of this exchanger by a protein kinase which is distinct from protein kinases A or C. Interestingly, Davis et al.[38] have shown recently in the single barnacle muscle fiber that external hypertonicity can activate an otherwise inactive Na^+/H^+ exchanger. This Na^+/H^+ exchanger activation by hypertonicity is not mimicked by cAMP or protein kinase C, but is blocked by cholera toxin, a G_S protein inhibitor. These results suggest that a G protein may be involved in transducing the shrinkage signal to the transporter(s) involved in cell volume regulation.

As noted above, the RVI response in renal medullary cells is unique in its requirement for vasopressin. We have shown recently in the mouse MTAL that vasopressin activates the basolateral Na^+/H^+ exchanger even under isotonic conditions.[32] In addition, external hypertonicity activates this transporter either in the presence or absence of vasopressin.[35] Thus, vasopressin and hypertonicity have an additive effect in stimulating the basolateral Na^+/H^+ exchanger under hypertonic conditions. The effect of vasopressin on the basolateral Na^+/H^+ exchanger in the mouse MTAL cannot be attributed to any hormone-induced cell shrinkage, since vasopressin causes no significant change in MTAL cell volume.[19] Future studies are required not only to study the mechanism of activation of the Na^+/H^+ exchanger but also to study the processes leading to activation of the basolateral Cl^-/HCO_3^- exchangers.

V. CELL VOLUME REGULATION: REGULATORY VOLUME DECREASE (RVD)

A. SOLUTE TRANSPORT MECHANISMS OF RVD AND THE INTERACTIONS BETWEEN RVD AND TRANSCELLULAR SALT TRANSPORT

Cell swelling may occur with an increase in cell solute entry or with an absolute or relative decrease in extracellular osmolality. Cell swelling activates an RVD response in which solute leak pathways are stimulated and solute with water exits cells, resulting in a decrease in cell volume. In most mammalian cells, RVD is mediated by passive KCl loss via KCl cotransporters, parallel K^+ and Cl^- channels, or parallel K^+/H^+ and Cl^-/HCO_3^- exchangers.[27,28,39] However, there are a number of exceptions.[9,40] For example, Strange[9] showed in the rabbit cortical collecting duct that depletion of intracellular K^+ and Cl^- had no effect on the rates or magnitude of RVD, thus arguing against a significant role of passive KCl efflux pathways in mediating RVD in this nephron segment.

The transport mechanisms involved in the RVD response of mouse MTAL or rat IMCD cells are unknown at present. However, several recent studies have implicated stretch-activated (SA) channels playing an important role in RVD in certain epithelia, including the MTAL.[41–43] SA channels represent a heterogeneous group of channels found in cell membranes that can be activated by stretching plasma membranes. In RVD, hypotonic cell swelling activates SA channels which, in turn, either directly or indirectly mediate ion effluxes, leading to cell volume regulation. In the renal proximal tubule, Sackin[43] has shown that hypotonic cell swelling induces membrane stretch and activates a SA-activated K^+ channel which mediates K^+ loss, leading to RVD. Alternatively, as shown in the choroid plexus epithelium obtained from *Necturus maculosus*,[42] hypotonic cell swelling opens a cation-selective, Ca^{2+} permeable SA channel with stretch causing an influx of Ca^{2+} which, in turn, activates a neighboring K^+ channel to produce K^+ efflux from the cell. Recent patch-clamp studies have identified both Ca^{2+}-activated large conductance (maxi)-K^+ channels[44] and Ca^{2+}-independent K^+ channels[45] in the apical membranes of mouse MTAL. The apical Ca^{2+}-dependent maxi-K^+ channel can be activated by membrane stretch in cell-detached patches and by hypotonic cell swelling in cell-attached patches.[44] In addition, swelling-induced activation of the maxi-K^+ channel only occurs when extracellular Ca^{2+} is present, suggesting that a membrane SA Ca^{2+} entry process is involved in mediating activation of the maxi-K^+ channel.[44] Taken together, the results of these recent studies suggest the scheme for the role of SA channel in RVD in the mouse MTAL shown in Figure 3–9. According to this hypothesis, cell swelling resulting either from increased solute transport or from a sudden reduction in osmolality of the media stretches the cell membrane, which (1) directly activates certain Ca^{2+}-independent K^+ channels, and/or (2) activates a cation channel that mediates Ca^{2+} entry leading to a rise in intracellular Ca^{2+} and stimulation of a Ca^{2+}-dependent maxi-K^+ channel. As a result, K^+ exits the cell and, if accompanied by loss of anion (presumably Cl^-), leads to a decrease in cell volume.

It is interesting to point out that organic osmolytes may play an important role in the rapid RVD process in renal medullary cells. In these cells, in contrast to a slow organic osmolyte accumulation during RVI, taking hours to days, organic osmolyte extrusion during RVD is relatively rapid. Bagnasco et al.[46] have shown that renal medullary cells transferred from hypertonic to isotonic media lose 50% of the sorbitol accumulated during RVI in 15 min. Of course, this does not exclude a role of rapid ion extrusion in RVD.

As discussed above, volume-regulatory mechanisms may play an important role in minimizing any cell volume changes resulting not only from changes in medium osmolality but also from imbalances between solute entry and exit processes. In mouse MTAL, therefore, the RVD response was assessed under three different states of salt absorption, shown in Figure 3–10. In the absence of vasopressin (Figure 3–10A), cells initially swell and then rapidly decrease their volume when exposed to a 50 mOsmol reduction in the bath osmolality. Upon returning to the isotonic medium cells shrink consistent with cell solute loss during the RVD response. When salt absorption is increased with vasopressin (Figure 3–10B), the RVD response is considerably slower. This hormone-associated impairment of RVD is in sharp contrast to the vasopressin requirement for RVI (either post-RVD RVI or hypertonic RVI). In contrast to the results shown in Figure 3–10A, the rate of RVD is increased fourfold when the vasopressin-mediated increase in salt absorption is abolished with luminal furosemide (Figure 3–10C). Therefore, the blunting of the RVD response observed in vasopressin-activated MTAL segments is most likely due to an effect of the hormone-mediated increase in NaCl absorption (enhanced transcellular solute movement) on the hypotonic-induced solute efflux mechanism. In addition, the results of Figure 3–10 indicate that increasing the rate of NaCl absorption decreases the rate of the RVD response that is

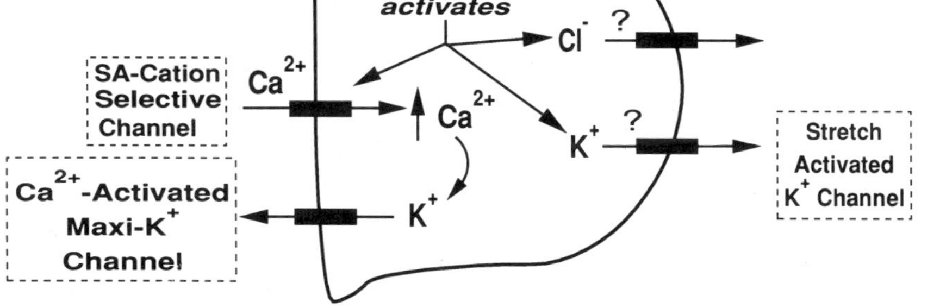

Figure 3–9 Model for the regulatory volume decrease, RVD, response in the mouse medullary thick ascending limb of Henle, MTAL. See text for discussion.

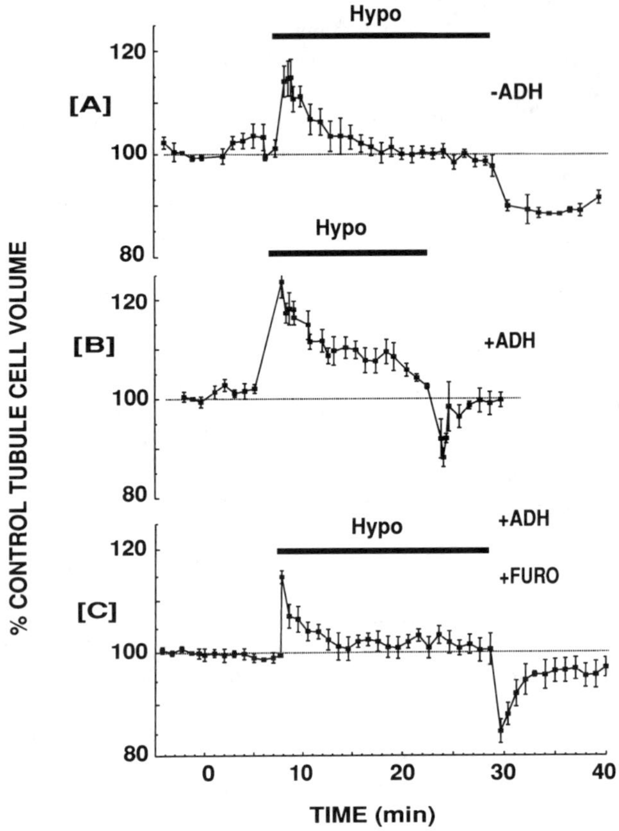

Figure 3–10 Effects of arginine vasopressin (antidiuretic hormone, ADH) on the rate of transcellular NaCl transport on the hypotonic RVD and post-RVD RVI responses in the *in vitro* perfused mouse MTAL. Where indicated by the solid bar the basolateral medium tonicity was reduced by 50 mOsmol/kg H_2O. (A) Tubule perfused in the absence of vasopressin (ADH); (B) tubule exposed to ADH (25 μU/ml) 20 min before measurement of cell volume; (C) as in (B), except luminal perfusate contained 10^{-4} *M* furosemide to block transcellular NaCl absorption. See text for discussion.

elicited by reducing extracellular osmolality, and therefore, suggests that both an increase in cell solute traffic and a decrease in extracellular osmolality activate similar mechanisms to enhance salt exit. Finally, these results support the notion that the RVD process may play an important role in minimizing any cell volume changes resulting from imbalances between solute entry and exit processes.

B. ACTIVATION MECHANISMS

It is not known how renal cells sense volume swelling and transduce this signal to the ion transporters. A number of factors have been implicated in regulating the function and activation of ion transporters: e.g., Ca^{2+}, calmodulin, cAMP, eicosanoids, phosphoinositide metabolism, proteinases, and cytoskeleton.[23] The roles of these factors in RVD will be discussed in Section III of this book, and therefore, will not be discussed here. It is to be noted that most of the studies assessing the roles of these regulatory factors in RVD have been performed in single cells and nonrenal epithelial cells. In renal epithelial cells, the role of intracellular Ca^{2+} in RVD has been studied in the proximal tubule.[47,48] These studies demonstrated that hypotonic cell swelling increases intracellular free Ca^{2+}. However, the activation of the RVD process appears to involve more than Ca^{2+}, since elevation of intracellular Ca^{2+} with Ca^{2+} ionophores did not induce cell shrinkage.[47,48] At present, there have been no studies examining the roles of Ca^{2+} and other regulatory factors in RVD in renal medullary cells.

VI. SUMMARY

In renal medullary cells, the dramatic changes in osmotic environment and rapid transcellular solute traffic tend to shrink or swell cells. Renal epithelial cells sense volume changes and initiate mechanisms to adjust intracellular osmolality, leading to return of their cell volume to normal. In both the mouse MTAL (an outer medullary segment) and rat IMCD (an inner medullary segment), the rapid RVI process is mediated by net NaCl uptake via basolateral coupled Na^+/H^+ and Cl^-/HCO_3^- exchangers. Vasopressin plays a central role in this process. The mouse MTAL is capable of a rapid RVD response. It is likely that RVD in these two nephron segments is mediated by KCl loss via separate K^+ and Cl^- channels. To a major extent the rate of RVD is determined by the rate of transcellular solute movement, since vasopressin increases the rates of NaCl absorption but concomitantly decreases the rate of RVD. The cellular mechanisms involved in activating the ion transporters involved in rapid cell volume regulatory responses in renal medullary cells are beginning to be elucidated. Future studies are required, however, not only to delineate how renal cells sense volume changes and transduce these signals to the solute transporters, but also how vasopressin affects these mechanisms.

REFERENCES

1. **Wirthensohn, G., Lefrank, S., Schmolke, M., and Guder, W. G.,** Regulation of organic osmolyte concentrations in tubules from rat renal inner medulla, *Am. J. Physiol.*, 256, F128, 1989.
2. **Bagnasco, S. M., Uchida, S., Balaban, R. S., Kador, P. F., and Burg, M. B.,** Induction of aldose reductase and sorbitol in renal inner medullary cells by elevated extracellular NaCl, *Proc. Natl. Acad. Sci. U.S.A.*, 84, 1718, 1987.
3. **Hebert, S. C.,** Hypertonic cell volume regulation in mouse thick limbs. II. Na-H and Cl-HCO_3 exchange in basolateral membranes, *Am. J. Physiol.*, 250, C920, 1986.
4. **Sun, A. M., Saltzberg, S. N., Kikeri, D., and Hebert, S. C.,** Mechanisms of cell volume regulation by the mouse medullary thick ascending limb of Henle, *Kidney Int.* 38, 1019, 1990.
5. **Sun, A. M. and Hebert, S. C.,** Rapid hypertonic cell volume regulation in the perfused inner medullary collecting duct, *Kidney Int.*, 36, 831, 1989.
6. **Gullans, S. R., Blumenfeld, J. D., Balshi, J. A., Brenner, R. M., Heilig, C. W., and Hebert, S. C.,** Dehydration increases organic osmolytes in the renal inner medulla, *Am. J. Physiol.*, 255, F626, 1988.
7. **Lohr, J. and Grantham, J. J.,** Isovolumetric regulation in anisotonic media: a new approach to study compensated cell volume regulation in isolated proximal S2 segments, *Kidney Int.*, 29, 419, 1986.
8. **Kirk, K. L., Dibona, D. L., and Schafer, J. A.,** Regulatory volume decrease in perfused proximal nephron: evidence for a dumping of cell K^+, *Am. J. Physiol.*, 252, F933, 1987.
9. **Strange, K.,** RVD in principal and intercalated cells of rabbit cortical collecting duct, *Am. J. Physiol.*, 255, C612, 1988.
10. **Sutherland, L., Cairnet, J., Elmore, M. J., Booth, I. R., and Higgins, C. F.,** Osmotic regulation of transcription: induction of the proU betaine transport gene is dependent on accumulation of intracellular potassium, *J. Bacteriol.*, 168, 805, 1986.
11. **Friedman, P. A. and Andreoli, T. E.,** CO_2-stimulated NaCl absorption in the mouse renal cortical thick ascending limb of Henle, *J. Gen. Physiol.*, 80, 683, 1982.
12. **Brezis, M., Rosen, S., Silva, P., and Epstein, F. H.,** Renal ischemia, *Kidney Int.*, 26, 375, 1984.
13. **Hebert, S. C. and Andreoli, T. E.,** Effects of antidiuretic hormone on cellular conductive pathways in mouse medullary thick ascending limbs of Henle. II. Determinants of the ADH-mediated increases in transepithelial voltage and in net Cl absorption, *J. Membr. Biol.*, 80, 221, 1984.
14. **Sun, A. M., Grossman, E. B., Lombardi, M. J., and Hebert, S. C.,** Vasopressin alters the mechanism of apical Cl^- entry from Na^+:Cl^- to Na^+:K^+:$2Cl^-$ cotransport in mouse medullary thick ascending limb, *J. Membr. Biol.*, 120, 83, 1991.
15. **Greger, R.,** Ion transport mechanisms in thick ascending limb of Henle's loop of mammalian nephrons, *Physiol. Rev.*, 65, 760, 1985.
16. **Leaf, A.,** Maintenance of concentration gradients and regulation of cell volume, *Ann. N.Y. Acad. Sci.*, 72, 396, 1959.
17. **Ussing, H. H.,** Active and passive transport of the alkali metal ions, in *The Alkali Metal Ions in Biology*, Ussing, H. H., Kruhoffer, P., Thaysen, J. H., and Thorn, N. A., Eds., Springer-Verlag, Berlin, 1960, 145.
18. **Tosteson, D. C. and Hoffman J. F.,** Regulation of cell volume by active cation transport in high and low potassium sheep red cells, *J. Gen. Physiol.*, 44, 169, 1960.
19. **Hebert, S. C.,** Hypertonic cell volume regulation in mouse thick limbs. I. ADH dependency and nephron heterogeneity, *Am. J. Physiol.*, 250, C9070, 1986.
20. **Schultz, S. G.,** Homocellular regulatory mechanisms in sodium-transporting epithelia: avoidance of extinction by "flush-through", *Am. J. Physiol.*, 241, F579, 1981.
21. **Hebert, S. C., Culpepper, R. M., and Andreoli, T. E.,** NaCl transport in mouse medullary thick ascending limbs. I. Functional nephron heterogeneity and ADH-stimulated NaCl cotransport, *Am. J. Physiol.*, 241, F412, 1981.
22. **Culpepper, R. M., Hebert, S. C., and Andreoli, T. E.,** The posterior pituitary and water metabolism, in *Textbook of Endocrinology*, Wilson, J. D. and Foster, D. W., Eds., W. B. Saunders, Philadelphia, 1985, 614.
23. **Hoffmann, E. K. and Simonsen, L. O.,** Membrane mechanisms in volume and pH regula-

tion in vertebrate cells, *Physiol. Rev.*, 69, 315, 1989.

24. **Rome, L., Lechene, C., Savin, V., and Grantham, J.,** Critical role of short-chain fatty acids in isovolumetric regulation of proximal S_2 segments in hypertonic media, *Kidney Int.*, 33, 439, 1988.
25. **Hebert, S. C. and Sun, A. M.,** Hypotonic cell volume regulation in mouse medullary thick ascending limb: effects of ADH, *Am. J. Physiol.*, 255, F962, 1988.
26. **Kirk, K. L., Schafer, J. A., and Dibona, D. L.,** Cell volume regulation in rabbit proximal straight tubule perfused in vitro, *Am. J. Physiol.*, 252, F922, 1987.
27. **Chamberlin, M. E. and Strange, K.,** Anisotonic cell volume regulation: a comparative view, *Am. J. Physiol.*, 257, C159, 1989.
28. **Eveloff, J. L. and Warnock, D. G.,** Activation of ion transport systems during cell volume regulation, *Am. J. Physiol.*, 252, F1, 1987.
29. **Light, D. B., McCann, F. V., Keller, T. M., and Stanton, B. A.,** Amiloride-sensitive cation channel in apical membrane of inner medullary collecting duct, *Am. J. Physiol.*, 255, F278, 1988.
30. **Stanton, B. A.,** Characterization of apical and basolateral membrane conductances of rat inner medullary collecting duct, *Am. J. Physiol.*, 256, F862, 1989.
31. **Kudo, L. H., Van Baak, A. A., and Rocha, A. S.,** Effect of vasopressin on sodium transport across inner medullary collecting duct, *Am. J. Physiol.*, 258, F1438, 1990.
32. **Sun, A. M., Kikeri, D., and Hebert, S. C.,** Vasopressin regulates apical and basolateral Na^+:H^+ antiporters in mouse medullary thick ascending limb, *Am. J. Physiol.*, 262, F241, 1992.
33. **Hebert, S. C., Culpepper, R. M., and Andreoli, T. E.,** NaCl transport in mouse medullary thick ascending limbs. III. Modulation of the ADH effect by peritubular osmolality, *Am. J. Physiol.*, 241, F443, 1981.
34. **Kikeri, D., Sun, A. M., Zeidel, M. L., and Hebert, S. C.,** Cell membranes impermeable to NH_3, *Nature*, 339, 478, 1989.
35. **Sun, A. M. and Hebert, S. C.,** Effect of hyperosmolality on the basolateral Na^+:H^+ exchanger in the mouse medullary thick ascending limb of Henle, *J. Am. Soc. Nephrol.*, 3, 799, 1992.
36. **Grinstein, S. and Rothstein, A.,** Mechanisms of regulation of the Na^+:H^+ exchanger, *J. Membr. Biol.*, 90, 1, 1986.
37. **Bianchini, L., Woodside, M., Sardet, C., Pouyssegur, J., Takai, A., and Grinstein, S.,** Okadaic acid, a phosphatase inhibitor, induces activation and phosphorylation of the Na^+:H^+ antiport, *J. Biol. Chem.*, 266, 15406, 1991.
38. **Davis, B. A., Hogan, E. M., and Boron, W. F.,** Role of G proteins in stimulation of Na-H exchange by cell shrinkage, *Am. J. Physiol.*, 262, C533, 1992.
39. **Hebert, S. C.,** Volume regulation in renal epithelial cells, *Semin. Nephrol.*, 7, 48, 1987.
40. **Parker, J. C., Giltelman, H. G., Glosson, P. S., and Leonard, D. L.,** Role of calcium in volume regulation by dog red blood cells, *J. Gen. Physiol.*, 65, 84, 1975.
41. **Taniguchi, J. and Guggino, W. B.,** Membrane stretch: a physiological stimulator of Ca^{2+}-activated K^+ channels in thick ascending limb, *Am. J. Physiol.*, 257, F347, 1989.
42. **Christensen, O.,** Mediation of cell volume regulation by Ca^{2+} influx through stretch-activated channels, *Nature*, 330, 66, 1987.
43. **Sackin H.,** A stretch-activated K^+ channel sensitive to cell volume, *Proc. Natl. Acad. Sci. U.S.A.*, 86, 1731, 1989.
44. **Guggino, S. E., Guggino, W. B., Green N., and Sacktor B.,** Ca^{2+}-activated K^+ channels in cultured medullary thick ascending limb cells, *Am. J. Physiol.*, 252, C121, 1987.
45. **Wang, W., White, S., Geibel, J., and Giebisch, G.,** A potassium channel in the apical membrane of rabbit thick ascending limb of Henle's loop, *Am. J. Physiol.*, 258, F244, 1990.
46. **Bagnasco, S. M., Murphy, H. R., Bedford, J. J., and Burg, M. B.,** Osmoregulation by slow changes in aldose reductase and rapid changes in sorbitol flux, *Am. J. Physiol.*, 254, C788, 1988.
47. **Beck, J. S., Breton, S., Laprade, R., and Giebisch, G.,** Volume regulation and intracellular calcium in the rabbit proximal convoluted tubule, *Am. J. Physiol.*, 260, F861, 1991.
48. **McCarty, N. A. and O'Neil, R. G.,** Calcium-dependent control of volume regulation in renal proximal tubule cells. I. Swelling-activated Ca^{2+} entry and release, *J. Membr. Biol.*, 123, 149, 1991.

Chapter 4

Clinical Significance of Cellular Osmoregulation

Michael L. McManus and Kevin B. Churchwell

TABLE OF CONTENTS

I. INTRODUCTION

Why is it important for the clinician to understand the mechanisms of cell volume regulation? As fundamental physiologic processes, those governing cell volume hold medical implications as far reaching as those of other fundamental cellular processes, such as protein synthesis, DNA replication, and mitochondrial respiration. Yet, like these other processes, the clinical ramifications of cell volume regulation become clear only as our basic understanding broadens.

Even superficial consideration of cell volume regulation and its place in medical practice, however, immediately yields three general questions. First, how do current medical interventions perturb cell volume and volume-regulatory processes? Osmotic dehydration of the brain, for example, is a time-honored component of therapy for cerebral edema, yet little is known of its cellular effects. How might understanding of the brain's volume-regulatory capacity improve this primitive therapeutic approach? Second, in pathologic states such as diabetes and renal failure where changes in plasma tonicity and cell volume are common, what is the best way to correct these derangements? What is the etiology of brain swelling in diabetic ketoacidosis or of "osmotic disequilibrium syndrome" in the dialysis patient? Why must therapy for chronic plasma hypo- or hypertonicity differ from that of acute osmotic disturbances? Third, and perhaps most significantly, what currently baffling disease states actually represent derangements of membrane transport and volume regulation? As outlined below, there is preliminary evidence to suggest that altered cell volume regulation may play a fundamental role in the pathogenesis of afflictions as diverse as essential hypertension, sickle cell anemia, and perhaps even cancer.

This chapter seeks to introduce briefly the concepts of cell volume regulation from a clinical perspective. While other chapters will detail the biochemistry, molecular biology, and cellular physiology of volume regulation, we will discuss these concepts as they relate to specific organ systems and the care of patients.

0-8493-4448-4/94/$0.00+$.50

II. CELL VOLUME REGULATION AND GENERAL OSMOTIC HOMEOSTASIS

As clinicians, the understanding of cell volume regulatory physiology adds insight to patient care on a variety of levels. The most obvious of these lies in the area of total body osmotic homeostasis. Although humans have evolved complex homeostatic mechanisms for maintenance of serum tonicity, pathophysiologic and iatrogenic derangements are common. Familiar pathologic hypertonic states include severe diarrhea, renal failure, diabetes mellitus and insipidus, osmotic diuresis with dehydration, water deprivation, inborn errors of metabolism, and certain poisonings (such as ethanol[1]). Iatrogenic causes of acute hypertonicity may include mannitol[2] or hypertonic saline[3] administration for cerebral edema, intravenous infusion of radiographic contrast material, and rapid infusion of certain medications (such as sodium bicarbonate in the setting of cardiac arrest). On occasion, chronic hypertonicity may be intentionally maintained through fluid restriction in the management of cardiopulmonary disease, capillary leak syndromes, or central nervous system injury.

Plasma hypotonicity, usually with hyponatremia, is seen clinically more frequently than plasma hypertonicity. Hyponatremia in the setting of low circulating blood volume is observed in malnutrition, congestive heart failure, nephrotic syndrome, and hepatic cirrhosis (where retained salt and water is effectively lost to edema formation). Hyponatremia due to simple sodium depletion is seen in disorders such as cerebral salt wasting[4] and the salt-wasting nephropathies. Hyponatremia with normal or expanded circulating blood volumes occurs in water intoxication and syndromes of inappropriate antidiuretic hormone secretion.[5] It is interesting to note that up to 20% of the deaths in schizophrenic patients under age 50 are related to the complications of self-induced water intoxication.[6]

Infants and children are particularly vulnerable to changes in serum tonicity due to their small size and large surface area-to-volume ratio. Disorders of both hypertonicity and hypotonicity are common. For example, while it is well known that fluid losses in pediatric diarrheal syndromes may rapidly produce profound dehydration, it is less commonly appreciated that hypertonicity is also frequently produced by inappropriate sodium loading via intravenous infusions. Conversely, since infants are dependent upon the judgment of others for their oral intake, water intoxication and hypertonicity due to inappropriate dilution of juices or infant formula preparations is quite common. Historically, hypotonic states produced in this manner have been a serious concern in undeveloped countries, yet it now appears that such problems are increasing in frequency in the U.S.[7] Finally, medical fluid management is sufficiently complicated in small, critically ill children that even meticulously controlled medical interventions, such as intravenous hyperalimentation, extracorporeal membrane oxygenation, and cardiopulmonary bypass, can be associated with rapid and wide shifts in serum tonicity.

In general, however, *acute* changes in plasma tonicity are well tolerated within the usual clinical ranges. Consequently, fluid and solute challenges are clinically given with relative impunity, while cell volume regulatory mechanisms work "in the background" to maintain general cellular homeostasis. Two such examples include rapid intravenous infusion of mannitol[8] for osmotic diuresis and transvenous absorption of nonelectrolyte irrigation fluids during prostate surgery.[9] When large volumes of potassium-free isosmotic electrolyte solutions are infused (such as during fluid resuscitation with normal saline), plasma potassium concentrations may occasionally fall precipitously.[10] However, when large volumes of an isosmotic permeant nonelectrolyte solution (such as glycine or mannitol) are infused, plasma potassium concentrations remain stable or may actually increase. Recent animal work by Hirose and colleagues[11] suggests that during the latter infusions, critical serum potassium levels are increased or stabilized largely by cell volume regulatory mechanisms. These authors note that hepatic and red blood cells, for example, are permeable to mannitol and may be expected to swell during such infusions. Potassium is then released from the intracellular space via regulatory volume decrease mechanisms. Further such studies will be required to provide the appreciation and quantitation of volume-regulatory processes necessary to permit a more enlightened approach to medical fluid administration.

In contrast, *chronic* changes in plasma tonicity are accompanied by longer-term adaptive mechanisms which substantially complicate therapy. While acute changes in tonicity and resultant cell volume are buffered primarily by rapid transport of electrolytes, chronic hypertonic stress evokes mechanisms that involve the slow accumulation of organic solutes. Classically termed "idiogenic osmoles",[12] these compounds are now known to include amino acids, polyols, methylamines, and similar small organic molecules that do not perturb cellular function at high concentrations. Accumulation of small organics such as taurine, *myo*-inositol, and betaine affords increased osmotic pressure without cell damage (see Chapter 5). While we are just beginning to appreciate the long-term effects of such adaptation, it is now clear that rapid medical correction of chronic hypo- or hypertonic states can be met with serious complications.[13–16]

III. VOLUME REGULATION IN THE CENTRAL NERVOUS SYSTEM

In the brain, small changes in volume may produce profound clinical sequelae. Current understanding of intracranial volume (V) relationships is predicated upon a four-compartment model: $V_{total} = V_{blood} + (V_{extracellular} + V_{intracellular}) + V_{cerebrospinal\ fluid}$. Because the rigid skull provides no avenue for expansion, increased volume of one compartment must be accompanied by a reciprocal decrease in the total volume of the remaining three. Although the cerebrospinal fluid (CSF), and perhaps extracellular volumes, may decrease significantly without detriment, loss of the blood volume ultimately leads to ischemia and injury. As intracranial pressure rises further, caudal displacement of brain parenchyma through the foramen magnum can result in death. Because of the critical nature of such pathology, decades of research have focused upon the individual behavior, general function, and interdependence of the blood, CSF, and extracellular compartments. It is interesting to consider then, that behavior of the largest intracranial volume compartment, $V_{intracellular}$, is poorly understood.

Pathologic changes in brain volume involving the intracellular compartment may be grouped into two broad categories: cytotoxic and anisotonic. Cytotoxic brain cell volume alterations involve cellular injury and include entities such as ischemic stroke, head trauma, and hypoxic/anoxic insults. These pathological conditions affect all age groups and account for substantial morbidity and mortality. Head trauma, for example, accounts for 75% of all trauma admissions to hospitals and for 70% of all deaths occurring within 48 h of admission for trauma.[17] In contrast, anisotonic volume disturbances in the brain result from changes in plasma tonicity and resultant fluid shifts without primary cellular injury. Though less common than cytotoxic injury, anisotonic volume disturbances produce significant morbidity and mortality. While such a distinction is convenient for purposes here, it may one day prove artificial as more is learned regarding the underlying mechanisms of injury.

As discussed above, anisotonic volume disturbances occur in settings such as the syndrome of inappropriate antidiuretic hormone secretion (SIADH), water intoxication, congestive heart failure, inappropriate diuretic use, dehydration, diabetes mellitus, and diabetes insipidus. Common to all is an altered extracellular osmotic environment and the need for brain cells to adapt in order to maintain constant volume. Associated clinical sequelae found in anisotonic volume disturbances include seizures, focal neurologic injury, and severe cerebral edema leading to coma and death. Despite the immense clinical significance of intracellular volume behavior during such stresses, little is known because of the *in vivo* inaccessibility and structural complexity of the brain. Since nervous tissue is composed of multiple cell types and physiologically distinct regions, complete understanding of the cellular and molecular mechanisms of cell volume regulation will require intense study of individual cell types alone and in various combinations. At the same time, *in vivo* work must constantly confirm and supplement *in vitro* observations to ultimately yield clinically applicable information. Thus, while primary cultures and transformed cell lines provide a suitable starting point, animal models are necessary to ultimately define important pathologic mechanisms.

In those disease states where acute decreases in plasma tonicity occur, resultant cell swelling elicits mechanisms for regulatory volume decrease (RVD). These mechanisms typically involve electrolyte transport, such as parallel K^+ and Cl^- channels, as seen by Kimmelburg and O'Connor[18] in primary culture of rat astrocytes. Additionally, loss of small organic molecules may occur, such as the efflux of taurine, aspartate, and glutamate that has been observed during RVD in cultured rat and mouse astrocytes swollen by exposure to elevated K^+ or hypotonic solutions.[19]

When the brain is exposed to elevated plasma tonicity, volume regulation occurs in two stages. First, a rapid regulatory volume increase (RVI) response is observed within minutes and is produced largely by an uptake of inorganic ions.[20] The response is sufficiently rapid that no change in intracellular water content could be observed by Cserr and colleagues in a rat model of plasma hypertonicity.[21] In these experiments, after 30 min of hypernatremia, brain extracellular water content decreased 27%, whereas intracellular water content remained unchanged. After 90 min, extracellular volume had decreased nearly 50%, still without changes in intracellular volume. This RVI of the intracellular space was associated with net cellular uptake of Na^+, Cl^-, and K^+. A similar response has been observed in the C6 glioma cell line. Chassande and co-workers[22] have demonstrated increased transport of Na^+, Cl^-, and K^+ in cells exposed to hypertonic conditions. We have observed rapid return of cell volume after hypertonic exposure by a process which is largely bumetanide-sensitive.[23] Using video microscopy, we have also observed similar volume regulation after hypertonic exposure in primary cultures of rat cortical neurons.[23a]

The aforementioned laboratory observations mirror and add insight to clinical experience. It has been long recognized that treatment of anisotonic conditions can produce significant neurologic problems, including severe cerebral edema, coma, and death.[14,24] Over time, it has become clear that such problems may be avoided by slow correction of pathologic anisotonic states. This is extremely critical for chronic hypo- and hypertonic disturbances (those present for >12 to 24 h) and less so for acute changes in plasma tonicity, which may be more rapidly corrected with few sequelae.[25]

In addressing one particular aspect of this problem, that of the differences in correcting acute and chronic hypertonic states, it appears that the pathophysiologic difference between the two may be related to the volume-regulatory mechanisms employed and how they are controlled by brain cells. Studies involving the rat brain show that during an acute exposure to a hypertonic stress RVI occurs by the uptake of inorganic ions.[21] Exposure then to a normal plasma tonicity causes a rapid loss of those accumulated electrolytes. This is supported by studies in the rat which reveal that electrolytes are lost rapidly from the brain following acute swelling.[26] Interestingly, when the hypertonic stress becomes prolonged (many hours to days), 40 to 60% of those accumulated electrolytes are exchanged for organic solutes, such as alanine, glycine, and inositol.[27–29] Unlike the situation with correction of acute hypertonicity, correction of chronic hypertonic stress results in a very slow loss of organic solutes from the cell. For example, Thurston et al.[30] showed that brain inositol levels increase during acclimation to hypertonicity and remain elevated for at least 4 h in young hypernatremic mice that have undergone rapid rehydration therapy. Lieny et al.[31] have made similar findings in the rat brain; the organic osmolytes betaine and glycerophosphorylcholine return to control levels 24 h after correction of chronic hypernatremia, whereas inositol remains elevated for at least 2 d. These investigations suggest that the slow loss of organic osmolytes may be involved in the development of the brain edema seen during rapid correction of chronic plasma hypertonicity.

Studies in culture cells also support the observations seen in the intact brain. Strange and Morrison[32] have shown that C6 glioma cells shrink when exposed to a hypertonic medium, then undergo complete RVI in 5 to 10 min. When these same cells are exposed to an isotonic medium, the cells swell but then rapidly undergo RVD. In contrast, when C6 cells are grown for 48 h in hypertonic media, 50 to 75% of the accumulated electrolytes used in the rapid RVI are replaced by inositol. This inositol accumulation is brought about by the upregulation of a Na^+-dependent uptake pathway.[32] When chronically hypertonic cells are exposed to an isotonic medium, they swell and exhibit little RVD. This lack of RVD correlates with a slow loss of inositol from the cell. This slow loss of inositol may be due in part to a slow downregulation of the Na^+-dependent uptake pathway.[32,33]

Unlike anisotonic volume changes, *cytotoxic* volume changes result from brain cell injury. Though originally ascribed to loss of oxygen delivery and disruption of cellular energetics, research over the past decade has demonstrated that extracellular mediators and neurotransmitters play a pivotal role in the pathophysiology of cytotoxic injury. One such mediator is the excitatory neurotransmitter glutamate. Glutamate excites virtually all central neurons and is normally present in nerve terminals at millimolar concentrations.[34] Glutamate interacts with at least three classes of membrane receptors as defined by their pharmacological agonist: *N*-methy-D-aspartate (NMDA), quisqualate, also called AMPA, and kainate, all of which are linked to cation channels. Glutamate itself is a potent neurotoxin causing cultured neurons to swell and die after as little as 5 min of exposure to 100 μM concentrations.[35]

The toxic changes seen with glutamate *in vitro* are biphasic and involve initial, acute swelling of neuronal dendrites and cell bodies followed by a more slowly evolving neuronal degeneration.[36] The molecular sequence of events in the initial phase of glutamate toxicity involve (1) binding to a specific receptor (e.g., NMDA), (2) opening of its associated membrane cation channel, leading to (3) Na^+ influx and membrane depolarization, followed by (4) secondary influx of Cl^- and water, thus producing the observed swelling.[35] The cell swelling seen *in vitro* is very dramatic and, *in vivo,* may be manifested by cerebral edema, neurologic dysfunction, coma, and death.

Although much current research into potential therapies for acute neuronal injury is focused upon antagonism of glutamate receptors, standard clinical therapy continues to revolve around osmotic opposition of cerebral swelling. After maximizing reductions in the cerebrospinal fluid and blood compartment volumes (via drainage and hyperventilation, respectively), routine practice calls for administration of osmotic diuretics such as glycerol or mannitol. This approach presumes that the extra- and/or intracellular compartments behave in an osmometric fashion and that excess water may be removed via creation of a brain-to-blood osmotic gradient. Although much data have accumulated demonstrating the rapid and transient reduction of intracranial pressure following intravenous infusion of agents such as mannitol, there is conflicting evidence as to whether this arises from hemodynamic, rheologic, or osmotic mechanisms.

If, as demonstrated by the experiments described above, brain cells possess active volume-regulatory mechanisms, the neuronal swelling associated with glutamate toxicity (and perhaps with all cytotoxic injuries), may be viewed as an instance of failure of these homeostatic systems. Whether this is due to inhibition, disruption, or supersaturation of these processes remains to be seen. It is clear, however, that basic understanding of normal cell volume regulatory processes in the brain is a prerequisite for the understanding of such pathophysiology.

Once understood, manipulation of volume-regulatory processes may form the basis of future therapies for cerebral edema. For example, it has long been known that "loop" and osmotic diuretics in combination act synergistically to reduce brain volume. This is important both for the care of patients with brain swelling and for reduction of brain bulk during neurosurgical procedures. While several mechanisms for such synergism have been proposed, it has only recently been noted that this may, in fact, be due to inhibition of volume-regulatory pathways (by "loop" diuretics) during osmotic shrinkage.[23]

IV. DIABETIC COMA AND FATAL CEREBRAL EDEMA

Diabetic coma presents a special case where the understanding of cell volume regulatory mechanisms in the brain is critical. Fatal cerebral edema during diabetic ketoacidosis (DKA) is an infrequent event, yet in some reports it is responsible for approximately 50% of the associated mortality.[37,38] When severe cerebral swelling and brainstem herniation does occur, it is usually seen after the start of medical intervention when other clinical features such as acidosis, hyperglycemia, and the state of hydration appear to be improving. There is even evidence to suggest that brain swelling occurs commonly in all patients with treated DKA[39,40] and that fatal edema is merely an exaggeration of this typical response.

Several cell volume regulatory mechanisms have been proposed as being responsible for this phenomenon. The first of these presumes that chronic cell volume regulation has occurred in these hypertonic patients and that medical rehydration creates an osmotic disequilibrium with shift of free water into the brain. Supporting this hypothesis is the presence of increased solute in the brains of chronically hyperglycemic animals,[24] the observation that spinal fluid from untreated patients is hypertonic relative to their plasma,[41] and the autopsy finding of cerebral edema in patients with ketoacidosis. Unfortunately, this simple explanation does not agree with clinical experience (reviewed in Reference 42). For example, in their review of 275 cases of uncontrolled diabetes, Carroll and Matz[43] found mortality to be more related to inadequate fluid resuscitation than to overzealous treatment and saw no deleterious effects of even massive fluid repletion, despite large osmolar gradients. Similarly, a review of 69 pediatric cases of diabetes and acute intracerebral complications could demonstrate no relationship between rehydration, falls in serum glucose, and neurologic outcome.[42] While this area of clinical investigation is complex and difficult to control, it remains true that a consistent pattern of osmolar injury with fluid administration has failed to emerge and that mortality still occurs with unchanged frequency despite slow correction and meticulous attention to fluid therapy.

Administration of insulin, deficient in such patients until medical intervention, has also been suggested as a possible etiology of brain swelling. Insulin is known to increase volume in several cell types (see below) through activation of the Na^+/H^+ exchanger and other electrolyte uptake paths. In the insulin-deficient state, these pathways may be relatively quiescent, yet suddenly activate when large replacement doses are administered. This consideration has prompted some to recommend low-dose insulin therapy for correction of DKA. It has not been demonstrated, however, that any specific deviation from standard medical care results in improved outcome.[42]

Finally, it has been proposed that acid-base shifts and the Na^+/H^+ exchanger (the primary regulator of intracellular pH) are largely responsible for cerebral edema in DKA.[44] According to this hypothesis, several important changes occur during DKA and its therapy which set the stage for brain cell swelling. Before therapy, both intra- and extracellular spaces are acidic due to accumulation of lactic and keto-acids. In such a state, low intracellular pH serves as a stimulus for Na^+/H^+ exchange, yet this is largely inhibited by extracellular competition between Na^+ and H^+. During treatment, however, extracellular pH normalizes rapidly and inhibition of Na^+/H^+ exchange is suddenly removed. An outward gradient for protons then exists favoring rapid uptake of Na^+ in exchange for H^+ and osmotic swelling results (see Figure 4–1). Thus, current therapy for DKA simultaneously produces (1) a rapid normalization of pH, (2) a rapid lowering of extracellular glucose levels, and (3) normalization of insulin levels after a period of insulin deficiency. All three of these factors may combine through complementary mechanisms to favor cell swelling (Figure 4–1). It is interesting to envision a future treatment strategy that would take these issues into consideration and thereby minimize the adverse effects of therapy. Potentially, such a strategy might include a combination of low-dose insulin, very slow correction of metabolic parameters, and early systemic administration of Na^+/H^+ exchange inhibitors.

V. DIABETES AND CELL VOLUME REGULATION OUTSIDE OF THE BRAIN

Cellular mechanisms involved in chronic volume regulation have been implicated in the peripheral neuropathy, retinopathy, and cataract formation familiar to diabetic patients. Chronically elevated levels of glucose and related compounds pose two challenges to the cell. The first is simply that of chronic hypertonicity, as previously discussed. The second is that of chronic concentration gradients favoring intracellular movement of carbohydrates.

It has been demonstrated that lens cells cultured in high galactose media accumulate polyols through the action of aldose reductase. This, in turn, results in cell swelling and the appearance of intracellular vacuoles similar to those seen in diabetic animals.[45] Polyol accumulation appears to be paralleled by both an increased efflux and decreased uptake of organic osmolytes such as *myo*-inositol.[46] Most of these effects may be prevented by exposure to the aldose reductase inhibitor, sorbinil, or by hypertonic prevention of cell swelling.

Similar observations have been made in cultured retinal cells, where accumulation of polyols, resultant swelling, and vacuolization also occur.[47] In a fashion similar to lens cells, after 72 h of culture in the presence of 30 m*M* galactose, intracellular taurine and inositol levels are diminished 80%, yet this response may be blocked by addition of an aldose reductase inhibitor.

Finally, comparable mechanisms appear to be at work in the peripheral nerve. Over the past decade it has become clear that many of the abnormalities in peripheral nerve conduction that are seen in animal models of diabetes are related to depletion of intracellular *myo*-inositol pools (reviewed in Reference 48). For example, both nerve conduction velocity and *myo*-inositol content fall within days in the sciatic nerves of rats with streptozocin-induced diabetes. These changes are unimproved by conventional insulin therapy, yet may be ameliorated by dietary supplements of *myo*-inositol or by administration of sorbinil.[49] Cultured neuroblastoma cells have been used as a model for the peripheral nerve and exhibit similar reduction in inositol uptake.[50] Study of this cell line has revealed that *myo*-inositol exists in discrete intracellular pools and its uptake is inhibited both competitively, by acute increases in extracellular glucose concentration, and noncompetitively, by chronic increases. As in the models described above, the chronically acquired noncompetitive inhibition could be blocked by addition of sorbinil.[51]

Burg and Kador[52] have demonstrated that synthesis of compounds such as sorbitol via upregulated aldose reductase activity is fundamental to hypertonic osmoregulation in renal medullary cells. They further argue that there is no such normal role for this behavior in other tissues. Thus, in the diabetic retina or peripheral nerve, accumulated polyols may themselves be toxic or may instead cause cell injury through disruption of normal osmoregulation. Strange et al.[53] have demonstrated that astroglial cells possess a swelling-activated efflux pathway for inositol, similar to those for sorbitol and taurine described in a number of other cell types. Further, they have demonstrated that downregulation of the *myo*-inositol uptake pathway occurs with swelling.[33] Swelling-induced downregulation may provide the basis for the noncompetitive inhibition of *myo*-inositol uptake described by Yorek and colleagues in chronically hyperglycemic neuroblastoma cells.[50]

Taken together, these data are exciting in that they suggest a common, volume-related mechanism for three of the major complications of diabetes (Figure 4–2). Constant exposure to pathologic concentrations of extracellular glucose, galactose, and related compounds may lead to cellular influx and subsequent conversion of these compounds to polyols via the action of aldose reductase. In nerve and related tissue, swelling caused by polyol accumulation leads to activation of inositol efflux pathways as a pathologic attempt toward volume regulation (RVD). Concurrently, swelling-induced downregulation of the *myo*-inositol uptake pathways may occur, resulting in severe refractory depletion of this compound. Depletion of *myo*-inositol, in turn, produces nerve dysfunction. While the mechanisms for this dysfunction remain unclear and, indeed, the entire analysis remains speculative, it is encouraging that elucidation of cell vol-

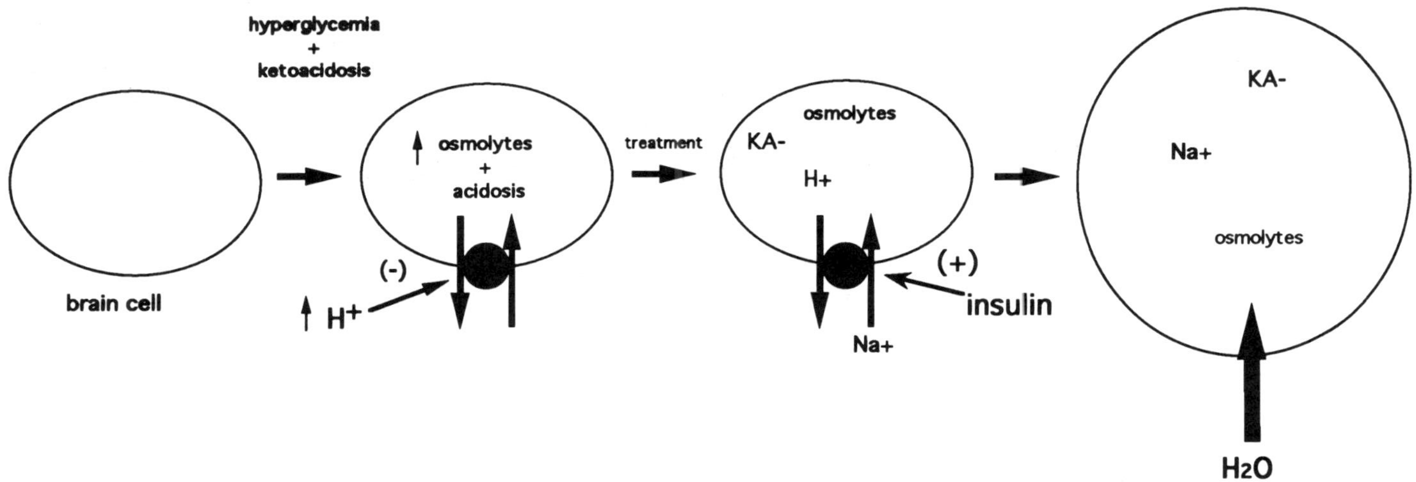

Figure 4–1 Proposed mechanism for acute cerebral edema during treatment of diabetic ketoacidosis. Brain cells exposed to acidic, ketotic, hyperglycemic, hypertonic extracellular fluid accumulate organic osmolytes to defend their volume. Intracellular acidosis develops due to diffusion of lactic and keto-acids (KA), yet Na^+/H^+ exchange is inhibited by extracellular acidosis. During medical therapy, extracellular pH is corrected, removing Na^+/H^+ inhibition. Additionally, insulin is administered and further stimulates Na^+/H^+ exchange. Net movement of Na^+ and water into the cell results, a process further exacerbated by medical correction of external (plasma) hypertonicity.

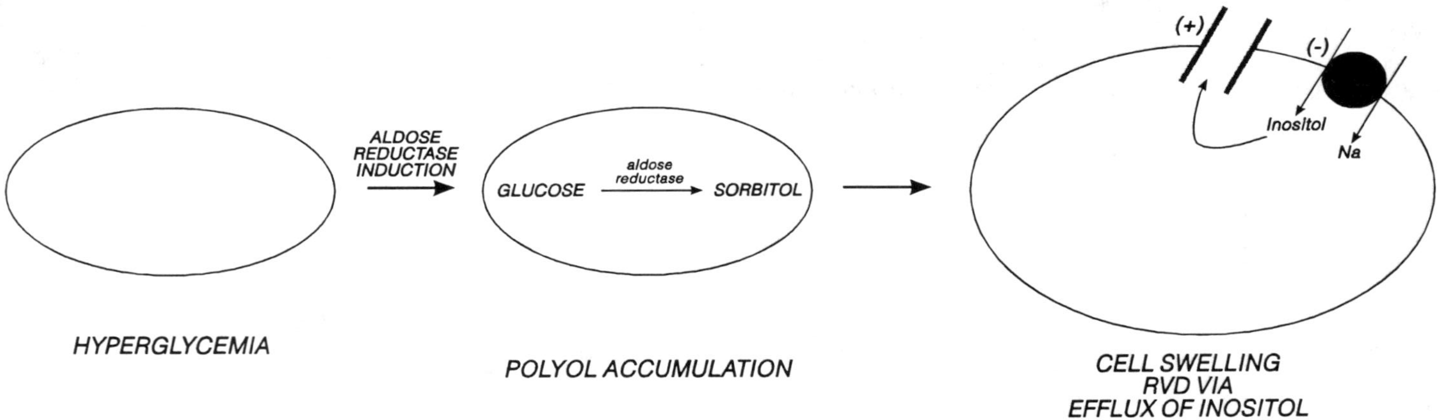

Figure 4–2 Proposed osmoregulatory mechanism for diabetic neuropathy. Chronic exposure to hyperglycemia leads to diffusion of glucose and related compounds into the cell. By the action of aldose reductase (AR), sugars are converted to polyols, which then accumulate intracellularly. Increased intracellular osmolarity secondary to polyol accumulation leads to cell swelling. In response, RVD mechanisms, such as the organic osmolyte efflux pathways, are activated, resulting in the loss of *myo*-inositol. Simultaneously, swelling-induced downregulation of the Na^+-inositol cotransporter also occurs. Ultimately, intracellular depletion of inositol leads to neuropathy via an incompletely understood mechanism.

ume regulatory processes now provides direction for research in an area where medical science otherwise currently offers little.

VI. CELL VOLUME REGULATION AND CARDIOVASCULAR FUNCTION

Because failure of cell volume regulation has been associated with severe damage after myocardial ischemia,[54,55] cardiac cell volume regulation has been the subject of several recent investigations. Using video microscopy, Drewnowski and Baumgarten[56] have demonstrated that the commonly used diuretics bumetanide and chlorothiazide cause myocyte shrinkage under isotonic conditions and attenuate volume-regulatory responses during hypertonic stress. These authors speculate that direct myocardial effects leading to reduced cardiac edema may account for some of the benefits of diuretic administration to patients with congestive heart failure.

In similar work, Clemo and colleagues[57] have identified and characterized a cGMP-mediated, atrial natriuretic factor (ANF)-sensitive $Na^+/K^+/2Cl^-$ cotransporter instrumental in regulation of volume in rabbit ventricular myocytes. Myocyte cell volume has implications for contractility through at least two mechanisms. First, overlap and spatial interaction of contractile elements are critically dependent upon intracellular distances. Thus, extremes of size may interfere with overlap and impair mechanical function. Second, volume changes involving Na^+ influx may indirectly effect contractile states through stimulation of Na^+/Ca^{2+} exchange and alteration of intracellular Ca^{2+}.

Like diuretics, sodium nitroprusside (SNP) has been shown to decrease myocyte volume, presumably through activation of guanylate cyclase.[57] While it is generally felt by clinicians that SNP enhances cardiac function through vasodilation and reduction of afterload, this new work suggests the possibility of other mechanisms. Indeed, in a preliminary communication,[58] Clemo and Baumgarten report interesting experiments in progress using a Langendorff preparation. In this model, the predicted macroscopic effects of cell shrinkage, such as improved compliance with SNP or bumetanide treatment, are demonstrable.

The newly discovered endocrine functions of the heart have also been linked to cell volume. In work with the $Na^+/K^+/2Cl^-$ cotransporter found in rabbit ventricular cells, Clemo and associates[57] suggest that cell volume changes may provide a local mechanism for feedback inhibition of ANF release. In such a model, ANF-stimulated cotransport may lead to cell shrinkage, which in turn inhibits further ANF release. Similarly, in atrial cells, osmotic stretch appears to be a major stimulus for release of atriopeptin, a key hormone involved in natriuresis, modulation of circulating blood volume, and blood pressure control.[59]

Some of the most intriguing of all work in the area of cell volume regulation and cardiovascular function is in the area of hypertension. As suggested above, the Na^+/H^+ exchanger is present in all cell types and functions in pH regulation, volume regulation (RVI), and Na^+ transport into cells and across epithelia. Because sodium intake, sodium metabolism, and intravascular volume status have been linked to essential hypertension, Na^+ transport in hypertensive individuals has been studied intensively (reviewed in Reference 60). Increased Na^+/H^+ exchange in red cells, platelets, and lymphocytes[61] has been identified consistently in patients with essential hypertension.

Increased Na^+/H^+ exchange is suspected of contributing to hypertension by two mechanisms. Increased renal absorption of Na^+ and water in the proximal tubule of the kidney clearly leads to expansion of circulating blood volume, elevation of filling pressures, increase in cardiac output, and resultant hypertension. Inappropriate excretion of dietary sodium loads mandates naturesis or restriction of intake for adequate blood pressure control. A second potential mechanism involves cell volume regulation directly. Increased Na^+/H^+ exchange in vascular smooth muscle cells may produce cell swelling (via entry of Na^+ and water) as well as increased intracellular pH. Cell swelling, in turn, may diminish vascular caliber directly or may lead to increased vascular tone through stimulation of Na^+/Ca^{2+} exchange. Increased intracellular pH may also engender cell proliferation and hypertrophy, producing the medial thickening usually seen in chronic hypertension. While much of the populational, genetic, and physiologic data regarding sodium exchange and hypertension remain to be sorted out, it has been tentatively suggested that inhibition of transport pathways involved in cell volume regulation may form the basis for future therapy of hypertension.[62]

VII. CELL VOLUME REGULATION AND SICKLE CELL DISEASE

Recently, a great deal of attention has been devoted toward understanding the mechanisms of cell volume regulation in erythrocytes of patients homozygous for the alleles of Hemoglobin S (HbS), Hemoglobin C (HbC), and patients heterozygous for Hb S and Hb C (SC disease). Abnormalities of hemoglobin structure clinically translate into dramatic alterations of red cell volume and morphology, which are in turn responsible for microvascular aggregation, thrombosis, and injury. Investigations into the relationship between cell volume regulation and sickle cell disease is predicated upon the observation that patients with sickle cell (SS) disease possess an abnormally high fraction of immature red cells and that such cells exhibit enhanced activity of several volume-regulatory ion transporters.[63,64] These observations have very important pathophysiologic implications, given that the rate and extent of hemoglobin polymerization following deoxygenation is strongly dependent on the initial hemoglobin concentration.[65] Ion transport processes that produce large volume reductions increase the mean corpuscular hemoglobin concentration (MCHC), increase hemoglobin polymerization, and favor sickling. Conversely, those processes producing volume increases and decreased MCHC have anti-sickling properties.[63]

The K^+/Cl^- cotransporter mediates RVD in SS, CC, and SC red cells. The high expression of this transport pathway is independent of genetically determined changes in the primary structure of hemoglobin.[63,66] The cotransporter is inhibited by deoxygenation but is stimulated by cell swelling induced by acidic pH or a hypotonic medium.[63,66]

Given these specific properties, the K^+/Cl^- cotransporter may play an important pathophysiologic role in the clinical sequelae related to sickle hemoglobinopathies. It is known that local or global tissue acidosis predisposes SS patients to painful sickle crises and exacerbates sickling injury. It is possible that the K^+/Cl^- cotransporter is activated in SS cells passing through acidotic vascular beds, which results in enhanced KCl efflux and cell shrinkage. Cell shrinkage in turn increases the MCHC, initiates hemoglobin polymerization, and produces further sickling, to the point where sickle forms predominate even in downstream areas of relatively normal plasma pH. A pictorial representation of this scenario is presented in Figure 4–3.

Other transport pathways that may play a role in red cell volume regulation and a pathologic role in hemoglobinopathies include the Na^+/H^+ antiporter and the K^+ selective channel controlled by cytostolic calcium. The actual cellular function of the K^+ selective channel is yet to be fully elucidated, but studies in SS cells have suggested that it may play a role in the deoxygenation-stimulated K^+ efflux associated with the formation of dense cells.[67] The potential pathophysiologic role played by the Ca^{2+}-activated K^+ channel relates to the increased cystolic

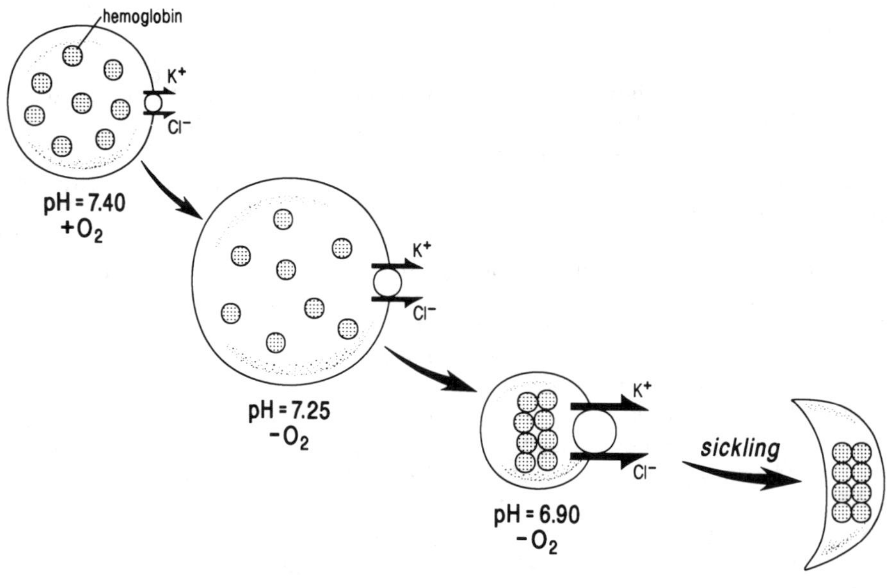

Figure 4–3 Effect of an acidic, deoxygenated vascular bed on red cells homozygous for hemoglobin S. As red cells flow from vascular beds with a normal pH and a normal supply of oxygen to vascular beds with very little oxygen and acidic pH, the red cell swells. This swelling induces increased activity of the K^+/Cl^- cotransporter,[66] causing red cell shrinkage. The decreased volume of the red cell increases the concentration of hemoglobin. This, along with deoxygenation, induces hemoglobin to polymerize further, causing sickling of the cell.

Ca^{2+} induced by deoxygenation.[68] In theory, local tissue deoxygenation and increased Ca^{2+} may cause opening of K^+ channels, rapid cell shrinkage, and irreversible sickling.[69]

The Na^+/H^+ antiporter has also been studied in SS and homozygous hemoglobin A (AA) red cells, both because of its age dependence and its responsiveness to cell shrinkage.[70] As compared to normal cells, high levels of expression of the Na^+/H^+ antiporter have been reported in SS cells and in AA reticulocytes.[70] The antiporter's physiologic role involves both pH regulation and, when coupled with anion exchange (Cl^-/HCO^-_3), volume regulation through net movement of NaCl and water. The *in vivo* consequences of increased Na^+/H^+ activity in sickle cell disease remain speculative, yet at least one study has demonstrated shrinkage activation of the Na^+/H^+ antiporter in SS cells.[70] If such Na^+/H^+ exchange is functional *in vivo,* it may help swell shrunken cells. Activation of this exchanger may therefore be clinically beneficial in the treatment of SS crisis.

VIII. CELL VOLUME REGULATION AND HEPATIC FUNCTION

To perform the basic functions of life, cells must accumulate solutes along or against electrochemical gradients. Many of these solutes become substrates for metabolism to smaller molecules and go on to further increase intracellular tonicity. Accumulation must be balanced by solute loss or result in cell swelling and death. It is the responsibility of volume-regulatory pathways to maintain volume equilibrium despite large solute fluxes. Clearly, the more metabolically active a cell, the more important these constant volume-regulatory mechanisms become. In the liver, it has been proposed that this association of volume and metabolism is so critical that cell volume changes actually form the basis for metabolic control (reviewed in Reference 71).

Volume-regulatory increase and decrease pathways are active in the liver and may even be provoked by the swelling induced by exposure of cells to amino acids at physiologic concentrations. Transporters involved in volume regulation are, in turn, modulated by the primary hormones of metabolism, insulin and glucagon, such that exposure to insulin produces swelling and exposure to glucagon causes shrinkage. Haussinger and Lang[71] suggest that, because osmotic swelling and shrinkage affect metabolism similarly, volume perturbation is actually the "second messenger" for metabolic control. They provide data that, in liver cells, the anabolic effects of insulin may be mimicked by hypotonic swelling and the catabolic effects of glucagon by hyperosmotic shrinkage.

Clinically, the implications of such understanding are highly speculative, yet exciting. For example, in starvation and protein deprivation states, does efflux of solutes occur causing cell shrinkage and metabolic shutdown? Do chronic regulatory mechanisms then come into play? If so, what is the best way to restore function once this has occurred? In critically ill patients, what is the nature of hyperalimentation hepatitis and is it in one way or another related to solute load? When planning nutrition or hyperalimentation protocols, might adjustment of tonicity and composition optimize substrate utilization and minimize liver injury? In liver transplantation, what is the best way to preserve and restore function in the harvested organ? Might osmotic manipulations during organ transport decrease metabolism and prolong storage time, and postoperative manipulations later enhance function in the recipient? Can metabolic activity be temporarily enhanced to hasten metabolism of poisons or drug overdoses? Obviously, further work is necessary to establish the validity of the model and determine its importance in the care of patients.

IX. CELL VOLUME REGULATION AND CONTROL OF CELLULAR PROLIFERATION

Continued unchecked proliferation of a particular cellular clone forms the basis of oncogenesis. The mechanisms by which particular cells lose the ability to differentiate and instead continue mitotic proliferation is, therefore, the essence of cancer research. Although changes in cell size precede and are necessary for division, this has generally been considered a secondary rather than primary phenomenon. Recently, however, there has been speculation that volume-sensitive pathways play a primary role in cellular proliferation.

Several studies have examined the role of different transport processes in cellular proliferation. Concurrent with the dramatic morphologic changes induced by mitogens in quiescent cells, there is a stimulation of monovalent ion fluxes. These fluxes represent the earliest observed phenomena in the sequence of events leading quiescent cells to proliferate.[72] It has been shown that there is an increase in Na^+ influx which stimulates Na^+/K^+ pump activity and a subsequent increase in cellular levels of K^+. Also present is a ouabain-insensitive furosemide-sensitive K^+ transport system that is activated early in cell transformation.[72]

The Na^+/H^+ antiporter, important for both pH regulation and volume control, may play a role in the initiation and maintenance of cellular proliferation. There is speculation, and some evidence in nonmammalian cells, that the intracellular alkalinization produced by stimulation of the Na^+/H^+ antiporter is an activator of the cellular machinery responsible for proliferation. There is further speculation that cell swelling caused by Na^+/H^+ exchange may, in fact, be part of the mitotic signaling pathway.[73] Research thus far has established that specific activation of the antiporter results in proliferation, yet only under a restrictive set of conditions in mammalian cells.[74] These conditions include an acidic environment or the absence of HCO^-_3. The physiologic significance of this effect remains unclear given that plasma pH varies relatively little and HCO^-_3 is always present at concentrations that support the operation of alternative pH regulatory systems.

Another example of a transport process having an effect on cellular proliferation is in the recent work of Meyer and co-workers.[75] These investigators examined the activity of the $Na^+/K^+/2Cl^-$ cotransporter in cells transfected with Ha-*ras*. In 3T3 fibroblasts transfected with the Ha-*ras* oncogene, the activity of a $Na^+/K^+/2Cl^-$ cotransporter was increased approximately 4-fold. This increased activity of the cotransporter was associated with a parallel increase in cell volume and an increased rate of cell proliferation. The increased activity of the $Na^+/K^+/2Cl^-$ cotransporter, the increase in cell volume, and interestingly, the increased rate of cellular proliferation were all blocked by furosemide. Cellular proliferation induced by the mitogen bombesin was also blocked by furosemide. These results suggest that activation of the cotransporter is an important component of the mitotic signaling pathway. Even more intriguing is the suggestion that the increase in cell volume associated with the cotransporter activity may also be a component in mitotic signaling.

Speculations have been offered regarding the possible role of ion transport pathways in normal and abnormal cellular proliferation. They include the idea that the increased salt and water content of cells resulting from stimulation of these pathways may be a necessary prelude to protein synthesis and subsequent cell division or transformation. Or, put another way, mitogen-induced swelling may be necessary to provide the intracellular space needed to accommodate newly synthesized macromolecules and organelles.[74] Finally, increased cell volume itself may produce membrane changes that serve as a positive feedback signal inducing further proliferative activity. Overall, preliminary experiments suggest a link between ion transport, cell volume perturbation, and mitotic proliferation. It remains to be seen whether such observations can be built upon to further our understanding of the mechanisms responsible for increasing cellular proliferation and how these mechanisms go awry in disease states such as cancer.

X. FUTURE DIRECTIONS

Space here does not permit discussion of the medical implications of cell volume regulation as it occurs in many other tissues, such as the lung, bronchial tree, pancreas, and hypothalamus. Indeed, nearly every cell type investigated to date exhibits volume-regulatory properties, which appear to be intimately related to overall cellular function. This is of little surprise, given that volume regulation is critical to survival and must have been one of evolution's oldest cellular homeostatic processes. It should come as no shock to physicians, then, that as more is learned regarding the structure and function of whole organs and organisms, the clinical implications of volume-regulatory physiology expand.

REFERENCES

1. **Robinson, A. G. and Loeb, J. N.,** Ethanol ingestion — commonest cause of elevated plasma osmolality?, *N. Engl. J. Med.*, 284, 1253, 1971.
2. **Wise, B. L. and Chater, N.,** The value of hypertonic mannitol solution in decreasing brain mass and lowering cerebrospinal-fluid pressure, *J. Neurosurg.*, 19, 1038, 1962.
3. **Fisher, B., Thomas, D., and Peterson, B.,** Hypertonic saline lowers raised intracranial pressure in children after head trauma, *J. Neurosurg. Anesthes.*, 4, 4, 1992.
4. **Weinland, M. E., O'Boynick, P. L., and Goetz, K. L.,** A study of serum antidiuretic hormone and atrial natriuretic peptide levels in a series of patients with intracranial disease and hyponatremia, *Neurosurgery*, 25, 781, 1989.
5. **Ho, M. S. and Carroll, H. J.,** Disorders of sodium metabolism: hypernatremia and hyponatremia, *Crit. Care Med.*, 20, 94, 1992.
6. **Andrew, R. D.,** Seizure and acute osmotic change: clinical and neurophysiological aspects, *J. Neurol. Sci.*, 101, 7, 1991.
7. **Keating, J. P., Shears, G. J., and Dodge, P. R.,** Oral water intoxication in infants: an American epidemic, *Am. J. Dis. Child.*, 145, 985, 1991.

8. **Moreno, M., Murphy, C., and Goldsmith, C.,** Increase in serum potassium resulting from the administration of hypertonic mannitol and other solutions, *J. Lab. Clin. Med.,* 71, 678, 1969.
9. **Hirose, M. and Tanaka, Y.,** Serum potassium change during the TURP syndrome by cell volume regulation, *Can. J. Anaesth.,* 39, 300, 1992.
10. **Malone, D. R., McNamara, R. M., Malone, R. S., Fleisher, G. R., and Spivey, W. H.,** Hypokalemia complicating emergency fluid resuscitation in children, *Pediatr. Emer. Care,* 6, 13, 1990.
11. **Hirose, M., Hashimoto, S., Nose, H., Morimoto, T., Itoh, T., Natsuyama, T., and Tanaka, Y.,** Mechanism underlying the changes in plasma potassium concentration during infusion of isosmotic nonelectrolyte solution, *Anesthesiology,* 77, 336, 1992.
12. **McDowell, M. E., Wolf, A. V., and Steer, A.,** Osmotic volumes of distribution: idiogenic changes in osmotic pressure associated with administration of hypertonic solutions, *Am. J. Physiol.,* 180, 545, 1955.
13. **Sterns, R. H., Riggs, J. E., and Schochet, S. S.,** Osmotic demyelination syndrome following correction of hyponatremia, *N. Engl. J. Med.,* 314, 1535, 1986.
14. **Sterns, R. H., Thomas, D. J., and Herndon, R. M.,** Brain dehydration and neurologic deterioration after rapid correction of hypernatremia, *Kidney Int.,* 35, 69, 1989.
15. **Morris-Jones, P. H., Houston, I. B., and Evans, R. C.,** Prognosis of the early complications of acute hypernatremia, *Lancet,* 2, 1385, 1967.
16. **Finberg, L.,** Hypernatremic (hypertonic) dehydration in infants, *N. Engl. J. Med.,* 289, 196, 1973.
17. **Davis, R. J., Dean, J. M., Goldberg, A. L., Carson, B. S., Rosenbaum, A. E., and Rogers, M. C.,** Head and spinal injury, in *Textbook of Pediatric Intensive Care,* Rogers, M. C., Ed., Williams & Wilkins, Baltimore, 1987, 650.
18. **Kimmelburg, H. K. and O'Connor, E.,** Swelling of astrocytes causes membrane potential depolarization, *Glia,* 1, 219, 1988.
19. **Kimmelburg, H. K., Goderie, S. K., Higman, S., Pang, S., and Waniewski, R. A.,** Swelling induced release of glutamate, aspartate, taurine, from cultured astrocytes, *J. Neurosci.,* 10, 1583, 1990.
20. **Cserr, H. F. and Patlak, C. S.,** Regulation of brain volume under isosmotic conditions, in *Advances in Comparative and Environmental Physiology,* Vol. 9, Gilles, F., Hoffman, E. K., and Boles, L., Eds., Springer-Verlag, Berlin, 1991, 61.
21. **Cserr, H. F., Depasquale, M., Nicholson, C., Patlak, C. S., Pettigrew, K. D., and Rice, M. E.,** Extracellular volume decreases while cell volume is maintained by ion uptake in rat brain during acute hypernatremia, *J. Physiol.,* 442, 277, 1991.
22. **Chassande, O., Frelin, C., Fafahifar, D., Thierry, J., and Lazdunski, M.,** The Na^+K^+/Cl^- cotransport in C6 glioma cells: properties and roles in volume regulation, *Eur. J. Biochem.,* 171, 425, 1988.
23. **McManus, M. L. and Strange, K.,** Acute volume regulation in brain cells exposed to hypertonic challenge, *Anesthesiology,* 78(6), 1132, 1993.
23a. **Churchwell, K. B., McManus, M. L., and Strange, K.,** unpublished observations.
24. **Arieff, A. I. and Kleeman, C. R.,** Cerebral edema in diabetic comas. II. Effects of hyperosmolarity, hyperglycemia and insulin in diabetic rabbits, *J. Clin. Endocrinol. Metab.,* 38, 1057, 1974.
25. **Elisa, F. M., Litou, H., and Siamoppoulos, K. C.,** Survival after severe iatrogenic hypernatremia, *Am. J. Kidney Dis.,* 14, 230, 1989.
26. **Melton, J. E., Patlak, C. S., Pettigrew, K. D., and Cserr, H. F.,** Volume regulatory loss of Na^+, Cl^-, and K^+ from rat brain during acute hyponatremia, *Am. J. Physiol.,* 252, F661, 1987.
27. **Lohr, J. W., Mcreynolds, J., and Grimaldi, T.,** Effect of acute and chronic hypernatremia on myo-inositol and sorbitol concentration in rat brain and kidney, *Life Sci.,* 43, 271, 1988.
28. **Helig, C. W., Stromski, M. E., Blemenfelf, J. D., Lee, J. P., and Gullans, S. R.,** Characterization of the major brain osmolytes that accumulate in salt-loaded rats, *Am. J. Physiol.,* 257, F1108, 1989.
29. **Pollack, A. S. and Arieff, A. I.,** Abnormalities of cell volume regulation and their functional consequences, *Am. J. Physiol.,* 239, F195, 1980.
30. **Thurston, J. H., Sherman, W. R., Hauhart, R. E., and Kloepper, R. F.,** Myo-inositol: a newly identified nonnitrogenous osmoregulatory molecule in mammalian brain, *Pediatr. Res.,* 26, 482, 1989.
31. **Lieny, H. H., Shapiro, J. I., and Chan, L.,** Study of brain electrolytes and organic osmolytes during correction of chronic hyponatremia, *J. Clin. Invest.,* 88, 303, 1991.

32. **Strange, K. and Morrison, R.,** Volume regulation during recovery from chronic hypertonicity in brain glial cells, *Am. J. Physiol.*, 263 (Cell Physiol. 32), C412, 1992.
33. **Paredes, A., McManus, M., Kwon, H. M., and Strange, K.,** Osmoregulation of Na^+-inositol cotransporter activity and mRNA levels in brain glial cells, *Am. J. Physiol.*, 263 (Cell Physiol. 32), C1282, 1992.
34. **Curtis, D. R. and Johnson, G. A. R.,** Amino acid transmitters in the mammalian central nervous system, *Ergeb. Physiol.*, 69, 98, 1974.
35. **Choi, D. W.,** Glutamate neurotoxicity and diseases of the central nervous system: a review, *Neuron*, 1, 623, 1988.
36. **Olney, J. W., Price, M. J., Samson, L., and Labruyere, J.,** The role of specific ions in glutamate neurotoxicity, *Neurosci. Lett.*, 65, 65, 1986.
37. **Young, E. and Bradley, R.,** Cerebral edema with irreversible coma in severe diabetic ketoacidosis, *N. Engl. J. Med.*, 12, 665, 1967.
38. **Scibila, J., Finegold, D., Dorman, J., Beker, D., and Drash, A.,** Why do children with diabetes die?, *Acta Endocrinol. Suppl.*, 279, 326, 1986.
39. **Krane, E. J., Rockoff, M. A., Wallman, J. K., and Wolfsdorf, J. I.,** Subclinical brain swelling in children during treatment of diabetic ketoacidosis, *N. Engl. J. Med.*, 312, 1147, 1985.
40. **Clements, R. S. J., Blumenthal, S. A., Morrison, A. D., and Winegrad, A. I.,** Increased cerebrospinal fluid pressure during treatment of diabetic ketosis, *Lancet*, 2, 657, 1971.
41. **Ohman, J. L., Marliss, E. B., Aoki, T. T., Munichoodappa, C. S., Khanna, V. V., and Kozak, G. P.,** The cerebrospinal fluid in diabetic ketoacidosis, *N. Engl. J. Med.*, 284, 283, 1971.
42. **Rosenbloom, A. L.,** Intracerebral crises during treatment of diabetic ketoacidosis, *Diabetes Care*, 13, 22, 1990.
43. **Carroll, P. and Matz, R.,** Uncontrolled diabetes mellitus in adults: experience in treating diabetic ketoacidosis and hyperosmolar nonketotic coma with low-dose insulin and a uniform treatment regimen, *Diabetes Care*, 6, 579, 1983.
44. **VanderMeulen, J. A., Klip, A., and Grinstein, S.,** Possible mechanism for cerebral edema in diabetic ketoacidosis, *Lancet*, i, 306, 1987.
45. **Lin, L. R., Reddy, V. N., Giblin, F. J., Kador, P. F., and Kinoshita, J. H.,** Polyol accumulation in cultured human lens epithelial cells, *Exp. Eye Res.*, 52, 93, 1991.
46. **Kawaba, T., Cheng, H.-M., and Kinoshita, J. H.,** The accumulation of myoinositol and rubidium ions in galactose-exposed rat lens, *Invest. Ophthalmol. Visual Sci.*, 27, 1522, 1986.
47. **Reddy, V. N., Lin, L. R., Giblin, F. J., Chakrapani, B., and Yokoyama, T.,** Study of the polyol pathway and cell permeability changes in human lens and retinal pigment epithelium in tissue culture, *Invest. Ophthalmol. Visual Sci.*, 33, 2334, 1992.
48. **Greene, D. A., Lattimer, S. A., and Sima, A. A. F.,** Sorbitol, phosphoinositides and sodium-potassium-ATPase in the pathogenesis of diabetic complications, *N. Engl. J. Med.*, 316, 599, 1987.
49. **Finegold, D., Lattimer, S. A., Nolle, S., Bernstein, M., and Greene, D.,** Polyol pathway activity and myo-inositol metabolism: a suggested relationship in the pathogenesis of diabetic neuropathy, *Diabetes*, 32, 988, 1983.
50. **Yorek, M. A., Dunlap, J. A., and Ginsberg, B. H.,** Effect of increased glucose levels on Na^+/K^+ pump activity in cultured neuroblastoma cells, *J. Neurochem.*, 51, 605, 1988.
51. **Yorek, M. A., Dunlap, J. A., and Stefani, M. R.,** Restoration of Na^+/K^+ pump activity and resting membrane potential by *myo*-inositol supplementation in neuroblastoma cells chronically exposed to glucose or galactose, *Diabetes*, 40, 240, 1991.
52. **Burg, M. B. and Kador, P. F.,** Sorbitol, osmoregulation, and the complications of diabetes, *J. Clin. Invest.*, 81, 635, 1988.
53. **Strange, K., Morrison, R., Shrode, L., and Putnam, R.,** Mechanism and regulation of swelling-activated inositol efflux in brain glial cells, *Am. J. Physiol.*, in press.
54. **Jennings, R. B.,** Cell volume regulation in acute myocardial ischemic injury, *Acta Med. Scand.*, 587, 83, 1975.
55. **Whalen, D. A., Hamilton, D. G., Ganote, C. E., and Jennings, R. B.,** Effect of transient period of ischemia on mycardial cells. I. Effects on cell volume regulation, *Am. J. Pathol.*, 74, 381, 1974.
56. **Drewnowska, K. and Baumgarten, C. M.,** Regulation of cellular volume in rabbit ventricular myocytes: bumetanide, chlorothiazide and ouabain, *Am. J. Physiol.*, 260, C122, 1991.
57. **Clemo, H. F., Feher, J. J., and Baumgarten, C. M.,** Modulation of rabbit ventricular cell volume and $Na^+/K^+/2Cl^-$ cotransport by cGMP and atrial natriuretic factor, *J. Gen. Physiol.*, 100, 89, 1992.

58. **Clemo, H. F. and Baumgarten, C. M.,** Nitroprusside shrinks cardiac cell volume by inhibiting $Na^+/K^+/2Cl^-$ cotransport and improves cardiac compliance, *Circulation,* 84, 168, 1991.
59. **Greenwald, J. E., Apkon, M., Hruska, K. A., and Needleman, P.,** Stretch-induced atriopeptin secretion in the isolated rat myocyte and its negative modulation by calcium, *J. Clin. Invest.,* 83, 1061, 1989.
60. **Huot, S. J. and Aronson, P. S.,** Na^+/H^+ exchanger and its role in essential hypertension and diabetes mellitus, *Diabetes Care,* 14, 521, 1991.
61. **Wehling, M., Kasmayr, J., and Theisen, K.,** The Na^+-H^+ exchanger is stimulated and cell volume increased in lymphocytes from patients with essential hypertension, *J. Hypertens.,* 9, 519, 1991.
62. **Denolle, T. and Plouin, P. F.,** The Na^+/H^+ antiport: a target for new antihypertensive agents, *Therapie,* 44, 215, 1989.
63. **Canessa, M.,** Red cell volume-related ion transport systems, *Hematol. Oncol. Clin. North Am.,* 5(3), 495, 1991.
64. **Cala, P. M.,** Volume regulation of red blood cells. Mechanism of ion transport, *Mol. Physiol.,* 4, 33, 1983.
65. **Eaton, W. A., Horfrichter, J., and Ross, P. D.,** Editorial hypothesis. Delay time of gelation: a possible determinant of clinical severity in sickle cell disease, *Blood,* 47, 621, 1976.
66. **Canessa, M., Fabry, M. E., and Nagel, R. L.,** Deoxygenation inhibits the volume-stimulated, Cl^- dependent K^+ efflux in SS and young AA cells: a cystolic Mg^{++} modulation, *Blood,* 70, 1861, 1987.
67. **Baroin, A. F., Garcia-Romeu, F., Lamarrer, T., and Motais, R.,** A transient sodium-hydrogen exchange system induced by catecholamines in erythrocytes of rainbow trout, Salmogairdeneri, *J. Physiol. (London),* 356, 21, 1984.
68. **Bookchin, R. M., Ortiz, O. E., and Lew, V. L.,** Activation of calcium-dependent potassium channels in deoxygenated sickled red cells, *Prog. Clin. Biol. Res.,* 240, 193, 1987.
69. **Balla, S. K., Larner, J., Smith, E. D., Surney, S., Schwartz, E., and Rappaport, E. F.,** The xerocytosis of Hb Sc disease, *Blood,* 69, 124, 1987.
70. **Canessa, M., Fabry, M. E. et al.,** Na^+/H^+ exchange is increased in sickle cell anemia and young normal red cells, *J. Membr. Biol.,* 116, 107, 1990.
71. **Haussinger, D. and Lang, F.,** Cell volume in the regulation of hepatic function: a mechanism for metabolic control, *Biochim. Biophys. Acta,* 1071, 331, 1991.
72. **Panet, R.,** Serum-induced net K^+ influx performed by the diuretic-sensitive transport system in quiescent NIH 3T3 mouse fibroblast, *Biochim. Biophys. Acta,* 813, 141, 1985.
73. **Saprirstein, V. S. and Benos, D. J.,** Activation of amiloride-sensitive sodium transport in C6 glioma cells, *J. Neurochem.,* 43, 1098, 1984.
74. **Grinstein, S., Rotin, D., and Mason, M. J.,** Na^+/H^+ exchange and growth factor induced cystolic pH changes. Role in cellular proliferation, *Biochim. Biophys. Acta,* 988, 73, 1989.
75. **Meyer, M., Maly, K., Uberall, F., Hoflacher, J., and Grunicke, H.,** Stimulation of K^+ transport systems by Ha-*ras, J. Biol. Chem.,* 266, 8230, 1991.

Part II
Organic Osmolytes

Chapter 5

Compatible and Counteracting Solutes

Paul H. Yancey

TABLE OF CONTENTS

0-8493-4448-4/94/$0.00+$.50

I. INTRODUCTION

One of the major discoveries in modern biology is the remarkable cellular and molecular similarity of life on Earth in all its diverse forms. Evolutionary similarities result from either convergence or conservation, patterns seen not only in the highly studied relationships among biological macromolecules, but also in the types and uses of small osmotic effector molecules — *osmolytes* — in cell volume regulation. The ability to regulate cell volume by inorganic ion transport, especially to prevent osmotic swelling, may be an ancient conserved trait found in cells from bacteria to mammals[1,2] (discussed in detail elsewhere in this volume). The occurrence of a variety of organic osmolytes, discovered initially in marine organisms almost a century ago,[3,4] was once thought to be a less common, diverse set of nonuniversal solutions to the problem of osmotic shrinkage during long-term water stress. Now it is known to occur in most major groups of organisms, including mammals (kidney and brain) and in desiccation and freezing as well as hypersaline stress,[5] with one generalization emerging: reduced cell volumes and/or volume adjustment by elevated inorganic ion concentrations may be tolerated in short-term exposures; however, in long-term adaptation organic osmolytes are preferred or essential (except in halophilic archaebacteria), with concentrations of inorganic ions either held relatively constant, or with K^+ increased considerably less than organic solutes.[5–9]

An illustration of this general pattern is given in Figure 5–1A, for the intertidal (eukaryotic) green alga, *Chlorococcum submarinum,* studied recently by Blackwell and Gilmour.[10] Following rapid transfer from 0.1 to 0.5 *M* NaCl medium, cells display an initial increase in concentrations of K^+, Na^+, and the polyhydric alcohol (polyol) glycerol, and a large decrease in photosynthesis rate. Later, concentrations of Na^+ decline and concentrations of glycerol rise much further, concomitant with restoration of photosynthesis rate (reaching 107 ± 20% of control

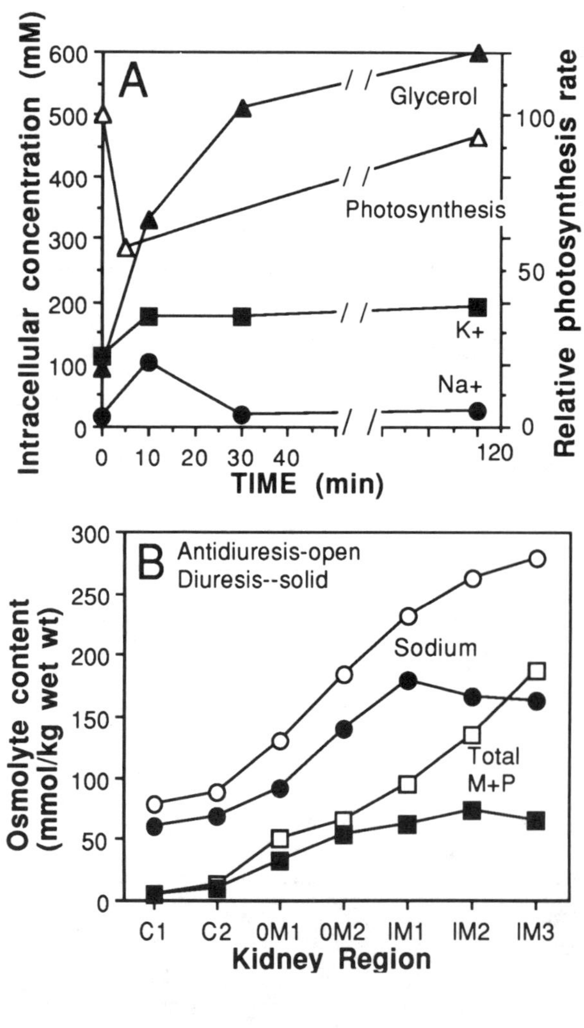

Figure 5–1 (A) Time course of intracellular osmolyte concentration and photosynthesis rate for the intertidal unicellular green alga, *Chlorococcum submarinum,* switched from 0.1 to 0.5 *M* NaCl (data from Reference 10). (B) Profiles of osmolyte contents in rabbit kidneys sliced in seven sections from outer cortex (C1) to outer medulla (OM) to inner medulla tip (IM3). Diuretic animals were given sugar water; antidiuretic animals had no water for 2 d. "Total M+P" (squares) = sum of betaine, GPC, sorbitol, and *myo*-inositol (Figure 5–2), dominant osmolytes in mammalian kidneys (data from Reference 13).

at 240 min). K^+ levels remain elevated, but to a much smaller degree than glycerol. Rather than being a rare adaptation of a primitive eukaryote to a unique environment, Figure 5–1A represents a widespread constrained pattern of evolutionary convergence and conservation from prokaryotes to higher eukaryotes. As examples, switching a euryhaline cyanobacterium from fresh water to 0.5 *M* NaCl medium results in a rapid NaCl influx to restore volume (2 min), exchange of Na^+ for K^+ (20 min), followed by replacement of KCl by sucrose and glucosylglycerol;[11] one line of cultured mammalian renal cells (chronically adapted to normal isosmotic medium), when exposed to hypersaline medium, initially accumulates high K^+, then over several days replaces this with sorbitol[12] (see also chapter on polyol regulation).

As a different example, Figure 5–1B shows the pattern of osmolytes found in mammalian kidney,[8,13] the medulla of which experiences high extracellular salt concentrations from the urine-concentrating mechanism. As the extracellular concentrations of salt rise in antidiuresis (open symbols, Figure 5–1B), or with increasing depth in the kidney, the total of several cellular organic osmolytes increases in parallel, while NaCl appears to remain largely extracellular.[8,14]

The constrained pattern of evolution represented in Figure 5–1 appears to be based on just a few physicochemical principles.[4–7] In general, varying and/or high concentrations of organic osmolytes, in contrast to disruptive effects of high inorganic ion levels, probably maintain a cellular environment "compatible"[15] with macromolecular function. In this chapter, these evolutionary patterns and principles for organic osmolytes will be discussed in detail. Chapters following this one will examine in detail the regulation of different classes of organic osmolytes.

II. ORGANIC OSMOLYTE TAXONOMY

Around 1900, L. Fredericq discovered that marine cartilaginous fishes (elasmobranchs) and invertebrates did not have enough inorganic ions in their cells to account for their approximately isosmotic balance with seawater (reviewed by Smith[3] and Clark[4]). This "osmotic gap" was almost immediately accounted for in elasmobranchs by the high quantities of the nitrogenous waste, urea, already known, and (in 1909) by trimethylamine oxide (TMAO).[3] However, not until after World War II were other organic osmolytes identified in organisms exposed to high osmolalities, e.g., free amino acids in marine invertebrates,[4] glycerophosphorylcholine (GPC) in the mammalian renal medulla,[8,16] polyols in osmophilic yeasts and marine algae,[15] proline and glycine betaine in marine plants.[17] While at first it seemed that very different adaptations had evolved in different phylogenetic groups, we now recognize that the types of organic osmolytes used are more widely distributed and universally restricted to about four major classes of solutes: free amino acids and derivatives, certain methylated compounds, sugars and polyols, and urea.[5]

Figure 5–2 shows the structures and major (kingdom) distributions of representative organic osmolytes in these categories; Table 1 provides a more detailed distribution.[21–70] Intracellular concentrations of these compounds range from a few millimolar to several molar in some microorganisms in concentrated brine habitats.[15] To qualify as an osmolyte, a cellular solute must be used in some significant way to maintain or adjust cell osmotic pressure and volume. This is clear when, following osmotic shrinkage, a tissue actively accumulates (by transport or production) a compound to tens or hundreds of millimolar. Is there a threshold level below which an osmotically regulated compound should not be considered an osmolyte? Recent studies in the mammalian brain, an organ that may be highly sensitive to slight osmotic changes,[18] show that amino acid, polyol, and methylamine solutes (each at only a few millimolar) are regulated to match small blood osmotic changes[18–20] (see also the following chapters). The question of a threshold level thus probably has no general answer.

A. SUGARS AND POLYOLS

A large number of different carbohydrate osmolytes are now known, including monosaccharides, disaccharides such as sucrose (including a new form, mannosucrose[31]), heterosides (sugar-glycerol combinations), linear polyols with three to six carbons, and cyclic polyols (cyclitols, Figure 5–2A). There are also reports of several methylated cyclitol osmolytes in plants (Table 1). In common are a large number of hydroxyl groups and a lack of charge. Carbohydrate osmolytes occur in the greatest variety in photosynthetic organisms and fungi,[33] but distribution of some types is scattered among distantly related groups, probably from evolutionary convergence. For example, *myo*-inositol and sorbitol are used as osmolytes in marine algae and mammals (Table 1, Figure 5–2A).

Sugars and polyols are also the most common solutes used in organisms adapting to freezing temperatures,[58,64] including the recent discovery of high glycerol levels in polar fishes.[63] In addition to lowering freezing points, they may serve as osmolytes, reducing cell shrinkage if extracel-

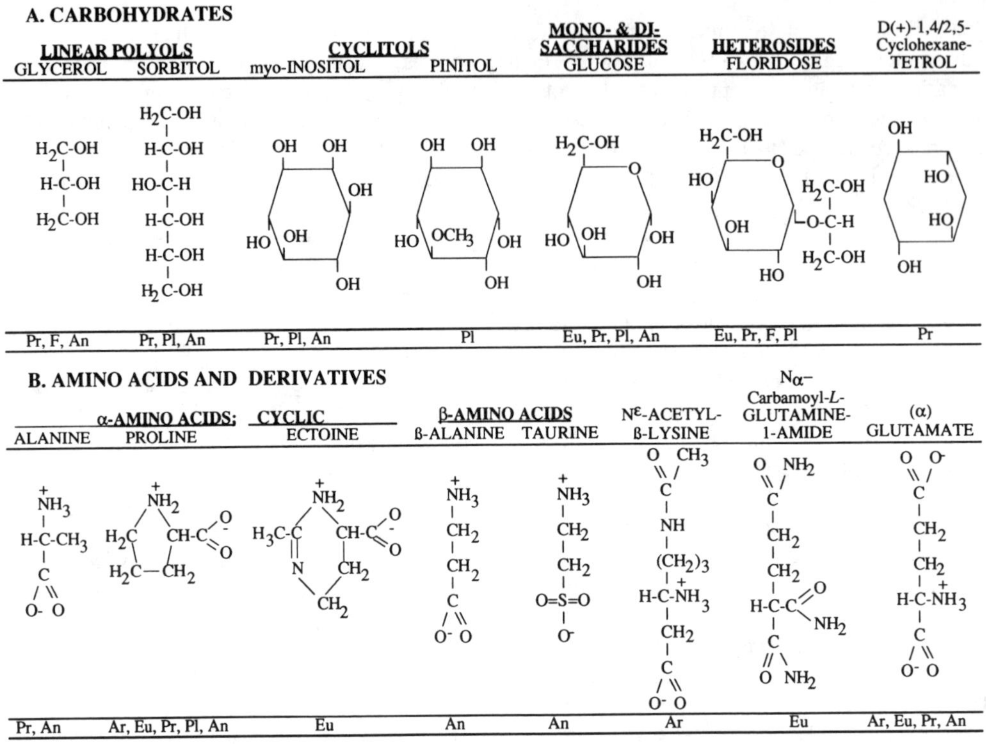

Figure 5–2 (A–C) Structures of major and/or newly discovered osmolytes and their distributions in organismal kingdoms: Ar, archaebacteria; Eu, eubacteria; Pr, Protista (including algae); F, fungi; Pl, true Plantae; An, Animalia; GPC, glycerophospho(ryl)choline; TMAO, trimethylamine oxide; DMSP, dimethylsulfoniopropionate (see Table 1 for references). (D) The Hofmeister series of inorganic ions ranked according to nearly universal stabilizing and destabilizing (kosmotropic, chaotropic[153]) effects in water-solute systems (see discussion of mechanisms in the text). (E) Molecular moieties ranked in order of hydrophilic properties as defined by Laughlin[159] (see text). Unless indicated, rankings within groups are not known.

lular fluids freeze and create high osmotic pressure. Disaccharides, most notably trehalose and sucrose, commonly build up in anhydrobiotic organisms, although in this situation the compounds probably act in different ways than osmotic stress osmolytes (below).[71]

B. FREE AMINO ACIDS AND DERIVATIVES

First discovered as osmolytes in several marine invertebrate phyla, the dominant free amino acid types are invariably small, polar zwitterionic ones, especially glycine, alanine, proline, glutamine, β-alanine, and the β-aminosulfonic acid, taurine.[4] These have since been found as osmolytes in a much greater diversity of organisms, including some subjected to freezing as well as to hyperosmotic stress (Table 1, Figure 5–2B). Charged amino acids are sometimes found as osmolytes; however, at least for β- and α-glutamate (Figure 5–2B) in bacteria, the excess charge is countered by elevated K^+ as an osmolyte.[9,22] In recent years, several other amino acids and derivatives have been found as definite or probable osmolytes in eubacteria and methanogenic archaebacteria (Table 1). These include other β-amino acids,[21,22] γ-amino butyrate, cyclic forms such as ectoine[29] (Figure 5–2B) and pipecolate,[25] N^{ε}-acetyl-β-lysine (an acetylated form that converts a basic form to a zwitterion; Figure 5–2B),[21] and a modified dipeptide, *N*-acetyl-glutaminylglutamine amide.[30]

C. METHYLATED AMMONIUM AND SULFONIUM COMPOUNDS

Another widespread type of compound is distinguished by having charged methylated nitrogen or sulfur atoms, with the former (also called methylamines) being far more common (Figure 5–2C).

C. METHYLAMMONIUM AND METHYLSULFONIUM OSMOLYTES

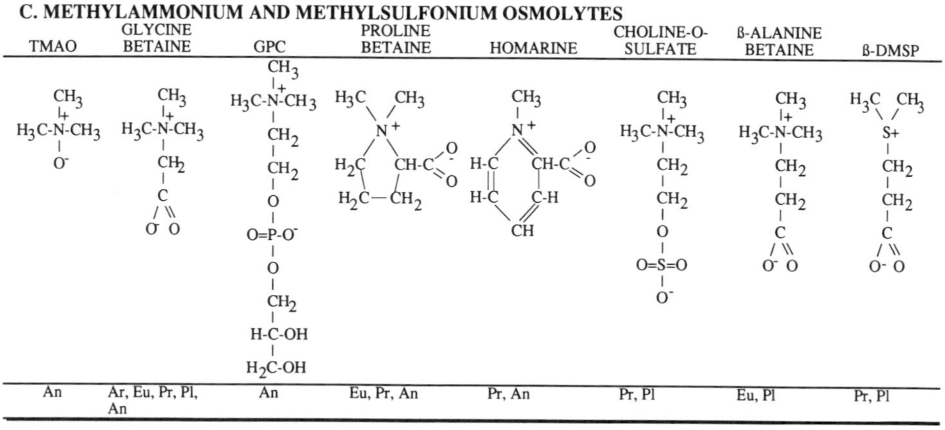

D. HOFMEISTER SERIES <---Stabilizing (Kosmotropes) Destabilizing (Chaotropes)--->

F^- PO_4^{3-} $SO_4^=$ CH_3COO^- Cl^- Br^- I^- SCN^-

$N^+(CH_3)_4$ $N^+(CH_3)_2H_2$ N^+H_4 K^+ Na^+ Cs^+ Li^+ Mg^{2+} Ca^{2+} Ba^{2+}

E. HYDROPHILICITY SERIES

Ionic	> Zwitterionic	> Semipolar	> Single bond
$-COO^-$ >> $-SO_3^-$ > $-SO_4^-$; $-PO_4^=$	$-N^+(CH_3)_2CH_2COO^-$	$-N(O)(CH_3)_2$	$-NH_2$
$-N^+(CH_3)_3$	$-N^+(CH_3)_2(CH_2)_2SO_3^-$	$-S(O)CH_3$	
$-S^+(O)(CH_3)_2$	$-OP(O)_2OCH_2CH_2N^+(CH_3)_3$	$-P(O)CH_3$	

Figure 5–2 (Continued)

All are also zwitterionic, remaining so over a wide pH range. Most are fully substituted and thus also termed quaternary ammonium or tertiary sulfonium compounds (QACs, TSCs); however, there are exceptions, such as sarcosine (*N*-methylglycine) in some cartilaginous fishes and *N,N*-dimethyltaurine in marine red algae (Table 1). By far the most widespread methylamine is betaine (Figure 5–2C) (*N,N,N,*-trimethylglycine), also called glycine betaine to distinguish it from other QACs which can also be termed betaines, such as proline betaine, β-alanine betaine (Figure 5–2C), and other types found mainly in bacteria (Table 1).[32]

D. UREA AND UREA WITH METHYLAMINES

Less widely distributed is the occurrence of high levels of the nitrogenous waste product urea, $O{=}C(NH_2)_2$ (not shown in Figure 5–2). First found as an osmolyte in marine elasmobranchs (at about 300 to 500 m*M*),[3,60,62] it is used similarly in coelacanths and some amphibians adapting to salinity,[72] and it accumulates to high levels in some estivating vertebrates and gastropods,[62] where it may help conserve body water. In the mammalian kidney, it is concentrated as a waste product, to several molar in some species such as desert rodents.[73] In nonestivating organisms, urea is accompanied by other osmolytes, usually methylamines[62] (indicated with an asterisk in Table 1). The significance of this pattern will be discussed later.

Table 1 and Figure 5–2 show that the use of a particular type of osmolyte sometimes correlates with phylogeny,[33] but not always. Indeed, some (e.g., α-amino acids, betaine) are found in only distantly related organisms, though whether a particular occurrence represents convergence or conservation of an ancient mechanism is not always clear. Some groups have seemingly unrelated osmolyte systems, (e.g., of three species of diatoms examined for salinity response in culture, all used glycine betaine and proline as osmolytes, but two species used homarine also, while the third used glycerol and dimethylsulfoniopropionate (DMSP) in addition[35]) (see Figure 5–2 for structures). Also, some organisms are found to have all categories of osmolytes in one tissue, e.g., the mammalian kidney, which along with urea has the polyols *myo*-inositol and sorbitol, the methylamines GPC and glycine betaine (Figure 5–1B), and the amino acid taurine.[8,70]

III. INORGANIC IONS

The major evolutionary question is, of course, why osmolytes are constrained by natural selection to a few solute types, mostly metabolically costly organic compounds, with the (presumably "cheaper") inorganic ions usually disfavored. Indeed, most organisms examined, independent of phylogeny and habitat, have broadly similar intracellular concentrations of K^+ (usually 100 to 200 m*M*) and Na^+ and Cl^- (typically <50 m*M*).[5,6,74] Why?

Table 1 **Distributions of Organic Osmolytes in Major Organismal Groups**

Organism	Sugars and Polyols	Amino Acids, Derivatives	Methylated Ammonium, Sulfonium Solutes	Ref.
Archaebacteria: methanogens		Glutamine, glutamate, proline, β-glutamate, β-glutamine, N^{ε}-acetyl-β-lysine	Glycine betaine	21, 22
Eubacteria	Trehalose, sucrose, mannitol, heteroside (glucosylglycerol), mannosucrose	Glutamine, glutamate, proline, ectoine, hydroxyectoine, N^{α}-carbamoyl-L-glutamate 1-amide, *N*-acetylglutaminyl-glutamine amide, pipecolate, GABA	Glycine betaine, choline-*O*-sulfate, proline betaine, taurine betaine, β-alanine betaine, pipecolate betaine, glutamate betaine, DMSP	23–32
Protista Algae	Glycerol, mannitol, sorbitol, volemitol, altritol, sucrose, mannose, glucose, heterosides (e.g., floridoside), cyclohexane	Proline, alanine, glycine; glutamate	Glycine betaine, proline betaine, homarine, choline-*O*-sulfate, DMSP, dimethyl-taurine, taurine betaine	32–45
Protozoa		Glycine, alanine, proline		46
Yeast, fungi	Glycerol, erythritol, arabitol, mannitol, etc.			33, 39, 47

Vascular plants	*Myo*-inositol, methyl-inositols (e.g., pinitol, ononitol), sorbitol, mannitol, glucose, sucrose	Proline	Glycine betaine, proline betaine, β-alanine betaine, choline-*O*-sulfate, DMSP	48–53
Animals				
Various aquatic invertebrates	Glycerol, trehalose *(Artemia)* Trehalose (insects)	Proline, serine (insects) Glycine, alanine, proline, serine, taurine (marine phyla)	Glycine betaine, TMAO, proline betaine (marine phyla), TMAO *(Artemia)*	4, 54–57
Mollusc (marine) & insects — freezing	Glycerol, sorbitol, erythritol, ribitol, threitol, sucrose (insects)	Taurine, strombine (mollusc)		58, 59
Agnathans		Glycine, alanine, proline, etc.	TMAO	60
Cartilaginous fishes[a]	*Myo*-inositol, *scyllo*-inositol	Taurine; β-alanine; glycine, alanine, etc.	Glycine betaine, TMAO, sarcosine	60–62
Bony fishes	Glycerol (antifreeze)	γ-amino butyrate, taurine (erythrocytes)	TMAO	54, 63
Amphibia	Glucose, glycerol (freezing)	Various α-amino acids	GPC[a]	62, 64, 65
Birds		Taurine (erythrocytes)		66
Mammals — kidney[a]	Sorbitol, *myo*-inositol	Taurine, others?	Glycine betaine, GPC	8, 19, 20, 67–70
— heart	*Myo*-inositol	Taurine		
— brain		Taurine, glutamine	Glycine betaine, GPC	

Note: See Figure 5–2 for structures and abbreviations.

[a] With high urea.

A. PERTURBATIONS OF BIOLOGICAL SYSTEMS BY INORGANIC IONS

Examination of a wide variety of biological processes *in vitro* have shown a widespread sensitivity to changes in inorganic ion concentration. In the case of NaCl, increased concentration usually leads to inhibition or disruption; while for K^+ salts, there may be an intermediate requirement coinciding with typical cellular K^+ levels, with perturbation at higher concentrations. Figure 5–3 shows a number of examples of these phenomena. KCl inhibition of malate dehydrogenases (MDH) from widely different organisms is basically similar (except for the archaebacteria; see below) (Figure 5–3A).[7,75] Rates of *in vitro* protein synthesis from salt-sensitive and tolerant plants are affected equally by K^+ salts (Figure 5–3B), but the type of K^+ salt used can make a large difference, implicating the anion as responsible for some of the inhibition, especially Cl^-.[76] Finally, K^+ salts inhibit muscle-fiber contraction force in a mammalian system (Figure 5–3C), again

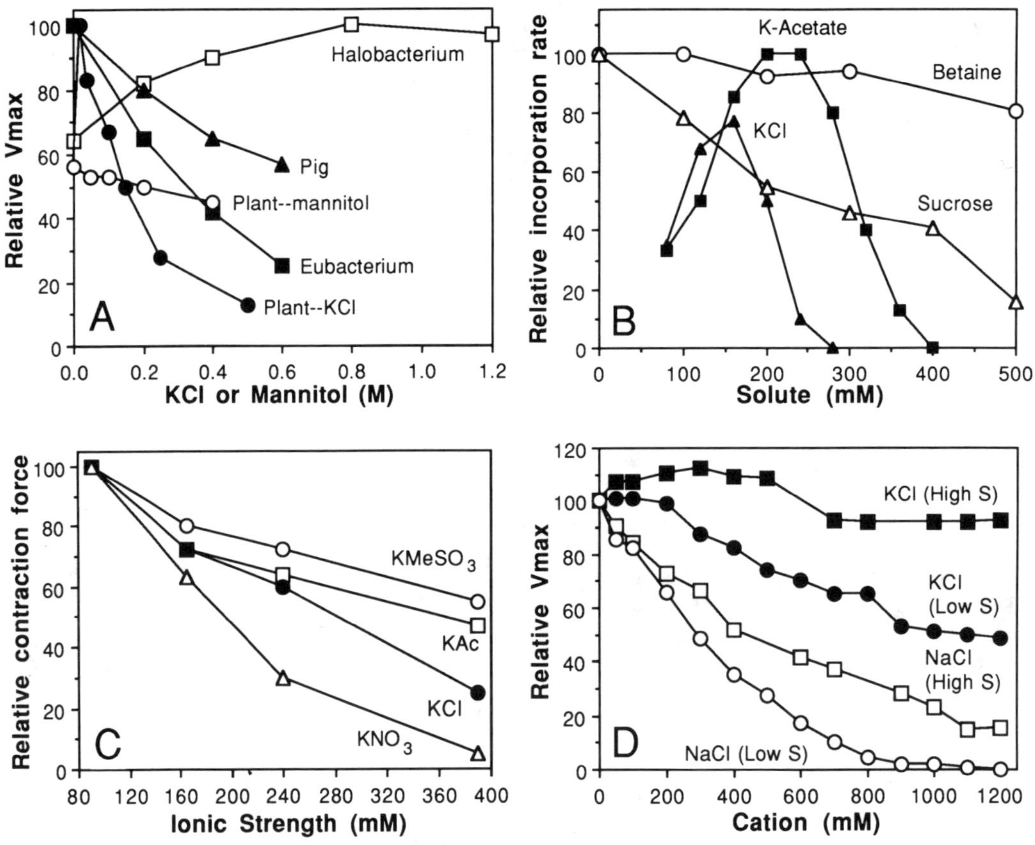

Figure 5–3 Effects of inorganic (and some organic) solutes on biological functions. (A) Maximal velocities of malate dehydrogenases (MDH) from the archaebacterium *Halobacterium halobium,* a marine manganese-oxiding eubacterium, and pig (from Somero, G. N., in *Water and Life: A Comparative Analysis of Water Relationships at the Organismic Cellular and Molecular Levels,* Somero, G. N., et al., Eds., Springer-Verlag, Berlin, 1992, 19. With permission.); and MDH from *Phaseolus vulgaris* (salt-sensitive plant). (From Greenway, H. and Osmond, C. B., *Plant Physiol.,* 49, 256, 1972. With permission of the American Society of Plant Physiologists.) Result with salt-tolerant plant MDH was similar. (B) Protein synthesis (methionine incorporation) rate in *in vitro* wheat germ translation system with RNAs from six plants (salt-tolerant and -sensitive) for K curves, and with wheat RNA for betaine, sucrose curves (from Gibson, T. S. et al., *Plant Cell Environ.,* 7, 579, 1984. With permission.) (C) Contractile force of skinned rabbit psoas muscle fibers. (From Andrews, M. A. W. et al., *J. Gen. Physiol.,* 98, 1105, 1991. With copyright permission of the Rockefeller University Press.) (D) Maximal velocity of pyruvate kinase from red alga *Porphyra umbilicalis;* High S, assay with 8 m*M* phosphoenolpyruvate (PEP) and 2.3 m*M* ADP; Low S, assay with 1 m*M* PEP, 0.3 m*M* ADP. (From Wiencke, C., *J. Plant Physiol.,* 116, 447, 1984. With permission.)

revealing the added role of the anion.[77] Muscle fibers from marine animals are also salt-sensitive (barnacle,[78] lobster[79]).

These examples are not meant to imply that all biological processes are salt-sensitive. Among 18 enzymes from *Dunaliella* species (halotolerant green algae), concentrations for 50% inhibition by NaCl range from 50 to near 400 m*M* for 13, but with 5 requiring as much as 3600 m*M*.[80] Also, NaCl is often more inhibitory than KCl. For example, in one study[81] concentrations for 10% inhibition of glutamate synthetases from freshwater, marine, and halotolerant cyanobacteria ranged from 160 to 920 m*M* for NaCl, and from 420 to >2000 m*M* for KCl. However, sensitivities showed no adaptive correlations with habitat. Also, lack of salt sensitivity in even a majority of enzymes would not protect hyperosmotic cells from suffering if a few critical proteins are affected.

Whether strong inhibitory effects *in vitro* can be extrapolated to significant effects *in vivo* is not certain. Cells subjected to hypersaline shock, during which intracellular ions (as well as other molecules) become concentrated from shrinkage, typically show metabolic and other system disturbances, e.g., reduced growth rate in bacteria,[82,83] reduced photosynthesis and respiration in algae[42] (Figure 5–1A), decreased metabolism in the crustacean *Artemia*,[56] gross damage in mammalian cells,[84] and reduced growth in mammalian renal cells (at low culture density).[85] However, intracellular inorganic ions have not been shown unequivocally to cause these changes and, in at least one case (DNA-protein binding in bacteria), salt inhibition may be relieved *in vivo* by molecular crowding[82,83] (see Section V, below). Nevertheless, *in vitro* effects on complex systems, such as protein synthesis and muscle fibers (Figures 3B and C), suggest some serious salt effects will occur in concentrated cells. Indeed, it has been proposed that the sensitivity of protein synthesis to ions (Figure 5–3B) may set the boundaries for cell ionic strength.[53]

B. EXCEPTIONS: HIGH-SODIUM OR POTASSIUM CELLS

Only one group of organisms appears to rely extensively on an inorganic ion as a major osmolyte for long periods and at very high concentrations. These organisms, the evolutionarily ancient halophilic archaebacteria, are restricted to marginal habitats, such as concentrated (brine) ponds.[32] Archaebacteria osmoconform to their environments by accumulating K^+, to as high as 7 *M* intracellularly, with the NaCl of the environment largely excluded.[86] Figure 5–3A (halobacterial MDH[7]) illustrates the radically different pattern that has evolved in these organisms' proteins: typically, halobacterial proteins actually require high salt levels for optimal activity. These proteins have large numbers of acidic amino acids (e.g., glutamate) which bind both water and K^+. In the absence of high cation concentrations, protein conformation is disrupted.[86]

Among eubacteria, most halophilic species have low intracellular ions compared to the environment, but there are some reports of cells with high salt contents,[32] e.g., *Vibrio costicola,* a moderate halophile in which intracellular Na^+ ions increase in hyperosmotic medium such that cell-associated Na and K salts reached about 2 *M*.[87] However, enzymes of this species are very salt-sensitive, and evidence suggests that cellular Na and Cl are bound to the cell envelope.[87] Similarly, in eukaryotes, many halophilic plants have high NaCl contents in their cells, but the ions appear to be sequestered in vacuoles, not in the cytoplasm.[88] One notable eukaryotic exception may be the intertidal red alga, *Porphyra umbilicalis,* which is reported to have high cytoplasmic K^+, up to 750 m*M* in 3.5-fold concentrated seawater.[89] Figure 5–3D illustrates salt effects on pyruvate kinase from this protist, showing a moderate Na but low K sensitivity. Whether this is a special adaption to K^+ is unknown (nor have other enzymes of this species been examined), but is not inherent in this enzymatic reaction in general: pyruvate kinases from a variety of marine and terrestrial animals have high KCl sensitivity (see below, Figure 5–4B).[90]

C. CONCLUSIONS ON SALT EFFECTS

Three broad conclusions follow from studies represented by those discussed above. First, inorganic ions, especially Na^+ and Cl^-, are not, in general, useful as long-term osmolytes except at relatively constant, low levels. Increases above these levels will likely inhibit some cellular functions, presumably through ionic interactions with macromolecular and substrate charges. Adaptations to prevent this by massive amino acid changes in proteins — i.e., the archaebacterial solution — result in systems that may be restricted to extremely saline environments, since their proteins tend to lose function with decreasing salt (e.g., Figure 5–3A). Second, salt sensitivity is phylogenetically widespread and not often different between species adapted to low or high environmental salt variability. Finally, intracellular K^+ without Cl^- may be tolerable over a limited (but not extreme) range in some cases, but this presumably entails a high energy cost in hypersaline conditions for ion pumping to keep K^+ in and Na^+ out.

IV. ORGANIC OSMOLYTE PROPERTIES

A. COMPATIBILITY AND STABILIZATION

1. Compatibility with Subcellular Function

The first major hypothesis for use of organic osmolytes resulted from the work of Brown and colleagues,[15,91] who found that, while inorganic ions (especially NaCl) at high concentrations disrupt protein functions in species of *Dunalliela* algae, glycerol (the major osmolyte in the halotolerant species) did not do so even at concentrations up to several molar. These effects were the same with enzymes from euryhaline and stenohaline species, leading to an important hypothesis: perturbing effects of high salt and nonperturbing effects of polyols are general features of protein-solute interactions, not specific adaptations of proteins to function in the presence of the solutes. Use of polyols should permit cellular functions without significant disruptions over a wide range of external salinites. Brown and Simpson[91] coined the term "compatibility" for the general nonperturbing properties of polyols on function.

Work in the laboratories of Clark, Somero, Wyn Jones and others extended this hypothesis to other osmolytes in addition to polyols, and to a variety of proteins from numerous species, with studies in numerous laboratories continuing to find supporting evidence. Examples of osmolyte effects are shown in Figures 5–3A, 5–3B, and 5–4. *In vitro* translation, inhibited by high KCl, is not altered much by glycine betaine (Figure 5–3B) or proline.[76] In Figure 5–4A are shown the effects of several amino solutes and salts on activity of an enzyme from a marine polychaete.[92] Clearly salts are again inhibitory, while amino acids in general are not, even at 0.5 *M* (increased activity with TMAO in Figure 5–4A will be discussed in Section IV.B). Figure 5–4B shows that similar patterns are seen with Michaelis constants (K_m), in this case for an enzyme from a marine crustacean. Inorganic salts are again detrimental, and basic amino acids (not used as osmolytes) are even more so. However, the types of neutral amino acid solutes used as osmolytes are fully innocuous up to 1 *M*. Similar differences are found for salt vs. organic osmolyte effects on catalytic velocity of this enzyme.[90] Figure 5–4C shows the effects of polyols and NaCl on several enzymes from a salt marsh plant (with sorbitol as its major osmolyte)[93] and on an enzyme from mammalian kidney (which uses sorbitol and *myo*-inositol as osmolytes, among others).[94] The pattern — inhibition by salt but not polyols — is consistent. Last, Figure 5–4D shows the effects on MDH from a marine plant of NaCl, proline, and the methyl-cyclitols pinitol (Figure 5–2A) and 1-D-1-*O*-methyl-muco-inositol, all known osmolytes from plants.[52]

2. Stabilization of Subcellular Structure

Independently of biologists working on functional compatibility of cell osmolytes, biochemists had found as early as 1929 that sugars and polyols (only later noted to be natural osmolytes) exhibit another widespread property: the ability to stabilize macromolecular structure under extreme conditions (for early studies, see references in Gerlsma[95]). Many studies have extended this phenomenon to a variety of compounds, systems, and disturbances.

a. High-temperature denaturation, inactivation

Figure 5–5A shows the effects of polyols on the transition melting temperatures of ovalbumin[96] and lysozyme.[97] In this and several other studies (bovine ribonuclease,[95] bovine chymotrypsinogen,[98] and invertase[99]), polyols consistently increase (in a concentration-dependent manner) protein thermal stability, with protection rising with number of carbons and cyclization: *myo*-inositol > sorbitol > five-carbon polyols > erythritol > glycerol > ethylene glycol.

Amino acids and methylamines, studied after their osmolyte functions were known, are also thermostabilizers. Figure 5–5B shows studies on bovine ribonuclease denaturation:[90,100] certain compounds (nonosmolyte amino acid arginine and urea) destabilize the protein, KCl and NaCl have little effect, and all organic osmolytes examined, except urea, are stabilizers. A new study has extended these observations to show that melting temperature continues to rise with osmolyte concentration up to 4 *M* betaine, 6 *M* dimethylglycine, and 8.2 *M* sarcosine.[101]

Numerous other studies find broadly similar results. Heat inactivations of wheat α-amylase, yeast glucose-6-phosphate dehydrogenase and hexokinase, and avian pyruvate carboxylase are alleviated by proline and glycine betaine about equally well.[102] Heat denaturation of three enzymes, MDH (Figure 5–5C), PEP carboxykinase, and nitrate reductase, from *S. soda,* a plant adapted to saline soils, showed similar trends; for the reductase, proline decreased stability slightly.[103] Figure 5–5D shows denaturation of a fish enzyme, with glycerol and DMSP (0.2 *M*)

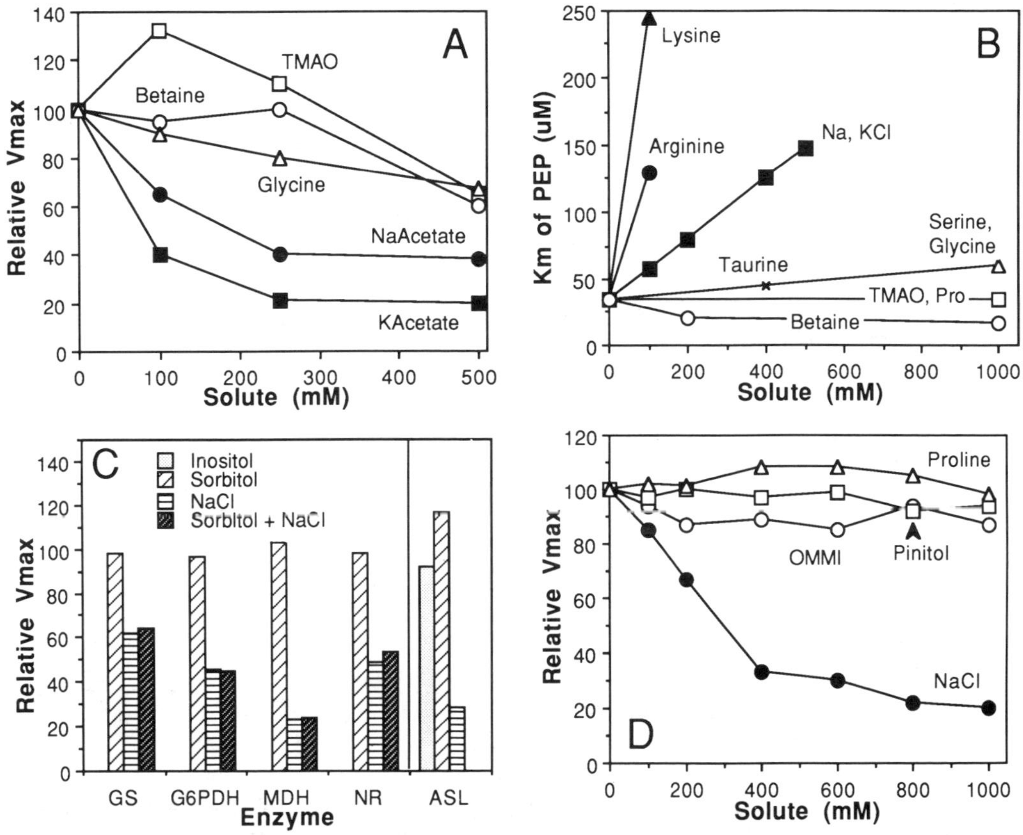

Figure 5–4 Effects of organic (and some inorganic) solutes on biological functions. (A) Maximal velocity of lactate dehydrogenase from a marine polychaete, *Nereis succinea.* (From Clark, M. E. and Zounes, M., *Biol. Bull.,* 153, 468, 1977. With permission.) (B) Apparent Michaelis constant (K_m) of PEP for pyruvate kinase from a marine crab *(Pachygrapsus crassipes).* Pro = proline. (From Bowlus, R. D. and Somero, G. N., *J. Exp. Zool.,* 208, 137, 1979. With permission of Wiley-Liss, a division of John Wiley and Sons, Inc., © 1979. (C) Maximal velocities of four enzymes from a marine halophyte, *Plantago maritima;* GS = glutamine synthetase; G6PDH = glucose-6-phosphate dehydrogenase; MDH = malate dehydrogenase; NR = nitrate reductase; and a mammalian renal enzyme, argininosuccinate lyase (ASL). (Halophyte data from Reference 93, renal data from Reference 94.) (D) Maximal velocity of MDH from a marine plant, *Rhizophora mangle;* OMMI = 1-D-1-*O* -methyl-muco-inositol. (From Sommer, C. et al., *Bot. Acta,* 103, 270, 1990. With permission.)

effectively stabilizing the protein[104] (for disturbance by 0.3 *M* DMSP, see Section VI, below).

Membranes are also stabilized by osmolytes, at least in initial studies done recently: betaine, proline, glycine, and trehalose all raise the pre- and transition temperatures of DCCP (dipalmitoylphosphatidylcholine) vesicles.[105]

b. Protein and membrane assembly

Stabilizing properties are also manifested during macromolecular assembly. Protein examples will be given in Section IV.B, below. A number of studies are aimed at improving membrane reconstitution — extraction from cells and reassembly of membrane proteins and lipids into functioning bilayers *in vitro.* Use of osmolytes in this procedure has been found to improve reconstitution of functioning prokaryotic and eukaryotic complexes, shown for one preparation in Figure 5–6A. Maloney and Ambudkar[106] have proposed that the stabilizing abilities of osmolytes are responsible for the effect. Another recent study[107] found that glycerol, glucose, and sucrose improve the ordering of the lipid bilayer and proper orientation of mammalian gonadotropin receptor in reconstituted proteoliposomes.

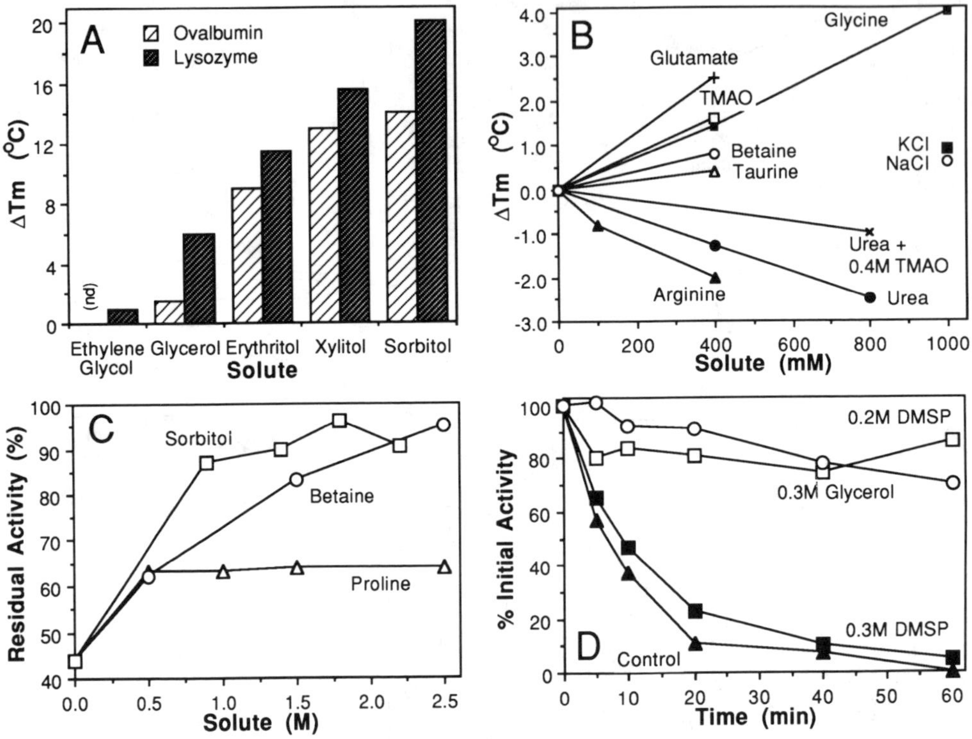

Figure 5–5 Effects of inorganic and organic solutes on biological structure. (A) Change in melting transition temperatures (T_m) of ovalbumin at pH 7 with 50% solute; nd, not determined (data from Back et al.[96]), and of lysozyme at pH 2 with 3 *M* solute (data from Fujita, Y. et al.[97]). (B) Melting temperature of bovine ribonuclease at pH 6.7 (From Bowlus, C. L. and Somero, G. N., *J. Exp. Zool.,* 208, 137, 1979. With permission of Wiley-Liss, a division of John Wiley and Sons, Inc., © 1979 and from Yancey, P. H. and Somero, G. N., *Biochem. J.,* 182, 317, 1979, by permission of The Biochemical Society, Portland Press.) (C) Remaining activity after 40 min at 45°C of MDH from a plant, *Salsola soda.* (From Niklolopoulos, D. and Manetas, Y., *Phytochemistry,* 30, 411, 1991. With permission.) (D) Activity loss over time at 50°C of barracuda lactate dehydrogenase. (From Nishiguichi, M. K. and Somero, G. N., *Cryobiology,* 29, 118, 1992. With permission.)

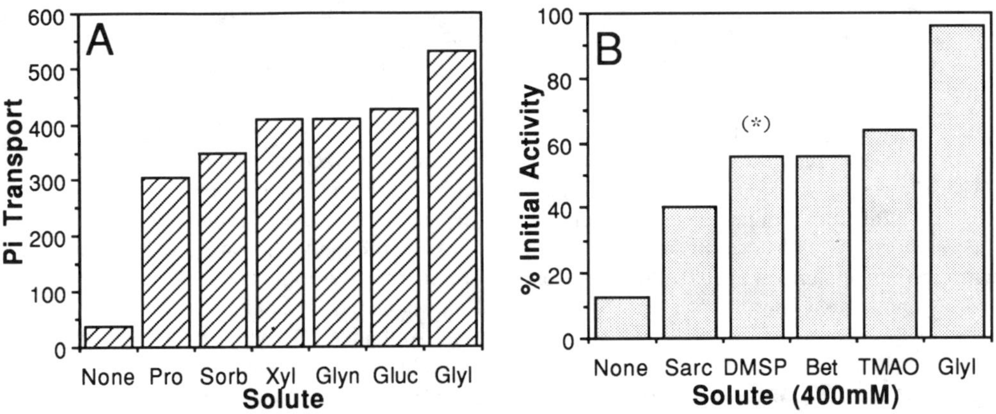

Figure 5–6 (A) Reconstitution of phosphate-linked antiport in lipid vesicles from *Streptococcus lactis* ($^{32}P_i$ transport in nmol/mg protein); Pro, 12% proline; Sorb, 20% sorbitol; Xyl, 20% xylitol; Glyn, 8% glycine; Gluc, 20% glucose; Glyl, 20% glycerol (data from Reference 106). (B) Protection of rabbit phosphofructokinase from cold inactivation after 1 h at 6°C; Sarc, sarcosine; Bet, betaine; *DMSP was 300 m*M* (data from References 104 and 109).

c. Cold inactivation

A number of proteins are labile in the cold (nonfreezing).[108] Several studies have found that organic osmolytes can slow or prevent loss of integrity, including protection of mammalian β-hydroxysteroid dehydrogenase by glycerol,[108] mammalian phosphofructokinase by several osmolytes (shown in Figure 5–6B),[104,109] and plant pyruvate-P_i dikinase by osmolytes, in the order TMAO > glycine betaine > proline > glycerol.[110]

d. Freeze-thaw, desiccation denaturation

Proteins are often denatured when subjected to a freeze-thaw cycle, and many organic osmolytes provide protection from this. For example, Higgins et al.[111] studied the leakage rate of streptomycin from freeze-thawed liposomes, with and without dimethylsulfoxide (DMSO) or betaine. Both solutes at 8% concentration reduced leakage from 30 to 10% (cooling rate of 1°C min^{-1}), while betaine was superior to DMSO at lower concentrations.

Desiccation (drying out) often disrupts membranes and proteins irreversibly. Sugar solutes can protect against this effectively.[71,112]

e. Salt, urea, and pH inhibition, denaturation

As noted above, high levels of inorganic ions inhibit and denature proteins. Hydrogen ions and urea[62] are also potent macromolecular destabilizers. Acid denaturation may be slowed by organic osmolytes.[53] Organic osmolytes can prevent some salt and urea effects also, to varying degrees, with the methylamine solutes showing the most consistency. Because urea and salt protection may occur *in vivo* in osmotic regulation, unlike most of the structural effects above, these will be discussed as "counteractions" (below).

3. Compatibility and Osmoprotection in Living Systems

Although tests *in vitro* with isolated proteins and membranes provide rationales for osmolyte evolution, they do not show whether such effects occur *in vivo*. The general benefit of osmolytes can be tested by defined manipulation of intracellular osmolyte contents. One testable prediction is that cells under hyperosmotic stress will suffer if prevented from accumulating (by mutation, or inhibitors) or may do better if provided with compatible osmolytes (or crucial precursors). Exogenous compounds capable of the latter are sometimes called *osmoprotectants*.[23]

a. Bacteria, yeast

In cultured cells, manipulation may be achieved by altering the composition of laboratory media. Most testing has been done with *Escherichia coli,* which can grow in elevated salinities (NaCl) but with growth slowing with increasing osmolality (ceasing in 0.8 *M*[113]). The cells use K^+, glutamate, and trehalose as their major osmolytes in this condition.[9,27,113] As shown in Figure 5–7A, addition of extracellular betaine (or its precursor choline), or other compatible osmolytes such as choline-*O*-sulfate or β-alanine betaine (plant osmolytes), greatly stimulates ability to grow,[114] as K^+ and glutamate are reduced and replaced with the osmoprotectant or intracellular osmolytes derived from it.[23,113,115]

Osmoprotectants also affect bacterial respiration. Halotolerant bacteria exposed to 1.8 *M* KCl have suppressed respiration, which can be restored by exogenous solutes with effectiveness in the order dimethylsulfonioacetate > glycine betaine > dimethylglycine > β-alanine betaine = sarcosine > glycine.[116]

Another approach to altering cell organic osmolytes is through mutation. *E. coli* mutants incapable of replacing K^+ and glutamate with trehalose are impaired in osmoadaptation,[9] while other mutants that overproduce proline or increase glycine betaine uptake have enhanced osmotolerance.[117,118] A halotolerant yeast, *Debaryomyces hansenii,* accumulates glycerol and arabitol when exposed to high NaCl medium. A mutant strain, which accumulates much less of these polyols, grows more slowly in that medium; addition of exogenous glycerol, arabitol, or mannitol (but not erythritol) restores growth.[119]

b. Vertebrate cells

For the mammalian renal medulla, tests have been done with aldose reductase inhibitors, a group of drugs that blocks the enzyme that converts glucose to sorbitol. In standard growth medium made hyperosmotic with NaCl, a line of rabbit inner medullary epithelial cells (PAP-HT25) relies mainly on sorbitol for cell volume maintenance, at least in normal growth medium. These cells grow well (while accumulating sorbitol) in hyperosmotic medium with 250 m*M* NaCl (compared to normal nonrenal blood at about 150 m*M*). However, if an aldose reductase inhibitor is added to this medium, cell growth drops in parallel with declining cell sorbitol content (Figure 5–7B, left) (as a control, the drug had no effect in medium at 150 m*M* NaCl, in which PAP cells use little or no sorbitol).[120] With the drug, the lack of sorbitol synthesis (with other organic solutes not increasing sufficiently to compensate) would result in higher concentrations of

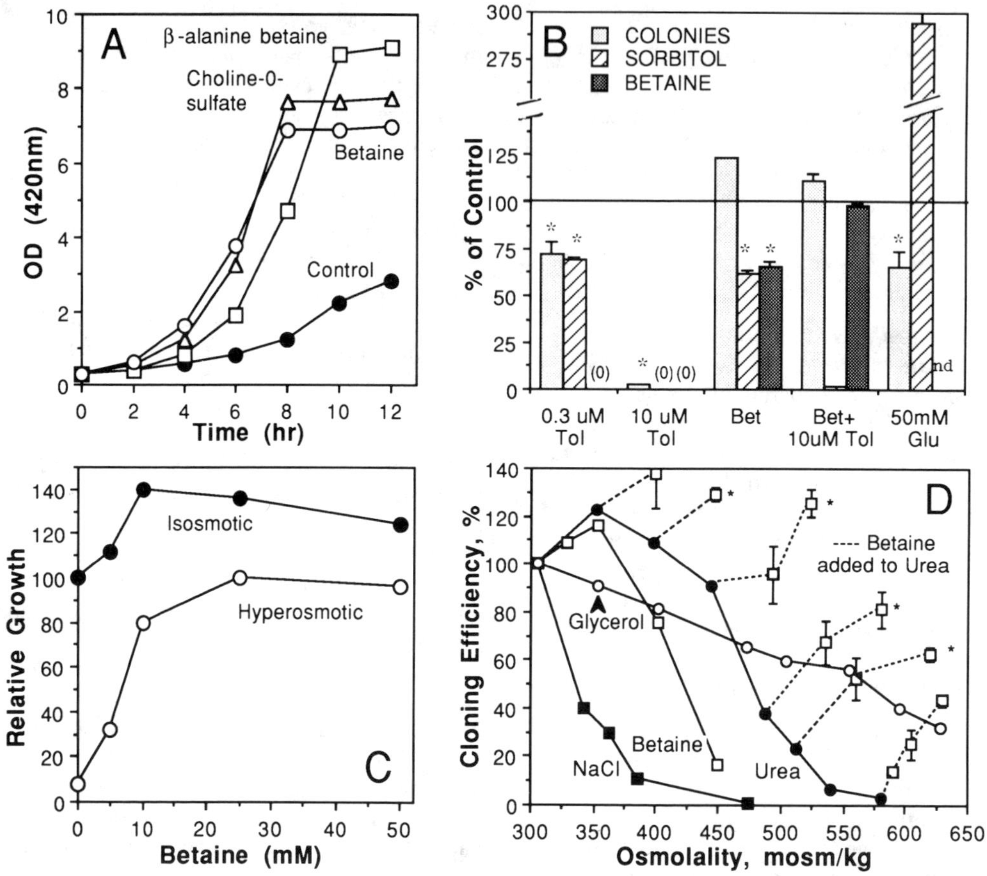

Figure 5–7 Compatibility of exogenous solutes with living cells. (A) Growth (by optical density) of *E. coli* at 37°C in 0.6 *M* NaCl, with or without 1 m*M* added solutes indicated. (From Hanson, A. D. et al., *Plant Physiol.,* 97, 1199, 1991. With permission from the American Society of Plant Physiologists.) (B) Effects of exogenous tolrestat (Tol, an aldose reductase inhibitor), betaine (Bet, 5 m*M*) or glucose (Glu) on number of successful colonies (from seeded individual cells) and osmolyte contents (mmol/kg protein normalized to control amount of sorbitol) of mammalian renal cells (PAP-HT25) after 7 to 14 d in culture, in 500 to 550 mOsmol media; *, significant change (data from References 120 and 142). (C) Cell density after 3 d of growth of SV-3T3 cells in 155 (isosmotic) or 250 m*M* NaCl, as percent of isosmotic value without betaine. (From Petronini, R. G. et al., *Biochem. J.,* 282, 69, 1992. With permission of The Biochemical Society Portland Press) (D) Number of successful colonies from seeded single mammalian renal (MDCK) cells after 4 to 5 d in culture, as percent of control with no added solutes; *, 2:1 urea/betaine. (From Yancey, P. H. and Burg, M. B., *Am. J. Physiol.,* 258, R198, 1990. With permission.)

noncompatible solutes, either by cell shrinkage or by ions entering from the medium, a clearly detrimental situation. Conversely, excessive quantities of osmolyte may be detrimental due to cell swelling: PAP cells exposed to high glucose overproduce sorbitol, and their growth suffers (Figure 5–7B, right). Indeed, this is thought to be one of the damaging side effects of diabetes mellitus and the use of aldose reductase inhibitors is being considered as possible treatment.[121]

These initial cell culture experiments were followed in two ways. First, *in vivo* testing was done by feeding an aldose reductase inhibitor to rats, and analyzing their kidneys for osmolytes. Like the PAP cells (Figure 5–7B), sorbitol content of the inner medulla declined greatly with drug treatment. Unlike the results with PAP cells, however, betaine contents rose in almost perfect compensation for the reduced sorbitol, such that there was no noticeable osmotic imbalance or damage.[122] Second, later studies on PAP cells confirmed the *in vivo* results: if PAP cells are provided betaine (normally absent) in growth medium, they will accumulate it when sorbitol production is suppressed by an aldose reductase inhibitor. The large decrease in cell viability previously noted[120] is largely reversed by betaine accumulation (Figure 5–7B, center).[123] Thus, betaine and sorbitol may serve as interchangeable compatible osmolytes in the kidney.

Other vertebrate cells have been examined for growth in hyperosmotic conditions. Mammalian blastocysts suffer *in vitro* when grown in elevated NaCl media. A recent report[124] shows that addition of glutamine restores development in proportion to the amount added to high-salt medium (while being inhibitory in optimal or low-salt media). The authors hypothesize that glutamine acts as a compatible osmolyte. SV-3T3 and chick fibroblast cells suffer reduced protein synthesis (>75%) and growth in 250 m*M* NaCl medium. Addition of some exogenous osmolytes reverses inhibition, shown for betaine in Figure 5–7C. For protein synthesis restoration, the order of effectiveness was betaine > proline > glycerol = taurine.[125]

Another prediction of the compatibility hypothesis is that penetrating compatible solutes should have less disruptive effects on cell function than other penetrating compounds. This has been tested on clonal colony growth of Madin-Darby canine kidney (MDCK) cells for two osmolytes that are relatively fast membrane penetrators — urea (noncompatible) and glycerol. With MDCK cells grown with these osmolytes in the medium, glycerol is clearly more compatible than urea at higher osmolalities (Figure 5–7D).[85]

4. Stabilization in Living Systems: Cryoprotection, Anhydrobiosis, Counteraction

The previous sections demonstrate general *compatibility* in living systems. However, they provide no evidence of actual macromolecular *stabilization*. Whether osmolytes are used widely *in vivo* to protect macromolecules against (nonfreezing) temperature stress, or to promote assembly, is unclear (e.g., a few studies suggest the former may occur in microorganisms[126]). But in three natural phenomena — freezing, anhydrobiosis, and perturbing osmolyte counteraction — it seems clear that stabilizing functions occur in addition to simple compatibility (nonperturbation). The first two phenomena are beyond the scope of this chapter; interested readers are directed to recent reviews.[58,64,71,112] In preventing urea and possibly salt perturbation of cell function during volume regulation, there is substantial evidence that certain cells use a combination of perturbing and stabilizing osmolytes to create a compatible intracellular environment.[100,127]

B. PERTURBATION AND COUNTERACTION

1. Perturbing Effects of Urea

In contrast to the usually benign effects of most organic osmolytes, the waste product urea is a well-known perturbant of macromolecules. At concentrations found in marine elasmobranchs and mammalian kidneys, a wide variety of biological functions are inhibited or disrupted, including alteration (usually inhibition) of enzyme kinetics (V_{max} or K_m), destabilization of protein folding and assembly, inhibition of muscle contraction and tissue respiration, death of exposed cells (e.g., Figure 5–7D) and organisms;[62] for example, bony fish exposed to 400 m*M* urea in seawater begin dying as their blood urea levels pass 200 m*M*.[128] Occasionally, major effects of low urea (e.g., 25 m*M*)[109] are seen on proteins, but most begin above 100 m*M*. At first this would seem to disprove the compatible osmolytes hypothesis; however, the cells of all nonestivating urea-accumulating animals examined so far (cartilaginous fishes, amphibia, mammalian kidneys) contain stabilizing osmolytes, mainly methylamines.[8,62,65]

2. Counteraction Between Urea and Methylamines: Subcellular Function

The occurrence of high concentrations (about 200 m*M*) of TMAO and betaine (Figure 5–2C) in most elasmobranchs (with urea at about 400 m*M*) prompted examination of these osmolytes.[100,127] Methylamines, when tested on enzyme functions, often do not exhibit simple compatibility, but enhance activity (e.g., Figure 5–4A). Figure 5–8A shows effects on an elasmobranch enzyme, in which methylamines and urea have opposite effects on K_m. These properties are additive, such that the two cancel out, most effectively at about a 2:1 urea/TMAO ratio (about the physiological average in elasmobranchs). This ratio also works well for several other enzymes (K_m or V_{max}), including elasmobranch and mammalian glutamate dehydrogenase;[127] elasmobranch and teleost actomyosin ATPase;[62] elasmobranch lactate dehydrogenase, argininosuccinase, creatine kinase,[127] and carbamoyl phosphate synthetase;[129] yeast alcohol dehydrogenase;[130] teleost glycine cleavage complex;[131] mammalian (renal) Ca-ATPase[132] and three enzymes shown in Figure 5–8B.[94,133,134]

Counteraction has also been found on more complex functions, including contraction of "skinned" shark muscle fibers[135] and various aspects of mitochondrial respiration. Less commonly observed is urea activation and TMAO inhibition of mitochondrial respiration[136,137] and a certain enzyme.[127] Per-

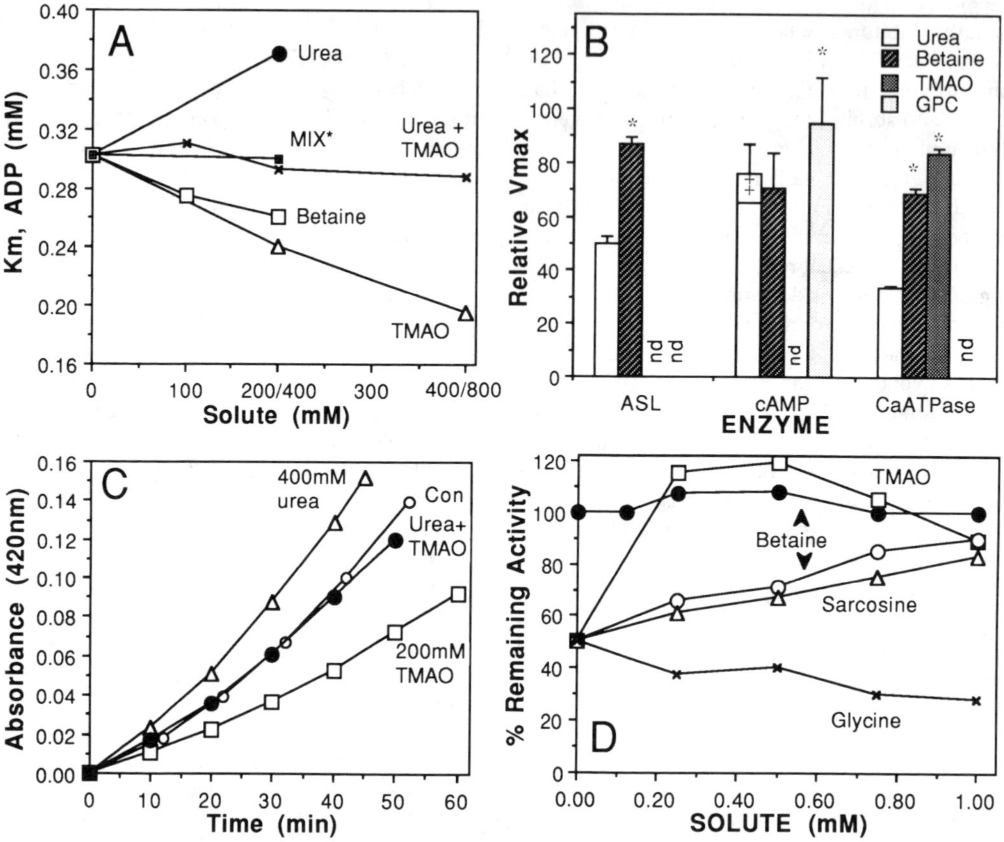

Figure 5–8 Counteracting effects of methylamines and perturbants on protein function and structure. (A) Apparent Michaelis constant (K_m) of ADP for stingray pyruvate kinase. Urea concentrations were twice that of other solutes (e.g., single urea point was 400 m*M;* rightmost urea + TMAO point was 800 m*M* urea + 400 m*M* TMAO). Mix = combination of TMAO, sarcosine, betaine and β-alanine totaling 200 m*M* plus 400 m*M* urea. (From Yancey, P. H. and Somero, G. N., *J. Exp. Zool.,* 212, 205, 1980. With permission of Wiley-Liss, a division of John Wiley and Sons, Inc., © 1980. (B) Reaction velocities (relative to control at 100%) with 1 *M* urea and 0.5 *M* methylamines of mammalian renal argininosuccinase (ASL, data from Reference 94), renal cAMP phosphodiesterase (cAMP, data from Reference 134), and muscle Ca-ATPase (data from Reference 133). nd, not determined; *, significantly higher than with urea only; ‡, two different control determinations. (C) Unfolding of bovine glutamate dehydrogenase at 22°C in reaction with thiol reagent 4-chloro-7-nitrobenzofurazan; Con, control; mixture is 400 m*M* urea + 200 m*M* TMAO. (From Yancey, P. H. and Somero, G. N., *Biochem. J.,* 182, 317, 1979. With permission of The Biochemical Society, Portland Press) (D) Activity of ribulose-1,5-bisphosphate carboxylase/oxygenase from cyanobacterium, *Aphanothece halophytica,* inhibited 50% by 250 m*M* KCl (open symbols) or without KCl (solid circles). (From Incharoensakdi, A. et al., *Plant Physiol.,* 81, 1044, 1986. With permission.)

haps TMAO makes some enzymes too rigid or too aggregated for proper function, while urea may restore flexibility or functional disaggregation.[5,127,130]

3. Counteraction between Urea and Methylamines: Subcellular Structure

Opposing effects are also seen on protein stability, including thermal denaturation of bovine ribonuclease (shown in Figure 5–5B),[100] catalase,[130] and lysozyme;[138] renaturation rate of elasmobranch and mammalian lactate dehydrogenase;[100] and unfolding of bovine glutamate dehydrogenase with a thiol reagent (Figure 5–8C). Recently, betaine at 2.5 *M* was found to reverse the large increase in the dielectric increment and relaxation time of bovine serum albumin in 5 *M* urea.[139] A non-natural urea counteractant may be DMSO, a commonly used cryoprotectant, found in one study to protect Na^+,K^+-ATPase from urea inactivation.[140]

4. Counteraction between Urea and Methylamines: Living Systems

The counteracting-osmolytes hypothesis suggests that a mixture is more beneficial to cells than urea or methylamine alone, since the latter might "overstabilize" proteins (above). Tests of this hypothesis *in vivo* are necessary for direct evidence. One study found that urea and betaine in growth medium will penetrate MDCK cells, such that intracellular contents can be manipulated.[85] As shown in Figure 5–7D, addition of urea or betaine alone to the medium greatly reduced cell colony growth, except for some stimulation at lower concentrations. However, addition of betaine with high urea partly or fully restored normal growth, in accordance with the counteracting hypothesis.

In another study, *E. coli* were found to use trace amounts of human urinary osmolytes (including glycine betaine) to grow effectively in the presence of salts and urea. Betaine alone did not provide protection, but it did so in combination with salts, sugars, or mannitol. This could play a role in urinary tract infections.[141]

Second, the hypothesis predicts that urea-counteracting osmolytes should be regulated specifically in response to intracellular urea and not external NaCl or osmotic pressure per se, to maintain a roughly constant ratio to urea inside cells. Marine elasmobranchs adapted to lower salinities reduce cellular urea and TMAO.[60,100] Mammalian kidneys have two methylamine osmolytes — glycine betaine and GPC (Figure 5–2C), which estimates suggest may be near a 1:2 ratio to urea.[94] *In vitro* experiments with both MDCK and PAP-HT25 mammalian renal cells lend support to the prediction of specific urea-methylamine regulation for GPC. Raising medium osmolality with NaCl triggers an increase in all intracellular organic osmolytes, including GPC; however, urea addition alone results in increased GPC only.[142] *In vivo,* the mammalian kidney also fits the prediction well for GPC, shown in tests with animals in which renal contents of urea were altered, largely independently of sodium through dietary intake.[143] In high-protein diets, renal urea (and to a lesser extent, sodium) contents increased, with only GPC increasing among the nonurea organic osmolytes (Figure 5–9, left). In low-protein diets, urea (but not sodium) contents decreased greatly, concomitant with decrease in GPC (and to a lesser extent, inositol) contents (Figure 5–9). Across all diets tested, only renal contents of urea and GPC maintained a strong linear correlation (Figure 5–9, inset). However, results with the other renal methylamine, glycine betaine, appear to contradict the hypothesis: betaine (and sorbitol) contents increased in apparent compensation for reduced GPC (and inositol) when medullary urea content is lowered *in vivo* by low dietary protein (Figure 5–9, middle).

Thus, although the link between renal urea and GPC may fit the counteracting hypothesis, a constant urea:methylamine (GPC plus betaine) ratio is clearly not maintained in all conditions. Betaine can counteract urea effects on enzymes and renal cells (above); if the counteracting-osmolytes hypothesis has validity, there are several possible reasons for the failure of total methylamines to follow urea, especially in low-protein diets (Figure 5–9).[143] First, it may be that GPC is a stronger (thus more critical) counteractant towards urea. Betaine is usually not as effective a stabilizer as TMAO, or as GPC in one preliminary study[131] (Figures 5–4A, 5–5B, 5–6B, 5–8B, and 5–8D). Adding that betaine is a widespread osmolyte in organisms with little or no urea (Table 1) and may functionally substitute for sorbitol as a compatible osmolyte (Figure 5–7B), suggests it may be more compatible than urea-counteracting, despite its methylamine structure. Second, it may be that a precise urea:methylamine ratio is not crucial for adequate protein functions at low urea levels, especially if betaine acts more as a compatible osmolyte. Finally, *myo*-inositol (which decreased in all low-urea conditions;[143] Figure 5–9) may be a urea-counteractant: among polyols, it is the strongest thermal stabilizer and preventer of urea denaturation of some proteins.[95,144] However, it could not offset urea inhibition of two mammalian renal enzymes.[94]

5. Salt Counteraction

The best case for another natural counteraction lies again in methylamines and their ability to offset functional inhibition and structural destabilization by inorganic salts, which could occur in conditions where elevated intracellular salts and organic osmolytes coexist (although salt counteractions are not yet found to be as consistent as urea counteraction). Pollard and Wyn Jones first showed (for barley MDH) the ability of glycine betaine to reverse NaCl inhibition of activity, dependent on methylation degree in the order betaine > dimethylglycine > sarcosine > glycine.[145] Other examples of partial or full counteraction of NaCl and/or KCl effects by methylamines include bacterial trehalase activity,[146] stability of spinach photosystem complex,[147] activity of a cyanobacterial enzyme, again with degree of methylation important, shown in Figure 5–8D.[148] In more complex systems, TMAO reverses the effects of salts on muscle fibers, namely disruption of myofilament architecture in barnacle preparation[78] and inhibi-

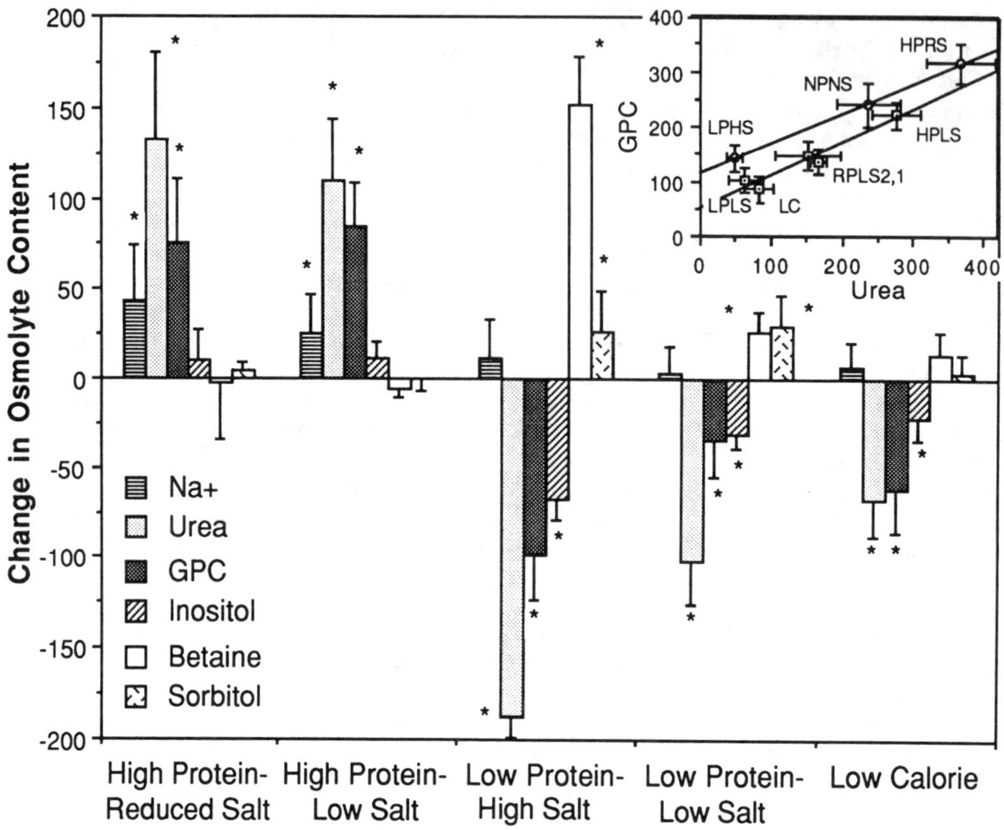

Figure 5–9 Main graph: changes (relative to control animals) in osmolyte contents (mmol/kg wet for urea and sodium, mmol/kg protein for others) in rat renal inner medullas, after 18 d on various dietary combinations of protein and salt. *, Significant change. Inset: correlation between GPC and urea contents, the only such relation found between individual osmolytes. Top line: experiment with low-protein, low-salt; normal-protein, normal-salt (control); and high-protein, reduced-salt diets. Lower line: two experiments with low-protein, low-salt; low-calorie, reduced-protein, low-salt (controls); and high-protein, low-salt diets. (From Peterson, D. P. et al., *Am. J. Physiol.,* 263, F594, 1992. With permission.)

tion of force development in rabbit (Figure 5–3C) and lobster fibers.[79]

Occasional counteraction also has been noted for other osmolytes, e.g., on the aggregation of chick erythrocyte chromatin induced by NaCl, with counteracting effectiveness in the order taurine > glycine > proline > mannitol > sorbitol > TMAO (with the latter, interestingly, having no effect).[149] Also, glycerol restores NaCl-inhibited activity of some *Dunaliella* (halophilic alga) enzymes but not others;[80] sorbitol reverses carp myosin denaturation and ATPase activity inhibition by NaCl and KCl.[150] A curious type of counteracting osmoprotection has been reported for primary cultures of mammalian renal medulla cells: explants survive in a medium with high urea and NaCl (two perturbants) better than with either alone.[151] This may not be true counteraction, however: explants undoubtedly start with intracellular methylamines and polyols matched to a urea-NaCl mixture *in vivo* (Figures 5–1B and 5–9), making them initially hyperosmotic to low-NaCl medium, or having excessive GPC relative to nonurea medium.

V. ORGANIC OSMOLYTE MECHANISMS AND CELL WATER

A fundamental premise of the compatible and counteracting hypotheses is that solute-macromolecule effects are in some sense universal, i.e., they rarely involve special adaptations in macromolecules (e.g., the solute effects seen in Figures 5–3A, 5–3B, 5–3C, 5–4B, and 5–7A are similar for proteins of high- and low-osmolyte species). How can this be? The answer probably involves universal water-solute-macromolecule interactions.

A. WATER-SOLUTE-MACROMOLECULE INTERACTIONS

1. The Hofmeister Series and Preferential Exclusion vs. Interaction

The initial explanation for osmolyte properties was proposed independently by Clark and Zounes[92] and Wyn Jones and co-workers.[6] Both groups realized that amino acid and methylamine solutes (Figures 5–2A to C) are structurally similar to the Hofmeister series, a ranking of anions and cations (shown in Figure 5–2D) which, since 1888, have been found to affect solution phenomena in the same way: ions to the left generally stabilize, salt out, and enhance catalysis of macromolecular systems, while those to the right do the opposite (e.g., Figure 5–3C).[152,153] Effects are algebraically additive, leading to counteraction if opposing ion types are paired [e.g., $(CH_3)_4NBr$], or if stabilizing salts are added with urea or other perturbants. K^+ and Na^+ rank in the middle, but K^+, the major cell cation, is consistently less destabilizing than Na^+. It now seems clear that the amino acid and methylated osmolytes, obvious Hofmeister homologs, operate by the same principles and probably mechanisms (details in Low,[154] Clark,[155] and Timasheff[156]). For instance, both methylated inorganic ions and organic osmolytes are typically the best overall stabilizers; GPC, the hypothesized main urea-counteractant in mammalian kidney, is a combination of methylamine cation, phosphate anion, and a diol (Figures 5–2C and D). As noted above (Figure 5–5A), polyols exhibit consistent rankings in their stabilizing effects.

Not fully understood, these universal properties have been attributed to the effects of (1) attractions of solutes to macromolecular moieties, and (2) solute effects on water structure.[152–156] Hofmeister ions in general, and organic perturbants, may be attracted to specific chemical groups on macromolecules. Alternatively, they may bind to water less well than water does to itself, resulting in sequestering of solute away from water towards the surfaces of macromolecules (Figure 5–10A[157]). If either effect dominates ("preferential interaction"[156]), then macromolecules will unfold, since this will maximize these favorable interactions. In contrast, stabilizing ions, and all organic stabilizers (except in desiccation[71]), plus some perturbants, exhibit an empirical tendency to be excluded from the protein hydration domain shell (documented extensively by Timasheff and colleagues[138,156]) (Figure 5–10B). If this effect dominates ("preferential exclusion"[156]), an entropically unfavorable order arises with neighboring regions of high and low solute (Figure 5–10B).[154] Proteins reduce this order if they minimize their exposed surface areas, by folding more compactly, by aggregation, or precipitation (salting out); i.e., they will be stabilized. If compact or aggregated forms are functionally better than loosened forms, and/or if exclusion drives substrates into active sites, enzymatic activity will also be enhanced.

2. Mechanisms of Exclusion

Preferential interaction from binding seems straightforward: charged solutes are attracted to charged protein groups, urea to hydrogen-bonding groups, hydrophobic solutes to hydrophobic groups. Preferential exclusion is less well understood, and there are several proposed mechanisms, in two broad categories (modified from References 154 to 156):

1. Steric exclusion/molecular crowding: some models use the fact that a bulky solute takes up more space than a water molecule, so will be less able to pack next to a protein than water; by simple geometry, water will be enriched relative to solute around the protein.[156] Bulky solutes will also increase protein activity by "crowding" proteins into a smaller space[158] (see Chapter 10, this volume). Any large solute not attracted to the macromolecule will have these effects, but also osmolytes like sugars with –OH groups that could form hydrogen bonds with proteins may have a poorer geometrical fit than water in the hydration lattice around the protein.[154]
2. Water structuring: stabilizers are thought to bind to water molecules better than to other molecules, and may actually enhance water-water interactions beyond their immediate hydration layer.[153] This water structure-making property is manifested in a reduced translational self-diffusion of water in osmolyte solutions[155] and their nearly universal tendency to raise water surface tension.[156] In one proposal, polar Hofmeister stabilizers with negative charges add to water structuring by charge distribution onto water oxygens, which readily take on a partial negative charge.[153] Stabilizers will be excluded from water at surfaces such as air and protein (Figure 5–10B), because less favorable (air-solute, protein-solute) interactions occur there, especially if hydration water around a protein is structured in a geometry unfavorable for hydrating the solute. Hydro-

phobic interactions that stabilize proteins are favored, since it becomes harder for them to interact with water structured by osmolytes.[156]

The unique ability of certain molecular types to interact with water has been thoroughly analyzed by Laughlin[159] in terms of "hydrophilicity". This involves a polar group X that confers association colloid behavior on a molecule RX, where R is a lipophilic group. In other words, attaching X to R makes an insoluble hydrophobe form structures such as micelles in water. Several types of X hydrophiles are listed in Figure 5–2E in order of effectiveness (where known). Diols, but not single hydroxyls, also work, so that all major osmolyte moieties have this property[159a] (not surprisingly, some of the groups are also the water-facing head groups of membrane phospholipids). Closely related groups [such as $-P(S)CH_3)_2$, $-NH(CH_3)$, $-N{=}O$] often do not work. Laughlin proposes that hydrophilicity arises from the groups having the correct binding geometry to substitute for water molecules in bulk water structure without major changes in that structure.[159]

Timasheff[156] concludes that most organic osmolytes work by water structuring, especially sugars and polyols with their multiple hydroxyl groups. However, methylamines remain problematical: betaine, though among the most strongly excluded solutes, does not raise surface tension[160] (see Clark[155] for analysis). It also did not fit the statistical crowding model at typical organismal concentrations in one study.[158] Possibly, the betaine methyl groups interact preferably with air over water but not with the polar protein surface.[160]

B. MODELS OF CELL WATER, SOLUTES, AND MACROMOLECULES

How do these mechanisms work on normal cell components in a cell undergoing hypertonic shrinkage, and how do osmolytes help at the mechanistic level? Can we extrapolate the typical disorganized, usually dilute-solution effects found *in vitro* (which provide most of the evidence cited) to the cell? In contrast to the old notion of the cytoplasm being an ordinary solution of enzymes and small solutes, currently there are two alternate views of cell organization.[161]

The crowded cytoplasm — In this view, cell solute content (especially protein) is so high as to cause the phenomenon of molecular crowding (see Chapter 10). Cayley et al.[82,83] show that, during shrinkage of *E. coli* in 1 *M* NaCl, cell K^+ activity rises from 0.14 to 0.76 *M*. Yet little disturbance of DNA-repressor binding occurs, despite the very strong inhibition of this by ionic strength *in vitro*. They attribute this to molecular crowding, which, during volume loss, will stabilize binding and macromolecular conformation in compensation for K^+ effects.[82] Nevertheless, growth ceases in these cells in direct proportion to increasing K^+ and decreasing volume of free water. They attribute this inhibition not to ion effects, but to crowding, which may slow diffusion or cause nonfunctional aggregations not stopped by K^+.[83] It is also possibly attributable to water loss, which could alter reactions in which water is involved (see Rand[162] for an overview of the large numbers of water molecules involved in protein reactions and assembly). Addition of osmoprotectants (betaine, proline) restores growth as they replace K^+ and restore cell water content: preferential exclusion makes them more osmotically active than K ions, which are partly bound to polyanions.[83] Thus, in their view, compatible solutes act to restore water and perhaps relieve unfavorable general crowding rather than deleterious ion effects. But K^+ and other ions could affect systems not studied, and their data cannot distinguish among these possibilities.[158]

The structured cytoplasm — In this view, championed by Clegg and co-workers,[163,164] eukaryotic cells have networks of cytoskeletal filaments to which most metabolic enzymes are bound, interspersed with a relatively uncrowded aqueous region (bulk water). Metabolism may take place within bound or "vicinal" water along the filaments, and bulk water is largely irrelevant. Clegg and co-workers[163] use sorbitol to shrink mammalian cells and show that, even after 85% water loss, glycolysis rate is not greatly altered. One might expect inhibition, as predicted by the ion-inhibition aspect of the compatible hypothesis, or strong activation, as should occur from simply concentrating the catalysts. They interpret the small effects as showing that the bulk water being lost is not necessary, consistent with the structured metabolism hypothesis. Interpretation is difficult, however; ion effects and increased concentration could work against each other. Alternatively, the use of a compatible solute for shrinkage could provide some compensation since effects on mammalian cells are much worse if a salt is used, rather than a compatible solute, for shrinkage[84] (e.g., Figure 5–7D). These issues and those raised by Cayley et al.[82,83] on cell water and ions require more study.

Given some evidence for the structured cytoplasm model, it is worth examining how osmolytes might fit in. Wiggins[165] has done so, as shown in Figures 5–10C and D. In Clegg's model, meta-

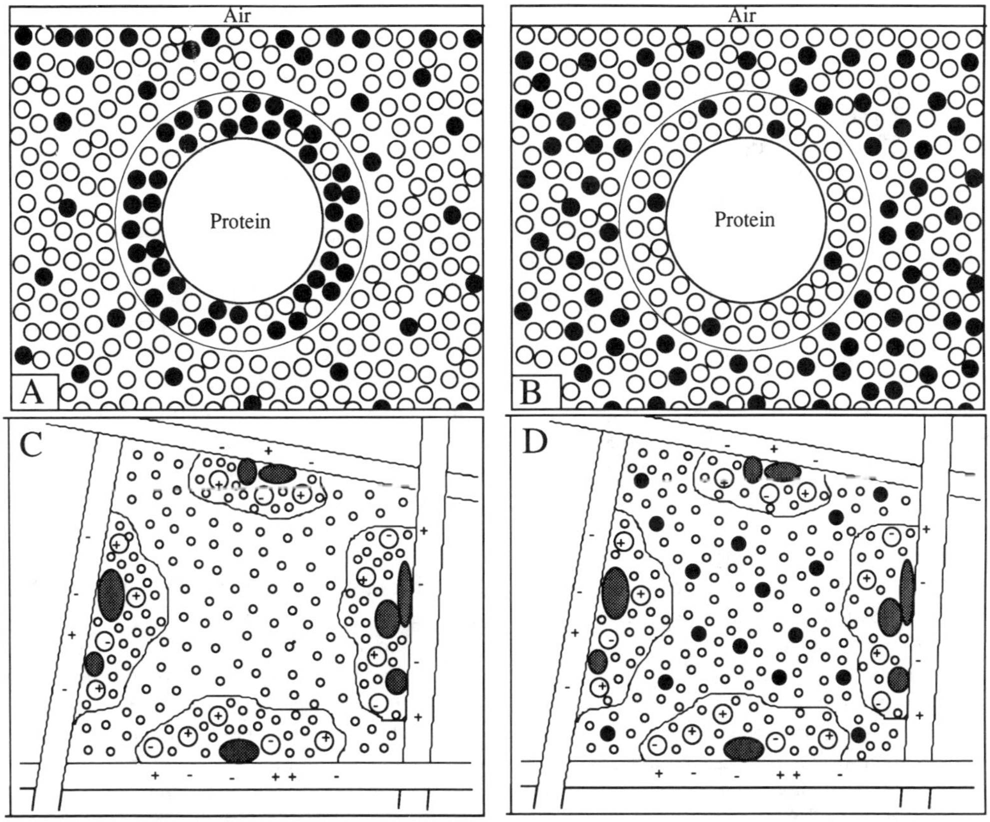

Figure 5–10 Simplified representations of water (small open circles) and test solutes (solid circles) at air and macromolecular interfaces. (A) Preferential interaction with destabilizer; (B) preferential exclusion with stabilizer. (Adapted from Na, G. C. and Timasheff, S. N., *J. Mol. Biol.,* 151, 165, 1981. With permission.) (C, D) Cell organization proposed by Clegg,[164] as modified by Wiggins,[165] with osmolytes (solid), rods representing F-actin filaments with bound enzymes (gray), attracted ions (+, –). See text for details. (Modified from Wiggins, P. M., *Microbiol. Rev.,* 54, 432, 1990. With permission.)

bolic enzymes are attached to cytoskeletal filaments. Due to protein charges, most cell ions (organic and inorganic) ought to be attracted to these areas also (Figure 5–10C). This would create areas of high osmolality around protein complexes, which could draw in bulk water. Wiggins proposes that the bound water must become denser while bulk water becomes less dense to counter this osmotic imbalance. Dense water, however, is weakly structured (since it cannot attain the open ice-like geometry for optimal hydrogen bonding), which could weaken protein hydration stabilization. Preferentially excluded osmolytes would accumulate in the bulk water and offset the osmolality differences that caused the problem (Figure 5–10D). Wiggins includes K^+ as a weakly compatible osmolyte, because it is more readily excluded from surfaces than is Na^+.

V. COMPARISON OF AND EXCEPTIONS TO COMPATIBILITY, STABILIZATION, AND COUNTERACTION

There is some confusion in the literature over the various osmolyte effects seen *in vitro,* the most common being the mistake of assuming compatibility and stabilization are the same thing. The strict concept of *compatibility* is that osmolytes are *nonperturbing* in steady-state active cells. Ability to *stabilize* may not come into play in most nonurea systems (though this is certainly unclear). Figure 5–8D shows an example: betaine, though able to restore function in high KCl, has no effect on this plant enzyme in the absence of deleterious salt levels, and whether betaine protects the protein from destabilizing forces (such as heat) is unknown. And, as Cayley et al.[83] show in bacteria, betaine increases in hyperosmotically stressed cells in proportion to K^+ decrease, so that counteraction against salt may never be needed. A similar K^+/osmolyte swap occurs in mammalian renal cells,[8,12] and it may be that "heat shock" stress proteins provide initial stabilization when proteins are disturbed following hyperosmotic shock (see Chapter 22). Many stabilizing effects on structure (Figures 5–5, 5–6, and 5–8) require osmolyte levels well above organismal values for significant effects, and in any case excess prolonged stabilization could impair function. Stabilization does occur *in vivo* in severe stress, such as freezing and desiccation, and in *counteractions* when perturbing solutes are involved (and perhaps in other extreme conditions yet to be clearly elucidated).

Caution must be applied to generalizing the ideas of compatibility and general stabilization. Glycerol, for example, is not always nonperturbing of function, e.g., in one study of 17 bacterial, yeast, and mammalian enzymes, some were inhibited (lower velocity and/or higher K_m), some activated, and some unaffected by 30% glycerol.[166] Glycerol above 0.5 *M* can also be toxic to mammalian cells through loss of membrane integrity[167] (see also Figure 5–7D). Possibly related to this or to glycerol retention, glycerol-using yeast grown in hyperosmotic medium substantially restructure their membranes.[168] Mono- and disaccharides tend to be less compatible than other solutes (e.g., Figure 5–3B); glucose has the additional problem of glycosylation of proteins, which may preclude its use as an osmolyte except at low temperature.[64] Other solutes show direct temperature- and concentration-dependent stabilization, e.g., DMSP in Figure 5–5D is effective at 0.2 but not 0.3 *M* at 50°C, yet works well at 0.3 *M* against cold inactivation (Figure 5–6B). Cryoprotectants such as DMSO show similar trends, a phenomenon attributed to increasing strength of hydrophobic interactions with temperature causing DMSO (and DMSP) to bind to proteins.[169] Thus, osmolytes with hydrophobic properties may not work at high temperature.[104]

Counteraction also does not always occur, i.e., a few urea-inhibited enzymes are not restored by TMAO.[94,109,127,130] This may require different adaptations, i.e., urea-insensitivity or urea-requiring activity in some elasmobranch proteins,[62] or perhaps some uncompensated inhibition is tolerable in nonlimiting reactions. Estivators that use urea may not need methylamines, since metabolism needs to be suppressed.[5,109]

VII. EVOLUTIONARY PERSPECTIVES, CONCLUSIONS

Remaining to be explained in detail is why certain osmolytes occur in certain groups, other than for compatibility or counteraction. Are, for example, betaine and sorbitol equivalent (as suggested by kidney experiments, above), with their selection among organisms due to other factors? The prevalence of sugars and polyols in photosynthesizers may reflect their primary energy pathway;[33] nitrogen limitation (especially in the oceans) may force use of DMSP and carbohydrates.[88] β-amino acids will not interfere with nonosmolyte functions of α-amino acids. Temperature effects on compatibility (above) may also be important, as may be metabolic roles of osmolytes (e.g., amino acids in protein synthesis, GPC in membrane phospholipid pathways[8]). Further study on possible unique or restricted properties of many osmolytes needs examination.

Despite these complexities, it seems likely that use of these osmolytes should speed up evolutionary adaptation relative to the alternative of altering most cell macromolecules to function in a dehydrated or concentrated ion solution.[5] Perhaps the ultimate test of this will be the transfer of the genetic system for osmolyte production from a salt-tolerant into a salt-intolerant species. This has been proposed as a logical step in genetic engineering of plants, e.g., adapting crops for saline soils or irrigation waters.[170,171] Work has begun with the successful transfer and expression of a bacterial mannitol dehydrogenase gene in tobacco, inducing greater salinity tolerance in the plant.[172]

A final topic is the practical use of knowledge of osmolytes. Agricultural uses, stabilization of biochemical preparations,[106,107] and understanding of diabetes[121] have already been mentioned. Stabilization of pharmaceuticals[173] and other medical uses are being explored. Zwitterionic detergents, capable of extracting membrane proteins while preserving function, are a fairly new group of sulfobetaine compounds with the formula (C_nH_{2n+1})-$N^+(CH_3)_2$-$(CH_2)_3$-SO^-_3.[174] The possible involvement of methylamines in human uremia and its treatment has been suggested,[175] and in an interesting extrapolation of the counteracting hypothesis, Kloiber et al.[176] have found that methylamine injections in rats provide strong protection against hyperammonemia. Finally, DMSP, a prevalent osmolyte in marine phytoplankton,[41] has been proposed to be a key regulator of the Earth's climate through dimethylsulfide gas production and the cloud formation it may trigger (part of the Gaia hypothesis).[177] Understanding osmolyte systems may have implications for the whole planet!

REFERENCES

1. **Schultz, S. G.,** Volume preservation: then and now, *News Physiol. Sci.,* 4, 169, 1989.
2. **Chamberlin, M. E. and Strange, K.,** Anisosmotic cell volume regulation: a comparative view, *Am. J. Physiol.,* 257, C159, 1989.
3. **Smith, H. W.,** The retention and physiological role of urea in the elasmobranchii, *Biol. Rev.,* 11, 49, 1936.
4. **Clark, M. E.,** The osmotic role of amino acids: discovery and function, in *Transport Processes, Iono- and Osmoregulation,* Gilles, R. and Gilles-Ballien, M., Eds., Springer-Verlag, Berlin, 1985, 412.
5. **Yancey, P. H., Clark, M. E., Hand, S. C., Bowlus, R. D., and Somero, G. N.,** Living with water stress: evolution of osmolyte systems, *Science,* 217, 1214, 1982.
6. **Wyn Jones, R. G., Storey, R., Leigh, R. A., Ahmad, N., and Pollard, A.,** A hypothesis on cytoplasmic osmoregulation, in *Regulation of Cell Membrane Activities in Plants,* Marre, E. and Ciferri, O., Eds., Elsevier, Amsterdam, 1977, 121.
7. **Somero, G. N.,** Adapting to water stress: convergence on common solutions, in *Water and Life: A Comparative Analysis of Water Relationships at the Organismic, Cellular, and Molecular Levels,* Somero, G. N., Osmond, C. B., and Bolis, C. L., Eds., Springer-Verlag, Berlin, 1992, 3.
8. **Garcia-Perez, A. and Burg, M. B.,** Renal medullary organic osmolytes, *Physiol. Rev.,* 71, 1081, 1991.
9. **Dinnibier, U., Limpinsel, E., Schmid, R., and Bakker, E. P.,** Transient accumulation of potassium glutamate and its replacement by trehalose during adaptation of growing cells of *Escherichia coli* K-12 to elevated sodium chloride concentrations, *Arch. Microbiol.,* 150, 348, 1988.
10. **Blackwell, J. R. and Gilmour, D. J.,** Physiological response of the unicellular green alga *Chlorococcum submarinum* to rapid changes in salinity, *Arch. Microbiol.,* 157, 86, 1991.
11. **Reed, R. H., Warr, S. R. C., Richardson, D. L., Moore, D. J., and Stewart, W. D. P.,** Multiphasic osmotic adjustment in a euryhaline cyanobacterium, *FEMS Microbiol. Lett.,* 28, 225, 1985.
12. **Uchida, S., Garcia-Perez, A., Murphy, H., and Burg, M. B.,** Signal for induction of aldose reductase in renal medullary cells by high external NaCl, *Am. J. Physiol.,* 256, C614, 1989.
13. **Yancey, P. H. and Burg, M. B.,** Distributions of major organic osmolytes in rabbit kidneys in diuresis and antidiuresis, *Am. J. Physiol.,* 257, F602, 1989.
14. **Beck, F., Dorge, A., Rick, R., and Thurau, K.,** Intra- and extracellular elemental concentrations of rat inner papilla in antidiuresis, *Kidney Int.,* 25, 397, 1984.
15. **Brown, A. D.,** Microbial water stress, *Bacteriol. Rev.,* 40, 803, 1976.
16. **Ullrich, K. J.,** Glycerophosphorylcholinumsatz und glycerophosphorylcholindiesterase in der Saugetier-Niere, *Biochem. Z.,* 331, 98, 1959.
17. **Stewart, G. R. and Lee, J. A.,** The role of proline accumulation in halophytes, *Planta,* 120, 279, 1974.
18. **Star, R. A.,** Hyperosmolar states, *Am. J. Med. Sci.,* 300, 402, 1990.
19. **Heilig, C. W., Stromski, M. E., Blumenfeld, J. D., Lee, J. P., and Gullans, S. R.,** Characterization of the major brain osmolytes that accumulate in salt-loaded rats, *Am. J. Physiol.,* 257, F1108, 1989.
20. **Lien, Y.-H. H., Shapiro, J. I., and Chan, L.,** Effects of hypernatremia on organic brain osmoles, *J. Clin. Invest.,* 85, 1427, 1990.
21. **Robertson, D. E. and Roberts, M. F.,** Organic osmolytes in methanogenic archaebacteria, *BioFactors,* 3, 1, 1991.
22. **Robertson, D. E., Noll, D., and Roberts, M. F.,** Free amino acid dynamics in marine methanogens. β-amino acids as compatible solutes, *J. Biol. Chem.,* 267, 14893, 1992.
23. **Csonka, L. N. and Hanson, A. D.,** Prokaryotic osmoregulation: genetics and physiology, *Annu. Rev. Microbiol.,* 45, 569, 1991.
24. **Galinski, E. A. and Oren, A.,** Isolation and structure of a novel compatible solute from the moderately halophilic purple sulfur bacterium *Ectothiorhodospira marismortue, Eur. J. Biochem.,* 198, 593, 1991.
25. **Gouesbet, G., Blanco, C., Hamelin, J., and Bernard, T.,** Osmotic adjustment in *Brevibacterium ammoniagenes:* pipecolic acid accumulation at elevated osmolalities, *J. Gen. Microbiol.,* 138, 959, 1992.
26. **Wohlfarth, A., Severin, J., and Galinski, E. A.,** The spectrum of compatible solutes in heterotrophic eubacteria of the family *Halomonadaceae, J. Gen. Microbiol.,* 136, 705, 1990.

27. **Larsen, P. I., Syndes, L. K., Landfald, B., and Strøm, A. R.,** Osmoregulation in *Escherichia coli* by accumulation of organic osmolytes: betaines, glutamic acid, and trehalose, *Arch. Microbiol.,* 147, 1, 1987.
28. **MacKay, M. A., Norton, R. S., and Borowitzka, L. J.,** Organic osmoregulatory solutes in cyanobacteria, *J. Gen. Microbiol.,* 130, 2177, 1984.
29. **Schuh, W. and Puff, H.,** The crystal structure of ectoine, a novel amino acid of potential osmoregulatory function, *Z. Naturforsch.,* 40c, 780, 1985.
30. **Smith, L. T. and Smith, G. M.,** An osmoregulated dipeptide in stressed *Rhizobium melilot, J. Bacteriol.,* 171, 4714, 1989.
31. **Smith, L. T., Smith, G. M., and Madkour, M. A.,** Osmoregulation in *Agrobacterium tumefaciens:* accumulation of a novel disaccharide is controlled by osmotic strength and glycine betaine, *J. Bacteriol.,* 172, 6849, 1990.
32. **Trüper, H. G. and Galinski, E. A.,** Concentrated brines as habitats for microorganisms, *Experientia,* 42, 1182, 1986.
33. **Borowitzka, L.,** Glycerol and other carbohydrate osmotic effectors, in *Transport Processes, Iono- and Osmoregulation,* Gilles, R. and Gilles-Ballien, M., Eds., Springer-Verlag, Berlin, 1985, 437.
34. **Brown, L. M. and Hellebust, J. A.,** The contribution of organic solutes to osmotic balance in some green and eustigmatophyte algae, *J. Phycol.,* 16, 265, 1980.
35. **Dickinson, D. M. J. and Kirst, G. O.,** Osmotic adjustment in marine eukaryotic algae: the role of inorganic ions, quaternary ammonium, tertiary sulphonium and carbohydrate solutes. I. Diatoms and a Rhodophyte, *New Phytol.,* 106, 645, 1987.
36. **Dickinson, D. M. J. and Kirst, G. O.,** Osmotic adjustment in marine eukaryotic algae: the role of inorganic ions, quaternary ammonium, tertiary sulphonium and carbohydrate solutes. II. Prasinophytes and Haptophytes, *New Phytol.,* 106, 657, 1987.
37. **Dickinson, D. M. J. and Kirst, G. O.,** The role of β-dimethylsulphoniopropionate, glycine betaine and homarine in the osmoacclimation of *Platymonas subcordiformis, Planta,* 167, 536, 1986.
38. **Edwards, D. M., Reed, R. H., and Stewart, W. D. P.,** Osmoaccumulation in *Enteromorpha intestinalis:* long-term effects of osmotic stress on organic solute accumulation, *Mar. Biol.,* 98, 467, 1988.
39. **Hellebust, J. A.,** Mechanisms of response to salinity in halotolerant microalgae, *Plant Soil,* 89, 69, 1985.
40. **Impellizzeri, G., Mangiafico, S., Oriente, G., Piattelli, M., and Scuito, S.,** Amino acids and low-molecular-weight carbohydrates of some marine red algae, *Phytochemistry,* 14, 1549, 1975.
41. **Keller, M. D.,** Dimethyl sulfide production and marine phytoplankton: the importance of species composition and cell size, *Biol. Oceanogr.,* 6, 375, 1988.
42. **Kirst, G. O.,** Salinity tolerance of eukaryotic marine algae, *Annu. Rev. Plant Physiol. Plant Mol. Biol.,* 40, 21, 1989.
43. **Mason, T. G. and Blunden, G.,** Quaternary ammonium and tertiary sulfonium compounds of algal origin as alleviators of osmotic stress, *Bot. Mar.,* 32, 313, 1989.
44. **Reed, R. H.,** Measurement and osmotic significance of β-dimethylsulphoniopropionate in marine macroalgae, *Mar. Biol. Lett.,* 4, 173, 1983.
45. **Wegmann, K.,** Osmoregulation in eukaryotic algae, *FEMS Microbiol. Rev.,* 39, 37, 1986.
46. **Kaneshiro, E. S., Holz, G. G., Jr., and Dunham, P. B.,** Osmoregulation in a marine ciliate, *Miamiensis avidus.* II. Regulation of intracellular free amino acids, *Biol. Bull.,* 137, 161, 1969.
47. **Beever, R. E. and Laracy, E. P.,** Osmotic adjustment in the filamentous fungus *Aspergillus nidulans, J. Bacteriol.,* 168, 1358, 1986.
48. **Evans, R. D., Black, R. A., Loescher, W. H., and Fellows, R. J.,** Osmotic relations of the drought-tolerant shrub *Artemisia tridentata* in response to water stress, *Plant Cell Environ.,* 15, 49, 1992.
49. **Ford, C. W.,** Accumulation of *o*-methyl-inositols in water-stressed *Vigna* species, *Phytochemistry,* 21, 1149, 1982.
50. **Gorham, J., Wyn Jones, R. G., and McDonnell, E.,** Some mechanisms of salt tolerance in crop plants, *Plant Soil,* 89, 15, 1985.
51. **Sacher, R. F. and Staples, R. C.,** Inositol and sugars in adaptation of tomato to salt, *Plant Physiol.,* 77, 206, 1985.
52. **Sommer, C., Thonke, B., and Popp, M.,** The compatibility of D-pinitol and 1D-1-O-methyl-muco-inositol with malate dehydrogenase activity, *Bot. Acta,* 103, 270, 1990.
53. **Wyn Jones, R. G.,** Phytochemical aspects of osmotic adaptation, *Rec. Adv. Phytochem.,* 18, 55, 1984.
54. **Burton, R. F.,** The composition of animal cells: solutes contributing to osmotic pressure

and charge balance, *Comp. Biochem. Physiol.*, 76B, 663, 1983.

55. **Garret, M. A. and Bradley, T. J.,** Extracellular accumulation of proline, serine and trehalose in the haemolymph of osmoconforming brackish-water mosquitoes, *J. Exp. Biol.*, 129, 231, 1987.
56. **Hand, S. C.,** Water content and metabolic organization in anhydrobiotic animals, in *Water and Life: A Comparative Analysis of Water Relationships at the Organismic, Cellular, and Molecular Levels,* Somero, G. N., Osmond, C. B., and Bolis, C. L., Eds., Springer-Verlag, Berlin, 1992, 104.
57. **Pierce, S. K., Edwards, S. C., Mazzocchi, P. H., Klinger, L. J., and Warren, M. K.,** Proline betaine: a unique osmolyte in an extremely euryhaline osmoconformer, *Biol. Bull.*, 167, 495, 1984.
58. **Duman, J. G., Wu, D. W., Xu, L., Tursman, D., and Olsen, T. M.,** Adaptations of insects to subzero temperatures, *Q. Rev. Biol.*, 66, 387, 1991.
59. **Loomis, S. H., Carpenter, J. F., and Crowe, J. H.,** Identification of strombine and taurine as cryoprotectants in the intertidal bivalve *Mytilus edulis, Biochim. Biophys. Acta,* 943, 113, 1988.
60. **King, P. A. and Goldstein, L.,** Organic osmolytes and cell volume regulation in fish, *Mol. Physiol.*, 4, 53, 1983.
61. **Sherman, W. R., Simpson, P. C., and Goodwin, S. L.,** *Scyllo*-inositol and *myo*-inositol levels in tissues of the skate *Raja erinacea, Comp. Biochem. Physiol.*, 59B, 201, 1978.
62. **Yancey, P. H.,** Organic osmotic effectors in cartilaginous fishes, in *Transport Processes, Iono- and Osmoregulation,* Gilles, R. and Gilles-Ballien, M., Eds., Springer-Verlag, Berlin, 1985, 424.
63. **Raymond, J. A.,** Glycerol is a colligative antifreeze in some northern fishes, *J. Exp. Zool.*, 262, 347, 1992.
64. **Storey, K. B. and Storey, J. M.,** Freeze tolerance in animals, *Physiol. Rev.*, 68, 27, 1988.
65. **Wilkie, D. R. and Wray, S.,** The response of the African frog *Xenopus laevis* to increased salinity demonstrated by ^{31}P-NMR spectroscopy, *J. Physiol.*, 381, 30P, 1986.
66. **Shihabi, Z. K., Goodman, H. O., Holmes, R. P., and O'Connor, M. I.,** The taurine content of avian erythrocytes and its role in osmoregulation, *Comp. Biochem. Physiol.*, 92A, 545, 1989.
67. **Avison, M. J., Rothman, D. L., Nixon, T. W., Long, W. S., and Siegel, N. J.,** ^{1}H NMR study of renal trimethylamine responses to dehydration and acute volume loading in man, *Proc. Natl. Acad. Sci. U.S.A.*, 88, 6053, 1991.
68. **Law, R. O.,** Amino acids as volume-regulatory osmolytes in mammalian cells, *Comp. Biochem. Physiol.*, 99A, 263, 1991.
69. **Trachtman, H., Barbour, R., Sturman, J. A., and Finberg, L.,** Taurine and osmoregulation: taurine is a cerebral osmoprotective molecule in chronic hypernatremic dehydration, *Pediatr. Res.*, 23, 35, 1988.
70. **Nakaniski, T., Uyama, O., and Sugita, M.,** Osmotically regulated taurine content in the rat renal inner medulla, *Am. J. Physiol.*, 261, F957, 1991.
71. **Crowe, J. H., Hoekstra, F. A., and Crowe, L. M.,** Anhydrobiosis, *Annu. Rev. Physiol.*, 54, 579, 1992.
72. **Katz, U.,** Strategies of adaptation to osmotic stress in anuran amphibia under salt and burrowing conditions, *Comp. Biochem. Physiol.*, 93A, 499, 1989.
73. **Yancey, P. H.,** Osmotic effectors in kidneys of xeric and mesic rodents: corticomedullary distributions and changes with water availability, *J. Comp. Physiol.*, 158B, 369, 1988.
74. **Steinbach, H. B.,** The prevalence of K, *Perspect. Biol. Med.*, 5, 338, 1962.
75. **Greenway, H. and Osmond, C. B.,** Salt responses of enzymes from species differing in salt tolerance, *Plant Physiol.*, 49, 256, 1972.
76. **Gibson, T. S., Speirs, J., and Brady, C. J.,** Salt-tolerance in plants. II. *In vitro* translation of m-RNAs from salt-tolerant and salt-sensitive plants on wheat germ ribosomes. Responses to ions and compatible organic solutes, *Plant Cell Environ.*, 7, 579, 1984.
77. **Andrews, M. A. W., Maughan, D. W., Nosek, T. M., and Godt, R. E.,** Ion-specific and general ionic effects on contraction of skinned fast-twitch skeletal muscle from the rabbit, *J. Gen. Physiol.*, 98, 1105, 1991.
78. **Clark, M. E., Hinke, J. A. M., and Todd, M. E.,** Studies on water in barnacle muscle fibres. II. Role of ions and organic solutes in swelling of chemically-skinned fibres, *J. Exp. Biol.*, 90, 43, 1981.
79. **Godt, R. E., Fogaca, R. T. H., Andrews, M. A. W., and Nosek, T. M.,** Influence of ionic strength on contractile force and energy consumption on skinned fibers from mammalian and crustacean striated muscle, in *Molecular Mechanisms of Muscle Contraction,* Sugi, H., Ed., Plenum Press, New York, 1992, in press.
80. **Ginzburg, M.,** *Dunaliella:* a green alga adapted to salt, *Adv. Bot. Res.*, 14, 93, 1987.

81. **Warr, S. R. C., Reed, R. H., and Stewart, W. D. P.,** The compatibility of osmotica in cyanobacteria, *Plant Cell Environ.,* 11, 137, 1988.
82. **Cayley, S., Lewis, B. A., Guttman, H. J., and Record, M. T.,** Characterization of the cytoplasm of *Escherichia coli* K-12 as a function of external osmolarity. Implications for protein-DNA interactions in vivo, *J. Mol. Biol.,* 222, 281, 1991.
83. **Cayley, S., Lewis, B. A., and Record, M. T.,** Origins of the osmoprotective properties of betaine and proline in *Escherichia coli* K-12, *J. Bacteriol.,* 174, 1586, 1992.
84. **Armitage, W. J., Moss, S. J., and Easty, D. L.,** Effects of osmotic stress on rabbit corneal endothelium, *Cryobiology,* 25, 425, 1988.
85. **Yancey, P. H. and Burg, M. B.,** Counteracting effects of urea and betaine on colony-forming efficiency of mammalian cells in culture, *Am. J. Physiol.,* 258, R198, 1990.
86. **Lanyi, J.,** Salt-dependent properties of proteins from extremely halophilic bacteria, *Bacteriol. Rev.,* 38, 272, 1974.
87. **Shindler, D. B., Wydro, R. W., and Kushner, D. J.,** Cell-bound cations of the moderately halophilic bacterium, *Vibrio costicola, J. Bacteriol.,* 130, 698, 1977.
88. **Hanson, A. D.,** Compatible solute synthesis and compartmentation in higher plants, in *Water and Life: A Comparative Analysis of Water Relationships at the Organismic, Cellular, and Molecular Levels,* Somero, G. N., Osmond, C. B., and Bolis, C. L., Eds., Springer-Verlag, Berlin, 1992, 52.
89. **Wiencke, C.,** The response of pyruvate kinase from the intertidal red alga *Porphyra umbilicalis* to sodium and potassium ions, *J. Plant Physiol.,* 116, 447, 1984.
90. **Bowlus, R. D. and Somero, G. N.,** Solute compatibility with enzyme function and structure: rationales for the selection of osmotic agents and end-products of anaerobic metabolism in marine invertebrates, *J. Exp. Zool.,* 208, 137, 1979.
91. **Brown, A. and Simpson, J.,** Water relations of sugar-tolerant yeasts: the role of intracellular polyols, *J. Gen. Microbiol.,* 72, 589, 1972.
92. **Clark, M. E. and Zounes, M.,** The effects of selected cell osmolytes on the activity of lactate dehydrogenase from the euryhaline polychaete, *Nereis succinea, Biol. Bull.,* 153, 468, 1977.
93. **Ahmad, I., Larher, F., and Stewart, G. R.,** Sorbitol, a compatible osmotic solute in *Plantago maritima, New Phytol.,* 82, 671, 1979.
94. **Yancey, P. H.,** Compatible and counteracting aspects of organic osmolytes in mammalian kidney cells in vivo and in vitro, in *Water and Life: A Comparative Analysis of Water Relationships at the Organismic, Cellular, and Molecular Levels,* Somero, G. N., Osmond, C. B., and Bolis, C. L., Eds., Springer-Verlag, Berlin, 1992, 19.
95. **Gerlsma, S. Y.,** Reversible denaturation of ribonuclease in aqueous solutions as influenced by polyhydric alcohols and some other additives, *J. Biol. Chem.,* 243, 957, 1968.
96. **Back, J. F., Oakenfull, D., and Smith, M. B.,** Increased thermal stability of proteins in the presence of sugars and polyols, *Biochemistry,* 18, 5191, 1979.
97. **Fujita, Y., Iwasa, Y., and Noda, Y.,** The effect of polyhydric alcohols on the thermal denaturation of lysozyme as measured by differential scanning calorimetry, *Bull. Chem. Soc. Jpn.,* 55, 1896, 1982.
98. **Gekko, K. and Morikawa, T.,** Thermodynamics of polyol-induced thermal stabilization of chymotrypsinogen, *J. Biochem.,* 90, 51, 1981.
99. **Combes, D. and Monsan, P.,** Effect of polyhydric alcohols on invertase stabilization, *Ann. N.Y. Acad. Sci.,* 434, 61, 1984.
100. **Yancey, P. H. and Somero, G. N.,** Counteraction of urea destabilization of protein structure by methylamine osmoregulatory compounds of elasmobranch fishes, *Biochem. J.,* 182, 317, 1979.
101. **Santoro, M. M., Liu, Y., Khan, S., M. A., Hou, L.-X., and Bolen, D. W.,** Increased thermal stability of proteins in the presence of naturally occurring osmolytes, *Biochemistry,* 31, 5278, 1992.
102. **Paleg, L. G., Douglas, T. J., van Daal, A., and Keech, D. B.,** Proline, betaine and other organic solutes protect enzymes against heat inactivation, *Aust. J. Plant Physiol.,* 8, 107, 1981.
103. **Nikolopoulos, D. and Manetas, Y.,** Compatible solutes and *in vitro* stability of *Salsola soda* enzymes: proline incompatibility, *Phytochemistry,* 30, 411, 1991.
104. **Nishiguchi, M. K. and Somero, G. N.,** Temperature- and concentration-dependence of compatibility of the organic osmolyte β-dimethylsulfoniopropionate, *Cryobiology,* 29, 118, 1992.
105. **Rudolph, A. S. and Goins, B.,** The effect of hydration stress solutes on the phase behavior of hydrated dipalmitoylphosphatidylcholine, *Biochim. Biophys. Acta,* 1066, 90, 1991.

106. **Maloney, P. C. and Ambudkar, S. V.,** Functional reconstitution of prokaryote and eukaryote membrane proteins, *Arch. Biochem. Biophys.*, 269, 1, 1989.

107. **Kolena, J., Matejčíková, K., and Šrenkelová, G.,** Osmolytes improve the reconstitution of luteinizing hormone/human chorionic gonadotropin receptors into proteoliposomes, *Mol. Cell. Endocrinol.*, 83, 301, 1992.

108. **Jarabak, J., Seeds, A. E., Jr., and Talalay, P.,** Reversible cold inactivation of a 17 β-hydroxysteroid dehydrogenase of human placenta: protective effect of glycerol, *Biochemistry*, 5, 1269, 1966.

109. **Hand, S. C. and Somero, G. N.,** Urea and methylamine effects on rabbit muscle phosphofructokinase, *J. Biol. Chem.*, 257, 734, 1982.

110. **Krall, J. P., Edwards, G. E., and Andreo, C. S.,** Protection of pyruvate, P_i dikinase from maize against cold lability by compatible solutes, *Plant Physiol.*, 89, 280, 1989.

111. **Higgins, J., Hodges, N. A., Olliff, C. J., and Phillips, A. J.,** A comparative investigation of glycinebetaine and dimethylsulphoxide as liposome cryoprotectants, *J. Pharm. Pharmacol.*, 39, 577, 1987.

112. **Carpenter, J. F. and Crowe, J. H.,** Modes of stabilization of a protein by organic solutes during desiccation, *Cryobiology*, 25, 459, 1988.

113. **Higgins, C. F., Cairney, J., Stirling, D. A., Sutherland, L., and Booth, I. R.,** Osmotic regulation of gene expression: ionic strength as an intracellular signal?, *Trends Biochem. Sci.*, 12, 339, 1987.

114. **Hanson, A. D., Rathinasabapathi, B., Chamberlin, B., and Gage, D. A.,** Comparative physiological evidence that β-alaninebetaine and choline-O-sulfate act as compatible osmolytes in halophytic *Limonium* species, *Plant Physiol.*, 97, 1199, 1991.

115. **Landfald, B. and Strøm, A. R.,** Choline-glycine betaine pathway confers a high level of osmotic tolerance in *Escherichia coli*, *J. Bacteriol.*, 165, 849, 1986.

116. **Skedy-Winkler, C. and Avi-Dor, Y.,** Betaine: induced stimulation of respiration of high osmolarities in halotolerant bacterium, *Biochem. J.*, 150, 219, 1975.

117. **Gowrishankar, J., Jayashree, P., and Rajkumari, K.,** Molecular cloning of an osmoregulatory locus in *Escherichia coli:* increased *proU* gene dosage results in enhanced osmotolerance, *J. Bacteriol.*, 168, 1197, 1986.

118. **Mahan, M. J. and Csonka, L. N.,** Genetic analysis of the *proBA* genes of *Salmonella typhimurium:* physical and genetic analyses of the cloned *proB+A+* genes of *Escherichia coli* and of a mutant allele that confers proline overproduction and enhanced osmotolerance, *J. Bacteriol.*, 156, 1249, 1983.

119. **Adler, L. and Gustafsson, L.,** Polyhydric alcohol production and intracellular amino acid pool in relation to halotolerance of the yeast *Debaryomyces hansenii*, *Arch. Microbiol.*, 124, 123, 1980.

120. **Yancey, P. H., Burg, M. B., and Bagnasco, S. M.,** Effects of NaCl, glucose and aldose reductase inhibitors on cloning efficiency of renal cells, *Am. J. Physiol.*, 258, C156, 1990.

121. **Burg, M. B. and Kador, P. F.,** Sorbitol, osmoregulation, and the complications of diabetes, *J. Clin. Invest.*, 81, 635, 1988.

122. **Yancey, P. H., Haner, R. G., and Freudenberger, T.,** Effects of an aldose reductase inhibitor on osmotic effectors in rat renal medulla, *Am. J. Physiol.*, 259, F733, 1990.

123. **Moriyama, T., Garcia-Perez, A., Olson, A. D., and Burg, M. B.,** Intracellular betaine substitutes for sorbitol in protecting renal medullary cells from hypertonicity, *Am. J. Physiol.*, 260, F494, 1991.

124. **Lawitts, J. A. and Biggers, J. D.,** Joint effects of sodium chloride, glutamine, and glucose in mouse preimplantation embryo culture media, *Mol. Reprod. Dev.*, 31, 189, 1992.

125. **Petronini, R. G., De Angelis, E. M., Borghetti, P., Borghetti, A. F., and Wheeler, K. P.,** Modulation by betaine of cellular responses to osmotic stress, *Biochem. J.*, 282, 69, 1992.

126. **Van Laere, A.,** Trehalose, reserve and/or stress metabolite?, *FEMS Microbiol. Rev.*, 63, 201, 1989.

127. **Yancey, P. H. and Somero, G. N.,** Methylamine osmoregulatory compounds in elasmobranch fishes reverse urea inhibition of enzymes, *J. Exp. Zool.*, 212, 205, 1980.

128. **Griffith, R. W., Pang, P. K. T., and Benedetto, L. A.,** Urea tolerance in the killifish, *Fundulus heteroclitus*, *Comp. Biochem. Physiol.*, 62A, 327, 1979.

129. **Anderson, P. M.,** Purification and properties of the glutamine- and N-acetyl-L-glutamate-dependent carbamoyl phosphate synthetase from liver of *Squalus acanthias*, *J. Biol. Chem.*, 256, 12228, 1981.

130. **Mashino, T. and Fridovich, I.,** Effects of urea and trimethylamine-N-oxide on enzyme activity and stability, *Arch. Biochem. Biophys.*, 258, 356, 1987.

131. **Moyes, C. D. and Moon, T. W.,** Solute effects on the glycine cleavage system of two

osmoconformers *Raja erinacea* and *Mya arenaria* and an osmoregulator *Pseudopleuronectes americanus*, *J. Exp. Zool.*, 242, 1, 1990.

132. **Vieyra, A., Caruso-Neves, C., and Meyer-Fernandes, J. R.,** ATP ↔ ^{32}P exchange catalyzed by plasma membrane Ca^{2+}-ATPase from kidney proximal tubules, *J. Biol. Chem.*, 266, 10324, 1991.

133. **de Meis, L. and Inesi, G.,** Effects of organic solvents, methylamines, and urea on affinity for P_i of the Ca^{2+}-ATPase of sarcoplasmic reticulum, *J. Biol. Chem.*, 263, 157, 1988.

134. **Subramaniam, S. and Jackson, B. A.,** Differential effects of trimethylamines TMA on enzyme activity in MDCK cells (Abstract), *FASEB J.*, 6, A958, 1992.

135. **Altringham, J. D., Yancey, P. H., and Johnston, I. A.,** The effects of osmoregulatory solutes on tension generation by dogfish skinned muscle fibres, *J. Exp. Zool.*, 96, 443, 1982.

136. **Anderson, P. M.,** Ketone body and phosphoenolpyruvate formation by isolated hepatic mitochondria from *Squalus acanthias* spiny dogfish, *J. Exp. Zool.*, 254, 144, 1990.

137. **Ballantyne, J. S. and Moon, T. W.,** The effects of urea, trimethylamine oxide and ionic strength on the oxidation of acyl carnitines by mitochondria isolated from the liver of the little skate *Raja erinacea*, *J. Comp. Physiol.*, 156B, 845, 1986.

138. **Arakawa, T. and Timasheff, S. N.,** The stabilization of proteins by osmolytes, *Biophys. J.*, 47, 411, 1985.

139. **Bateman, J. B., Evans, G. F., Brown, P. R., Gabriel, C., and Grant, E. H.,** Dielectric properties of the system bovine albumin:urea:betaine in aqueous solution, *Phys. Med. Biol.*, 37, 175, 1992.

140. **Mirsalikhova, N. M.,** Stabilization of Na^+,K^+-adenosine triphosphatase by dimethyl-sulfoxide in the case of inactivation by urea, *Biokhimiya*, 43, 43, 1978.

141. **Chambers, S. T. and Kunin, C. M.,** The osmoprotective properties of urine for bacteria: the protective effect of betaine and human urine against low pH and high concentrations of electrolytes, sugars and urea, *J. Infect. Dis.*, 152, 1308, 1985.

142. **Moriyama, T., Garcia-Perez, A., and Burg, M. B.,** Factors affecting the ratio of different organic osmolytes in renal medullary cells, *Am. J. Physiol.*, 259, F847, 1990.

143. **Peterson, D. P., Murphy, K. M., Ursino, R., Streeter, K., and Yancey, P. H.,** Effects of dietary protein and salt on rat renal osmolytes: co-variation in urea and GPC contents, *Am. J. Physiol.*, 263, F594, 1992.

144. **Shifrin, S. and Parrott, C. L.,** Influence of glycerol and other polyhydric alcohols on the quaternary structure of an oligomeric protein, *Arch. Biochem. Biophys.*, 166, 426, 1975.

145. **Pollard, A. and Wyn Jones, R. G.,** Enzyme activities in concentrated solutions of glycinebetaine and other solutes, *Planta*, 144, 291, 1979.

146. **Herzog, R. M., Galinski, E. A., and Trüper, H. G.,** Degradation of the compatible solute trehalose in *Ectothiorhodospira halochloris*: isolation and characterization of trehalase, *Arch. Microbiol.*, 153, 600, 1990.

147. **Murata, N., Mohanty, P. S., Hayashi, H., and Papageorgiou, G. C.,** Glycinebetaine stabilizes the association of extrinsic proteins with the photosynthetic oxygen-evolving complex, *FEBS Lett.*, 296, 187, 1992.

148. **Incharoensakdi, A., Takabe, T., and Akazawa, T.,** Effect of betaine on enzyme activity and subunit interaction of ribulose-1,5-bisphosphate carboxylase/oxygenase from *Aphanothece halophytica*, *Plant Physiol.*, 81, 1044, 1986.

149. **Buche, A., Colson, P., and Houssier, C.,** Organic osmotic effectors and chromatin structure, *J. Biomol. Struc. Dynam.*, 8, 601, 1990.

150. **Kumazawa, Y. and Arai, K.,** Suppressive effect of sorbitol on denaturation of carp myosin B induced by neutral salts, *Nippon Suisan Gakkaishi*, 56, 679, 1990.

151. **Woolverton, W., Githens, S., O'Dell-Smith, R., and Bartell, C.,** Rat renal papillary tissue explants survive and produce epithelial monolayers in culture media made hyperosmotic with sodium chloride and urea, *J. Exp. Zool.*, 256, 189, 1990.

152. **Von Hippel, P. H. and Schleich, T.,** The effects of neutral salts on the structure and conformational stability of macromolecules in solution, in *Structure and Stability of Biological Macromolecules*, Timasheff, S. N. and Fasman, G. D., Eds., Marcel Dekker, New York, 1969, 417.

153. **Collins, K. D. and Washabaugh, M. W.,** The Hofmeister effect and the behavior of water at interfaces, *Q. Rev. Biophys.*, 18, 323, 1985.

154. **Low, P. S.,** Molecular basis of the biological compatibility of nature's osmolytes, in *Transport Processes, Iono- and Osmoregulation*, Gilles, R. and Gilles-Ballien, M., Eds., Springer-Verlag, Berlin, 1985, 469.

155. **Clark, M. E.,** Non-Donnan effects of organic osmolytes in cell volume changes, *Curr. Top. Membr. Transp.,* 30, 251, 1987.

156. **Timasheff, S. N.,** A physicochemical basis for the selection of osmolytes by nature, in *Water and Life: A Comparative Analysis of Water Relationships at the Organismic, Cellular, and Molecular Levels,* Somero, G. N., Osmond, C. B., and Bolis, C. L., Eds., Springer-Verlag, Berlin, 1992, 70.

157. **Na, G. C. and Timasheff, S. N.,** Interaction of calf brain tubulin with glycerol, *J. Mol. Biol.,* 151, 165, 1981.

158. **Winzor, C. L., Winzor, D. J., Paleg, L. G., Jones, G. P., and Naidu, B. P.,** Rationalization of the effects of compatible solutes on protein stability in terms of thermodynamic nonideality, *Arch. Biochem. Biophys.,* 296, 102, 1992.

159. **Laughlin, R. G.,** HLB, from a thermodynamic perspective, *J. Soc. Cosmet. Chem.,* 32, 371, 1981.

159a. **Laughlin, R. G.,** personal communication.

160. **Arakawa, T. and Timasheff, S. N.,** Preferential interactions of proteins with solvent components in aqueous amino acid solutions, *Arch. Biochem. Biophys.,* 224, 169, 1983.

161. **Fulton, A. B.,** How crowded is the cytoplasm?, *Cell,* 30, 345, 1982.

162. **Rand, R. P.,** Raising water to new heights, *Science,* 256, 618, 1992.

163. **Clegg, J. S., Jackson, S. A., and Fendl, K.,** Effects of reduced cell volume and water content on glycolysis in L-929 cells, *J. Cell. Physiol.,* 142, 386, 1990.

164. **Clegg, J. S.,** Properties and metabolism of the aqueous cytoplasm and its boundaries, *Am. J. Physiol.,* 246, R133, 1984.

165. **Wiggins, P. M.,** Role of water in some biological processes, *Microbiol. Rev.,* 54, 432, 1990.

166. **Myers, J. S. and Jakoby, W. B.,** Effect of polyhydric alcohols on kinetic parameters of enzymes, *Biochem. Biophys. Res. Commun.,* 51, 631, 1973.

167. **Armitage, W. J. and Mazur, P.,** Toxic and osmotic effects of glycerol on human granulocytes, *Am. J. Physiol.,* 247, C382, 1984.

168. **Hosono, K.,** Effect of salt stress on lipid composition and membrane fluidity of the salt-tolerant yeast *Zygosaccharamyces rouxii, J. Gen. Microbiol.,* 138, 91, 1992.

169. **Arakawa, T., Carpenter, J. F., Kita, Y. A., and Crowe, J. H.,** The basis for toxicity of certain cryoprotectants: a hypothesis, *Cryobiology,* 27, 401, 1990.

170. **LeRudulier, D., Strom, A. R., Dandekar, A. M., Smith, L. T., and Valentine, R. C.,** Molecular biology of osmoregulation, *Science,* 224, 1064, 1984.

171. **McCue, K. F. and Hanson, A. D.,** Drought and salt tolerance: towards understanding and application, *Trends Biotechnol.,* 8, 358, 1990.

172. **Tarczynski, M. C., Jensen, R. G., and Bohnert, H. J.,** Stress protection of transgenic tobacco by production of the osmolyte mannitol, *Science,* 259, 508, 1993.

173. **Arakawa, T., Kita, Y., and Carpenter, J. F.,** Protein-solvent interactions in pharmaceutical formulations, *Pharm. Res.,* 8, 285, 1991.

174. **Navarrette, R. and Serrano, R.,** Solubilization of yeast plasma membranes and mitochondria by different types of non-denaturing detergents, *Biochim. Biophys. Acta,* 728, 403, 1983.

175. **Lee, J. A., Lee, H. A., and Sadler, P. J.,** Uraemia: is urea more important than we thought?, *Lancet,* 338, 1438, 1991.

176. **Kloiber, O., Banjac, B., and Drewes, L. R.,** Protection against acute hyperammonemia: the role of quaternary amines, *Toxicology,* 49, 83, 1988.

177. **Charlson, R. J., Lovelock, J. E., Andreae, M. O., and Warren, S. G.,** Oceanic phytoplankton, atmospheric sulphur, cloud albedo and climate, *Nature,* 326, 655, 1987.

Chapter 6

Regulation of Cellular Amino Acid Levels

J. S. Ballantyne and M. E. Chamberlin

TABLE OF CONTENTS

I. OVERVIEW OF THE PHYSIOLOGICAL ROLES AND PHYSICOCHEMICAL PROPERTIES OF AMINO ACIDS IN CELLS

Many organisms maintain high concentrations of intracellular free amino acids to offset high osmolarity in the environment or extracellular fluid. The role of these solutes in cell volume regulation has been the subject of study and review for several decades.[1–8] Because virtually all amino acids used as intracellular solutes also have important metabolic functions, amino acid concentration and turnover must be strictly regulated so that their various roles may be fulfilled. It is our goal to explore the regulation of amino acid levels in a variety of cell types within the context of cell volume regulation.

Although amino acids may be important in mediating the regulation of cell volume in many taxa, the quantitative importance of amino acids in cell volume regulation is greatest in euryhaline osmoconforming marine invertebrates and lower vertebrates (hagfish, elasmobranchs, and coelacanths). These animals experience the greatest changes in external osmolarity and maintain some of the highest intracellular levels of amino acids of all animal taxa. These animals, in particular, make excellent systems for the investigation of amino acid regulation during cell volume stress and studies on these animals will be emphasized in this chapter.

0-8493-4448-4/94/$0.00+$.50

There are in excess of 800 amino acids found in nature.[9] Nevertheless, only a few (alanine, glycine, taurine, glutamate, β-alanine, and proline) are of quantitative importance in cell volume regulation. The consistent use of one or more of these select amino acids by so many taxa suggests that these amino acids have special characteristics compatible with their role in regulating cell volume. These characteristics, which are likely a combination of their physicochemical properties and their suitability for the physiological processes, are discussed in the following sections.

A. PHYSIOLOGICAL ROLES OF AMINO ACIDS

1. Osmolytes

When exposed to anisosmotic media, cells change in volume. In many cases adjustments in intracellular amino acid concentration may be involved in restoring the cell to near its original volume. The loss of amino acids during regulatory volume decrease (RVD) or the gain of amino acids during regulatory volume increase (RVI) may be due to activation of specific transport pathways,[10–16] changes in amino acid oxidation,[7,17,18] or changes in the turnover of intracellular[19–22] or extracellular[23,24] macromolecules. Because of their important role in cellular metabolism, the use of amino acids as osmolytes presents a potential conflict with their role as metabolic intermediates.

2. Metabolic Substrates

Although they play a central role in metabolism, not all amino acids can be synthesized by animal cells. Those amino acids that cannot be synthesized are termed "essential" and must be obtained from the external environment. Dietary requirements for L-amino acids vary with species, growth rate, and age.[25] Ten amino acids, arginine, histidine, isoleucine, leucine, lysine, methionine, phenylalanine, threonine, tryptophan, and valine, are considered essential in most animals.[25] The essential amino acids of many marine invertebrates have not been determined because many of these organisms harbor intracellular symbionts capable of synthesizing the "essential" amino acids. The symbiotic associations (bacterial and algal) are common in marine organisms that employ amino acids as osmolytes (e.g., molluscs) and, therefore, the contribution of each partner in the symbiotic association to the amino acid pools needs to be determined. The changes in the amino acid pools of the hosts, as well as of the symbionts, during osmotic stress would be an interesting area for investigation.

Essential amino acids are generally found in low concentration in cells and do not play a major role in cell volume recovery. Conservation of these amino acids for protein synthesis and other metabolic functions is undoubtedly a higher priority than simply using them as osmotically active solutes. Any changes in essential amino acid levels during cell volume regulation or salinity acclimation are probably due to altered rates of protein synthesis rather than a specific, volume-sensitive modulation of their levels.

Taurine is an amino acid not used in protein synthesis but is a very important intracellular solute in many animal groups.[15] By using this amino acid for cell volume regulation, animals avoid the potentially conflicting use of amino acids in protein synthesis and volume regulation. Interestingly, taurine is an essential amino acid for some animals that use taurine as an osmolyte,[15] but these are primarily carnivores (e.g., cats, elasmobranchs) whose dietary intake far exceeds the non-osmoregulatory needs for this amino acid.

Nonessential amino acids are synthesized by cells and the cost of synthesizing different types varies considerably (Table 1). In general, the amino acids used as osmolytes (e.g., alanine and glycine) require less energy to produce *de novo* than other amino acids. Clearly, metabolic cost is not the only criterion important in the selection of a solute; otherwise, glycine would be universally employed. One factor that might override the cost of synthesis is the abundance of a particular amino acid, or its precursor, in the diet, thus removing the requirement for *de novo* synthesis. This would be the case for marine osmoconformers that consume other amino acid-rich marine osmoconformers.

Amino acids serve as important oxidative substrates in liver tissues[26,27] of virtually all animals examined, in muscle tissues of many invertebrates[28–30] and lower vertebrates,[31,32] and in the excretory organs of molluscs,[33] insects,[34] fish,[35] reptiles,[36] birds,[37] and mammals.[38] Directly oxidizing amino acids is more efficient than using amino acid carbon to synthesize lipid and subsequently using the lipid as an energy source. In cells with high turnover rates (e.g., cultured cells, enterocytes, and cells of the immune system of mammals) oxidation of amino acids provides both energy and the precursors for purine synthesis.[39]

With the exception of leucine and lysine, all amino acids of metabolic importance can be used to synthesize glucose or glycogen.[40] Alanine is an important gluconeogenic amino acid in the liver and many other tissues. In the mammalian hepatocyte the transport of alanine into the cell exerts significant control over alanine utilization by the cell[41] and the importance of this uptake process would increase during starvation when amino acids

Table 1 **Physicochemical Properties of Biologically Important L-Amino Acids**

Solute	Solubility[52] (mmol/l)	Enthalpy[99] (kcal/mol)	log K_d[52]		Isoelectric Point[100]	Partition Coefficient[101]
			Ca^{2+}	Mg^{2+}		
Alanine	1,867	387	1.2	2.0	6.00	520
β-Alanine	6,117				6.89[15]	
Arginine	861	894	2.2	1.3	10.76	
Asparagine	226	461			5.41	
Aspartate	38	383	1.6	2.4	2.77	
Betaine	13,396					
Cysteine	395		<4		5.07	
Cystine	<1	725			5.06[102]	
Glutamate	58	537	1.4	1.9	3.22	
Glutamine	291	614			5.65	
Glycine	3,329	231	1.4	1.3	5.97	991
Histidine	268				7.59	330
Isoleucine	314	856			6.02	48.5
Leucine	185	856			5.98	33.0
Lysine					9.74	
Methionine	665				5.74	74.5
Phenylalanine	179	1,111			5.48	22.5
Proline	14,101				6.30	
Sarcosine	4,804					
Serine	2,379	348	1.4		5.68	
Taurine	837		1.6[103]		5.16[15]	
Threonine	491				5.60	
Tryptophan	56	1,345			5.89	
Tyrosine	2	1,062	1.5	2.0	5.66	
Valine	756	698			5.96	

References for values are given at the top of the column unless otherwise indicated in the body of the table. Log K_d is the log of the dissociation (stability) constant.[52]

are released during increased protein catabolism. As a result, starvation may interfere with amino acid-mediated cell volume regulation in tissues such as the liver and kidney, in which significant levels of gluconeogenesis occur. The interplay between starvation, gluconeogenesis, and cell volume regulation has not been investigated.

3. Neurotransmitters

In addition to their role in intermediary metabolism, nervous tissues employ amino acids to transmit information from one cell to another. Eight amino acids have been identified as important neurotransmitters in the vertebrate central nervous system (glutamate, aspartate, cysteic acid, homocysteic acid, γ-aminoisobutyric acid, glycine, taurine, and β-alanine)[42] and some may be important in the nervous systems of invertebrates as well.[43] Use of an amino acid by a particular neuron may preclude the use of this amino acid for volume regulatory purposes in that cell type or may require strict partitioning of its metabolism. The nerve cells of osmoconforming invertebrates that use amino acids as neurotransmitters and osmolytes[7] may be the best systems for studying this problem.

B. PHYSICOCHEMICAL PROPERTIES OF AMINO ACIDS

1. Stereospecificity

Whereas L-amino acids are considered to predominate in most eukaryotic cells, D-amino acids have been reported in tissues of organisms from many phyla: Porifera, Cnidaria, Annelida, Arthropoda, Bryozoa, Echinodermata, Mollusca, Rhynchocoela, and Arthropoda.[44–46] The availability of chiral columns for high-performance liquid chromatography (HPLC) separation of stereoisomers now permits the enantiomeric composition of the amino acid pools of organisms to be examined. The origin of the D-amino acids is uncertain, but diet or transport from the external environment must certainly play an important role in the case of some D-amino acids. Recently, Matsushima and Yamada[47] demonstrated that the gills of *Corbicula japonica* (a brackish-water bivalve mollusc) transport D- and L-alanine competitively. Levels in some marine invertebrates are high enough to be considered osmotically important[45] and Matsushima et al.[48] have found that D-alanine levels increase in *C. japonica* upon acclimation to seawater. Conversion of D-amino acids to L- forms may occur by the action of D-amino acid oxidase with deamination followed by reamination to the L- form or by the action of a racemase, directly converting the D- to the L- form.[49] Matsushima et al.[48] have reported an alanine racemase in tissues of *C. japonica* and ^{14}C from ^{14}C-D-alanine can be found in TCA-ethanol precipitated fractions (presumably proteins) apparently after conversion of D-alanine to L-alanine via an alanine racemase.[16]

Somero[50] has argued that D-amino acids may be better compatible solutes than the L-forms, since their kinetic interference with enzymes responsible for the metabolism of L-amino acids would be minimal (for discussion of compatible solutes see Chapter 5). By using D-amino acids as osmolytes, many of the problems of integrating cell volume regulation processes with "mainstream" metabolism can be circumvented. In light of the potential volume-regulatory importance of D-amino acids, further studies are needed regarding the role of the kinetics and compartmentation of amino acid racemases and oxidases.

2. Isoelectric Point of Amino Acids

The electrical charge of amino acids in the cell plays an important role in their transport, solubility, and metabolism. All of the amino acids important in cell volume regulation have isoelectric points below 7.00 (Table 1). At physiological pH these amino acids would have a slight negative charge. This negative charge would decrease the lipophilicity of the amino acid and diminish the membrane's permeability to the amino acid. This can be seen in the high water:oil partition coefficients of amino acids (Table 1). This relative impermeability contributes to the maintenance of large concentration gradients of some amino acids, such as taurine.[15]

Arginine has the highest pK of all biologically important amino acids and can be found in high concentration in some invertebrate cells. This does not, however, reflect a volume-regulatory role for this amino acid. Arginine phosphate is an important high-energy compound in these animals. Breakdown of arginine phosphate to arginine and ATP may occur during volume-regulatory stress. High levels of arginine are more likely caused by anoxia or exercise when ATP hydrolysis exceeds ATP production.

3. Solubility

The zwitterionic nature of amino acids makes them quite soluble in water, although the solubility of L-amino acids in pure water varies more than three orders of magnitude (Table 1). Data on the solubility of pure amino acids in water should be used with caution when predicting their solubility in the cytosol. Because ionic strength, pH, and temperature all affect solubility, their solubility in the cytoplasm may be considerably different from that in pure water. Nevertheless, those amino acids most soluble in pure water are also those that are important osmolytes in animal cells: alanine, β-alanine, betaine, glycine, proline, and taurine. Glutamate, although not very soluble in pure water, is far more soluble in the presence of ions.[9] Perhaps this explains why glutamate is accumulated in the cells of some marine osmoconformers that have high levels of intracellular inorganic ions (e.g., oysters).[51]

4. Chelation of Calcium and Magnesium

All amino acids chelate calcium to some extent and this action could interfere with the role of this divalent ion in signal transduction. The dissociation constants (Table 1), however, indicate binding strengths far below those of chelating agents such as EGTA (11.0) or ethylene diamine tetraacetate (EDTA, 10.6), and lower than similar complexes of ADP and ATP,[52] and far lower than those of the high-affinity calcium-binding proteins in cells (log K_d = 6 to 10).[53] Dissociation constants for magnesium are slightly higher than those for calcium, where comparisons can be made (Table 1). The binding of magnesium by amino acids is lower than that for ADP and ATP.[52] The relatively low affinity of amino acids for both calcium and magnesium indicates modulating amino acid levels during cell volume regulation would not interfere with normal intracellular signaling mechanisms and pathways involving Mg^{2+} or Ca^{2+}. This physicochemical characteristic, along with the others mentioned in the previous sections, may have favored selection of amino acids as osmolytes.

II. METABOLISM OF AMINO ACIDS DURING OSMOTIC STRESS

A major source of amino acids during RVI may be proteins. These proteins may be found extracellularly in protozoa,[22] crustaceans,[23] and molluscs[24] or intracellularly in molluscs.[20,21,54] The importance of protein as a source of amino acids for RVI may explain why taurine (not found in protein) does not increase in concentration during hyperosmotic stress.[54]

Another macromolecule, glycogen, may also serve as a precursor for amino acids generated during RVI. Baginski and Pierce[19] have suggested that alanine is synthesized from glycogen by *Modiolus demissus* (ribbed mussel, now called *Guekensia demissa*).

An increase in amino acid oxidation during hypoosmotic stress may help mediate RVD. Several studies have demonstrated that amino acid oxidation is osmotically sensitive. Oxidation of proline by isolated mitochondria from the heart of *Mercenaria mercenaria* (quahog) and gill of the oyster *(Crassostrea virginica)*[55] is increased in hypoosmotic conditions. Similarly, sarcosine oxidation by elasmobranch hepatocytes[56] and isolated mitochondria[27,56] is stimulated under dilute conditions.

III. REGULATION OF AMINO ACID METABOLISM

Many studies, including those described above, have shown *in vitro* effects consistent with changes in amino acid concentrations during cell volume regulation. Nevertheless, few studies have combined measurements of cell volume with detailed studies on the regulation of amino acid metabolism during volume recovery. Metabolic regulation can be studied in a variety of ways. In this section, techniques and approaches that may be useful in studying volume-mediated changes in amino acid metabolism are discussed.

A. ANALYSIS OF METABOLIC CONTROL

A major goal of most metabolic investigations is the identification of sites of regulation. Measurement of maximal enzyme or transporter activities can provide initial clues to these sites. Enzymes or transporters with low maximal activities are "bottlenecks" that potentially limit flux. Changes in maximal activities of these sites, induced by a variety of signals, allow flux to increase or decrease.

Further confirmation of the "regulatory" site can be obtained by examining metabolite concentrations before and after changes in flux. Upstream from a regulatory site, metabolite levels increase or decrease as flux changes while downstream metabolite levels change in the opposite manner. By measuring all the metabolites in a pathway the regulatory site is identified by the position of the changeover from increasing to decreasing metabolite levels. This method of analysis, termed "crossover theory",[57] lends some insight into regulation, but it may be an inadequate approach to most systems.

Although many quantitative measurements of metabolite concentrations are made, the extent of control exerted by the regulatory site is not quantified. Seldom do regulatory pathways have a single regulatory site and, indeed, the sites of regulation shift as conditions change. It has become widely accepted that the quantitative contribution of a site in a pathway is needed to understand the metabolic regulation of a given process. Such a method of analysis was first introduced independently, and virtually simultaneously, by Kacser and Burns[58] and Heinrich and Rapoport.[59,60] This methodology, called "control strength theory", assigns a fractional value, control strength (CS), to each site in a metabolic sequence with the sum of values for all sites equalling one. In its simplest application, determination of the value for a site relies on measuring the fractional change (dF) in flux (F) induced by a fractional change (dE) in the amount of enzyme (E) (i.e., $dE/E \cdot dF/F = CS$). In its original application modulation of the amount of enzyme could only be achieved through the use of specific irreversible inhibitors or by genetic manipulation. The limited availability of irreversible inhibitors limited the application of control strength theory to well-characterized mammalian systems, such as liver mitochondria.[61] A more general method of application of control strength theory has been developed by Brown and co-workers.[62] This "top-down" approach does not require specific knowledge of the mode of action of the inhibitor used. This approach has been used to study hormonal effects on isolated mitochondria and cells[63] and has considerable potential for the analysis of metabolic regulation during cell volume regulation.

B. REGULATION OF MITOCHONDRIAL FUNCTION

Mitochondria are responsible for most of the ATP production in the cell and contain significant portions of many metabolic pathways, including those of amino acid metabolism. There are several reasons why isolated mitochondria are particularly useful systems for the study of metabolic regulation. First, isolated mitochondria contain the transporters and enzymes of metabolic pathways in their proper proportions and are second only to isolated cells in their value for the study of metabolic regulation. Second, there are established methods for isolating mitochondria from tissues of osmoconformers, such as molluscs,[26,29,30,55,64] crustaceans,[65,66] and elasmobranchs,[32,56,67,68] as well as osmoregulators, such as bony fishes,[31,68–70] amphibians,[71,72] reptiles,[73] birds,[74,75] mammals,[73,76–78] and insects.[34,79–82] Third, the electron transport chain of mitochondria is quite conservative, such that inhibitors known to work in rat liver mitochondria usually work in mitochondria isolated from more obscure species. Finally, mitochondria are affected by their osmotic and solute environment and, therefore, it is possible to osmotically perturb them and monitor the changes in flux through a pathway. Mitochondrial studies have revealed that the osmotically sensitive site of action varies depending on the species or the amino acid being studied. For example, osmolarity modulates the mitochondrial glycine carrier in elasmobranch liver mitochondria,[27] whereas monovalent cations affect the electron transport chain during glutamate oxidation by mollusc gill mitochondria.[83]

C. REGULATION OF ENZYME AND TRANSPORTER ACTIVITY

1. Protein Synthesis

One way to increase flux through a metabolic pathway is to increase the activity of the enzyme or transporter responsible for a bottleneck. This type of study is performed by measuring the V_{max} activity of key enzymes in animals exposed to changes in external osmolarity. Short-term changes in some enzymes of amino acid metabolism have been demonstrated in marine molluscs[84] but in the long term such changes cannot be demonstrated.[85] Studies of this type, although relatively easy to perform, do not determine the nature of change in enzyme activity. There are several ways in which enzyme activity may be altered. These include amplification of the production of a single type of the enzyme or transporter, production of different variants of the enzyme (isozymes or allozymes), or modulation of existing enzymes or carriers (e.g., phosphorylation/dephosphorylation). These mechanisms are the subject of the following sections.

a. Amplification

Suprisingly, there are few examples proving enhanced production of an enzyme in response to osmotic stress. Enhanced sorbitol production due to increased synthesis of aldose reductase has been demonstrated in hyperosmotically stressed, cultured mammalian kidney cells.[86] Induction of the aldose reductase gene is apparently due to increase in intracellular potassium and sodium or ionic strength. Similar studies aimed at the enzymes of amino acid metabolism would considerably enhance our understanding of the regulation of amino acid metabolism during cell volume regulation.

b. Isozymes

In some cases production of alternate forms of the enzyme or transporter may occur. Isozymes are variants of the catalyst produced by different genes. In multimeric enzymes or transporter, gradations of the properties of the separate gene products can be produced by mixing subunits from one gene with those of another gene. Although there are studies demonstrating the use of isozymes during temperature adaptation, at this time there are no studies identifying this strategy in association with osmotic stress.

c. Allozymes

In populations of organisms, the genes associated with the production of enzymes may differ from one individual to another. The products of such genes are termed allozymes and, although they catalyze the same chemical reaction, the kinetics differ. The kinetic differences form the basis for adaptation to new conditions. Such a case associated with cell volume regulation has been identified in euryhaline molluscs.[87] Leucine aminopeptidase (Lap) has five alleles in wild populations of *Mytilus edulis* (blue mussels). The enzyme catalyzes the breakdown of protein to amino acids and is thought to provide amino acids for cell volume regulation at high salinities. Individuals with one allozyme, Lap,[98] produce amino acids at higher rates than those individuals with other forms of Lap. At low salinities this allele is maladaptive because the individuals possessing it also lose amino acids at higher rates. The loss of nitrogen and carbon results in enhanced mortality in individuals with Lap[98] at low salinities. Populations of mussels in low salinities, therefore, have lower frequencies of Lap[98] than those found at higher salinities.

2. Allosteric Modulation

Allosteric modification of existing enzymes or transporters is a widely used strategy for changing their kinetic properties. Such changes are rapid, readily reversible, and do not involve costly synthesis of new proteins. The following sections outline the ways proteins are allosterically modified in the context of cell volume regulation.

a. Phosphorylation and dephosphorylation

Polites and Pierce[88] report phosphorylation of erythrocyte membranes from the mollusc *Noetia ponderosa* (blood clam) exposed to lower salinities. The identity of the proteins was not established, but the requirement for calcium indicates calcium may activate a protein kinase responsible for the phosphorylation. Exposing skate erythrocytes to the calcium ionophore, A23187, or a phorbol ester causes the release of taurine. Loss of taurine is also observed when these red blood cells are exposed to hypotonic conditions.[89] These results indicate that intracellular calcium and the inositol phosphate signal transduction pathway may be involved in modulating the phosphorylation status of the taurine carrier. Clearly, more studies into the role of protein phosphorylation in controlling amino acid metabolism and transport during volume stress are warranted.

b. Organic ligands and signals

There are few studies demonstrating the role of organic ligands in mediating changes in amino acid metabolism during cell volume regulation. This may be because metabolic systems exposed to large fluctuations in organic solutes may be relatively insensitive to these solutes. For example, Moyes and Moon[90] have shown the glycine cleavage system is not sensitive to urea or methylamines in elasmobranchs, in which large changes in urea and methylamine levels occur in response to changing environmental salinity. This system, however, is sensitive to urea in organisms that do not use urea as a solute (e.g., teleosts).

Amino acids, themselves, have been shown to modulate the activity of enzymes associated with amino acid metabolism in osmoconformers and osmoregulators.[90] In addition, they are involved in regulating flux through gluconeogenic,[91,92] glycolytic,[93] and lipogenic pathways.[91] Rarely has this been attributed to allosteric modulation of an enzyme.

c. Ionic strength or specific ions

Several studies have implicated specific ions or ionic strength in modulating the activity of enzymes or transporters in the context of cell volume regulation. In some marine osmoconformers changes in external osmolarity can change intracellular levels of potassium, chloride, and sodium. Ion effects on amino acid oxidation have been suggested based on the effects of ionic strength on glutamate dehydrogenase in Cnidarians.[94] In crustaceans, Gilles[95] has found the activities of enzymes of amino acid metabolism, in particular glutamate dehydrogenase, are altered by changing salt concentrations. This is not the case in marine molluscs,[83] in which potassium modulates glutamate oxidation by affecting the electron transport chain rather than glutamate dehydrogenase. The glycine cleavage system is also sensitive to potassium chloride in the physiological range in an elasmobranch, a marine teleost, and a marine mollusc.[90] Interestingly, enzymes in other pathways not specific to amino acid catabolism (e.g., citrate synthase), are insensitive to ionic strength. The same enzymes of osmoregulators, however, are sensitive to ionic strength.[96] All of these studies have used either isolated mitochondria or enzymes. Confirmation of the importance of ionic strength or specific ions (other than divalent cations) in regulating amino acid metabolism during cell volume regulation must await studies of intact cells.

d. Membrane lipids

Membrane phospholipids are important for maintaining membrane permeability as well as modulating the activity of membrane-bound transporters and enzymes. Rudolph et al.[97] have demonstrated changes in the physical properties of artificial membranes in response to changing concentrations of proline and betaine. If amino acids have a similar effect on biological membranes, then the wide swings in intracellular amino acid concentration during volume perturbation and recovery may affect the functioning of the membrane and warrant further investigation.

There are few studies of the role of membrane lipids in modulating amino acid transport in animal cells. A study of amino acid transport (diffusion) in fish acclimated to temperatures ranging from 8 to 30°C, a maneuver which changes membrane lipid composition, showed that transport of less lipophilic amino acids were affected.[98] Changing the fatty acid composition of phosphatidylcholine liposomes altered the rate of efflux of all amino acids but the more lipophilic amino acids were the least affected.[98]

IV. SUMMARY

The use of amino acids as osmolytes in widely divergent taxa indicates that this role may have been important in the earliest organisms. The role of amino acids as osmolytes evolved with their other functions (protein synthesis, neurotransmitters, energy sources, etc.). In this chapter we have tried to highlight the physicochemical and biological properties of amino acids that allow them to fulfill their roles as both osmolytes and metabolic intermediates. Studies of isolated enzymes or mitochondria indicate that changes in monovalent ions or ionic strength alters amino acid metabolism in a manner consistent with observations of osmotically stressed cells. Extension of these studies to look for other signals and sites of action within intact cells is clearly warranted.

An important feature of amino acid metabolism is compartmentation, especially within membrane-bound organelles. These organelles are subject to volume changes when the cell is exposed to anisosmotic media, yet our understanding of the effect of such volume perturbations on organelle function, including amino acid metabolism, is limited. Clearly, more research into the compartmentation of amino acid metabolism and the effects of osmotic stress on subcellular organelles are warranted.

Although many taxa use amino acids in volume regulation, we have emphasized studies performed on marine osmoconformers. These organisms are excellent systems for studying volume-mediated changes in amino acid metabolism, because many of these animals routinely experience wide fluctuations in external osmolarity and modulate the concentration of their large intracellular free amino acid pool to effect volume regulation. Research employing these organisms has revealed, and will continue to reveal, mechanisms involved in regulating cellular levels of amino acids. Similar studies can be applied to osmoregulators, which experience much smaller swings in osmolarity.

REFERENCES

1. **Schoffeniels, E.,** Role des acides amines dans la regulation de la pression osmotique du milieu interieur des insectes aquatiques, *Arch. Int. Physiol. Biochim.*, 68, 507, 1960.
2. **Schoffeniels, E.,** Adaptations with respect to salinity, *Biochem. Soc. Symp.*, 41, 179, 1976.
3. **Forster, R. P. and Goldstein, L.,** Amino acids and cell volume regulation, *Yale J. Biol. Med.*, 52, 497, 1979.
4. **Pierce, S. K.,** Invertebrate cell volume control mechanisms: a coordinated use of intracellular amino acids and inorganic ions as osmotic solute, *Biol. Bull.*, 163, 405, 1982.
5. **Clark, M. E.,** The osmotic role of amino acids: discovery and function, in *Transport Processes, Iono- and Osmoregulation,* Gilles, R. and Gilles-Baillien, M., Eds., Springer-Verlag, New York, 1985, 412.
6. **Clark, M. E.,** Non-Donnan effects or organic osmolytes in cell volume regulation, in *Current Topics in Membranes and Transport,* Vol. 30, Kleinzeller, A., Ed., Academic Press, New York, 1987, 251.
7. **Gilles, R.,** Volume regulation in cells of euryhaline invertebrates, in *Cell Volume Control: Fundamental and Comparative Aspects,* Kleinzeller, A., Ed., Academic Press, New York, 1987, 205.
8. **Chamberlin, M. E. and Strange, K.,** Anisosmotic cell volume regulation: a comparative view, *Am. J. Physiol.*, 257, C159, 1989.
9. **Eggum, B. O. and Sorensen, H.,** Chemistry and analysis of amino acids, in *Absorption and Utilization of Amino Acids,* Friedman, M., Ed., CRC Press, Boca Raton, FL, 1989, 265.
10. **Amende, L. M. and Pierce, S. K.,** Cellular volume regulation in salinity stressed molluscs: the response of *Noetia ponderosa* (Arcidae) red blood cells to osmotic variation, *J. Comp. Physiol.*, 138B, 283, 1980.
11. **Costa, C. J., Pierce, S. K., and Warren, M. K.,** The intracellular mechanism of salinity tolerance in polychaetes: volume regulation by isolated *Glycera dibranchiata* red coelomocytes, *Biol. Bull.*, 159, 626, 1980.
12. **Fincham, D. A., Wolowyk, M. W., and Young, J. D.,** Volume-sensitive taurine transport in fish erythrocytes, *J. Membr. Biol.*, 96, 45, 1987.
13. **Goldstein, L. and Brill, S.,** Isosmotic swelling of skate *(Raja erinacea)* red blood cells causes a volume regulatory release of intracellular taurine, *J. Exp. Zool.*, 253, 132, 1990.
14. **Ballatori, N. and Boyer, J. L.,** Taurine transport in skate hepatocytes. II. Volume activation, energy, and sulfhydryl dependence, *Am. J. Physiol.*, 262, G451, 1992.
15. **Huxtable, R. J.,** Physiological actions of taurine, *Physiol. Rev.*, 72, 101, 1992.
16. **Matsushima, O. and Hayashi, Y. S.,** Metabolism of D-alanine and L-alanine and regulation of intracellular free amino acid levels during salinity stress in a brackish-water bivalve *Corbicula japonica, Comp. Biochem. Physiol.*, 102A, 465, 1992.
17. **Gilles, R. and Pequeux, A.,** Cell volume regulation in crustaceans: relationship between mechanisms for controlling the osmolarity of extracellular and intracellular fluids, *J. Exp. Zool.*, 215, 351, 1981.
18. **Ballantyne, J. S. and Chamberlin, M. E.,** Adaptation and evolution of mitochondria: osmotic and ionic considerations, *Can. J. Zool.*, 66, 1028, 1988.
19. **Baginski, R. M. and Pierce, S. K.,** A comparison of amino acid accumulation during high salinity adaptation with anaerobic metabolism in the ribbed mussel, *Modiolus demissus demissus, J. Exp. Zool.*, 203, 419, 1978.
20. **Bishop, S. H., Greenwalt, D. E., and Burcham, J. M.,** Amino acid cycling in ribbed mussel tissues subjected to hyperosmotic shock, *J. Exp. Zool.*, 215, 277, 1981.
21. **Deaton, L. E.,** Hyperosmotic cellular volume regulation in the ribbed mussel *Geukensia demissa:* inhibition by lysosomal and proteinase inhibitors, *J. Exp. Zool.*, 244, 375, 1987.

22. **Cronkite, D. L. and Pierce, S. K.,** Free amino acids and cell volume regulation in the euryhaline ciliate *Paramecium calkinsi, J. Exp. Zool.,* 251, 275, 1989.

23. **Gilles, R.,** Effects of osmotic stresses on the proteins concentration and pattern of *Eriocheir sinensis* blood, *Comp. Biochem. Physiol.,* 56A, 109, 1977.

24. **Matsushima, O., Khan, H. R., and Saleuddin, A. S. M.,** Haemolymph protein as a source for free amino acid accumulation in the pulmonate *Helisoma trivolvis* during hypertonic stress, *J. Exp. Zool.,* 257, 141, 1991.

25. **Morris, J. G.** Nutrition, in *Environmental and Metabolic Animal Physiology,* Prosser, C. L., Ed., Wiley-Liss, New York, 1991, 231.

26. **Ballantyne, J. S. and Moon, T. W.,** Hepatopancreas mitochondria from *Mytilus edulis:* substrate preference and effects of pH and osmolarity, *Mar. Biol.,* 87, 239, 1985.

27. **Moyes, C. D., Moon, T. W., and Ballantyne, J. S.,** Osmotic effects on amino acid oxidation in skate liver mitochondria, *J. Exp. Biol.,* 125, 181, 1986.

28. **Weeda, E., de Kort, C. A. D., and Beenakers, A. M. T.,** Fuels for energy metabolism in the Colorado potato beetle, *Leptinotarsa decemlineata* Say, *J. Insect Physiol.,* 25, 951, 1979.

29. **Ballantyne, J. S., Mommsen, T. P., and Hochachka, P. W.,** Studies of the metabolism of the migratory squid, *Loligo opalescens:* enzymes of tissues and heart mitochondria, *Mar. Biol. Lett.,* 2, 75, 1981.

30. **Ballantyne, J. S. and Storey, K. B.,** Mitochondria from the ventricle of the marine clam, *Mercenaria mercenaria:* substrate preferences and effects of pH and salt concentration on proline oxidation, *Comp. Biochem. Physiol.,* 76B, 133, 1983.

31. **Chamberlin, M. E., Glemet, H. C., and Ballantyne, J. S.,** Glutamine metabolism in a Holostean fish *(Amia calva)* and a teleost *(Salvelinus namaycush), Am. J. Physiol.,* 260, R159, 1991.

32. **Chamberlin, M. E. and Ballantyne, J. S.,** Glutamine metabolism in elasmobranch and agnathan muscle, *J. Exp. Zool.,* in press.

33. **Hochachka, P. W. and Fields, J. H. A.,** Arginine, glutamate, and proline as substrates for oxidation and for glycogenesis in cephalopod tissues, *Pac. Sci.,* 36, 325, 1982.

34. **Chamberlin, M. E. and Phillips, J. E.,** Oxidative metabolism in the locust rectum, *J. Comp. Physiol.,* 151B, 191, 1983.

35. **Stieber, S. F. and Cvancara, V. A.,** Tissue deamination of L-amino acids in the teleost *Stizostedion vitreum* (Mitchill), *Comp. Biochem. Physiol.,* 56B, 285, 1977.

36. **Lemieux, G., Craan, A. G., Quenneville, A., Lemieux, C., Berkofsky, J., and Lewis, V. S.,** Metabolic machinery of the alligator kidney, *Am. J. Physiol.,* 247, F686, 1984.

37. **Brosnan, J. T., Lowry, M., Vinay, P., Gougoux, A., and Halperin, M. L.,** Renal ammonium production — une vue canadienne, *Can. J. Physiol. Pharmacol.,* 65, 489, 1987.

38. **Goldstein, L.,** Glutamine transport by mitochondria isolated from normal and acidotic rats, *Am. J. Physiol.,* 229, 1027, 1975.

39. **Newsholme, E. A., Crabtree, B., and Ardawi, M. S. M.,** The role of high rates of glycolysis and glutamine utilization in rapidly dividing cells, *Biosci. Rep.,* 5, 393, 1985.

40. **Newsholme, E. A. and Leech, A. R.,** *Biochemistry for the Medical Sciences,* John Wiley & Sons, New York, 1983, 450.

41. **Christensen, H. N.,** Hypothesis: control of hepatic utilization of alanine by membrane transport by cellular metabolism, *Biosci. Rep.,* 3, 905, 1983.

42. **Copper, J. R, Bloom, F. E., and Roth, R. H.,** *The Biochemical Basis of Neuropharmacology,* Oxford University Press, New York, 1991, chap. 7.

43. **Garcia, U., Garcia-Colunga, J., and Arechiga, H.,** Differential effects of gamma-aminobutyric acid, glycine and taurine on central synapses of crustaceans, *Proc. West. Pharmacol. Soc.,* 34, 215, 1991.

44. **Corrigan, J. J.,** D-amino acids in animals, *Science,* 164, 142, 1969.

45. **Preston, R. L.,** Occurrence of D-amino acids in higher organisms: a survey of the distribution of D-amino acids in marine invertebrates, *Comp. Biochem. Physiol.,* 87B, 55, 1987.

46. **Preston, R. L.,** Sodium amino acid cotransport systems in marine invertebrates, in *Comparative Aspects of Sodium Cotransport Systems,* Kinne, R. K. H., Ed., S. Karger, Basel, 1990, 1.

47. **Matsushima, O. and Yamada, A.,** Uptake of L- and D-alanine by a brackish-water bivalve, *Corbicula* japonica, with special reference to their transport pathways and with salinity effect, *J. Exp. Zool.,* 263, 8, 1992.

48. **Matsushima, O., Katayama, H., Yamada, K., and Kado, Y.,** Occurrence of free D-alanine and alanine recemase activity in bivalve molluscs with special reference to intracellular osmoregulation, *Mar. Biol. Lett.,* 5, 217, 1984.

49. **Borg, B. S. and Wahlstrom, R. C.,** Species and isometric variation in the utilization of

amino acids, in *Absorption and Utilization of Amino Acids,* Friedman, M., Ed., CRC Press, Boca Raton, FL, 1989, 155.

50. **Somero, G. N.,** Protons, osmolytes and fitness of internal milieu for protein function, *Am. J. Physiol.,* 251, R197, 1986.
51. **Prosser, C. L.,** Inorganic Ions, in *Comparative Physiology,* Prosser, C. L., Ed., W. B. Saunders, Philadelphia, 1973, 79.
52. **Dawson, R. M. C., Elliot, D. C., Elliot, W. H., and Jones, K. M.,** *Data for Biochemical Research,* Oxford Science Publications, New York, 1986, 1.
53. **Rasmussen, H. and Barrett, P. Q.,** Calcium messenger system: an integrated view, *Physiol. Rev.,* 64, 938, 1984.
54. **Deaton, L. E., Hilbish, T. J., and Koehn, R. K.,** Protein as a source of amino nitrogen during hyperosmotic volume regulation in the mussel *Mytilus edulis, Physiol. Zool.,* 57, 609, 1984.
55. **Moyes, C. D. and Ballantyne, J. S.,** Amino acid transport in bivalve mollusc mitochondria: the role of the proton motive force and monovalent cations in proline oxidation by oyster *(Crassostrea virginica)* gill mitochondria, *J. Exp. Zool.,* 243, 1, 1987.
56. **Ballantyne, J. S., Moyes, C. D., and Moon, T. W.,** Osmolarity affects oxidation of sarcosine by isolated hepatocytes and mitochondria from a euryhaline elasmobranch, *J. Exp. Zool.,* 238, 267, 1986.
57. **Rolleston, F. S.,** A theoretical background to the use of measured concentrations of intermediates in study of the control of intermediary metabolism, *Curr. Top. Cell. Regul.,* 5, 47, 1972.
58. **Kacser, H. and Burns, J. A.,** The control of flux, *Symp. Soc. Exp. Biol.,* 32, 65, 1973.
59. **Heinrich, H. and Rapoport, T. A.,** A linear steady-state treatment of enzymatic chains. Critique of the crossover theorem and a general procedure to identify interaction sites with an effector, *Eur. J. Biochem.,* 42, 97, 1974.
60. **Heinrich, R. and Rapoport, T. A.,** A linear steady-state treatment of enzymatic chains. General properties, control and effector strength, *Eur. J. Biochem.,* 42, 89, 1974.
61. **Tager, J. M., Wanders, R. J. A., Groen, A. K., Kunz, W., Bohnensack, R., Kuster, U., Letko, G., Bohme, G., Duszynski, J., and Wojtczak, L.,** Control of mitochondrial respiration, *FEBS Lett.,* 151, 1, 1983.
62. **Brown, G. C., Hafner, R. P., and Brand, M. D.,** A "top-down" approach to the determination of control coefficients in metabolic control theory, *Eur. J. Biochem.,* 188, 321, 1990.
63. **Brown, G. C.,** Control of respiration and ATP synthesis in mammalian mitochondria and cells, *Biochem. J.,* 284, 1, 1992.
64. **Moyes, C. D., Moon, T. W., and Ballantyne, J. S.,** Glutamate catabolism in mitochondria from *Mya arenaria* mantle: effects of pH on the role of glutamate dehydrogenase, *J. Exp. Zool.,* 236, 293, 1985.
65. **Chen, C. and Lehninger, A. L.,** Respiration and phosphorylation by mitochondria from the hepatopancreas of the blue crab *(Callinectes sapidus), Arch. Biochem. Biophys.,* 154, 449, 1973.
66. **Doumen, C. D. and Ellington, W. R.,** Substrate preferences of the heart mitochondria of the horseshoe crab *Limulus polyphemus, Comp. Biochem. Physiol.,* 93B, 883, 1989.
67. **Lea, M. and Hillman, S.,** Effects of osmolality and solutes on performance of shark heart mitochondria, *J. Exp. Zool.,* 255, 9, 1990.
68. **Ballantyne, J. S., Chamberlin, M. E., and Singer, T. D.,** Oxidative metabolism in thermogenic tissues of the swordfish and mako shark, *J. Exp. Zool.,* 261, 110, 1992.
69. **Ballantyne, J. S. and Moon, T. W.,** Solute effects on mitochondria from an elasmobranch *(Raja erinacea)* and a teleost *(Pseudopleuronectes americanus), J. Exp. Zool.,* 239, 319, 1986.
70. **Ballantyne, J. S., Flannigan, D., and White, T. B.,** The effects of temperature on the oxidation of fatty acids, acyl carnitines and ketone bodies by mitochondria isolated from the liver of the lake charr, *Salvelinus namaycush, Can. J. Fish. Aquat. Sci.,* 46, 950, 1989.
71. **Skoog, C., Kromer, U., Mitchell, R. W., and Stephens, N. L.,** Characterization of frog muscle mitochondria, *Am. J. Physiol.,* 234, C1, 1978.
72. **Hillman, S. S., Lea, M. S., and Duerr, J. M.,** Osmotic effects on mitochondria from two species of amphibians, *Bufo marinus* and *Rana catesbeiana, Physiol. Zool.,* 64, 1552, 1991.
73. **Brand, M. D., Couture, P., Else, P. L., Withers, K. W., and Hulbert, A. J.,** Evolution of energy metabolism. Proton permeability of the inner membrane of liver mitochondria is greater in a mammal than in a reptile, *Biochem. J.,* 275, 81, 1991.
74. **Suarez, R. K., Brown, G. S., and Hochachka, P. W.,** Metabolic sources of energy for hummingbird flight, *Am. J. Physiol.,* 251, R537, 1986.
75. **Ballantyne, J. S., John, T. M., and George, J. C.,** The effects of glucagon on hepatic mito-

chondrial metabolism in the pigeon, *Columbia livia*, *Gen. Comp. Endocrinol.*, 72, 130, 1988.

76. **Harris, S. I., Balaban, R. S., Barret, L., and Mandel, L. J.,** Mitochondrial respiratory capacity and Na^+- and K^+-dependent adenosine triphosphatase-mediated ion transport in the intact renal cell, *J. Biol. Chem.*, 256, 10319, 1981.
77. **Hutson, S. M.,** Branched chain α-keto acid oxidative decarboxylation in skeletal muscle mitochondria. Effect of isolation procedure and mitochondrial delta pH, *J. Biol. Chem.*, 261, 4420, 1986.
78. **Takeuchi, Y., Morii, H., Tamura, M., Hayaishi, O., and Watanabe, Y.,** A possible mechanism of mitochondria dysfunction during cerebral ischemia: inhibition of mitochondria respiration activity by arachidonic acid, *Arch. Biochem. Biophys.*, 289, 33, 1991.
79. **Hansford, R. G.,** Some properties of mitochondria isolated from the flight muscle of the periodical cicada, *Magicicada septendecim*, *Biochem. J.*, 121, 771, 1971.
80. **Bygrave, F. L., Daday, A. A., and Doy, F.,** Evidence for a calcium-ion-transport system in mitochondria isolated from flight muscle of the developing sheep blowfly *Lucilia cuprina*, *Biochem. J.*, 146, 601, 1975.
81. **Hansford, R. G. and Johnson, R. N.** Some aspects of the oxidation of pyruvate and palmitoylcarnitine by moth *(Manduca sexta)* flight muscle mitochondria, *Comp. Biochem. Physiol.*, 55B, 543, 1976.
82. **Chamberlin, M. E.** Enzyme activities and mitochondrial substrate oxidation in tobacco hornworm midgut, *J. Comp. Physiol.*, 157B, 643, 1987.
83. **Ballantyne, J. S. and Moyes, C. D.,** The role of divalent cations and ionic strength in the osmotic sensitivity of glutamate oxidation in oyster gill mitochondria, *J. Exp. Biol.*, 130, 203, 1987.
84. **Dupaul, W. D. and Webb, K. L.,** Salinity-induced changes in the alanine and aspartic aminotransferase activity in three marine bivalve molluscs, *Arch. Int. Physiol. Biochim.*, 82, 817, 1974.
85. **Ballantyne, J. S. and Berges, J. A.,** Enzyme activities of gill, hepatopancreas, mantle and adductor muscle of the oyster (*Crassostrea virginica*, Gemlin, 1791) after changes in diet and salinity, *Can. J. Fish. Aquat. Sci.*, 48, 1117, 1991.
86. **Garcia-Perez, A. and Burg, M. B.,** Renal medullary organic osmolytes, *Physiol. Rev.*, 71, 1081, 1991.
87. **Hilbish, T. J., Deaton, L. E., and Koehn, R. K.,** Effect of an allozyme polymorphism on regulation of cell volume, *Nature*, 298, 688, 1982.
88. **Politis, A. D. and Pierce, S. K.,** Specific protein phosphorylation occurs in molluscan red blood cell ghosts in response to hypoosmotic stress, *J. Membr. Biol.*, 124, 169, 1991.
89. **Leite, M. V. and Goldstein, L.,** Ca^{2+} ionophore and phorbol ester stimulate taurine efflux from skate erythrocytes, *J. Exp. Zool.*, 242, 95, 1987.
90. **Moyes, C. D. and Moon, T. W.,** Solute effects on the glycine cleavage system of two osmoconformers (*Raja erinacea* and *Mya arenaria*) and an osmoregulator *(Pseudopleuronectes americanus)*, *J. Exp. Zool.*, 242, 1, 1987.
91. **Lavoinne, A., Baquet, A., and Hue, L.,** Stimulation of glycogen synthesis and lipogenesis by glutamine in isolated rat hepatocytes, *Biochem. J.*, 248, 429, 1987.
92. **Baquet, A., Hue, L., Meijer, A. J., van Woerkom, G. M., and Plomp, P. J. A. M.,** Swelling of rat hepatocytes stimulates glycogen synthesis, *J. Biol. Chem.*, 265, 955, 1990.
93. **Kayne, F. J. and Price, N. C.,** Conformational changes in the allosteric inhibition of muscle pyruvate kinase by phenylalanine, *Biochemistry*, 11, 4415, 1972.
94. **Male, K. B. and Storey, K. B.,** Kinetic characterization of NADP-specific glutamate dehydrogenase from the sea anemone, *Anthopleura xanthogrammica:* control of amino acid biosynthesis during osmotic stress, *Comp. Biochem. Physiol.*, 76B, 823, 1983.
95. **Gilles, R.,** Effect of various salts on the activity of enzymes implicated in amino-acid metabolism, *Arch. Int. Physiol. Biochim.*, 77, 441, 1969.
96. **Sarkissian, I. V.,** Regulation by salt of activity of citrate synthases from osmoregulators and osmoconformers, *Trans. N.Y. Acad. Sci.*, 36, 775, 1974.
97. **Rudoplh, A. S., Crowe, J. H., and Crowe, L. M.,** Effects of three stabilizing agents — proline, betaine, and trehalose — on membrane phospholipids, *Arch. Biochem. Biophys.*, 245, 134, 1986.
98. **Smith, M. W.,** Temperature adaptation of transport properties in poikilotherms, in *Hibernation and Hypothermia: Perspectives and Challenges*, South, F. E., Ed., Elsevier, Amsterdam, 1972, 219.
99. **Sober, H. A., Ed.,** *Handbook of Biochemistry*, CRC Press, Boca Raton, FL, 1968, B8.

100. **Lide, D. R., Ed.,** *Handbook of Chemistry and Physics,* CRC Press, Boca Raton, 1990, chap. 7.
101. **Klein, R. A., Moore, M. J., and Smith, M. W.,** Selective diffusion of neutral amino acids across lipid bilayers, *Biochim. Biophys. Acta,* 233, 420, 1971.
102. **Mahler, H. R. and Cordes, E. H.,** *Biological Chemistry,* Harper and Row, New York, 1971, chap. 3.
103. **Irving, C. S., Hammer, B. E., Danyluk, S. S., and Klein, P. D.,** Coordination and binding of taurine as determined by nuclear magnetic measurements on ^{13}C-labelled taurine, in *Taurine in Nutrition and Neurology,* Huxtable, R. J. and Pasantes-Morales, H., Eds., Plenum Press, New York, 1982, 5.

Chapter 7

The Role and Regulation of Methylamines in the Response of Cells to Osmotic Stress

Julia Dragolovich and Sidney K. Pierce

TABLE OF CONTENTS

I. INTRODUCTION

In response to volume changes produced by exposure to anisosmotic conditions, most cells attempt to recover volume by regulating intracellular osmolyte concentrations.[1,2] A rapid mechanism of volume recovery is most critical for cells undergoing hypoosmotic shock. The dilute environment initiates cellular swelling, which must be rapidly controlled to avoid cell lysis. Most cells utilize a two-step coordinated efflux of both inorganic ions and organic osmolytes to produce volume reduction.[3] Conversely, following a hyperosmotic challenge, cells shrink as they lose osmotically obligated water. To offset this volume reduction, many cells increase their total osmolyte pool by accumulating both inorganic ions and low-molecular-weight organic compounds. This response is also crucial to cell survival, but since lysis is not an issue, it is often relatively slower than the hypoosmotic response. High intracellular salt levels can disrupt protein structure and impair metabolic processes. Therefore, most cells must limit the amount of inorganic ions accumulated and, above a certain concentration, rely on the increased organic osmolytes, which do not affect protein stability and/or actually counteract the deleterious effects of the protein-denaturing solutes.[4,5] However, many cell types do not completely rehydrate following hyperosmotic stress. Under dehydrating conditions, volume recovery may not be as important as maintaining a balance between intracellular salt and organic osmolyte concentrations that is favorable for continued metabolic processes.[6]

In the response to both hyper- and hypoosmotic stress, cells use several types of organic osmolytes, predominantly polyols, free amino acids, and methylamines. In this chapter, we will describe the regulation of the methylamines proline betaine, trimethylamine oxide, and glycine betaine by osmotically stressed animal cells. The osmotically controlled regulation of glycerophosphorylcholine, another methylamine, has been extensively reviewed recently.[7,8] Although the methylamines are important osmolytes in a wide variety of organisms, including bacteria, higher plants, invertebrates, and mammalian kidney cells, our knowledge of the mechanisms that regulate their intracellular concentrations remains relatively limited, especially in animal cells.[7–11]

0-8493-4448-4/94/$0.00+$.50

II. PROLINE BETAINE

A. HYPEROSMOTIC RESPONSE

Proline betaine is a relatively rare methylamine, originally identified in several higher plants and subsequently in some marine algae.[12,13] Its occurrence in animal cells may be limited. Its regulation has been described in *Elysia chlorotica,* an extremely euryhaline sea slug that has a symbiotic relationship with chloroplasts from the green alga *Vaucheria.*[14] *Elysia* can survive for at least 2 weeks in salinities ranging from a very dilute 24 mOsmol to an extremely concentrated 2422 mOsmol. Throughout this range, the animal is an osmoconformer, with blood osmolality slightly hyperosmotic to that of the external environment.[15] When placed in exceedingly hyperosmotic medium (1950 mOsmol), the slugs rapidly lose body weight, as much as 36% of the initial weight within 4 h of exposure. This large dehydration suggests that the cells are experiencing both substantial water loss and increases in intracellular salt concentrations. Over a 2-week period of hyperosmotic acclimation, the tissue does not regain any water, indicating an inability for volume regulation in *Elysia* subjected to extremely high salinities. Despite the absence of volume increase, the animals survive in elevated salinities for extended periods of time. After 7 d of hyperosmotic stress, proline betaine begins to accumulate and new steady-state levels are attained following 2 weeks of hyperosmotic acclimation.[16]

The intracellular concentrations of proline betaine can be very high in hyperosmotically adapted *Elysia* (190 μmol/g dry weight in animals adapted to 87 mOsmol, compared to 1041 μmol/g dry weight in 1834 mOsmol acclimated animals).[14] The mechanism(s) responsible for the hyperosmotically stimulated proline betaine accumulation in animal cells is not known, although in *Elysia,* proline betaine is produced by the animal cells and not the symbiotic algal chloroplasts.[16] Further, the 7-d delay in proline betaine accumulation suggests that synthetic regulation may be under genetic control rather than a modification of existing biochemical pathways.[16] Alternatively, if proline betaine is a metabolic intermediate under nonstress conditions, hyperosmotic stress may inhibit proline betaine utilization, resulting in increased accumulation of the osmolyte. This type of control, decrease in degradative enzyme activity, has been reported in the accumulation of glycerophosphorylcholine by mammalian kidney cells.[7]

In salt-stressed plants, it appears that proline betaine is not synthesized directly from proline, but the metabolic source is not known.[17]

B. HYPOOSOMOTIC RESPONSE

Proline betaine is also regulated in *Elysia* transferred to dilute medium. In whole animal experiments, exposure to hypoosmotic conditions produces swelling, which corresponds to the severity of the osmotic shift. Regardless of the magnitude of the nonlethal hypoosmotic shock, the animals at least partially recover volume. This volume-regulatory response is associated with an efflux of tissue proline betaine. Animals acclimated to 920 mOsmol and transferred to 150 mOsmol lose approximately 50% of the tissue proline betaine (715 to 388 μmol/g dry weight) within 48 h. As the osmolyte concentrations fall, cell volume is recovered as osmotically obligated water leaves the cells. In hypoosmotically stressed *Elysia,* proline betaine is excreted intact, an event common to most methylamines.[18] A response similar to that of the whole animal has been recorded in isolated cells (Parker cells) of the abdominal ganglia from *Elysia* circumesophageal nerve rings. The volume-regulatory response is much faster in the isolated cells than the whole animal, but is also dependent on the efflux of proline betaine. In addition to proline betaine regulation, a significant Na^+ efflux from hypoosmotically stressed Parker cells occurs, which is also involved in cell volume recovery.[19,20]

III. TRIMETHYLAMINE OXIDE

A. HYPEROSMOTIC RESPONSE

Trimethylamine oxide (TMAO) is an osmolyte found predominantly in urea-concentrating organisms, such as the elasmobranchs. Many of these animals use urea as an extracellular osmolyte to achieve plasma osmolalities comparable to those of the external environment. The amount of urea accumulated can be substantial. Urea concentrations as high as 630 mmol/l have been recorded in the plasma of the stingray, *Dasyatis americana.*[21] Since cell membranes are highly permeable to urea, intracellular concentrations closely parallel those of the plasma. Even at low concentrations, urea is a potent protein denaturant. Therefore, many hyperosmotically stressed elasmobranchs utilize increased intracellular TMAO concentrations to counteract the potentially deleterious effects of urea. Accumulation of total methylamines, including TMAO, in a ratio of 2:1 to urea is favorable for continued enzyme function, even in the presence of elevated urea concentrations.[4] Intracellular TMAO in *Squalus acanthias* rectal gland slices also permits the maintenance of intracellular concentrations of sodium and potassium below those found in the plasma, thereby further reducing the intracellular ionic strength, which also reduces the chance of protein perturbation.[22] In addition to TMAO, glycine betaine is also accumulated by some hyperosmotically stressed elasmobranchs, as well as some cultured mammalian cells, and like TMAO serves both as an osmolyte and osmoprotectant.[4,23]

The counteracting effects of TMAO have been described in a variety of elasmobranch cell types; however, the mechanism(s) regulating the accumulation of this osmolyte following hyperosmotic stress remains unclear.

B. HYPOOSMOTIC RESPONSE

Plasma membranes are relatively impermeable to TMAO under isosmotic and hyperosmotic conditions. However, hypoosmotic stress results in an increased membrane permeability to TMAO and the osmolyte exits cells unaltered. Kleinzeller and colleagues[24–26] suggest that the efflux of organic osmolytes, including TMAO, from shark rectal gland cells following hypoosomotic stress results from a reorganization (possibly depolymerization) of cytoskeletal F-actin. A few other studies have implicated the cytoskeleton in osmolyte efflux and cell volume recovery.[27,28] In addition to the cytoskeletal changes, organic osmolyte efflux may be mediated by volume-sensitive channels and/or exchangers.[29–33] Although the research in this area has focused on the efflux of taurine, these pathways may be accessible to TMAO and the other methylamines.

Despite the tremendous interest in the involvement of the cytoskeleton and volume-activated chloride channels, the mechanism(s) controlling hypoosmotically induced loss of intracellular TMAO is unknown.

IV. GLYCINE BETAINE

A. HYPEROSMOTIC RESPONSE

Glycine betaine is a very common and important methylamine used in cellular osmoregulation. Although its occurrence has been documented in an ever-expanding list of diverse organisms, the mechanisms controlling intracellular accumulation following hyperosmotic stress have been described for only a few species of bacteria and plants, and more recently, mammalian kidney cells and invertebrate heart tissue. Cells accumulate glycine betaine by either intracellular synthesis or transport from extracellular pools.

1. Intracellular Synthesis

In those cells that rely on osmotically stimulated synthesis for the accumulation of glycine betaine, the osmolyte is the end-product of a two-step oxidation of choline (choline → betaine aldehyde → glycine betaine). The first reaction is catalyzed by choline dehydrogenase (oxidase) while the second and terminal reaction is regulated by betaine aldehyde dehydrogenase (BADH) activity. In the few species studied, BADH is an enzyme that is substrate (betaine aldehyde)-specific and NAD^+-dependent. In addition, in all eukaryotes studied, the glycine betaine synthetic pathway is located in subcellular organelles — chloroplasts in plants and mitochondria in both mammalian liver and kidney cells, as well as horseshoe crab heart cells.[34–37]

The most extensively studied eukaryotic organisms that synthesize glycine betaine in response to hyperosmotic stress are spinach and barley.[38–40] In these plants, BADH activity is the primary regulatory step in osmotically induced increases in glycine betaine levels. Following 24 h of exposure to hyperosmotic conditions, the amount of intracellular BADH protein in both spinach and barley increases.[38,40] This event results in an increase in BADH activity and elevation of intracellular glycine betaine levels. Accumulation of the protein, hence increased activity, continues in plants chronically stressed for 1 to 3 weeks. In spinach, the increase in the amount of protein is genetically regulated as the amount of translatable BADH mRNA increases two- to threefold in response to hyperosmotic stress.[38]

Many marine invertebrates accumulate extremely high intracellular concentrations of glycine betaine when acclimated to hyperosmotic conditions. However, relatively little is known about the mechanisms that control this accumulation. Heart cells of the horseshoe crab *(Limulus polyphemus)* increase glycine betaine concentrations by 125 µmol/g dry weight as early as 12 h following transfer to hyperosmotic conditions. Over the next 48 h of acclimation, *Limulus* heart cells continue to accumulate glycine betaine.[41]

Limulus cardiac cells synthesize glycine betaine from the precursor, choline, in the mitochondrial matrix. Isolated heart mitochondria take up choline and convert over 80% of it into glycine betaine. Both the uptake of choline and its conversion to glycine betaine are stimulated approximately fivefold by increased osmolality.[42]

Unlike the regulatory control in spinach, the rate of choline uptake, rather than modulation of BADH activity, appears to be the rate-limiting step in glycine betaine synthesis by isolated *Limulus* mitochondria. Choline enters isolated mitochondria at a rate that is much slower than the rate at which BADH converts betaine aldehyde to glycine betaine. Choline uptake is stimulated by increases in both the medium ionic and osmotic concentrations. More than one type of choline transporter may be present in *Limulus* heart mitochondria.[37] The possible existence of multiple choline transporters with varying degrees of osmotic sensitivity has also been reported for the bacterium *Rhizobium meliloti*.[43]

It appears that the mitochondria are the site of glycine betaine synthesis in animal cells. However, its osmotic effect is exerted in the cytoplasm. The efflux of glycine betaine from liver mitochondria is controlled by extramitochondrial monovalent cation concentrations and mitochondrial states of respiration. High intracellular salt concentrations (in particular K^+) and "state 3" respiration inhibit mitochondrial glycine betaine efflux. On the other hand, isosmotic media with lowered salt concentrations and uncoupled respiration promote mitochondrial loss of glycine betaine.[44] This influence of cation concentrations on mitochondrial glycine betaine efflux seems to be in opposition to the events of intracellular osmolyte regulation. For example, as the extracellular osmotic concentration increases, intracellular inorganic ion levels increase as water leaves. If the elevated intracellular ionic concentrations inhibit osmolyte release from the mitochondria, then newly synthesized glycine betaine would not appear in the cytoplasmic pool. The signal(s) modulating glycine betaine efflux from mitochondria in osmotically stressed cells is not known, but it appears that its release into the cytoplasm may be influenced differently or controlled by factors other than those described in liver cells under isosmotic conditions.

2. Na^+-Dependent Transport

Glycine betaine is an important osmolyte in the mammalian kidney. Cells from the cortex, outer medulla, and inner medulla are able to synthesize glycine betaine; however, only the cortical oxidation of choline is responsive to osmotic perturbations.[36,45] Inner medullary mitochondrial choline dehydrogenase activity does not increase following hyperosmotic challenge, although glycine betaine concentrations increase dramatically in the inner medulla.[36] Therefore, the ability of hyperosmotically stressed inner medullary cells to accumulate high levels of glycine betaine relies on the active transport of the osmolyte from extracellular sources, possibly supplied from cortical synthesis, rather than local synthesis.

The mechanism regulating glycine betaine uptake has been extensively studied in the Madin Darby canine kidney cell (MDCK) line.[7,8] These cells do not synthesize glycine betaine when exposed to choline. However, in the presence of extracellular glycine betaine, MDCK cells accumulate the osmolyte at concentrations 1000-fold greater than the medium concentration.[46] In MDCK cells, glycine betaine uptake is mediated by both low- and high-affinity Na^+-dependent betaine transporters with a K_m of 6 and 0.1 m*M*, respectively, in isosmotic medium (315 mOsmol). The betaine transporters are localized predominantly in the basolateral regions of the cells.[47] In MDCK cells chronically (1 year) exposed to hyperosmotic conditions (615 mOsmol), glycine betaine is elevated by a tenfold increase in V_{max} of the betaine transporters, while K_m remains unaffected by the increased osmolality.[48] Both the ionic (increased NaCl) and osmotic (addition of mannitol) components of the stress are responsible for the increased rates of glycine betaine accumulation. In addition, short-term exposure (up to 1 week) of MDCK cells to hyperosmotic conditions also results in a Na^+-dependent accumulation of glycine betaine. A maximal increase in glycine betaine uptake is observed 1 d after transfer to hyperosmotic medium.[46,47]

The hyperosmotically stimulated increase in glycine betaine accumulation in MDCK cells is produced by an increase in the number of transporters. Hyperosmotic stress elevates the V_{max} of betaine uptake, but does not alter substrate affinity. In addition, injecting *Xenopus* oocytes with mRNA isolated from hyperosmotically stressed cells results in an increase in Na^+-dependent glycine betaine uptake, while mRNA taken from isosmotically maintained cells has no effect on oocyte betaine transport.[49] It appears that hyperosmotic stress triggers an increase in transcription of the gene responsible for encoding the betaine transporter. The betaine transporter (BGT-1) has been cloned from MDCK cells. Injection of mRNA transcribed *in vitro* from a cDNA screened from a MDCK cell library into *Xenopus* oocytes induces Na^+- (and Cl^-)-dependent uptake of glycine betaine similar to the response in hyperosmotically stressed MDCK cells. The cDNA product is a single protein containing 614 amino acids and appears to be a member of the GABA family of transporters first identified in brain tissue. In fact, the renal medullary betaine transporter has a greater affinity for γ-amino-*n*-butyric acid (GABA) than for glycine betaine, but GABA plasma levels are well below those of glycine betaine.[50] Therefore, glycine betaine is the substrate transported by this carrier in the kidney medulla. Although the signal for increased transcription of the BGT-1 gene in MDCK cells is not yet known, these results show that the regulation of glycine betaine is under genetic control.

The gills of the mussel, *Mytilus californianus,*

also accumulate high levels of glycine betaine by transport rather than synthesis. Glycine betaine and taurine contribute 50% of the total osmolyte pool and both are taken up directly from the seawater. The transport of glycine betaine is strictly Na^+-dependent. Removal of Na^+ from the bathing medium results in a 97% decrease in uptake rate. In this tissue, it appears that glycine betaine transport is mediated by the alanine-proline (AP) carrier in the apical membrane. The AP pathway transports other neutral amino acids and glycine betaine uptake is inhibited significantly when the gill is incubated in medium containing L-proline, L-alanine, and other structural analogues. Glycine betaine transport is stimulated by increasing osmolality and the transport rate is directly proportional to the Na^+ concentration of the bathing solution up to 425 m*M* (100% artificial seawater).[51]

Bivalves are not limited to active transport as the only strategy for hyperosmotically stimulated accumulation of osmolytes. In fact, since the concentration of taurine and glycine betaine in the external medium is a most unreliable source for an osmolyte, certain populations of Atlantic Coast oysters *(Crassostrea virginica),* for example, synthesize glycine betaine when exposed to hyperosmotic conditions. Following high salinity adaptation, the amount of glycine betaine in adductor muscle tissue increased approximately 1.5-fold, and in gill tissue accumulation was nearly 2-fold greater than in low salinity-adapted animals.[52] As in horseshoe crab heart tissue, the glycine betaine synthetic pathway in oyster gill is localized in the mitochondria and choline uptake by isolated mitochondria requires Na^+. The rates of both choline uptake and glycine betaine synthesis are increased in mitochondria isolated from high salinity-adapted animals as compared to low salinity-adapted oysters from the same population (Figures 7–1 and 7–2). The time course of accumulation has not been established and more research is necessary to define the regulatory pathway of hyperosmotically induced glycine betaine synthesis.

Bivalve mitochondria also synthesize taurine.[53] In the ribbed mussel *(Modiolus demissus),* mitochondrial activity of the taurine biosynthetic pathway is much greater than that of the cytoplasmic fraction, suggesting that the mitochondria contribute, albeit slowly, to the overall accumulation of this important osmolyte.[54]

B. HYPOOSMOTIC RESPONSE

Accumulation of glycine betaine following hyperosmotic stress takes hours and even days to occur, often longer than the increases in intracellular amino acids. However, in most cell types (isolated or in culture), the release of glycine betaine following hypoosmotic stress is similar to the very rapid efflux of free amino acids. For example, 30% of the intracellular glycine betaine is lost to the medium within 15 min of returning hyperosmotically stressed MDCK cells to the initial osmotic conditions. In addition to the glycine betaine efflux, the uptake rate of betaine is quickly reduced in these cells. Three hours after returning the cells to the original lower osmotic medium, the hyperosmotically stimulated betaine uptake has decreased by 25%. Within 1 d, the rate has declined further to 50% of the hyperosmotically stimulated level. Control levels (as determined in cells kept in isosmotic medium throughout the time course) of betaine transport are reached after 2 d of exposure to the reduced osmolality.[55]

The efflux of glycine betaine from kidneys in intact animals displays a time course similar to that of cultured renal cells, although there is some variation in time courses between species and diuretic treatments. For example, kidney tissue from the vole, *Microtus montanus,* accumulates glycine betaine, as well as other organic osmolytes, when subjected to extended dehydration (4 to 6 d). After 24 h of rehydration, a 68% drop in medullary betaine content occurs. Within 48 h, a new steady-state level of medullary betaine at 38% of the dehydrated controls is achieved.[56] In Wistar rats, 4 h of water loading leads to a 56% decrease in papillary base glycine betaine content. In the papillary tip, the amount of glycine betaine is reduced by 39% within 1 h of diuresis and 62% within 4 h.[57] Rabbit kidneys also rapidly lose glycine betaine under furosemide-induced diuresis. After only 20 min of treatment, the inner medullary level of glycine betaine is 35% of control and within 45 min a new steady-state level has been established at approximately 20% of control.[58] Whether these steep reductions in tissue glycine betaine are a result of the combined effects of osmolyte efflux and/or downregulation of betaine transporter remains to be established.

Not all tissues exhibit such a speedy osmolyte response to acute hypoosmotic stress. Horseshoe crab heart tissue experiences a 40% increase in wet weight within 2 h of exposure to hypoosmotic conditions. After 3 h of incubation in dilute medium, the tissue wet weight begins to decrease, suggesting that cell volume regulation is occurring. This reduction in cell water is produced by the rapid efflux of Na^+ and Cl^-. In intact animals, heart tissue glycine betaine levels do not begin to drop until 24 h after transfer to dilute seawater. Tissue glycine betaine amounts continue to decline slowly over a 7-d period, at which time 50% of the initial glycine betaine

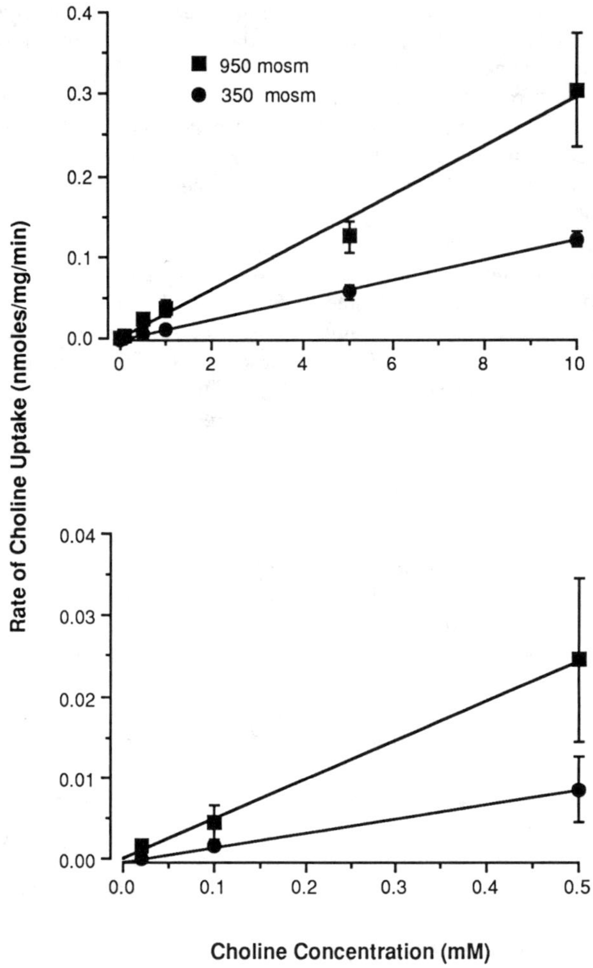

Figure 7–1 The effect of salinity adaptation on choline uptake by isolated oyster mitochondria. Mitochondria isolated from the gills of high (950 mOsmol) salinity-adapted oysters *(Crassostrea virginica)* have a much higher rate of choline uptake than their low (350 mOsmol) salinity-adapted counterparts. The uptake rate is linear for mitochondria in both salinities. The range of extramitochondrial choline concentrations used in these experiments was substantial and the uptake rates measured in the lower choline concentrations have been expanded (bottom).

remains in the tissue. Therefore, glycine betaine appears to be more important in long-term adaptation to hypoosmotic environments in this species.[59]

Despite our knowledge of the time courses of glycine betaine efflux from cells and tissue in response to hypoosmotic stress, the underlying mechanism(s) that controls the efflux is unknown.

V. CONCLUDING REMARKS

Methylamines are very common osmolytes which are involved in cellular responses to both hyper- and hypoosmotic stress. The strategies for the accumulation of methylamines include synthetic processing of specific precursors, uptake of final products, and inhibition of degradative processes. Hypoosmotically triggered elimination of methylamines occurs most often by efflux of the nonmetabolized osmolyte. In some cell types, increased osmolyte permeability coincides with a decrease in the rate of osmolyte uptake. The regulatory steps of these physiological responses may be genetically and/or biochemically controlled; however, more research is needed to fully under-

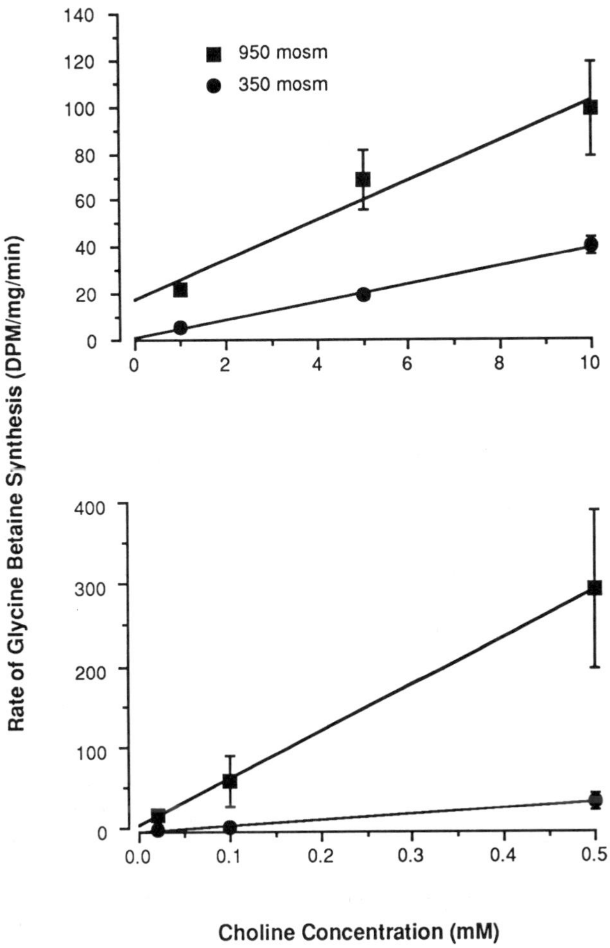

Figure 7–2 The effect of salinity adaptation and choline concentration on glycine betaine synthesis by isolated oyster mitochondria. Similar to choline uptake (see Figure 7–1), glycine betaine synthesis increased in a linear fashion as extramitochondrial choline increased. Adaptation to high (950 mOsmol) salinity stimulates glycine betaine synthesis in isolated gill mitochondria. Top: synthetic rate as measured over the range of choline concentrations from 0 to 0.5 m*M*. Bottom: synthetic rate measured over the range of choline concentrations from 1.0 to 10 m*M*.

stand the regulatory mechanisms of these osmotically stimulated processes.

REFERENCES

1. **MacKnight, A. D. C.,** Principles of cell volume regulation, *Renal Physiol. Biochem.*, 3–5, 114, 1988.
2. **Chamberlin, M. E. and Strange, K.,** Anisomotic cell volume regulation: a comparative view, *Am. J. Physiol.*, 257, C159, 1989.
3. **Pierce, S. K.,** Invertebrate cell volume control mechanisms: a coordinated use of intracellular amino acids and inorganic ions as osmotic solute, *Biol. Bull.*, 163, 405, 1982.
4. **Yancey, P. H., Clark, M., E., Hand, S. C., Bowlus, R. D., and Somero, G. N.,** Living with water stress: evolution of osmolyte systems, *Science*, 217, 1214, 1982.
5. **Arakawa, T. and Timasheff, S. N.,** The stabilization of proteins by osmolytes, *Biophys. J.*, 47, 411, 1985.

6. **Somero, G. N.,** Protons, osmolytes, and fitness of internal milieu for protein function, *Am. J. Physiol.,* 251, R197, 1986.
7. **Garcia-Perez, A. and Burg, M. B.,** Renal medullary organic osmolytes, *Physiol. Rev.,* 71, 1081, 1991.
8. **Burg, M. B.,** Molecular basis for accumulation of compatible osmolytes in mammalian cells, in *Water and Life: Comparative Analysis of Water Relationships at the Organismic, Cellular, and Molecular Levels,* Somero, G. N. et al., Eds., Springer-Verlag, Berlin, 1992, chap. 3.
9. **Csonka, L. N.,** Physiological and genetic responses of bacteria to osmotic stress, *Microbiol. Rev.,* 53, 121, 1989.
10. **Storey, R. and Wyn Jones, R. G.,** Quaternary ammonium compounds in plants in relation to salt resistance, *Phytochemistry,* 16, 447, 1977.
11. **Beers, J. R.,** The species distribution of some naturally-occurring quaternary ammonium compounds, *Comp. Biochem. Physiol.,* 21, 11, 1967.
12. **Cornforth, J. W. and Henry, A. J.,** The isolation of L-stachydrine from the fruit of *Capparis tomentosa, J. Chem. Soc.,* Part 1, 601, 1952.
13. **Blunden, G., Gordon, S. M., McClean, W. F. H., and Guiry, M. D.,** The distribution and possible taxonomic significance of quaternary ammonium and other Dragendorff-positive compounds in some genera of marine algae, *Bot. Mar.,* 25, 563, 1982.
14. **Pierce, S. K., Edwards, S. C., Mazzocchi, P. H., Klinger, L. J., and Warren, M. K.,** Proline betaine: a unique osmolyte in an extremely euryhaline osmoconformer, *Biol. Bull.,* 167, 495, 1984.
15. **Pierce, S. K., Warren, M. K., and West, H. H.,** Non-amino acid mediated volume regulation in an extreme osmoconformer, *Physiol. Zool.,* 56, 445, 1983.
16. **Pierce, S. K. and Rowland, L. M.,** Proline betaine and amino acid accumulation in sea slugs *(Elysia chlorotica)* exposed to extreme hyperosmotic conditions, *Physiol. Zool.,* 61, 205, 1988.
17. **Robertson, A. V. and Marion, L.,** The biosynthesis of alkaloids. XXIII. The role of proline and ornithine in the synthesis of stachydrine, *Can. J. Chem.,* 37, 1197, 1959.
18. **Rowland, L. M. and Pierce, S. K.,** The utilization and fate of quaternary ammonium compounds during low salinity adaptation in *Elysia chlorotica, Physiol. Zool.,* 58, 149, 1985.
19. **Parker, H. T. and Pierce, S. K.,** Comparative electrical properties of identified neurons in *Elysis chlorotica* before and after low salinity acclimation, *Comp. Biochem. Physiol.,* 82A, 367, 1985.
20. **Quinn, R. H. and Pierce, S. K.,** The ionic basis of the hypo-osmotic depolarization in neurons from the opisthobranch mollusc *Elysia chlorotica, J. Exp. Biol.,* 163, 169, 1992.
21. **Forster, R. P. and Goldstein, L.,** Intracellular osmoregulatory role of amino acids and urea in marine elasmobranchs, *Am. J. Physiol.,* 230, 925, 1976.
22. **Kleinzeller, A.,** Trimethylamine oxide and the maintenance of volume of dogfish rectal gland cells, *J. Exp. Zool.,* 236, 11, 1985.
23. **Yancey, P. H. and Burg, M. B.,** Counteracting effects of urea and betaine in mammalian cells in culture, *Am. J. Physiol.,* 258, R198, 1990.
24. **Kleinzeller, A. and Ziyadeh, F. N.,** Cell volume regulation in epithelia — with emphasis on the role of osmolytes and the cytoskeleton, in *Comparative Physiology,* Vol. 4, *Cell Volume Regulation,* Beyenbach, K. W., Ed., S. Karger, Basel, 1990, 59.
25. **Kleinzeller, A. and Mills, J. W.,** K^+-induced swelling of the dogfish shark *(Squalus acanthias)* rectal gland cells is associated with changes in the cytoskeleton, *Biochim. Biophys. Acta,* 1014, 40, 1989.
26. **Ziyadeh, F. N., Mills, J. W., and Kleinzeller, A.,** Hypotonicity and cell volume regulation in shark rectal gland: role of organic osmolytes and F-actin, *Am. J. Physiol.,* 262, F468, 1992.
27. **Pierce, S. K. and Politis, A. D.,** Ca^{2+}-activated cell volume recovery mechanisms, *Annu. Rev. Physiol.,* 52, 27, 1990.
28. **Foskett, J. K. and Spring, K. R.,** Involvement of calcium and cytoskeleton in gallbladder epithelial cell volume regulation, *Am. J. Physiol.,* 248, C27, 1985.
29. **Goldstein, L. and Brill, S. R.,** Volume-activated taurine efflux from skate erythrocytes: possible band 3 involvement, *Am. J. Physiol.,* 260, R1014, 1991.
30. **Goldstein, L., Brill, S. R., and Freund, E. V.,** Activation of taurine efflux in hypotonically stressed elasmobranch cells: inhibition by stilbene disulfonates, *J. Exp. Zool.,* 254, 114, 1990.
31. **Kirk, K., Ellory, J. C., and Young, J. D.,** Transport of organic substrates via a volume-activated channel, *J. Biol. Chem.,* 267, 23475, 1992.
32. **Grunder, S., Thiemann, A., Pusch, M., and Jentsch, T. J.,** Regions involved in the open-

ing of the ClC-2 chloride channel by voltage and cell volume, *Nature*, 360, 759, 1992.

33. **Uchida, S., Sasaki, S., Furukawa, T., Hiraoka, M., Imai, T., Hirata, Y., and Marumo, F.,** Molecular cloning of a chloride channel that is regulated by dehydration and expressed predominantly in kidney medulla, *J. Biol. Chem.*, 268, 3821, 1993.
34. **Hanson, A. D., May, A. M., Grumet, R., Bode, J., Jamieson, G. C., and Rhodes, D.,** Betaine synthesis in chenopods: localization in chloroplasts, *Proc. Natl. Acad. Sci. U.S.A.*, 82, 3678, 1985.
35. **Rothchild, H. A. and Guzmann Barron, E. S.,** The oxidation of betaine aldehyde by betaine aldehyde dehydrogenase, *J. Biol. Chem.*, 209, 511, 1954.
36. **Grossman, E. B. and Hebert, S. C.,** Renal inner medullary choline dehydrogenase activity: characterization and modulation, *Am. J. Physiol.*, 256, F107, 1989.
37. **Dragolovich, J. and Pierce, S. K.,** Characterization of partially purified betaine aldehyde dehydrogenase from horseshoe crab *(Limulus polyphemus)* cardiac mitochondria, *J. Exp. Zool.*, in press.
38. **Weretilnyk, E. A. and Hanson, A. D.,** Betaine aldehyde dehydrogenase from spinach leaves: purification, *in vitro* translation of the mRNA, and regulation by salinity, *Arch. Biochem. Biophys.*, 271, 56, 1989.
39. **Arakawa, K., Katayama, M., and Takabe, T.,** Levels of betaine and betaine aldehyde dehydrogenase activity in the green leaves, and etiolated leaves and roots of barley, *Plant Cell Physiol.*, 31, 797, 1990.
40. **Arakawam K., Mizuno, K., Kishitani, S., and Takabe, T.,** Immunological studies of betaine aldehyde dehydrogenase in barley, *Plant Cell Physiol.*, 33, 833, 1992.
41. **Dragolovich, J. and Pierce, S. K.,** Comparative time courses of inorganic and organic osmolyte accumulation as horseshoe crabs *(Limulus polyphemus)* adapt to high salinity, *Comp. Biochem. Physiol.*, 102A, 79, 1992.
42. **Dragolovich, J.,** The Control of Glycine Betaine Synthesis and its Role in the Cellular Adaptation to Hyperosmotic Stress: a Horseshoe Crab Model, Ph.D. dissertation, University of Maryland, College Park, MD, 1991.
43. **Pocard, J.-A., Bernard, T., Smith, L. T., and Le Rudulier, D.,** Characterization of three choline transport activities in *Rhizobium meliloti:* modulation by choline and osmotic stress, *J. Bacteriol.*, 171, 531, 1989.
44. **De Ridder, J. J. M. and Van Dam, K.,** The efflux of betaine from rat-liver mitochondria, a possible regulating step in choline oxidation, *Biochim. Biophys. Acta*, 291, 557, 1973.
45. **Lohr, J. and Acara, M.,** Effect of dimethylaminoethanol, an inhibitor of betaine production, on the disposition of choline in the rat kidney, *J. Pharmacol. Exp. Ther.*, 252, 154, 1989.
46. **Nakanishi, T., Balaban, R. S., and Burg, M. B.,** Survey of of osmolytes in renal cell lines, *Am. J. Physiol.*, 255, C181, 1988.
47. **Yamauchi, A., Kwon, H. M., Uchida, S., Preston, A. S., and Handler, J. S.,** *Myo*-inositol and betaine transporters regulated by tonicity are basolateral in MDCK cells, *Am. J. Physiol.*, 261, F197, 1991.
48. **Nakanishi, T., Turner, R. J., and Burg, M. B.,** Osmoregulation of betaine transport in mammalian renal medullary cells, *Am. J. Physiol.*, 258, F1061, 1990.
49. **Robey, R. B., Ellis, D. E., Kwon, H. M., Handler, J. S., Garcia-Perez, A., and Burg, M. B.,** Expression of Na^+-dependent betaine uptake in *Xenopus* oocytes injected with mRNA from osmotically stressed MDCK cells, *J. Am. Soc. Nephrol.*, 1, 705, 1990.
50. **Yamauchi, A., Uchida, S., Kwon, H. M., Preston, A. S., Robey, R. B., Garcia-Perez, A., Burg, M. B., and Handler, J. S.,** Cloning of a Na^+- and Cl^--dependent betaine transporter that is regulated by hypertonicity, *J. Biol. Chem.*, 267, 649, 1992.
51. **Wright, S. H., Wunz, T. M., and Silva, A. L.,** Betaine transport in the gill of a marine mussel, *Mytilus californianus*, *Am. J. Physiol.*, 263, R226, 1992.
52. **Pierce, S. K., Rowland-Faux, L. M., and O'Brien, S., M.,** Different salinity tolerance mechanisms in Atlantic and Chesapeake Bay conspecific oysters: glycine betaine and amino acid pool variations, *Mar. Biol.*, 113, 107, 1992.
53. **McCormick, A., Paynter, K. T., Brodey, M. M., and Bishop, S. H.,** Aspartate aminotransferases from ribbed mussel gill tissue: reactivity with b-L-cysteinesulfinic acid and other properties, *Comp. Biochem. Physiol.*, 84B, 163, 1986.
54. **Bishop, S. H., Ellis, L. L., and Burcham, J. M.,** Amino acid metabolism in molluscs, in *The Mollusca, Metabolic Biochemistry and Molecular Biomechanics*, Vol. 1, Hochachka, P. W., Ed., Academic Press, New York, 1983, chap. 6.
55. **Nakanishi, T. and Burg, M. B.,** Osmoregulatory fluxes of myo-inositol and betaine in renal cells, *Am. J. Physiol.*, 257, C964, 1989.

56. **Yancey, P. H.,** Osmotic effectors in kidneys of xeric and mesic rodents: corticomedullary distributions and changes with water availability, *J. Comp. Physiol. B,* 158, 369, 1988.
57. **Beck, F. X., Schmolke, M., Guder, W. G., Dorge, A., and Thurau, K.,** Osmolytes in renal medulla during rapid changes in papillary tonicity, *Am. J. Physiol.,* 262, F849, 1992.
58. **Wolff, S. D., Stanton, T. S., James, S. L., and Balaban, R. S.,** Acute regulation of the predominant organic solutes of the rabbit renal inner medulla, *Am. J. Physiol.,* 257, F676, 1989.
59. **Warren, M. K. and Pierce, S. K.,** Two cell volume regulatory systems in the *Limulus* myocardium: an interaction of ions and quaternary ammonium compounds, *Biol. Bull.,* 163, 504, 1982.

Chapter 8

Regulation of Intracellular Polyols and Sugars in Response to Osmotic Stress

Jeff M. Sands

TABLE OF CONTENTS

I. INTRODUCTION

Organic osmolytes can be divided into two categories, compatible osmolytes and counteracting osmolytes (see Chapter 5 for definition of these terms). This chapter will review the regulation of production, transport, and intracellular concentration of polyols and sugars, the major compatible osmolytes. The polyols include sorbitol, inositol, arabitol, mannitol, and heterosides, while the sugars include glucose, mannose, fructose, sucrose, and trehalose. The polyols are involved in cell volume regulation in organisms as diverse as cyanobacteria, algae, fungi, plants, insects, crustaceans, and mammals.[1] To date, no role for sugars in volume regulation has been reported in mammalian cells. Sugars are involved in cell volume regulation in cyanobacteria, diatoms, fungi, eubacteria, algae, plants, insects, and crustaceans.[1]

II. SORBITOL

A. METABOLIC PATHWAYS

Sorbitol, or D-glucitol, is the polyol produced from D-glucose in a NADPH-dependent reduction reaction catalyzed by the enzyme aldose reductase (low K_m aldehyde reductase, ALR2, EC 1.1.1.21).[2]

$$\text{D-glucose} + \text{NADPH} \leftrightarrow \text{Sorbitol} + \text{NADP}^+$$

Sorbitol can be oxidized to form D-fructose in a NAD^+-dependent reaction catalyzed by sorbitol dehydrogenase (L-iditol dehydrogenase, EC 1.1.1.14).[2]

$$\text{Sorbitol} + \text{NAD}^+ \leftrightarrow \text{D-fructose} + \text{NADH}$$

The polyol, or sorbitol, pathway is the combination of these two reactions and results in the conversion of D-glucose to D-fructose.

0-8493-4448-4/94/$0.00+$.50

This pathway has at least two physiologic roles and is pathologically important. Physiologically, this pathway is important for generating sorbitol as a compatible osmolyte in the mammalian renal inner medulla[3–7] and for generating fructose in sheep placentas and seminal vesicles.[4] Pathologically, the polyol pathway has been implicated in the development of diabetic retinopathy and neuropathy in humans.[5]

B. INTRARENAL DISTRIBUTION

Within the mammalian renal inner medulla, osmolality increases from the inner-outer medullary border to the papillary tip (Figure 8–1). In both rats and rabbits, tissue sorbitol concentration also increases from the inner-outer medullary border to the papillary tip, in parallel with the increase in tissue osmolality, whereas only small amounts of sorbitol are found in the renal cortex (Figure 8–1).[3,8–16] In microdissected rat kidney tubules, high concentrations of sorbitol are found in cells of the inner medullary collecting duct (IMCD), especially the terminal IMCD, and the thin limbs of the loop of Henle.[17,18] Thus, sorbitol measurements in both tissue slices and in microdissected tubules show that sorbitol concentration parallels variations of osmolality in the inner medulla, as would be expected for a compound involved in volume regulation.

Inner medullary osmolality varies with the state of hydration: increasing during antidiuresis and decreasing during diuresis. Sorbitol accumulates during antidiuresis (water deprivation) in the inner medulla of all mammals studied to date, including rat, rabbit, vole, deer mouse, pocket mouse, sheep, and dog.[10–14,19–23] Similarly, inner medullary sorbitol increases when Brattleboro rats are treated with vasopressin.[16,20,21,24]

The antidiuresis-induced increase in inner medullary sorbitol occurs slowly. In two studies of Sprague-Dawley rats, 3 d of water deprivation (with food available *ad libitum*) were required before a significant increase in inner medullary sorbitol was detected[12,22] (in one study there was no increase in sorbitol after 3 d of water deprivation[21]). In Brattleboro rats, 3 d of vasopressin are required to increase inner medullary sorbitol.[16] However, sorbitol continues to increase for up to 12 d after vasopressin administration is begun.[20]

During diuresis, the concentration of sorbitol in the inner medulla generally decreases as osmolality decreases in the inner medulla. Sorbitol concentration has been demonstrated to decrease during water diuresis[11–13,19,21,22,25] and during furosemide diuresis,[12,21,25,26] as does inner medullary sodium.[21,25,26] On the other hand, when the sodium concentration of the inner medulla was increased in rats given hypertonic NaCl for 5 d, inner medullary

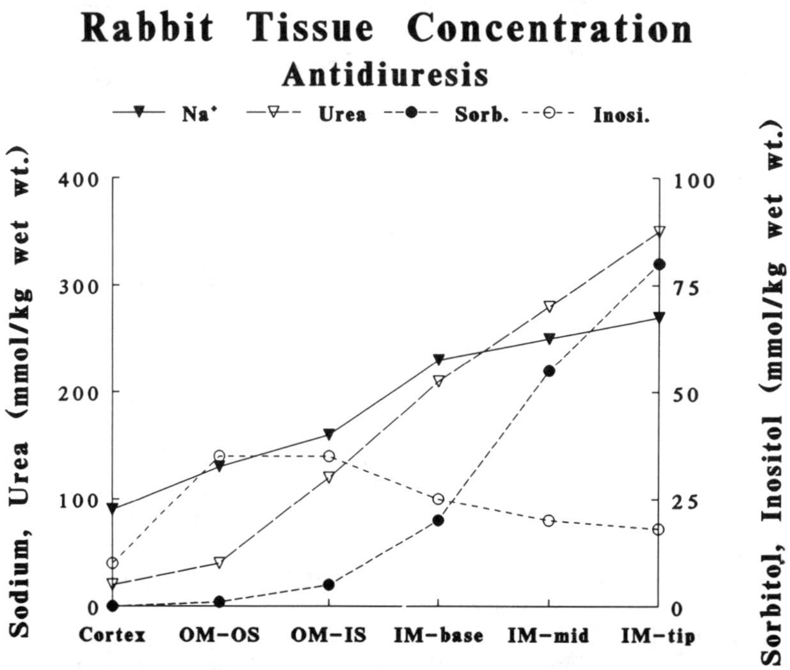

Figure 8–1 Sodium, urea, sorbitol, and inositol content in six regions of antidiuretic rabbit kidney. OM, outer medulla; OS, outer stripe; IS, inner stripe; IM, inner medulla. (Data from Reference 11.)

sorbitol rose even though urine flow increased,[27] as did inner medullary sodium.[28] Taken together, these results have led to the hypothesis that sorbitol accumulates in inner medullary cells to balance the osmotic effects of extracellular NaCl.

Unlike the 3 d required to increase inner medullary sorbitol during antidiuresis, sorbitol can decrease within hours of inducing a water diuresis.[12,21,25,26] The difference in the time course between loss of sorbitol during diuresis and the increase during antidiuresis suggests that inner medullary sorbitol concentration must be regulated by multiple factors. Evidence in kidney tubules supports three regulatory pathways for cellular sorbitol concentration: (1) sorbitol production by aldose reductase; (2) sorbitol degradation by sorbitol dehydrogenase; and (3) sorbitol efflux by sorbitol permease.

C. ALDOSE REDUCTASE

Aldose reductase catalyzes the production of sorbitol from D-glucose. Although its major physiologic role is the generation of sorbitol in the renal inner medulla, aldose reductase is ubiquitous; it is found in lens, nerve, eye, adrenal, muscle, brain, liver, pancreas, intestine, ovary, salivary glands, blood vessels, and erythrocytes.[4] Its function in these other tissues is unclear; however, aldose reductase has been linked to the pathogenesis of diabetic retinopathy and neuropathy.[5]

1. Intrarenal Distribution

The intrarenal distribution of aldose reductase has been determined by immunocytochemistry, enzyme activity, northern blotting, and reverse transcription-polymerase chain reaction (RT-PCR). There is universal agreement that aldose reductase is present in the medulla but absent in the cortex.[9,21,23,29–37] In the older literature it was suggested that aldose reductase was present in the proximal tubule, but four recent studies have clarified this issue. Sands et al.[32] measured aldose reductase in individual microdissected rat nephron segments using a fluorescent microassay and showed that aldose reductase was present in initial and terminal IMCDs, thin ascending limbs, and the inner medullary portions of the thin descending limb. Although proximal tubules also displayed reductase activity, this reductase activity was shown to be aldehyde, not aldose, reductase.[32] Terubayashi et al.[33] corroborated these results when they found immunostaining for aldose reductase in the inner medulla. They also found low levels of aldose reductase in glomeruli, but only when large numbers of glomeruli were isolated by glomerular sieving and immunostained.[33]

The intrarenal distribution of aldose reductase mRNA shows a similar pattern. Using *in situ* hybridization, Bondy and Lightman[35] found aldose reductase mRNA only in the inner medulla. Moriyama et al.[34] used RT-PCR to identify aldose reductase mRNA in microdissected nephron segments and found it in IMCD subsegments, but not in proximal tubules. Interestingly, they could detect aldose reductase mRNA in glomeruli only if they used six or more glomeruli. They interpreted these results as evidence either for low abundance of the mRNA or that aldose reductase was only present in some cell types within the glomerulus.[34] Thus, aldose reductase mRNA, immunoreactive protein, and enzyme activity are located in all nephron segments within the inner medulla but not in cortical nephron segments. (Aldose reductase may be present in low abundance in the glomerulus.)

2. Regulation in Kidney Tubules

Aldose reductase activity increases with depth into the inner medulla in rats receiving food and water *ad libitum*,[7,32,38] in parallel with the increase in tissue sorbitol concentration (Figure 8–1). This pattern can be assigned to an increase in aldose reductase activity in all inner medullary nephron segments: thin descending limbs, thin ascending limbs, and IMCDs.[32]

a. Regulation by hydration status

If inner medullary sorbitol were regulated by production, aldose reductase activity should vary with hydration status. In general, this is true (Figure 8–2).[21,32] When a water diuresis is induced in rats for 3 d by feeding them 10% glucose in water, aldose reductase activity and mRNA abundance decrease.[7,37–39] Aldose reductase activity remains unchanged at 24 h,[7,38] whereas mRNA begins to decrease within 2 h,[21] because of the 6-d half-life.[40] These results suggest that during 3 d of water diuresis, transcription of the aldose reductase gene is downregulated. However, during an acute diuresis (<1 d), aldose reductase activity is unchanged. Thus, the ability to produce sorbitol is preserved during an acute diuresis.

Two models have been used to induce antidiuresis, administering vasopressin to Brattleboro rats and water deprivation of Sprague-Dawley rats. When Brattleboro rats are given vasopressin for 7 d, aldose reductase mRNA, protein, and activity all increase.[21] Depriving Sprague-Dawley rats of water for 1 d causes no change in aldose reductase activity;[7,38] however, the response to 3 d of water deprivation is controversial. Sands

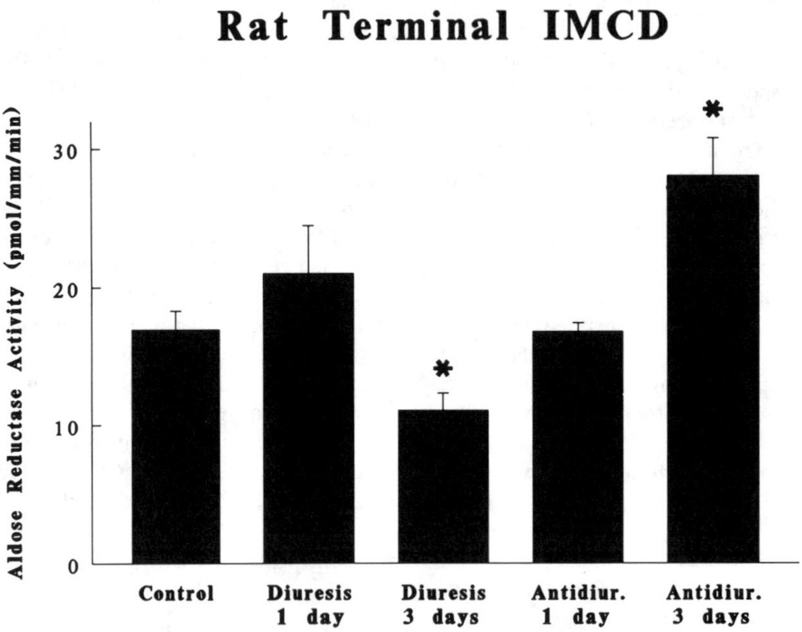

Figure 8–2 Aldose reductase activity in microdissected terminal IMCDs from water diuretic or water-restricted (antidiur.) rats compared to control rats receiving water *ad libitum.* Asterisk, $P < .05$ vs. control rats; $N = 6$ in all groups. (Data from Reference 7.)

and Schrader[7,38] found that 3 d of water deprivation caused a significant increase in aldose reductase activity in the deepest half of the IMCD (Figure 8–2). However, Cowley et al.[21] found that 3 d of water deprivation caused no change in aldose reductase mRNA, protein, or activity. One difference between these studies is that Cowley et al.[21] used inner medullary homogenates, whereas Sands and Schrader[7,38] used microdissected IMCDs. Recently, two preliminary studies have reported that 3 d of water deprivation does increase aldose reductase mRNA in inner medullary homogenates.[37,39] Thus, the majority of the data suggests that during a chronic (3 d) antidiuresis, aldose reductase mRNA abundance and enzyme activity are upregulated in the inner medulla. Presumably, the increase in aldose reductase results in an increase in sorbitol production.

It is important that aldose reductase remains functional over a wide range of intracellular ionic or osmotic compositions as may occur during changes in hydration (see Part I of this volume). Using permeabilized rat terminal IMCDs, Sands and Schrader simulated changes in intracellular ionic composition, osmolality, and pH, and measured aldose reductase activity.[41] Aldose reductase activity did not vary with concentration of Na^+, K^+, or Cl^-, or with osmolality, but did vary with pH (6.50 vs. 7.50) and with the concentration of sulfate.[41] Thus, the aldose reductase enzyme remains active over a wide range of simulated intracellular compositions and, presumably, during wide variation in hydration status.

b. Regulation during post-natal development

Newborn animals are unable to concentrate their urine.[42–45] Rats normally develop the ability to concentrate urine at 18 to 21 d of age (Figure 8–3).[42,46–48] With the development of urinary concentrating ability and of a hypertonic inner medulla, the question of whether the hypertonic inner medulla stimulates osmolyte production, or whether osmolyte production is induced prior to concentrating ability, can be addressed.

Recently, Schwartz et al.[42] investigated this problem by determining whether aldose reductase was induced before or after the appearance of urinary concentrating ability in rats. They found that aldose reductase mRNA abundance (by RT-PCR) was absent in newborns (<12 h old) but was detectable at 3 d of age and was maximal between 8 and 20 d; mRNA abundance decreased in adult rats (Figure 8–3).[42] Similar results were obtained by pooling tissue from several rat pups and performing northern analysis.[35,42]

To determine whether aldose reductase mRNA was transcribed into functional protein, Schwartz et al.[42] measured aldose reductase activity in microdissected terminal IMCDs from the same rat

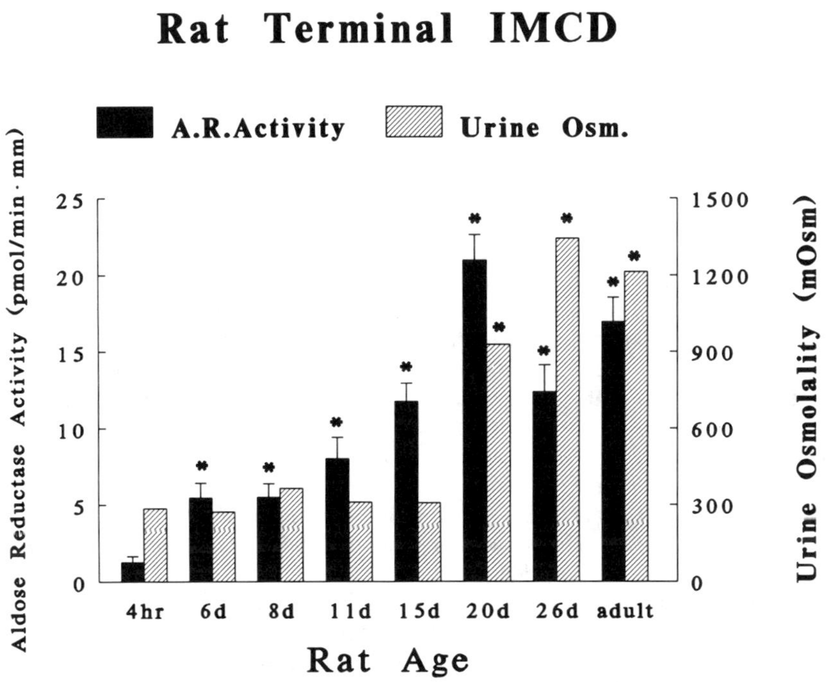

Figure 8–3 Aldose reductase activity in microdissected terminal IMCDs and urine osmolality (from the same rats) in neonatal rats receiving water *ad libitum*. Asterisk, $P < .05$ vs. control rats. (Data from Reference 42.)

pups. They found that enzyme activity was absent in newborns (<12 h old) but was detectable in 3-d-old rats and increased progressively up to 20 d of age; enzyme activity decreased in adult rats (Figure 8–3).[42] Thus, during development, aldose reductase is induced prior to concentrating ability, ensuring that at least one osmoprotective osmolyte, sorbitol, is present before the rat pup's inner medulla becomes hypertonic.

3. Osmotic Induction of Sorbitol in Cultured Cells

Cultured cells have been used widely to study the regulation and genetic control of sorbitol production. In the following section, insights gained from cultured cells will be reviewed, excluding studies dealing with genetic regulation, as this topic is covered elsewhere in this book.

PAP-HT25 cells are a cell line derived from rabbit papilla. They can grow in culture media in which the osmolality has been increased to 600 mOsmol by the addition of NaCl.[5] Increasing media osmolality from 300 to 600 mOsmol by adding NaCl induced sorbitol accumulation; 4 d were required before maximal sorbitol concentrations were achieved.[40,49,50] Sorbitol accumulation was also induced by adding raffinose, but not by adding urea or glycerol.[50] Decreasing the culture media osmolality from 600 to 300 mOsmol resulted in rapid sorbitol efflux from the cultured cells, with approximately 50% of the intracellular sorbitol appearing in the culture media within 15 min.[40]

The sorbitol accumulation in response to hypertonic culture media in PAP-HT25 cells is mediated by increases in aldose reductase activity.[40,49,50] The abundance of aldose reductase mRNA increased after 6 h in hypertonic media, reached a maximum at 18 to 24 h, then gradually decreased.[29] Although synthesis of aldose reductase protein increased in parallel with the increase in aldose reductase mRNA,[51] increased aldose reductase activity was not detectable until 1 d after increasing media osmolality[50] and intracellular sorbitol concentration was not increased for 2 d.[50] Aldose reductase activity reached a maximum after 3 d, whereas sorbitol concentration increased for up to 1 week.[50] When sorbitol synthesis was inhibited by adding the aldose reductase inhibitor Tolrestat, the rate of aldose reductase synthesis continued to increase beyond 24 h, consistent with the conclusion that an increased intracellular sorbitol concentration contributes to downregulating the rate of aldose reductase protein synthesis.[51]

PAP-HT25 cells are not the only cell line in which hypertonicity induces sorbitol accumulation via aldose reductase. Similar responses have been found in Chinese hamster ovary (CHO) cells,[52] rat kidney mesangial cells,[52] and dog lens epithelial cells.[53] Interestingly, MDCK cells, a dog kidney cell line thought to originate from a distal nephron cortical segment, did not accumulate sorbitol in response to hypertonic media.[54]

D. SORBITOL DEHYDROGENASE

Sorbitol dehydrogenase catalyzes the degradation of sorbitol to D-fructose. In rats receiving food and water *ad libitum,* sorbitol dehydrogenase activity was found to be quite low in all medullary collecting duct subsegments, but was 20-fold higher in proximal tubule segments.[7,32] Sands et al.[7,38] investigated whether sorbitol dehydrogenase might contribute to the regulation of intracellular sorbitol by measuring changes in the activity of the enzyme in response to manipulation of the hydration status in rats. When rats were given sugar water for 3 d to induce a water diuresis, sorbitol dehydrogenase activity doubled in terminal IMCDs; there was no change after only 1 d of water diuresis (Figure 8–4).[7,38] When water was restricted for 3 d, there was no change in sorbitol dehydrogenase activity (Figure 8–4).[7,38] This result was not surprising, considering the low levels of sorbitol dehydrogenase activity present in IMCDs from rats receiving food and water *ad libitum.*[7,38]

It is interesting to compare the time course of changes in aldose reductase and sorbitol dehydrogenase in microdissected rat terminal IMCDs. The activities of both enzymes were unchanged after 1 d of water diuresis or antidiuresis, but the activity of both enzymes changed after 3 d of water diuresis.[7,38] The interesting observation is that the changes were in opposite directions.[7,38] Following 3 d of water restriction, aldose reductase activity increased, whereas sorbitol dehydrogenase activity remained quite low. Thus, there is coordinated regulation of aldose reductase and sorbitol dehydrogenase during changes in hydration status *in vivo.*

PAP-HT25 cells have barely detectable levels of sorbitol dehydrogenase, and reducing media osmolality does not increase sorbitol dehydrogenase activity.[40] Thus, while sorbitol degradation increased during diuresis in the renal inner medulla, an appropriate cultured cell line for studying its regulation *in vitro* has not been reported.

E. SORBITOL EFFLUX

Sorbitol permeability, hence sorbitol efflux from cells, can change rapidly in response to decreases in extracellular osmolality, unlike the slower time course of changes in sorbitol production and degradation. When PAP-HT25 cells were maintained at constant osmolality, sorbitol permeability was quite low.[40] However, sorbitol permeability increased 80-fold within 15 min following reduction

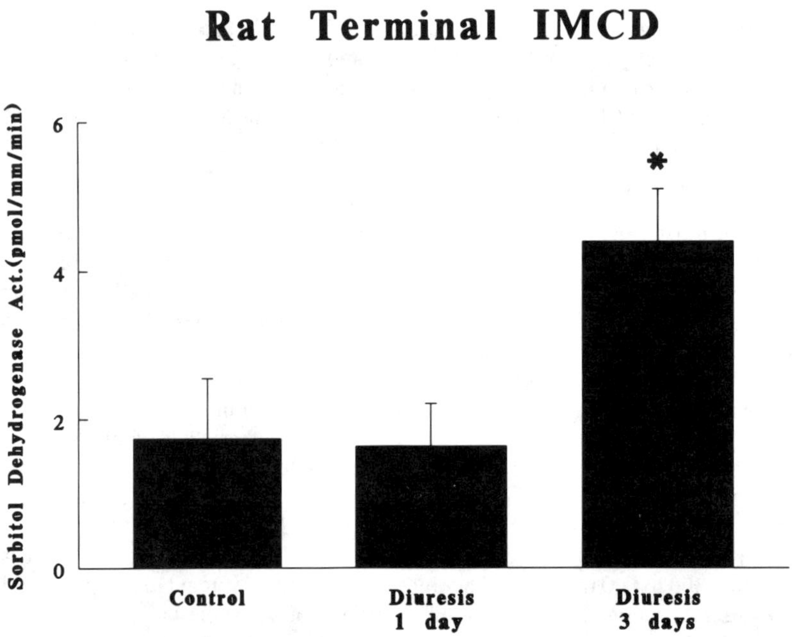

Figure 8–4 Sorbitol dehydrogenase activity in microdissected terminal IMCDs from water diuretic rats compared to control rats receiving water *ad libitum.* Asterisk, $P < .05$ vs. control rats. $N = 6$ in all groups. (Data from Reference 7.)

in media osmolality, resulting in the loss of 50% of the intracellular sorbitol.[40,55] Sorbitol permeability increased within 30 s after decreasing media osmolality, and decreased within 30 s of restoration of the original osmolality.[55]

Sorbitol transport is mediated by sorbitol permease.[55,56] This transporter is located in the apical membrane of PAP-HT25 cells and does not display saturation kinetics at sorbitol concentrations up to 315 m*M;* higher concentrations have not been studied.[55,56] Sorbitol permease activity was inhibited by quinidine, eicosa-5,8,11,14-tetraynoic acid, or cytochrome P-450 inhibitors (ketoconazole and proadifen).[57]

Suspensions of rat IMCDs and of papillary cells display changes in sorbitol permeability similar to those in PAP-HT25 cells.[12,58,59] Calcium and calmodulin are apparently involved in the sorbitol efflux pathway of IMCDs, since removing extracellular calcium while adding EGTA and trifluoperazine from solutions containing suspensions of IMCDs inhibited sorbitol efflux, while A23187 stimulated sorbitol efflux.[59] In PAP-HT25 cells, removal of calcium from the media plus addition of EGTA inhibited sorbitol efflux; however, many of the cells were lifted off the culture dish under these conditions.[55] Trifluoperazine and A23187 had no effect on sorbitol efflux in PAP-HT25 cells.[2] Whether this discrepancy between rat IMCDs and cultured cells is due to more than one transporter, to differences in the second messenger pathways, or to differences between kidney tubules and cultured cells, remains to be determined.

F. PHYSIOLOGIC ROLE

Terrestrial animals do not consume water continuously, so their kidneys should function chronically to conserve water but have the capacity to respond quickly to water ingestion. The sorbitol regulatory mechanisms appear to fit these requirements. The long half-life of aldose reductase, coupled with the time required to change either sorbitol production or degradation, ensures that sorbitol is continuously synthesized. When water is ingested, sorbitol can be transported quickly into the urine by sorbitol permease. When water ingestion ends, sorbitol efflux can be stopped quickly, and intracellular sorbitol rapidly regenerated as sorbitol production and degradation would not change. However, if the change in hydration status is prolonged for several days, then coordinated regulation of aldose reductase and sorbitol dehydrogenase can occur to change sorbitol production and degradation and regulate cell sorbitol concentration.

This model does not provide a mechanism for rapidly regenerating sorbitol following chronic water loading, as would be necessary in a patient with psychogenic polydipsia who stops ingesting water. Such patients are unable to concentrate their urine for several days following cessation of water ingestion due to "medullary washout" even though their glomerular filtration rate is normal. One may speculate that, as in neonatal development of urinary concentrating ability, the kidney is "programmed" to delay concentrating ability until sorbitol synthetic capacity is reestablished to prevent osmotic damage.

III. INOSITOL

A. METABOLIC PATHWAYS

Inositol, or *myo*-inositol, is synthesized intracellularly from D-glucose or from inositol phospholipids.[2] The reaction using D-glucose involves three enzymes: hexokinase, L-*myo*-inositol-1-phosphate synthase, and *myo*-inositol-1-phosphatase, whereas the reaction using inositol phospholipids only involves the enzyme *myo*-inositol-1-phosphatase.[2] Inositol is degraded by *myo*-inositol oxygenase to form D-glucuronic acid.[2]

Cells take up inositol via an electrogenic sodium-inositol cotransporter.[60] The cDNA for this transporter was cloned recently[61] and will be discussed in more detail in Chapter 24. Secondary active transport of inositol occurs in kidney, red blood cells, intestine, nerve, pancreas, and brain,[2] permitting these cells to accumulate inositol against a concentration gradient.

B. INTRARENAL DISTRIBUTION

Tissue inositol concentration is higher in the renal medulla than cortex (Figure 8–1).[3,11,24] Unlike sorbitol, the maximal inositol concentration is found in outer medullary tissue and decreases progressively in inner medullary tissue from the base to the papillary tip (Figure 8–1).[3,11,24] However, when inositol concentration was measured in microdissected tubules, inositol concentration increased progressively along the collecting duct from cortex to inner medulla.[24] The medullary thick ascending limb also had high concentrations of inositol,[24] which may account for outer medullary tissue having a higher tissue inositol concentration than inner medullary tissue.

Despite higher concentrations of inositol in outer medullary tissue, inositol concentration varies with hydration status only in inner medullary tissue. Following acute water diuresis in dogs, inner medullary, but not outer medullary, inositol concentration decreased significantly.[62] In vasopressin-treated Brattleboro rats, inner medullary inositol concentration increased after 7 to 12 d of vasopressin, but not after 3 d of vasopressin.[16,20,21,24] Outer medullary inositol concentration also increased by 30%, but the change was not statistically significant.[16] Finally, neither inner nor outer medullary tissue inositol concentration changed after hypertonic saline diuresis.[27]

In summary, tissue inositol concentration was higher in medulla than in cortex, and it changed with hydration status in the inner medulla in most studies. Its distribution and response to changes in hydration were quite different from other osmolytes, such as sorbitol (described above), betaine and glycerylphosphorylcholine (see previous chapters), raising the possibility that inositol is involved in more than osmoregulation. Currently, there are no data regarding what these other roles or regulators could be.

C. ROLE IN CENTRAL NERVOUS SYSTEM

Inositol plays an important role in the response of brain cells to hypernatremia (dehydration) and is one of the so-called "idiogenic osmoles".[63] Unlike sorbitol, inositol is accumulated in rat brain cells in response to chronic hypernatremia.[63–67] Inositol is accumulated and released slowly, thereby preventing brain cells from releasing inositol quickly if hypernatremia is rapidly corrected by rehydration.[66,67] For example, in hypernatremic rats and mice, brain inositol was retained for over 4 h after rapid rehydration.[66,67] Retention of inositol and other "idiogenic osmoles" is thought to play a role in the rebound cerebral edema that follows overly rapid correction of hypernatremia.

During hyponatremia (water overload), there is loss of ions and organic osmolytes, including inositol, from brain cells.[63,68,69] During rapid correction of hyponatremia, brain cells accumulate ions, as there is insufficient time to regenerate inositol and other "idiogenic osmoles".[63,68,69] Reaccumulation of organic osmolytes following hyponatremia does not occur with a uniform time course within the brain.[70] The pons, midbrain, and striatum reaccumulate inositol more slowly than other regions of the brain.[70] This slower recovery may explain the syndrome of central pontine myelinolysis which is observed in patients whose hyponatremia is corrected too rapidly.[70]

D. OSMOTIC INDUCTION OF INOSITOL IN CULTURED CELLS

Cultured cells have been a useful model system for studying the regulation of inositol and for expression cloning of the sodium-inositol cotransporter,[61] which will be discussed in detail in Chapter 24. Cultured C6 glioma cells have been useful for studying inositol accumulation in the brain during hypernatremia.[63,71,72]

When the culture medium was made hypertonic by adding NaCl, inositol concentration increased in the following cultured cell lines: MDCK cells,[54,73–75] PAP-HT25 cells,[76] LLC-PK_1 and LLC-PK_3 cells,[54] GRB-MAL1 cells,[54] rat IMCD cells,[77] and C6 glioma cells.[71,72,78] Inositol concentration increased in MDCK cells when osmolality was increased by adding NaCl, NaCl plus urea, raffinose, sorbitol, or dextrose, but not when urea alone or glycerol was added.[54,73,79]

When either MDCK or PAP-HT25 cells were switched from isotonic culture media to culture media made hypertonic by the addition of NaCl, they accumulated inositol slowly, requiring ≥1 week before maximal intracellular inositol concentrations were achieved.[76,80] However, when these same cells were switched from hypertonic to isotonic culture media, inositol concentration decreased rapidly, with about 15% of cell inositol leaving within the first 15 min after decreasing media osmolality.[55,72,76] Thus, the time course of change in intracellular inositol concentration was similar to that for sorbitol.

The increase in intracellular inositol concentration only occurred when inositol was present in the culture medium, consistent with the conclusion that intracellular inositol concentration was increased by transport rather than by synthesis.[73] In MDCK cells exposed to hypertonic culture media, sodium-inositol cotransport increased across the basolateral, but not the apical, membrane, due to changes in V_{max}, rather than K_m, of the cotransporter.[73,81] As intracellular inositol concentration increased, inositol transport decreased.[73,81] Thus, inositol exerted negative feedback upon its own transport.

Cultured C6 glioma cells also accumulated inositol when culture medium osmolality was increased by adding NaCl.[71,72,78] Intracellular inositol content began to increase after 10 h in hypertonic media and reached a maximum after 24 h.[71] This increase was due to a three- to fourfold increase in inositol transport.[71] Interestingly, C6 glioma cells,[72] unlike cultured renal cells,[55,76] lose inositol slowly when placed in a hypotonic culture medium. This slow loss of inositol appeared to be the rate-limiting step in cell volume regulation in C6 glioma cells.[72] Thus, this cultured brain cell line behaves like brain cells regarding the time course of changes in inositol content following changes in osmolality. Future studies using C6 glioma cells may provide interesting insights into mechanisms by which the brain responds to changes in hydration.

ACKNOWLEDGMENTS

This work was supported by National Institutes of Health grant R29-DK41707. JMS performed this work during the tenure of an Established Investigatorship from the American Heart Association. The author thanks Drs. W. E. Mitch and S. R. Price for their critical reading of this manuscript.

REFERENCES

1. **Chamberlin, M. E. and Strange, K.,** Anisosmotic cell volume regulation: a comparative view, *Am. J. Physiol.*, 257, C159, 1989.
2. **Garcia-Perez, A. and Burg, M. B.,** Renal medullary organic osmolytes, *Physiol. Rev.*, 71, 1081, 1991.
3. **Bagnasco, S. M., Balaban, R. S., Fales, H. M., Yang, Y. M., and Burg, M. B.,** Predominantly osmotically active organic solutes in rat and rabbit renal medullas, *J. Biol. Chem.*, 261, 5872, 1986.
4. **Burg, M. B.,** Role of aldose reductase and sorbitol in maintaining the medullary intracellular milieu, *Kidney Int.*, 33, 635, 1988.
5. **Burg, M. B. and Kador, P. F.,** Sorbitol, osmoregulation, and the complications of diabetes, *J. Clin. Invest.*, 81, 635, 1988.
6. **Balaban, R. S. and Burg, M. B.,** Osmotically active organic solutes in the renal inner medulla, *Kidney Int.*, 31, 562, 1987.
7. **Sands, J. M. and Schrader, D. C.,** Coordinated response of renal medullary enzymes regulating net sorbitol production in diuresis and antidiuresis, *J. Am. Soc. Nephrol.*, 1, 58, 1990.
8. **Chauncey, B., Leite, M. V., and Goldstein, L.,** Renal sorbitol accumulation and associated enzyme activities in diabetes, *Enzyme*, 39, 231, 1988.
9. **Corder, C. N., Collins, J. G., Brannan, T. S., and Sharma, J.,** Aldose reductase and sorbitol dehydrogenase distribution in rat kidney, *J. Histochem. Cytochem.*, 25, 1, 1977.
10. **Oates, P. J. and Goddu, K. J.,** A sorbitol gradient in rat renal medulla, *Kidney Int.*, 31 (Abstr.), 448, 1987.
11. **Yancey, P. H. and Burg, M. B.,** Distribution of major organic osmolytes in rabbit kidneys in diuresis and antidiuresis, *Am. J. Physiol.*, 257, F602, 1989.
12. **Wirthensohn, G., LeFrank, S., Schmolke, M., and Guder, W. G.,** Regulation of organic osmolyte concentrations in tubules from rat renal inner medulla, *Am. J. Physiol.*, 256, F128, 1989.
13. **Wirthensohn, G., LeFrank, S., Guder, W. G., and Beck, F. X.,** Studies on the role of glycerophosphorylcholine and sorbitol in renal osmoregulation, in *Molecular Nephrology. Biochemical Aspects of Kidney Function*, Kovacevic, Z. and Guder, W. G., Eds., Walter de Gruyter, Berlin, 1987, 321.
14. **Schmolke, M., Beck, F. X., and Guder, W. G.,** Effect of antidiuretic hormone on renal organic osmolytes in Brattleboro rats, *Am. J. Physiol.*, 257, F732, 1989.
15. **Hutton, J. C., Schofield, P. J., Williams, J. F., and Hollows, F. C.,** The localization of sorbitol pathway activity in the rat renal cortex and its relationship to the pathogenesis of the renal complications of diabetes, *Aust. J. Biol. Med. Sci.*, 53, 49, 1975.
16. **Schmolke, M., Beck, F., and Guder, W.,** Effect of antidiuretic hormone on renal organic osmolytes in Brattleboro rats, *Am. J. Physiol.*, 257, F732, 1989.
17. **Schmolke, M. and Guder, W. G.,** Metabolic regulation of organic osmolytes in tubules from rat renal inner and outer medulla, *Renal Physiol. Biochem.*, 12, 347, 1989.
18. **Schmolke, M., Bornemann, A., and Guder, W. G.,** Polyol determination along the rat nephron, *Biol. Chem. Hoppe-Seyler*, 371, 909, 1990.
19. **Yancey, P. H.,** Osmotic effectors in kidneys of xeric and mesic rodents: corticomedullary distribution and changes with water availability, *J. Comp. Physiol. B*, 158, 369, 1988.
20. **Blumenfeld, J. D., Hebert, S. C., Heilig, C. W., Balschi, J. A., Stromski, M. E., and Gullans, S. R.,** Organic osmolytes in inner medulla of Brattleboro rat: effects of ADH and dehydration, *Am. J. Physiol.*, 256, F916, 1989.
21. **Cowley, B. D., Ferraris, D., Carper, D., and Burg, M. B.,** *In vivo* osmoregulation of aldose reductase mRNA, protein, and sorbitol in renal medulla, *Am. J. Physiol.*, 258, F154, 1990.
22. **Gullans, S. R., Blumenfeld, J. D., Balschi, J. A., Kaleta, M., Brenner, R. M., Heilig, C. W., and Hebert, S. C.,** Accumulation of major organic osmolytes in renal inner medulla in dehydration, *Am. J. Physiol.*, 255, F626, 1988.
23. **Boulanger, Y., Legault, P., Tejedor, A., Vinay, P., and Theriault, Y.,** Biochemical characterization and osmolytes in papillary collecting ducts from pig and dog kidneys, *Can. J. Physiol. Pharmacol.*, 66, 1282, 1988.
24. **Guder, W. G., Beck, F. X., and Schmolke, M.,** Regulation and localization of organic osmolytes in mammalian kidney, *Klin. Wochenschr.*, 68, 1091, 1990.

25. **Wolff, S. D., Stanton, T. S., James, S. L., and Balaban, R. S.,** Acute regulation of the predominant organic solutes of the rabbit renal inner medulla, *Am. J. Physiol.,* 257, F676, 1989.
26. **Beck, F. X., Schmolke, M., Guder, W. G., Dörge, A., and Thurau, K.,** Osmolytes in renal medulla during rapid changes in papillary tonicity, *Am. J. Physiol.,* 262, F849, 1992.
27. **Heilig, C. W., Stromski, M. E., and Gullans, S. R.,** Methylamine and polyol responses to salt loading in renal inner medulla, *Am. J. Physiol.,* 257, F1117, 1989.
28. **Wald, H., Scherzer, P., and Popovtzer, M. M.,** Inhibition of thick ascending limb Na^+-K^+-ATPase activity in salt loaded rats by furosemide, *Am. J. Physiol.,* 256, F549, 1989.
29. **Garcia-Perez, A., Martin, B., Murphy, H. R., Uchida, S., Murer, H., Cowley, B. D., Handler, J. S., and Burg, M. B.,** Molecular cloning of cDNA coding for kidney aldose reductase. Regulation of specific mRNA accumulation by high NaCl-mediated osmotic stress, *J. Biol. Chem.,* 264, 16815, 1989.
30. **Kern, T. S. and Engerman, R. L.,** Immunohistochemical distribution of aldose reductase, *Histochem. J.,* 14, 507, 1982.
31. **Ludvigson, M. A. and Sorenson, R. L.,** Immunohistochemical localization of aldose reductase. Rat eye and kidney, *Diabetes,* 29, 450, 1980.
32. **Sands, J. M., Terada, Y., Bernard, L. M., and Knepper, M. A.,** Aldose reductase activities in microdissected rat renal tubule segments, *Am. J. Physiol.,* 256, F563, 1989.
33. **Terubayashi, H., Sato, C., Nishimura, C., Kador, P. F., and Kinoshita, J. H.,** Localization of aldose and aldehyde reductase in the kidney, *Kidney Int.,* 36, 843, 1989.
34. **Moriyama, T., Murphy, H. R., Martin, B. M., and Garcia-Perez, A.,** Detection of specific mRNAs in single nephron segments by use of the polymerase chain reaction, *Am. J. Physiol.,* 258, F1470, 1990.
35. **Bondy, C. A. and Lightman, S. L.,** Developmental and physiological regulation of aldose reductase mRNA expression in renal medulla, *Mol. Endocrinol.,* 3, 1409, 1989.
36. **Gabbay, K. H. and O'Sullivan, J. B.,** The sorbitol pathway. Enzyme localization and content in normal and diabetic nerve and cord, *Diabetes,* 17, 239, 1968.
37. **Price, S. R., Hoban, C. A., Isozaki, T., and Sands, J. M.,** Effects of hydration on sorbitol dehydrogenase and aldose reductase mRNAs in rat kidney, *J. Am. Soc. Nephrol.,* 3 (Abstr.), 334, 1992.
38. **Sands, J. M. and Schrader, D. C.,** Coordinated response of aldose reductase and sorbitol dehydrogenase regulates net sorbitol production in the renal inner medulla, in *Cellular and Molecular Biology of the Kidney, Contrib. Nephrol.,* Vol. 95, Koide, H., Endou, H., and Kurokawa, K., Eds., S. Karger, Basel, 1991, 272.
39. **Nakanishi, T., Uyama, O., Yamamoto, S., Ueda, N., and Sugita, M.,** In vivo renal medullary aldose reductase mRNA is regulated by Na and K, *J. Am. Soc. Nephrol.,* 2 (Abstr.), 765, 1991.
40. **Bagnasco, S. M., Murphy, H. R., Bedford, J. J., and Burg, M. B.,** Osmoregulation by slow changes in aldose reductase and rapid changes in sorbitol flux, *Am. J. Physiol.,* 254, C788, 1988.
41. **Sands, J. M. and Schrader, D. C.,** Acute changes in intracellular ions or pH and regulation of aldose reductase activity, *J. Am. Soc. Nephrol.,* 2, 212, 1991.
42. **Schwartz, G. J., Zavilowitz, B. J., Radice, A. D., Garcia-Perez, A., and Sands, J. M.,** Maturation of aldose reductase expression in the neonatal rat inner medulla, *J. Clin. Invest.,* 90, 1275, 1992.
43. **Calcagno, P. L., Rubin, M. I., and Weintraub, D. H.,** Studies on the renal concentrating and diluting mechanisms in the premature infant, *J. Clin. Invest.,* 33, 91, 1954.
44. **Edelmann, C. M., Jr., Barnett, H. L., and Troupkou, V.,** Renal concentrating mechanisms in newborn infants. Effect of dietary protein and water content, role of urea, and responsiveness to anti-diuretic hormone, *J. Clin. Invest.,* 39, 1062, 1960.
45. **Winberg, J.,** Determination of renal concentrating capacity in infants and children without renal disease, *Acta Paediatr.,* 48, 318, 1959.
46. **Falk, G.,** Maturation of renal function in infant rats, *Am. J. Physiol.,* 181, 157, 1955.
47. **Heller, H.,** Effects of dehydration on adult and newborn rats, *J. Physiol.,* 108, 303, 1949.
48. **McCrory, W. W., Jr.,** *Developmental Nephrology,* Harvard University Press, Cambridge, MA, 1972, 123.
49. **Bagnasco, S. M., Uchida, R. S., Balaban, R. S., Kador, P. F., and Burg, M. B.,** Induction of aldose reductase and sorbitol in renal inner medullary cells by elevated extracellular NaCl, *Proc. Natl. Acad. Sci. U.S.A.,* 84, 1718, 1987.
50. **Uchida, S., Garcia-Perez, A., Murphy, H., and Burg, M.,** Signal for induction of aldose reductase in renal medullary cells by high external NaCl, *Am. J. Physiol.,* 256, C614, 1989.

51. **Moriyama, T., Garcia-Perez, A., and Burg, M. B.,** Osmotic regulation of aldose reductase protein synthesis in renal medullary cells, *J. Biol. Chem.,* 264, 16810, 1989.
52. **Kaneko, M., Carper, D., Nishimura, C., Millen, J., Bock, M., and Hohman, T. C.,** Induction of aldose reductase expression in rat kidney mesangial cells and Chinese hamster ovary cells under hypertonic conditions, *Exp. Cell Res.,* 188, 135, 1990.
53. **Carper, D., Kaneko, M., Stark, H., and Hohman, T.,** Increase in aldose reductase mRNA in dog lens epithelial cells under hypertonic conditions, *Exp. Eye Res.,* 50, 743, 1990.
54. **Nakanishi, T., Balaban, R. S., and Burg, M. B.,** Survey of osmolytes in renal cell lines, *Am. J. Physiol.,* 255, C181, 1988.
55. **Siebens, A. W. and Spring, K. R.,** A novel sorbitol transport mechanism in cultured renal papillary epithelial cells, *Am. J. Physiol.,* 257, F937, 1989.
56. **Garty, H., Furlong, T. J., Ellis, D. E., and Spring, K. R.,** Sorbitol permease: an apical membrane transporter in cultured renal papillary epithelial cells, *Am. J. Physiol.,* 260, F650, 1991.
57. **Furlong, T. J., Moriyama, T., and Spring, K. R.,** Activation of osmolyte efflux from cultured renal papillary epithelial cells, *J. Membr. Biol.,* 123, 269, 1991.
58. **Grunewald, R. W. and Kinne, R. K. H.,** Intracellular sorbitol content in isolated rat inner medullary collecting duct cells. Regulation by extracellular osmolarity, *Pfluegers Arch.,* 414, 178, 1989.
59. **Bevan, C., Theiss, C., and Kinne, R. K. H.,** Role of Ca^{2+} in sorbitol release from rat inner medullary collecting duct (IMCD) cells under hypoosmotic stress, *Biochem. Biophys. Res. Commun.,* 170, 563, 1990.
60. **Hammerman, M. R., Sacktor B., and Daughaday, W. H.,** Myo-inositol transport in renal brush border vesicles and its inhibition by D-glucose, *Am. J. Physiol.,* 239, F113, 1980.
61. **Kwon, H. M., Yamauchi, A., Uchida, S., Preston, A. S., Garcia-Perez, A., Burg, M. B., and Handler, J. S.,** Cloning of the cDNA for a Na^+/myo-inositol co-transporter, a hypertonicity stress protein, *J. Biol. Chem.,* 267, 6297, 1992.
62. **Cohen, M. A., Hruska, K. A., and Daughaday, W. H.,** Free myo-inositol in canine kidney: selective concentration in the renal medulla, *Proc. Soc. Exp. Biol. Med.,* 169, 380, 1982.
63. **Strange, K.,** Regulation of solute and water balance and cell volume in the central nervous system, *J. Am. Soc. Nephrol.,* 3, 12, 1992.
64. **Heilig, C. W., Stromski, M. E., Blumenfeld, J. D., Lee, J. P., and Gullans, S. R.,** Characterization of the major brain osmolytes that accumulate in salt-loaded rats, *Am. J. Physiol.,* 257, F1108, 1989.
65. **Lohr, J. W., McReynolds, J., Grimaldi, T., and Acara, M.,** Effect of acute and chronic hypernatremia on myoinositol and sorbitol concentration in rat brain and kidney, *Life Sci.,* 43, 271, 1988.
66. **Lien, Y.-H. H., Shapiro, J. L., and Chan, L.,** Effects of hypernatremia on organic brain osmoles, *J. Clin. Invest.,* 85, 1427, 1990.
67. **Thurston, J. H., Sherman, W. R., Hauhart, R. E., and Kloepper, R. F.,** Myo-inositol: a newly identified nonnitrogenous osmoregulatory molecule in mammalian brain, *Pediatr. Res.,* 26, 482, 1989.
68. **Verbalis, J. G. and Gullans, S. R.,** Hyponatremia causes large sustained reductions in brain content of multiple organic osmolytes in rats, *Brain Res.,* 567, 274, 1991.
69. **Lien, Y.-H. H., Shapiro, J. I., and Chan, L.,** Study of brain electrolytes and organic osmolytes during correction of chronic hyponatremia, *J. Clin. Invest.,* 88, 303, 1991.
70. **Lai, L.-W. and Lien, Y.-H.,** Regional analysis of brain organic osmolytes: effect of hyponatremia, *J. Am. Soc. Nephrol.,* 2 (Abstr.), 764, 1991.
71. **Strange, K., Morrison, R., Heilig, C. W., DiPietro, S., and Gullans, S. R.,** Upregulation of inositol transport mediates inositol accumulation in hyperosmolar brain cells, *Am. J. Physiol.,* 260, C784, 1991.
72. **Strange, K. and Morrison, R.,** Volume regulation during recovery from chronic hypertonicity in brain glial cells, *Am. J. Physiol.,* 263, C412, 1992.
73. **Nakanishi, T., Turner, R. J., and Burg, M. B.,** Osmoregulatory changes in myo-inositol transport by renal cells, *Proc. Natl. Acad. Sci. U.S.A.,* 86, 6002, 1989.
74. **Heilig, C. W., Brenner, R. M., Yu, A. S. L., Kone, B. C., and Gullans, S. R.,** Modulation of osmolytes in MDCK cells by solutes, inhibitors, and vasopressin, *Am. J. Physiol.,* 259, F653, 1990.
75. **Shayman, J. A. and Wu, D.,** Myo-inositol does not modulate PI turnover in MDCK cells under hyperosmolar conditions, *Am. J. Physiol.,* 258, F1282, 1990.

76. **Moriyama, T., Garcia-Perez, A., and Burg, M. B.,** Factors affecting the ratio of different organic osmolytes in renal medullary cells, *Am. J. Physiol.,* 259, F847, 1990.
77. **Veis, J. H., Molitoris, B. A., Teitelbaum, I., Mansour, J. A., and Berl, T.,** *Myo*-inositol uptake by rat cultured inner medullary collecting tubule cells: effect of osmolality, *Am. J. Physiol.,* 260, F619, 1991.
78. **Lien, Y.-H. H.,** Differential regulation of organic osmolytes in brain cells cultured in isotonic and hypertonic media, *Clin. Res.,* 40 (Abstr.), 112A, 1992.
79. **Nakanishi, T., Turner, R. J., and Burg, M. B.,** Osmoregulation of betaine transport in mammalian renal medullary cells, *Am. J. Physiol.,* 258, F1061, 1990.
80. **Nakanishi, T. and Burg, M. B.,** Osmoregulatory fluxes of myo-inositol and betaine in renal cells, *Am. J. Physiol.,* 257, C964, 1989.
81. **Yamauchi, A., Kwon, H. M., Uchida, S., Preston, A. S., and Handler, J. S.,** *Myo*-inositol and betaine transporters regulated by tonicity are basolateral in MDCK cells, *Am. J. Physiol.,* 261, F197, 1991.

Part III
Volume Sensing and Signal Transduction

Chapter 9

Cellular Signal Transduction Mechanisms

Thomas Force and Joseph V. Bonventre

TABLE OF CONTENTS

I. INTRODUCTION

Hypertonic or hypotonic challenge produces marked changes in the activity of various ion transport systems involved in cell volume control. The signal transduction mechanisms activated by a change in cell volume, which, in turn, produce the regulatory volume decrease (RVD) or regulatory volume increase (RVI) responses, are not well defined. Mechanisms postulated to play a role include increases in cytosolic free [Ca^{2+}] ([Ca^{2+}]$_i$), Ca^{2+}-induced calmodulin binding to a regulatory protein, changes in cAMP concentration, production of various eicosanoids, release of inositol phosphates by phospholipase C, activation of several protein kinases, and rearrangements of components of the cytoskeleton (reviewed in Reference 1). In most cases, although production of one or more of these signal-transducing molecules is correlated with the change in cell volume produced by hypotonic or hypertonic challenge, the mechanisms by which the change in cell volume initiates activation of a signaling pathway and how that pathway results in activation of the appropriate volume-regulatory mechanism, are unknown.

0-8493-4448-4/94/$0.00+$.50

Less dramatic, but clearly important, changes in the activity of the various ion transport systems involved in cell volume regulation also occur when growth factors or hormones bind to their cell surface receptors. The signal transduction pathways activated by agonist binding are, in general, much better understood than those activated by hypotonic or hypertonic challenge. In many cases, however, the link between these ligand-activated pathways and the volume-regulatory mechanisms are unknown.

It is the purpose of this chapter to explore the signal transduction mechanisms that may be important for cellular volume regulation. These signaling pathways may be activated by hypotonic or hypertonic challenge. We will draw heavily from studies designed to characterize and examine the regulation of the pathways activated by ligand-receptor interactions. Similar paradigms are likely to characterize signaling events important for cell volume regulation. We will first classify cell surface receptors into one of five broad categories based on their proximal signal transduction mechanisms. We will then discuss in more detail the early signal transduction events of two of the classes, the G protein-linked receptor and the growth factor receptor with intrinsic tyrosine kinase activity. We will explore the role of small-molecular-weight GTP binding proteins and heterotrimeric G proteins in the propagation of the signal from both classes of receptors. Finally, we will explore more distal signal transduction mechanisms activated by these two classes of receptors, involving enzymes such as phospholipases C, phospholipases A_2 and the eicosanoids, the serine/threonine protein kinases, in particular, protein kinase C and the mitogen-activated protein (MAP) kinase cascade, and the p85/phosphatidylinositol-3-kinase. We will discuss how the various signaling pathways activated by the G protein-linked hormones and the growth factors converge and diverge. We will attempt to link these pathways to the various volume control mechanisms of the cell, but in most cases, the discussion will, of necessity, highlight areas of uncertainties rather than provide definitive answers. Because of limitations in space, we will not discuss very distal steps in the signaling cascades — steps that result in modulation of transcription factor activity, so important for complex cellular responses, such as proliferation and differentiation.

II. CLASSIFICATION AND FUNCTION OF RECEPTORS

When a ligand binds to its receptor on the cell surface, the resulting conformational change in the receptor initiates a series of intracellular events that culminate in a characteristic response by the cell. That response is often ligand-specific and may be cell type-specific. Ligand binding is translated into a signal transduction "profile" in the cell. That profile is determined by the effectors with which the receptor interacts.

We have divided cell surface receptors into five broad classes based on their structure and proximal mechanisms of signal transduction (Table 1). The first class, growth factor receptors with intrinsic tyrosine kinase activities, are characterized by a cytoplasmic tyrosine kinase domain that is activated by autophosphorylation when ligand binding induces receptor dimerization (reviewed in Reference 2). Activation of this kinase initiates signal transduction in the cell.

A large number of agonists bind to a second class of receptors that are linked to guanine nucleotide-binding regulatory proteins (G proteins). More than 100 different G protein-linked receptors have been identified.[3] Signal transduction is initiated when hormone binding to the receptor causes activation of the α-subunit of the heterotrimeric G protein,[3] which then activates intracellular effectors including a variety of enzymes and ion channels (see below).

The third class of receptors are those for transforming growth factor (TGF)-β and the activins.[4,5] These receptors signal through a novel mechanism. The receptors, like the growth factor receptors, contain a cytoplasmic kinase domain, but the kinase appears to be a serine/threonine kinase. Full activation of the TGF-β signaling pathways appears to require association of the type II receptor, containing the kinase domain, with another TGF-β binding protein, called the type I receptor.[4]

The hematopoietic receptor family comprises the fourth class of receptors. The ligands for these receptors are all growth and/or differentiation factors for cells of hematopoietic lineage and include interleukin (IL)-2, IL-3, IL-4, IL-6, IL-7, granulocyte-macrophage colony-stimulating factor (GM-CSF), and erythropoietin. For some of these receptors (IL-2, IL-3, IL-6, GM-CSF), the ligand-binding domain associates with a second protein (e.g., IL-2 β-chain and GP130 for IL-6) which then transduces the signal. Although these receptors lack intrinsic tyrosine kinase activity, a key component of the proximal signal transduction mechanisms for many of them is the association with, and activation of, one or more members of the *src* family of nonreceptor tyrosine kinases (reviewed in Reference 6) and activation of the Raf-1 serine/threonine kinase, which is critical to the mitogenic response to many growth factors and oncogenes.[7–9]

Table 1 **Classification of Receptors**

Receptor Type	Examples	Proximal Signal Transduction Mechanism
1. Growth factor receptor	PDGF, EGF, FGF, CSF-1, insulin	Intrinsic tyrosine kinase. Interaction with multiple cytosolic proteins mediated via SH_2 domains. Activate Ras.
2. G protein-linked receptor	β-ARs, α-ARs, AII, AVP, thrombin, PTH, muscarinic, dopamine, thromboxane A_2	Seven-transmembrane-spanning configuration. Signal transduction mediated by activation of one or more heterotrimeric G proteins.
3. TGF-β, activins	TGF-β_1 and -β_2	Single transmembrane domain and cytosolic serine/threonine kinase. Multiple types of receptors may be required for full activation of signal transduction. Activation of Ras may play role.
4. Hematopoietic growth factors	IL-2, IL-3, GM-CSF, erythropoietin	No intrinsic tyrosine kinase but associate with and activate one or more nonreceptor TKs possibly via activation of CD45 tyrosine phosphatase. Many have 2 subunits; both required for full activation. Many activate Ras.
5. Immunoglobulin superfamily	T Cell AgR	No intrinsic tyrosine kinase but activate nonreceptor tyrosine kinases. Activation of CD45 may play critical role. May also activate Ras.

How ligand binding initiates activation of these tyrosine kinases is unclear, but the tyrosine phosphorylation that results is highly correlated with the proliferative response.[6] In the case of the IL-2 receptor, the *src* family member, p56^{lck}, associates with the IL-2 receptor β-chain and is activated by IL-2 binding.[10] The mechanism of activation may involve dephosphorylation of the tyrosine kinase's regulatory carboxy terminal tyrosine residue since T cells lacking the membrane-associated tyrosine phosphatase, CD45, do not respond normally with proliferation or tyrosine phosphorylation of cellular substrates (reviewed in Reference 11).

The fifth class of receptors is the immunoglobulin superfamily. Representative members include IL-1 and the T cell antigen receptor. Like the hematopoietic growth factors, a prominent and very early step in the signal transduction pathways activated by the T cell receptor is activation of two nonreceptor tyrosine kinases, p56^{lck} and p60^{fyn}, also probably occurring via activation of CD45 (reviewed in Reference 11).

A. GROWTH FACTOR RECEPTORS

In this group, we include those receptors with intrinsic protein tyrosine kinase activity. These receptors have an extracellular ligand-binding domain, a single transmembrane region, and a cytosolic region that includes a tyrosine kinase catalytic domain (Figure 9–1; reviewed in Reference 2). Prior to ligand binding the receptors may be monomeric (EGF, PDGF, FGF, CSF-1)* or heterotetrameric (insulin, insulin-like growth factor [IGF]-1). Ligand-receptor interaction produces conformational changes in the extracellular domain, which induces receptor dimerization in the case of the monomeric receptors. This leads to interactions between cytoplasmic domains, which enhances receptor protein tyrosine kinase activity. Subsequent autophosphorylation events occur which increase kinase activity and keep the kinase in the activated state even when ligand is no longer bound to the receptor.

Tyrosine residues of the receptor that have no direct effect on kinase activity are also autophosphorylated. Some of these play a critical role in directing the association of cytosolic signaling molecules with the receptor, which can then phosphorylate the signaling molecules.[12–14] In general, the particular cytosolic signaling proteins that as-

*EGF, Epidermal growth factor; PDGF, platelet-derived growth factor; FGF, fibroblast growth factor; CSF-1, colony-stimulating factor-1.

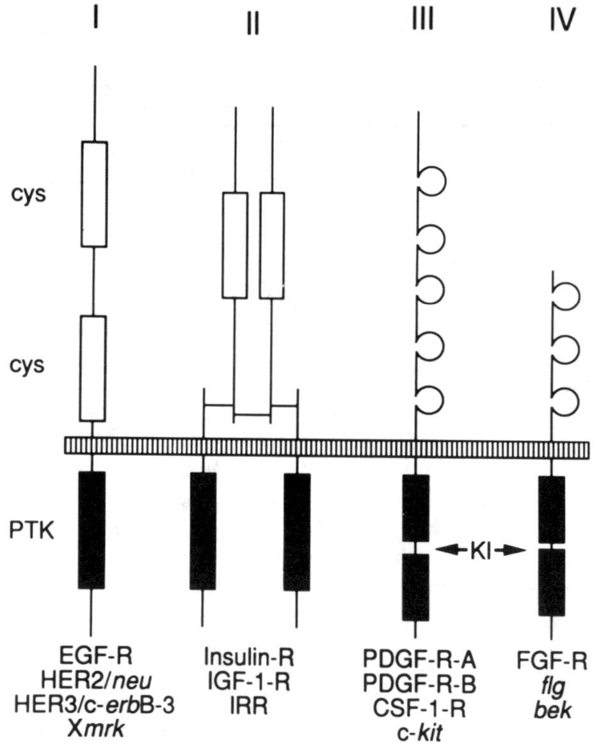

Figure 9–1 The four receptor tyrosine kinase subclasses, based on structural similarities. Subclass I has two cysteine-rich sequences in the extracellular domain and is monomeric. Subclass II receptors are disulfide-linked heterotetramers with two α and two β chains. Subclasses III and IV have kinase insert sequences (KI, see text) with either five (subclass III) or three (subclass IV) immunoglobulin-like repeats in the extracellular domain. PTK, protein tyrosine kinase; KI, kinase insert. (From Ullrich, A. and Schlessinger, J., *Cell,* 61, 203, 1990. With permission.)

sociate with a specific growth factor receptor determine the signal transduction "profile" of the growth factor in that cell. Autophosphorylation triggers a cascade of events in which multiple cytosolic signal transducing proteins interact with the phosphotyrosine residues of the receptor via sequences called SH2 (for *src* homology 2) domains on these cytosolic proteins. The SH2 domain, originally identified in the nonreceptor tyrosine kinase $pp60^{c\text{-}src}$, is approximately 100 amino acids in length and is found in a large number of both signal-transducing proteins with enzymatic activity (e.g., PLCγ, Ras-GAP, and $pp60^{c\text{-}src}$[15]) and "adapter" proteins that appear to have no intrinsic enzymatic activity but allow other signaling molecules to interact with the receptor (e.g., p85, which allows activation of the 110-kDa subunit of the PI-3-kinase, GRB-2 [for growth factor receptor-bound protein-2], possibly mediating Ras activation, and IRS-1, the insulin receptor substrate-1.[16–20])

The tyrosine phosphorylation of the signaling molecule by the receptor in some cases (e.g., PLCγ) activates the signaling molecule.[21] The role of tyrosine phosphorylation of the signaling molecule by the receptor is less clear in the case of the other major catalytically active growth factor receptor substrates (the Ras GTPase-activating protein [GAP] and PI-3-kinase). There is no direct evidence that the phosphorylation alters catalytic activity of either of these enzymes. Rather, the phosphorylation may be important for downstream effector or cofactor binding, or for the correct cellular localization of the proteins to allow maximal activation.

Another consequence of activation of some growth factor receptors is activation of one or more of the nonreceptor tyrosine kinases of the *src* family. Following PDGF stimulation, for example, three of these tyrosine kinases ($pp60^{c\text{-}src}$, $p56^{lck}$, and $p62^{c\text{-}yes}$) associate with the PDGF receptor and are activated.[22]

The signaling pathways that may be activated by one or more of the growth factor receptors include activation of PLC, PI-3-kinase, Na^+/H^+ exchange, nonreceptor tyrosine kinases, the Raf-1/MAP kinase serine/threonine kinase cascade, protein kinase C, and PLA_2. Each of these signaling pathways, how they are activated, and the consequences of that activation on cell volume regulation, if known, will be discussed below.

B. G PROTEIN-LINKED RECEPTORS

G protein-coupled receptors have a variably sized ligand-binding domain, a common seven-transmembrane-spanning configuration and a variable-length cytoplasmic tail (Figure 9–2). Specific sequences within the receptor are believed to interact with one of four major classes of heterotrimeric G proteins, G_s,

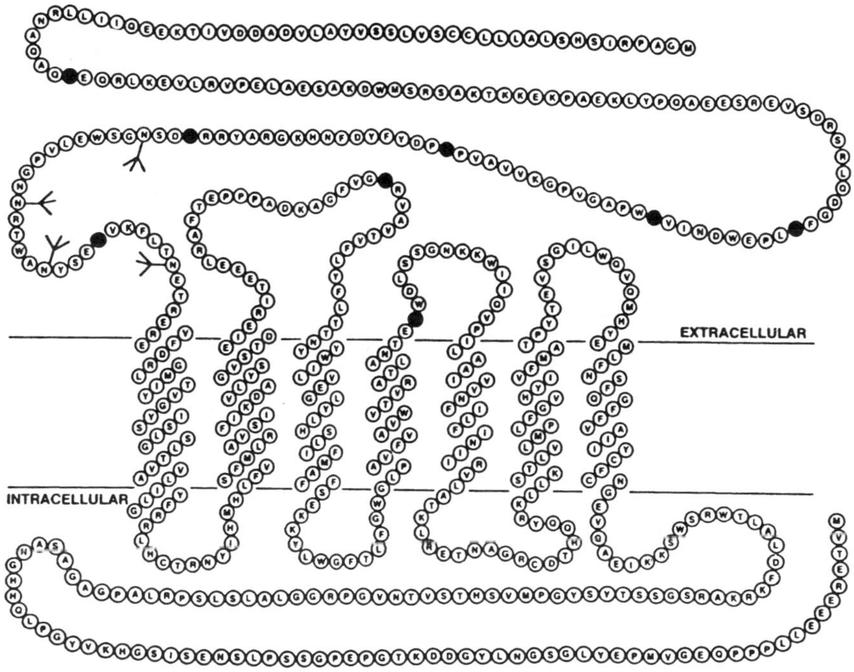

Figure 9–2 G protein-linked receptor. Amino acid sequence of the opossum kidney parathyroid hormone/parathyroid hormone-related peptide (PTH/PTHrP) receptor. This receptor demonstrates the typical seven-transmembrane-spanning configuration of the G protein-linked receptors. Potential glycosylation sites (ψ) and conserved cysteine residues (filled circles) are noted. (From Segre, G. V. et al., Second Int. Conf. New Actions of PTH, Pisa, Italy, 1991.)

G_i/G_o, $G_{q/11}$, or G_{12} (see below).[3,23] The profile of G proteins with which the receptor interacts determines in large part the cell's response to the hormone. The signal transduction pathways activated by these receptors will, therefore, be discussed in detail in the section covering the heterotrimeric G proteins.

One example will illustrate the progress made in identifying determinants of how the receptor interacts with and controls activation of specific G proteins. Specific sequences within the β_2-adrenergic receptor appear to interact with different G proteins. A 15-amino acid segment of the third cytoplasmic loop of the β_2-adrenergic receptor has been found to activate directly G_s, the G protein that couples to and stimulates adenylyl cyclase.[24] This region of the receptor also weakly stimulates G_i, the G protein inhibiting adenylyl cyclase. One result of adenylyl cyclase activation is activation of the cAMP-dependent protein kinase (protein kinase A). This kinase phosphorylates a serine residue of this 15-amino acid regulatory segment and this phosphorylation markedly reduces G_s activation by this region of the receptor. In addition, this phosphorylation appears to increase activation of G_i by the peptide. These findings may help explain the desensitization of the β-adrenergic receptor (via uncoupling of the receptor from G_s) and suggest an autoregulatory mechanism.[24]

III. SIGNAL TRANSDUCERS

A. GUANINE NUCLEOTIDE-BINDING PROTEINS

1. Small-Molecular-Weight G Proteins

The best-studied families of the large group of small-molecular-weight G proteins are the *ras, rho,* and *rab* families. Ras activation is critical for the mitogenic response of cells to serum,[25,26] many growth factors, including PDGF, EGF, insulin, and CSF-1,[25,26] phorbol esters,[25] and certain oncogenes, including v-*src* and *abl*.[27,28] Ras is also critical to nerve growth factor (NGF)-induced (and probably FGF- and IL-6-induced) neuronal differentiation of the PC12 adrenal pheochromocytoma cell line[29–31] and TGF-β-induced growth inhibition of epithelial cells.[32] The role of Ras in these processes has been demonstrated in experiments utilizing microinjection of antibodies to Ras or a dominant inhibitory mutant of Ras (see below), both of which abolish the response.

The *rho* family, which also includes Rac, has been shown recently to play a critical role in serum and growth factor-induced actin cytoskeletal reorganization.[33,34] Specifically, Rac appears to induce actin membrane ruffling and Rho stimulates stress fiber formation and focal adhesions.[33–35] Recently, a link between signal transduction pathways of the *ras* and *rho* families was suggested when the 190-kDa protein that complexes with the GTPase-activating protein for Ras was shown to be a GTPase-activating protein for *rho* family members.[36] This provides a potential connection between Ras and its many cellular functions, including transmission of the mitogenic signal, and the *rho* family, which modulates cytoskeletal reorganization. The *rab* family appears to be involved in vesicular transport (reviewed in Reference 37).

In general, G proteins are inactive when in the GDP-bound form and active when in the GTP-bound form. Not surprisingly, many of the agonists (e.g., PDGF, EGF, insulin, IL-2, IL-3, erythropoietin, NGF, etc.) that induce the Ras-mediated responses noted above have been demonstrated to increase the proportion of Ras in the GTP-bound form, and in many cases, the relative amount of Ras bound to GTP appears to be dependent upon the tyrosine phosphorylation of cellular proteins that occurs as a result of receptor or nonreceptor tyrosine kinase activation (reviewed in Reference 38).

The proportion of Ras in the GTP-bound form is modulated by the relative activities of (at least) two types of proteins — one a GTPase-activating protein or GAP, increasing GTP hydrolysis by Ras, and the other a nucleotide exchange factor, exchanging GDP bound to Ras for GTP (Figure 9–3) (reviewed in References 39 and 40).

A number of exchange proteins and GAPs have been identified, but whether and how their activity is modulated by activated growth factor receptors remains unclear. Mammalian nucleotide exchange factors analogous to yeast and *Drosophila* factors have been cloned recently and at least one possesses exchange activity.[41] However, receptor modulation of its activity remains to be demonstrated. Nucleotide exchange factor activity in crude cell lysates does appear to be modulated somewhat by growth factors,[42,43] but whether this activity is accounted for by the recently cloned exchange factors and what proportion of the increase in the amount of Ras in the GTP-bound state is due to an increase in exchange factor activity is unknown.

GAP activity in crude lysates is clearly modulated by various activated receptors. Two mammalian Ras GTPase-activating proteins have been identified — Ras-GAP and the neurofibromatosis gene product, NF-1. PDGF and EGF stimulate binding of Ras-GAP to their respective receptors and subsequent tyrosine phosphorylation of Ras-GAP, as do some viral oncogenes (v-*src* and v-*fps*).[13,44–46]

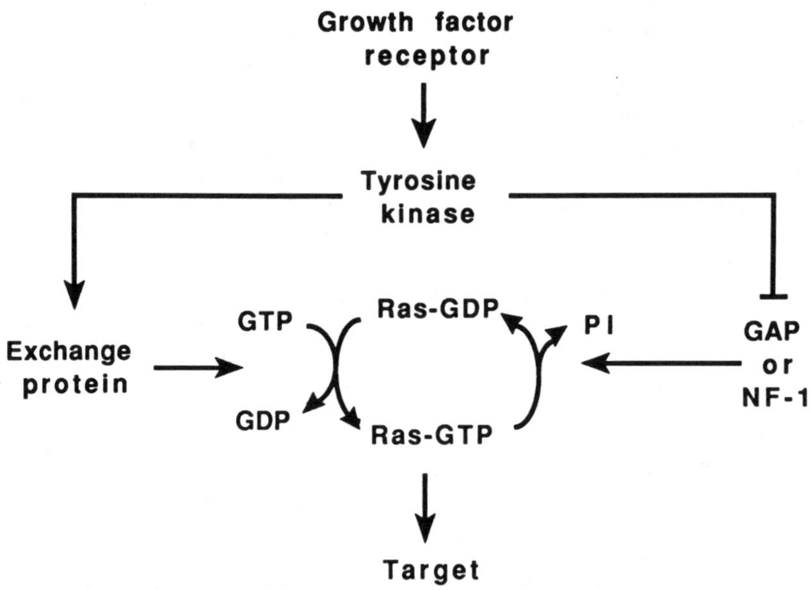

Figure 9–3 Two possible mechanisms of activation of Ras. Growth factors may activate a guanine nucleotide exchange protein which enhances exchange of GDP bound to Ras for GTP, and/or inhibit activity of a GTPase-activating protein (e.g., GAP or NF-1). Either will increase the proportion of Ras in the GTP-bound or active state. (From Satoh, T. et al., *J. Biol. Chem.*, 267, 24149, 1992. With permission.)

While tyrosine phosphorylation appears to be important for Ras activation, suggesting this phosphorylation may modulate GAP activity, tyrosine phosphorylation has not been shown convincingly to modulate GAP activity directly. Furthermore, activation of protein kinase C in T cells appears to activate Ras by decreasing activity of GAP, but this does not appear to occur through phosphorylation of GAP.[47] Finally, mutant PDGF receptors without GAP binding or phosphorylating activity can still activate Ras.

Modulation of Ras-GAP activity has been demonstrated for arachidonic acid and several of its metabolites, suggesting GAP activity may be allosterically modulated.[48–50] Also possible roles of novel proteins such as GRB-2, which binds to activated growth factor receptors and appears to lie between the receptors and Ras in the signal transduction cascade,[17] remain to be clarified. In summary, both GAP and exchange factors may modulate Ras activity in growth factor-stimulated cells, but the mechanisms are unclear. The role, if any, of allosteric protein or lipid modulators and of proteins that inhibit the GTPase activity of Ras[51] or catalyze the conversion of GDP bound to Ras to GTP[52,53] is unknown.

Downstream effectors of Ras and Ras-related proteins have, until recently, eluded identification. In the yeast, *Saccharomyces cerevisiae,* the mammalian homolog to Ras, RAS, when bound to GTP, stimulates adenylyl cyclase by acting on a cyclase-associated protein (CAP) (reviewed in Reference 54). In mammalian cells, Ras clearly plays no role in modulating adenylyl cyclase activity. However, experiments using a dominant inhibitory mutant of Ras demonstrated that Ras was essential for NGF-induced activation of the serine-threonine protein kinase cascade that includes Raf-1, mitogen-activated protein kinases (MAP kinases), and the ribosomal S6 kinase (Rsk)[55,56] (see below). Although it is clear that activation of this cascade is dependent upon Ras, and that the cascade may be the primary downstream target of Ras, it is not clear how Ras activates the cascade. Since Raf-1 appears to be the most proximal kinase in this cascade,[57] and no Raf-1 kinase that activates Raf-1 has been identified, it is conceivable that activated Ras, with or without an additional modulatory protein(s) or lipid(s), allosterically activates Raf-1.

Activation of at least some downstream effectors appears to require the same domain of Ras that binds to GAP.[58] These and other data suggest that GAP might not only inactivate Ras, but also the Ras-GAP complex may interact with and activate downstream effectors. Recently, GAP (when bound to Ras) was found to inhibit directly the muscarinic atrial K^+ channel, and this inhibition appeared to be mediated by the SH2 and/or SH3 domains of the protein.[59]

Despite the central role of Ras in the signal transduction mechanisms of growth factors and cytokines,[60] and the role of Rac and Rho in cytoskeletal reorganization, the role of these small-molecular-weight G proteins, if any, in cell volume regulation is unknown. Studies to determine the role of Ras in cell volume regulation could make use of two experimental approaches. First, Ras "activity" can be assayed by determining the percentage of ras in the GTP- vs. GDP-bound form in immunoprecipitates.[61] An increase in the percentage of Ras in the GTP-bound form is taken as evidence of activation of Ras by the stimulus. Second, a dominant inhibitory Ras mutant has been used with great success to characterize the role of Ras in various signal transduction pathways. This mutant contains an asparagine substitution at position 17 and binds GDP with an affinity significantly higher than that for GTP.[28,58] The mutant appears to compete with normal cellular Ras for upstream activators yet cannot propagate a signal downstream.[56,58] Study of cell volume regulation mechanisms using a cell line expressing this mutation might allow one to determine what role, if any, Ras plays in this vital physiological function of cells.

2. Heterotrimeric G Proteins

The major proximal signal transducing mechanism employed by the large family of serpentine receptors, in the process of transmission of ligand-receptor binding into activation of a number of processes in the interior of the cell, is stimulation of a heterotrimeric G protein with one subunit having intrinsic GTPase activity (reviewed in References 3 and 23). Like their small-molecular-weight relatives, these G proteins are also conceived of as molecular switches that can be in one of two states, depending on whether they are bound to GDP or GTP.[39] These proteins are composed of α-, β-, and γ-subunits. The α-subunits have been found to be responsible for many of the processes that transduce the signal to cellular effectors when hormone binds to cell surface receptor, although the $\beta\gamma$-subunits may also play a direct role in signaling (see below).

a. The G protein cycle

When a ligand binds to a cell surface receptor, the ligand-receptor complex acts as an exchange factor and causes GDP to dissociate from the $\alpha\beta\gamma$ complex (Figure 9–4). This allows GTP to bind to the α-subunit, which causes dissociation of the $\alpha\beta\gamma$

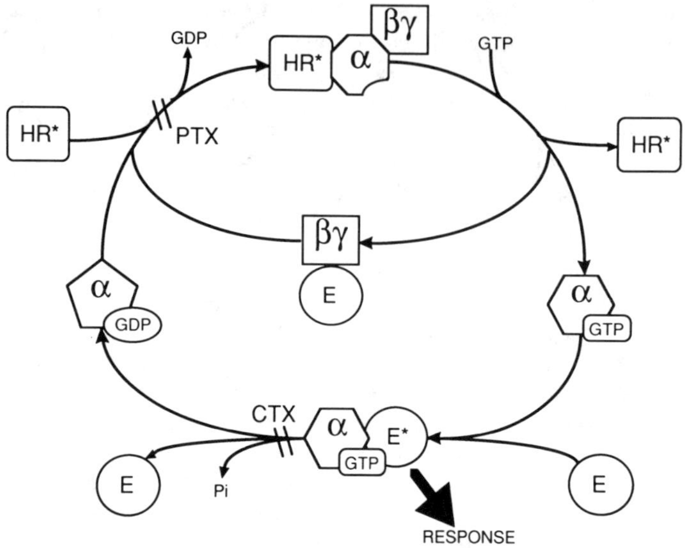

Figure 9–4 The heterotrimeric G-protein cycle. Hormone binding to receptor (HR*) triggers the exchange of GDP bound to the α-subunit of the G protein for GTP. The α- and βγ-subunits then dissociate. While α-GTP interactions with, and activation of, effectors (E and E*) are the predominant response arm of the heterotrimeric G proteins, βγ-subunits may also interact with effectors (see text). The α-subunit then hydrolyzes GTP to form the inactive α-GDP which then reassociates with βγ and is available for interaction with another hormone-receptor complex. Also shown are the sites of action of pertussis toxin (PTX), which uncouples the hormone-receptor complex from the G protein, and cholera toxin (CTX), which prevents the hydrolysis of GTP and "locks" the G protein in the "on" configuration. (Adapted from Bourne, H. R. et al., *Nature,* 348, 125, 1990. With permission.)

from the receptor and of βγ from α-GTP. α-GTP then activates an effector, either an enzyme (adenylyl cyclase, phospholipase C, phosphodiesterase, phospholipase A_2) or an ion channel. The α-subunit has intrinsic GTPase activity that hydrolyzes GTP to GDP, terminating the interaction with the effector. α-GDP then reassociates with βγ and the αβγ-GDP complex is available for interaction with another activated receptor. Until recently, unlike the small-molecular-weight GTPases, α-subunit GTPase activity was not thought to be regulated by a GAP. However, the demonstration that phospholipase C-β_1 acts as a GAP for $G_{q/11}$, the G protein that regulates PLC-β activation,[62] raises the possibility that other effectors can serve as GAPs and terminate the response.

The roles of the free βγ-subunits generated when α and βγ dissociate are less clear. At least four different β- and four γ-subunits have been found.[3] βγ-subunits in yeast activate the pheromone signaling pathway directly[63] and some studies suggest that βγ can directly transmit signals in higher organisms as well. βγ-subunits have been reported to modulate a K^+ channel, possibly via activation of phospholipase A_2 with subsequent production of eicosanoids which directly mediate the effect.[64,65] βγ-subunits have been shown also to activate retinal phospholipase A_2.[66] βγ, dissociated from α-GTP, appears to target the β-adrenergic receptor kinase (βARK) to the membrane where it phosphorylates (and desensitizes) the β-adrenergic receptor.[67,68] This would provide a means whereby receptor activation, which leads to dissociation of α-GTP and βγ, is self-limited. Finally, although β-subunits have extensive sequence similarity, suggesting they may be interchangeable in many αβγ complexes, specific β-subunits are required for the coupling between certain receptors and effectors. Coupling of the muscarinic receptor and the somatostatin receptor via G_o to inhibition of the voltage-sensitive Ca^{2+} channel depends on the presence of specific β-subunits.[69] It is likely that many more instances of β or γ specificity of interactions of receptors with effectors will be found.

The heterotrimeric G proteins perform two major functions — amplification of the signal and sorting of information.[39] G proteins amplify the initial signal of hormone/receptor binding. Each receptor can activate multiple α-subunits, each of which can then activate an effector. The sorting function refers to the integration by the G protein of signals coming from potentially a huge number of receptors and conversion of these into a specific intracellular response. The G proteins accomplish this largely by α-subunit specificity for both receptor and effectors. Each type of G protein α-subunit interacts with a specific set of receptors and then, in turn, with a specific set of effectors. This α-subunit specificity for receptor and effector determines the response of a particular cell type to a ligand.

Based on amino acid sequence similarity, Simon et al.[3,70] have defined four classes of α-subunits — G_s, G_i, G_q, and G_{12}. Although this is a structural classification, functional roles tend to segregate by structural class as well. Table 2 is a partial list of G protein α-subunits, the receptors and effectors with which the G proteins interact, and the cellular response produced. In the majority of cases (α_s stimulation of adenylyl cyclase, α_i opening of atrial potassium channels and $G_{q/11}$ activation of PLCβ [but not PLCγ or PLCδ] being exceptions),[71–75] direct modulation of the effector by reconstituted G proteins has not been demonstrated. Rather, a role for G proteins in activating various effectors has been suggested based on a response that is lost after pretreatment with pertussis toxin. Pertussis toxin (PTX) ADP ribosylates a cysteine residue four amino acids from the carboxy terminus of the α-subunit. This modification uncouples the receptor from the G protein. The α_i class (including α_o) contains this cysteine, and the abolition of various intracellular responses after PTX treatment suggests that activation of several of the effectors/ responses listed in Table 2 for the α_i group are indeed activated by α_i. Importantly, the $\alpha_{q/11}$ group does not contain this cysteine residue and is probably the PTX-insensitive G protein responsible for activating phospholipase C in response to a host of agonists, including bradykinin, angiotensin II, vasopressin, and histamine.[76]

One type of receptor often couples to more than one type of G protein, and the pattern is, presumably, dependent on specific sequences within the receptor.[77] Representative combinations include coupling via G_{α_i} to inhibition of adenylyl cyclase and activation of K^+ currents, and via G_{α_o} to inhibition of voltage-dependent Ca^{2+} channels (α_2 adrenergic, D_2 dopaminergic, somatostatin, and M_2 muscarinic receptors) and activation via G_{α_s} of adenylyl cyclase and activation, probably via $G_{q/11}$, of phospholipase C (calcitonin, parathyroid hormone/parathyroid hormone-related peptide).[78,79]

b. *Regulation of ion channels by G proteins*

One effect of activated G proteins that is likely to be critical to cell volume regulation is direct modulation of ion channels. A large variety of ion channels, including K^+, Na^+, and Ca^{2+} channels, appear to be controlled or modulated by a direct interaction of G proteins with the channel (reviewed in Reference 80). The G proteins may be *obligatory* for opening of the channel (muscarinic atrial K^+ channel and somatostatin-activated K^+ channel in a pituitary cell line controlled by G_{α_i}) or they may *modulate* channel opening induced, for example, by membrane potential (decrease in voltage-gated dihydropyridine-sensitive Ca^{2+} channel current by G_{α_o} in response to opioids, dopamine, α_2-adrenergic receptor activation, etc.). Other examples of direct modulation by G proteins include G_{α_s} stimulation of Ca^{2+} channels in cardiac cells and $G_{\alpha_{i-3}}$ enhancement of an amiloride-sensitive cation chan-

Table 2 **G Protein Subunit Receptor and Effector Specificity**

G protein family	Receptors	Effectors	Response
α_s	β-adrenergic, PTH calcitonin, ACTH, glucagon, H_2 histamine, TSH	Adenylyl cyclase Ca^{2+} channel	↑cAMP ↑Ca^{2+} influx
α_q	Angiotensin II, vasopressin, bradykinin, histamine, thromboxane A_2, muscarinic	PLC-β	↑IP_3 (and Ca^{2+}) ↑DAG
$\alpha_{i/o}$	α_2-adrenergic, thrombin, D_2-dopamine, M_2-muscarinic, somatostatin, serotonin, angiotensin II	Adenylyl cyclase K^+ channels Na^+ channel PLA_2 ? PLC	↓cAMP membrane polarization ↑Na^+ current Production of AA ↑IP_3/Ca^{2+}, DAG
$\alpha_{12/13}$	?	?	?

nel in inner medullary collecting duct cells.[74,80–82] The α-subunits most frequently implicated in modulation of ion channels are the α_o and α_i class and are PTX-sensitive.

B. PHOSPHATIDYLINOSITOL-3-KINASE

The type I phosphatidylinositol kinase, or phosphatidylinositol-3-kinase, phosphorylates the inositol ring at the D3 position as opposed to the type II kinase, which phosphorylates it at the D4 position.[83] The PI-3-kinase is composed of two "subunits", one 85 kDa and one 110 kDa. The 85-kDa subunit has no catalytic activity, contains one SH3 and two SH2 domains, and binds to a variety of growth factor receptors. In so doing, it couples the 110-kDa catalytic subunit to these receptors. After stimulation with growth factors (EGF, PDGF, β-FGF, NGF, CSF-1, and insulin),[15,84] the 85/110 protein associates with the receptor. This association, following stimulation with PDGF-β, is dependent on phosphorylation of tyrosine 751 in the kinase insert region of the PDGF-β receptor. It is not clear how the PI-3-kinase catalytic activity is modulated, but possibilities include direct phosphorylation of the 110-kDa subunit, modulation by the 85-kDa subunit, or relocalization of the catalytic subunit to the membrane, directed by p85, which may allow increased access of the catalytic subunit to its substrate. Activation of PI-3-kinase is not restricted to growth factors or to cells transformed by viral oncogenes, but is also increased by thrombin in platelets and fMLP in neutrophils,[85,86] two agonists that act via G protein-linked receptors. The mechanisms of activation of PI-3-kinase by these agonists are also unclear, but may involve activation of a nonreceptor tyrosine kinase of the *src* family, since oncogenically activated *src* (either v-*src* or middle T antigen-c-*src* complex) is associated with increased PI-3-kinase activity.[87]

The PI-3-kinase may produce any of three products in the intact cell — PI-3-P, PI-3,4-P_2, or, presumably, PI-3,4,5-P_3, none of which are substrates for phospholipase C. The physiologically relevant product(s) is not clear, but PIP_3 levels appear to be tightly controlled in cells (lower at baseline, rising proportionally higher with stimulation, and returning to baseline more quickly).[85,88]

The physiologic role of the 3′ phosphorylated lipids is also not clear. A protein related to the 110-kDa subunit, vps34p in *S. cervisiae,* is involved with the targeting of proteins to the yeast vacuole and possibly vesicular transport.[89] In mammalian cells, some evidence points to a role of 3′ phosphorylated lipids in cytoskeletal reorganization and cell morphology. In neutrophils, the rise and fall in the level of PIP_3 correlates with F-actin polymerization and depolymerization.[85] In addition to these data, a cell line expressing a PDGF-β receptor mutant lacking the kinase insert region (to which the p85 subunit binds) failed to undergo actin rearrangement in response to PDGF.[90] In a similar mutant involving the kinase insert region of the PDGF-α receptor, however, receptor-associated PI-3-kinase activity was abolished but chemotactic signal transduction was not.[12]

There is also evidence that implicates a PI-3-kinase product in the mitogenic response. The ability of PI-3-kinase activity to associate with the middle T antigen-pp60$^{c\text{-}src}$ complex correlates with the ability of middle T antigen to transform cells.[91] Similarly, a kinase insert mutant of the PDGF-β receptor similar to that described above was ineffective at transmitting a mitogenic signal,[92] but the PDGF-α receptor mutant transmitted the mitogenic signal normally.[12] In summary, although the 3′ phosphorylated lipids probably play a role in cytoskeletal reorganization and vesicle trafficking in some cells, and in the mitogenic response to some growth factors in others, their precise role remains unclear.

Two methods have been used to study PI-3-kinase activity in cells. The first involves immunoprecipitation of PI-3-kinase activity with antibodies to either phosphotyrosine or the p85 subunit (commercially available). One can then measure PI-3-kinase activity as ^{32}P-phosphorylation of PI, PI-4-P, or PI-4,5-P_2 after separation of the lipids by thin-layer chromatography (TLC; methods thoroughly reviewed in Reference 93). Since PI-4-P and PI-3-P as well as PI-4,5-P_2 and PI-3,4-P_2 comigrate on TLC, one has to confirm the identity of the products with anion exchange high-performance liquid chromatography (HPLC) after deacylation. One can also examine production of 3-phosphorylated lipids in intact cells by labeling with either ^3H-inositol or $^{32}PO_4$ and then analyzing by anion exchange HPLC after extraction and deacylation. With the cloning of both subunits of the PI-3-kinase and the availability of larger quantities of these lipids for studies *in vitro,* it becomes more feasible to define the possible roles of these proteins and their lipid products in cell volume regulation.

C. PHOSPHOLIPASE C

Hormone- and growth factor-dependent increases in inositol phosphates are mediated by phosphatidylinositol-specific phospholipase C (PLC). This phospholipase C hydrolyzes PI-4,5-P_2 (in preference to PI or PI-4-P) to produce two second messengers, inositol 1,4,5 trisphosphate

(IP_3), both cyclic and noncyclic forms, and diacylglycerol. The catalytic activity of all PLC subtypes described to date are dependent on the [Ca^{2+}]. Three subtypes, beta (150 to 154 kDa), gamma (145 to 148 kDa), and delta (85 to 88 kDa), each with multiple isoforms, have been characterized and cDNAs isolated and expressed[94] (Figure 9–5). All subtypes have two similar regions, designated X and Y, which are presumably involved with catalytic functions (hydrolysis, lipid/Ca^{2+}-binding, etc.). Between the X and Y regions, the γ-subtypes have two SH2 and one SH3 domain, which are critical for interaction with growth factor receptors (see below). The δ-subtype has an extremely short carboxy terminal section after the Y region. There is significant homology within the carboxy and amino-terminal regions between subtype isoforms (e.g., 45 to 50% identity within the amino-terminal region), but there is little homology between different subtypes.

It has been known for some time that the mechanisms of growth factor-induced and hormone-induced PLC activation were different. For example, vasopressin-induced increases in IP_3 appeared to be modulated by a G protein(s) (increased by GTPγS and decreased by GDPβS, stable analogs of GTP and GDP, respectively), whereas the growth factor-induced increase was not (minimally affected by the thio analogues of the guanine nucleotides). PLCγ is an effector for many growth factor receptors (PDGF, EGF, NGF, and FGF but not CSF-1 or insulin).[95] These receptors tyrosine phosphorylate PLCγ. The association of PLCγ with these receptors is dependent on the SH2 domains of PLCγ which interact with phosphotyrosines on the autophosphorylated receptor. For example, association of PLCγ with the EGF receptor is dependent on phosphorylation of Tyr 992 in the carboxy terminus of the EGF receptor[96] and association with the FGF receptor is dependent on phosphorylation

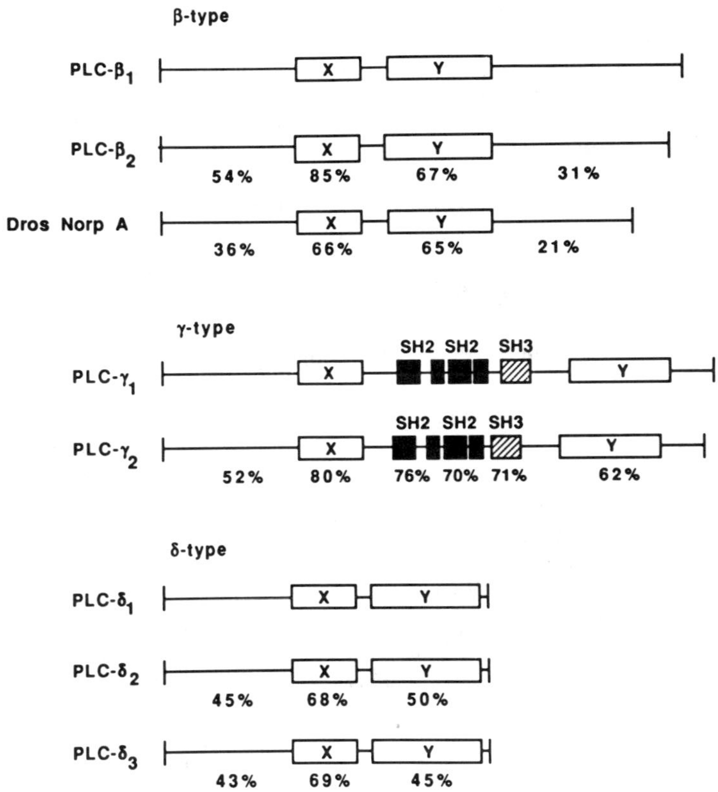

Figure 9–5 Schematic of the similarities and differences between the three types of phospholipase C enzymes (α, β, γ) and subtypes (designated by Arabic numeral subscripts). Also shown is the *Drosophila* Norp A protein which belongs to the PLC-β group. X and Y regions, and SH2 and SH3 domains are discussed in the text. Amino acid sequence identity (in percent) between subtypes are shown for X, Y, SH2, SH3, amino terminal (left) and carboxy terminal (right) regions. (From Rhee, S. G., *Trends Biochem. Sci.*, 16, 297, 1991. With permission.)

of Tyr 766.[97] PLCγ-1 may, in turn, be phosphorylated at three predominant sites (tyrosine 771, 783, and 1254) and EGF or PDGF-stimulated PI-4,5-P_2 hydrolysis is dependent on phosphorylation of two of these (783 and 1254).[98] Site-directed mutagenesis of these tyrosines to phenylalanine residues prevents activation of PLCγ-1, indicating that the tyrosine phosphorylation is crucial in the activation of PLCγ-1. *In vitro,* however, phosphorylated and unphosphorylated PLCγ-1 display similar activity.[98] This suggested that PLCγ-1 activity in cells might be regulated by a negative modulator. Candidate modulators include profilin, an actin-binding protein that binds four or five molecules of PIP_2 with high affinity and inhibits hydrolysis by unphosphorylated PLCγ.[99] The tyrosine phosphorylation of PLCγ-1 by the growth factor receptors appears to allow PLCγ-1 to hydrolyze PI-4,5-P_2 even in the presence of profilin. Of note, the release of profilin from the hydrolyzed PIP_2 could then alter actin polymerization and induce changes in cell shape and motility. Other candidate modulators of PLCγ include glucosphingolipid or any of several proteins that co-immunoprecipitate with PLCγ. Although it is clear that the phosphorylation of tyrosines 783 and 1254 eliminate this negative modulation and allow PLCγ-1 to hydrolyze PIP_2, what intramolecular changes in PLCγ-1 occur as a result of the tyrosine phosphorylation are unknown.

PLCγ-1 tyrosine phosphorylation and activation can also occur following stimulation of the T cell antigen receptor, presumably via the latter's activation of the nonreceptor tyrosine kinase $p60^{fyn}$ and/or $p56^{lck}$.[100] Tryptic mapping showed that the phosphorylation sites of PLCγ in these cells were identical to the sites phosphorylated by the EGF and PDGF receptors.

Feedback inhibition of PLCγ occurs in cells following stimulation of either protein kinase C or protein kinase A, although how this is achieved in most instances is not clear. One target of these serine/threonine kinases is Ser 1248 of PLCγ-1. It appears that phosphorylation of this residue reduces tyrosine phosphorylation of PLCγ-1 (and phospholipase activity), possibly by altering the association of PLCγ-1 with the tyrosine kinase phosphorylating it.[101]

The β- and δ-subtypes of PLC are not activated by tyrosine phosphorylation. PLCβ-1 is activated by the α-subunits of the PTX-insensitive G_{α_q} protein family, which includes α_q, α_{11}, α_{14}, and α_{16}.[75] Receptors for several agonists are known to be coupled to α_q and α_{11}, including thromboxane A_2, vasopressin, angiotensin II, histamine, acetylcholine (muscarinic), and bradykinin,[76,102] and it is speculated that most of the ligands stimulating PLC via PTX-insensitive mechanisms will do so via a G_q family member. There is some specificity in the interactions between PLCβ isoforms and α_q family members, since α_q and α_{11} are more effective at stimulating PLCβ-1 while only α_{16} effectively activates PLCβ-2.[72,73] This is not surprising since the carboxy terminal region of PLCβ appears to direct the interaction with α_q, and PLCβ-1 and PLCβ-2 are only 31% identical in this region.

Although the identity of the PTX-insensitive G proteins mediating activation of PLC (G_q family), and the subtype of PLC activated (PLCβ) have been identified, the identity of the G protein and PLC subtype mediating the PTX-sensitive hormone-induced response remains unclear. There is speculation that PLCδ may mediate the PTX-sensitive response, based largely on the inability of the G_q members to activate it. However, neither coupling of PLCδ to any receptor nor activation of PLCδ by a G protein has been demonstrated as yet.

The site of action of inositol 1,4,5 trisphosphate, once released by PLC, is an intracellular Ca^{2+} channel with sequence homology to the ryanodine receptor of the sarcoplasmic reticulum. This process, as well as activation of protein kinase C by diacylglycerol, the other product of PIP_2 hydrolysis, will be discussed below.

D. CALCIUM

The understanding of intracellular calcium homeostasis is critical to the evaluation of cell volume regulatory phenomena since calcium may be involved at various steps of the process. A number of studies have indicated that RVD is dependent upon extracellular Ca^{2+},[103,104] whereas other reports, on different cell types, suggest that intracellular release of Ca^{2+} is primarily responsible for RVD. In many cells, including epithelial cells, Ca^{2+}-dependent channels are present. Ca^{2+}-dependent Cl^- and K^+ channels may be important for cell volume regulation.[105,106] Ca^{2+}-dependent K^+ channels are present in a number of different cells, including the apical membrane of the cortical collecting duct,[107] the thick ascending limb cell,[108] Madin-Darby canine kidney (MDCK) cells,[109] and opossum kidney cells.[110] Ca^{2+}-dependent cation channels are also present in the cortical thick ascending limb basolateral membrane[111] and the proximal convoluted tubule apical membrane.[112] Calmodulin has been implicated in the Ca^{2+}-dependent regulation of some of these Ca^{2+}-dependent ion channels.[105,106,113]

Ca^{2+} can also be involved in other ways in cell volume regulation. Many cytoskeletal proteins bind Ca^{2+} and these cytoskeletal proteins may play important roles in membrane channel activity or other cellular phenomena important for volume regulation.[114] Ca^{2+} can also activate phospholipases and phosphatases. The activation of phospholipase A_2 leads to the generation of free fatty acids, including arachidonic acid, and ultimately to various eicosanoid metabolites of arachidonic acid that have been implicated in membrane channel regulation and volume regulation, as described below. The implications of alterations in phospholipase and phosphatase activity are also described below.

Cytosolic free [Ca^{2+}] can increase in response to a perturbation such as ligand-receptor interaction or volume change due to release of Ca^{2+} from intracellular pools and/or enhanced entry of Ca^{2+}. The inositol trisphosphate (IP_3), produced as a result of phosphatidylinositol-specific phospholipase C activation, interacts with a specific receptor on a subcompartment of the endoplasmic reticulum, which has been purified[115] and cloned.[116] Upon reconstitution into lipid vesicles, IP_3 stimulates Ca^{2+} release from these vesicles, indicating that the binding protein also has Ca^{2+} channel properties.[117]

This IP_3-induced release of Ca^{2+} from intracellular stores likely plays an important role in the oscillatory response characteristic of the changes in cytosolic free [Ca^{2+}] observed with many perturbations both in excitable and nonexcitable cells[118–126] (Figure 9–6). There are various hypotheses which have been presented to account for these oscillatory changes in Ca^{2+} concentration, with no consensus reached at present. One hypothesis proposes oscillations in IP_3 levels. Another possible explanation is that the effects of IP_3 are diminished when ambient levels of Ca^{2+} are increased, leading to desensitization of the receptor, thus allowing for temporal changes in IP_3-induced release.[127] In the presence of permissive levels of IP_3 it is possible that the spontaneous release of Ca^{2+} from intracellular IP_3-sensitive stores that occurs once the stores are overloaded with Ca^{2+} may be periodic in nature.[128] It has also been proposed[1,129] that oscillations may occur as a result of the Ca^{2+}-dependent conversion of IP_3 to inositol tetrakisphosphate (I-1,3,4,5-P_4). According to this theory as cytosolic [Ca^{2+}] increases, IP_3 will be converted to IP_4, thus reducing the IP_3-induced release of Ca^{2+} from intracellular stores and decreasing cytosolic [Ca^{2+}], while increasing the effects of IP_4 to potentiate Ca^{2+} entry into the cell and ultimately into the stores, thus replenishing them. Space constraints do not permit a detailed evaluation of the various hypotheses.

This phenomena of IP_3-induced release of Ca^{2+} is self-limited due to the fact that the Ca^{2+} stores are limited in size. In many cases of phospholipase C-induced IP_3 generation and release of Ca^{2+} from intracellular stores, this release is associated with enhanced Ca^{2+} entry into the cell. It is believed by many investigators that IP_4 is important in this Ca^{2+}

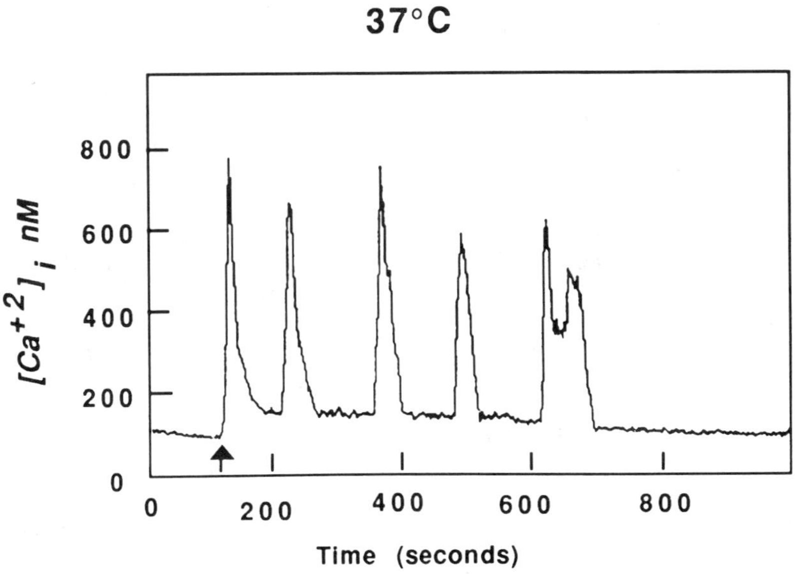

Figure 9–6 Pattern of oscillations in intracellular free Ca^{2+} in response to vasopressin in renal mesangial cells. Vasopressin (1 n*M*) was added at the time indicated by the arrow. The experiment was performed at 37°C. (From Hajjar, R. J. and Bonventre, J. V., *J. Biol. Chem.*, 266, 21589, 1991.)

entry.[130] Exactly how IP_4 participates in the process remains controversial and the reader is referred to a recent review for the current status of this discussion.[130] Recently, Theibert and colleagues[131] have isolated two distinct high-affinity IP_4 binding proteins from rat brain membranes.

E. PROTEIN KINASES AND PHOSPHATASES

A number of reports have suggested that a protein kinase(s) and phosphatase(s) are involved in controlling RVI via regulation of the $Na^+/K^+/2Cl^-$ cotransporter and Na^+/H^+ exchange. Although most of these studies have been limited to activating or blocking activation of a particular response with phosphatase inhibitors (sodium orthovanadate, okadaic acid, sodium fluoride) or kinase inhibitors (e.g., staurosporine), the data suggest that a kinase(s) and a phosphatase(s) may be key modulators of RVI and, possibly, volume decreases as well as changes in cell volume in response to various agonists. Experiments utilizing tyrosine and/or serine/threonine phosphatase inhibitors suggest phosphatases may terminate $Na^+/K^+/2Cl^-$ transport activity and Na^+/H^+ exchange that had been maintained in an active form by a kinase.[132,133] The identities of the kinases and phosphatases involved in modulating activity of these volume-regulatory pathways are unknown. The cAMP-dependent protein kinase (protein kinase A) can activate $Na^+/K^+/2Cl^-$ cotransport in some systems, but it is probably not the physiologically relevant kinase controlling the RVI.[133–135] Since the same volume-regulatory pathways are activated by catecholamines, it is possible that there are multiple kinase activators, including protein kinase A.

The role of kinases and phosphatases in modulating RVD is much less clear. Phosphatase inhibitors have been reported to reduce activation of KCl cotransport induced by cell swelling,[134,136,137] suggesting dephosphorylation activates KCl cotransport and potentiates RVD. In summary, although kinases and phosphatases appear to be involved in modulating RVI and probably RVD, the identity of the kinases or phosphatases, their targets, mechanisms of activation, and the signals leading to their activation are unknown.

Growth factors and G protein-linked peptides produce marked increases in phosphorylation of multiple intracellular proteins via activation of tyrosine, serine/threonine, or bifunctional kinases. These kinases control or modulate a host of cellular responses. In this section we will briefly review some of the major kinases involved in cellular signaling which may have prominent roles in cell volume regulation.

a. Protein kinase C

Protein kinase C is critical to many of the signal transduction pathways activated by growth factors, many G protein-linked hormones, and neurotransmitters. The protein kinase C family is made up of at least seven genes encoding nine or more polypeptides.[138] The "conventional" isoforms (α, β1, β2, and γ) are Ca^{2+}- and phospholipid-dependent and require diacylglycerol or phorbol ester for activation *in vitro*. Furthermore, there is evidence that these isoforms are activated in intact cells by an allosteric mechanism involving diacylglycerol produced when agonists activate phospholipase C. These isoforms have four conserved regions, a lipid/phorbol ester-binding region (C1), containing two cysteine-rich zinc finger-like motifs, a probable Ca^{2+}-binding region (C2), and two kinase domains (C3 and C4).[139] Four other "novel" isoforms have also been identified: δ, ε, ζ, and n(L). All are Ca^{2+}-independent, but δ, ε, and n(L) require diacylglycerol or phorbol ester for activity *in vitro*. The ζ isoform is not activated by diacylglycerol or phorbol esters. It is activated by acidic phospholipids, including 3′ phosphorylated phosphoinositides and unsaturated fatty acids *in vitro*, and this may be the mechanism of activation in intact cells.[139]

The existence of several isoforms has raised speculation that different targets exist and the different modes of regulation of the isoforms suggest different pathways of activation. Indeed, PKCα, PKCδ, and PKCβ1 (the latter of which differs from PKCβ2 only in the C-terminal 50 amino acids) phosphorylate glycogen synthase kinase-3β (which phosphorylates and inactivates the transcription factor c-Jun) whereas PKCβ2 and ε do not.[140]

Despite the wealth of information on regulation of PKC activity *in vitro*, very few physiological targets have been identified. PKC does phosphorylate threonine 654 of the EGF receptor which is responsible for transmodulation of the receptor. This phosphorylation reduces the ability of the receptor to phosphorylate and activate PLCγ-1.[141] In addition, exposure of cells to phorbol esters activates two kinase cascades, the p70 S6 kinase cascade and the Raf-1/MAP kinase cascade, and may activate a third involving GSK-3β and c-Jun. For the p70 S6 kinase and Raf-1/MAP kinase cascades, however, it is not clear what the PKC target is and, thus, how PKC activates the cascades. Finally, PKC phosphorylates the α-subunit of the Na^+,K^+-ATPase *in vitro*.[142] Activation of PKC in intact cells leads to an increase in phosphorylation of this subunit as well as a downregulation of Na^+,K^+-ATPase activity. This might be a fifth physiologic target for PKC.

Techniques to study the role of PKC in the intact cell have largely relied on one of two methods — downregulation or inhibitors. Downregulation refers to the effect of prolonged exposure of cells to high concentrations of phorbol esters to markedly reduce the Ca^{2+}-phospholipid-dependent protein kinase activity in the cell. While abolition of a particular response in downregulated cells strongly suggests involvement of PKC in the generation of that response, failure to abolish the response does not, since PKCζ does not undergo phorbol ester-induced downregulation. Furthermore, cells differ in their ability to undergo downregulation and the efficacy of any protocol should be confirmed by an assay of PKC activity and by immunoblotting.

Use of inhibitors is more problematic in the intact cell since most, including the widely used staurosporine and the isoquinoline, H7, are not specific inhibitors. These inhibitors bind to the ATP-binding site of the catalytic domain and there is significant sequence homology within the ATP-binding site between PKC and other serine/threonine kinases (reviewed in Reference 143).

The most specific PKC inhibitors are peptide pseudosubstrates based on the pseudosubstrate sequence in the regulatory domain of PKC. These inhibitors cannot, however, be used in intact cells but may be used for *in vitro* kinase assays or in permeabilized cells.[144,145]

b. Raf-1/MAP kinase cascade

The activity of many protein kinases is modulated by phosphorylation of that kinase by a "kinase kinase". These kinase cascades with sequential activation of several kinases provide a mechanism for dramatically amplifying a signal initiated at the cell membrane. One extensively studied cascade which is activated by many different classes of agonists is the MAP (*m*itogen-*a*ctivated *p*rotein) kinase or ERK (*e*xtracellular signal-*r*egulated *k*inase) kinase cascade. The MAP kinases are a family of serine/threonine kinases including the extensively studied 42- and 44-kDa kinases.[146,147] This cascade appears to be of critical importance in transmitting mitogenic signals to the nucleus. In addition, however, and particularly important for cell volume regulation, it may activate other signal transduction pathways not primarily involved with mitogenesis. For example, MAP kinase phosphorylates and activates *in vitro* the large-molecular-weight PLA_2 activated by growth factors (see below).

The MAP kinase cascade (Figure 9–7) is activated by growth factors, many cytokines and oncogenes, and at least some agonists that signal via G protein-linked receptors, including endothelin and thrombin.[148] Raf-1 is the most proximal kinase in the cascade that has been identified to date.[57,149,150] Raf-1 is a ubiquitous serine/threonine kinase which is critical to the mitogenic response to serum.[7] It appears to be downstream of Ras based on experiments utilizing a dominant inhibitory mutant of *ras* to block signal transduction in NGF-stimulated PC12 cells.[55,56] The mechanism of activation of Raf-1 is not clear. Although Raf-1 is phosphorylated in response to mitogens, no Raf-1 kinase which is capable of activating Raf-1 *in vitro* has been identified, and it is conceivable that Raf-1 is allosterically activated by a lipid or protein cofactor, analogous to PKC, since Raf-1 also has a zinc finger motif in its amino terminal regulatory domain.[151] The multiple classes of agonists which activate the MAP kinase cascade and the proximal position of Raf-1 in the cascade suggest Raf-1 is an integrator of mitogenic signals from multiple sources.

The activated Raf-1 appears to phosphorylate and activate the MAP kinase kinase or MEK (*M*AP or *E*RK *K*inase),[152] a bifunctional kinase that in turn phosphorylates MAP kinase on a threonine and a tyrosine residue and activates it. MAP kinase, in turn, phosphorylates and activates the ribosomal S6 kinase, $p90^{rsk}$.[153] In addition, p42/44 MAP kinases (and the related p54 MAP kinase) phosphorylate transcription factors including $p62^{TCF}$ and c-Jun, and these phosphorylations could directly control gene expression.[154,155]

Important to cell volume regulation is the fact that ERK-1, but not p54MAP kinase, has been shown to phosphorylate and activate the large-molecular-weight PLA_2 *in vitro*.[156] Given the demonstrated involvement of arachidonic acid and its metabolites in cell volume regulatory pathways (see below), the Raf-1/MAP kinase cascade may be of critical importance to maintenance of cell volume.

c. Protein phosphatases

Protein phosphatases are relatively specific for either phosphotyrosine or phosphoserine/threonine (reviewed in References 157 and 158). Given the data suggesting involvement of serine/threonine phosphatases in cell volume regulation, we will largely direct our comments to this group of phosphatases. To date, over 20 serine/threonine phosphatases have been cloned and sequenced. They are generally divided into type 1 and type 2, with the latter being subdivided into 2A, 2B, and 2C. The divisions are based on sensitivity to two thermostable protein inhibitors (inhibitor-1 and inhibitor-2) and on the specificity for the protein kinase A sites on the subunits of phosphorylase kinase, and the subdivisions are based on dependence on

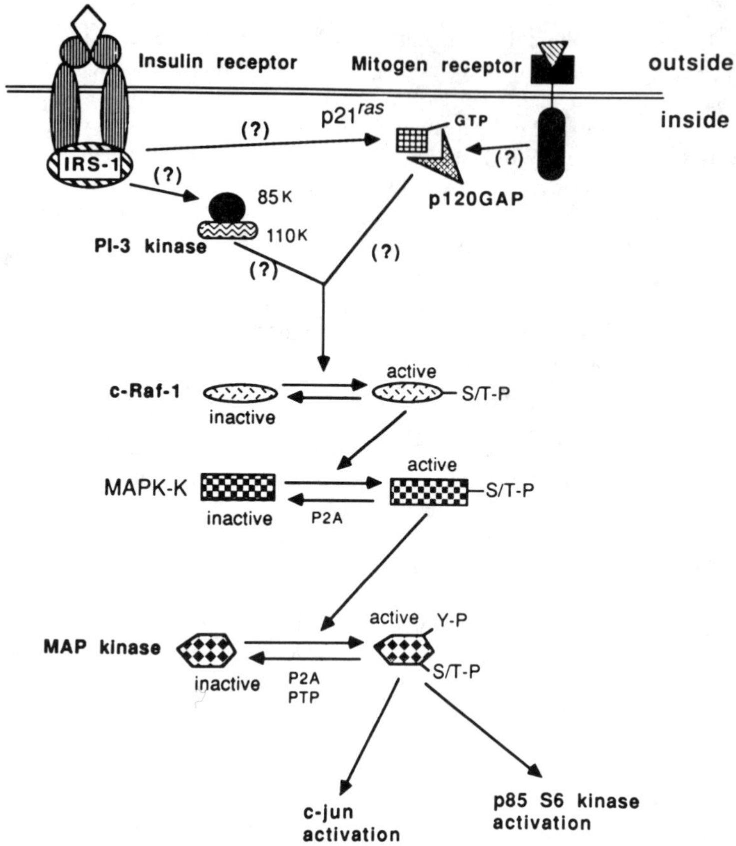

Figure 9–7 Model for the mitogen-activated protein kinase (MAP kinase) cascade. For discussion, see text. Abbreviations used: S/T-P, phosphoserine/phosphothreonine; Y-P, phosphotyrosine; IRS-1, insulin receptor substrate-1; GAP, GTPase-activating protein; PI-3 kinase, phospatidylinositol-3 kinase. Uncertainties over the activator (s) of Raf-1 are highlighted as question marks. (From Kyriakis, J. M. et al., *Nature,* 358, 417, 1992. With permission.)

Ca^{2+} vs. Mg^{2+} (Table 3).[158] Protein phosphatase-1 (PP-1) and protein phosphatase-2A (PP-2A) are the predominant serine/threonine phosphatases in eukaryotic cells.[159] These phosphatases have catalytic and regulatory subunits, both of which confer substrate specificity.

The role of the serine/threonine phosphatases in signal transduction and in cell volume regulation has been explored using several marine and algal toxins (Table 3). Okadaic acid is a cell-permeable tumor promoter from sponges that inhibits PP-2A much more than PP-1. Calyculin A, another tumor promoter, inhibits PP-1 significantly more than PP-2A. Microcystin is an hepatotoxin from cyanobacteria that is not cell permeable and thus is of use only in *in vitro* assays. It potently inhibits both PP-2A and PP-1. The use of phosphatase inhibitors in intact cells demonstrates the wide range of processes modulated by these enzymes, including relaxation of smooth muscle, prevention of deactivation of Ca^{2+} channels, triggering of meiotic maturation, and regulation of $p34^{cdc2}$, the cell cycle kinase (reviewed in Reference 158).

Regulation of the activity of these phosphatases is often complex. PP-2B is a Ca^{2+}-dependent calmodulin-stimulated enzyme, although its 19-kDa subunit subserves calmodulin's function in the absence of the latter. Targets of PP-2B appear to include several substrates of protein kinase A.

Substrate specificity of PP-2A, the Ca^{2+}-independent phosphatase, is modulated by regulatory subunits, but these do not acutely modulate activity of the phosphatase.[159] PP-2A appears to be inactivated by phosphorylation of a tyrosine residue catalyzed by such diverse tyrosine kinases as $pp60^{src}$, $p56^{lck}$, the insulin receptor, and the EGF receptor. Phosphatase activity is then regained by an autodephosphorylation reaction.[159] This may be a critically important observation, since several serine/threonine kinases (including most of the members

Table 3 **Protein Serine/Threonine Phosphatases**

Phosphatase	Dephosphorylates Phosphorylase Kinase	Sensitivity to Inhibitors -1 and -2	Divalent Cation Dependence	Sensitivity to Okadaic Acid	Sensitivity to Other Inhibitors
PP-1	β-subunit	Yes	No	Somewhat	Calyculin A > microcystin
PP-2	α-subunit	No			
PP-2A			No	Very	Microcystin > calyculin A
PP-2B			Ca^{2+}	None	
PP-2C			Mg^{2+}	None	

of the MAP kinase cascade and p54 MAP kinase) are inactivated *in vitro* by PP-2A. This inactivation of PP-2A by receptor and nonreceptor tyrosine kinases might then allow full activation of critical serine/threonine kinases involved in cellular signal transduction.

F. PHOSPHOLIPASE A_2

Since a number of studies have implicated products of arachidonic acid metabolism in the regulation of ion channel activity and cell volume control we will devote the following sections to a summary of what is known about the production of arachidonic acid, its metabolites, and the effects of these substances on volume regulation.

Phospholipases A_2 comprise a family of enzymes that are important for many intracellular signaling events. These enzymes act on the acyl bond at the *sn*-2 position of naturally occurring L-phospholipids to release a free fatty acid and a lysophospholipid. When 5,8,11,14 eicosatetraenoic acid (arachidonic acid) is the fatty acid at the *sn*-2 position and is released by a PLA_2 it may be metabolized to a number of eicosanoids, some of which have been implicated in the regulation of cell volume. While arachidonic acid is present in various other lipid molecules of the cell besides phospholipids, including diglycerides, triglycerides, and cholesterol esters, and other signaling pathways not involving PLA_2 may result in enhanced release of arachidonic acid, as depicted in Figure 9–8,[160] it is generally felt that, in most cells, PLA_2 activation is responsible for most of the arachidonic acid released in response to cell activation by various stimuli. The reader is also referred to a recent review written by one of the authors which covers in

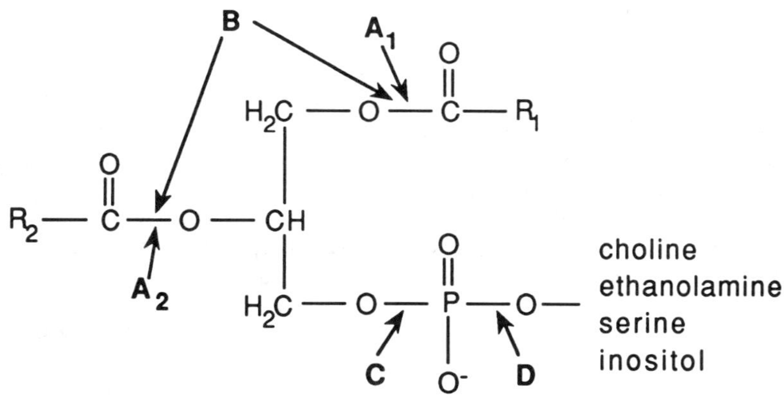

Figure 9–8 Schematic representation of the sites of hydrolysis of phospholipids by various phospholipases. Phospholipase A_1 (A_1) hydrolyzes the lipid molecule at the 1-acyl ester linkage. Phospholipase B acts at both the 1- and 2-acyl ester bonds. Phospholipase A_2 (A_2) acts at the *sn*-2 position. Phospholipase C (C) and D are phosphodiesterases. Phospholipase C cleaves the phosphoglycerol bond. Phospholipase D (D) acts beyond the phosphate to release phosphatidic acid and a base. R_1 and R_2 represent fatty acids. (From Bonventre, J. V., *J. Am. Soc. Nephrol.*, 3, 128, 1993.)

greater detail many of the topics presented in this section.[161]

PLA_2s are present in a wide range of tissues and there is a growing number of forms of the enzyme. There are secretory and intracellular forms of PLA_2. Because of the more accessible nature of the secretory forms from snake venom and pancreas these smaller (13 to 15 kDa)-molecular-weight forms were well characterized a number of years ago.[162] These secretory forms require millimolar concentrations of calcium (extracellular levels) for activation,[163] are very stable, and have seven disulfide bonds.

Secretory PLA_2s have been categorized into two groups according to the positions of cysteine pairs:[164] group I (present in elapid and hydrophid snake venom and mammalian pancreas[165]) and group II (present in crotalid and viperid snake venom and a wide range of mammalian tissues[166–171]). Large amounts of group II PLA_2 are found in inflammatory exudates,[169] and administered group II PLA_2 is a potent inflammatory stimulus.[172,173] Since levels of calcium concentration required to activate secretory PLA_2s are higher than those present in the cell under physiological conditions of activation or volume perturbation even at sites of limited calcium diffusion, and since the secretory enzymes do not demonstrate any significant selectivity among fatty acids at the *sn*-2 position of phospholipids,[174] and are compartmentalized into secretory granules,[175] it is unlikely that these secretory enzymes play important roles in cell signal transduction.

Our laboratory, as well as others, has characterized a number of membrane-associated microsomal and mitochondrial PLA_2s[171,175–180] that range in size from 13 to 18.5 kDa. These forms have characteristics which place them in the group II class of PLA_2s, although many have not been characterized to the point where sequence information is known. The mitochondrial and synaptosomal forms we have characterized from the kidney and brain, respectively, have alkaline pH optima and are Ca^{2+} dependent, requiring calcium concentrations greater than 1 μmol for activation.[179,180] Since mitochondrial calcium concentrations[181] and pH are higher than cytosolic concentrations, it is possible that these mitochondrial forms play an important role in mitochondrial physiology.

Recently we and others have argued for the importance of a 100- to 110-kDa form of PLA_2 found in the cytosolic fraction of kidney,[182] platelets,[183] human monocytic cell line, U937, and rat macrophage cell line, RAW 264.7.[184] This enzyme is activated at physiological levels of intracellular free calcium and is stably activated in cells treated with vasopressin, phorbol esters, or EGF.[185,186] The enzyme has a preference for arachidonic acid at the *sn*-2 position of phospholipids. This form of PLA_2 has been cloned recently by two groups from human U937 cells.[184,187] The cDNA sequence encodes a protein with a predicted molecular weight of 85.2 kDa, with no sequence homology to other forms of PLA_2.

Other distinct intracellular forms of PLA_2 have been characterized recently. In the heart there is a 40-kDa form which has high activity toward plasmalogens and, unlike most forms of PLA_2, is independent of Ca^{2+} for activity.[188] Calcium-independent forms of the enzyme have also been reported to be present in cytosolic fractions in a number of organs, including brain, lung, liver, and spleen of the rat.[189]

1. Regulation of PLA_2

When examining the regulation of PLA_2 activity within the cell one encounters a number of interacting signaling systems. Thus, a number of signal transducers affect PLA_2 activation and hence eicosanoid production. To complicate the system further, however, many of the products of arachidonic acid, as well as arachidonic acid itself, affect intracellular enzymes involved in signaling as well as membrane channel activities. Hence, it is clear that this multifaceted interacting signaling system has the potential to be involved in many cellular responses, including the volume-regulatory responses.

In this section we will summarize the various factors believed to be involved in regulation of intracellular PLA_2 activity. Then we will discuss some of the many possible ways that arachidonic acid and its products can alter cellular function.

Calcium — As discussed above, many, but not all, of the PLA_2 enzymes are activated by increases in cytosolic free Ca^{2+} concentration. In the case of the large-molecular-mass (100 to 110 kDa by SDS-PAGE and 85.2 kDa based on predicted amino acid sequence) intracellular form of PLA_2 ($cPLA_2$), one half maximal activation of the enzyme occurs by increasing the cytosolic free Ca^{2+} concentration from 100 to approximately 500 n*M*.[182] This is similar to the calcium sensitivity of eicosanoid production in permeabilized mesangial cells.[190] Thus, the Ca^{2+} dependency of the "physiological read-out" of increased eicosanoid production paralleled nicely the results found with the isolated enzyme. Calcium also regulates PLA_2 activity of the cytosolic form by mediating the translocation of the enzyme from the cytosol to the membrane, thus promoting the apposition of the enzyme to its substrate in the membrane.[185,191] The amino-terminal, 138-amino acid fragment of this form of PLA_2 translocates to natural membrane vesicles in a Ca^{2+}-dependent fashion.[184] Furthermore, this part of the protein contains a 45-amino acid region that is homologous to a number of enzymes, including protein kinase C isoforms, which translocate to the membrane in a Ca^{2+}-dependent fashion; the synaptic vesicle protein, p65;[192] the GTPase-activating protein (GAP);[193] and phospholipase C.[194] Thus, this 45-amino acid region may determine Ca^{2+}-dependent phospholipid binding.

pH — With the exception of the 40-kDa calcium-independent cardiac form[188] and the lysosomal form of PLA_2,[195] which have acid pH optima at 4.5 and 6.4, all the other forms of PLA_2 for which pH dependency has been determined have neutral or alkaline pH optima.[196] Although optimal activities of the kidney and brain forms of the enzyme occur at pH of approximately 8.5 *in vitro,* there is regulated activity at physiological pH.[179,180,185,186,190] Thus, an increase in pH, as might occur by increased Na^+/H^+ exchange, would be expected to increase PLA_2 activity.

Protein kinases — It was reported from our laboratory[190] that protein kinase C enhances the activity of PLA_2 in permeabilized mesangial cells when intracellular free calcium concentration was fixed at levels achieved normally after stimulation with agonists such as vasopressin or endothelin.[197] Protein kinase C activation also potentiated Ca^{2+} ionophore (A23187)-stimulated PLA_2 activity in nonpermeabilized mesangial cells. Protein kinase C activation alone was not sufficient to increase PLA_2 activity if cytosolic free calcium concentrations were not raised above baseline levels. In another set of experiments PLA_2 activity in cytosolic extracts from cells stimulated with phorbol esters was determined and found to be increased over that in control cell extracts when assayed under saturating Ca^{2+} concentrations. Under these conditions the activity measured was primarily the cytosolic high-molecular-weight form of the enzyme.[185] Prolonged treatment with phorbol esters to downregulate protein kinase C resulted in loss of PMA-induced activation as well as vasopressin-induced activation.[186] These studies led to the conclusion that the cytosolic PLA_2 might be regulated by phosphorylation mediated in part by protein kinase C.

Protein kinase C and the p42MAP kinase have been reported by Nemenoff and colleagues[156] to directly phosphorylate and activate the cytosolic PLA_2. The two kinases phosphorylated different amino acid residues of the PLA_2 molecule. $cPLA_2$ was not a substrate for the ribosomal S6 kinase or p54 MAP kinase. In lysed synaptosomes Ca^{2+}/calmodulin-dependent protein kinase II inhibited PLA_2 activity,[198] whereas cAMP-dependent protein kinase and casein kinase II increased PLA_2 activity. If phosphorylation represents an important mechanism of activation of PLA_2 then phosphatases active against these phosphorylation sites may also be involved in regulation of enzymatic activity.[199]

G Proteins — Clearly, G proteins are important in activation of PLA_2 via the phospholipase C-IP_3-Ca^{2+} cascade. This involves G proteins of the G_q[75,200] and G_i[201] families. In addition, however, there is evidence for PLA_2 activation without, or prior to, activation of phospholipase C.[202] With some agonists these pathways may involve direct coupling of G proteins to PLA_2 activation. GTP analogues, when added to permeabilized cells, increase PLA_2 activity independently of phospholipase C activation.[203] When the α_o subunit cDNA is transfected into a murine adrenal cell line there is increased GTPγS-stimulatable release of arachidonic acid.[204] In addition to involvement of α subunits in the activation of PLA_2, there have been reports that $\beta\gamma$ subunits may be involved in some cells.[64,66]

Inhibitory and activating proteins — A family of proteins, referred to as lipocortins, were implicated over a number of years in the inhibition of PLA_2 that occurs secondary to glucocorticoids.[205,206] Two proteins that had PLA_2 inhibitory activity were cloned and termed lipocortin I and lipocortin II.[207,208] These compounds, however, inhibit PLA_2 activity by competing with the enzyme for interaction with the phospholipid substrate, rather than by directly inhibiting the enzyme.[209] A PLA_2-activating protein (PLAP) has been cloned recently.[210] LTD_4 increases synthesis of this protein in both smooth muscle and endothelial cells. This protein is not universally involved in PLA_2 regulation[210] and its role in PLA_2 regulation remains unclear.

Metabolic product inhibition — Unesterified *cis*-unsaturated, but not *trans*-unsaturated, fatty acids inhibit PLA_2 and may function as endogenous inhibitors of PLA_2 activity.[211,212] Oxidation of the *cis*-fatty acids destroys their inhibitory action. Lipoxygenase products of arachidonic acid, 5-hydroxy-6,8,11,15-eicosatetraenoic acid (5-HETE), 12-hydroxy-5,8,10,14-eicosatetraenoic acid (12-HETE), and 15-hydroxy-5,8,11,13-eicosatetraenoic acid (15-HETE), inhibit platelet PLA_2.[213]

Transcriptional regulation — While there is some information regarding transcriptional regulation of the expression of the secretory forms of PLA_2,[214–219] little is known about potential transcriptional regulation of intracellular forms of the enzyme. Interleukin 1-α (Il-1α) addition to a human lung fibroblast cell line, WI-38, causes a large increase in production of cytosolic PLA_2.[220] Dexamethasone blocks this effect on PLA_2.

G. ARACHIDONIC ACID AND ITS METABOLITES AS INTRACELLULAR MESSENGERS

Arachidonic acid and its metabolites have been demonstrated to be intracellular messengers in a number of eukaryotic systems. Some of these effects may be related directly to processes important for cell volume regulation, since they include the modification of the activities of membrane channels, phospholipase C, intracellular calcium release, guanylate cyclase, GTP-ase-activating protein (GAP), and protein kinase C.

1. Arachidonic Acid

Arachidonic acid, as well as other fatty acids that are not substrates for eicosanoid production, can directly modulate the activity of a number of different types of membrane channels.[221] They activate K^+ channels in smooth muscle cells.[222] They activate the ATP-insensitive K^+ channel in rat atrial[64] and ventricular[65] cells, but inhibit an ATP-dependent K^+ channel in rat ventricular cells.[65] They activate the muscarinic-gated K^+ channel in atrial cells.[64] Arachidonic acid, and other *cis*-unsaturated fatty acids, block Cl^- channels in membrane patches from airway epithelial cells.[223,224] These channels are implicated in the pathophysiology of cystic fibrosis.[225]

Arachidonic acid also affects a number of enzyme systems involved in intracellular signaling. It alters cellular guanylate cyclase activity[226] and may participate in the coupling of receptors to soluble guanylate cyclase.[227] Arachidonic acid stimulates phospholipase C in GH_3 cells[228] and human placental cells.[229] Arachidonic acid inhibits the Ca^{2+}/calmodulin-dependent protein kinase II purified from brain.[230] Arachidonic acid inhibits the GTPase-activating protein, GAP,[49] decreasing the GAP-stimulated rate at which the cellular H-Ras protein converts bound guanosine triphosphate (GTP) to guanosine diphosphate (GDP), thus enhancing Ras activity. Arachidonic acid also may regulate cellular Ras activity by stimulation of a GTPase-inhibitory protein.[51] Thus, by inhibiting a GTPase-activating protein and stimulating a GTPase-inhibitory protein, GTPase activity will be reduced and cellular Ras activity will increase.

Arachidonic acid activates protein kinase C,[231,232] activating the γ and ζ isoforms[139] of protein kinase C but not the α or β forms.[233] In addition to this effect of the fatty acid alone, arachidonic acid potentiates the effects of diacylglycerol on all subspecies of protein kinase C, increasing the apparent affinity of protein kinase C for Ca^{2+}. We have reported that arachidonic acid can release Ca^{2+} from intracellular stores of mesangial cells, although this release is not as great as that seen with some lipoxygenase and P450 mixed function oxidase products.[234]

2. Arachidonic Acid Metabolites

Arachidonic acid is metabolized along three primary pathways: (1) PGH synthases (with cyclooxygenase and peroxidase function), followed by specific synthases (isomerases) or reductases resulting in the formation of prostanoids or thromboxanes. (2) Lipoxygenases introduce oxygen to form hydroperoxyeicosatetraenoic acids (HPETEs). HPETEs can then be converted to either: (a) diHETEs, by lipoxygenation at another site on the aliphatic chain; (b) HETEs, by reduction of the hydroperoxy group; or (c) leukotrienes, by dehydration of the hydroperoxy group to an epoxy bond followed by ring opening in the presence of GSH and attachment of GSH via a sulfoether linkage. (3) Epoxygenases acting upon arachidonic acid result in the formation of epoxyeicosatrienoic acids (EETs), which can then be metabolized to dihydroxyeicosatrienoic acids (DHETs) by hydrolases. These pathways are depicted schematically in Figure 9–9.

A number of these metabolites of arachidonic acid have been found to have properties suggesting roles in intracellular signaling events. PGH synthase products have been reported to inhibit atrial muscarinic K^+ channels.[235] Prostaglandin E_2 inhibits RVD in Ehrlich ascites tumor cells when the hypotonic medium contains Na^+, but not in the absence of Na^+, leading to the conclusion that PGE_2 increases Na^+ permeability.[1] Lipoxygenase metabolites increase the open time of S channels and thereby increase net outward K^+ current in *Aplysia* sensory neurons.[236] Lipoxygenase products alter K^+ channel activity linked to muscarinic receptors in the heart.[64] Cantiello et al.[237] reported that $G_{\alpha i\text{-}3}$, GTPγS, or mellitin, an activator of PLA_2, increased amiloride-sensitive Na^+ channel activity in a manner inhibitable by quinacrine or NDGA, in inside-out patches of A6 renal epithelial cells. 5-HPETE or LTD_4 overcame the effect of NDGA. Chloride channels may also be regulated by lipoxygenase products. In the bovine chromaffin cell, GTPγS activates chloride channels in a manner inhibited by a PLA_2 inhibitor or by NDGA.[238] Thus, a mechanism of chloride channel activation may involve G protein-linked activation of PLA_2 with subsequent production of a lipoxygenase product.

During RVD, which is rate limited by K^+ efflux in Erlich ascites tumor cells, the synthesis of leukotrienes is enhanced and the synthesis of pros-

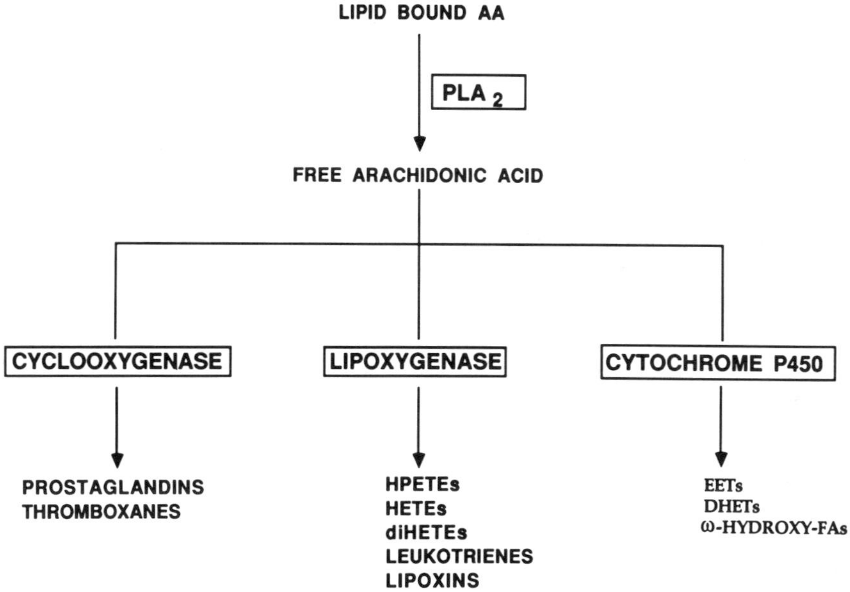

Figure 9–9 Pathways of arachidonic acid (AA) metabolism. Prostaglandins and thromboxanes are the products of the metabolism of arachidonic acid by PGH synthase, one component of which is a cyclooxgenase enzyme. Alternatively, arachidonic acid can be metabolized by lipoxygenases to hydroperoxyeicosatetraenoic acids (HPETEs). HPETEs can be hydroxylated to hydroxyeicosatetraenoic acids (HETEs) or acted upon by other lipoxygenases to form diHETEs. Leukotrienes are the products of the 5′-lipoxygenase and lipoxins are trihydroxylated products of the 5′- and 15′-lipoxygenases. The cytochrome P-450 monooxygenase system metabolizes arachidonic acid to epoxyeicosatrienoic acids (EETs), which can then be metabolized to dihydroxyeicosatrienoic acids (DHETs) by epoxide hydrolases. This enzyme system can also result in the production of ω-hydroxylated fatty acids. (From Bonventre, J. V., *J. Am. Soc. Nephrol.,* 3, 128, 1993.)

taglandins is reduced.[239] Addition of LTD_4 enhanced RVD and addition of an inhibitor of lipoxygenase activity (NDGA) inhibited RVD. Lambert et al.[239] proposed that hypotonic cell swelling enhances leukotriene production, which in turn increases K^+ and Cl^- transport. Cl^- conductance is important, as demonstrated by the fact that RVD was inhibited when NDGA was used even in the presence of gramacidin, which would ensure high K^+ permeability. It was further proposed that leukotrienes might mediate the well-known Ca^{2+}-calmodulin dependence of RVD since LTD_4 overcame the inhibition of RVD induced by the calmodulin inhibitor, pimozide.[240]

Lipoxygenase products can also modulate activity of soluble guanylate cyclase,[241] stimulate protein kinase C activity,[232] and inhibit Ca^{2+}/calmodulin-dependent protein kinase from rat brain cortex.[230]

In rat mesangial cells we have found that 15-HPETE, and to a lesser extent, PGE_2 and 15-HETE released Ca^{2+} from intracellular storage sites.[234] Thus, after an agonist interacts with a cell to increase phospholipase C and polyphosphoinositide turnover, enhancing PLA_2 activity, subsequent lipoxygenase product production will potentiate the inositol trisphosphate-induced release of Ca^{2+}.

3. Epoxygenase Products

There is little information regarding potential roles for epoxygenase products in cell volume regulatory events. It should be recognized, however, that NDGA, which is widely used as a "lipoxygenase" inhibitor, also has inhibitory activity toward the cytochrome P450 system, which is involved in the generation of epoxide metabolites of arachidonic acid.[242] Therefore, it is possible that some of the effects attributable to lipoxygenase products may actually be mediated by epoxygenase products.

It has been suggested that the P450 mixed function oxidase system plays an important role in S-K^+ channel regulation in *Aplysia* neurons.[243] We have reported that SKF-525A, an inhibitor of the P450 system, and NDGA inhibit the vasopressin-induced increase in cytosolic free [Ca^{2+}] ([Ca^{2+}]$_i$) in glomerular mesangial cells without altering the vasopressin-induced increase in inositol phosphates.[244] This inhibition was overcome by addition of 14,15-EET (100 n*M*) without an associated increase in phospholipase C activity. 14,15-EET increased [Ca^{2+}]$_i$ when added to intact single mesangial cells, and also induced Ca^{2+} release from intracellular stores of permeabilized mesangial cells. Madhun et al. have reported that 5,6-EET enhances Ca^{2+} influx through voltage-sensitive channels of rabbit proximal tubule epithelial cells.[245] An epoxygenase product inhibits microsomal Na^+,K^+-ATPase activity in rabbit kidney medullary thick ascending limb cells.[246]

H. ADENYLYL CYCLASE

Cyclic AMP is involved in the regulation of many cell membrane ionic conductances as well as transport pathways such as Na^+/H^+ and Cl^-/HCO_3^- exchange processes. Its potential role in cell volume regulatory processes remains controversial.

The classical mechanism of adenylyl cyclase activation involves a receptor coupled to one or more G proteins, and the catalytic unit of the cyclase. Adrenergic, parathyroid hormone, calcitonin, and PGE_2 receptors are examples of receptors that activate the adenylyl cyclase enzyme. The G_s and G_i families of G proteins confer stimulatory or inhibitory specificities in this signaling pathway.

In addition to receptor activation, however, it has been recognized that mechanical deformation can activate adenylyl cyclase. This is of importance in the volume-regulatory responses which may be triggered by cell deformation induced by hypoosmotic or hyperosmotic stimuli, such as swelling of S49 mouse lymphoma cells,[247] and turkey erythrocytes.[248] Deformation-related increases in cAMP in the S49 cells were blocked by miconazole, an inhibitor of adenylyl cyclase which does not affect G protein function.[249] Increased levels of cAMP can enhance Na^+,K^+,$2Cl^-$ cotransport in avian red cells.[250,251] This cotransporter is believed to be important in the RVI response. It does not, however, appear that cAMP is involved in the RVI since adenylyl cyclase is not activated by cell shrinkage.[252]

I. CELLULAR pH

Intracellular pH is regulated by three primary processes: the Cl^-/HCO_3^- exchanger, the Na^+-dependent Cl^-/HCO_3^- exchanger, and the Na^+/H^+ exchanger. Na^+/H^+ exchangers are found in the plasma membranes of a large variety of cells. There are at least two, and possibly three, pharmacologically distinct types of Na^+/H^+ exchangers.[253] One class is present on nonepithelial cells and basolateral membranes of many epithelial cells. This class is inhibited by low concentrations of 5-amino-substituted amiloride derivatives. A second class, found in the apical membrane of the proximal tubule, is inhibited only by high concentrations of the amiloride derivatives. A possible third class of Na^+-dependent acid extruders, found in nonepithelial cells, is not inhibited by amiloride compounds.[253]

Many factors that cause cell deformation also cause an activation of Na^+/H^+ exchange and intracellular alkalinization. The shape change in platelets that occurs shortly after epinephrine activation is associated with activation of the Na^+/H^+ exchanger and the subsequent metabolic consequences of activation, including aggregation, and activation can be blocked by amiloride derivatives.[254] Shrinkage of cells by osmotic stimuli results in activation of Na^+/H^+ exchange in a large number of cell types.[255–258] The mechanism of activation of the exchanger has been proposed by Grinstein and Rothstein[256] to involve an alkaline shift in the pH dependence of the exchanger. This change may be due to phosphorylation of the exchanger by a process independent of protein kinase C.[259]

Growth factors and mitogenic peptides also have effects on the exchangers regulating cellular pH. It has been shown in some cells that vasopressin and EGF activate all three exchangers involved in pH regulation with a similar time course.[253] In red cells and lymphocytes, but not in ascites tumor cells,[260] the RVI response appears to involve the coupling of an amiloride-sensitive Na^+/H^+ exchange to Cl^-/HCO_3^-.

IV. CONCLUSIONS

We have attempted to summarize the basic features of general cellular signaling pathways. We have only scratched the surface and each area discussed could easily serve as the topic for a chapter of this length. We have also left a good deal out. We have ignored the events that translate the cytosolic signaling into nuclear events and we have also left out intranuclear events critical to gene transcription and therefore critical to many responses of cells to ligands. Nevertheless, we have covered many signaling processes that are likely to be involved in the rapid cellular responses characteristic of volume-regulatory behavior in the setting of perturbation of environmental conditions. While many studies have been reported on changes in membrane channel conductances the regulatory processes involved in these changes are not well understood. We hope that this chapter and those that follow will give the investigator who has not wandered into the complicated arena of cellular signals the courage to venture forth into this morass in search of answers to the very fundamental questions dealing with how cells respond to perturbations in their environment leading to change in their volume.

ACKNOWLEDGMENTS

The authors are supported by National Institutes of Health grants DK 39773, DK38452, NS 10828, and DK01986.

REFERENCES

1. **Hoffmann, E. K. and Ussing, H. H.,** Membrane mechanisms in volume regulation in vertebrate cells and epithelia, in *Membrane Transport in Biology,* Vol. 5, Schafer, J. A., Ussing, H. H., Kristensen, P., and Giebisch, G. H., Eds., Springer-Verlag, Berlin, 1992, 317.
2. **Ullrich, A. and Schlessinger, J.,** Signal transduction by receptors with tyrosine kinase activity, *Cell,* 61, 203, 1990.
3. **Simon, M. I., Strathmann, M. P., and Gautam, N.,** Diversity of G proteins in signal transduction, *Science,* 252, 802, 1991.
4. **Wrana, J. L., Attisano, L., Carcamo, J., Zentella, A., Doody, J., Laiho, M., Wang, X., and Massague, J.,** TGFβ signals through a heteromeric protein kinase receptor complex. *Cell,* 71, 1003, 1992.
5. **Matthews, L. S. and Vale, W. W.,** Expression cloning of an activin receptor, a predicted transmembrane serine kinase, *Cell,* 65, 973, 1991.
6. **Miyajima, A., Kitamura, T., Harada, N., Yokota, T., and Arai, K.,** Cytokine receptors and signal transduction, *Annu. Rev. Immunol.,* 10, 295, 1992.
7. **Kolch, W., Heidecker, G., Lloyd, P., and Rapp, U. R.,** Raf-1 protein kinase is required for growth of NIH/3T3 cells, *Nature,* 349, 426, 1991.
8. **Turner, B., Rapp, U., App, H., Greene, M., and Dobashi, K.,** IL-2 induces tyrosine phosphorylation and activation of p72–74 Raf-1 kinase in a T-cell line, *Proc. Natl. Acad. Sci. U.S.A.,* 88, 1227, 1991.
9. **Carroll, M. P., Spivak, J. L., McMahon, M., Weich, N., Rapp, U. R., and May, W. S.,** Erythropoietin induces Raf-1 activation and Raf-1 is required for erythropoietin-mediated proliferation, *J. Biol. Chem.,* 266, 14964, 1991.
10. **Hatakeyama, M., Kono, T., Kobayashi, N., Kawahawa, A., Levin, S. D., Perlmutter, R. M., and Taniguchi, T.,** Interaction of the IL-2 receptor with the *src*-family kinase $p56^{lck}$: identification of novel intermolecular association, *Science,* 252, 1523, 1991.
11. **Samelson, L. E. and Klausner, R. D.,** Tyrosine kinases and tyrosine-based activation motifs, *J. Biol. Chem.,* 267, 24913, 1992.
12. **Yu, J., Heidaran, M. A., Pierce, J. H., Gutkind, J. S., Lombardi, D., Ruggiero, M., and Aaronson, S. A.,** Tyrosine mutations within the α platelet-derived growth factor receptor kinase insert domain abrogate receptor-associated phosphatidylinositol-3 kinase activity without affecting mitogenic or chemotactic signal transduction, *Mol. Cell. Biol.,* 11, 3780, 1991.
13. **Kazlauskas, A., Ellis, C., Pawson, T., and Cooper, J. A.,** Binding of GAP to activated PDGF receptors, *Science,* 247, 1578, 1990.
14. **Margolis, B., Rhee, S. G., Felder, S., Mervic, M., Lyall, R., Levitzki, A., Ullrich, A., Zilberstein, A., and Schlessinger, J.,** EGF induces tyrosine phosphorylation of phospholipase C-II: a potential mechanism for EGF receptor signalling, *Cell,* 57, 1101, 1989.
15. **Cantley, L. C., Auger, K. R., Carpenter, C., Duckworth, B., Graziani, A., Kapeller, R., and Soltoff, S.,** Oncogenes and signal transduction, *Cell,* 64, 281, 1991.
16. **Sun, X. J., Miralpeix, M., Myers, M. G., Jr., Glasheen, E. M., Backer, J. M., Kahn, C. R., and White, M. F.,** Expression and function of IRS-1 in insulin signal transmission, *J. Biol. Chem.,* 267, 22662, 1992.
17. **Lowenstein, E. J., Daly, R. J., Batzer, A. G., Li, W., Margolis, B., Lammers, R., Ullrich, A., Skolnik, E. Y., Bar-Sagi, D., and Schlessinger, J.,** The SH2 and SH3 domain-containing protein GRB2 links receptor tyrosine kinases to ras signaling, *Cell,* 70, 431, 1992.
18. **Margolis, B., Silvennoinen, O., Comoglio, F., Roonprapunt, C., Skolnik, E., Ullrich, A., and Schlessinger, J.,** High-efficiency expression/cloning of epidermal growth factor-receptor-binding proteins with Src homology 2 domains, *Proc. Natl. Acad. Sci. U.S.A.,* 89, 8894, 1992.
19. **Matuoka, K., Shibata, M., Yamakawa, A., and Takenawa, T.,** Cloning of ASH, a ubiquitous protein composed of one Src homology region (SH) 2 and two SH3 domains, from human and rat cDNA libraries, *Proc. Natl. Acad. Sci. U.S.A.,* 89, 9015, 1992.
20. **Pelicci, G., Lanfrancone, L., Grignani, F., McGlade, J., Cavallo, F., Forni, G., Nicoletti, I., Grignani, F., Pawson, T., and Pelicci, P.**

G., A novel transforming protein (SHC) with an SH2 domain is implicated in mitogenic signal transduction, *Cell,* 70, 93, 1992.

21. **Nishibe, S., Wahl, M. I., Hernandez-Sotomayor, S. M. T., Tonks, N. K., Rhee, S. G., and Carpenter, G.,** Increase of the catalytic activity of phospholipase C-γ1 by tyrosine phosphorylation, *Science,* 250, 1253, 1990.
22. **Kypta, R. M., Goldberg, Y., Ulug, E. T., and Courtneidge, S. A.,** Association between the PDGF receptor and members of the *src* family of tyrosine kinases, *Cell,* 62, 481, 1990.
23. **Gilman, A. G.,** G proteins: transducers of receptor-generated signals, *Annu. Rev. Biochem.,* 56, 615, 1987.
24. **Okamoto, T., Murayama, Y., Hayashi, Y., Inagaki, M., Ogata, E., and Nishimoto, I.,** Identification of a G_S activator region of the β2-adrenergic receptor that is autoregulated via protein kinase A-dependent phosphorylation, *Cell,* 67, 723, 1991.
25. **Mulcahy, L. S., Smith, M. R., and Stacey, D. W.,** Requirement for *ras* proto-oncogene function during serum-stimulated growth of NIH 3T3 cells, *Nature,* 313, 241, 1985.
26. **Cai, H., Szeberenyi, J., and Cooper, G. M.,** Effect of a dominant inhibitory Ha-*ras* mutation on mitogenic signal transduction in NIH 3T3 cells, *Mol. Cell. Biol.,* 10, 5314, 1990.
27. **Smith, M. R., DeGudicibus, S. J., and Stacey, D. W.,** Requirement for c-*ras* proteins during viral oncogene transformation, *Nature,* 320, 540, 1986.
28. **Feig, L. A. and Cooper, G. M.,** Inhibition of NIH 3T3 cell proliferation by a mutant *ras* protein with preferential affinity for GDP, *Mol. Cell. Biol.,* 8, 3235, 1988.
29. **Hagag, N., Halegoua, S., and Viola, M.,** Inhibition of growth factor-induced differentiation of PC12 cells by microinjection of antibody to *ras* p21, *Nature,* 319, 680, 1986.
30. **Nakafuku, M., Satoh, T., and Kaziro, Y.,** Differentiation factors, including nerve growth factor, fibroblast growth factor, and interleukin-6, induce an accumulation of an active Ras-GTP complex in rat pheochromocytoma PC12 cells, *J. Biol. Chem.,* 267, 19448, 1992.
31. **Szeberenyi, J., Cai, H., and Cooper, G. M.,** Effect of a dominant inhibitory Ha-*ras* mutation on neuronal differentiation of PC12 cells, *Mol. Cell. Biol.,* 10, 5324, 1990.
32. **Mulder, K. M., and Morris, S. L.,** Activation of $p21^{ras}$ by transforming growth factor β in epithelial cells, *J. Biol. Chem.,* 267, 5029, 1992.
33. **Ridley, A. J., Paterson, H. F., Johnston, C. L., Diekmann, D., and Hall, A.,** The small GTP-binding protein rac regulates growth factor-induced membrane ruffling, *Cell,* 70, 401, 1992.
34. **Ridley, A. J. and Hall, A.,** The small GTP-binding protein rho regulates the assembly of focal adhesions and actin stress fibers in response to growth factors, *Cell,* 70, 389, 1992.
35. **Downward, J.,** Rac and Rho in tune, *Nature,* 359, 273, 1992.
36. **Settleman, J., Albright, C. F., Foster, L. C., and Weinberg, R. A.,** Association between GTPase activators for Rho and Ras families, *Nature,* 359, 153, 1992.
37. **Balch, W. E.,** Small GTP-binding proteins in vesicular transport, *Trends Biochem. Sci.,* 15, 473, 1990.
38. **Satoh, T., Nakafuku, M., and Kaziro, Y.,** Function of Ras as a molecular switch in signal transduction, *J. Biol. Chem.,* 267, 24149, 1992.
39. **Bourne, H. R., Sanders, D. A., and McCormick, F.,** The GTPase superfamily: a conserved switch for diverse cell functions, *Nature,* 348, 125, 1990.
40. **Simon, M. A., Bowtell, D. D., Dodson, G. S., Laverty, T. R., and Rubin, G. M.,** Ras1 and a putative guanine nucleotide exchange factor perform crucial steps in signaling by the sevenless protein tyrosine kinase, *Cell,* 67, 701, 1991.
41. **Shou, C., Farnsworth, C. L., Neel, B. G., and Feig, L. A.,** Molecular cloning of cDNAs encoding a guanine-nucleotide-releasing factor for Ras p21, *Nature,* 358, 351, 1992.
42. **Li, B., Kaplan, D., Kung, H., and Kamata, T.,** Nerve growth factor stimulation of the Ras-guanine nucleotide exchange factor and GAP activities, *Science,* 256, 1456, 1992.
43. **Zhang, K., Papageorge, A. G., and Lowy, D. R.,** Mechanistic aspects of signaling through Ras in NIH 3T3 cells, *Science,* 257, 671, 1992.
44. **Molloy, C. J., Bottaro, D. P., Fleming, T. P., Marshall, M. S., Gibbs, J. B., and Aaronson, S. A.,** PDGF induction of tyrosine phosphorylation of GTPase activating protein, *Nature,* 342, 711, 1989.
45. **Kaplan, D. R., Morrison, D. K., Wong, G., McCormick, F., and Williams, L. T.,** PDGF β-receptor stimulates tyrosine phosphorylation of GAP and association of GAP with a signaling complex, *Cell,* 61, 125, 1990.
46. **Ellis, C., Moran, M., McCormick, F., and Pawson, T.,** Phosphorylation of GAP and GAP-associated proteins by transforming and

mitogenic tyrosine kinases, *Nature,* 343, 377, 1990.

47. **Downward, J., Graves, J. D., Warne, P. H., Rayter, S., and Cantrell, D. A.,** Stimulation of $p21^{ras}$ upon T-cell activation, *Nature,* 346, 719, 1990.
48. **Han, J., McCormick, F., and Macara, I.,** Regulation of Ras-GAP and the neurofibromatosis-1 gene product by eicosanoids, *Science,* 252, 576, 1991.
49. **Tsai, M.-H., Yu, C.-L., Wei, F.-S., and Stacey, D. W.,** The effect of GTPase activating protein upon Ras is inhibited by mitogenically responsive lipids, *Science,* 243, 522, 1989.
50. **Yu, C., Tsai, M., and Stacey, D. W.,** Serum stimulation of NIH 3T3 cells induces the production of lipids able to inhibit GTPase-activating protein activity, *Mol. Cell. Biol.,* 10, 6683, 1990.
51. **Tsai, M.-H., Yu, C.-L., and Stacey, D. W.,** A cytoplasmic protein inhibits the GTPase activity of H-Ras in a phospholipid-dependent manner, *Science,* 250, 982, 1990.
52. **Ruggieri, R. and McCormick, F.,** Ras and the *awd* couple, *Nature,* 353, 390, 1991.
53. **Teng, D. H. F., Engele, C. M., and Venkatesh, T. R.,** A product of the *prune* locus of *Drosophila* is similar to mammalian GTPase-activating protein, *Nature,* 353, 437, 1991.
54. **Downward, J.,** The ras superfamily of small GTP-binding proteins, *Trends Biochem. Sci.,* 15, 469, 1990.
55. **Thomas, S. M., DeMarco, M., D'Arcangelo, G., Halegoua, S., and Brugge, J. S.,** Ras is essential for nerve growth factor- and phorbol ester-induced tyrosine phosphorylation of MAP kinases, *Cell,* 68, 1031, 1992.
56. **Wood, K. W., Sarnecki, C., Roberts, T. M., and Blenis, J.,** *ras* mediates nerve growth factor receptor modulation of three signal-transducing protein kinases: MAP kinase, Raf-1, and RSK, *Cell,* 68, 1041, 1992.
57. **Kyriakis, J. M., App, H., Zhang, X., Banerjee, P., Brautigan, D. L., Rapp, U. R., and Avruch, J.,** Raf-1 activates MAP kinase-kinase, *Nature,* 358, 417, 1992.
58. **Farnsworth, C. L., Marshall, M. S., Gibbs, J. B., Stacey, D. W., and Feig, L. A.,** Preferential inhibition of the oncogenic form of Ras^{H} by mutations in the GAP binding/"effector" domain, *Cell,* 64, 625, 1991.
59. **Martin, G. A., Yatani, A., Clark, R., Conroy, L., Polakis, P., Brown, A. M., and McCormick, F.,** GAP domains responsible for Ras p21-dependent inhibition of muscarinic atrial K+ channel currents, *Science,* 255, 192, 1992.
60. **Satoh, T., Nakafuku, M., Miyajima, A., and Kaziro, Y.,** Involvement of *ras* p21 protein in signal-transduction pathways from interleukin 2, interleukin 3, and granulocyte/macrophage colony-stimulating factor, but not from interleukin 4, *Proc. Natl. Acad. Sci. U.S.A.,* 88, 3314, 1991.
61. **Gibbs, J. B., Marshall, M. S., Scolnick, E. M., Dixon, R. A. F., and Vogel, U. S.,** Modulation of guanine nucleotides bound to Ras in NIH3T3 cells by oncogenes, growth factors, and the GTPase activating protein (GAP), *J. Biol. Chem.,* 265, 20437, 1990.
62. **Berstein, G., Blank, J. L., Jhon, D., Exton, J. H., Rhee, S. G., and Ross, E. M.,** Phospholipase C-β1 is a GTPase-activating protein for $G_{q/11}$, its physiologic regulator, *Cell,* 70, 411, 1992.
63. **Whiteway, M., Hougan, L., Dignard, D., Thomas, D. Y., Bell, L., Saari, G. C., Grant, F. J., O'Hara, P., and MacKay, V. L.,** The *STE4* and *STE18* genes of yeast encode potential β and γ subunits of the mating factor receptor-coupled G protein, *Cell,* 56, 467, 1989.
64. **Kim, D., Lewis, D. L., Graziadei, L., Neer, E. J., Bar-Sagi, D., and Clapham, D. E.,** G-protein βγ-subunits activate the cardiac muscarinic K^+ channel via phospholipase A_2, *Nature,* 337, 557, 1989.
65. **Kim, D. and Duff, R. A.,** Regulation of K^+ channels in cardiac myocytes by free fatty acids, *Circ. Res.,* 67, 1040, 1990.
66. **Jelsema, C. L. and Axelrod, J.,** Stimulation of phospholipase A_2 activity in bovine rod outer segments by the βγ subunit of transducin and its inhibition by α subunit, *Proc. Natl. Acad. Sci. U.S.A.,* 84, 3623, 1987.
67. **Pitcher, J. A., Inglese, J., Higgins, J. B., Arriza, J. L., Casey, P. J., Kim, C., Benovic, J. L., Kwatra, M. M., Caron, M. G., and Lefkowitz, R. J.,** Role of βγ subunits of G proteins in targeting the β-adrenergic receptor kinase to membrane-bound receptors, *Science,* 257, 1264, 1992.
68. **Inglese, J., Koch, W. J., Caron, M. G. and Lefkowitz, R. J.,** Isoprenylation in regulation of signal transduction by G-protein-coupled receptor kinases, *Nature,* 359, 147, 1992.
69. **Kleuss, C., Scherubl, H., Hescheler, J., Schultz, G., and Wittig, B.,** Different β-subunits determine G-protein interaction with transmembrane receptors, *Nature,* 358, 424, 1992.

70. **Strathmann, M. P. and Simon, M. I.,** Gα12 and Gα13 subunits define a fourth class of G protein α subunits, *Proc. Natl. Acad. Sci. U.S.A.*, 88, 5582, 1991.
71. **Jones, D. T., Masters, S. B., Bourne, H. R., and Reed, R. R.,** Biochemical characterization of three stimulatory GTP-binding proteins, *J. Biol. Chem.*, 265, 2671, 1990.
72. **Lee, C. H., Park, D., Wu, D., Rhee, S. G., and Simon, M. I.,** Members of the G_q α subunit gene family activate phospholipase C β isozymes, *J. Biol. Chem.*, 267, 16044, 1992.
73. **Park, D., Jhon, D., Kriz, R., Knopf, J., and Rhee, S. G.,** Cloning, sequencing, expression, and G_q-independent activation of phospholipase C-β2, *J. Biol. Chem.*, 267, 16048, 1992.
74. **Yatani, A., Mattera, R., Codina, J., Graf, R., Okabe, K., Padrell, E., Iyengar, R., Brown, A. M., and Birnbaumer, L.,** The G protein-gated atrial K^+ channel is stimulated by three distinct $G_i\alpha$-subunits, *Nature*, 336, 680, 1988.
75. **Smrcka, A. V., Hepler, J. R., Brown, K. O., and Sternweis, P. C.,** Regulation of polyphosphoinositide-specific phospholipase C activity by purified G_q, *Science*, 251, 804, 1991.
76. **Gutowski, S., Smrcka, A., Nowak, L., Wu, D., Simon, M., and Sternweis, P. C.,** Antibodies to the αq-subfamily of guanine nucleotide-binding regulatory protein α subunits attenuate activation of phophatidylinositol 4,5-bisphosphate hydrolysis by hormones, *J. Biol. Chem.*, 266, 20519, 1991.
77. **Surprenant, A., Horstman, D. A., Akbarali, H., and Limbird, L. E.,** A point mutation of the α2-adrenoceptor that blocks coupling to potassium but not calcium currents, *Science*, 257, 977, 1992.
78. **Abou-Samra, A.-B., Juppner, H., Force, T., Freeman, M. W., Kong, X.-F., Schipani, E., Urena, P., Richards, J., Bonventre, J. V., Potts, J. T., Kronenberg, H. M., and Segre, G. V.,** Expression cloning of a PTH/PTHrP receptor from rat osteoblast-like cells: a single receptor stimulates intracellular accumulation of both cAMP and inositol trisphosphates and increases intracellular free calcium, *Proc. Natl. Acad. Sci. U.S.A.*, 89, 2732, 1992.
79. **Force, T., Bonventre, J. V., Flannery, M. R., Gorn, A. H., Yamin, M., and Goldring, S.,** A cloned porcine renal calcitonin receptor couples to adenylyl cyclase and phospholipase C, *Am. J. Physiol.*, 262, F1110, 1992.
80. **Brown, A. M. and Birnbaumer, L.,** Ionic channels and their regulation by G protein subunits, *Annu. Rev. Physiol.*, 52, 197, 1990.
81. **Light, D. B., Ausiello, D. A., and Stanton, B. A.,** Guanine nucleotide-binding protein α_{1-3}i directly activates a cation channel in rat renal inner medullary collecting duct cells, *J. Clin. Invest.*, 84, 352, 1989.
82. **Yatani, A., Codina, J., Imoto, Y., Reeves, J. P., Birnbaumer, L., and Brown, A. M.,** Direct regulation of mammalian cardiac calcium channels by a G protein, *Science*, 238, 1288, 1987.
83. **Whitman, M., Downes, C. P., Keeler, M., Keller, T., and Cantloy, L.,** Type 1 phosphatidylinositol kinase makes a novel inositol phospholipid phosphatidylinositol-3-phosphate, *Nature*, 332, 644, 1988.
84. **Raffioni, S. and Bradshaw, R. A.,** Activation of phosphatidylinositol 3-kinase by epidermal growth factor, basic fibroblast growth factor, and nerve growth factor in PC12 pheochromocytoma cells, *Cell Biol.*, 89, 9121, 1992.
85. **Eberle, M., Traynor-Kaplan, A. E., Sklar, L. A., and Norgauer, J.,** Is there a relationship between phosphatidylinositol triphosphate and F-actin polymerization in human neutrophils?, *J. Biol. Chem.*, 265, 16725, 1990.
86. **Kucera, G. L., and Rittenhouse, S. E.,** Human platelets form 3-phosphorylated phosphoinositides in response to α-thrombin, U46619, or GTPγS, *J. Biol. Chem.*, 265, 5345, 1990.
87. **Talmage, D. A., Freund, R., Young, A. T., Dahl, J., Dawe, C. J., and Benjamin, T. L.,** Phosphorylation of middle T by pp60$^{c\text{-}src}$: a switch for binding of phosphatidylinositol 3-kinase and optimal tumorigenesis, *Cell*, 59, 55, 1989.
88. **Majerus, P. W., Ross, T. S., Cunningham, T. W., Caldwell, K. K., Jefferson, A. B., and Bansal, V. S.,** Recent insights in phosphatidylinositol signaling, *Cell*, 63, 459, 1990.
89. **Hiles, I. D., Otsu, M., Volinia, S., Fry, M. J., Gout, I., Dhand, R., Panayotou, G., Ruiz-Larrea, F., Thompson, A., Totty, N. F., Hsuan, J. J., Courtneidge, S. A., Parker, P. J., and Waterfield, M. D.,** Phosphatidylinositol 3-kinase: structure and expression of the 110 kd catalytic subunit, *Cell*, 70, 419, 1992.
90. **Severinsson, L., Ek, B., Mellstrom, K., Claesson-Welsh, L., and Heldin, C. H.,** Deletion of the kinase insert sequence of the platelet-derived growth factor β-receptor affects receptor kinase activity and signal transduction, *Mol. Cell. Biol.*, 10, 801, 1990.
91. **Kaplan, D. R., Whitman, M., Schaffhausen, B., Pallas, D. C., White, M., Cantley, L., and**

Roberts, T. M., Common elements in growth factor stimulation and oncogenic transformation: 85kd phosphoprotein and phosphatidylinositol kinase activity, *Cell,* 50, 1021, 1987.

92. **Escobedo, J. A. and Williams, L. T.**, A PDGF receptor domain essential for mitogenesis but not for many other responses to PDGF, *Nature,* 335, 85, 1988.
93. **Serunian, L. A., Auger, K. R., and Cantley, L. C.**, Identification and quantification of polyphosphoinositides produced in response to platelet-derived growth factor stimulation, *Meth. Enzymol.,* 198, 78, 1991.
94. **Rhee, S. G. and Choi, K. D.**, Regulation of inositol phospholipid-specific phospholipase C isozymes, *J. Biol. Chem.,* 267, 12393, 1992.
95. **Meisenhelder, J., Suh, P., Rhee, S. G., and Hunter, T.**, Phospholipase C-γ is a substrate for the PDGF and EGF receptor protein-tyrosine kinases *in vivo* and *in vitro, Cell,* 57, 1109, 1989.
96. **Rotin, D., Margolis, B., Mohammadi, M., Daly, R. J., Daum, G., Li, N., Fischer, E. H., Burgess, W. H., Ullrich, A., and Schlessinger, J.**, SH2 domains prevent tyrosine dephosphorylation of the EGF receptor: identification of Tyr992 as the high-affinity binding site for SH2 domains of phospholipase C-2, *EMBO J.,* 11, 559, 1992.
97. **Mohammadi, M., Honegger, A. M., Rotin, D., Fischer, R., Bellot, F., Li, W., Dionne, C. A., Jaye, M., Rubinstein, M., and Schlessinger, J.**, A tyrosine-phosphorylated carboxy-terminal peptide of the fibroblast growth factor receptor (Flg) is a binding site for the SH2 domain of phospholipase C-γ1, *Mol. Cell. Biol.,* 11, 5068, 1991.
98. **Kim, H. K., Kim, J. W., Zilberstein, A., Margolis, B., Kim, J. G., Schlessinger, J., and Rhee, S. G.**, PDGF stimulation of inositol phospholipid hydrolysis requires PLC-γ1 phosphorylation on tyrosine residues 783 and 1254, *Cell,* 65, 435, 1991.
99. **Goldschmidt-Clermont, P. J., Kim, J. W., Machesky, L. M., Rhee, S. G., and Pollard, T. D.**, Regulation of phospholipase C-γ1 by profilin and tyrosine phosphorylation, *Science,* 251, 1231, 1991.
100. **Park, D. J., Rho, H. W., and Rhee, S. G.**, CD3 stimulation causes phosphorylation of phospholipase C-γ1 on serine and tyrosine residues in a human T-cell line, *Proc. Natl. Acad. Sci. U.S.A.,* 88, 5453, 1991.
101. **Park, D. J., Min, H. K., and Rhee, S. G.**, Inhibition of CD3-linked phospholipase C by phorbol ester and by cAMP is associated with decreased phosphotyrosine and increased phosphoserine contents of PLC-γ1, *J. Biol. Chem.,* 267, 1496, 1992.
102. **Berstein, G., Blank, J. L., Smrcka, A. V., Higashijima, T., and Sternweis, P. C.**, Reconstitution of agonist stimulated phosphatidylinositol 4,5-biphosphate hydrolysis using purified m1 muscarinic receptor, $G_{q/11}$, and phospholipase C-β1, *J. Biol. Chem.,* 267, 8081, 1992.
103. **Wong, S. M. E. and Chase, H. S., Jr.**, Role of intracellular calcium in cellular volume regulation, *Am. J. Physiol.,* 250, C841, 1986.
104. **Davis, C. W. and Finn, A. L.**, Cell volume regulation in frog urinary bladder, *Fed. Proc.,* 44, 2520, 1985.
105. **Hoffmann, E. K., Lambert, I. H., and Simonsen, L. O.**, Separate Ca^{2+}-activated K^+ and Cl^- transport pathways in Ehrlich ascites tumor cells, *J. Membr. Biol.,* 91, 227, 1986.
106. **Hoffmann, E. K., Simonsen, L. O., and Lambert, I. H.**, Volume-induced increase of K^+ and Cl^- permeabilities in Ehrlich ascites tumor cells. Role of internal Ca^{2+}, *J. Membr. Biol.,* 78, 211, 1984.
107. **Hunter, M., Lopes, A. G., Boulpaep, E. L., and Giebisch, G. H.**, Single channel recordings of calcium-activated potassium channels in the apical membrane of rabbit cortical collecting tubules, *Proc. Natl. Acad. Sci. U.S.A.,* 81, 4237, 1984.
108. **Guggino, S. E., Guggino, W. B., Green, N., and Sacktor, B.**, Ca^{2+}-activated K^+ channels in cultured medullary thick ascending limb cells, *Am. J. Physiol.,* 252, C121, 1987.
109. **Lang, F., Friedrich, F., Paulmichl, M., Schobersberger, W., Jungwith, A., Ritter, M., Steidl, M., Weiss, H. W. E., Tshernko, E., Paulmichl, R., and Hallbrucker, C.**, Ion channels in Madin-Darby canine kidney cells, *Renal Physiol. Biochem.,* 13, 82, 1990.
110. **Ubl, J., Murer, H., and Kolb, H. A.**, Hypotonic shock evokes opening of Ca^{2+}-activated K channels, *Pfluegers Arch.,* 412, 551, 1988.
111. **Teulon, J., Paulais, M., and Boutheir, M.**, A Ca^{2+}-activated cation-selective channel in the basolateral membrane of the cortical thick ascending limb of Henle's loop of the mouse, *Biochim. Biophys. Acta,* 905, 125, 1987.
112. **Merot, J., Bidet, M., Gachot, B., LeMaout, S., Tauc, M., and Poujeol, P.**, Patch clamp study on primary culture of isolated proximal convoluted tubules, *Pfluegers Arch.,* 413, 51, 1988.
113. **Grinstein, S., DuPre, A., and Rothstein, A.**, Volume regulation by human lymphocytes.

Role of calcium, *J. Gen. Physiol.*, 79, 849, 1982.

114. **Cantiello, H. F., Prat, A. G., Bonventre, J. V., Cunningham, C. C., Hartwig, J. H., and Ausiello, D. A.,** Actin-binding protein contributes to cell volume regulatory ion channel activation in melanoma cells, *J. Biol. Chem.*, 268, 4596, 1993.
115. **Supattapone, S., Worley, P. F., Baraban, J. M., and Snyder, S.,** Solubilization, purification, and characterization of an inositol triphosphate receptor, *J. Biol. Chem.*, 263, 1530, 1988.
116. **Furuichi, T., Yoshikawa, S., Miyawaki, A., Wada, K., Maeda, N., and Mikoshiba, K.,** Primary structure and functional expression of the inositol 1,4,5-triphosphate-binding protein P_{400}, *Nature*, 342, 32, 1989.
117. **Ferris, C. D., Huganir, R. L., Supattapone, S., and Snyder, S. H.,** Purified inositol 1,4,5-triphosphate receptor mediates calcium flux in reconstituted lipid vesicles, *Nature*, 342, 87, 1989.
118. **Ambler, S. K., Poenie, M., Tsien, R. Y., and Taylor, P.,** Agonist-stimulated oscillations and cycling of intracellular free calcium in individual cultured muscle cells, *J. Biol. Chem.*, 263, 1952, 1988.
119. **Cuthbertson, K. S. R. and Cobbold, P. H.,** Phorbol ester and sperm activate mouse oocytes by inducing sustained oscillations in cell Ca^{2+}, *Nature*, 316, 541, 1985.
120. **Gray, P. T. A.,** Oscillations of free cytosolic calcium evoked by cholinergic and catecholaminergic agonists in rat parotid acinar cells, *J. Physiol. (London)*, 406, 35, 1988.
121. **Hajjar, R. J. and Bonventre, J. V.,** Oscillations of intracellular calcium induced by vasopressin in individual fura-2-loaded mesangial cells. Frequency dependence on basal calcium concentration, agonist concentration, and temperature, *J. Biol. Chem.*, 266, 21589, 1991.
122. **Holl, R. W., Thorner, M. O., Mandell, G. L., Sullivan, J. A., Sinha, Y. N., and Leony, D. A.,** Spontaneous oscillations of intracellular calcium and growth hormone secretion, *J. Biol. Chem.*, 263, 9682, 1988.
123. **Jacob, R., Merritt, J. E., Hallam, T. J., and Rink, T. J.,** Repetitive spikes in cytoplasmic calcium evoked by histamine in human endothelial cells, *Nature*, 335, 40, 1988.
124. **Ueda, S., Oiki, S., and Okada, Y.,** Oscillations of cytoplasmic concentrations of Ca^{2+} and K^{2+} in fused L cells, *J. Membr. Biol.*, 91, 65, 1986.
125. **Prentki, M., Glennon, M. C., Thomas, A. P., Morris, R. L., Matschinksy, F. M., and Corkey, B. E.,** Cell-specific patterns of oscillating free Ca^{2+} in carbomylcholine-stimulated insulinoma cells, *J. Biol. Chem.*, 263, 11044, 1988.
126. **Lipscombe, D., Madison, D. V., Poenie, M., Reuter, H., Tsien, R. Y., and Tsien, R. W.,** Spatial distribution of calcium channels and cytosolic calcium transients in growth cones and cell bodies of sympathetic neurons, *Proc. Natl. Acad. Sci. U.S.A.*, 85, 2398, 1988.
127. **Worley, P. F., Baraban, J. M., Supattapone, S., Wilson, V. S., and Snyder, S. H.,** Characterization of inositol triphosphate receptor binding in brain: regulation by pH and calcium, *J. Biol. Chem.*, 262, 12132, 1987.
128. **Missiaen, L., Taylor, C. W., and Berridge, M. J.,** Spontaneous calcium release from inositol trisphosphate sensitive calcium stores, *Nature*, 352, 241, 1991.
129. **Yule, D. I. and Gallacher, D. V.,** Oscillations of cystolic calcium in single pancreatic acinar cells stimulated by acetylcholine, *FEBS Lett.*, 239, 358, 1988.
130. **Irvine, R. F.,** Inositol phosphates and Ca^{2+} entry: toward a proliferation or a simplification?, *FASEB J.*, 6, 3085, 1992.
131. **Theibert, A. B., Estevez, V. A., Mourey, R. J., Marecek, J. F., Barrow, R. K., Prestwich, G. D., and Snyder, S. H.,** Photoaffinity labeling and characterization of isolated inositol 1,3,4,5-tetrakisphosphate- and inositol hexakisphosphate-binding proteins, *J. Biol. Chem.*, 267, 9071, 1992.
132. **Altamirano, A. A., Breitweiser, G. E., and Russell, J. M.,** Vanadate and fluoride effects on Na-K-2Cl cotransport in squid giant axon, *Am. J. Physiol.*, 254, C582, 1988.
133. **Pewitt, B. E., Hedge, R. S., Haas, M., and Palfrey, H. C.,** The regulation of Na-K-2Cl cotransport and bumetanide binding in avian erythrocytes by protein phosphorylation and dephosphorylation, *J. Biol. Chem.*, 265, 20747, 1990.
134. **Jennings, M. L., and Schulz, R. K.,** Okadaic acid inhibition of KCl cotransport. Evidence that protein dephosphorylation is necessary for activation of transport by either cell swelling or N-ethylmaleimide, *J. Gen. Physiol.*, 97, 799, 1991.
135. **Bianchini, L. M., Woodside, M., Sardet, C., Pouyssegur, J., Takai, A., and Grinstein, S.,** Okadaic acid, a phosphatase inhibitor, induces activation and phosphorylation of the Na/H antiport, *J. Biol. Chem.*, 266, 15406, 1991.

136. **Jennings, M. L., and Al-Rohil, N.,** Kinetics of activation and inactivation of swelling-stimulated K/Cl transport. The volume sensitive parameter is the rate constant for inactivation, *J. Gen. Physiol.,* 95, 1021, 1990.

137. **Kaji, D. M., and Tsukitani, Y.,** Role of protein phosphatase in activation of KCl cotransport in human erythrocytes, *Am. J. Physiol.,* 260, C178, 1991.

138. **Nishizuka, Y.,** Studies and prospectives of the protein kinase C family for cellular regulation, *Cancer,* 63, 1892, 1989.

139. **Nakanishi, H. and Exton, J. H.,** Purification and characterization of the ζ isoform of protein kinase C from bovine kidney, *J. Biol. Chem.,* 267, 16347, 1992.

140. **Goode, N., Hughes, K., Woodgett, J. R., and Parker, P. J.,** Differential regulation of glycogen synthase kinase-3β by protein kinase C isotypes, *J. Biol. Chem.,* 267, 16878, 1992.

141. **Decker, S. J., Ellis, C., Pawson, T., and Velu, T.,** Effects of substitution of threonine 654 of the epidermal growth factor receptor on epidermal growth factor-mediated activation of phospholipase C, *J. Biol. Chem.,* 265, 7009, 1990.

142. **Chibalin, A. V., Vasilets, L. A., Hennekes, H., Pralong, D., and Geering, K.,** Phosphorylation of Na, K-ATPase α-subunits in microsomes and in homogenates of *Xenopus* oocytes resulting from the stimulation of protein kinase A and protein kinase C, *J. Biol. Chem.,* 267, 22378, 1992.

143. **Ruegg, U. T., and Burgess, G. M.,** Staurosporine, K-252 and UCN-01: potent but non-specific inhibitors of protein kinases, *Trends Pharm. Sci.* 10, 218, 1989.

144. **Alexander, D. R., Granes, J. D., Lucas, S., Cantrell, D. A., and Crumpton, M. J.,** A method for measuring protein kinase C activity in permeabilized T lymphocytes by using peptide substrates, *Biochem. J.,* 268, 303, 1990.

145. **House, C. and Kemp, B. E.,** Protein kinase C contains a pseudosubstrate prototype in its regulatory domain, *Science,* 238, 1726, 1987.

146. **Boulton, T. G., Yancopoulos, G. D., Gregory, J. S., Slaughter, C., Moomaw, C., Hsu, J., and Cobb, M. H.,** An insulin-stimulated protein kinase similar to yeast kinases involved in cell cycle control, *Science,* 249, 64, 1990.

147. **Boulton, T. G., Nye, S. H., Robbins, D. J., Ip, N. Y., Radziejewska, E., Morgenbesser, S. D., DePinho, R. A., Panayotatos, N., Cobb, M. H., and Yancopoulos, G. D.,** ERKs: A family of protein-serine/threonine kinases that are activated and tyrosine phosphorylated in response to insulin and NGF, *Cell,* 65, 663, 1991.

148. **Kyriakis, J. M., Force, T. L., Rapp, U. R., Bonventre, J. V., and Avruch, J.,** Mitogen regulation of c-Raf-1 protein kinase activity toward mitogen-activated protein kinase kinase, *J. Biol. Chem.,* in press.

149. **Dent, P., Haser, W., Haystead, T. A. J., Vincent, L. A., Roberts, T. M., and Sturgill, T. W.,** Activation of mitogen-activated protein kinase kinase by v-Raf in NIH 3T3 cells and in vitro, *Science,* 257, 1404, 1992.

150. **Howe, L. R., Leevers, S. J., Gomez, N., Nakielny, S., Cohen, P., and Marshall, C. J.,** Activation of the MAP kinase pathway by the protein kinase raf, *Cell,* 71, 335, 1992.

151. **Pelech, S. L., and Sanghera, J. S.,** MAP kinases: charting the regulatory pathways, *Science,* 257, 1355, 1992.

152. **Crews, C. M., Alessandrini, A., and Erikson, R. L.,** The primary structure of MEK, a protein kinase that phosphorylates the *ERK* gene product, *Science,* 258, 478, 1992.

153. **Sturgill, T. W., Ray, L. B., Erikson, E., and Maller, J. L.,** Insulin-stimulated MAP-2 kinase phosphorylates and activates ribosomal protein S6 kinase II, *Nature,* 334, 715, 1988.

154. **Pulverer, B., Kyriakis, J. M., Avruch, J., Nikolakaki, E., and Woodgett, J. R.,** Phosphorylation of c-*jun* mediated by MAP kinases, *Nature,* 353, 670, 1991.

155. **Gille, H., Sharrocks, A. D., and Shaw, P. E.,** Phosphorylation of transcription factor $p62^{TCF}$ by MAP kinase stimulates ternary complex formation at c-*fos* promoter, *Nature,* 358, 414, 1992.

156. **Nemenoff, R. A., Johnson, G. L., and Heasley, L. E.,** Phosphorylation and activation of a high molecular weight form of phospholipase A_2 by p42MAP kinase and protein kinase C, *J. Am. Soc. Nephrol.,* 3, 503, 1992.

157. **Fischer, E. H., Charbonneau, H., and Tonks, N. K.,** Protein tyrosine phosphatases: a diverse family of intracellular and transmembrane enzymes, *Science,* 253, 401, 1991.

158. **Cohen, P. and Cohen, P. T. W.,** Protein phosphatases come of age, *J. Biol. Chem.,* 264, 21435, 1989.

159. **Chen, J., Martin, B. L., and Brautigan, D. L.,** Regulation of protein serine-threonine phosphatase type-2A by tyrosine phosphorylation, *Science,* 257, 1261, 1992.

160. **Bonventre, J. V. and Nemenoff, R.,** Renal tubular arachidonic acid metabolism, *Kidney Int.,* 39, 438, 1991.

161. **Bonventre, J. V.,** Phospholipase A_2 and signal transduction, *J. Am. Soc. Nephrol.,* 3, 128, 1993.

162. **Dennis, E. A.,** Phospholipases, *Enzymes,* 16, 307, 1983.

163. **Waite, M.,** Approaches to the study of mammalian cellular phospholipases, *J. Lipid Res.,* 26, 1379, 1985.

164. **Hendrikson, R. L., Kreuger, E. T., and Keim, P. S.,** Amino acid sequence of phospholipase A_2-α from the venom of *Crotalus adamanteus, J. Biol. Chem.,* 252, 4913, 1977.

165. **deHaas, G. H., Postema, N. M., Nieuwenhuizen, W., and van Deenen, L. L. M.,** Purification and properties of phospholipase A from porcine pancreas, *Biochim. Biophys. Acta,* 159, 103, 1968.

166. **Stoner, C. R., Reik, L. M., Donohue, M., Levin, W., and Crowl, R. M.,** Human group II phospholipase A_2, *J. Immunol. Meth.,* 145, 127, 1991.

167. **Verger, R., Ferrato, F., Mansbach, C. M., and Pieroni, G.,** Novel intestinal phospholipase A_2: purification and some molecular characteristics, *Biochemistry,* 21, 6883, 1982.

168. **Ono, T., Tojo, H., Kuramitsu, S., Kagamiyama, H., and Okamoto, M.,** Purification and characterization of a membrane-associated phospholipase A_2 from rat spleen, *J. Biol. Chem.,* 263, 5732, 1988.

169. **Frost, S., Weiss, J., Elsbach, P., Maraganore, J. M., Reardon, I., and Heinrikson, R. L.,** Structural and functional properties of phospholipase A_2 purified from an inflammatory exudate, *Biochemistry,* 25, 8381, 1986.

170. **Tojo, H., Ono, T., Kuramitsu, S., Kagamiyama, H., and Okamoto, M.,** A phospholipase A_2 in the supernatant of rat spleen, *J. Biol. Chem.,* 263, 5724, 1988.

171. **Hayakawa, M., Kudo, I., Tomita, M., and Inoue, K.,** Purification and characterization of membrane-bound phospholipase A_2 from rat platelets, *J. Biochem.,* 103, 263, 1988.

172. **Vadas, P., Pruzanski, W., Kim, J., and Fornasier, V.,** The proinflammatory effect of intra-articular injection of soluble human and venom phospholipase A_2, *Am. J. Pathol.,* 134, 807, 1989.

173. **Murakami, M., Kudo, I., Nakamura, H., Yokoyama, Y., Mori, H., and Inoue, K.,** Exacerbation of rat adjuvant arthritis by intradermal injection of purified mammalian 14-kDa group II phospholipase A_2, *FEBS Lett.,* 268, 113, 1990.

174. **Schalkwijk, C., Marki, F., and Van den Bosch, H.,** Studies on the acyl-chain selectivity of cellular phospholipases A_2, *Biochim. Biophys. Acta,* 1044, 139, 1990.

175. **Aarsman, A. J., de Jong, J. G. N., Arnoldussen, E., Neys, F. W., van Wassenaar, P. D., and Van den Bosch, H.,** Immunoaffinity purification, partial sequence, and subcellular localization of rat liver phospholipase A_2, *J. Biol. Chem.,* 264, 10008, 1989.

176. **Kramer, R. M., Wuthrich, C., Bollier, C., Allegrini, P. R., and Zahler, P.,** Isolation of phospholipase A_2 from sheep erythrocyte membranes in the presence of detergents, *Biochim. Biophys. Acta,* 507, 381, 1978.

177. **Crowl, R., Stoner, C., Stoller, T., Pan, Y.-C., and Conroy, R.,** Isolation and characterization of cDNA clones from human placenta coding for phospholipase A_2, *Adv. Exp. Med. Biol.,* 279, 173, 1990.

178. **Van den Bosch, H., De Jong, J. G. N., and Aarsman, A. J.,** Phospholipase A_2 from rat liver mitochondria, *Meth. Enzymol.,* 197, 365, 1991.

179. **Nakamura, H., Nemenoff, R. A., Gronich, J. H., and Bonventre, J. V.,** Subcellular characteristics of phospholipase A_2 activity in rat kidney. Enhanced cytosolic, mitochondrial, and microsomal phospholipase A_2 enzymatic activity after renal ischemia and reperfusion, *J. Clin. Invest.,* 87, 1810, 1991.

180. **Rordorf, G., Uemura, Y., and Bonventre, J. V.,** Characterization of phospholipase A_2 (PLA_2) activity in gerbil brain. Enhanced activities of cytosolic, mitochondrial and microsomal forms after ischemia and reperfusion, *J. Neurosci.,* 11, 1829, 1991.

181. **Hansford, R. G. and Castro, F.,** Intramitochondrial and extramitochondrial free calcium concentrations of suspensions of heart mitochondria with very low, plausibly physiological, contents of total calcium, *J. Bioeng. Biomembr.,* 14, 361, 1982.

182. **Gronich, J. H., Bonventre, J. V., and Nemenoff, R. A.,** Purification of a high molecular mass phospholipase A_2 from rat kidney activated at physiological calcium concentrations, *Biochem. J.,* 271, 37, 1990.

183. **Kim, D. K., Kudo, I., and Inoue, K.,** Purification and characterization of rabbit platelet cytosolic phospholipase A_2, *Biochim. Biophys. Acta,* 1083, 80, 1991.

184. **Clark, J. D., Lin, L.-L., Kriz, R. W., Ramesha, C., Sultzman, L. A., Lin, A. Y., Milona, N., and Knopf, J. L.,** A novel arachidonic acid-selective cytosolic PLA_2 contains a Ca^{2+}-dependent translocation domain with homology to PKC and GAP, *Cell,* 65, 1043, 1991.

185. **Gronich, J. H., Bonventre, J. V., and Nemenoff, R. A.,** Identification and characterization of a hormonally regulated form of phospholipase A_2 in rat renal mesangial cells, *J. Biol. Chem.,* 263, 16645, 1988.
186. **Bonventre, J. V., Gronich, J. H., and Nemenoff, R. A.,** Epidermal growth factor enhances glomerular mesangial cell soluble phospholipase A_2 activity, *J. Biol. Chem.,* 265, 4934, 1990.
187. **Sharp, J. D., White, D. L., Chiou, X. G., Goodson, T., Gamboa, G. C., McClure, D., Burgett, S., Hoskins, J., Skatrud, P. L., Sportsman, J. R., Becker, G. W., Kang, L. H., Roberts, E. F., and Kramer, R. M.,** Molecular cloning and expression of human Ca^{2+}-sensitive cytosolic phospholipase A_2, *J. Biol. Chem.,* 266, 14850, 1991.
188. **Hazen, S. L., Stuppy, R. J., and Gross, R. W.,** Purification and characterization of canine myocardial cytosolic phospholipase A_2, *J. Biol. Chem.,* 265, 10622, 1990.
189. **Pierik, A. J., Nijssen, J. G., Aarsman, A. J., and Van den Bosch, H.,** Calcium-independent phospholipase A_2 in rat tissue cytosols, *Biochim. Biophys. Acta,* 962, 345, 1988.
190. **Bonventre, J. V. and Swidler, M.,** Calcium dependency of prostaglandin E_2 production in rat glomerular mesangial cells. Evidence that protein kinase C modulates the Ca^{2+}-dependent activation of phospholipase A_2, *J. Clin. Invest.,* 82, 168, 1988.
191. **Channon, J. Y. and Leslie, C. C.,** A calcium-dependent mechanism for associating a soluble arachidonyl-hydrolyzing phospholipase A_2 with membrane in the macrophage cell line RAW 264.7, *J. Biol. Chem.,* 265, 5409, 1990.
192. **Perin, M. S., Fried, V. A., Mignery, G. A., Jhan, R., and Sudhof, T. C.,** Phospholipid binding by a synaptic vesicle protein homologous to the regulatory region of protein kinase C, *Nature,* 345, 260, 1990.
193. **Vogel, U. S., Dixon, R. A. F., Schaber, M. D., Diehl, R. E., Marshall, M. S., Scolnick, E. M., Sigal, I. S., and Gibbs, J. B.,** Cloning of bovine GAP and its interaction with oncogenic ras p21, *Nature,* 335, 90, 1988.
194. **Stahl, M. L., Ferenz, C. R., Kelleher, K. L., Kriz, R. W., and Knopf, J. L.,** Sequence similarity of phospholipase C with non-catalytic region of src, *Nature,* 332, 269, 1988.
195. **Franson, R., Waite, M., and LaVia, M.,** Identification of phospholipase A_1 and A_2 in the soluble fraction of rat liver lysosomes, *Biochemistry,* 10, 1942, 1971.
196. **Waite, M.,** *The Phospholipases,* Plenum Press, New York, 1987, 332.
197. **Bonventre, J. V., Skorecki, K. L., Kreisberg, J. I., and Cheung, J. Y.,** Vasopressin increases cytosolic free calcium concentration in glomerular mesangial cells, *Am. J. Physiol.,* 251, F94, 1986.
198. **Piomelli, D. and Greengard, P.,** Bidirectional control of phospholipase A_2 activity by Ca^{2+}/calmodulin-dependent kinase II, cAMP-dependent protein kinase, and casein kinase II, *Proc. Natl. Acad. Sci. U.S.A.,* 88, 6770, 1991.
199. **Zor, U., Her, E., Braquet, P., Ferber, E., and Reiss, N.,** A novel mechanism of glucocorticosteroid action in suppression of phospholipase A_2 activity stimulated by Ca^{2+} ionophore A23187: induction of protein phosphatases, *Adv. Prost. Thromb. Leukotriene Res.,* 21, 265, 1990.
200. **Blank, J. L., Ross, A. H., and Exton, J. H.,** Purification and characterization of two G-proteins that activate the β1 isozyme of phosphoinositide-specific phospholipase C, *J. Biol. Chem.,* 266, 18206, 1991.
201. **Birmbaumer, L., Abramowitz, J., and Brown, A. M.,** Receptor-effector coupling by G proteins, *Biochim. Biophys. Acta,* 1031, 163, 1990.
202. **Bonventre, J. V., Weber, P. C., and Gronich, J. H.,** PAF and PDGF increase cytosolic $[Ca^{2+}]$ and phospholipase activity in mesangial cells, *Am. J. Physiol.,* 254 (Renal Fluid Electrolyte Physiol. 23), F87, 1988.
203. **Burch, R. M., Luini, and A., Axelrod, J.,** Phospholipase A_2 and phospholipase C are activated by distinct GTP-binding proteins in response to α_1-adrenergic stimulation in FRTL5 thyroid cells, *Proc. Natl. Acad. Sci. U.S.A.,* 83, 7201, 1986.
204. **Bloch, D. B., Bonventre, J. V., Neer, E. J., and Seidman, J. G.,** The G protein α_o subunit alters morphology, growth kinetics, and phospholipid metabolism of somatic cells, *Mol. Cell. Biol.,* 9, 5434, 1989.
205. **Hirata, F.,** The regulation of lipomodulin phospholipase inhibitory protein in rabbit neutrophils by phosphorylation, *J. Biol. Chem.,* 256, 7730, 1981.
206. **Hirata, F., Notsu, Y., Yamada, R., Ishihara, Y., Wano, Y., Kunos, I., and Kunos, G.,** Isolation and characterization of lipocortin (lipomodulin), *Agents Actions,* 17, 263, 1985.
207. **Pepinsky, R. B., Sinclair, L. K., Browning, J. L., Mattaliano, R. J., Smart, J. E., Chow, E. P., Falbel, T., Ribolini, A., Garwin, J. L., and Wallner, B. P.,** Purification and partial

sequence analysis of a 37-kDa protein that inhibits phospholipase A_2 activity from rat peritoneal exudates, *J. Biol. Chem.*, 261, 4239, 1986.

208. **Huang, K. S., Wallner, B., Mattaliano, R. J., Tizard, R., Burne, C., Frey, A., Hession, C., McGray, P., Sinchair, L. K., Chow, E. P., Browning, J. L., Ramachandran, K. L., Tang, J., Smart, J. E., and Pepsinsky, R. B.,** Two human 35 kd inhibitors of phospholipase A_2 are related to substrates of pp60$^{v\text{-}src}$ and of the epidermal growth factor receptor/kinase, *Cell*, 46, 191, 1986.

209. **Davidson, F. F., Dennis, E. A., Powell, M., and Glenney, J. R.,** Inhibition of phospholipase A_2 by "lipocortins" and calpactins, *J. Biol. Chem.*, 262, 1698, 1987.

210. **Clark, M. A., Ozgur, L. E., Conway, T. M., Dispoto, J., Crooke, S. T., and Bomalaski, J. S.,** Cloning of a phospholipase A_2-activating protein, *Proc. Natl. Acad. Sci. U.S.A.*, 88, 5418, 1991.

211. **Ballou, L. R. and Cheung, W. Y.,** Inhibition of human platelet phospholipase A_2 activity by unsaturated fatty acids, *Proc. Natl. Acad. Sci. U.S.A.*, 82, 371, 1985.

212. **Franson, R., Raghupathi, R., Fry, M., Saal, J., Vishwanath, B., Ghosh, S. S., and Rosenthal, M. D.,** Inhibition of human phospholipase A_2 by *cis*-unsaturated fatty acids and oligomers of prostaglandin B1, *Adv. Exp. Med. Biol.*, 279, 219, 1990.

213. **Chang, J., Blazek, E., Kreft, A. F., and Lewis, A. J.,** Inhibition of platelet and neutrophil phospholipase A_2 by hydroxyeicosatetraenoic acids (HETES), *Biochem. Pharmacol.*, 34, 1571, 1985.

214. **Ishizaki, J., Ohara, O., Nakamura, E., Tamaki, M., Ono, T., Kanda, A., Yoshida, N., Teraoka, H., Tojo, H., and Okamoto, M.,** cDNA cloning and sequence determination of rat membrane-associated phospholipase A_2, *Biochem. Biophys. Res. Commun.*, 162, 1030, 1989.

215. **Sedor, J. R., Sedlak, T. W., Hornbuckle, K., Herman, W. H., and Konieczkowski, M.,** Inflammatory cytokines induce expression of a non-pancreatic (type II) mesangial cell phospholipase A_2 mRNA in a highly regulated manner, *JASN*, 1, 481, 1990.

216. **Nakazato, Y., Simonson, M. S., Herman, W. H., Konieczkowski, M., and Sedor, J. R.,** Interleukin-1α stimulates prostaglandin biosynthesis in serum-activated mesangial cells by induction of a non-pancreatic (type II) phospholipase A_2, *J. Biol. Chem.*, 266, 14119, 1991.

217. **Muhl, H., Geiger, T., Pignat, W., Marki, F., van den Bosch, H., Vosbeck, K., and Pfeilschifter, J.,** PDGF suppresses the activation of group II phospholipase A_2 gene expression by interleukin 1 and forskolin in mesangial cells, *FEBS Lett.*, 291, 249, 1991.

218. **Schalkwijk, C., Vervoordeldonk, M., Pfeilschifter, J., Marki, F., and van den Bosch, H.,** Cytokine- and forskolin-induced synthesis of group II phospholipase A_2 and prostaglandin E_2 in rat mesangial cells is prevented by dexamethasone, *Biochem. Biophys. Res. Commun.*, 180, 46, 1991.

219. **Schalkwijk, C., Pfeilschifter, J., Marki, F., and van den Bosch, H.,** Interleukin-1β, tumor necrosis factor and forskolin stimulate the synthesis and secretion of group II phospholipase A_2 in rat mesangial cells, *Biochem. Biophys. Res. Commun.*, 174, 268, 1991.

220. **Lin, L.-L., Lin, A. Y., and DeWitt, D. L.,** Interleukin-1α induces the accumulation of cytosolic phospholipase A_2 and the release of prostaglandin E_2 in human fibroblasts, *J. Biol. Chem.*, 267, 23451, 1992.

221. **Ordway, R. W., Singer, J. J., and Walsh, J. V., Jr.,** Direct regulation of ion channels by fatty acids, *Trends Neurosci.*, 14, 96, 1991.

222. **Ordway, R. W., Walsh, J. V., Jr., and Singer, J. J.,** Arachidonic acid and other fatty acids directly activate potassium channels in smooth muscle cells, *Science*, 244, 1176, 1989.

223. **Hwang, T.-C., Guggino, S. E., and Guggino, W. B.,** Direct modulation of secretory chloride channels by arachidonic and other cis unsaturated fatty acids, *Proc. Natl. Acad. Sci. U.S.A.*, 87, 5706, 1990.

224. **Anderson, M. P. and Welsh, M. J.,** Fatty acids inhibit apical membrane chloride channels in airway epithelia, *Proc. Natl. Acad. Sci. U.S.A.*, 87, 7334, 1990.

225. **Riordan, J. R., Rommens, J. M., Kerem, B.-S., Alon, N., Rozmahel, R., Grzelczak, Z., Zielenski, J., Lok, S., Plavsic, N., Chou, J.-L., Drumm, M. L., Iannuzzi, M. C., Collins, F. S., and Tsui, L.-C.,** Identification of the cystic fibrosis gene: cloning and characterization of complementary DNA, *Science*, 245, 1066, 1989.

226. **Gerzer, R., Brash, A. R., and Hardman, J. G.,** Activation of soluble guanylate cyclase by arachidonic acid and 15-lipoxygenase products, *Biochim. Biophys. Acta*, 886, 383, 1986.

227. **Murad, F., Leitman, D., Waldman, S., Chang, C.-H., Hirata, M., and Kohse, K.,** Effects of nitrovasodilators, endothelin-dependent vasodilators, and atrial peptides on cGMP,

Cold Spring Harbor Symp. Quant. Biol., 53, 1005, 1988.

228. **Kolesnick, R. N., Musacchio, I., Thaw, C., and Gershengorn, M. C.,** Arachidonic acid mobilizes calcium and stimulates prolactin secretion from GH_3 cells, *Am. J. Physiol.*, 246, E458, 1984.

229. **Zeitler, P. and Handwerger, S.,** Arachidonic acid stimulates phosphoinositide hydrolysis and human placental lactogen release in an enriched fraction of placental cells, *Mol. Pharmacol.*, 28, 549, 1985.

230. **Piomelli, D., Wang, J. K. T., Sihra, T. S., Nairn, A. C., Czernik, A. J., and Greengard, P.,** Inhibition of Ca^{2+}/calmodulin-dependent protein kinase II by arachidonic acid and its metabolites, *Proc. Natl. Acad. Sci. U.S.A.*, 86, 8550, 1989.

231. **McPhail, L. C., Clayton, C. C., and Snyderman, R.,** A potential second messenger role for unsaturated fatty acids. Activation of Ca^{2+}-dependent protein kinase, *Science*, 224, 662, 1984.

232. **Hansson, A., Serhan, C. N., Haeggström, J., Ingelman-Sundberg, M., and Samuelsson, B.,** Activation of protein kinase C by lipoxin A and other eicosanoids. Intracellular action of oxygenation products of arachidonic acid, *Biochem. Biophys. Res. Commun.*, 134, 1215, 1986.

233. **Shearman, M. S., Naor, Z., Sekiguchi, K., Kishimoto, A., and Nishizuka, Y.,** Selective activation of the γ-subspecies of protein kinase C from bovine cerebellum by arachidonic acid and its lipoxygenase metabolites, *FEBS Lett.*, 243, 177, 1989.

234. **Force, T., Hyman, G., Hajjar, R., Sellmayer, A., and Bonventre, J. V.,** Noncyclooxygenase metabolites of arachidonic acid amplify the vasopressin-induced Ca^{2+} signal in glomerular mesangial cells by releasing Ca^{2+} from intracellular stores, *J. Biol. Chem.*, 266, 4295, 1991.

235. **Scherer, R. W. and Breitwiesser, G. E.,** Arachidonic acid metabolites alter G protein-mediated signal transduction in heart. Effects on muscarinic K^+ channels, *J. Gen. Physiol.*, 96, 735, 1990.

236. **Piomelli, D., Volterra, A., Dale, N., Siegelbaum, S. A., Kandel, E. R., Schwartz, J. H., and Belardetti, F.,** Lipoxygenase metabolites of arachidonic acid as second messengers for presynaptic inhibition of *Aplysia* sensory cells, *Nature*, 328, 38, 1987.

237. **Cantiello, H. F., Patenaude, C. R., Codina, J., Birnbaumer, L., and Ausiello, D. A.,** $G_{\alpha i\text{-}3}$ regulates epithelial Na^+ channels by activation of phospholipase A_2 and lipoxygenase pathways, *J. Biol. Chem.*, 265, 21624, 1990.

238. **Doroshenko, P.,** Second messengers mediating activation of chloride current by intracellular GTPγS in bovine chromafin cells, *J. Physiol.*, 436, 725, 1991.

239. **Lambert, I. H., Hoffmann, E. K., and Christensen, P.,** Role of prostaglandins and leukotrienes in volume regulation by Ehrlich ascites tumor cells, *J. Membr. Biol.*, 98, 247, 1987.

240. **Lambert, I. H.,** Activation of the volume- or ionophore A23187 plus Ca^{2+}-induced Cl^- transport pathway in Ehrlich ascites tumor cells is mediated via leucotriene-D4, *J. Physiol. (London)*, 394, 85P, 1987.

241. **Graff, G., Stephenson, J. H., Glass, D. B., Haddox, M. K., and Goldberg, N. D.,** Activation of soluble splenic cell guanylate cyclase by prostaglandin endoperoxides and fatty acid hydroperoxides, *J. Biol. Chem.*, 253, 7662, 1978.

242. **Capdevila, J., Gil, L., Orellana, M., Marnett, L. J., Mason, J. I., Yadagiri, P., and Falck, J. R.,** Inhibitors of cytochrome P-450-dependent arachidonic acid metabolism, *Arch. Biochem. Biophys.*, 261, 257, 1988.

243. **Weiss, R., Arnold, J., and Estabrook, R.,** Transformation of an arachidonic acid hydroperoxide into epoxyhydroxy and trihydroxy fatty acids by liver microsomal P-450, *Arch. Biochem. Biophys.*, 252, 334, 1987.

244. **Force, T., Hyman, G., Sellmayer, A., and Bonventre, J. V.,** Cytochrome P-450 monooxygenase products modulate the calcium response to vasopressin in glomerular mesangial cells, *Clin. Res.*, 38, 371A, 1990.

245. **Madhun, Z. T., Goldthwait, D. A., McKay, D., Hopfer, U., and Douglas, J. G.,** An epoxygenase metabolite of arachidonic acid mediates angiotensin II-induced rises in cytosolic calcium in rabbit proximal tubule epithelial cells, *J. Clin. Invest.*, 88, 456, 1991.

246. **Schwartzman, M., Ferreri, N. R., Carroll, M. A., Songu-Mize, E., and McGiff, J. C.,** Renal cytochrome P450-related arachidonate metabolite inhibits (Na^+ + K^+) ATPase, *Nature*, 314, 620, 1985.

247. **Watson, P. A.,** Accumulation of cAMP and calcium in S49 mouse lymphoma cells following hypoosmotic swelling, *J. Biol. Chem.*, 264, 14735, 1989.

248. **Morgan, H. E., Xenophontos, X. P., Haneda, T., McGlauphlin, S., and Watson, P. A.,** Stretch-anabolism transduction, *J. Appl. Cardiol.*, 4, 415, 1989.

249. **Watson, P. A.,** Function follows form: generation of intracellular signals by cell deformation, *FASEB J.*, 5, 2013, 1991.

250. **Kregenow, F. M.,** Osmoregulatory salt transporting mechanisms: control of cell volume in anisotonic media, *Annu. Rev. Physiol.*, 43, 493, 1981.

251. **Palfrey, H. C. and Rao, M. C.,** Na/K/Cl cotransport and its regulation, *J. Exp. Biol.*, 106, 43, 1983.

252. **Kregenow, F. M., Robbie, D. E., and Orloff, J.,** Effect of norepinephrine and hypertonicity on K influx and cyclic AMP in duck erythrocytes, *Am. J. Physiol.*, 231, 306, 1976.

253. **Boron, W. F.,** Control of intracellular pH, in *The Kidney: Physiology and Pathophysiology*, Vol. 1, Seldin, D. W. and Giebisch, G., Eds., Raven Press, New York, 1992, 219.

254. **Sweat, J. D., Blair, I. A., Cragoe, E. J., and Limbird, L. E.,** Inhibitors of Na^+/H^+ exchange block epinephrine- and ADP-induced stimulation of human platelet phospholipase C by blockade of arachidonic acid release at a prior step, *J. Biol. Chem.*, 261, 8660, 1986.

255. **Parker, J. C.,** Volume-responsive sodium movements in dog red blood cells, *Am. J. Physiol.*, 244, C324, 1983.

256. **Grinstein, S. and Rothstein, A.,** Mechanisms of regulation of the Na^+/H^+ exchanger, *J. Membr. Biol.*, 90, 1, 1986.

257. **Moolenaar, W. H.,** Effects of growth factors on intracellular pH regulation, *Annu. Rev. Physiol.*, 48, 363, 1986.

258. **Borgese, F., Garcia-Romeu, F., and Motais, R.,** Control of cell volume and ion transport by β-adrenergic catecholamines in erythrocytes of rainbow trout, *Salmo gairdneri*, *J. Physiol. (London)*, 382, 123, 1987.

259. **Grinstein, S., Goetz-Smith, J. D., Stewart, D., Beresford, B. J., and Mellors, A.,** Protein phosphorylation during activation of Na^+/H^+ exchange by phorbol esters and by osmotic shrinking. Possible relation to cell pH and volume regulation, *J. Biol. Chem.*, 261, 8009, 1986.

260. **Doppler, W., Maly, K., and Grunike, H.,** Role of Na^+/H^+ antiport in the regulation of the internal pH of Ehrlich ascites tumor cells in culture, *J. Membr. Biol.*, 91, 147, 1986.

261. **Segre, G. V., Abou-Samra, A. B., Juppner, H., Schipani, E., Force, T., Urena, P., Freeman, M., Kong, X. F., Kolakowski, L. F., Jr., Hock, J., Bonventre, J. V., Potts, J., Jr., and Kronenberg, H. M.,** Characterization of cloned PTH/PTHrP receptors, Second International Conference on New Actions of PTH, Pisa, Italy, 1991.

262. **Rhee, S. G.,** Inositol phospholipid-specific phospholipase C: interaction of the γ_1 isoform with tyrosine kinase, *Trends Biochem. Sci.*, 16, 297, 1991.

Chapter 10

Influence of Macromolecular Crowding on Intracellular Association Reactions: Possible Role in Volume Regulation

Allen P. Minton

TABLE OF CONTENTS

I. INTRODUCTION

The maintenance of cell volume in the face of osmotic stress necessitates the existence of a feedback loop containing at least two functional elements: a sensor that detects transient changes in volume, and a sensor-activated effector that operates to restore the original volume. Of these two elements, the effector is the better understood. Numerous molecular mechanisms for the alteration of cell volume via synthesis or transport of osmolytes have been identified and characterized.[1] Considerably less is known about the sensor. We know that it is sensitive: cellular volume changes of only a few percent have been shown to elicit tenfold or greater increases in transporter-mediated ion fluxes.[2,3] We shall refer to this phenomenon as *signal amplification*.

Two schools of thought regarding the nature of the sensing mechanism have emerged.[4] It has been suggested that certain membrane transporters are activated by mechanical deformations of the membrane, such as stretching, that may accompany cellular volume changes.[5] A stretch-sensitive transporter would seem, in principle, to be capable of volume regulation only under rather special circumstances (viz., transporter activation via swelling of a spherical cell), and even under those circumstances it is not clear how the transporter could provide the requisite signal amplification.[1]

A more general hypothesis for the sensing of volume changes is that a change in intracellular volume is associated with a corresponding change in the concentrations of all impermeant intracellular species, one or more of which act as regulators of membrane transporters. Molecular models for swelling-activated ion flux based upon this hypothesis[2,6] can account qualitatively for the observed activation of K-Cl cotransporters by small increases in red cell volume. These models postulate reversible associations between soluble regulatory proteins, or associations between at least one soluble regulatory protein and a membrane protein (the transporter). In order to characterize such associa-

0-8493-4448-4/94/$0.00+$.50

tions correctly within an intact cell it is necessary to take into account the effects of volume occupancy by macromolecules, often referred to as "macromolecular crowding".[7]

The purpose of this chapter is twofold: first, to indicate how macromolecular crowding profoundly affects macromolecular associations in an intact cell; and second, to show how extremely simple (but not necessarily realistic) models incorporating principles of macromolecular crowding can account for signal amplification of both swelling-activated and shrinkage-activated ion flux.

II. MACROMOLECULAR CROWDING AND MACROMOLECULAR ASSOCIATIONS

It is common practice to express chemical equilibria in terms of ratios of concentrations of reactants. For example, given the hypothetical reaction for reversible association of two macromolecular species A + B $\rightleftarrows$ AB, one frequently sees the following expression for an equilibrium association constant:

$$K = \frac{c_{AB}}{c_A c_B} \tag{1}$$

where c_i denotes the concentration (molar or w/v) of the *i*th species. We shall show below that in order to describe macromolecular equilibria correctly in a biological system it is essential to keep in mind that the quantity K defined by Equation 1 is not a true constant at constant temperature and pressure, and only becomes constant in certain limiting cases. The true thermodynamic equilibrium constant is given by

$$K_o = \frac{a_{AB}}{a_A a_B} = \frac{\gamma_{AB} c_{AB}}{\gamma_A c_A \, \gamma_B c_B} \tag{2}$$

where a_i denotes the thermodynamic activity and γ_i denotes the activity coefficient of species *i* (see any textbook of physical chemistry). The activity coefficient of solute species i is a measure of the interactions between a molecule of species i and all other solute molecules present, as will be discussed in more detail below. It follows from Equations 1 and 2 that

$$K = K_o \Gamma \tag{3}$$

where Γ is a correction factor equal to $\gamma_A\gamma_B/\gamma_{AB}$. Equation 3 is true for all equilibria, but the expression for Γ depends upon the particular reaction being considered.[7,8]

Living cells contain a large total concentration of macromolecules. Taken together, the macromolecular species occupy 25 to 50% of the total volume of the cell interior.[9] Because the macromolecules are mutually impenetrable, any arbitrarily selected individual macromolecule (designated the *probe molecule*) is excluded from the volume collectively occupied by all of the other macromolecules (designated the *background molecules*), as depicted schematically in Figures 1A and B. Consequently, the translational entropy of the probe molecule is lower, and its chemical potential (denoted by μ) is higher than would be the case in the absence of the background macromolecules, i.e., in an ideal solution. The difference between the actual chemical potential of the *i*th species and the chemical potential in ideal solution is called the *nonideal* part of the chemical potential, and is related to the thermodynamic activity coefficient γ_i as follows:[7]

$$\mu_i^{\text{nonideal}} \equiv \mu_i - \mu_i^{\text{ideal}} = RT \ln \gamma_i \tag{4}$$

where R denotes the molar gas constant and T the absolute temperature. Statistical thermodynamics provides a formal relation between the activity coefficient (or chemical potential) of the *i*th solute species and the potential of effective interaction between a molecule of the *i*th species and one or more other solute molecules in the solution.[7,10] If the effective interaction is predominantly attractive, then μ_i will be less than μ_i^{ideal}, and $\gamma_i < 1$. If the effective interaction is predominantly repulsive (or purely repulsive, as in the case of molecules that interact only via volume exclusion), then μ_i will exceed μ_i^{ideal}, and γ_i will exceed 1.

In a pure volume exclusion system, the thermodynamic activity coefficient of a probe molecule of species i is simply related to the volume accessible to the center of mass of that species, called the available or accessible volume:[11]

$$\gamma_i = V/V_{a,i} \tag{5}$$

where V denotes the total volume of the fluid and $V_{a,i}$ denotes the volume accessible to the *i*th solute species. Comparison of Figures 1A and 1B reveals that in a volume-occupied solution, the value of $V_{a,i}$ decreases substantially (and hence γ_i increases substantially) as the size of the probe increases relative to that of the predominant background species.

For the purpose of estimating activity coefficients due to volume exclusion, it has proven useful to model proteins by hard (i.e., impenetrable) particles of simple shape, such as spheres, ellipsoids of rotation, or cylindrical rods.[7,8,12–14] For example, the experimentally measured activity coefficient of hemoglobin in solution is well described

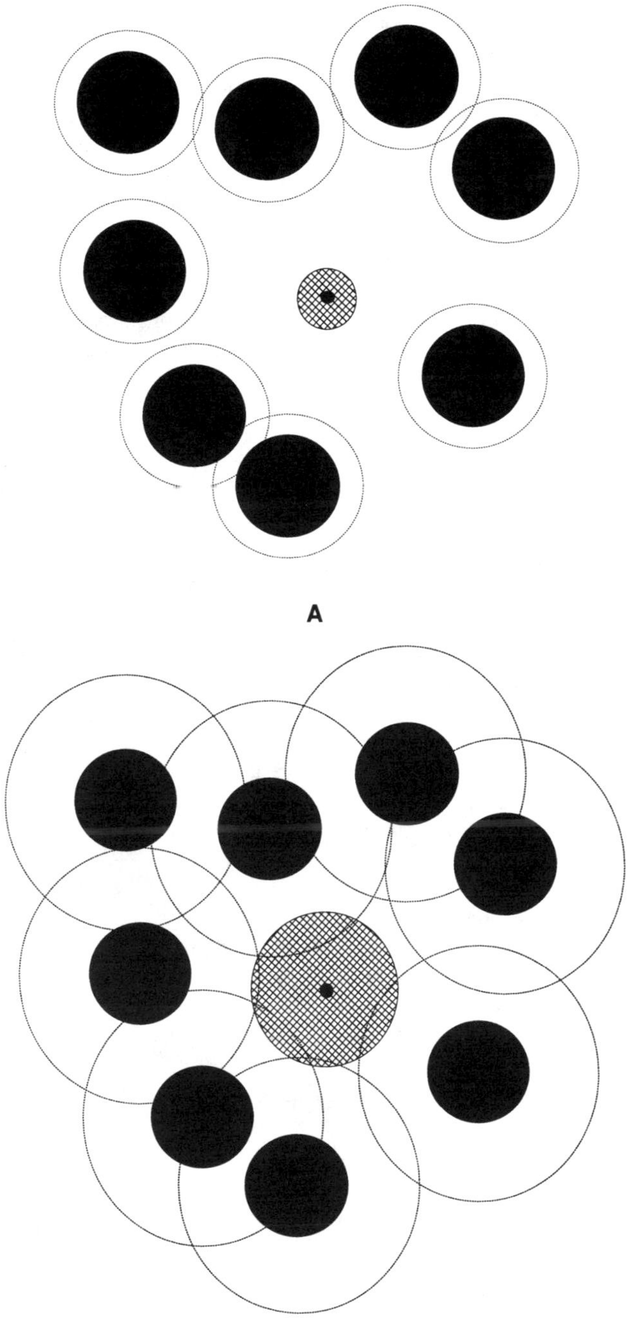

A

B

Figure 10–1 Schematic illustration of volume available to a probe particle (cross-hatched sphere) in a solution containing background particles (solid spheres). Each background particle lies at the center of a spherical region, indicated by a dashed surface, the radius of which is equal to the sum of the radii of background and probe particles. This region is inaccessible to the center of the probe particle, indicated by a black dot. The available volume, or volume accessible to probe, is thus that part of the total solution volume that lies exterior to all dashed surfaces. (A) Probe molecule smaller than background molecules; (B) probe molecule larger than background molecules.

as a function of protein concentration over the entire range of measurement (up to about 400 g/l) by theoretical relations derived for a suspension of hard spherical particles, where the size and shape of the particle is close to that of a hemoglobin molecule viewed at low resolution.[12] According to the scaled particle theory of fluid mixtures,[15–17] abbreviated SPT, the activity coefficient of the *i*th solute species may be expressed as a finite power series in the characteristic dimension R of the equivalent hard particle (for example, the radius of a sphere) representing a molecule of this species:

$$\ln \gamma_i = A_0 + A_1 R_i + A_2 R_i^2 + A_3 R_i^3 \qquad \textbf{(6)}$$

where the coefficients A_j are positively valued functions of the number density, size, and shape of the background particles. As a simple example of crowding, we consider an erythrocyte which, for the present purpose, may be considered to be a solution of hemoglobin at a concentration of around 350 g/l. Equation 6 was used to estimate the effect of macromolecular crowding upon the activity coefficient of a spherical probe molecule in such a solution. The results, plotted in Figure 10–2, indicate that crowding has little effect upon the activity of test solutes that are small relative to the predominant background species. However, the activity of the probe molecule begins to increase strongly as the size of the probe approaches that of the predominant background species.

With this background we can now examine the effect of crowding upon two types of association reactions that are likely to be of importance in cellular volume regulation. The first is the association of two soluble macromolecules to form a soluble complex. In Figure 10–3 the logarithm of Γ for the association A + B $\rightleftarrows$ AB, calculated using Equations 3 and 6, is plotted as a function of the concentration of background protein (hemoglobin) for different relative sizes of A and B. The results shown in curves 1a–e indicate that when the concentration of background molecules becomes similar to that found in a variety of intracellular fluids (>200 g/l), the equilibrium constant for formation of a heterodimer may be enhanced by a factor of ten, a hundred, or even more, depending upon conditions. For a fixed size of A and a fixed concentration of background molecules, the enhancement of association due to crowding increases with increasing size of B.

The second type of association likely to be of importance in a molecular model for volume regulation is that between a soluble macromolecule (P) and a specific, independent site located on the cell membrane (most likely a regulatory site on a transporter):

$$\text{P} + \text{site} \rightleftarrows \text{P} \cdot \text{site}$$

If it is assumed that bound P is immobilized, then one can readily show[18] that the fractional saturation of sites with P is given by:

$$y \equiv \frac{n_{\text{occupied sites}}}{n_{\text{occupied sites}} + n_{\text{unoccupied sites}}} = \frac{K_o a_P}{1 + K_o a_P} \qquad \textbf{(7)}$$

where K_o is the thermodynamic equilibrium constant for association of P to the site. Equation 7 may be rewritten as the conventional independent binding site isotherm:

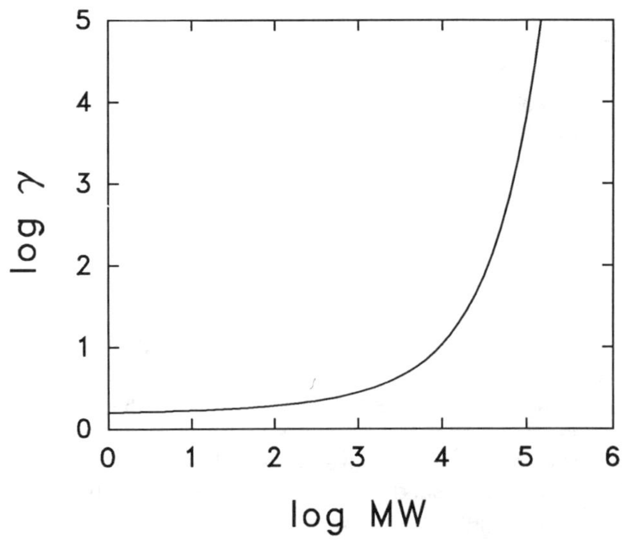

Figure 10–2 Activity coefficient of a probe molecule in a 350 g/l solution of hemoglobin (M_{Hb} = 68,000), plotted as a function of the molecular weight of the probe. Activity coefficients used to calculate the results shown here and in subsequent figures were obtained via SPT relations presented in Reference 17; probe and hemoglobin molecules are modeled by equivalent hard spheres with a specific volume of 1 ml/g.

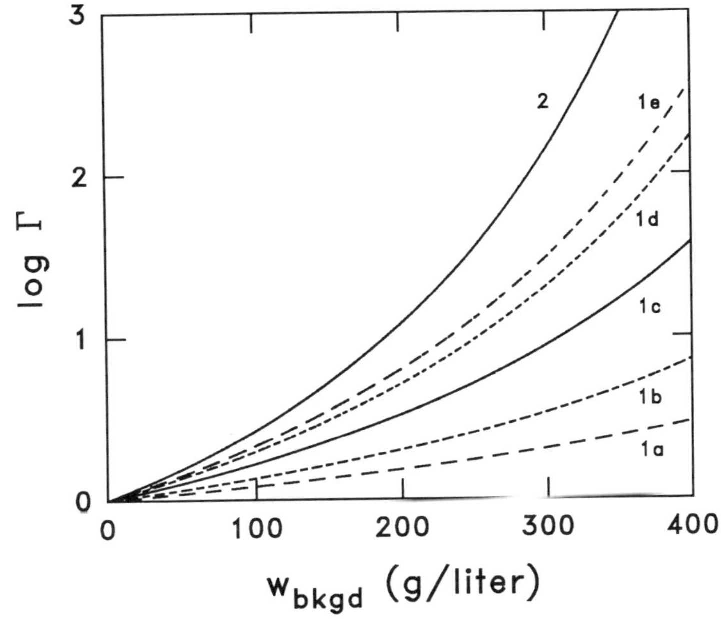

Figure 10–3 Equilibrium correction factor for association reactions, plotted as a function of the concentration of background protein (hemoglobin). Curves 1a–e: A + B ⇄ AB. $M_A/M_{Hb} = 1$, $M_B/M_{Hb} = 0.25$ (curve a); 0.5 (curve b); 1 (curve c); 2 (curve d); and 4 (curve e). Curve 2: A + site ⇄ A · site, $M_A/M_{Hb} = 1$.

$$y = \frac{Kc_P}{1+Kc_P} \quad (8)$$

where $K = K_o\gamma_P$. Thus, for this association, $\Gamma \equiv K/K_o = \gamma_P$. Curve 2 plotted in Figure 10–3 illustrates the dependence of log Γ upon the concentration of background protein, when the size of P is equal to the size of background protein. Just as in the case of heteroassociating macromolecules in solution, in a crowded medium the equilibrium constant for association of a macromolecular ligand with a surface site can exceed the dilute solution value by factors of up to several orders of magnitude.

III. MODELS FOR VOLUME-REGULATED TRANSPORTER-MEDIATED ION FLUX

In the following section we shall formulate and examine two simple models for volume regulation of the activity of a membrane transporter. The first model is the simplest model invoking reversible associations that is capable of exhibiting volume regulation. The second model is slightly more complex and perhaps slightly more realistic as well. The reader should keep in mind that the intent of this chapter is not to propose an actual mechanism of volume regulation, but rather to demonstrate how the concept of macromolecular crowding is incorporated into particular models and affects the results obtained from such models. We shall assume throughout that the membrane transporter can exist in one of two states: fully active (A), in which osmolytes are allowed to traverse the membrane, and inactive (I), in which osmolytes are prevented from traversing the membrane. The addition of partially activated states of the transporter will complicate models but add no insight.

We define here two model-independent functions characterizing the behavior of the system. The values of these functions may be evaluated within the context of any particular model. The first function is the fractional activity of transporter, denoted by $\mathcal{F}$. For two-state models, such as those considered here, the fractional activity of transporter is simply equal to the fraction of transporter in the active state:

$$\mathcal{F} \equiv \frac{A}{I+A} \quad (9)$$

The second function is the response function, denoted by $\Re$, equal to the ratio of the fractional activities of transporter after and before a volume change, or equivalently, the ratio between the steady-state transporter-mediated ion flux following volume activation and the baseline flux prior to activation:

$$\Re \equiv \frac{\mathcal{F}(V_{final})}{\mathcal{F}(V_{initial})} = \frac{\mathcal{F}(Vx)}{\mathcal{F}(V)} \quad (10)$$

where V is the initial volume, and $x \equiv V_{final}/V_{initial}$.

A. MODEL I: THE MEMBRANE TRANSPORTER REVERSIBLY BINDS AN INTRACELLULAR REGULATORY PROTEIN

We shall assume, again for simplicity, that the membrane transporter reversibly binds regulatory protein at a single site, with an equilibrium association constant K. The inclusion of additional sites and site heterogeneity or site-site interaction would add complexity to the model without a proportional increase in understanding of the fundamental phenomenon.

This model may be formulated to describe activation by swelling and activation by shrinkage. We treat each case in turn.

1. Model IA: Swelling-Activated Transport

In the absence of regulatory protein the transporter is active, and binding of a protein inhibitor P renders the transporter inactive.

$$A + P \rightleftarrows I$$

In this model, the fractional activity of the transporter is equal to the fractional *unsaturation* of the transporter, i.e., the fraction of transporter to which the inhibitor P is *not* bound. It follows from Equation 7 that:

$$\mathcal{F} = 1 - y = \frac{1}{1 + K_o a_P} = \frac{1}{1 + K_o \gamma_P c_P} \tag{11}$$

where y is the fractional saturation defined in Equation 7. The response of the system is given by:

$$\Re = \frac{1 + K_o \gamma_P(V) c_P(V)}{1 + K_o \gamma_P(Vx) c_P(Vx)} \tag{12}$$

where $\gamma_P(V)$ and $c_P(V)$, respectively, denote the activity coefficient of P and the concentration of P at volume V. The maximum response will be obtained when y is close to 1, i.e., when $\mathcal{F}$ (= 1 – y $\ll$1) is approximately linear in the thermodynamic activity of regulatory protein. In this limit, $K_o\gamma_P c_P \gg 1$, and Equation 12 becomes:

$$\Re \approx \frac{\gamma_P(V)\, c_P(V)}{\gamma_P(Vx)\, c_P(Vx)} = \frac{\gamma_P(V)\, x}{\gamma_P(Vx)} \tag{13}$$

In Figure 10–4, the value of $\Re$, calculated using Equation 13 with activity coefficients derived from SPT, is plotted as a function of percentage volume increase for putative inhibitors of several molecular weights, in an erythrocyte containing hemoglobin at an initial concentration of 350 g/l. The dashed line in this figure represents the *maximum* response obtainable with this model when the effect of macromolecular crowding on the thermodynamic activity of regulatory protein P is neglected, i.e., when $\Re$ = x.

2. Model IB: Shrinkage-Activated Transport

In the absence of regulatory protein, the transporter is inactive, and binding of a protein regulator renders the transporter active.

$$I + P \rightleftarrows A$$

The fractional activation of transporter is equal to the fraction of transporter that binds regulator P:

$$\mathcal{F} = y = \frac{K_o a_P}{1 + K_o a_P} = \frac{K_o \gamma_P c_P}{1 + K_o \gamma_P c_P} \tag{14}$$

The response function of this model is given by:

$$\Re = \frac{K_o \gamma_P(Vx)\, c_P(Vx)}{K_o \gamma_P(V)\, c_P(V)} \times \frac{1 + K_o \gamma_P(V)\, c_P(V)}{1 + K_o \gamma_P(Vx)\, c_P(Vx)} \tag{15}$$

The reader may readily verify that once again, the maximum response will be obtained from this model when $\mathcal{F} \ll 1$. However, in the present instance this condition corresponds to $y \ll 1$, and $K_o\gamma_P c_P \ll 1$. Equation 15 thus simplifies to:

$$\Re = \frac{\gamma_P(Vx)}{\gamma_P(V)\, x} \tag{16}$$

In Figure 10–5, the value of $\Re$, calculated using Equation 16 with activity coefficients derived from SPT, is plotted as a function of percentage volume decrease for putative activators of several molecular weights, in an erythrocyte containing hemoglobin at an initial concentration of 350 g/l. The dashed line in this figure represents the *maximum* response obtainable with this model when the effect of macromolecular crowding on the activity of regulatory protein P is neglected, i.e., when $\Re$ = 1/x.

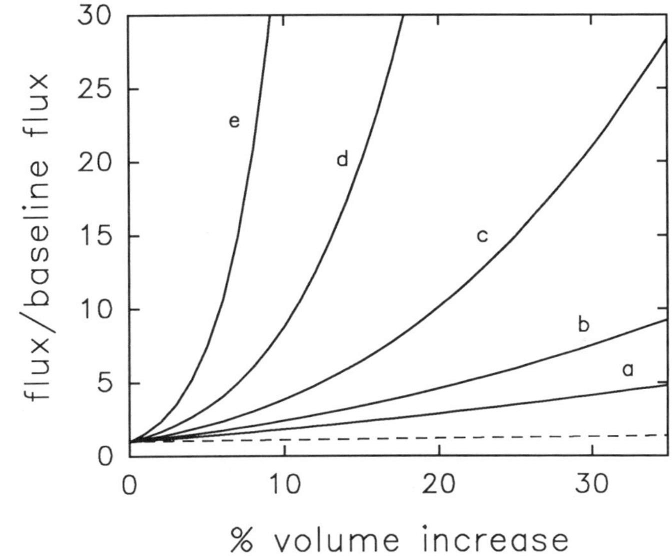

Figure 10–4 Response of Model IA, plotted as a function of fractional increase in cellular volume. Initial hemoglobin concentration = 350 g/l. M_PM_{Hb} = 0.25 (curve a); 0.5 (curve b); 1 (curve c); 2 (curve d); and 4 (curve e).

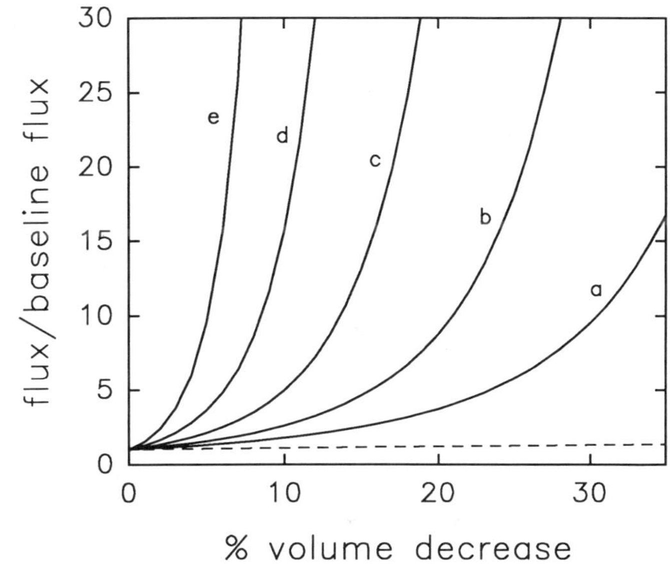

Figure 10–5 Response of Model IB, plotted as a function of fractional decrease in cellular volume. Conditions as described in the legend to Figure 10–4.

B. MODEL II: TRANSPORTER IS CONVERTED FROM INACTIVE TO ACTIVE FORM BY AN INTRACELLULAR ENZYME AND CONVERTED FROM ACTIVE TO INACTIVE FORM BY A SECOND INTRACELLULAR ENZYME

It is assumed that the inactive and active states of the transporter are not in equilibrium with each other. One intracellular enzyme, called α, catalyzes a reaction that converts inactive transporter to active, and a second intracellular enzyme, called β, catalyzes a complementary reaction that converts active transporter to inactive.

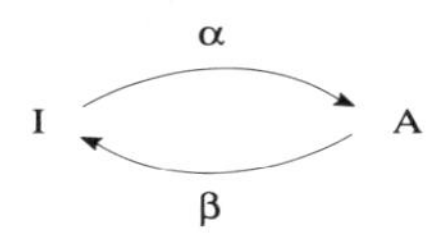

The two enzymes are assumed to act in accordance with Michaelis-Menten kinetics, i.e.,

$$\text{rate}_{I \to A} = \alpha_{tot}\, k_{cat}^{\alpha} \frac{K_{\alpha} n_I}{1 + K_{\alpha} n_I} \qquad \textbf{(17a)}$$

$$\text{rate}_{A\to I} = \beta_{tot}\, k_{cat}^{\beta} \frac{K_\beta n_A}{1+K_\beta n_A} \quad \textbf{(17b)}$$

where ε_{tot} ($\varepsilon = \alpha$ or β) is the total *amount* (not concentration) of enzyme ε, $k_{cat}{}^{\varepsilon}$ is the specific turnover (catalytic) rate for fully saturated enzyme ε, K_ε is the quasi-equilibrium constant for association of enzyme ε with its substrate (equal to the inverse of the Michaelis constant), and n_A and n_I, respectively, denote the amounts of active and inactive transporter. The expressions $K_\alpha n_I/(1+K_\alpha n_I)$ and $K_\beta n_A/(1+K_\beta n_A)$ in Equations 17a and 17b represent, respectively, the fractional saturation of α with inactive transporter and the fractional saturation of β with active transporter. By the same reasoning that leads to Equations 7 and 8, one can demonstrate that $K_\alpha = K^o_\alpha\gamma_\alpha$ and $K_\beta = K^o_\beta\gamma_\beta$, where K^o_x is the thermodynamic quasi-equilibrium constant for binding of substrate to enzyme x. At steady state, the rates of conversion of I to A and of A to I are equal, by definition to:

$$\alpha_{tot}\, k_{cat}^{\alpha} \frac{K^o_\alpha \gamma_\alpha n_I}{1+K^o_\alpha \gamma_\alpha n_I} = \beta_{tot}\, k_{cat}^{\beta} \frac{K^o_\beta \gamma_\beta n_A}{1+K^o_\beta \gamma_\beta n_A} \quad \textbf{(18)}$$

Equation 18 is complex, and predicts many different types of behavior, depending upon the relative values of the various parameters. It is instructive to examine two limiting cases in some detail, as they provide simple expressions for activity and response functions.

1. Model IIA: Both Enzyme α and Enzyme β Bind Weakly to Transporter

Under this set of conditions, Equation 18 reduces to:

$$\alpha_{tot}\, k_{cat}^{\alpha} K^o_\alpha \gamma_\alpha n_I = \beta_{tot}\, k_{cat}^{\beta} K^o_\beta \gamma_\beta n_A \quad \textbf{(19)}$$

Letting $n_I = (1 - \mathcal{F})\, T_{tot}$ and $n_A = \mathcal{F} T_{tot}$ (where T_{tot} represents the total amount of transporter), and grouping all concentration-independent terms together, Equation 19 may be rewritten as:

$$\mathcal{F} = \frac{J_\alpha \gamma_\alpha}{J_\alpha \gamma_\alpha + J_\beta \gamma_\beta} \quad \textbf{(20)}$$

where $J_\alpha = \alpha_{tot}\, k^\alpha_{cat}\, K^o_\alpha$ and $J_\beta = \beta_{tot}\, k^\beta_{cat}\, K^o_\beta$. According to this model, no volume regulation is expected when enzymes α and β are the same size, since γ_α would be approximately equal to γ_β for all V, and Equation 20 would reduce to:

$$\mathcal{F} = \frac{J_\alpha}{J_\alpha + J_\beta} \quad \textbf{(21)}$$

independent of V. There are two special cases of interest here:

1. $J_\alpha \approx J_\beta$. Equation 20 reduces to:

$$\mathcal{F} = \frac{\gamma_\alpha}{\gamma_\alpha + \gamma_\beta} \quad \textbf{(22)}$$

According to Equation 22, no volume regulation would be expected when enzyme α is significantly larger than enzyme β, since under crowded conditions, γ_α would be much larger than γ_β and $\mathcal{F}$ would be approximately equal to 1 over a wide range of volumes.

2. $J_\alpha\gamma_\alpha \ll J_\beta\gamma_\beta$. Equation 20 reduces to:

$$\mathcal{F} = \frac{J_\alpha \gamma_\alpha}{J_\beta \gamma_\beta} \quad \textbf{(23)}$$

$$\Re = \frac{\gamma_\alpha(Vx)}{\gamma_\alpha(V)} \times \frac{\gamma_\beta(V)}{\gamma_\beta(Vx)} \quad \textbf{(24)}$$

In Figure 10–6, the value of $\Re$, calculated using Equation 24, is plotted as a function of fractional volume change. When M_α (the molecular weight of α) $< M_{Hb} < M_\beta$, the transporter is activated by shrinkage (deactivated by swelling). Conversely, when $M_\alpha > M_{Hb} > M_\beta$, the transporter is activated by swelling (deactivated by shrinkage). In both cases, signal amplification increases with increasing difference between the sizes of α and β.

2. Model IIB: Enzyme α Is Tightly Bound to Transporter, Enzyme β Is Weakly Bound

Under this set of conditions, Equation 18 reduces to:

$$\alpha_{tot}\, k_{cat}^{\alpha} = \beta_{tot}\, k_{cat}^{\beta} K^o_\beta \gamma_\beta n_A \quad \textbf{(25)}$$

Recalling that $\mathcal{F} = n_A/T_{tot}$, Equation 25 may be rewritten:

$$\mathcal{F} = Q / \gamma_\beta \quad \textbf{(26)}$$

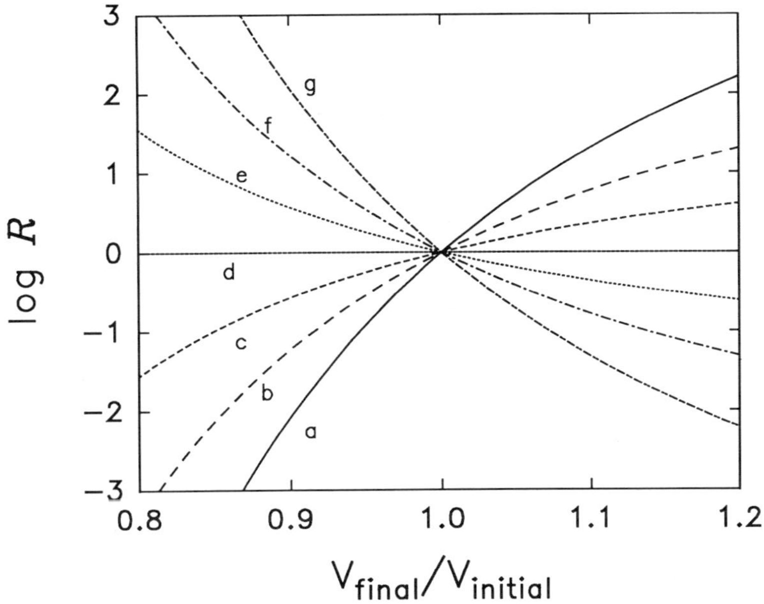

Figure 10–6 Response of Model IIA, calculated as a function of fractional volume change, for $M_\alpha/M_{Hb} = M_{Hb}/M_\beta$ = 1/4 (curve a); 1/2.5 (curve b); 1/1.6 (curve c); 1 (curve d); 1.6 (curve e); 2.5 (curve f); and 4 (curve g).

where

$$Q = \frac{\alpha_{tot}\, k^{\alpha}_{cat}}{T_{tot}\beta_{tot}\, k^{\beta}_{cat} K^{o}_{\beta}}$$

In this model the response function has the especially simple form:

$$R = \frac{\gamma_\beta(V)}{\gamma_\beta(Vx)} \qquad \textbf{(27)}$$

It is evident that model IIB can account for activation of transporter via swelling but not via shrinkage. Since Equation 27 differs from Equation 13 by only a numerically small factor of x, the dependence of R upon percentage volume increase, calculated for various molecular weight of β, is quite similar to that calculated using Equation 13 and plotted in Figure 10–4.

IV. DISCUSSION

The models presented here are highly simplified, and probably much simpler than actual mechanisms of volume regulation. However, even these simplified models are capable of exhibiting activation of transporters via swelling and shrinkage, with signal amplification as great as that observed experimentally. Recently, model IIB was considered as a possible model for the experimentally observed response of the K-Cl cotransporter to swelling of dog erythrocytes.[6] In the context of this particular system, the active form of the transporter was taken to be dephosphorylated and the inactive form phosphorylated. Thus, enzyme α was postulated to be a phosphatase and β to be a kinase. According to this model, swelling-induced activation of the K-Cl cotransporter is due to the dissociation of kinase from the cotransporter (i.e., inhibition of the inactivating enzyme). This conclusion is qualitatively consistent with the finding that swelling-induced activation of K-Cl cotransport in rabbit erythrocytes is due to a decrease in the rate of inactivation of the active form of the cotransporter.[19]

The primary purpose of this chapter is not to advocate particular models for volume regulation. The author has learned from personal experience that elegant, simple, well-crafted, and apparently comprehensive models can be easily rendered irrelevant by experimental disclosure of previously unsuspected complexity. The important lesson to be learned here is that small fractional volume changes of a cell can, via alteration of macromolecular crowding, influence association equilibria and enzyme-catalyzed reaction rates by a far greater extent than expected on the basis of simplistic considerations of mass action. This fact must be taken into account when formulating any model for volume regulation that involves the formation or dissociation of macromolecular complexes, including transient complexes formed between enzymes and macromolecular substrates.

ACKNOWLEDGMENTS

The author thanks Steven Zimmerman, National Institutes of Health, and John Parker, University of North Carolina, for helpful comments on a preliminary draft of this chapter.

REFERENCES

1. **Sarkadi, B. and Parker, J. C.,** Activation of ion transport pathways by changes in cell volume, *Biochim. Biophys. Acta,* 1071, 407, 1991.
2. **Jennings, M. L. and Schulz, R. K.,** Swelling-activated KCl cotransport in rabbit red cells: flux is determined mainly by cell volume rather than shape, *Am. J. Physiol.,* 259, C960, 1990.
3. **Colclasure, G. C. and Parker, J. C.,** Cytosolic protein concentration is the primary volume signal in dog red cells, *J. Gen. Physiol.,* 98, 881, 1991.
4. **Cossins, A. R.,** A sense of cell size, *Nature,* 352, 667, 1991.
5. **Watson, P. A.,** Function follows form: generation of intracellular signals by cell deformation, *FASEB J.,* 5, 2013, 1991.
6. **Minton, A. P., Colclasure, G. C., and Parker, J. C.,** Model for the role of macromolecular crowding in regulation of cellular volume, *Proc. Natl. Acad. Sci. U.S.A.,* 89, 10504, 1992.
7. **Minton, A. P.,** The effect of volume occupancy upon the thermodynamic activity of proteins: some biochemical consequences, *Mol. Cell. Biochem.,* 55, 119, 1983.
8. **Minton, A. P.,** Excluded volume as a determinant of macromolecular structure and reactivity, *Biopolymers,* 20, 2093, 1981.
9. **Fulton, A. B.,** How crowded is the cytoplasm?, *Cell,* 30, 345, 1982.
10. **Hill, T. L.,** *Introduction to Statistical Thermodynamics,* Addison-Wesley, Reading, MA, 1960, chap. 19.
11. **Giddings, J. C., Kucera, E., Russell, C. P., and Myers, M. N.,** Statistical theory for the equilibrium distribution of rigid molecules in inert porous networks, *J. Phys. Chem.,* 72, 4397, 1968.
12. **Ross, P. D. and Minton, A. P.,** Analysis of nonideal behavior in concentrated hemoglobin solutions, *J. Mol. Biol.,* 112, 437, 1977.
13. **Nichol, L. W., Ogston, A. G., and Wills, P. R.,** Effect of inert polymers on protein self-association, *FEBS Lett.,* 126, 18, 1981.
14. **Ralston, G. B.,** Effects of "crowding" in protein solutions, *J. Chem. Educ.,* 67, 857, 1990.
15. **Lebowitz, J. L., Helfand, E., and Praestgaard, E.,** Scaled particle theory of fluid mixtures, *J. Chem. Phys.,* 43, 774, 1965.
16. **Gibbons, R. M.,** The scaled particle theory for particles of arbitrary shape, *Mol. Phys.,* 17, 81, 1969.
17. **Chatelier, R. C. and Minton, A. P.,** Sedimentation equilibrium in macromolecular solutions of arbitrary concentration. I. Self-associating proteins (Appendix), *Biopolymers,* 26, 522, 1987.
18. **Hill, T. L.,** *Cooperativity Theory in Biochemistry,* Springer-Verlag, New York, 1985, chap. 2.
19. **Jennings, M. L. and Al-Rohil, N.,** Kinetics of activation and inactivation of swelling-induced KCl transport, *J. Gen. Physiol.,* 95, 1021, 1990.

Chapter 11

How an Epithelial Cell Swells Is a Determinant of the Signaling Pathways that Activate RVD

R. John MacLeod

TABLE OF CONTENTS

I. INTRODUCTION

How does a cell sense changes in volume and transduce those signals into a regulatory response? Several recent reviews have implicated known second messengers, such as Ca^{2+}, phosphoinositides, membrane stretch, cytoskeletal changes, changes in intracellular pH (pH_i), cytosolic ionic strength or protein content as transducing mechanisms.[1-4] Invariably, these mechanisms are invoked to explain the control of volume regulation in cells that have undergone substantial volume increases induced by exposure to large, and in most cases, unphysiological hypotonic dilutions. In epithelial cells, however, physiological volume changes may be small relative to those induced in most experimental models. Furthermore, epithelial cell volume changes are commonly brought about by changes in transcellular solute transport. For example, in the small intestine, amino acids and glucose are reabsorbed by Na^+-dependent cotransport processes.[5,6] Normal changes in the activity of these cotransporters can alter intestinal cell volume due to increases or decreases in net, apical membrane solute influx. Consequently, it is reasonable to ask if the signaling of regulatory volume decrease (RVD) after hypotonic dilution is different compared with that induced by changes in Na^+-dependent amino acid or glucose transport. This chapter will focus on studies which suggest that the regulation of K^+ and Cl^- channel activation in jejunal villus enterocytes is dependent on both the extent of cell swelling, as well as the mechanism by which cell volume is increased.

II. Na^+-NUTRIENT UPTAKE AND RVD

We first determined the volume response of guinea pig jejunal villus cells exposed to solutions diluted either 30 or 50%.[7] The pharmacological sensitivity of the subsequent RVD, as well as the generation of secondary volume changes induced by anion gradients, were consistent with volume regulation being mediated by separate K^+ and Cl^- conductances. Next, we sought to determine whether these cells would swell when exposed to elevated concentra-

0-8493-4448-4/94/$0.00+$.50

tions of L-alanine or D-glucose.[8] Using a Coulter/channelyzer apparatus, we observed volume increases when 20 to 25 m*M* of the solute was added to cells suspended in isotonic medium. Volume increases reached peak values 1 to 5 min after solute addition. Once the maximal volume was reached, the cells began to shrink, completing RVD by 2 min (D-glucose) or 3 min (L-alanine). These volume changes were compared with those occurring after addition of the non-Na^+ cotransported isomer (D-alanine or L-glucose). Immediately after the addition of these solutes, the cells shrunk approximately 6% and remained shrunken. Addition of the same amount of the impermeant solute, mannitol, also generated immediate shrinkage. While villus cells manifest a brisk regulatory volume increase (RVI) following hypertonic shrinkage due to the activation of $NaKCl_2$ cotransport,[9] these cells must first be depleted of KCl with a round of RVD before the cotransporter can be activated. The volume increase caused by L-alanine or D-glucose was prevented in Na^+-free medium; without extracellular Na^+ the cells shrank immediately. Furthermore, cell swelling induced by α-CH_3-glucoside was prevented by phlorizin, an inhibitor of Na-glucose cotransport. Together these experiments suggested that the increase in villus cell volume was due to the influx of Na^+ with either L-alanine or D-glucose, together with water.

Different classes of K^+ and Cl^- channel blockers prevented RVD. For example, 5 min after Na^+-nutrient addition, the relative volume of the cells treated with anthracene-9-carboxylate (A-9C) was larger ($1.05 \pm .02$ vs. $0.91 \pm .01$, $p < .001$) compared with cells in the absence of the Cl^- channel blocker. A-9C alone had no effect on the isotonic volume of the cells while the stilbene-derivative DIDS had no effect on RVD following cell swelling induced by Na^+-nutrient cotransport.

An additional experimental approach also suggested that a Cl^- conductance was activated for RVD. In these studies, cells were permeabilized to Na^+ by exposing them to gramicidin D. If a Cl^- conductance has been activated for RVD, the provision of a parallel conductive pathway for Na^+ entry (gramicidin D) should result in 2° cell swelling. As expected, striking 2° swelling was observed in Cl^--containing medium, whereas in gluconate-containing (Cl^--free) medium, no swelling was observed. In addition, A-9C attenuated the 2° swelling observed in the Cl^--containing medium. Together, the pharmacology of RVD and the generation of 2°, Cl^--dependent volume changes suggested that RVD following Na^+-nutrient cotransport swelling was due to the activation of both K^+ and Cl^--conductive pathways.

Hudson and Schultz showed that 10 m*M* glycine caused a slow volume increase of 17% in Ehrlich ascites cells.[10] The volume increase was prevented in Na^+-free medium consistent with cell swelling being due to the uptake of Na^+ together with glycine. The Ehrlich cells, however, showed no RVD and remained swollen throughout the course of the experiment. Single-channel patch-clamp analysis after cell swelling induced by glycine addition or a 50% hypotonic dilution indicated that a Cl^- channel with a conductance of 23 pS was activated under these conditions. Christensen and Hoffmann[11] have also shown that swelling activates a Cl^- conductance in this cell line. The nature of the signaling pathways responsible for the activation of these Cl^- channels following hypotonic and Na^+-nutrient swelling remain to be determined.

III. PKC INHIBITORS AND RVD

The first evidence that different signaling mechanisms were responsible for activation of the RVD conductive pathways following hypotonic or Na^+-nutrient swelling came from volume experiments using protein kinase C (PKC) inhibitors.[12] In the presence of 1-(5-isoquinolinylsulfonyl)-2-methylpiperazine (H-7), the RVD following cell swelling stimulated by L-alanine was prevented (Figure 11–1B). However, the same concentration of H-7 had no effect on RVD after the villus cells were swollen by hypotonic dilutions of 5 or 10%, to mimic the volume changes caused by L-alanine (Figure 11–1C), or 50% (Figure 11–1A).

Since H-7 will also inhibit protein kinase A, the more selective and potent PKC inhibitor (K_i = 3 n*M*), staurosporine, was assessed for its effect on these volume changes. With maximal effect at 10 n*M*, staurosporine prevented RVD of cells swollen by L-alanine absorption. In contrast, staurosporine (0.1 to 100 n*M*) had no effect on RVD in cells swollen by hypotonic dilutions of 5 or 50%. When the ^{36}Cl efflux rate after hypotonic dilution (50%) was determined, H-7 was found to have no effect on either the isotonic rate of ^{36}Cl loss or the increased rate after hypotonic dilution. However, A-9C was found to attenuate the increased efflux rate caused by hypotonic dilution. The addition of L-alanine to these cells caused biphasic ^{36}Cl efflux; the rapid rate, occurring over the first minute, was inhibited by H-7. The increased rate of ^{36}Cl efflux that was stimulated by L-alanine was also inhibited by A-9C. Thus, the volume response to the PKC inhibitor, as well as to the Cl^- channel blocker was mirrored by Cl^- loss. This suggested that PKC was

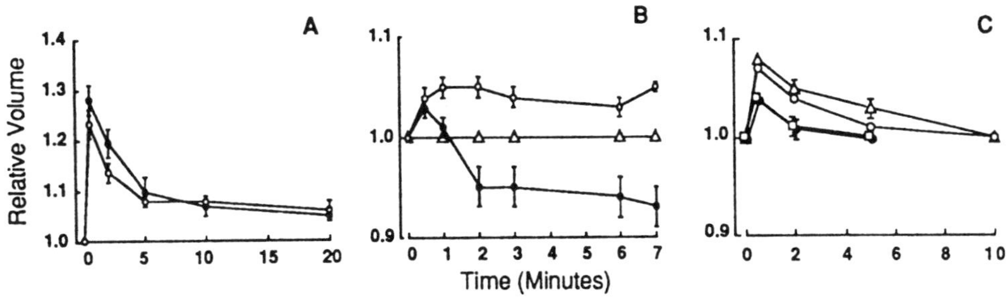

Figure 11–1 Effect of 1-(5-isoquinolinylsulfonyl)-2-methylpiperazine (H-7) on regulatory volume decrease (RVD). (A) Hypotonic medium: ○, 0.5 × isotonic; •, 0.5 × isotonic with 100 μ*M* H-7. (B) Volume changes after addition of 25 m*M* L-Ala: ●, control; ○, 100 μ*M* H-7 and 25 m*M* L-Ala; Δ, 100 μ*M* H-7 (n = 7). (C) Hypotonic medium: Δ, 0.9 × isotonic; ○, 0.9 × isotonic with 100 μ*M* H-7; □, 0.95 × isotonic; ▼, 0.95 × isotonic with 100 μ*M* H-7 (n = 5). Volume measured electronically, expressed relative to isotonic control. (From MacLeod, R. J. et al., *Am. J. Physiol.,* 262, C950, 1992. With permission.)

required to activate the Cl^- conductance necessary for RVD when villus cells were swollen because of Na^+-nutrient uptake, but not for RVD following passive hypotonic dilution.

Additional evidence for the PKC sensitivity of the Cl^- conductance activated for Na^+-nutrient RVD came from volume experiments using gramicidin. Striking 2° swelling was observed in cells permeabilized with gramicidin added when they were maximally swollen by L-alanine absorption in Cl^--containing medium. This 2° swelling was attenuated by staurosporine. These experiments suggested that during cell swelling stimulated by L-alanine cotransport, PKC was required to activate the Cl^- conductance for RVD. Accordingly, it should be possible to activate a PKC-sensitive Cl^- conductance without causing cell swelling by exposing cells to PKC activators, such as 12-tetradecanoyl-phorbol-13-acetate (TPA) or synthetic diacylglycerol, under isotonic conditions. Previously, we had shown that guinea pig jejunal villus cells permeabilized with gramicidin under isotonic conditions swell and that this volume increase was rate-limited by the anion permeability of the cells.[7] Consequently, the volume changes stimulated by TPA in gramicidin-permeabilized cells under isotonic conditions were compared to cells challenged with the inactive 4-α-phorbol didecanoate (Figure 11–2). TPA (10^{-8} *M*) caused both faster and greater cell swelling, which was prevented by staurosporine. Synthetic diacylglycerol (OAG) was also found to stimulate both the rate and extent of villus cell swelling and this stimulation was prevented by staurosporine. An important finding from this experiment was that staurosporine alone had no impact on the volume changes occurring after gramicidin permeabilization, suggesting that the depolarization caused by Na^+ entry was not activating the PKC-sensitive Cl^- conductance.

IV. Ca^{2+} MEDIATION OF HYPOTONIC AND Na^+-NUTRIENT RVD

The second line of evidence that the signaling mechanisms responsible for activating K^+ and Cl^- conductances was dependent on the mode of cell volume increase came from studies of the Ca^{2+} dependence of volume regulation.[13] The RVD following passive, hypotonic dilution absolutely required extracellular Ca^{2+}. Even when villus cells were hypotonically swollen to a volume comparable to the one generated by L-alanine absorption (0.95 × isotonic), RVD was prevented by Ca^{2+} removal (Figures 3A and C). In striking contrast, the volume changes caused by L-alanine addition were not affected by Ca^{2+}-free medium (Figure 11–3B). When putative increases in intracellular Ca^{2+} were buffered by loading these cells with the Ca^{2+} chelator, BAPTA, the RVD was prevented in both models of villus cell swelling. These experiments suggested that for the hypotonic model of RVD, the Ca^{2+} required was extracellular, consistent with studies in other epithelia and symmetrical cells (reviewed in Chapter 15 of this volume). However, the volume response of the cells swollen by Na^+-alanine cotransport suggested that intracellular Ca^{2+} was mobilized. In preliminary experiments using Indo-1, we have found that intracellular Ca^{2+} increases from 79 to 263 n*M* within 4 s after addition of L-alanine. The source of this Ca^{2+} rise is currently being investigated using the sesquiterpene lactone, thapsigargin. By inhibiting microsomal Ca^{2+}-ATPase, thapsigargin causes a rapid release

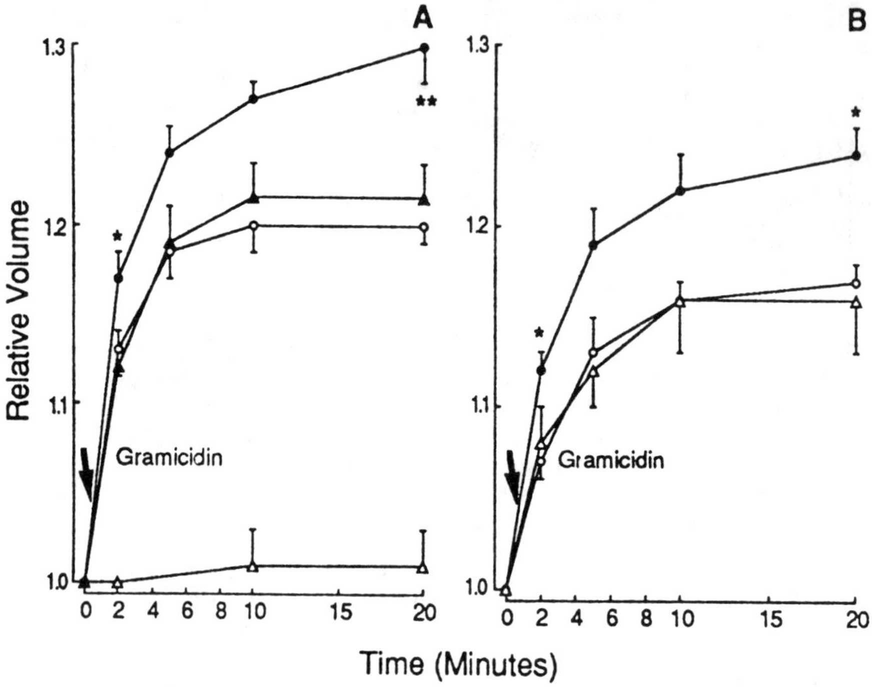

Figure 11–2 Effect of phorbol ester, synthetic diacylglycerol, and staurosporine on isotonic villus cell volume. (A) Gramicidin (0.5 μ*M*) and phorbol esters added immediately after isotonic volume measurement: ●, 12-tetradecanoyl-phorbol-13-acetate (TPA, 10um^{-8} *M*); ▲, staurosporine (10 n*M*) + TPA; ○, 4 α-phorbol 12, 13-didecanoate (10^{-7} *M*); Δ, gramicidin in sodium gluconate medium. *, $p < .05$, ** $p < .005$ compared with staurosporine + TPA (n = 5). (B) 1-Oleoyl-2-acetylglycerol (2.5 μg/ml) added together with gramicidin: ●, OAG; ○, staurosporine (10 n*M*) + OAG; Δ, staurosporine. *, $p < .05$ compared with OAG + staurosporine (n = 5). Volume was measured electronically and expressed relative to control before gramicidin or PKC activator addition. (From MacLeod, R. J. et al., *Am. J. Physiol.,* 262, C950, 1992. With permission.)

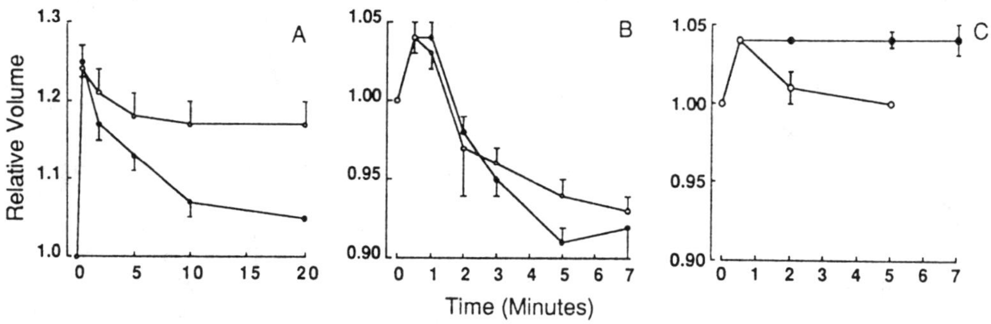

Figure 11–3 Effect of extracellular Ca^{2+} chelation on RVD. (A) Hypotonic medium: ●, 0.5 × isotonic (1 m*M* Ca^{2+}); ○, 0.1 m*M* EGTA in Ca^{2+}-free medium; n = 8. (B) Volume changes after addition of 25 m*M* L-Ala: ●, isotonic (1 m*M* Ca^{2+}); ○, 0.1 m*M* EGTA in Ca^{2+}-free medium; n = 20. (C) ○, 0.95 × isotonic in Ca^{2+}-containing medium; ●, 0.95 × isotonic in Ca^{2+}-free medium; n = 5. Volume was measured electronically, expressed relative to isotonic control. (From MacLeod, R. J. et al., *J. Membr. Biol.,* 130, 23, 1992. With permission.)

of Ca^{2+} from an IP_3-sensitive compartment.[14] In a variety of cell types, addition of thapsigargin also results in elevation of membrane Ca^{2+} permeability. The addition of thapsigargin to villus cells causes a rapid decrease in cell volume, presumably due to a transient increase in intracellular Ca^{2+} followed by the activation of K^+ and Cl^- conductances.[15] However, the addition of L-alanine to the thapsigargin-treated cells caused cell swelling, but subsequent RVD was absent (Figure 11–4). Thapsigargin-treated cells swollen by exposure to media diluted 5 or 50% showed complete RVD in Ca^{2+}-containing medium. In contrast, RVD was prevented in the thapsigargin-treated cells hypotonically swollen in Ca^{2+}-free medium. We have also found that villus cells treated with ryanodine (20 μ*M*) show RVD after swelling caused by L-alanine, but consistent with observations from an intestinal cell line,[16] RVD after hypotonic dilution in Ca^{2+}-containing medium was prevented by ryanodine. Our volume data suggest that the Ca^{2+} required for RVD after L-alanine swelling is derived from an IP_3-sensitive pool, while the Ca^{2+} required for RVD following hypotonic swelling is derived from a ryanodine-sensitive pool.

The volume response of hypotonically swollen villus cells to both organic and inorganic Ca^{2+} channel blockers differed in comparison with villus cells swollen by Na^+-nutrient absorption.[13] RVD in hypotonically swollen cells was blocked by the Ca^{2+} channel blockers niguldipine (0.1 μ*M*), nifedipine (5 μ*M*), diltiazem (10 to 100 μ*M*), Ni^{2+}, and Co^{2+} (1 m*M*). In contrast, these Ca^{2+} channel blockers had no effect on RVD following L-alanine addition. The volume response of the hypotonically swollen cells (0.5 or 0.95 × isotonic) in the presence of either organic or inorganic Ca^{2+} channel blockers is consistent with the interpretation that extracellular Ca^{2+} enters the cells through a Ca^{2+} channel.

The Ca^{2+} dependency of RVD following hypotonic or Na^+-nutrient swelling also differs in two additional respects. Charybdotoxin (CTX)[17,18] prevents the RVD of villus cells swollen because of the uptake of D-glucose.[13] In contrast, the RVD following large hypotonic cell swellings was not blocked by the toxin. In addition, the response to calmodulin antagonists differed in these two models of cell swelling. The napthalene sulfonamide derivates W-13 or W-7 had no effect on RVD after cell swelling caused by L-alanine. In contrast, these agents blocked RVD following hypotonic dilution. The inhibitory effect on volume regulation was bypassed by adding gramicidin, suggesting that the K^+ conductance was the target of calmodulin. Thus, in contrast with RVD after hypotonic dilution, RVD after swelling induced by Na^+-nutrient uptake was not affected by calmodulin antagonists, but was prevented by CTX.

We have also observed a role for the Ca^{2+}/calmodulin-dependent protein kinase in the activation of RVD pathways. RVD after small

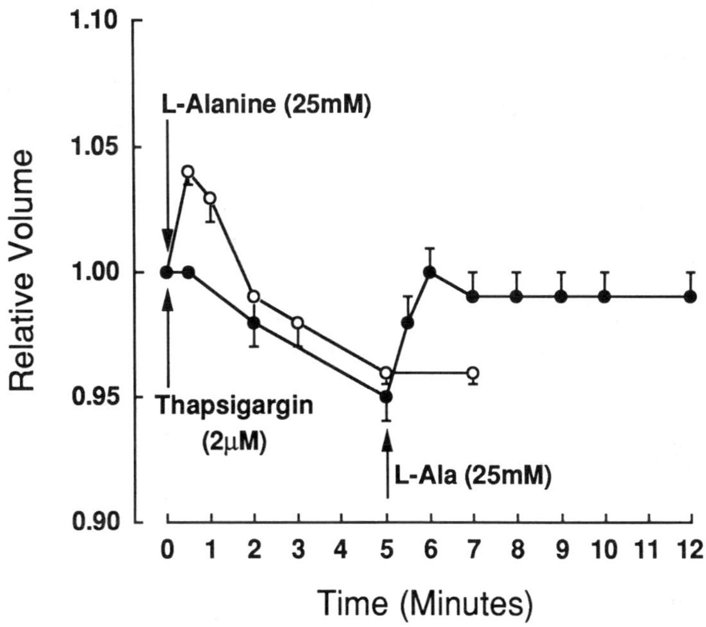

Figure 11–4 Effect of thapsigargin on RVD after L-alanine-induced swelling: ○, control cells treated with L-alanine (25 m*M*); ●, cells treated with thapsigargin (2 μ*M*) then treated with L-alanine. Data shown are means ± SE of 10 experiments. Volume measured electronically, expressed relative to isotonic control.

hypotonically induced volume increases was not blocked by the isoquinoline, KN-62, which is a reputed specific inhibitor of this enzyme.[27] In contrast, RVD after large hypotonic swellings was blocked by KN-62.

V. A MODEL

The results of the studies described above suggest that the mode of cell swelling determines the nature of the signaling pathways that are responsible for activating volume regulatory K^+ and Cl^- conductances. More importantly, these studies suggest a different way of thinking about the function of Na^+-nutrient cotransporters.[19,20] A model of how the Na^+-nutrient cotransporter activates the K^+ and Cl^- conductances necessary for RVD must account for (1) the CTX sensitivity of the K^+ conductance, (2) the mobilization of intracellular Ca^{2+} from a thapsigargin-sensitive source, and (3) the PKC requirement of the Cl^- conductance. At the risk of having one's hands amputated with Occam's razor, the simplest model would require the Na^+-nutrient cotransporter to be coupled to an effector pathway that would stimulate inositol lipid hydrolysis and diacylglycerol production. This model requires that the Na^+-nutrient cotransporter stimulate phospholipase C activity and, by implication, requires coupling to the lipase in the same manner as a polypeptide or adrenergic hormone receptor is coupled to its effector pathways, via a G protein. Accordingly, we would like to know, in intact cells and in isolated brush border membranes, if D-glucose or L-alanine addition stimulate PLCase. Is a GTPase stimulated? If diacylglycerol is produced in intact cells by L-alanine-induced swelling, which isozyme of PKC is translocated to which membrane? This is an important consideration, since earlier work found that L-alanine increased Cl^- permeability of the apical membrane of enterocytes of *Necturus* small intestine.[21] When the villus cell swells 5 to 7% because of Na^+-alanine uptake, why isn't a dihydropyridine-sensitive Ca^{2+} influx pathway activated? Answering these questions would explain a great deal about how the Na^+-nutrient cotransporter activates second messengers in order to activate the conductances required for volume regulation.

VI. CHANGES IN INTRACELLULAR pH DURING HYPOTONIC OR L-ALANINE-INDUCED SWELLING

We found that the addition of either L-alanine or D-glucose to BCECF-loaded villus cells caused a rapid alkalinization of approximately 0.1 pH units, which was prevented by 1 μM 5-*N*-methyL-isobutyl amiloride (MIA). However, addition of the stereoisomers of these solutes or 20 m*M* mannitol, which do not cause a Na^+-dependent increase in cell volume, all caused the same pH_i changes, suggesting that it was a response to the hyperosmolarity of the media. This was not a surprising finding since, in parallel studies, it was established that epithelial cells isolated from the crypt compartment undergo RVI after hypertonic challenge by activating Na/H and Cl/HCO_3 exchangers.[22] Since the villus enterocytes differentiate from cells from the crypt, but mediate RVI by activating bumetanide-sensitive $NaKCl_2$ cotransport,[9] an osmotically responsive Na/H exchanger on villus cells did not seem unreasonable. The real surprise came when these villus cells were hypotonically diluted 5 to 7% to mimic the volume changes stimulated by L-alanine absorption (Figure 11–3C). We found that villus cells hypotonically diluted 5 or 7% first acidified approximately 0.03 pH units, then alkalinized by 0.05 to 0.07 pH units over 3 min. In both cases, 1 μM MIA inhibited the alkalinization. In contrast, villus cells hypotonically swollen 30% continued to acidify over the same time period. We therefore assessed the MIA sensitivity of RVD after small hypotonic swellings and found that the RVD was blocked by this drug.[23] When the villus cells were hypotonically swollen 5 or 7% in Na^+-free medium, the subsequent RVD was prevented. Again in contrast, cells swollen by large hypotonic dilutions (30 or 50%) underwent complete RVD in the Na^+-free medium (Figure 11–5A). Together these results suggested that the RVD following small hypotonically induced swellings was dependent on activation of Na^+/H^+ exchange. This Na^+/H^+ exchanger appears to be located on the basolateral membrane of these cells.[23]

These results seemed counterintuitive. Why would an epithelial cell take up Na^+ while it was losing K^+ and Cl^- for volume reduction? Indeed, in symmetrical cells Na^+/H^+ exchange has been shown to be activated subsequent to the acidification found with large hypotonic swelling.[24,25] But in these cases, amiloride accelerates RVD, consistent with the concept that solute uptake would offset solute loss during RVD. Since it has been shown that the affinity for Ca^{2+} of the K^+ conductance is exquisitely sensitive to pH, with affinity increasing at more alkaline pH_i,[26] we speculated that the "target" of the resulting alkalinization occurring when the basolateral Na^+/H^+ exchange was activated was the K^+ conductance. To test this idea, we swelled cells by exposure to a medium diluted 50%. As discussed above, the RVD activated following such

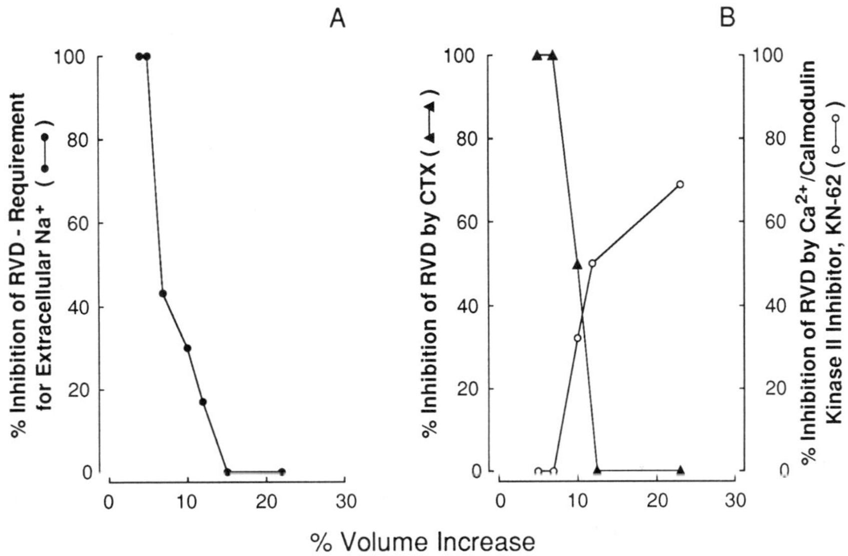

Figure 11–5 Relationship between extracellular Na^+, charybdotoxin (CTX), and KN-62 on RVD as a function of percent volume increase. (A) Effect of Na^+-free medium on RVD. Na^+ isotonically replaced with *N*-methyl D-glucamine. Villus cells were hypotonically diluted 5 to 50%. Data are means from eight experiments performed in duplicate. (B) Effect of charybdotoxin (50 n*M*), ▲-▲, or KN-62 (50 μ*M*), O-O, on RVD after hypotonic dilutions from 5 to 50%. Data are means of five experiments performed in duplicate. Volume measured electronically, expressed relative to isotonic control.

volume increases is CTX-*insensitive*. When these same cells were alkalinized, CTX-*sensitive* volume changes were observed.[23] These results suggest that intracellular alkalinization increases the affinity of the K^+ conductance for Ca^{2+}, resulting in activation of this pathway for RVD.

The relationships between extracellular Na^+, CTX-sensitive and Ca^{2+}-calmodulin kinase II-sensitive RVD as a function of amount of volume increase are illustrated in Figure 11–5. While small volume increases absolutely require extracellular Na^+ for any RVD, larger volume increases show no requirement for extracellular Na^+. Intermediate amounts of cell swelling show partial volume recovery in Na^+-free medium. The CTX-sensitive RVD (Figure 11–5B) is superimposable on the extracellular Na^+ relationship and the response of RVD to the Ca^{2+}-calmodulin kinase inhibitor is a mirror image of CTX sensitivity. RVD following small volume increases is not affected by the kinase inhibitor, but the larger the villus cell gets, the greater the inhibition of volume regulation. These results are consistent with the interpretation that cell swelling (by passive hypotonic dilution) will activate different conductive pathways contingent on the extent of the cell swelling. Whether this applies to other Na^+-absorbing epithelial cells or symmetrical cells that are capable of Na^+-solute cotransport is not known. However, it may be that some of the signaling pathways responsible for RVD when a cell swells because of Na^+-nutrient absorption can be mimicked by a small hypotonic dilution.

VII. OTHER EXAMPLES OF DIFFERENTIALLY REGULATED RVD PATHWAYS

RVD in trout red blood cells following hypotonic swelling is mediated by both KCl and taurine loss. When catecholamines are added to these cells in isotonic medium they swell due to NaCl uptake mediated by activation of Na^+/H^+ and Cl^-/HCO^-_3 exchangers. In response to the catecholamine-induced swelling, KCl loss is activated, but amino acid loss is not.[28] Motais et al.[29] compared cytoplasmic ionic strength in these two models of cell swelling in a very elegant study. Isotonic urea-containing saline was found to induce osmolyte loss that was the same as hypotonic swelling, and isotonic NH_4Cl-containing saline induced the same osmolyte loss as that stimulated by catecholamines. Motais et al.[29] found that both the hypotonic swelling and the urea-induced, isotonic swelling decreased internal ionic strength, whereas the catecholamine and the NH_4Cl isotonic swellings increased internal ionic strength. The relationship of amino acid loss as a function of cytoplasmic ionic strength was shown to be a gradual, continuous function, in contrast to the abrupt changes of the volume "set points" illustrated for villus epithelial cells in Figure 11–5B. These authors speculated how the volume-sensitive KCl cotransporter could be modulated by internal ionic strength using microfilaments as the volume sensor.

Ouabain will induce villus cell swelling when added under isotonic conditions.[7] We presume that preventing Na^+ efflux via Na^+,K^+-ATPase while still allowing Na^+ and Cl^- influx accounts for this volume increase. If the swollen cells are resuspended in isotonic medium that is ouabain-free, these cells undergo RVD (Figure 11–6A). The volume decrease is indeed RVD and not just due to loss of Na^+ via the reactivated Na^+,K^+-ATPase, since Ba^{2+} (Figure 11–6B) and A-9C (Figure 11–6C) prevent the subsequent volume recovery. Furthermore, a dihydropyridine Ca^{2+} channel blocker also prevented RVD (Figure 11–6D). Consistent with the data involving the Na^+-nutrient RVD, the PKC inhibitor staurosporine had no effect on RVD (Figure 11–6C). Together, the data shown in Figure 11–6 suggest that ouabain-induced swelling induces RVD by the activation of a dihydropyridine-sensitive Ca^{2+} influx pathway as well as the activation of K^+ and Cl^- channels similar to that seen with hypotonic swelling. These results contrast with findings of ouabain-induced swelling of CCT principal cells where anisosmotic RVD is unaffected by Cl^- transport inhibitors or cellular Cl^- depletion,[30] but RVD induced by Na^+ pump inhibition clearly requires Cl^- and is prevented by Cl^- channel blockers.[31]

Table 1 summarizes some of the characteristics of RVD in villus cells following swelling induced by three different experimental maneuvers. In all

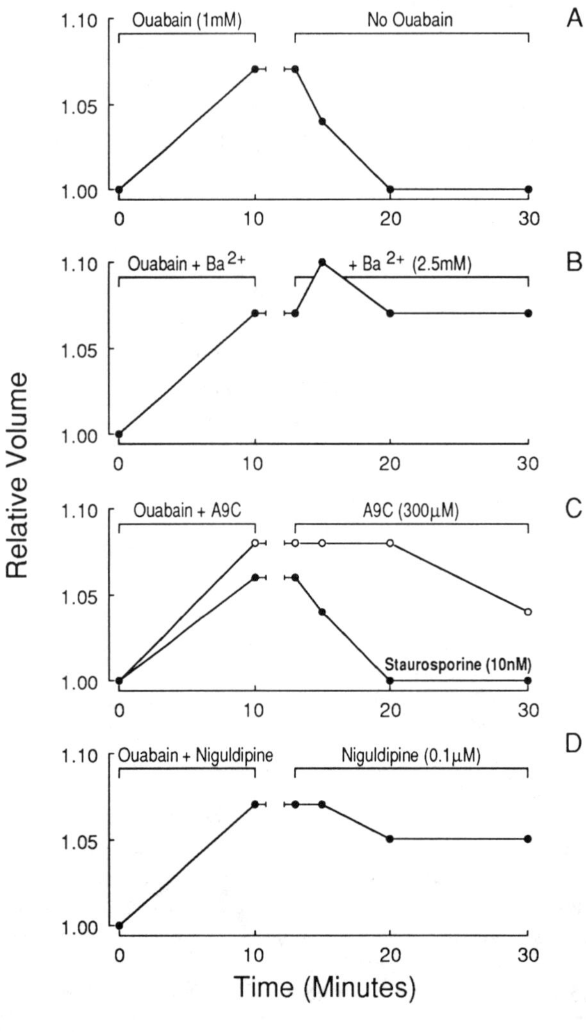

Figure 11–6 Effect of ouabain on isotonic villus cell volume. (A) Ouabain (1 m*M*) added to cells in isotonic medium, at 10 min cells centrifuged and resuspended in isotonic, ouabain-free medium. (B) Effect of Ba^{2+} (2.5 m*M*). (C) Effect of anthracene-9-carboxylate (A-9C) or staurosporine (10 n*M*). (D) Effect of niguldipine (0.1 μ*M*). Data shown are means of six experiments performed in duplicate. Volume measured electronically, expressed relative to isotonic control.

Table 1 **Characteristics of RVD in Villus Cells[a]**

	Hypotonic Dilution	Ouabain (1 m*M*)	L-alanine (20 m*M*)
Extent of volume increase	5–7%	5–7%	5–7%
Rate of volume increase	Fast	Slow	Fast
Dihydropyridine-sensitive RVD	Yes	Yes	No
Staurosporine-sensitive RVD	No	No	Yes
Methylisobutyl amiloride-sensitive RVD	Yes	No	Yes
Charybdotoxin-sensitive RVD	Yes	Yes	Yes
Cytoskeletal requirement for RVD	Yes	Yes	Yes

[a] Following swelling induced by three different experimental designs.

cases, the K^+ conductance that is activated is CTX-sensitive and RVD can be prevented by cytochalasin B (25 μ*M*). MIA (1 μ*M*) prevents RVD after hypotonic or L-alanine swelling, but not ouabain-induced swelling. While intracellular ionic strength probably decreases with hypotonic swelling, we expect it to increase with ouabain swelling. Is there a causal relationship between the rate of swelling, ionic strength and the activation of basolateral Na^+/H^+ exchange? Dihydropyridine Ca^{2+} channel blockers prevent both the hypotonic- and ouabain-induced RVD, yet have no effect on L-alanine-induced RVD. Since the membrane is stretching a comparable amount with the L-alanine- or D-glucose-induced swelling, it is difficult to imagine why a Ca^{2+} influx pathway is not activated. The PKC inhibitor staurosporine only prevents RVD when the cells are swollen by uptake of an osmotically active solute.

VIII. CONCLUSIONS

The studies described above suggest strongly that the way in which cell volume is perturbed is a determinant of the signal transduction pathways that activate subsequent volume-regulatory mechanisms. In addition, the nature of the volume perturbation may also determine the type of transporters activated for RVD. Such findings have important implications for understanding how cells sense their size. These results also indicate that a complete understanding of homocellular regulation[6] in Na^+-transporting epithelia will require the study of volume changes induced by changes in transcellular transport.

ACKNOWLEDGMENTS

I thank Peter Lembessis for his expert technical assistance, Dick Hamilton for his critical encouragement and interest, and Hugh Bennett for valuable discussions and comments on the manuscript. Our work is funded by the Medical Research Council of Canada.

REFERENCES

1. **Hoffman, E. K. and Simonsen, L. O.,** Membrane mechanisms in volume and pH regulation in vertebrate cells, *Physiol. Rev.*, 69, 315, 1989.
2. **Sarkadi, B. and Parker, J. C.,** Activation of ion transport pathways by changes in cell volume, *Biochim. Biophys. Acta*, 1071, 407, 1991.
3. **Lewis, S. A. and Donaldson, P.,** Ion channels and cell volume regulation: chaos in an organized system, *News Physiol. Sci.*, 5, 112, 1990.
4. **Minton, A. P., Colclasure, C. G., and Parker, J. C.,** Model for the role of macromolecular crowding in regulation of cellular volume, *Proc. Natl. Acad. Sci. U.S.A.*, 89, 10504, 1992.
5. **Schultz, S. G., Fuisz, R. E., and Curran, P. F.,** Amino acid and sugar transport in rabbit ileum, *J. Gen. Physiol.*, 49, 849, 1966.
6. **Schultz, S. G. and Hudson, R. L.,** How do sodium-absorbing cells do their job and survive?, *News Physiol. Sci.*, 1, 175, 1986.
7. **MacLeod, R. J. and Hamilton, J. R.,** Separate K^+ and Cl^- transport pathways are activated for regulatory volume decrease in jejunal villus cells, *Am. J. Physiol.*, 260, G405, 1991.
8. **MacLeod, R. J. and Hamilton, J. R.,** Volume regulation initiated by Na^+-nutrient cotransport in isolated mammalian villus enterocytes, *Am. J. Physiol.*, 260, G26, 1991.
9. **MacLeod, R. J. and Hamilton, J. R.,** Regulatory volume increase in mammalian jejunal villus cells is due to bumetanide-sensitive $NaKCl_2$ cotransport, *Am. J. Physiol.*, 258, G665, 1990.
10. **Hudson, R. L. and Schultz, S. G.,** Sodium-coupled glycine uptake by Ehrlich ascites tumor cells results in an increase in cell volume

and plasma membrane channel activities, *Proc. Natl. Acad. Sci. U.S.A.*, 85, 279, 1988.

11. **Christensen, O. and Hoffmann, E. K.,** Cell swelling activates K^+ and Cl^- channels as well as nonselective, stretch-activated cation channels in Ehrlich ascites tumor cells, *J. Membr. Biol.*, 129, 13, 1992.
12. **MacLeod, R. J., Lembessis, P., and Hamilton, J. R.,** Effect of protein kinase C inhibitors on Cl^- conductance required for volume regulation after L-alanine cotransport, *Am. J. Physiol.*, 262, C950, 1992.
13. **MacLeod, R. J., Lembessis, P., and Hamilton, J. R.,** Differences in Ca^{2+}-mediation of hypotonic and Na^+-nutrient regulatory volume decrease in suspensions of jejunal enterocytes, *J. Membr. Biol.*, 130, 23, 1992.
14. **Takemura, H., Hughes, A. R., Thastrup, O., and Putney, J. W.,** Activation of calcium entry by the tumor promoter, thapsigargin, in parotid acinar cells, *J. Biol. Chem.*, 264, 12266, 1989.
15. **MacLeod, R. J., Hamilton, J. R., Bateman, A., Belcourt, D., Hu, J., Bennett, H. P. J., and Solomon, S.,** Corticostatic peptides cause nifedipine-sensitive volume reduction in jejunal villus enterocytes, *Proc. Natl. Acad. Sci. U.S.A.*, 88, 552, 1991.
16. **Hazama, A. and Okada, Y.,** Involvement of Ca^{2+}-induced Ca^{2+} release in the volume regulation of human epithelial cells exposed to a hypotonic medium, *Biochem. Biophys. Res. Commun.*, 167, 287, 1990.
17. **Gimenez-Gallego, G., Navia, M., Reuben, J., Katz, G., and Garcia, M.,** Purification, sequence and model structure of charybdotoxin, a potent selective inhibitor of calcium-activated K^+ channels, *Proc. Natl. Acad. Sci. U.S.A.*, 85, 3329, 1988.
18. **Grinstein, S. and Smith, J. D.,** Calcium-independent cell volume regulation in human lymphocytes. Inhibition by charybdotoxin, *J. Gen. Physiol.*, 95, 97, 1990.
19. **Maenz, D., Chenn, C., Breton, S., and Berteloot, A.,** pH-dependent heterogeneity of acidic amino acid transport in rabbit jejunal brush border membrane vesicles, *J. Biol. Chem.*, 267, 1510, 1992.
20. **Parent, L., Supplisson, S., Loo, D. D. F., and Wright, E. M.,** Electrogenic properties of the cloned Na^+/glucose cotransport. I. Voltage-clamp studies, *J. Membr. Biol.*, 125, 49, 1992.
21. **Giraldez, F. and Sepulveda, F. V.,** Changes in the apparent chloride permeability of Necturus enterocytes during the sodium-coupled transport of alanine, *Biochim. Biophys. Acta*, 898, 248, 1987.
22. **MacLeod, R. J., Lembessis, P., and Hamilton, J. R.,** Na^+/H^+ exchange in isolated crypt cells is activated by hyperosmolarity and intracellular acidification, *Clin. Invest. Med.*, 14, A48, 1991.
23. **MacLeod, R. J., Lembessis, P., and Hamilton, J. R.,** Hypotonic activation of Na^+/H^+ antiport is required for small volume RVD in villus epithelial cells, *Physiologist*, 35(5), A18, 1992.
24. **Livine, A., Grinstein, S., and Rothstein, A.,** Volume-regulating behavior of human platelets, *J. Cell. Physiol.*, 131, 354, 1987.
25. **Livine, A. and Hoffmann, E. K.,** Cytoplasmic acidification and activation of Na^+/H^+ exchange during regulatory volume decrease in Ehrlich ascites tumor cells, *J. Membr. Biol.*, 114, 153, 1990.
26. **Chang, D., Kushman, N. L., and Dawson, D. C.,** Intracellular pH regulates basolateral K^+ and Cl^- conductances in colonic epithelial cells by modulating Ca^{2+} activation, *J. Gen. Physiol.*, 98, 183, 1991.
27. **Tokomitsu, H., Chijiwa, T., Hagiwara, M., Mizutani, M., Terrasawa, M., and Hidaka, H.,** KN-62, 1-[N-O-bis(5-isoquinoline-sulfonyl)-N-methyl-L-tyrosyl-4-phe–nyl–piperazine, a specific inhibitor of Ca^{2+}/cal–modulin-dependent protein kinase II, *J. Biol. Chem.*, 265, 4315, 1990.
28. **Garcia-Romen, R., Cossins, A. R., and Motais, R.,** Cell volume regulation by trout erythrocytes: characteristics of the transport systems activated by hypotonic swelling, *J. Physiol. (London)*, 440, 547, 1991.
29. **Motais, R., Guizouarn, H., and Garcia-Romen, F.,** Red cell volume regulation: the pivotal role of ionic strength in controlling swelling-dependent transport systems, *Biochim. Biophys. Acta*, 1075, 169, 1991.
30. **Strange, K.,** RVD in principal and intercalated cells of rabbit cortical collecting duct, *Am. J. Physiol.*, 255, C612, 1988.
31. **Strange, K.,** Volume regulatory Cl^- loss after Na^+ pump inhibition in CCT principal cells, *Am. J. Physiol.*, 260, F225, 1991.

Chapter 12

Protein Phosphorylation Control in the Activity of Volume-Sensitive Transport Systems

H. Clive Palfrey

TABLE OF CONTENTS

I. INTRODUCTION

The internal osmotic environment is closely monitored in vertebrates, yet most cells have retained volume-sensitive transport mechanisms that tend to return cell volume towards normal after perturbation.[1] In addition to transient alterations in the tonicity of the extracellular fluid experienced by certain epithelial cells, it is likely that the operation of various plasma membrane transport systems, in concert with passive Donnan forces, is the primary contributor to cell volume fluctuations in an isosmotic environment. While there is an abundance of empirical data attesting to the existence of volume-sensitive transporters in multiple cell types, the transduction of the "volume signal" remains mysterious. The situation is complicated by the fact that different cells use distinct transport pathways to control cell volume. For example, regulatory volume decrease (RVD) can proceed by the loss of amino acids (through largely uncharacterized pathways) as well as K and Cl (through separate K and Cl channels of various types or by KCl cotransport).[2] Each cell may control these transport systems through different mechanisms and it is probable that multiple intracellular messengers are involved. This review will focus on the possibility that a shared characteristic of volume-regulatory transport systems may be their modulation by protein phosphorylation.

The hypothesis that protein phosphorylation is involved in volume regulation is hardly revolutionary, given the primacy of this mechanism in the control of a plethora of biological phenomena.[3] In fact, modulation of the activity of various ion channels by protein phosphorylation seems to be a virtually universal characteristic of this class of passive transporters.[4] Protein phosphorylation and dephosphorylation would also appear to be a logical mechanism for volume regulation from a kinetic standpoint as volume-induced changes occur rapidly and reversibly, characteristics of phosphorylation-mediated processes. However, it is still too early to define the chain of events that link changes in cell volume to protein phosphorylation events. We do not know how cell size controls the activity of the enzymes involved in protein phos-

0-8493-4448-4/94/$0.00+$.50

phorylation-dephosphorylation cycles. This review will focus on recent studies linking protein phosphorylation to the activity of volume-sensitive transporters such as the Na/K/2Cl cotransporter, Na/H exchanger, and KCl cotransporter. We start with a brief overview of the enzymes involved in protein phosphorylation systems as well as comments on the use of permeant inhibitors of this process that may be useful in defining components that are regulated by cell volume.

II. PROTEIN KINASE AND PHOSPHATASE FAMILIES

Protein phosphorylation and dephosphorylation are controlled by the activity of two large families of enzymes: the protein kinases and protein phosphatases. The reader is referred to several excellent reviews for detailed information.[5–9] Protein kinases have been suggested to comprise the largest family of proteins in eukaryotic cells; thus far over 170 distinct enzymes have been identified.[10] While the enzymes that have grown most familiar to biologists are those kinases that are regulated by the classical second messengers, cAMP, cGMP, Ca, and diacylglycerol,[5] these comprise only a small fraction of the total number of kinases. Moreover, certain kinases that were previously thought to be "constitutive" in their activity (e.g., "casein" kinases) are now known to be responsive to extracellular factors.[11] Recently, novel categories of kinase that are themselves regulated entirely by phosphorylation due to other kinases have been identified.[12–16] Frequently, these enzymes participate in kinase cascades that are initiated through activation of receptor tyrosine kinases at the plasma membrane. For example, "MAP kinases" (originally so named because they could phosphorylate microtubule-associated protein-2 *in vitro*) constitute a group of enzymes[12,13] that are activated in a variety of cells via different receptor tyrosine kinases (e.g., insulin receptor, nerve growth factor [NGF] receptor, epidermal growth factor [EGF] receptor). The MAP kinases are thought to be quiescent prior to receptor binding, but rapidly (<10 min) activate on binding of the polypeptide hormone to its receptor. Though incompletely understood, it is thought that MAP kinases are activated by tyrosine and threonine phosphorylations, brought about by "MAP kinase kinases"[15,16] that seem to be members of a novel family of dual-specificity enzymes capable of phosphorylating both tyrosine and ser/thr residues on target substrates.[17] Earlier work had suggested that ser/thr kinases and tyr kinases belonged to entirely separate families, but this dogma is now broken. MAP kinase kinase may itself be phosphorylated by Raf,[14,16] a ser/thr kinase that may prove to be the missing link between the receptor tyrosine kinases and the ensuing kinase cascade.

Second messenger-dependent enzymes seem to be particularly important with respect to the function of volume-regulatory transporters. For example, cAMP-dependent (cAMP-PK) and cGMP-dependent protein kinases, as well as protein kinase C (PKC) and possibly calmodulin-dependent protein kinase II, are thought to be involved in the regulation of Na/K/2Cl cotransport by hormonal stimuli in various tissues.[18–20] Current evidence suggests, however, that these enzymes are not intermediaries in the effects of cell volume on the activity of transport systems, and it may well be that some new category of kinase could be involved in such processes.

Though protein phosphatases (PPases) appear to be a less complex group of enzymes than the protein kinases, it has become evident that substantial diversity exists not only in the molecular structure of this class of enzyme, but also in their regulation. Both structural and functional studies of ser/thr phosphatases has progressed rapidly in the last several years.[8,9] The classification of these enzymes into types 1, 2A, 2B, and 2C is deceptively simple. For example, though the catalytic subunits of the type 1 enzymes are very similar and evolutionarily well conserved, the nature of the binding or regulatory subunits is probably diverse. One function of the regulatory subunits may be in targeting the holoenzyme to distinct regions of the cell. For example, in skeletal muscle the 160-kDa PP-1_G subunit anchors the catalytic subunit to glycogen particles.[8,9] Membrane forms of PPase-1 have been identified and may have distinct anchoring subunits. As indicated below it appears that PPase-1 may be particularly important in regulating the activity of both Na/K/2Cl and KCl cotransporters. The regulatory subunits of the type 2A enzyme are less well understood and may exhibit some degree of tissue specificity. The role of each subunit will need to be evaluated biochemically before a complete picture of PPase-2A function emerges. It seems likely that the majority of ser/thr dephosphorylation in many cells is accomplished by the catalytic subunits of PPases-1 and -2A. The functions of PPases-2B and -2C seem to be more restricted than those of -1 and -2A.[8] Molecular cloning suggests that the protein tyrosine phosphatases comprise a large family of enzymes, some of which are transmembrane molecules, suggesting that extracellular signals may regulate the activity of this class of enzyme directly. Studies on this group of enzymes are advancing rapidly.

III. USE OF KINASE AND PHOSPHATASE INHIBITORS IN INTACT CELLS

The availability of permeant and specific inhibitors of particular enzyme systems generally leads to advances in our understanding of the roles of those enzymes in physiological processes. More often than not, however, initial claims of specificity for novel compounds are overblown and it sometimes takes several years for the realization that side-effects are inevitable to sink in. Protein kinase inhibitors have proven to be no exception to this rule, as virtually all classes of these compounds have been shown either to have actions unrelated to kinase-mediated effects or to inhibit a wider range of kinases than first suspected. As of this writing, there are no permeant inhibitors of specific protein kinases that can be used to establish definitively the role of any single member of this family of enzymes in a biological process. (A single exception is (Rp)-cAMPS, a permeant cyclic nucleotide derivative that can specifically inhibit cAMP-PKs in intact cells.[21]) For example, many studies were conducted using the compound H7, a member of the H-series of isoquinoline sulfonamides,[22] under the supposition that it was a specific inhibitor of PKC. This conclusion was probably reached on the basis of early experiments on intact platelets showing that H7 inhibited the secretagogue-induced phosphorylation of a 40-kDa PKC substrate but not the 20-kDa myosin light chain,[23,24] a substrate of the calmodulin-dependent enzyme, myosin light chain kinase. However, most studies ignored the original publication on the family of H compounds (e.g., H7, H8, and H9) that showed inhibition of a number of protein kinases at very similar concentrations. Though H7, for example, might discriminate well between PKC and myosin light chain kinase (K_is *in vitro* of 6 and 97 μM, respectively), it would not discriminate between PKC and cAMP- or cGMP-dependent protein kinase (K_is *in vitro* for the latter two enzymes are 3 and 5.8 μM, respectively). Similarly, sphingosine was touted as a specific inhibitor of PKC,[25] but later proved to be an equally effective inhibitor of calmodulin-dependent protein kinase II.[26] Yet another example is the staurosporine-K252a group of inhibitors.[27] These potent alkaloids affect a number of protein kinases, by competing at the nucleotide binding site, as is also the case with the H-compounds. While staurosporine appears to inhibit PKC more potently than other kinases *in vitro,* a number of other effects of this alkaloid unrelated to inhibition of PKC have been described. K252a is also a potent inhibitor of certain tyrosine kinases.[28] The situation with this and other inhibitors becomes even more difficult to interpret because some have been shown to have actions that clearly do not involve protein kinase inhibition. Nevertheless, used judiciously, such inhibitors can provide circumstantial evidence of possible roles for kinases in complex biological phenomena, but only when combined with corroborating information from other approaches.

The most selective inhibitors of protein kinases have proven to be substrate analogue peptides that lack a phosphorylatable residue.[7] Unfortunately, these are impermeant and therefore ineffective, unless access to the cell interior can be gained. Such access can be provided by patch pipette or other means of microinjection, methods that have been extensively employed to analyze the role of protein kinases in ion channel modulation.[4] They have not yet been employed in analyses of volume regulation.

The situation with respect to protein phosphatases is remarkably different, largely due to the quite-recent identification of two marine toxins that appear to be highly specific and potent antagonists of PPases-1 and -2A catalytic subunits. These inhibitors, okadaic acid (OkA) and calyculin A (CalA), bind stoichiometrically to the catalytic subunits of these two enzymes and block their activity.[29] While OkA shows a preference for PPase-2A ($IC_{50} \sim 0.05$ nM) over PPase-1 ($IC_{50} \sim 10$ nM), CalA inhibits both enzymes at similar subnanomolar concentrations. These two agents have proven to be quite specific thus far; direct effects on kinases have not been found and they only weakly inhibit PPase-2B[29] (that can be specifically antagonized by immunophilins such as FK506). PPase-2C and the protein tyrosine phosphatases as well as other non-protein-type phosphatases are not affected. The tyrosine phosphatases are inhibited by vanadyl ions that do not affect ser/thr phosphatases to a significant extent.[30] One caveat that should always be borne in mind with respect to the use of OkA and CalA is that the massive increases in protein phosphorylation incurred upon treatment of intact cells with these agents are often toxic.

IV. THE REGULATION OF Na/K/2Cl COTRANSPORT BY PHOSPHORYLATION

The Na/K/2Cl cotransporter has emerged as a key element in transcellular Cl transport in a wide variety of epithelia, but exists in many other nonepithelial cell types where its function is not so well understood (for reviews see References 18 to 20). The physiological and pharmacological characteristics of this transport system are by now well known. Suffice it to say that the properties of this system in tissues as diverse as the shark rectal gland, avian red blood cell, mammalian kidney, and parotid gland are quite similar, leading to the conclusion that this system is probably highly conserved. This transporter is the locus of action of the "loop" diuretics furosemide and bumetanide, and labeled forms of the latter have been used to reversibly and irreversibly label the putative cotransport protein.[19,31] Current evidence suggests that the diuretic-binding component is comprised of a 150- to 200-kDa glycoprotein, depending on the tissue of origin.[19] Cloning and expression studies suggest that this protein may be the sole component necessary for transport function.[32,33]

The Na/K/2Cl cotransporter is of interest in the context of volume regulation because of its ability to both regulate cell volume, through its capacity to mediate net salt transport, and be regulated by cell volume. Thus, regulatory volume increase (RVI) after hypertonic perturbation has been demonstrated to be bumetanide-sensitive and to possess other characteristics of Na/K/2Cl cotransport in several instances.[20] This is not to say, however, that possession of a cotransport system necessarily renders cells capable of RVI or that RVI invariably involves Na/K/2Cl cotransport (see below). Most cells have a basal level of cotransport that is up- or down-modulated by a variety of signals.[20] A common feature of the many first and second messengers that affect cotransport is their ability to activate protein kinases. Consequently, a reasonable hypothesis is that the cotransporter is itself a phosphoprotein with perhaps several sites in the cytoplasmic domain for different protein kinases (some of which would be stimulatory and some inhibitory). At present, evidence is consolidating in support of such a notion.

The regulation of the avian erythrocyte cotransport system has been well studied and provides a useful and unique model, in that it exhibits very low activity under normal conditions *in vitro.* A variety of stimuli increase cotransport rate; these can be grouped into cAMP-dependent and cAMP-independent classes.[34] A great deal of evidence supports the notion that cAMP-dependent stimuli work by activation of cAMP-PK and phosphorylation of the cotransporter:

1. The sole cAMP receptors in these cells are the regulatory subunits of cAMP-PK[34]
2. Permeant cAMP analogues are as effective as hormones that raise cAMP, such as β-catecholamines
3. Cotransport activity correlates very well with phosphorylation of plasma membrane proteins at specific cAMP-PK sites
4. cAMP-stimulated cotransport is inhibited by permeant protein kinase inhibitors such as K252a and H9[35]

Though we have shown that the putative cotransport protein in avian erythrocytes is phosphorylated in the presence of isoproterenol[35] it has not been possible to date to quantitate this phosphorylation, due to the low abundance of the protein. As discussed below, this has been achieved in the case of forskolin-stimulated shark rectal gland.

The non-cAMP-dependent stimuli include cell shrinkage, NaF, and deoxygenation.[34] What links these seemingly unrelated stimuli is presently unclear. The protein phosphatase inhibitors OkA[35] and CalA[36] may be included in this group for reasons discussed below. Figure 12–1 shows a comparison of the stimulation of Na/K/2Cl cotransport in duck erythrocytes by the two phosphatase inhibitors; CalA is about 100 times more effective than OkA. Based on the *in vitro* inhibitory potency of these inhibitors against the purified catalytic subunits of PPases-1 and -2B (see above), these results suggest that PPase-1 is more likely to be responsible for cotransport stimulation in avian erythrocytes than PPase-2A.[36] Unfortunately, little is known about the protein phosphatases in avian red blood cells; thus, it is difficult to speculate at present as to the structure and function of the relevant enzymes. Nevertheless, the fact that a phosphatase inhibitor *alone* is able to stimulate the system strongly suggests that some endogenous kinase must be present that phosphorylates and activates the cotransporter. It should be pointed out that the relevant endogenous kinase may itself be "constitutively active". Alternatively, the kinase could be turned on by phosphatase inhibition, implying the existence of a "kinase kinase"; as indicated above, such kinase cascades do exist in diverse cell types. Whatever the situation, the kinase operates very rapidly as turn-on of cotransport with either OkA or CalA is rapid ($t_{1/2}$ ~2 min). The prevailing evidence argues against the involvement of cAMP-PK in the response to phosphatase inhibitors: (1) K252a inhibits OkA-stimulated cotransport only at concentrations an order of magnitude greater than those inhibiting cAMP-stimulated cotransport (Figure 12–2), and (2) OkA does not lead to phosphorylation of the membrane skeletal protein goblin at its cAMP-PK sites.[35] K252a was also found to inhibit cotransport after stimulation by either NaF (Figure 12–2) or cell shrinkage.[35] The dose-inhibition relationship was similar to that obtained with cells stimulated by OkA but different from that obtained with cAMP, suggesting that perhaps

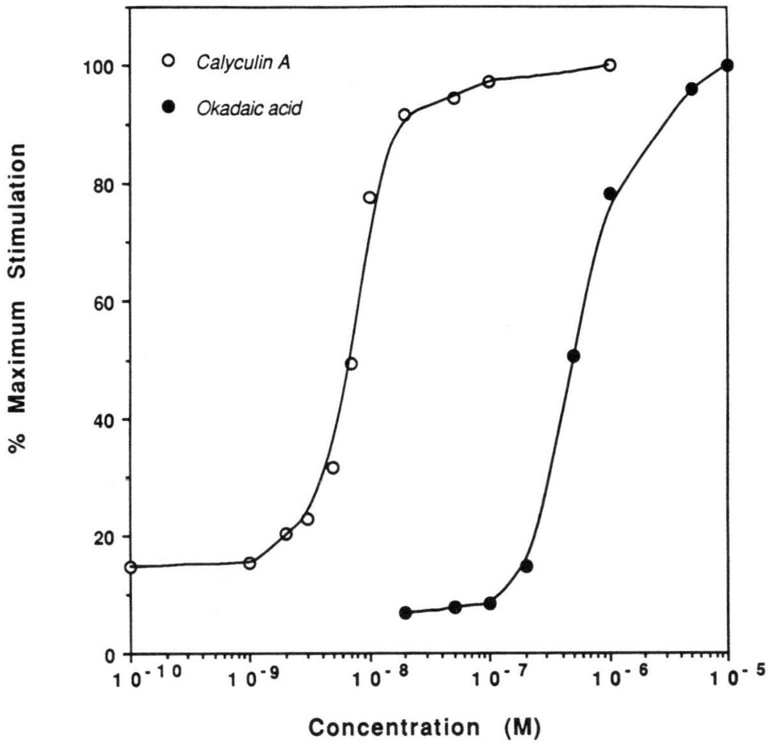

Figure 12–1 Stimulation of Na/K/2Cl cotransport in duck erythrocytes by okadaic acid and calyculin A. Assays of cotransport activity in duck erythrocytes were carried out as described.[35] The two phosphatase inhibitors were preincubated with cells for 15 min prior to addition of ^{86}Rb; bumetanide-sensitive influx was assayed for 1 min. The EC_{50} for CalA was ~10 n*M* and for OkA ~600 n*M*. Cotransport rates in these experiments were 80 to 100 mmol/l · cells/h. (Adapted from References 35 and 36.)

the same non-cAMP-dependent kinase is involved in the effects of NaF, hypertonicity, and phosphatase inhibitors. This interpretation assumes that K252a is only inhibiting the activity of a kinase in this system which, as mentioned above, is not necessarily true. Hopefully, this question can be explored further when it is possible to analyze phosphorylation of the cotransporter directly in these cells.

A requirement for phosphorylation in Na/K/2Cl cotransport function may extend to other tissues that exhibit high "basal" activity of this system. This raises the important question as to whether the cotransporter must be in a phosphorylated form to operate in any cell. Cotransport activity has been found to be dependent on metabolic state in a variety of cell types. Inhibition of energy metabolism with consequent depletion of ATP generally depresses cotransport rates,[36–39] though it is probable that the system does not consume ATP directly.[37] Studies in perfused squid axon, where it is feasible to specifically remove ATP by dialysis, indicate a requirement for this nucleotide for cotransport operation that cannot be substituted by nonhydrolyzable analogues.[40] The possibility that ATP is needed to maintain the cotransporter in a phosphorylated state has received support from recent experiments with phosphatase inhibitors in avian red cells.[36] Stimulation of cells with CalA was found to render them resistant to the effects of subsequent ATP depletion by the mitochondrial uncoupler antimycin A, whereas in cells that were stimulated by cAMP, cotransport rate fell precipitously once the [ATP] dropped below a critical value (Figure 12–3). Similar results were obtained when $[Mg^{2+}]_i$ was depleted by incubation of cells in A23187 and ethylene diamine tetraacetate (EDTA). Starke[41] and Flatman[42] have both shown that $[Mg^{2+}]_i$ modulates cotransport in duck and ferret erythrocytes, respectively, but neither investigator provided a mechanism for the effect. We found that incubation of duck red cells in A23187 plus EDTA for 30 min (conditions known to deplete $[Mg^{2+}]_i$ to undetect-

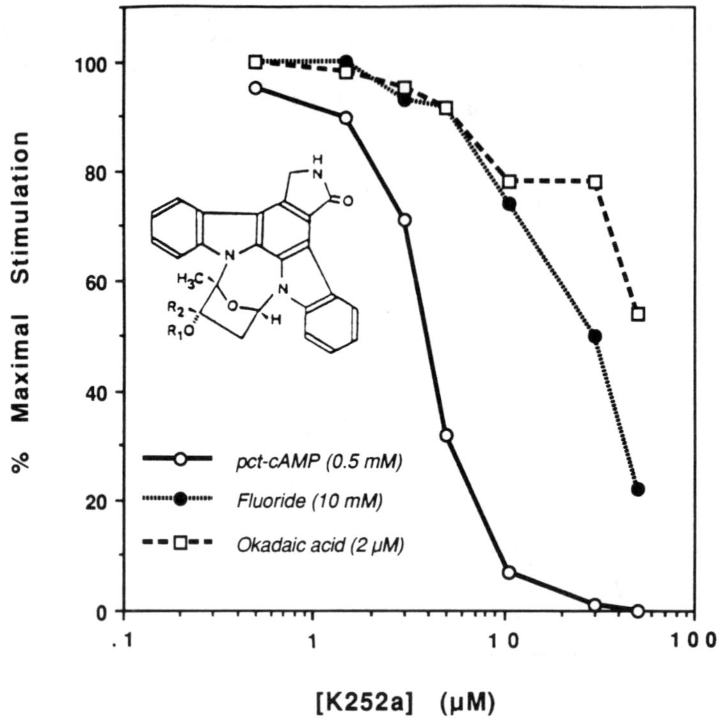

Figure 12–2 Inhibition of Na/K/2Cl cotransport in duck erythrocytes by K252a. Cells were preincubated with various concentrations of K252a prior to treatment with the indicated cotransport-activating stimuli for 15 min. Assays of cotransport activity were carried out as previously described.[35] The IC_{50}s for K252a in these experiments were 4.1 μ*M* (pct-cAMP), 30 μ*M* (NaF), and >50 μ*M* (okadaic acid). The structure of K252a is shown (inset): R_1 = H and R_2 = $COOCH_3$. (Modified from Reference 35.)

able levels) resulted in cotransport inhibition when cells were stimulated with cAMP but not when cells were stimulated with CalA.[36] The most straightforward explanation for these results is that when dephosphorylation of the transporter is prevented by CalA there is no requirement for sustained activity of protein kinases, hence depletion of either Mg^{2+} or ATP has a negligible effect on cotransport rate. On the basis of such a hypothesis one would predict that if either ATP or Mg^{2+} were depleted *before* the addition of activator then both cAMP and CalA stimulation would be blocked; this turns out to be true.[36]

The possibility that the hypertonic stimulation of cotransport operates via a phosphorylation mechanism has been directly proven in the shark rectal gland. This tissue is a typical salt-secreting epithelium with basolateral Na/K/2Cl cotransporters and apical Cl channels. Secretion is elicited by agents that elevate cAMP and by hypertonicity, probably by processes that generate increased activity of both the cotransporter and channel, and is inhibited by serosally applied loop diuretics in a similar manner to the avian red cell.[43] Lytle and Forbush[44] have shown recently that both forskolin and hypertonicity lead to increased phosphorylation of the ~195-kDa rectal gland cotransport protein at both ser and thr residues (Figure 12–4). With both stimuli, phosphorylation was temporally correlated with increased [^{3}H]-benzmetanide binding, and could be inhibited by K252a and H8. The nature of the putative hypertonically stimulated kinase is obscure at present, but stimulation of phosphorylation by forskolin and hypertonicity was not additive, suggesting that similar sites were phosphorylated on the cotransporter in the two cases. This is the first demonstration of volume-induced phosphorylation of a specific transport molecule. An alternative explanation for the hypertonic effect on cotransporter phosphorylation could be that hypertonicity inhibits protein phosphatase activity. Some support for this idea comes from experiments on avian erythrocytes involving ATP or Mg^{2+} depletion. As discussed above, ATP or Mg^{2+} depletion after cell stimulation differentially affects cells activated by cAMP or CalA. It turns out that

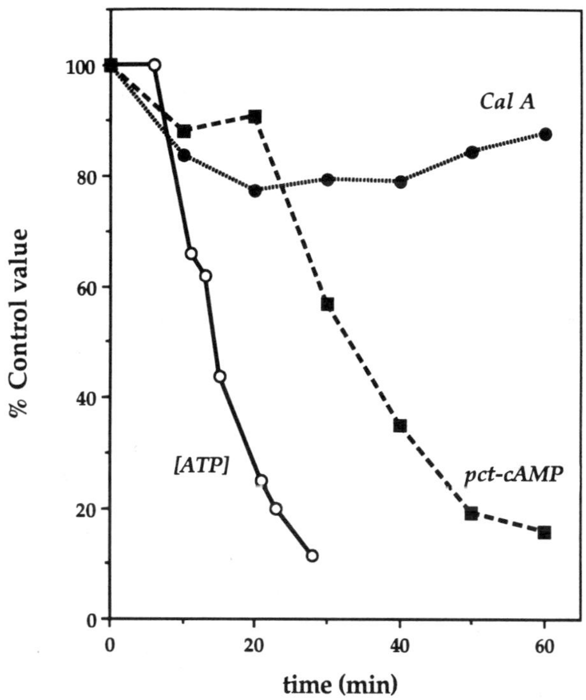

Figure 12–3 ATP depletion leads to inhibition of Na/K/2Cl cotransport following cAMP stimulation but not after calyculin A stimulation of duck erythrocytes. Cells were pretreated with either cpt-cAMP (0.5 m*M*) or calyculin A (Cal A; 200 n*M*) for 15 min, then aliquots were either left as controls or further treated with antimycin A (10 μ*M*) for the ensuing 60 min to deplete cellular ATP; cells were suspended in a glucose-free medium to minimize ATP resynthesis in depleted batches. At the times indicated samples were removed for determination of cotransport rate as described.[35] Aliquots were also taken for determination of total cell ATP by the luciferase procedure. The results are expressed as percent control values. In this experiment control cAMP-stimulated cotransport rate was 72 mmol K/l · cells/h, CalA-stimulated cotransport rate was 82 mmol K/l · cells/h and cellular [ATP] was 3.3 m*M;* these values did not change significantly in cells not subjected to ATP depletion. Note that pct-cAMP-stimulated cotransport rate began to decline only after cellular [ATP] was <20% of control and that CalA-stimulated rates were largely unaffected by the decline in [ATP]. (Adapted from Reference 36.)

hypertonically activated cells are also relatively resistant to these manipulations, rendering them similar in this respect to CalA-treated cells.[36]

V. KCl COTRANSPORT

The characterization of hypotonically induced KCl cotransport has proceeded simultaneously in erythrocytes from a number of different species and appears to be a mechanism utilized by several cell types for RVD.[2] Certain epithelia may also employ a similar basolaterally disposed transporter that participates in Cl reabsorption. The relatively narrow anion specificity, activation by sulfhydryl reagents at isotonic volume, and insensitivity to loop diuretics compared to the Na/K/2Cl cotransporter are characteristics shared by the system in different red cells.[45] On the basis of these studies it is unlikely that KCl cotransport is a specialized mode of operation of the Na/K/2Cl cotransporter, but the two systems may ultimately turn out to be related at the structural level. Of primary importance to the current discussion are findings that the system is inhibited by permeant protein phosphatase inhibitors such as OkA and CalA.[46–49] This has led to speculation that KCl cotransport may be regulated by its phosphorylation state (i.e., the more phosphorylation, the greater the inactivation of the system) and that the most likely mechanism of hypotonic activation of cotransport might be by inhibition of the relevant protein kinase (dubbed "V kinase").[50] The latter conclusion was reached by Jennings and Al-Rohil[50] on kinetic grounds. There is a distinct lag between osmotic swelling and activation of KCl cotransport in rab-

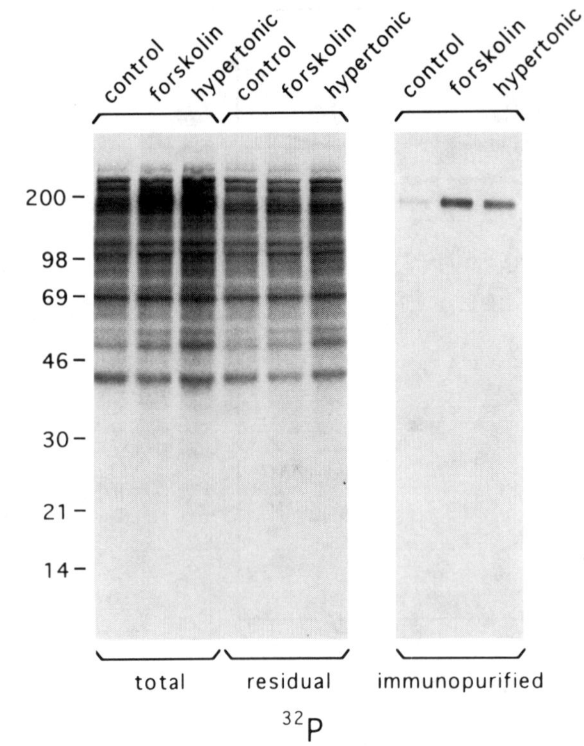

Figure 12–4 Phosphorylation of the Na/K/2Cl cotransport protein in isolated shark rectal gland tubules. Isolated rectal gland tubules were prepared by enzymatic dissociation from slices of gland and incubated for 1 h in ^{32}P-orthophosphate, they were then washed free of extracellular isotope and incubated in effectors (forskolin: 10 μ*M* or hypertonicity: +580 m*M* sucrose) for 15 min. Samples were solubilized and the cotransporter immunoprecipitated using a monoclonal antibody immunomatrix. Proteins were separated by SDS-PAGE and the gel autoradiographed. The right-hand panel shows the ^{32}P-labeled, immunopurified cotransporter; in this experiment forskolin led to a 4.3-fold and hypertonicity a 3.3-fold increase in phosphorylation, respectively. The left-hand panel shows the original solubilized samples ("total") and the supernatants following immunoprecipitation ("residual"). (From Lytle, C. Y. and Forbush, B., III, *J. Biol. Chem.*, 267, 25438, 1992. With permission.)

bit,[50] dog,[51] and avian[41] red cells, whereas no lag in deactivation is seen on shrinking the cells back to their normal volume. Thus, in a simple two-state model ($R \underset{k_{12}}{\overset{k_{21}}{\rightleftarrows}} A$), where R is the phosphorylated inactive form and A is the active dephosphorylated form, k_{12} (the kinase rate constant) > k_{21} (the phosphatase rate constant) at isotonic volume because the system is deactivated and therefore phosphorylated. Swelling-induced activation involves net dephosphorylation that could be either a result of a diminution in k_{12} or an increase in k_{21}, such that $k_{21} > k_{12}$. As the rate constant for deactivation (phosphorylation) is rapid, implying a fast kinase, while the activation lag implies a slow phosphatase, only inhibition of a kinase explains the slow kinetics of activation on swelling. If the phosphatase rate was accelerated such that it exceeded that of (an already-fast) kinase there would be no reason to expect a lag in activation (the rate would be equal to or greater than that observed on shrinkage-induced inactivation). This hypothesis might hold up if the simple scenario outlined above is correct. On the other hand one could easily imagine, for example, that the rate of the phosphatase is accelerated but that the lag is due to a relatively slow generation of active phosphatase (e.g., release of catalytic subunits from regulatory subunits). Clearly, the activity of the relevant enzymes must be determined directly before firm conclusions can be reached, and this may necessitate the development of tools to study the phosphorylation of the transporter itself. The molecular structure of KCl cotransporters is entirely unknown, thus it is not possible to confirm at present whether the cotransport protein(s) themselves are phosphorylated.

Another finding that tends to support the phosphorylation hypothesis is that activation of KCl cotransport can be obtained at isotonic volumes by Mg^{2+} depletion (which, by analogy with the hypothesis advanced above for the inhibitory effect of this manipulation on Na/K/2Cl cotransport, should diminish kinase activity) in sheep and dog red cells.[45,46] However, the situation is made more complex by reports that ATP depletion actually inhibits KCl cotransport whether the latter is stimulated by hypotonicity or by *N*-ethylmaleimide (NEM) at isotonic volume.[45] This result does not conform in a readily understandable manner with the phosphorylation hypothesis, as such a manipulation would be expected to lower kinase activity and stimulate the transporter. Speculative hypotheses to explain this paradox, such as a requirement for phosphorylation in the action of a phosphatase, have been proposed, but supporting evidence is lacking.

Some recent results from this laboratory on the duck red cell KCl cotransporter replicate and extend observations made on other systems. The system is transiently activated by cell swelling; after a lag there is a peak of cotransport activity at 10 to 20 min, then cotransport gradually declines as the cells lose KCl and shrink back to their normal volume. The system can be activated at isotonic volume by sulfhydryl reagents such as NEM, ethacrynic acid, and diamide (all of which, incidentally, inhibit the Na/K/2Cl cotransporter). The magnitude of the effect of -SH reagents is comparable to that of cell swelling and is also switched off as the cells shrink. We have compared the ability of OkA and CalA to inhibit both the NEM- and hypotonically activated KCl cotransport system in this cell type. As shown in Figure 12–5, titration of these two inhibitors reveals that CalA is considerably more potent (IC_{50} ~20 n*M*) than OkA (IC_{50} ~500 n*M*), suggesting that, as with Na/K/2Cl cotransport in the same cells (Figure 12–1), PPase-1 is likely to be the enzyme responsible for regulating this system. Similar results have been obtained in rabbit red cells[49] (though it is interesting to note that the IC_{50} values are somewhat lower than those found in avian erythrocytes). For reasons given in the next section it is intriguing that inhibition of what appears to be the same phosphatase activates Na/K/2Cl cotransport and inhibits KCl cotransport in the same cell. In other studies we failed to find any effect of kinase inhibitors or ATP depletion using similar manipulations that lead to inhibition of the Na/K/2Cl cotransporter in duck erythrocytes. Thus, K252a, for example, had no effect on KCl cotransport in isotonic cells and failed to modify the responses to NEM or hypotonicity. It may well be that a different kinase from that responsible for activating Na/K/2Cl cotransport affects KCl cotransport in these cells.

VI. COORDINATE REGULATION OF VOLUME-SENSITIVE COTRANSPORT SYSTEMS

The diametrically opposite regulation of shrinkage- and swelling-induced transporters in various red cell types led to the proposal, first enunciated explicitly by Cossins[52] and covered in more detail elsewhere in this tome,[53] that a common regulatory process may turn RVI- and RVD-induced transport systems on and off as a function of cell volume. The nature of this process has been proposed to be phosphorylation,[46,52] based largely on the effects of the permeant phosphatase inhibitors discussed above. The avian erythrocyte has the advantage that both RVD and RVI are present, the former mediated by KCl cotransport and the latter by Na/K/2Cl cotransport. Starke[41] showed that a volume set-point existed, emanating from which either the KCl cotransporter was activated by cell swelling or Na/K/2Cl cotransport was activated by cell shrinkage. This set point could be shifted by various treatments, such as norepinephrine or Mg^{2+} depletion, but no situation could be found where the activity of the two cotransporters was increased simultaneously. For example, when cells were osmotically swollen in the presence of norepineph-

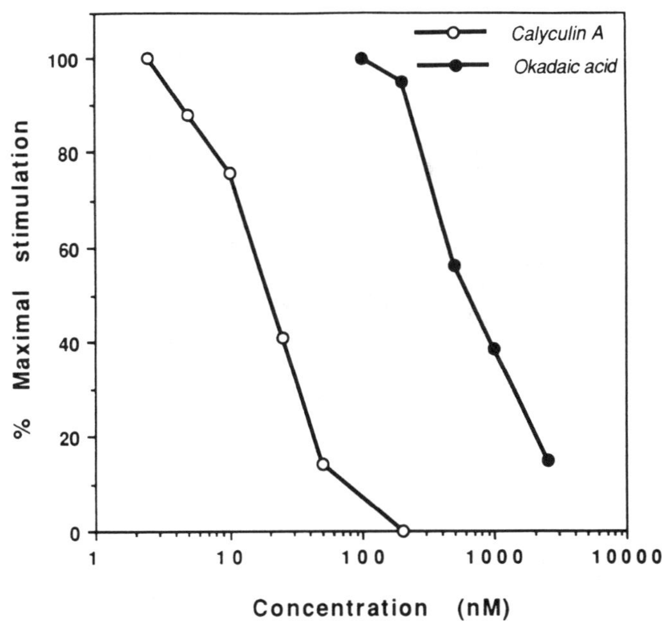

Figure 12–5 Inhibition of KCl cotransport in duck erythrocytes by protein phosphatase inhibitors. Cells were preincubated with either CalA or OkA at the indicated concentrations for 10 min prior to exposure to hypotonic conditions (190 mOsmol) for 10 min. Cells were then assayed for KCl cotransport by measurement of bumetanide (20 μ*M*)-insensitive ^{86}Rb uptake in a Na-free medium containing ouabain. The difference between influx under isotonic and hypotonic conditions was taken as KCl cotransport; control values of 9 to 10 mmol K/l · cells/h were obtained. The IC_{50} values for CalA and OkA in this experiment were ~20 and ~500 n*M*, respectively.

rine, KCl cotransport and RVD were suppressed and Na/K/2Cl cotransport continued. We have carried the comparison further and a summary of some of the data is depicted in Figure 12–6. It is apparent that any manipulation that turns on one system switches off the other. Similar data have been obtained for dog red cells where shrinkage accelerates Na/H exchange, whereas swelling increases KCl cotransport.[46] While this reciprocity may be coincidental, it strongly suggests the presence of a common regulatory system. However, it does not necessarily indicate that the "regulator" is some mysterious third molecule that inhibits one system and stimulates the other. It may be, for example, that the Na/K/2Cl and KCl cotransporters possess cytoplasmic domains that are capable of being modified by common kinases and phosphatases. Phosphorylation of one cotransporter by a kinase may activate it, but inhibit the activity of the other transport system. Though it is too early to provide a molecular model of how similar phosphorylations could up- and downregulate transport activity, it is known that the same second messenger may activate or inhibit the *same* cotransporter in different cells.[18] For example, not only does cAMP activate Na/K/2Cl cotransport in avian red cells yet inhibit the system in human and ferret red cells, but it activates KCl cotransport in pig red cells[54] while inhibiting this process in avian red cells![41] These results may indicate species-specificity in phosphorylation sites that could translate into opposite responses to phosphorylation of the molecules in question.

VII. PHOSPHORYLATION IN OTHER VOLUME-REGULATORY TRANSPORT SYSTEMS

The concept that cell shrinkage or swelling affects protein phosphorylation systems that modulate the activity of appropriate transporters has been invoked in a number of other instances. Grinstein and coworkers showed that shrinkage of lymphocytes led to an increased phosphorylation of a protein of ~60 kDa that was also phosphorylated in phorbol ester-treated cells.[55] These cells undergo RVI by activation of a Na/H exchanger; the exchanger is also activated by phorbol esters in a manner consistent with the involvement of PKC in the process.[56] However, cell shrinkage did not lead to an accumulation of inositol phosphates, suggesting that phosphatidylinositol turnover was not accelerated under these conditions. Hence, it was concluded that hypertonicity activated Na/H exchange by a process involving a kinase similar to, but not identical with, PKC.[55] More recently, it has been shown that OkA can increase the activity of the lymphocyte exchanger in the absence of other stimuli, leading to a conclusion much the same as that reached in work on the Na/K/2Cl cotransporter alluded to above, viz., that a constitutively activated kinase is responsible for phosphorylating the exchanger under conditions of phosphatase inhibition.[57] Osmotic activation of the exchanger is also diminished by prior ATP depletion, an effect that could be due to substrate deprivation of a necessary kinase.[55]

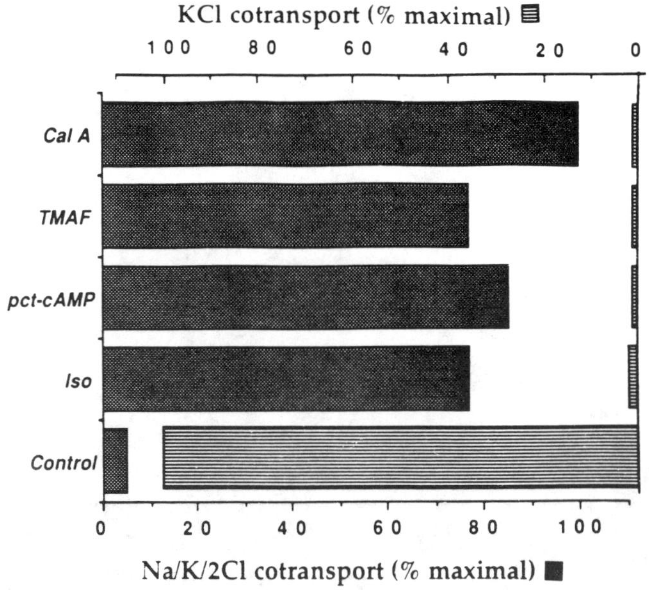

Figure 12–6 Opposite responses of Na/K/2Cl cotransport and hypotonically induced KCl cotransport to various effectors. Measurements of the two transport systems were made as described (Reference 35 and Figure 5 legend). Effectors were generally added 15 min prior to measurement of transport rates. The stimuli used were isoproterenol (10 μ*M*), pct-cAMP (0.5 m*M*), tetramethylammonium fluoride (10 m*M*), and calyculin A (200 n*M*). The maximal transport rates observed in these experiments were 114.6 mmol K/l · cells/h (Na/K/2Cl cotransport) and 9.6 mmol K/l · cells/h (KCl cotransport).

Recent cloning of the fibroblast exchanger (NHE1),[58] as well as related genes from other tissues (NHE2-4),[59,60] indicates that this transport family is diverse. Examination of the putative cytoplasmic tails from these molecules suggests that each has a distinct set of regulatory phosphorylation sites. For example, NHE1 has consensus sites for PKC and CaM-PK II but not cAMP-PK while NHE3 has consensus sites for all three kinases. Expression of truncated transporters demonstrates that the cytoplasmic domain is essential for growth factor activation of the system, though the basic exchange function of the molecule remains unaltered.[61] Development of immunological tools has allowed a direct evaluation of the phosphorylation of the Na/H exchanger in a variety of systems.[57,58] Good evidence has accumulated in platelets, fibroblasts, and lymphocytes that the antiporter is indeed regulated by phosphorylation, the result of which is to shift the pH_i sensitivity of the transport system to the left and render it more active at physiological pH.[58] Because stimulation with different agents in fibroblasts appears to lead to phosphorylation of the same set of ser-containing tryptic peptides, it has been suggested that an early signal-integration system impinging on the relevant kinase exists.[58] Candidates for such a kinase could be the MAP kinases or Raf-1 referred to earlier in this chapter. Clearly, identification of the specific ser residues in the exchanger by site-directed mutagenesis or sequence analysis would aid in the identification of this kinase. While there is little evidence for cAMP regulation of Na/H exchange in fibroblasts, the trout erythrocyte exchanger is markedly activated by cAMP. Cloning of the trout exchanger (βNHE) reveals a divergent cytoplasmic region containing two consensus cAMP-PK phosphorylation sites. Deletion of this region abrogated the ability of the exchanger to be regulated by cAMP in transfected fibroblasts.[62] To date, no reports of shrinkage-induced phosphorylation of the exchanger have appeared, though a recent report suggests that this is not the case in several cell lines.[63] Nevertheless, it would seem well worth investigating this possibility in other cells in which hypertonicity activates this system (e.g., dog, frog, salamander, and trout red cells, as well as mammalian lymphocytes).

Many cells use the transfer of organic osmolytes in addition to ions to accomplish volume regulation (usually RVD).[64] In skate and clam red cells RVD is accomplished in part by release of amino acids, specifically taurine.[65,66] It is of interest to note that taurine is also found in mammalian cells and may participate in RVD. Recent reports suggest that swelling of skate and clam cells results in changes in protein phosphorylation.[65,66] In skate red cells an increase in ^{32}P content of proteins of 99, 65, and 32 kDa was found on cell swelling.[65] The 99-kDa protein was identified as the skate equivalent of "band III" or capnophorin, the anion exchanger, and DIDS was found to inhibit the swelling-induced loss of taurine from these cells. Immunoprecipitation of band III revealed that swelling led to a ~50% increase in phosphorylation, leading to the suggestion that phosphorylation of the anion exchanger may explain the hypotonic effect on taurine efflux. Hypotonic conditions also increased phosphorylation of 63- and 34-kDa bands in clam red cells.[66] The relationship of these unidentified proteins to the volume response is not known. The second messenger involved was hypothesized to be Ca^{2+} acting via calmodulin, based on Ca^{2+} removal experiments and calmodulin inhibition by phenothiazines.

VIII. PROSPECT

While it seems likely that cell volume controls the activity of transporters, at least in part, by reversible phosphorylation, elucidation of the molecular mechanisms involved are still in their infancy. Molecular details about some of the key transporters are just beginning to emerge but it will be some time before the specific phosphorylation sites that up- and down-modulate the activity of these systems are identified and catalogued. Determination of the structural events that alter transport activity as a function of addition of phosphate to amino acid side chains will clearly take much longer. We also know little about the "sensors" and transduction mechanisms involved in converting information about cell size into transport activity, and these processes may vary among different tissues. Are specific second messengers generated or are concepts such as "macromolecular crowding" sufficient to explain the activation and inactivation of the enzymes involved? Identification of the relevant kinases and phosphatases will be aided by the availability of molecular tools to identify the transporters involved. For example, the putative hypertonically activated kinase of shark rectal gland could be assayed and possibly purified using a peptide fragment of the Na/K/2Cl cotransporter containing the appropriate phosphorylatable amino acid(s). Experiments of this nature should be feasible in the next few years.

ACKNOWLEDGMENTS

I thank Drs. C. Lytle, B. Forbush, L. Starke, M. Jennings, and M. Musch for access to prepublication manuscripts.

REFERENCES

1. **Chamberlin, M. E. and Strange, K.,** Anisosmotic cell volume regulation: a comparative view, *Am. J. Physiol.,* 257, C159, 1989.
2. **Sarkadi, B. and Parker, J. C.,** Activation of ion transport pathways by changes in cell volume, *Biochim. Biophys. Acta,* 1071, 407, 1991.
3. **Greengard, P.,** Phosphorylated proteins as physiological effectors, *Science,* 196, 146, 1978.
4. **Levitan, I. B.,** Modulation of ion channels in neurons and other cells, *Annu. Rev. Neurosci.,* 11, 119, 1988.
5. **Edelman, A., Blumenthal, D. A., and Krebs, E. G.,** Protein serine/threonine kinases, *Annu. Rev. Biochem.,* 56, 567, 1987.
6. **Hanks, S. K., Quinn, A. M., and Hunter, T.,** The protein kinase family: conserved features and deduced phylogeny of the catalytic domains, *Science,* 241, 42, 1988.
7. **Kemp, B.,** Ed., *Peptides and Protein Phosphorylation,* CRC Press, Boca Raton, FL, 1990.
8. **Cohen, P.,** The structure and regulation of protein phosphatases, *Annu. Rev. Biochem.,* 58, 453, 1989.
9. **Shenolikar, S. and Nairn, A. C.,** Protein phosphatases: recent progress, *Adv. Second Messenger Phosphoprotein Res.,* 23, 1, 1990.
10. **Hunter, T.,** Protein kinase classification, *Meth. Enzymol.,* 200, 3, 1992.
11. **Tuazon, P. T. and Traugh, J. A.,** Casein kinase I and II — multipotential serine protein kinases: structure, function and regulation, *Adv. Second Messenger Phosphoprotein Res.,* 23, 123, 1990.
12. **Sturgill, T. and Wu, T.,** Recent progress in characterization of protein kinase cascades for phosphorylation of ribosomal protein S6, *Biochim. Biophys. Acta,* 1092, 350, 1991.
13. **Boulton, T. G.** et al., ERKs: a family of protein-serine/threonine kinases that are activated and tyrosine phosphorylated in response to insulin and NGF, *Cell,* 65, 663, 1991.
14. **Li, P., Wood, K., Mamon, H., Harer, W., and Roberts, T. M.,** Raf-1: a kinase currently without a cause but not lacking in effects, *Cell,* 64, 479, 1991.
15. **Nakielny, S., Cohen, P., Wu, J., and Sturgill, T.,** MAP kinase activator from insulin-stimulated skeletal muscle is a protein threonine/tyrosine kinase, *EMBO J.,* 6, 2123, 1992.
16. **Pelech, S. L. and Sanghera, J. S.,** MAP kinases: charting the regulatory pathways, *Science,* 257, 1355, 1992.
17. **Lindberg, R. A., Quinn, A. M., and Hunter, T.,** Dual-specificity protein kinases, *Trends Biochem. Sci.,* 17, 114, 1992.
18. **O'Grady, S. M., Palfrey, H. C., and Field, M.,** Na/K/Cl cotransport in epithelial tissues, *Am. J. Physiol.,* 253, C177, 1987.
19. **Haas, M.,** Properties and diversity of (Na-K-Cl) cotransporters, *Annu. Rev. Physiol.,* 51, 443, 1989.
20. **Palfrey, H. C. and O'Donnell, M. E.,** Characteristics and regulation of the Na/K/2Cl cotransporter, *Cell. Physiol. Biochem.,* 2, 293, 1992.
21. **Parker-Botelho, L. H., Rothermel, J. D., Coombs, R. V., and Jastorff, B.,** cAMP analog antagonists of cAMP action, *Meth. Enzymol.,* 159, 159, 1988.
22. **Hidaka, H., Watanabe, M., and Kobayashi, R.,** Properties and use of H-series compounds as protein kinase inhibitors, *Meth. Enzymol.,* 201, 328, 1991.
23. **Inagaki, M., Kawamoto, S., and Hidaka, H.,** Serotonin secretion from human platelets may be modified by Ca^{2+}-activated phospholipid-dependent myosin phosphorylation, *J. Biol. Chem.,* 259, 14321, 1984.
24. **Kawamoto, S. and Hidaka, H.,** 1-(5-isoquinolinesulfonyl)-2-methylpiperazine (H7) is a selective inhibitor of protein kinase C in rabbit platelets, *Biochem. Biophys. Res. Commun.,* 125, 258, 1984.
25. **Hannun, Y. A., Loomis, C. R., Merrill, A. H., and Bell, R. M.,** Sphingosine inhibition of protein kinase C activity and of phorbol dibutyrate binding in vitro and in human platelets, *J. Biol. Chem.,* 261, 12604, 1986.
26. **Jefferson, A. B. and Schulman, H.,** Sphingosine inhibits calmodulin-dependent enzymes, *J. Biol. Chem.,* 263, 15241, 1988.
27. **Ruegg, U. T. and Burgess, G. M.,** Staurosporine, K252 and UCN-01: potent but nonspecific inhibitors of protein kinases, *Trends Pharmacol. Sci.,* 10, 218, 1989.
28. **Tapley, P., Lamballe, F., and Barbacid, M.,** K252a is a selective inhibitor of the tyrosine kinase activity of the *trk* family of oncogenes and neurotrophin receptors, *Oncogene,* 7, 371, 1992.
29. **Cohen, P., Holmes, C. F. B., and Tsukitani, Y.,** Okadaic acid: a new probe for the study of cellular regulation, *Trends Biochem. Sci.,* 15, 99, 1990.
30. **Brautigan, D. L. and Shriner, C. L.,** Methods to distinguish various types of protein phosphatase activity, *Meth. Enzymol.,* 159, 339, 1988.

31. **Forbush, B., III and Palfrey, H. C.,** [³H]Bumetanide binding to membranes from dog kidney outer medulla. Relationship to Na,K, 2Cl cotransport, *J. Biol. Chem.,* 258, 11787, 1983.
32. **Burnham, C. E., Kidd, J., and Palfrey, H. C.,** Expression of the Na/K/2Cl cotransporter in *Xenopus* oocytes, *Am. J. Physiol.,* 259, F383, 1990.
33. **Xu, J., Lytle, C., Zhu, T., Benz, E. J., and Forbush, B., III,** submitted.
34. **Palfrey, H. C. and Greengard, P.,** Hormone-sensitive ion transport systems in erythrocytes as models for epithelial ion pathways, *Ann. N.Y. Acad. Sci.,* 372, 291, 1981.
35. **Pewitt, E. B., Hegde, R. S., Haas, M., and Palfrey, H. C.,** The regulation of Na/K/2Cl cotransport and bumetanide binding by protein phosphorylation and dephosphorylation, *J. Biol. Chem.,* 265, 20747, 1990.
36. **Palfrey, H. C. and Pewitt, E. B.,** The ATP and Mg dependence of Na/K/2Cl cotransport reflect a requirement for protein phosphorylation: studies using calyculin A, *Eur. J. Physiol.,* in press.
37. **Geck, P. and Heinz, E.,** The Na-K-2Cl cotransport system, *J. Membr. Biol.,* 91, 97, 1986.
38. **Ikehara, T., Yamaguchi, H., Hosokawa, K., and Miyamoto, H.,** Kinetic mechanism of ATP action in Na^+-K^+-Cl^- cotransport of HeLa cells determined by Rb^+ influx studies, *Am. J. Physiol.,* 258, C599, 1990.
39. **Flatman, P. W.,** The effects of metabolism on Na-K-Cl cotransport in ferret red cells, *J. Physiol.,* 437, 495, 1991.
40. **Russell, J. M.,** Role of K and ATP in Na,K,Cl cotransport by the squid giant axon, *Regulation of K Transport Across Biological Membranes,* Reuss, L., Russell, J. M., and Szabo, G., Eds., University of Texas Press, Austin, TX, 1990, 265.
41. **Starke, L. C.,** Regulation of [Na-K-2Cl] and [K-Cl] Cotransport in Duck Red Cells, Ph.D. thesis, Duke University, Durham, NC, 1989.
42. **Flatman, P. W.,** The effects of magnesium on potassium transport in ferret red cells, *J. Physiol.,* 397, 471, 1988.
43. **Palfrey, H. C., Silva, P., and Epstein, F. H.,** Sensitivity of cAMP-stimulated salt secretion in the shark rectal gland to "loop" diuretics, *Am. J. Physiol.,* 246, C242, 1984.
44. **Lytle, C. Y. and Forbush, B., III,** The Na-K-Cl cotransport protein of shark rectal gland. II. Regulation by direct phosphorylation, *J. Biol. Chem.,* 267, 25438, 1992.
45. **Lauf, P. K.,** KCl cotransport: sulfhydryls, divalent cations and the mechanism of volume activation in a red cell, *J. Membr. Biol.,* 88, 1, 1985.
46. **Parker, J. C., Colclasure, G. C., and McManus, T. J.,** Coordinated regulation of shrinkage-induced Na/H exchange and swelling induced KCl cotransport in dog red cells, *J. Gen. Physiol.,* 98, 869, 1991.
47. **Jennings, M.-L. and Schulz, R. K.,** Okadaic acid inhibition of KCl cotransport, *J. Gen. Physiol.,* 97, 799, 1991.
48. **Kaji, D. M. and Tsukitani, Y.,** Role of protein phosphatase in activation of KCl cotransport in human erythrocytes, *Am. J. Physiol.,* 260, C178, 1991.
49. **Starke, L. C. and Jennings, M. L.,** KCl cotransport in rabbit red cells: further evidence for regulation by protein phosphatase type 1, *Am. J. Physiol.,* 264, C118, 1993.
50. **Jennings, M. L. and Al-Rohil, N.,** Kinetics of activation and inactivation of swelling-stimulated KCl cotransport, *J. Gen. Physiol.,* 95, 1021, 1990.
51. **Parker, J. C., McManus, T. J., Starke, L. C., and Gitelman, H. J.,** Coordinated regulation of Na/H exchange and KCl cotransport in dog red cells, *J. Gen. Physiol.,* 96, 1141, 1990.
52. **Cossins, A. R.,** A sense of cell size, *Nature* 352, 667, 1991.
53. **McManus, T. J. and Parker, J. C.,** Coordinated regulation of volume-activated transport pathways, Chapter 18, this volume.
54. **Kim, H. D., Sergeant, S., Forte, L., Sohn, D. H., and Im, J. H.,** Activation of Cl-dependent K flux by cAMP in pig red cells, *Am. J. Physiol.,* 256, C772, 1989.
55. **Grinstein, S., Goetz-Smith, J. D., Stewart, D., Beresford, B. J., and Mellors, A.,** Protein phosphorylation during activation of Na^+-H^+ exchange by phorbol esters and by osmotic shrinking, *J. Biol. Chem.,* 261, 8009, 1986.
56. **Grinstein, S., Cohen, S., Goetz, J. D., and Rothstein, A.,** Osmotic and phorbol ester-induced activation of Na^+/H^+ exchange: possible role of protein phosphorylation in lymphocyte volume regulation, *J. Cell Biol.,* 101, 269, 1985.
57. **Bianchini, L., Woodside, M., Sardet, C., Pouyssegur, J., Takai, A., and Grinstein, S.,** Okadaic acid induces activation and phosphorylation of Na/H antiport, *J. Biol. Chem.,* 266, 15406, 1991.
58. **Wakabayashi, S., Sardet, C., Fafournoux, P., Counillon, L., Meloche, S., Pages, G., and Pouyssegur, J.,** Structure function of the

growth-factor-activatable Na^+/H^+ exchanger (NHE1), *Rev. Physiol. Biochem. Pharmacol.*, 119, 157, 1992.

59. **Orlowski, J., Kandasamy, R. A., and Shull, G. E.,** Molecular cloning of putative members of the Na/H exchanger gene family, *J. Biol. Chem.*, 267, 9331, 1992.
60. **Tse, C.-M., Brant, S. R., Walker, M. S., Pouyssegur, J., and Donowitz, M.,** Cloning and sequencing of a rabbit cDNA encoding an intestinal and kidney-specific Na^+/H^+ exchanger isoform (NHE-3), *J. Biol. Chem.*, 267, 9340, 1992.
61. **Wakabayashi, S., Fafournoux, P., Sardet, C., and Pouyssegur, J.,** The Na^+/H^+ antiporter cytoplasmic domain mediates growth factor signals and controls "H^+-sensing", *Proc. Natl. Acad. Sci. U.S.A.*, 89, 2424, 1992.
62. **Borgese, F., Sardet, C., Cappadoro, M., Pouyssegur, J., and Motais, R.,** Cloning and expression of a cAMP-activated Na/H exchanger: evidence that the cytoplasmic domain mediates hormonal regulation, *Proc. Natl. Acad. Sci. U.S.A.*, 89, 6765, 1992.
63. **Grinstein, S., Woodside, M., Sardet, C., Pouyssegur, J., and Rotin, D.,** Activation of the Na/H antiporter during cell volume regulation, *J. Biol. Chem.*, 267, 23823, 1992.
64. **Yancey, P. H., Clark, M. E., Hand, S. C., Bowlus, R. D., and Somero, G. N.,** Living with water stress: evolution of osmolyte systems, *Science*, 217, 1214, 1982.
65. **Musch, M., Leffingwell, T. R., and Goldstein, L.,** Band III modulation in hypotonic-stimulated taurine efflux in skate erythrocytes, *Am. J. Physiol.*, in press.
66. **Politis, A. D. and Pierce, S. K.,** Specific protein phosphorylation occurs in molluscan red blood cell ghosts in response to hypoosmotic stress, *J. Membr. Biol.*, 124, 169, 1991.

Chapter 13

Stretch-Activated Ion Channels

Henry Sackin

TABLE OF CONTENTS

I. INTRODUCTION

Mechanosensitive or stretch-activated (SA) channels respond to membrane stress by changes in open probability. These channels exist in auditory cells, stretch receptors, muscle spindles, vascular endothelium, and other neurosensory tissues where their physiological function seems readily apparent. It is less obvious why nonexcitable cells, such as those of blood and epithelial tissues, need channels that respond to mechanical stimuli. Clearly, all cells must cope with the dual problems of volume regulation and electrolyte homeostasis. Since the primary function of epithelia is salt and water transport, these cells face both extracellular and intracellular osmotic challenges. For example, Na-transporting epithelia in the intestine and kidney must accommodate significant variations in net solute uptake without suffering destructive changes in cell volume, caused by slight discrepancies between influx and efflux.[1]

Since volume is a physical property of the cell, it can only be sensed physically, or mechanically. Although cell swelling may dilute the concentration of certain impermeants that could conceivably function as intracellular volume sensors, a volume controller that relies on chemical sensors is inher-

0-8493-4448-4/94/$0.00+$.50

ently unstable because chemical concentration is affected by multiple factors within the cell. A reliable feedback system for short-term volume regulation ultimately requires some type of *mechanical* sensor to convey information about cell size. Similarly, regulation of cell growth may also require specific mechanotransducers that detect physical changes in cell size and shape. Conversely, the abnormal growth of cancer cells could involve a breakdown of such a mechanotransduction system.

Early studies implied that mechanosensitive channels belonged to a fairly homogeneous family.[2–4] Their ion selectivity, single channel conductance, kinetics, and stretch sensitivity were surprisingly similar, despite diverse tissues of origin. This led to the concept of a primitive, mechanosensitive channel type that persisted through the course of evolution. Perhaps it started by conferring the important property of volume regulation to its one-celled host, and was then retained in higher animals as a multipurpose channel, serving as sensory transducer in one tissue and volume regulator in another.

Although the concept of a single SA channel with multiple functions was appealing from a theoretical point of view, it soon became apparent that SA channels were sufficiently different to defy description by a single model. Nonetheless, one can still define a family of mechanosensitive channels, with common properties, that are clearly distinguishable from the many channel types that are not stretch activated. In an attempt to group SA channels into logical categories, a functional rather than a phylogenetic scheme will be stressed, with emphasis on volume regulation and electrolyte homeostasis. Next, the properties that characterize SA channels will be considered, emphasizing the various hypotheses for electromechanical transduction. Finally, the important question of how SA channels might be involved in the physiological regulation of transport and volume will be addressed. This treatment of SA channels is not intended to be all-inclusive and the reader is referred to reviews that deal more extensively with the growing family of mechanosensitive channels.[5–7]

II. FUNCTIONAL CLASSIFICATION OF SA CHANNELS

A. STRETCH-ACTIVATED CATION CHANNELS

The original family of mechanosensitive, cation-selective, ion channels was discovered (probably by accident), in chick skeletal muscle[2] and embryonic *Xenopus* muscle,[8] while applying pipette suction during seal formation. Subsequently, these stretch-activated cation channels were identified in choroid plexus,[9] corneal epithelium,[10] neuroblastoma cells,[11] osteoblasts,[12] Ehrlich ascites tumor cells,[13] opossum kidney cells,[14] *Xenopus* oocytes,[15] aortic endothelial cells,[16] amphibian smooth muscle cells,[17] frog diluting segments,[18] basolateral membrane of frog proximal tubule,[19] and apical membrane of *Necturus* renal proximal tubule.[20]

The channels were consistently cation selective, sometimes with a slight preference for K over Na; hence, the nomenclature *SA-cat* will be used. Their single-channel conductance ranged between 25 and 35 pS (in normal bathing solutions). The trivalent lanthanide, gadolinium (Gd), was found to block the channel completely at low concentrations.[15,20] Although this blocker is a useful tool for identifying SA channels and relating them to macroscopic currents (see Section IV), it is not 100% specific for the SA-cat channel, since it also blocks some calcium channels[21] and endplate channels.[15] Fortunately, this is not a serious drawback for epithelial studies.

The significant calcium permeability of the SA-cat channel raised the possibility that Ca might function as a second messenger for translating mechanical stress into regulation of ion transport. This would be particularly appropriate for volume regulation. An example is the choroid plexus response to hypotonic shock, where opening of SA-cat channels presumably increases cytosolic Ca sufficiently to activate Ca-dependent maxi-K channels.[9] This type of coupling seemed to establish a clear function for the SA-cat channel as a regulator of cytosolic calcium.

The SA-cat channel was subsequently implicated as a mediator for two types of Ca-dependent, K channels, both of which were activated by hypotonic shock: (1) a small-conductance (15 pS), Ca-dependent K channel in opossum kidney cells[22] and (2) a large-conductance (150 pS) K channel in cultured MTAL cells.[23] Stretch and osmotic activation of both these channels appeared to depend on extracellular Ca. The presumption was that mechanical deformations of the membrane were turning on SA-cat channels and increasing cytosolic Ca, which then gated the Ca-dependent channels.

Actually, the details of this cascade may be more complex than originally proposed. Studies on both opossum kidney cells and cultured TALH indicate that release of Ca from internal stores may be as important as extracellular Ca for the hypotonic regulatory volume decrease (RVD) response.[24] Fluorescence measurements of cytosolic Ca in rabbit proximal tubule during cell swelling suggested that the changes in free intracellular Ca during swelling were too small to activate Ca-dependent maxi-K channels.[25] It is still possible that SA-cat channels could establish unstirred layers of elevated Ca concentration adjacent to the interior of the cell membrane, and that these (micromolar?) levels of Ca could gate Ca-dependent channels. In any case, calcium appears to be important for normal RVD in a number of epithelial preparations.[26–28] In experiments on isolated cells, waves of Ca were produced by mechanically prodding heart cells (possessing SA-cat channels) with a smooth pipette.[29] Presumably, this calcium is extracellular in origin and enters via SA-cat channels. Final evaluation of the SA-cat calcium-influx hypothesis must await higher resolution techniques for calcium imaging in volume-regulating cells.

Figure 13–1 illustrates what may be a subgroup of the SA cat family.[14] These channels are seen in the opossum kidney cell line, and resemble the SA-cat channel in conductance (22 pS) and permeability to divalents like Ca. However, they differ from most SA-cat channels by displaying a significant permeability to anions like Cl. The exact nature of anion permeation through this channel is unclear. Furthermore, it appears that the cation conductance of the channel is affected by the presence of chloride.

As indicated in Figure 13–1, these opossum kidney channels are activated equally well by either pipette suction or hypotonic shock. This makes them possible mediators of the renal RVD response, provided they allow significant amounts of Ca to enter the cell during swelling. A small-conductance channel (15 pS), which is gated by calcium,[22] may be functionally coupled to this SA channel since it is active during hyposmotic shock. In contrast to maxi-K channels, the 15-pS Ca-dependent K channel activates at lower levels of intracellular Ca (at normal membrane potential), making it a better candidate than the maxi-K for eliminating K from swollen cells. Another case of functional coupling between ion-selective channels and the SA-cat channel seems to occur in Ehrlich ascites tumor cells, in which cell swelling activates independent K and Cl fluxes via channels that are not themselves mechanosensitive.[13]

An obvious problem with the SA-cat channel and volume regulation is that cation-selective channels also allow Na to enter the cell down its electrochemical gradient. The influx of osmotically active Na (and anion) would offset the volume-regulatory effect of Ca-dependent K efflux. One way around this problem would be for the cell to use Ca as a biological amplifier, so that a relatively small influx of Ca would gate a large efflux of K. The steep Ca dependence of the maxi-K channel makes this a possible candidate for an amplification scheme. However, such a volume control system would also require Na levels in the cell to be under tight regulation by an independent system. This could be tested experimentally by assessing the dependence of RVD on the Na pump. In fact, renal tubules, whose basement membranes have been removed by collagenase, lose the ability to volume regulate if the Na pump is blocked by ouabain.[30] However, if the basement membrane and the cell cytoskeleton remain intact, these tubules can still volume regulate after pump inhibition.[31] Hence, there may be redundant systems for RVD, some of which seem to be independent of Na transport.

The SA-cat channel may also be involved in mechanotransduction of shear stress forces in blood vessels. Studies on cultured endothelial cells from pig aorta reveal a 19-pS SA-cat channel, similar to what is seen in cultured chick muscle cells. Ca fluxes through these channels appear as unitary currents of about 1 to 2 pA that are reversibly activated by 15-mmHg pipette suction.[16] This raises the possibility that changes in blood pressure and flow rate produce vasodilatation via Ca entry through mechanically activated SA-cats.

A mechanosensitive channel has also been seen in amphibian lens preparations.[10] Its ion selectivity is similar to SA-cat but it has a slightly higher single-channel conductance, 50 pS, compared to the 20–30 pS of other SA-cats. Since it is the only channel in the lens that appears to be stretch-sensitive, it is distinguished by the label CAT-50. This channel could play a role in cataract formation since any increase in lens pressure would activate CAT-50s and allow abnormal amounts of sodium, calcium, and water to enter the cell. The influx of Na would also depolarize the lens, producing a greater driving force for K exit. The resulting increase in Na and decrease in K is observed in many types of cataracts.

An interesting variant of the SA-cat channel is seen in membrane patches of the fungus *Uromyces*.[32] This channel is cation selective, permeable to divalent ions like Ca, blocked by low concentrations of gadolinium, but has a single-channel conductance of 600 pS, more than 20 times the conductance of most SA-cat channels in animal cells. This mechanosensitive channel does not seem to be involved in volume regulation but does seem to be essential for allowing the fungus to gain entry into wheat and bean plants. Stretch-activated plant channels seem to be more involved in transducing topographical and geotropic signals than in maintaining cell homeostasis.

As a final caveat, maxi-K channel involvement in RVD does not absolutely require the presence of

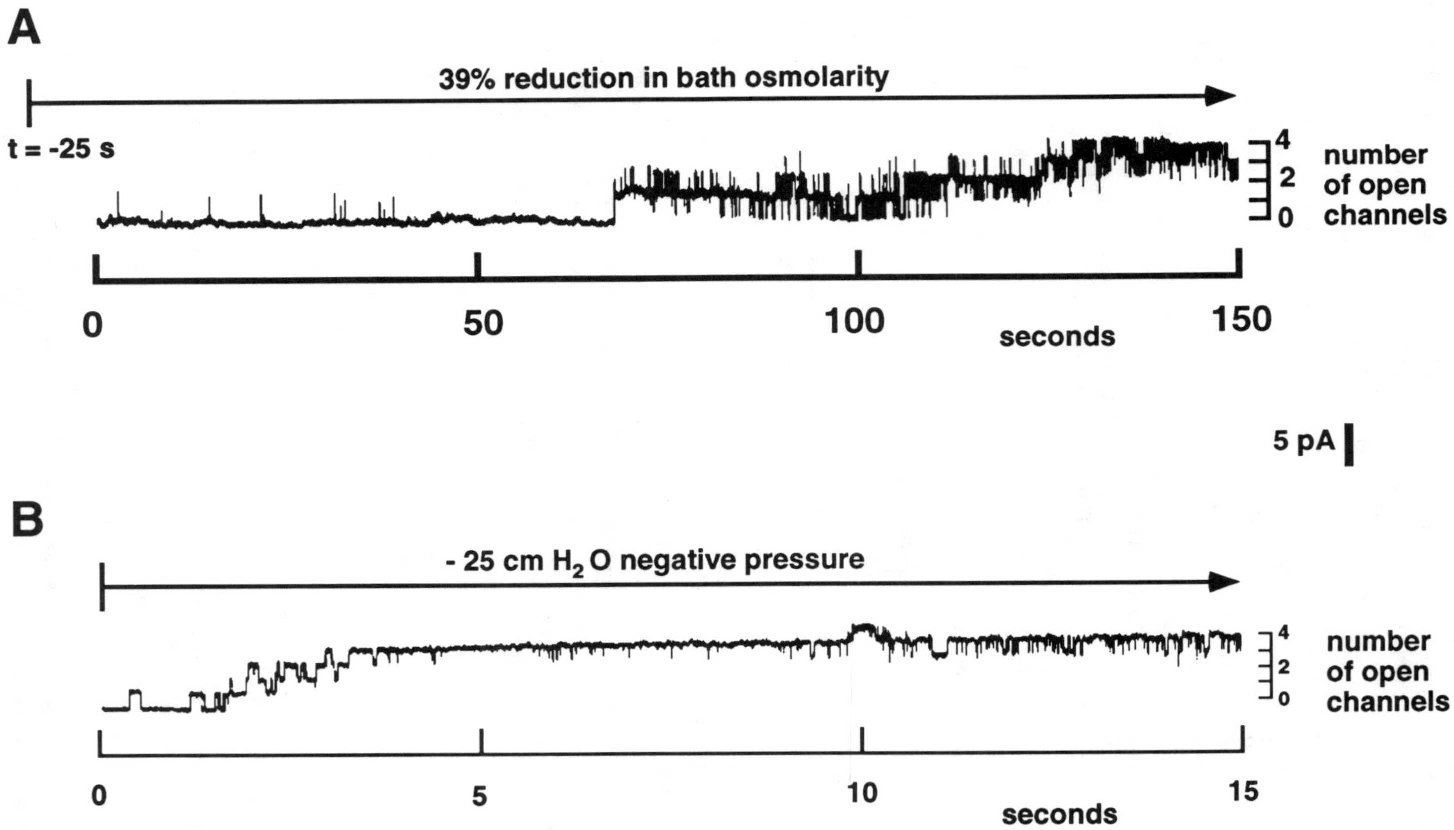

Figure 13–1 Activation of a channel population in opossum kidney cells by hyposmotic shock (A) and pipette suction (B). The channels are similar to the SA-cat channel. (Redrawn from Figure 13–4 of Ubl, J. et al., *J. Membr. Biol.*, 104, 223, 1988.)

SA-cat channels. For example, in cultured rabbit proximal tubule cells a 25% hypotonic bath elevated cytosolic Ca from 96 to 468 n*M* and dramatically activated Ca-dependent maxi-K channels.[33,34] In these experiments, it was estimated that 84% of the Ca increase arose from G protein-mediated Ca influx. The rest of the rise in Ca during swelling presumably occurred via $IP_{3,4}$ release of Ca from intracellular stores. In support of this hypothesis are experiments with atrial tissues which indicate that atrial stretch increases levels of IP_3.[35]

In rabbit cortical collecting tubule (CCT) cultures, hypotonic solutions also increase activity of a maxi-K channel at the apical membrane of principal cells.[36] This response can be mimicked by arachidonic acid or prostaglandin (PGE_2), which presumably release Ca from intracellular stores. The lag between hypotonic exposure and channel activation suggests that intracellular intermediates (e.g., eicosanoids), rather than SA-cat channels, may be producing the rise in cytosolic Ca. Precedents for such a stretch-induced increase in prostaglandin synthesis come from studies in vascular endothelium.[37] Calcium-dependent maxi-K channels were also implicated in the RVD after ouabain-induced swelling of perfused rabbit CCTs. Again the results could be mimicked by exogenous PGE_2, suggesting that volume regulation via maxi-K channels may not require direct influx of Ca from the bath.[38]

B. STRETCH-ACTIVATED POTASSIUM CHANNELS

In addition to the large SA-cat family of mechanosensitive channels, there are smaller families of mechanosensitive channels that are not as widely distributed among different species but are more highly selective for particular ions. Those SA channels displaying a predominant selectivity for K will be labeled SAKs and can be grouped into two broad classes based on their sensitivity to Ca.

1. Ca-Insensitive SAKs

A flickery K-selective SA channel with two open and three closed states has been reported in molluscan heart cells.[39] The spontaneous activity of this ventricular SAK, in the absence of applied stretch, raises the possibility that this channel could contribute to the normal resting potential. Application of pipette suction up to 25 mmHg dramatically increased the P_o of this channel without altering its selectivity or its 33-pS single-channel conductance. The kinetics and density of distribution of this channel are similar to the less selective SA-cat channel described above. However, the characteristic K selectivity of this SA channel makes it a better candidate for mediating RVD than the SA-cat channel. Since hypotonic volume regulation depends on an efflux of cellular cations, any channel with a significant selectivity for K over Na would be useful in restoring cell volume.

In many cells, short-term volume regulation after exposure to hypotonic medium is accompanied by efflux of solutes and water out of the cell.[7] In epithelia, this volume-regulatory phase (or RVD) seems to occur via a barium-sensitive K efflux,[40,41] together with a chloride flux,[42] where K and Cl ions leave the cell via separate channels.[43] Although K channels definitely seem to be involved in short-term epithelial volume regulation, there is no unified theory for how this occurs. In mammalian renal cells, K channels seem to be regulated indirectly by Ca, pH, and possibly ATP levels. In contrast, many of the K-selective channels found in amphibian cells are directly activated by stretch.

Two types of SAK channels were found at the basolateral membrane of *Necturus* proximal tubule: (1) a short open time, flickery SAK with conductance of about 45 pS and mean open time on the order of 1.5 msec,[44,45] and (2) a longer open time SAK with conductance of about 30 pS and mean open time between 40 and 50 msec.[45,46] In addition to being activated by pipette suction, these K channels were also sensitive to osmotic swelling. Hyposmotic solutions increased P_o, without changes in conductance or channel selectivity, in both K depolarized cells,[47] as well as cells maintained in Na-Ringer solutions (Figure 13–2).[45] In the latter case, measurements of electrical potential during hyposmotic shock indicated that the accompanying changes in membrane potential were not large enough to affect the relatively weak voltage dependence of this channel. The similarity in the kinetic response to both swelling and pipette suction[47] suggests that cell swelling activates these SAK channels via mechanical forces rather than by release of some intracellular mediator. The role of chemical signaling during RVD may be more important in mammalian cells than in amphibian cells.

2. Ca-Sensitive SAKs

There are now a number of preparations in which what appear to be maxi-K channels also display a direct sensitivity to stretch. The channels retain their characteristic steep dependence on voltage and Ca sensitivity, but also respond directly to increases in membrane tension, even when Ca concentrations are tightly controlled. Maxi-K channels have been identified at the apical membrane of rat and rabbit CCT that react independently to the three variables: voltage, cytosolic Ca, and mem-

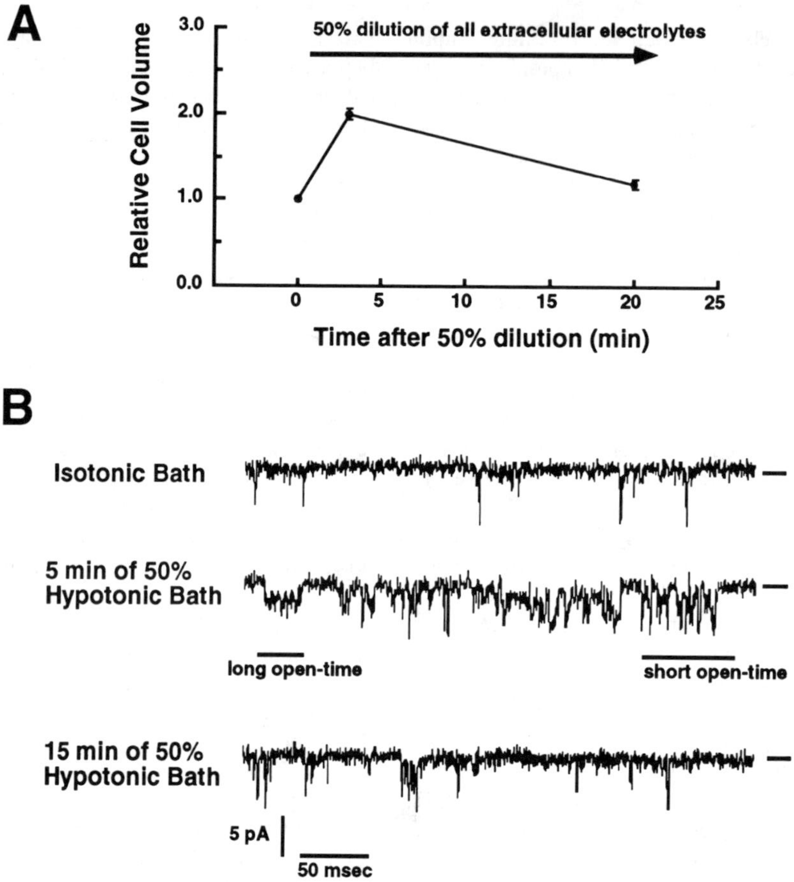

Figure 13–2 Effect of bath dilution on cell volume and channel activity. All experiments on isolated, polarized *Necturus* proximal cells, suspended by patch pipettes in flowing solution. (A) Relative cell volume as a function of time after a 50% dilution of extracellular electrolytes (n = 10 animals). (B) Basolateral single channel currents during one experiment. Isotonic bath, sodium chloride Ringer solution; 50% dilution, half sodium chloride Ringer; pipette, potassium chloride Ringer. Transpatch potential = 0 mV. Downward deflections from the closed state (horizontal bar) denote inward (K) current. (From Filipovic, D. and Sackin, H., *Am. J. Physiol.,* 262, F857, 1992. With permission.)

brane stretch.[48] Surprisingly, the density of maxi-K channels on intercalated cells (IC) was much higher than on principal cells (PC). The example of Figure 13–3 is taken from an excised (inside-out) patch on an intercalated cell of rabbit CCT where the calcium at the cytoplasmic side of the patch was buffered to 10^{-7} *M,* and there was no Ca added to the NaCl pipette solution.[48]

Successive application of negative pipette pressure dramatically increased the mean number of open channels in this excised patch (Figure 13–3). Upward deflections denote outward K movement from a 140 m*M* KCl cytoplasmic side solution into the Na-containing pipette. Although these channels are sensitive to elevations in cytosolic side Ca, it is unlikely that the observed increase in P_o is due to a change in calcium. With excised patches formed on steeply tapered pipettes, the patch remains close to the pipette tip so that the Ca at the inside face of the membrane should be effectively buffered by BAPTA in the external bath.

Although there appeared to be no obvious effect of apical hypotonicity in these cells, experiments on everted rat CCTs indicate that reducing the osmolarity at the basolateral side of the cell activates an apical, maxi-K channel that might be involved in RVD.[49] However, it is not known whether these channels responded to increases in cell volume via changes in membrane tension or via some intracellular mediator.

In addition to possibly being involved in hypotonic RVD, the stretch-sensitive class of maxi-K

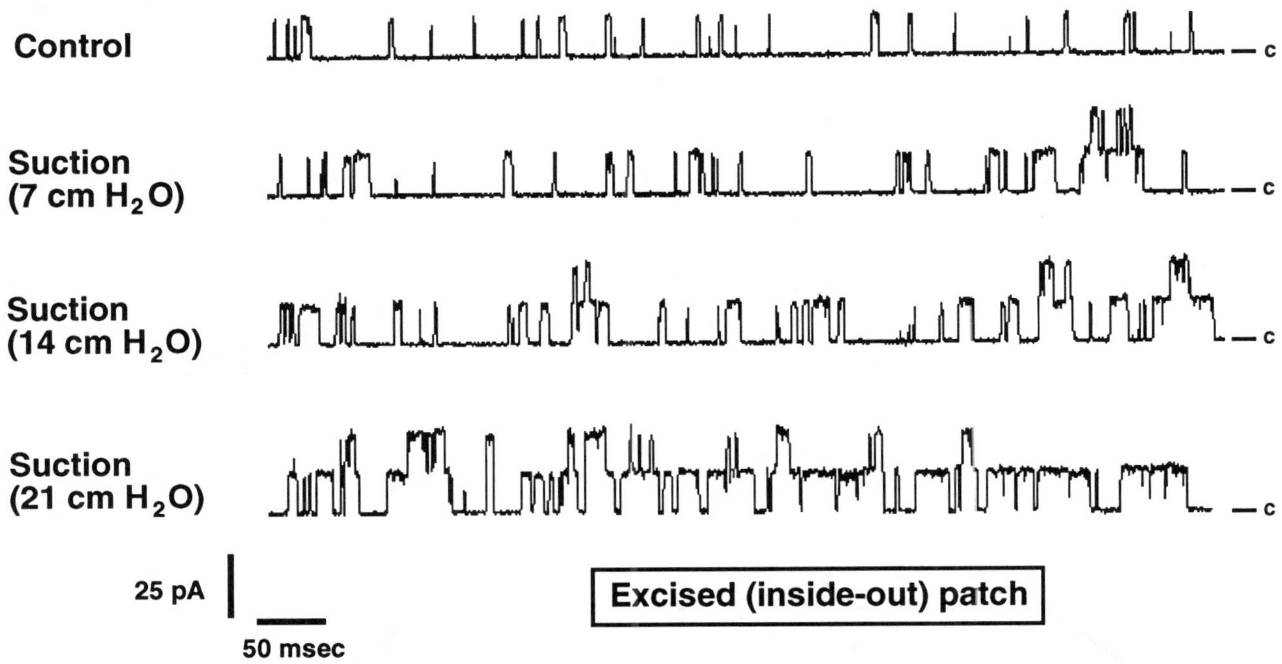

Figure 13–3 Effect of pipette suction on single-channel currents in an excised (inside-out) patch on an intercalated cell from rabbit cortical collecting tubule. Pipette contained 50 m*M* NaCl, but no added Ca. Bath (cytosolic side) contained 140 m*M* KCl with Ca maintained at pCa = 7 with BAPTA. Holding potential was –40 mV, interior side of the membrane positive. (From Pacha, J. et al., *Am. J. Physiol.*, 261, F696, 1991. With permission.)

channels could be important for flow-dependent K secretion in distal tubule and CCT. Since increases in flow rate (distal delivery) could increase CCT intraluminal pressure by as much as 6 or 7 mmHg, SA maxi-K channels, at the apical membrane of intercalated (and principal) cells, might contribute to the flow-dependent K secretion observed in this nephron segment.

Stretch-sensitive maxi-K channels have also been reported in pulmonary vascular smooth muscle cells.[50] In these studies four distinct types of stimuli were effective at increasing the P_o of this large-conductance K channel; cytosolic Ca, cell depolarization, membrane stretch, and exogenous fatty acids, where the effect of stretch does not seem to be mediated by calcium. This was elegantly demonstrated by applying pipette suction to a series of excised (inside-out) patches with 5 m*M* EGTA and no added Ca on both sides of the patch. It is interesting that these channels could also be activated by fatty acids via a mechanism that did not seem to involve formation of biologically active metabolites, phosphorylation, or Ca. This raises the possibility that increases in membrane tension could stimulate a membrane-bound phospholipase to release endogenous fatty acids from the membrane, which then activate the channel. This mechanism, and its reverse, are discussed further in Section III.B.4.

C. STRETCH-ACTIVATED ANION CHANNELS

Mechanosensitive anion channels are not as prevalent as SA-cat or SAK channels. With the exception of chloride channels in Ehrlich ascites tumor cells, most mechanosensitive anion channels (SA-an) are large (>300 pS) and often subject to regulation by specific chemical mediators. For example, a mechanosensitive anion channel found in *Escherichia coli* spheroblasts has a conductance between 650 and 970 pS[51] and is readily activated by amphipathic compounds via a mechanism that may (or may not) be distinct from its mechanosensitive activation.[52] A mechanosensitive anion channel of about 100 pS has been observed in tobacco protoplasts that is consistently stretch activated in both cell-attached and excised patches.[53] It is not known whether this channel responds to changes in osmolarity.

A 305-pS mechanosensitive anion-selective channel has also been described in a cell line (RCCT-28A), which has the phenotype of α-intercalated renal CCT cells. Both pipette suction and hyposmotic solutions activated this channel, which suggests that it might be involved in hyposmotic volume regulation.[54] In addition, exposing these cells to dihydrocytochalasin B dramatically increased the stretch sensitivity of the channel.[55] This effect is similar to what has been seen with SA-cat channels.[2] However, an important difference is that the SA-an channel is not only activated by stretch/swelling, but is also activated by an inositol lipid cascade involving a membrane-bound phospholipase C, diacylglycerol, protein kinase C, and a G protein.[56] The net result is the opening of a significant pathway for Cl efflux during RVD. Direct stretch activation of protein kinase C has also been reported for rat skeletal muscle.[57]

In Ehrlich ascites tumor cells, cell swelling activates a 23-pS conductance channel with an 11-fold preference for chloride over potassium.[58] Exposure to 10 m*M* glycine produced a slow increase in the volume of Ehrlich ascites tumor cells that was due to accumulation of osmotically active Na and glycine within the cells. Activation of chloride channels in cell-attached patches occurred at the same time as the increase in cell volume. These Cl channels could also be activated by reducing the osmolarity of the external solution. This suggests that the effect of glycine on channel activity may be mediated by mechanical forces associated with cell swelling.

Hoffman and co-workers[7,13] have reported a volume-sensitive, small-conductance (7 pS) chloride channel in Ehrlich ascites tumor cells, which can be activated both by swelling and by increases in cytosolic calcium. Interestingly, these Cl channels cannot be activated directly by pipette suction. This suggests that they may function in concert with SA-cat channels that are also present in ascites tumor cells. In this scenario, stretch or swelling would activate SA-cat channels, thereby allowing sufficient Ca to enter the cell and gate the small-conductance Cl channels, permitting exit of this anion during hyposmotic RVD. This would be similar to the virtual coupling between the SA-cat channel and the maxi-K channel that was discussed in the previous section.

Since many chloride channels are activated by cAMP, it is interesting that cAMP production is sometimes modulated by membrane tension.[37] This many constitute a novel "indirect" form of anion channel stretch activation. In fact, both mechanical deformation and hyposmotic swelling increase the production of cAMP in S49 mouse lymphoma cells.[59] The possibility that membrane tension activates adenylate cyclase in much the same way that it activates certain ion channels is reinforced by how closely the secondary and tertiary structure of adenylate cyclase resembles the multiple membrane-spanning domains of many ion channels.[60] Furthermore, disruption of actin microfilaments by cytochalasin B increased cAMP content of S49 cells;[59] however, cells that were pretreated with cytochalasins exhibited no further increase in cAMP after osmotic swelling.[37] Hence, an intact cytoskeleton may be required for mechanostransduction systems of this type. Presumably, mechanical forces are focused via cytoskeletal filaments onto the membrane-bound adenylate cyclase, perturbing its structure in a way that increases catalytic activity.

D. STRETCH-ACTIVATED NONSELECTIVE CHANNELS AND STRETCH-INACTIVATED CHANNELS

Truly nonselective channels (SA-non), which exhibit little discrimination between anions and cations, are unusual. Perhaps the best example is the 36-pS nonselective channel in yeast plasma membrane, which has been studied both at the whole cell and single-channel levels.[61] Although this yeast channel is blocked by 10 μM gadolinium, its inability to select between cations and anions and its kinetic response to suction distinguish it from the SA-cat family, and even from the slightly nonselective channel of opossum kidney cells. Presumably, the physiological function of SA-non and SA-cat channels are similar; namely, to elevate intracellular divalents (Ca) during changes in cell volume or membrane tension.

Stretch-inactivated channels (SI) are still relatively rare and have little application to volume regulation. Details of SIs have been described extensively by Morris[5] and Morris and Sigurdson,[62] and their role as "notch filters" for excitable cells will not be elaborated on here. Nonetheless, an interesting implication of SI channels is that their very existence refutes the notion that mechanosensitivity is just a subtle form of channel enlargement. The fact that SI channels shut down during increases in membrane tension proves that stretch is a true gating process, rather than a mechanically induced leakage current.

III. THE MECHANICS OF STRETCH ACTIVATION

A. PROPERTIES COMMON TO SA CHANNELS

1. Increase in Open Probability

Despite profound differences in kinetics, ion selectivity, and stretch sensitivity, there are a surprising number of properties common to the different classes of SA channels. In all SA channels studied to date, mechanical stimuli induce an increase in open probability, without significant changes in either singe-channel conductance or ion selectivity. This preservation of selectivity and conductance reinforces the notion that membrane stretch gates the channel. "Breakdown currents" do occur at high levels of suction, but these can be distinguished readily from discrete openings that characterize channel currents.

2. A Membrane-Specific Phenomenon

Stretch activation is a membrane (or membrane-cytoskeletal) phenomenon. It does not seem to involve soluble cytosolic messengers or insertion of preexisting cytoplasmic channels. The principal evidence for this comes from the repeated observation that mechanosensitive channels can be stretch activated in cell free excised patches. This effectively rules out mediators like Ca, ATP, etc., provided the excised patch is formed near the tip of the pipette. Nonetheless, a variable amount of cytoskeleton often remains attached to an excised patch, and the stimulus for an increase in P_o may be more complicated than simply membrane tension.

Stretch activation in excised patches also rules out any mechanism dependent on insertion of channels from the cytoplasm. This has been proposed to explain osmotically induced changes in urinary bladder conductance.[63] Rapid reversibility of channel activation is another argument against channel insertion as an explanation of mechanosensitivity. Many SA channels reversibly activate on a time scale that is much faster than membrane fusion processes. Finally, SA channels can be turned on by both positive and negative pipette pressures, even though it is technically more difficult to stabilize the seal during application of positive pressure. This finding rules out the possibility that pipette suction is somehow mechanically drawing channels or channel components into the patch. If this were the case, positive and negative pipette pressures should not have the same effect on P_o.

3. A Common Kinetic Scheme

In addition to gross similarities in their general mode of activation, mechanosensitive channels also display subtle similarities in the kinetics of their activation. Most SA channels activate in bursts of flickery openings, although the extent of flicker seems to be greater in SAKs than SA-cats. Figures 13–4 and 13–5 illustrate the effect of suction on K-selective and cation-selective SA channels, respectively. Figure 13–4 is taken from a study on molluscan heart cells[39] and Figure 13–5 from chick skeletal muscle.[2] Despite differences in conductance, selectivity, and kinetics, suction appears to increase the frequency of bursts to the open state, in both types of channels.

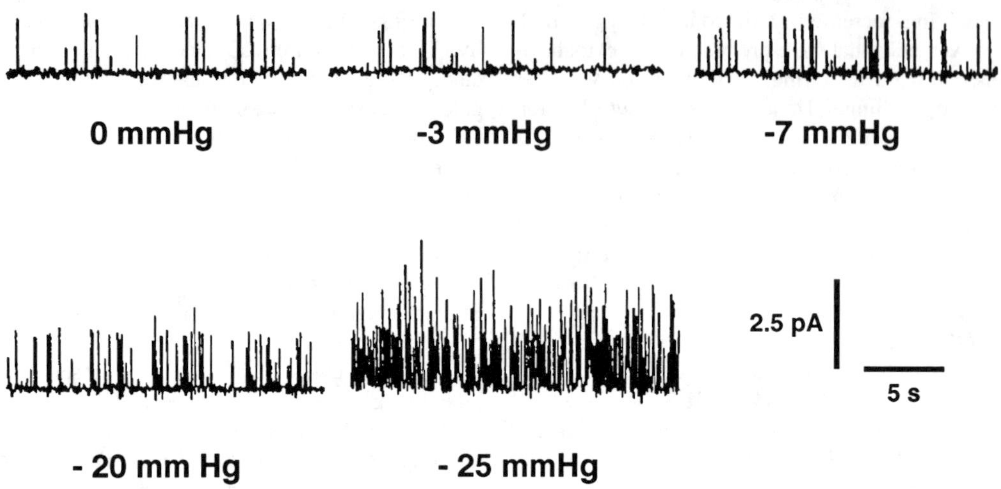

Figure 13–4 Effect of pipette suction on K channel activity in molluscan heart cells. Cell-attached patch with 30 m*M* K in the pipette and a pipette holding potential of +40 mV relative to bath. Higher negative pressures produced increased flickering to the open state, with no apparent change in mean open time. (From Sigurdson, W. J. et al., *J. Exp. Biol.,* 127, 191, 1987. With permission.)

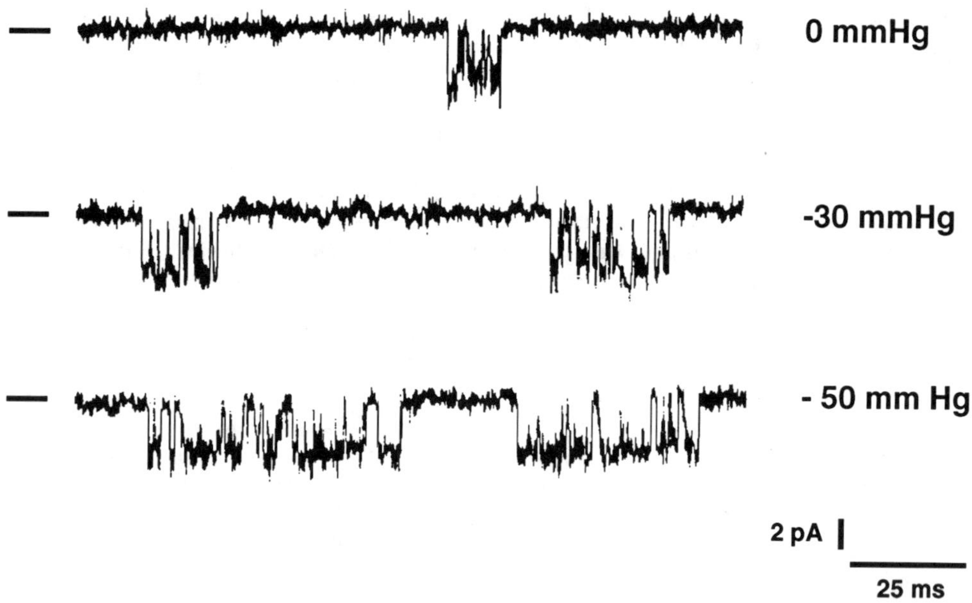

Figure 13–5 Effect of pipette suction on cation-selective channel in embryonic chick skeletal muscle. Inward currents (down deflections) from an excised (inside-out) patch with 150 m*M* K in the pipette and normal saline in the bath. Pipette potential was +50 mV with respect to the bath. (From Guharay, F. and Sachs, F., *J. Physiol. (London),* 352, 685, 1984. With permission.)

Some of these qualitative features of stretch-activation kinetics were described in models proposed by Sachs and Guharay[2] (Figure 13–6A). In this scheme there is one open state, with a mean open time that is relatively unaffected by stretch.[2] Three closed states are arranged in a linear fashion, where only the rate constant governing transitions out of the longest closed state is stretch sensitive. This model (Figure 13–6A) is consistent with data on SA channels from molluscan heart cells[39] and

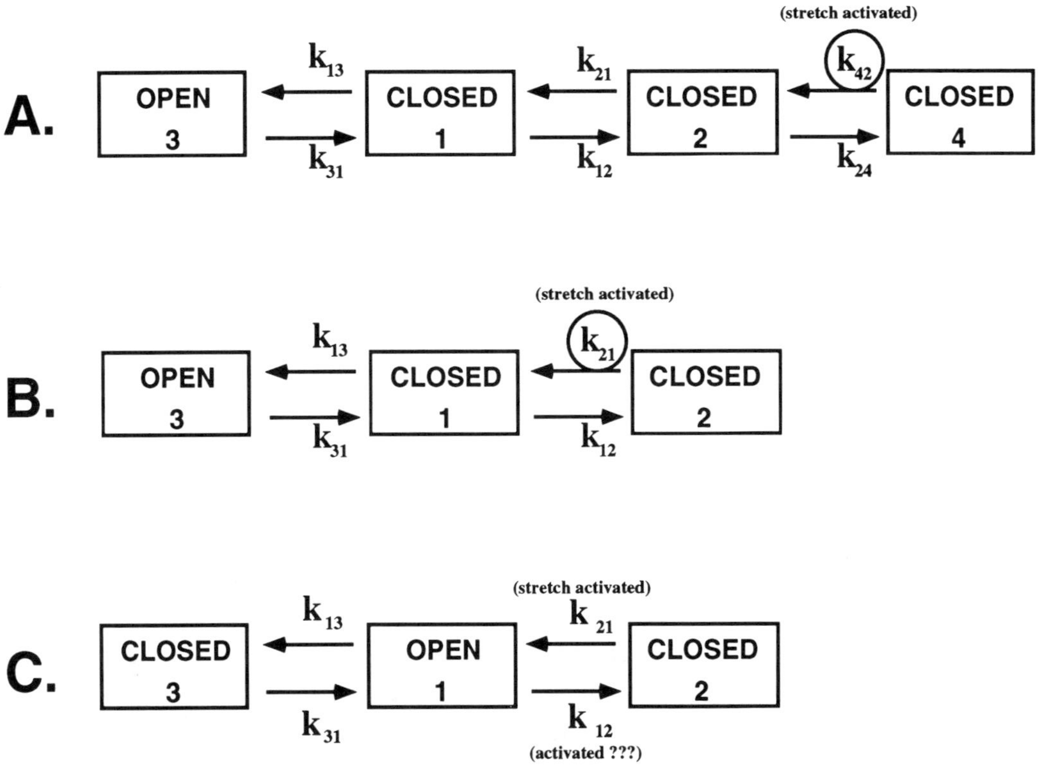

Figure 13–6 Possible kinetic schemes for stretch-activated channels. Model A was originally proposed for SA channels in chick skeletal muscle cells. The mechanosensitive rate constant governs transitions out of the longest closed state. Model B is a variation of this, appropriate for amphibian preparations, in which the mechanosensitive rate constant governs transitions between the interburst closed state (2) and the flickery closed state (1). In model C, the open state is surrounded by two closed states. Rate constants (Table 2) derived from model C are counterintuitive to the qualitative behavior of currents during suction.

chick skeletal muscle.[2] Mechanosensitive channels from amphibian proximal tubule have a similar kinetic scheme (Figure 13–6B), except that only two distinct closed states can be resolved.[47] As with the first model, the predominant effect of pipette suction is to increase the rate constant (k_{21}) governing transitions out of the longest closed state.

4. Alternative Kinetic Schemes

The linear kinetic scheme with a single open state and contiguous closed states is based to some extent on exclusion of other possible models. For example, models B and C were compared using single-channel data from patches on the basolateral membranes of *Necturus* proximal tubules and equations developed by Colquohoun and Hawkes.[64] In model B (contiguous closed states) both suction and hypotonic swelling significantly increased the rate constant, k_{21}, governing transitions between the interburst closed state and the flickery closed state (Table 1). Pipette suction also produced a small increase in the rate constant, k_{13}, compared to control; however, the change in k_{13} was not significant during hypotonic swelling. Even if stretch activation involves more than one rate constant, the direction of the changes are consistent with the general appearance of SA currents during stretch.

If the same data are fit to the model of Figure 13–3C with a single open state flanked by two closed states, the predicted changes in rate constants are counterintuitive to the process of stretch activation (Table 2). Although suction and swelling still increase k_{21}, which governs transitions between the interburst state and the open state, suction now increases k_{12} by 37%. Since k_{12} is the rate constant for transitions between the open state and the longer closed state, it is hard to reconcile an increase in k_{12} with the observation that suction always *decreases* the interburst closed time of the channel. This does

Table 1 **Rate Constants for Linear Kinetic Model with Two Contiguous Closed States**

	Open Time Rate Constant $(msec)^{-1}$	Closed Time Rate Constants $(msec)^{-1}$		
	k_{31}	k_{13}	k_{12}	k_{21}
Control	1.3 ± 0.2	0.19 ± 0.04	0.31 ± 0.04	0.06 ± 0.004
Suction (–6 cm H_2O)	1.2 ± 0.05	0.29 ± 0.06*	0.22 ± 0.04	0.14 ± 0.02*
Cell swelling (95 mOsmol hypotonic)	1.1 ± 0.1	0.23 ± 0.05	0.17 ± 0.02*	0.20 ± 0.03*

Data for this table were obtained on K-selective SAKs at the basolateral membrane of *Necturus* proximal tubules. Rate constants were determined from model B (Figure 13–6), using equations developed by Colquhoun and Hawkes (*Proc. R. Soc. London Ser. B.* 211, 205, 1981); * denotes significantly different from control values.

Table 2 **Rate Constants for Linear Kinetic Model with Open State Surrounded by Two Closed States**

	Open Time Rate Constant $(msec)^{-1}$			Closed Time Rate Constant $(msec)^{-1}$	
	k_{13}	k_{12}	$k_{13} + k_{12}$	k_{21}	k_{31}
Control	0.91 ± 0.2	0.38 ± 0.04	1.3 ± 0.2	0.02 ± 0.003	0.54 ± 0.07
Suction (–6 cm H_2O)	0.68 ± 0.1	0.52 ± .04*	1.2 ± 0.1	0.07 ± 0.01*	0.59 ± 0.10
Cell swelling (95 mOsmol hypotonic)	0.77 ± 0.2	0.36 ± 0.06	1.1 ± 0.2	0.07 ± 0.01*	0.54 ± 0.06

Data for this table were obtained on K-selective SAKs at the basolateral membrane of *Necturus* proximal tubules. Rate constants were determined from model C (Figure 13–6), using equations developed by Colquhoun and Hawkes (*Proc. R. Soc. London Ser. B,* 211, 205, 1981); * denotes significantly different from control values.

not prove that Figure 13–6C is *incorrect,* but comparisons of this type can provide an intuitive basis for selecting one particular kinetic scheme over another.

5. Sigmoidal Dependence of Open Probability

Early models for mechanosensitive gating relied on a strict application of Hooke's law to the elastic elements of the membrane, with the result that the free energy of gating was dependent on the square of applied tension.[2,5,65] This model predicted a sigmoid dependence similar to that of Equation 1, where K is a pressure-independent constant, θ is the stretch sensitivity of the membrane, and T is the applied tension. Although there is a good empirical basis for Equation 1 and it is usually possible to obtain a "good" fit to this equation, the underlying assumptions of a T^2 dependence are incomplete.[66]

$$P_0 = \frac{P_{max}}{1 + K \cdot \exp[-\theta \cdot T^2]} \quad (1)$$

A better representation for open probability can be derived by assuming that the free energy available for gating is linearly[67] (rather than quadratically) related to applied tension T. Open probability, P_o, will still be a sigmoid function of tension, but the curve will have a slightly different shape and obey the equation:

$$P_0 = \frac{P_{max}}{1 + k_{eq} \cdot \exp\left[\frac{-T\Delta A - q \cdot V}{kT_k}\right]} \quad (2)$$

where k_{eq} is the equilibrium constant for channel opening in the absence of stretch, ΔA is the increase in channel cross-sectional area during opening ($\Delta A > 0$), q is the total gating charge (in coulombs) moving across the potential field during channel opening ($q > 0$), k is the Boltzman constant, and T_k is the absolute temperature in degrees Kelvin and V is voltage.

Sachs and Lecar have considered the most general case in which both the linear and quadratic terms are included,[66,68] resulting in an expression of the form:

$$P_0 = \frac{P_{max}}{1 + k_{eq} \cdot \exp\left[\dfrac{-T^2\left(\dfrac{1}{K_s} - \dfrac{1}{K_0}\right)\cdot\dfrac{1}{2} - T\Delta A - q\cdot V}{kT_k}\right]} \quad (3)$$

The additional quadratic term (T^2) in Equation 3 comes from elastic energy considerations and depends on the elasticity coefficients for the open state (K_o) and the closed state (K_s). In a formal sense, one can relate the old stretch-sensitivity parameter, θ, to the elasticity coefficients, according to Equation 4:

$$\theta = \left(\frac{1}{K_s} - \frac{1}{K_0}\right)\cdot\left(\frac{1}{2kT_k}\right) \quad (4)$$

In many cases, the stiffness of the open state is much greater than that of the closed state ($K_o >> K_s$), so that Equation 3 becomes:

$$P_0 = \frac{P_{max}}{1 + k_{eq} \cdot \exp\left[\dfrac{-\dfrac{T^2}{2K_s} - T\Delta A - q\cdot V}{kT_k}\right]} \quad (5)$$

If the detailed dynamic parameters of the channel are unknown (as is often the case), Equation 5 can be rewritten in terms of four independent parameters (k_{eq}, a, b, c) that would define a fit to the data (see Equation 6). Without prior information about the values of these parameters, a standard fitting algorithm will result in many equally good fits to a given data set. In all of these, P_o will be sigmoid functions of applied tension (T).

$$P_0 = \frac{P_{max}}{1 + k_{eq} \cdot \exp\left[-a\cdot T^2 - b\cdot T - c\cdot V\right]} \quad (6)$$

For example, Figure 13–7 illustrates data for the SA-cat channel of the crayfish stretch receptor.[65] When P_o is plotted as a function of pipette suction (in cm H_2O), the data can be fit slightly better by an equation that assumes that the free energy is linearly related to tension (thick gray line in Figure 13–7, fit by Equation 2) than by an equation that assumes that free energy varies as the square of the tension (thin dark line in Figure 13–7, fit by Equation 1). However, the difference in "goodness" of the fit is relatively minor. Slightly more scatter in the data would make it impossible to distinguish the two models on an empirical basis. The pressure

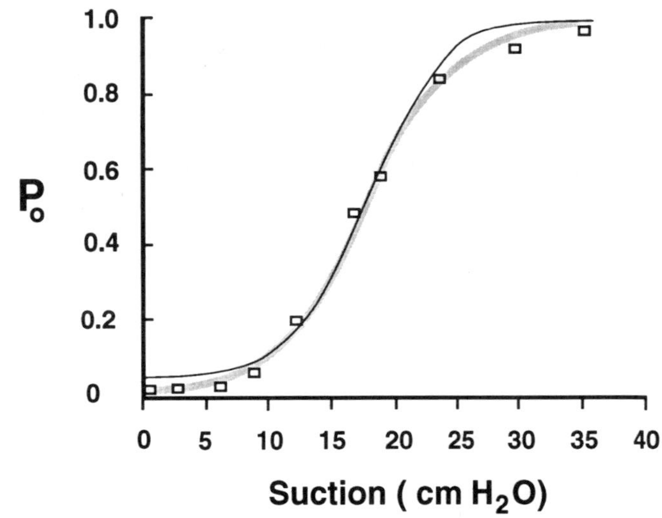

Figure 13–7 Effect of pipette suction on crayfish abdominal stretch receptor neuron. Single-channel open probability, P_o, is sigmoidally dependent on pipette suction, in which the exponential term is either a quadratic function of tension (thin dark line, Equation 1) or a linear function of tension (thick gray line, Equation 2). (From Erxleben, C., *J. Gen. Physiol.*, 94, 1071, 1989. With permission.)

dependence of the SAK channel of *Necturus* proximal tubule is illustrated in Figure 13–8. Both the quadratic and linear dependence models fit the data equally well.

The sigmoid curve in Figure 13–8 was fit using Equation 2 with an effective stretch sensitivity of 0.62 (cm $H_2O)^{-1}$ or 0.84 $(mmHg)^{-1}$. This value is between two and three times larger than the stretch sensitivity of the crayfish stretch receptor (Figure 13–7), and is consistent with the half maximal suction being much less for the *Necturus* SAK than for the crayfish SA-cat channel.

B. THEORIES OF MECHANOSENSITIVE CHANNEL ACTIVATION

1. Tension vs. Pressure

Early studies of mechanosensitive channels often contained tacit assumptions about the equivalence of hydrostatic pressure and membrane tension. In fact these two forces are orthogonal to each other, with pressure being exerted perpendicular to the plane of the membrane and tension being a force *in* the plane of the membrane. In thin-walled spheres, hydrostatic pressure (P) and membrane tension (T) are related according to Laplace's law, where d_c is the diameter of the cell or sphere:

$$T = \frac{P \cdot d_c}{4} \quad (7)$$

The issue of pressure vs. tension was addressed in whole cell clamps on yeast spheroblasts of various sizes.[61] As indicated by the data of Figure 13–9, plots of mechanosensitive current vs. tension were independent of spheroblast diameter, whereas plots of current vs. pressure exhibited a clear dependence on diameter. These studies indicate that membrane *tension* and not pressure is the relevant parameter that controls SA channel activity.

2. Swelling and Membrane Tension

The discovery of SA channels in nonexcitable cells immediately raised the possibility that these channels were somehow involved in cell volume regulation. The theoretical basis for the relation between cell volume changes and membrane tension is deceptively simple, and depends primarily on application of Laplace's law (Equation 7) to an ideally elastic system. However, the assumptions of the model are less obvious, and reflect our basic lack of information. For example, it is usually assumed that increases in cell volume are associated with significant changes in membrane tension. However, most epithelial cells possess multiple membrane infoldings that may change shape during swelling without producing a real change in membrane tension. Since we cannot visualize dynamic changes in cell surface morphology during swelling of living cells, we can only guess at the changes in membrane structure that are occurring.

Nonetheless, it is instructive to examine the simplest model, consisting of a spherical cell with no membrane infolding. In this idealized case, the membrane tension on the patch (T_p) that is associ-

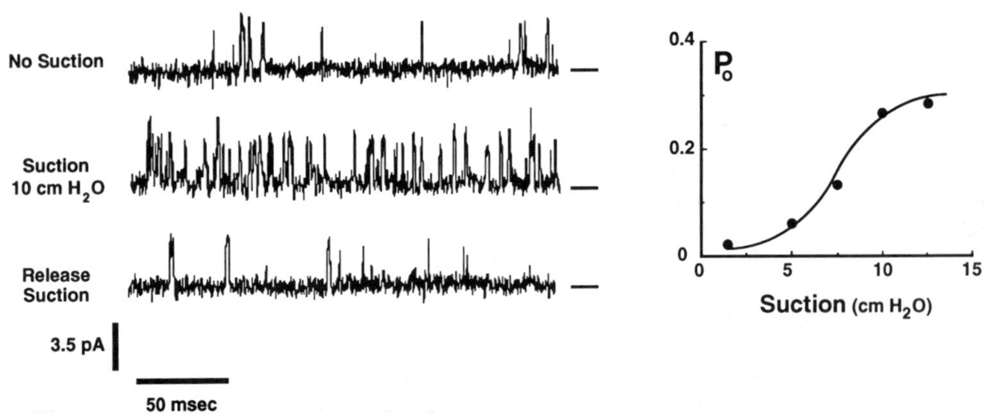

Figure 13–8 Effect of pipette suction on the short open time, stretch-activated K channel at the basolateral membrane of *Necturus* proximal tubule. Cell-attached patch on an isolated tubule. Basement membrane was manually removed without the use of enzymes. Pipette, Na Ringer; Bath, K Ringer, zero applied potential. The open probability, P_o, is a sigmoid function of applied tension. The curve was fit to Equation 2 of the text using an effective stretch sensitivity of 0.6 (cm $H_2O)um^1$.

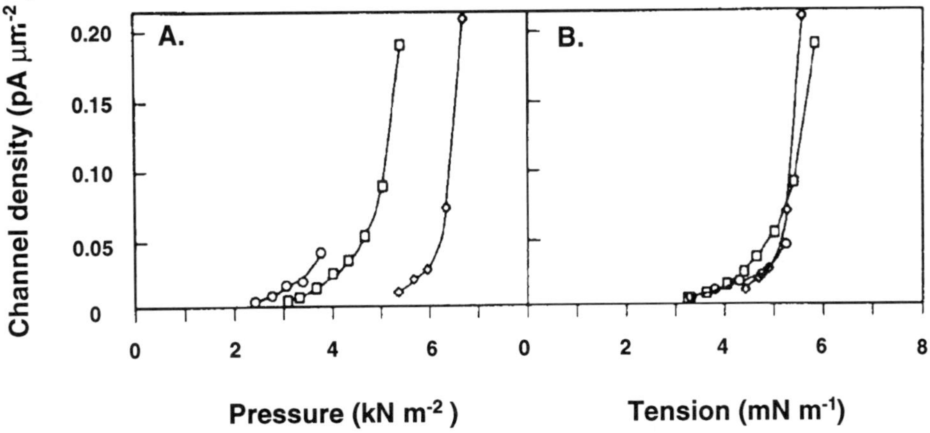

Figure 13–9 Current density through mechanosensitive channels of three cells of different sizes plotted against the applied pressure or tension. (A) Pressure dependence of SA channels in three whole cell recordings. Cell diameters were 5.7 μm (○), 4.3 μm (□), and 3.3 μm (◇). (B) Tension dependence of SA channels in the same three cells. Tension was calculated from Laplace's law. Applied voltage = –60 mV. (From Gustin, M. C. et al., *Science,* 242, 762, 1988. With permission.)

ated with a specific increase in cell surface area, ΔA, can be described by a two-dimensional form of Hooke's law:[2]

$$T_p = K_A(\Delta A/A)_p \qquad \textbf{(8)}$$

where $(\Delta A/A)_p$ is the fractional increase in membrane patch area, and K_A is the area elasticity coefficient. This latter parameter can be estimated from the measured hydrostatic pressure (P*) and the associated increase in relative area (ΔA/A)* that produces membrane breakdown (i.e., lysis, denoted by *). Using Equations 7 and 8:

$$K_A = \frac{T_p}{\Delta A/A} = \frac{P^* \cdot d_p}{4(\Delta A/A)^*} \qquad \textbf{(9)}$$

If the relative increase in surface area associated with *cell* swelling is $(\Delta A/A)_c$, then the membrane tension in a normal (unclamped) cell is T_c where:

$$T_c = K_A(\Delta A/A)_c \qquad \textbf{(10)}$$

The elasticity coefficient, K_A is determined empirically from Equation 9, and $(\Delta A/A)_c$ is geometrically related to the relative change in cell volume, ΔV/V, according to Equation 11:

$$\left(\frac{\Delta A}{A}\right)_c = \left(1+\frac{\Delta V}{V}\right)^{2/3} - 1 \qquad \textbf{(11)}$$

which follows directly from the equations for surface area and volume of an ideal sphere.

Combining Equations 10 and 11, the cell tension associated with a particular relative increase in cell volume (ΔV/V) is

$$T_c = K_A \cdot \left[\left(1+\frac{\Delta V}{V}\right)^{2/3} - 1\right] \qquad \textbf{(12)}$$

Even though Equation 12 represents the cell as an ideal sphere, we can still use it to estimate the increase in cell volume that would generate sufficient membrane tension, T_c, to activate SA channels in an unclamped cell. If a 30 μm diameter amphibian cell is hypotonically swollen by 1% (i.e., ΔV/V = 0.01), the relative volume factor in Equation 12 becomes:

$$\left[\left(1+\frac{\Delta V}{V}\right)^{2/3} - 1\right] = (1.01)^{2/3} - 1 = 0.007 \qquad \textbf{(13)}$$

The elasticity coefficient, K_A, for the "unclamped" cell can be estimated by patching the cell and noting the maximum suction that can be sustained by the membrane. In general, amphibian proximal tubule cells can withstand 50 cm H_2O in a cell-attached patch of about 0.5 μm diameter. Assuming that cell lysis occurs after a 3% increase in real area,[69,70] the K_A for the whole cell can be computed by inserting the above values into Equation 9:

$$K_A = \frac{0.5 \cdot 10^{-4}\,\text{cm} \cdot 30 \cdot 980.6\ \text{dyne cm}^{-2}}{4 \cdot 0.03} = 12.2\ \text{dyne/cm} \quad \textbf{(14)}$$

This elasticity coefficient can then be used to estimate the membrane tension, T_c, that would occur in an "unclamped" cell that was subjected to a 1% increase in cell volume. Combining Equations 12, 13, and 14, the membrane tension produced by the increase in cell volume is

$$T_c = 12.2\ \text{dyne/cm} \cdot 0.007 = 0.09\ \text{dyne/cm} \quad \textbf{(15)}$$

It turns out that this tension is sufficient to activate at least some classes of SA channels. For example, SAK channels at the basolateral membrane of amphibian proximal tubule can be activated by 6 cm H_2O negative pressure in patches of 0.5 μm diameter. Applying Laplace's law to a hemispherical patch, 6 cm H_2O negative pressure is equivalent to a cell-attached patch tension of:

$$T = \frac{P \cdot d}{4} = \frac{6 \cdot 980\ \text{dyne/cm}^{-2} \cdot 0.05 \cdot 10^{-4}\,\text{cm}}{4} = 0.07\ \text{dyne/cm} \quad \textbf{(16)}$$

Hence, the cell membrane tension (T_c) of 0.09 dyne/cm that is generated by only a 1% increase in cell volume (Equation 15) could be sufficient to turn on SA channels in this preparation.

If this is the case, why is it usually necessary to swell cells by about 50% in order to detect significant increases in P_o? This apparent paradox arises from the mismatch between the diameter of the cell-attached patch and the diameter of the cell, and is illustrated schematically in Figure 13–10. If it were possible to form gigaohm seals on half the cell (Pipette B, Figure 13–10), a 1% increase in cell volume *would* produce sufficient tension in this "super patch" to activate mechanosensitive channels.

In practice, pipette tips larger than 1 μm fail to form stable high-resistance seals in most tissues. Hence, channel currents are usually recorded from patches of relatively small area. Under these conditions, the mechanical constraints at the tight seal reduce the tension in the cell-attached patch (T_p) in proportion to the ratio of patch to cell diameter (d_p/d_c). This result, expressed by Equation 17, follows directly from Laplace's law and the requirement that both the unclamped region of the cell and the section of membrane within the pipette must experience the same hydrostatic pressure (P).

$$P = \frac{T_c}{d_c} = \frac{T_p}{d_p} \quad \textbf{(17)}$$

Combining Equations 12 and 17, the actual tension that a cell-attached patch experiences from a 60% increase in relative volume, $\Delta V/V$, is given by Equation 18:

$$T_p = \frac{d_p}{d_c} K_A \cdot \left[\left(1 + \frac{\Delta V}{V}\right)^{2/3} - 1\right] = \frac{0.5}{30} \cdot 12.2\ \text{dyne/cm}\left[(1+0.6)^{2/3} - 1\right] = 0.08\ \text{dyne/cm} \quad \textbf{(18)}$$

In other words, a 60% increase in relative cell area is required to produce sufficient tension (>0.07

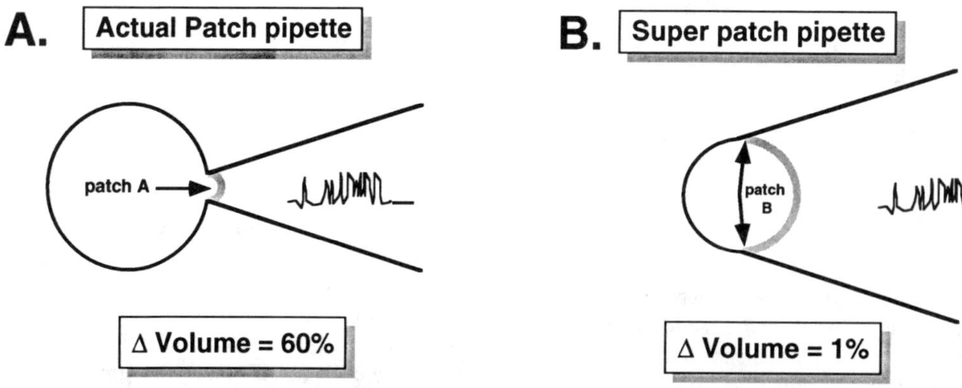

Figure 13–10 Cell-attached patches are mechanically as well as electrically clamped. (A) Small patch pipettes require a large change in volume to produce sufficient tension in a cell-attached patch to activate mechanosensitive channels. (B) If the cell-attached patch were equal to the diameter of the cell, SA channel activity could be recorded with relatively small changes in cell volume.

dyne/cm) in a cell-attached patch to turn on SA channels to the same extent as a direct application of 6 cm H_2O suction (see Equation 16).

3. Nonideal Membrane Mechanics

Although the above calculations are instructive, it is a considerable simplification to assume that the shape of the membrane at the tip of the patch pipette is an ideal hemisphere. The spherical portion of the patch often occurs quite far from the pipette tip so that the patch assumes an "omega shape" with a long straight section of membrane in close contact with the glass wall of the pipette.[71] Presumably, this straight section is the site of the gigaohm seal. Light microscopic observations,[71] as well as electron microscopy,[72] reveal cytoplasmic and cytoskeletal material within the patch pipette (Figure 13–11). This material often persists even after the patch is excised.

In some cases, suction completely distorted the shape of the membrane; whereas in other cases, suction simply decreased the radius of curvature of the domed portion of membrane spanning the inside of the pipette, while the straight section of membrane, attached to the glass, remained unchanged.[73] Hence, some of the basic concepts of ideal hemispherical patches considered in the previous section could still be applied to the spherical portion of membrane spanning the pipette. However, the presence of a viscous cytoplasm within the pipette and different estimates of K_A resulting from the omega shape of the patch, could alter some of the predictions about how cell swelling affects membrane tension. Furthermore, the observation that the patches of membrane held in long-tipped pipettes are stretched flat at zero applied pressure[73] would mean that unstretched patches have a radius of curvature similar to that of the cell. In this case, both the cell and the patch could experience the same tensions during osmotic stress. Consequently, in certain cases, small changes in cell volume could produce measurable channel activity in cell-attached patches.

Difficulties in reproducing the stretch sensitivity of a particular channel type may be due partially to an inability to control patch geometry. Variability in the "omega" shape of a patch makes it difficult to reproduce specific membrane tensions with the same pipette pressures.

High-resolution video microscopy indicates that pipette suction proportionately increases both apparent patch area and patch capacitance so that the *specific capacitance* of the patch remains relatively unchanged.[71] Hence, patch area is increased at nearly constant membrane thickness. Consequently, the increase in dome area (Figure 13–11) does not arise from membrane thinning but from an influx of lipid recruited from along the walls of the patch pipette. Release of suction causes retraction of the patch, probably arising from elastic restoring forces of the underlying cytoskeleton.

Despite movement of lipid along the wall into the dome region, the membrane itself remains rigidly attached to the glass in both excised and cell-attached patches.[71] This suggests that the membrane-glass attachment involves specific proteins that bind to the glass, while surface glycoproteins provide a high-resistance "sucrose gap" between the outer membrane leaflet and the glass pipette. In

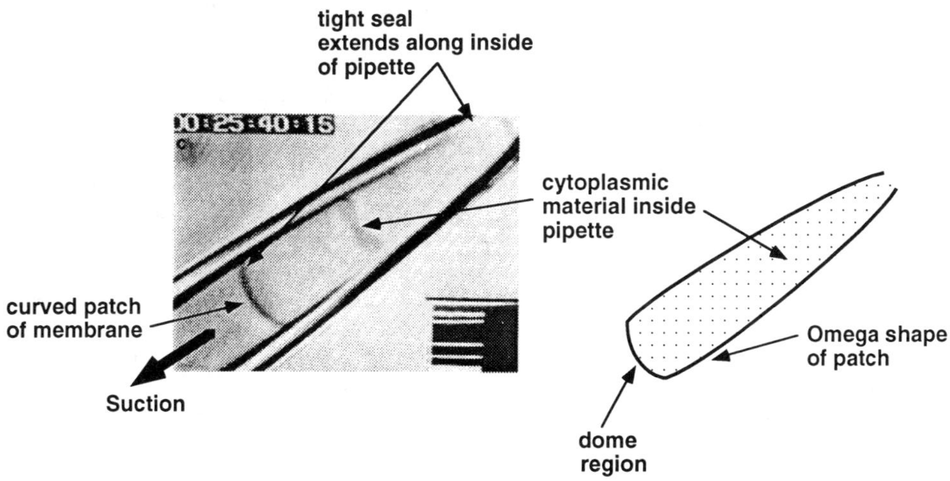

Figure 13–11 Cell-attached patch on cultured chick skeletal muscle cell. Applied pressure on the patch was −10 mmHg. Note that membrane curvature (domed region) can occur up to 10 μm from the tip of the pipette. (From Sokabe, M. et al., *Biophys. J.,* 59, 722, 1991. With permission.)

such a model, suction could produce real increases in patch area, via lipid flow, without pulling new channel proteins into the patch.

4. Lipid-Soluble Mediators and Membrane Deformation

The process of stretch activation may mimic normally occurring physiological regulators. For example, endogenous fatty acids seem to augment the open probability of certain SA channels in a fashion similar to membrane stretch. In Ehrlich ascites tumor cells, arachidonic acid metabolites, such as the leukotrienes, activate a K conductance pathway that seems to be essential for normal volume regulation.[74] It is also becoming apparent that fatty acids can activate channels directly by processes that do not depend on the arachidonic acid pathways. For example, recent studies on vascular smooth muscle cells have demonstrated a large-conductance K channel that is independently regulated by four types of stimuli: calcium, voltage, stretch, and free fatty acids.[50] The demonstration of mechanosensitive channels with multiple agonists reinforces the possibility that membrane deformation might be an important clue to a more general process of channel gating. Dual activation by both free fatty acids and stretch was also observed for a lower-conductance K channel, isolated from smooth muscle cells of toad stomach.[17,75,76]

One hypothesis for fatty acid mediation of mechanotransduction is that physical deformation of the patch activates membrane-bound phospholipases. These release fatty acids from the bilayer, which could then directly gate the channel protein.[77] The mechanism may involve binding of the fatty acid to the channel via a process similar to fatty acid activation of purified protein kinase C.[78–80] It is equally possible that endogenous fatty acids, released by stretch or cell swelling, indirectly affect P_o by altering either membrane fluidity or the lipid environment of the channel.[77]

A clue to the process of fatty acid channel activation may be the finding that lipid-soluble amphipaths activate mechanosensitive channels of *E. coli,* by a mechanism that appears to alter the curvature of the bilayer.[52] Following a model originally proposed by Sheetz and Singer to explain drug-erythrocyte interactions,[81] Martinac et al.[52] suggested that amphipathic compounds (molecules having both hydrophobic and hydrophilic characteristics) would preferentially insert into the inner or outer leaflet of the bilayer. This would expand one monolayer relative to the other and produce a microscopic bending of the bilayer that transmits tension to mechanosensitive receptors on the channel protein or cytoskeleton (Figure 13–12). In this model, hydrophobic *cations* would insert into the relatively negatively charged inner leaflet, causing a "cup-shaped" bending (as seen in red cells, Figure 13–12). Conversely, hydrophobic *anions* would insert into the less negative outer leaflet, expanding it and crenating a local region of the cell membrane (also seen in red cells, Figure 13–12).

Interestingly, hydrophobic cations and anions were equally effective at increasing the P_o of mechanosensitive channels.[52] In the simplest model (Figure 13–12), amphipathic increases in bilayer tension could distort channel proteins in a way that increases their open probability. Alternatively, the changes in bilayer curvature could be communicated to the cytoskeletal network, which may then gate the channel. Both these mechanisms are consistent with the experimental observation that most SA channels can be activated by both positive and negative pipette pressures.

Hence we are left with two alternative theories to explain fatty acid activation of mechanosensitive channels. In the first model, the primary process is fatty acid activation of the channel. Pipette suction, swelling, and membrane stretch all act to release endogenous fatty acids from the bilayer, which then move to their active site on the channel. In the second (or bilayer couple) model, a microscopic change in membrane tension is the primary process that gates the channel. Exogenous free fatty acids mimic stretch by preferentially inserting into the inner or outer leaflet of the bilayer, thereby mechanically deforming the membrane. Although this is an intriguing hypothesis, the time course of fatty acid activation of SA channels in gastric smooth muscle cells[75] and pulmonary vascular smooth muscle[50] is much faster than what is seen with amphipathic channel activators in *E. coli*.[52] This implies that the bilayer-couple model may not be a universal mechanism for the stretch activation process.

5. The Cytoskeleton and Stretch Activation

Given the intimate association between channels and the cytoskeleton,[82,83] it would not be too surprising if suction forces were coupled to SA channels via the cytoskeletal network. Early studies by Guharay and Sachs found dramatic increases in the stretch sensitivity (θ) of mechanosensitive channels after treatment with a class of cytochalasins known to disrupt actin filaments.[2] At the time, these results were interpreted as indicating a network of cytoskeletal strands funneling membrane strain energy into widely spaced nodal points (Figure 13–13).[4] The increased sensitivity to stretch after cytochalasin would then be consistent with disruption of the elastic elements in parallel with the channel, since this would increase the effective stress on the individual channel proteins.

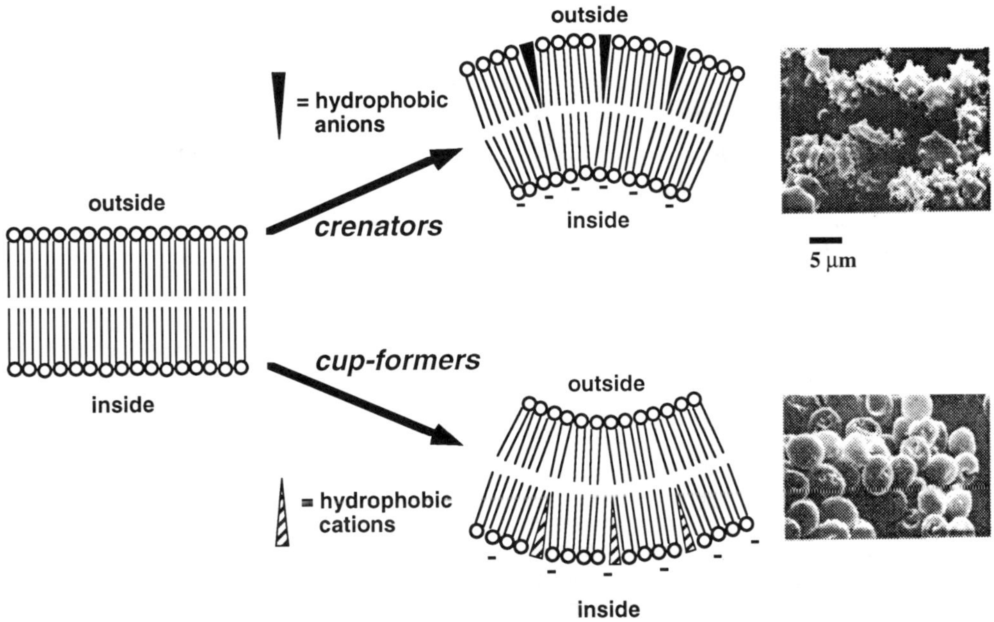

Figure 13–12 The bilayer couple model. The binding of amphipathic cations into the more negatively charged inner leaflet causes it to expand relative to the extracellular side, thereby bending the bilayer and locally distorting the membrane into a cup shape. Conversely, amphipathic anions would bind to the less negatively charged outer leaflet, expanding it relative to the inside monolayer and crenating local regions of the cell. (From Sheetz, M. P. and Singer, S. J., *Proc. Natl. Acad. Sci. U.S.A.*, 71, 4457, 1974. With permission.)

More recent studies in which elastic constants (K_A) of individual membrane patches were determined from direct video imaging indicated no effect of cytochalasins on K_A, and agents that disrupted tubulin and actin did not block SA channel activity.[71] This finding does not completely exclude the model of Figure 13–13, but it does imply that the elastic elements are not actin based, although they could consist of spectrin. Alternatively, it may be that the model of Figure 13–13 is too simple a representation of how the cytoskeletal network actually interacts with membrane channels.

The proximity of an actin-based cytoskeleton to Na channels in A6 cells[84] suggests that the cytoskeletal network may be essential for the proper regulation of at least some types of channels. This structural data was reinforced by observations that anything which increased the relative amount of short actin filaments (cytochalasin D, exogenous actin + ATP, or gelsolin) increased Na channel open probability in A6 cells.[84] The relationship between SA channels and cytoskeleton is indirectly strengthened by work with *E. coli* mutants lacking Braun's lipoprotein, which normally links the bacterial cell wall with the stretch-sensitive membrane.[85] Bacteria lacking this lipoprotein exhibit major alterations in their mechanosensitive channels. The possibility of producing selective mutations in cytoskeletal elements of *E. coli* makes the study of bacterial SA channels a very promising avenue of exploration.

There are well-known links between the cytoskeleton and volume regulation in a variety of tissues. Disruption of cytoskeletal elements by cytochalasin B inhibits volume regulation in *Necturus* gallbladder,[86] rabbit proximal tubule,[31] PC12 cells,[87] and Ehrlich ascites tumor cells.[88] This establishes a tentative link between the cytoskeleton and membrane ion channels, at least in those cases where hypotonic volume regulation is known to depend on electrolyte efflux through channels. In more recent experiments, cytochalasin B not only disrupted volume regulation but also activated mechanosensitive channels and decreased cellular content of K and Cl.[7] One possibility is that microfilaments are somehow involved in sensing changes in cell volume and transmitting this information to specific ion channels. Disruption of this network could have two simultaneous effects: (1) a loss of ability to transmit volume or stretch information to the ion channels, and (2) an increase in stretch sensitivity to the point where the channel spends much more time in the open state.

Another line of evidence supporting cytoskeletal-channel interactions comes from ultrastructural observations of PC12 cells after hyposmotic swell-

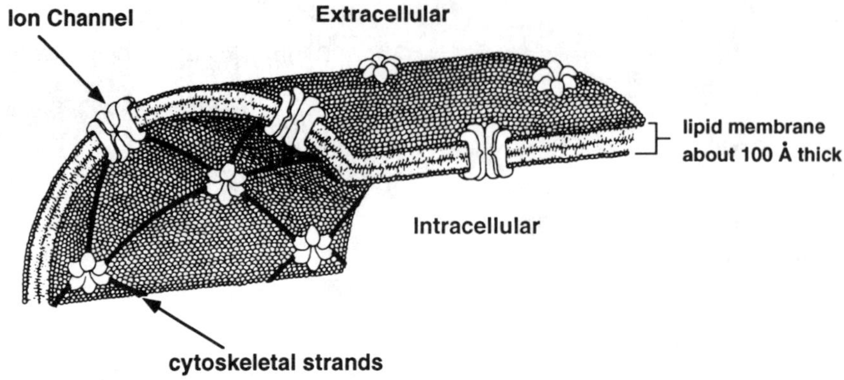

Figure 13–13 Diagram of the hypothetical linkage between ion channels and submembranous cytoskeleton. (From Sachs, F., *Fed. Proc.,* 46, 12, 1987. With permission.)

ing.[87] During hyposmotic shock, homogeneously distributed microfilaments (but not microtubules) become concentrated around the nucleus, with radial extensions to the cell periphery.[87] These cytoskeletal changes take place on a time scale coincident with the time required for RVD in normal PC12 cells.

In summary, the association between SA channels and the cytoskeleton is intriguing, but unclear at this time. Disruption of cytoskeletal elements by cytochalasins has prevented volume regulation in some, but not all, of the tissues in which it was examined. Morphological changes in the cytoskeleton during swelling and volume regulation suggest that these elements may function as volume transducers, but experiments with fixed cells cannot resolve whether cytoskeletal changes precede or follow volume regulation.

IV. DO MECHANOSENSITIVE CHANNELS MEDIATE CELLULAR CURRENTS?

A. SA-cat, SA-non CHANNELS AND MACROSCOPIC CURRENTS

It has been proposed recently that the phenomenon of stretch activation may be dependent on the manner in which the high-resistance gigaohm seal is formed.[89] This hypothesis arose from experiments that showed no correlation between activation of single mechanosensitive channels and changes in macroscopic currents.[89] Morris and Horn[89] generalized this hypothesis to suggest that the process of gigaseal formation artificially sensitizes channels to mechanical stimuli. Although this issue is still in dispute, it has provided an important stimulus for reexamining mechanosensitive phenomena using a variety of tools.

A number of investigators have begun to compare single-channel and whole cell currents in the same preparation. Some of these results constitute important counter-examples to the hypothesis of Morris and Horn. In one case, Gustin et al.[90] reported that inflating yeast spheroblasts with positive pressure elicited macroscopic currents of a magnitude that was consistent with the expected total current from individual SA channels in these spheroblasts. Furthermore, the ion selectivity and sensitivity of these channels to inhibitors, such as gadolinium, were identical for individual single channels and macroscopic currents. This seems to clearly establish (at least for yeast) that SA ion channels underlie the whole cell currents produced by a mechanical deformation.

Experiments in a number of other preparations also support the notion that SA channels form the basis for whole cell currents. Stretched smooth muscle cells exhibited a net inward current that could be attributed to microscopic SA channels.[91] In liver cells, hyposmotic stress activated Ca-permeable, SA channels on the same time scale as observed increases in cytosolic calcium.[92] Hyposmotically stressed intestinal cells (I407 line) manifested large gadolinium-sensitive inward currents that were correlated with Ca influx,[93] as did *Xenopus* myocytes.[94]

In nystatin-permeabilized,[95] whole cell recordings from isolated opossum kidney cells, stepwise reductions in bath osmolarity simultaneously increased both cell size, inward current, and total conductance.[96] However, these whole cell currents were not completely blocked by gadolinium, although the cells possessed SA channels that were readily activated by either pipette suction or hyposmotic stress.[7] Since the opossum SA channels are different from most SA-cat channels (by virtue of their anion permeability), they also may not be directly blockable by gadolinium.

Alternatively, hyposmotic swelling may have more than simply a mechanical effect on SA channels. For example, there is some evidence that swelling alters the cytoskeleton in a way that activates a membrane-bound phospholipase C. The IP_3 produced from this process would mobilize Ca from intracellular stores, and the diacylglycerol produced would elevate protein kinase C, which could, in turn, stimulate specific ion channels.[56] Diacylglycerol could also function as a precursor in arachidonic acid metabolism. Some of the end products of this pathway, such as the leukotrienes, may also gate specific K or Cl channels essential for RVD.

B. SAK CHANNELS AND MACROSCOPIC CURRENTS

The absence of highly specific blockers for the different classes of SA channels makes it difficult to attribute macroscopic currents to SA channels with absolute certainty. The experiments by Gustin et al. may be the most convincing (to date) that a direct relationship exists between mechanosensitive channels and volume- or stress-induced macroscopic currents.[90] In a somewhat different approach, we have sought to circumvent the absence of specific blockers for SAK channels by working with the amphibian proximal tubule, where all of the demonstrable K conductance seems to be mechanosensitive. In this preparation, a relatively unspecific blocker like barium can be used to assess the amount of macroscopic current directly attributable to potassium channels.

In these experiments, isolated frog proximal tubule cells were voltage clamped and the effects of glucose (GLU) and phenylalanine (PHE) on cell volume and whole cell current were examined. Consistent with previous studies, both GLU and PHE were cotransported with Na into proximal tubule cells. The principal new finding was that cotransport of GLU and PHE increased both cell volume and barium-sensitive current on about the same time scale.[97] In 11 conventional whole cell clamps, isosmotic addition of 40 m*M* glucose to the bathing solution increased cell volume by 23 ± 4% and increased barium-sensitive (i.e., K) conductance by 40 ± 10%. Similar results were obtained for isosmotic addition of PHE, and for addition of cotransported, but nonmetabolized, glucose analogues.

Figure 13–14 illustrates the current voltage relations for one of the glucose experiments.[97] Only the segment of the curve in the range of negative cell potential (left of the ordinate) is meaningful because of voltage-dependent barium block at positive cell potentials (bending over of the curves to the right of the ordinate). The linear portion of the current-voltage relation depicts both inward and outward currents in the range of physiological membrane potentials. Although the glucose-induced increase in barium-sensitive conductance was relatively modest (33% in this case), each cell was used as its own control and the differences were highly significant ($P < .005$). The swelling-associated change in conductance also occurred without significant shifts in the barium-sensitive reversal potential, indicating that addition of GLU and PHE does not alter channel selectivity.

A priori, the temporal association between cell swelling and increased barium-sensitive conductance does not prove that the increase in conductance is caused by mechanosensitive channels.

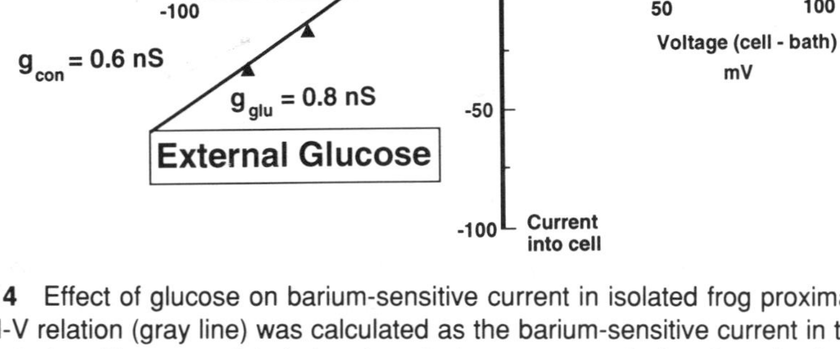

Figure 13–14 Effect of glucose on barium-sensitive current in isolated frog proximal tubule cells. The *control* I-V relation (gray line) was calculated as the barium-sensitive current in the absence of external substrates. The curve labeled *external glucose* was calculated as the barium-sensitive current after isosmotic addition of 40 m*M* external glucose. Each cell served as its own control. The terms g_{con} and g_{glu} refer, respectively, to the linear portions of the barium-sensitive current in the absence and presence of glucose. In all cases, the pipette contained high K Ringer.

However, one can propose the following argument. Since all K channels observed in amphibian proximal tubule cells are stretch-activated and barium sensitive, all barium-sensitive macroscopic currents can be attributed to SAKs. Use of the conventional whole cell technique in these experiments, rather than the permeabilized patch procedure,[95] effectively washes out diffusable intracellular mediators that might link substrate transport to increased K conductance (Figure 13–15). Therefore, the substrate-induced increase in G_K probably arises from a swelling-associated deformation of the submembrane cytoskeleton or a direct change in membrane tension. In this regard mechanosensitive K channels would be the microscopic correlates of the macroscopic change in G_K.

In some of the experiments, cells were specifically loaded with high levels of ATP to exclude this substance as a possible mediator. Therefore, the observed increase in macroscopic barium-sensitive (K) conductance must be mediated by a "membrane-linked" mechanism that does not depend on subtle changes in cytosolic pH or intracellular metabolites (Figure 13–15). This mechanism could involve changes in P_o produced by the physical forces of membrane deformation[66,68] (see Equation 6), or by subtle changes in the cytoskeleton that communicates cell size information to specific ion channels. There is ample evidence that even excised patches contain viable cytoskeletal elements. In any case, the above results strongly suggest that SAKs underlie the increases in macroscopic K conductance that accompany GLU and PHE cotransport. This does not exclude other parallel processes that may also increase K conductance in a way that does not depend on SA channels. The whole cell experiments described above were designed to maximize cell-pipette equilibration and, therefore, provide no information about chemically mediated conductance signals.

C. CONCLUDING REMARKS

The Morris and Horn hypothesis[89] that the patch-clamp technique hypersensitizes channels to stretch raises a more general concern; namely, is it ever possible to design a totally noninvasive experiment? Intracellular electrodes, cell-attached patches, excised patches, whole cell clamps, nystatin-permeabilized whole cell clamps, all perturb the cell to some degree from its normal state; not to mention the very process of cell isolation that traumatically wrenches a cell away from its neighbors. Clearly, gigaseal formation is not without its hazards to the cell. In video-micrographs, the formation of a cell-attached patch is often accompanied by much initial distortion of the membrane and possible modification of the cytoskeleton.[71,73]

There are already indications that the amount of pipette suction required to form the initial seal can affect the kinetic properties of the channels being

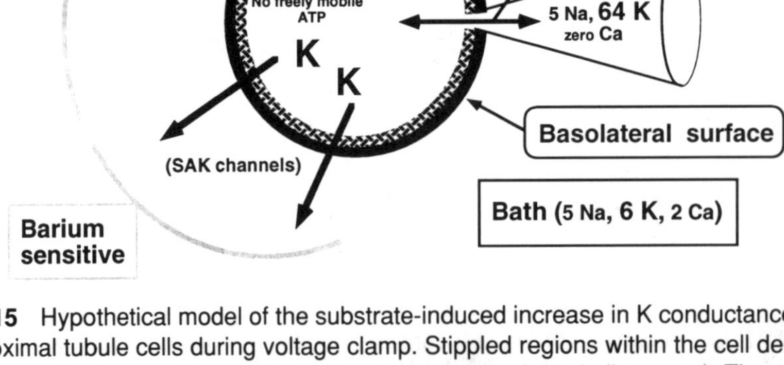

Figure 13–15 Hypothetical model of the substrate-induced increase in K conductance observed in isolated, proximal tubule cells during voltage clamp. Stippled regions within the cell denote possible unstirred layers where osmolarity transiently exceeds that of the bulk cytosol. The cross-hatched region denotes a submembranous cytoskeletal network that may be altered by cell swelling. SA-cat refers to stretch-activated, cation-selective channels found at the apical membrane. Maxi-K refers to Ca-dependent, high-conductance K channels at the apical surface. SAK refers to stretch-activated K channels.

studied. Certain channels, which appear to be linked to the cytoskeletal network, exhibit adaptive responses to suction that are only seen with "gentle" sealing.[98] Seals formed with more vigorous suction abolish the adaptation and possibly increase the stretch sensitivity of the channel. Hence, it is conceivable that SA currents are not always seen in the whole cell mode because these currents would adapt (shut down) quickly, whereas cell-attached patches formed with enough suction to destroy adaptation would remain sensitive to membrane tension.

Ultimately, one is left with a sort of "biological uncertainty principle" where the process of measurement itself disturbs the system (even when dealing on a scale much larger than the wavelength of light). It may be inevitable that the method of forming gigaseals, which are necessary to observe single channels, increases the stretch sensitivity of those channels. Even if this turns out to be the case, the pessimism of Morris and Horn is unjustified.[89] Much useful information and insight can still be gleaned from discovering how the patching process itself alters channel gating. As more data become available on mechanosensitive whole cell currents and more selective blockers are utilized, the function of stretch-activated channels will undoubtedly be clarified.

ACKNOWLEDGMENTS

The author gratefully acknowledges the assistance and advice of Professor Larry G. Palmer (Cornell University Medical College) and Professor Fred Sachs (Department of Biophysics, State University of New York, Buffalo). This work was supported in part by a grant from the New York Affiliate of the American Heart Association.

REFERENCES

1. **Schultz, S. G.,** Homocellular regulatory mechanisms in sodium-transporting epithelia: avoidance of extinction by "flush-through", *Am. J. Physiol.,* 241, F579, 1981.
2. **Guharay, F. and Sachs, F.,** Stretch-activated single ion channel currents in tissue-cultured embryonic chick skeletal muscle, *J. Physiol. (London),* 352, 685, 1984.
3. **Sachs, F.,** Biophysics of mechanoreception, *Membr. Biochem.,* 6, 173, 1986.
4. **Sachs, F.,** Baroreceptor mechanisms at the cellular level, *Fed. Proc.,* 46, 12, 1987.
5. **Morris, C. E.,** Mechanosensitive ion channels, *J. Membr. Biol.,* 113, 93, 1990.
6. **Sachs, F.,** Ion channels as mechanical transducers, in *Cell Shape: Determinants, Regulation and Regulatory Role,* Bonner, F. and Stein, W., Eds., Academic Press, New York, 1989.
7. **Hoffmann, E. K. and Kolb, H.-A.,** Mechanisms of activation of regulatory volume responses after cell swelling, in *Comparative and Environmental Physiology,* Gilles, R., Hoffmann, E. K., and Bolis, L., Eds., Springer-Verlag, Berlin, 1991.
8. **Brehm, P., Kullberg, R., and Moody-Corbett, F.,** Properties of non-junctional acetylcholine receptor channels in innervated muscle of *Xenopus* larvae, *J. Physiol.,* 350, 631, 1984.
9. **Christensen, O.,** Mediation of cell volume regulation by Ca influx through stretch-activated channels, *Nature (London),* 330, 66, 1987.
10. **Cooper, K. E., Tang, J. M., Rae, J. L., and Eisenberg, R. S.,** A cation channel in frog lens epithelia responsive to pressure and calcium, *J. Membr. Biol.,* 93, 259, 1986.
11. **Falke, L. C. and Misler, S.,** Activity of ion channels during volume regulation by clonal N1E115 neuroblastoma cells, *Proc. Natl. Acad. Sci. U.S.A.,* 86, 3919, 1989.
12. **Duncan, R. and Misler, S.,** Voltage-activated and stretch-activated Ba conducting channels in an osteoblast-like cell line (umR 106), *FEBS Lett.,* 251, 17, 1989.
13. **Christensen, O. and Hoffmann, E. K.,** Cell swelling activates K and Cl channels as well as non-selective stretch-activated cation channels in ehrlich ascites tumor cells, *J. Membr. Biol.,* 129, 13, 1992.
14. **Ubl, J., Murer, H., and Kolb, H. A.,** Ion channels activated by osmotic and mechanical stress in membranes of opossum kidney cells, *J. Membr. Biol.,* 104, 223, 1988.
15. **Yang, X. and Sachs, F.,** Block of stretch-activated ion channels in *Xenopus* oocytes by gadolinium cna calcium ions, *Science,* 243, 1068, 1989.
16. **Lansman, J. B.,** Single stretch-activated ion channels in vascular endothelial cells as mechanotransducers, *Nature,* 325, 811, 1987.
17. **Kirber, M. T., Walsh, J. V., and Singer, J. J.,** Stretch-activated ion channels in smooth muscle: a mechanism for the initiation of stretch-induced contraction, *Pfluegers Arch.,* 412, 339, 1988.
18. **Hurst, A. M. and Hunter, M.,** Stretch-activated channels in single early distal tubule cells of the frog, *J. Physiol. (London),* 430, 13, 1990.
19. **Hunter, M.,** Stretch-activated channels in the basolateral membrane of single proximal cells of frog kidney, *Pfluegers Arch.,* 416, 448, 1990.

20. **Filipovic, D. and Sackin, H.,** A calcium-permeable stretch-activated cation channel in renal proximal tubule, *Am. J. Physiol.,* 260, F119, 1991.
21. **Lansman, J. B.,** Blockage of current through single calcium channels by trivalent lanthanide cations. Effect of ionic radius on the rates of ion entry and exit, *J. Gen. Physiol.,* 95, 679, 1990.
22. **Ubl, J., Murer, H., and Kolb, H.-A.,** Hypotonic shock evokes opening of Ca-activated K channels in opossum kidney cells, *Pfluegers Arch.,* 412, 551, 1988.
23. **Taniguchi, J. and Guggino, W. B.,** Membrane stretch: a physiological stimulator of Ca-activated K channels in thick ascending limb, *Am. J. Physiol.,* 257, F347, 1989.
24. **Montrose-Rafizadeh, C. and Guggino, W. B.,** Role of intracellular calcium in volume regulation by medullary thick ascending limb cells, *Am. J. Physiol.,* 260, 402, 1991.
25. **Beck, J. S., Breton, S., Laprade, R., and Giebisch, G.,** Volume regulation and intracellular calcium in the rabbit proximal convoluted tubule, *Am. J. Physiol.,* 260, F861, 1991.
26. **Okada, Y. and Hazama, A.,** Volume-regulatory ion channels in epithelial cells, *NIPS,* 4, 238, 1989.
27. **Hazama, A. and Okada, Y.,** Ca sensitivity of volume-regulatory K and Cl channels, *J. Physiol.,* 402, 687, 1988.
28. **Wong, S. M., DaBell, M. C., and Chase, H., Jr.,** Cell swelling increases intracellular free [Ca] in cultured toad bladder cells, *Am. J. Physiol.,* 258, F292, 1990.
29. **Sigurdson, W. S., Ruknudin, A., and Sachs, F.,** Calcium imaging of mechanically induced fluxes in tissue-cultured chick heart: role of stretch-activated ion channels, *Am. J. Physiol.,* 262, H1110, 1992.
30. **Linshaw, M. A. and Grantham, J. J.,** Effect of collagenase and ouabain on renal cell volume in hypotonic media, *Am. J. Physiol.,* 238, F491, 1980.
31. **Linshaw, M., Fogel, C. A., Downey, G. P., Koo, E. W. Y., and Gotlieb, A. I.,** Role of cytoskeleton in volume regulation of rabbit proximal tubule in dilute medium, *Am. J. Physiol.,* 262, F144, 1992.
32. **Zhou, X., Stumpf, M., Hoch, H., and Kung, C.,** A mechanosensitive channel in whole cells and in membrane patches of the fungus *Uromyces, Science,* 253, 1415, 1991.
33. **Suzuki, M., Kawahara, K., Ogawa, A., Morita, T., Kawaguchi, Y., Kurihara, S., and Sakai, O.,** Ca rises via G protein during regulatory volume decrease in rabbit proximal tubule cells, *Am. J. Physiol.,* 258, F690, 1990.
34. **Kawahara, K., Ogawa, A., and Suzuki, M.,** Hyposmotic activation of Ca-activated K channels in cultured rabbit kidney proximal tubule cells, *Am. J. Physiol.,* 260, F27, 1991.
35. **Harsdorf, R. V., Lang, R., Fullerton, M., Smith, A., and Woodcock, E. A.,** Right atrial dilation increases inositol-(1,4,5) triphosphate accumulation, *FEBS Lett.,* 233, 201, 1988.
36. **Ling, B., Webster, C., and Eaton, D.,** Eicosanoids modulate apical Ca-dependent K channels in cultured rabbit principal cells, *Am. J. Physiol.,* 263, F116, 1992.
37. **Watson, P.,** Function follows form: generation of intracellular signals by cell deformation, *FASEB J.,* 5, 2013, 1991.
38. **Strange, K.,** Volume regulation following Na pump inhibition in CCT principal cells: apical K loss, *Am. J. Physiol.,* 258, F732, 1990.
39. **Sigurdson, W. J., Morris, C. E., Brezden, B. L., and Gardner, D. R.,** Stretch activation of a K channel in molluscan heart cells, *J. Exp. Biol.,* 127, 191, 1987.
40. **Welling, P. A., Linshaw, M. A., and Sullivan, L. W.,** Effect of barium on cell volume regulation in rabbit proximal straight tubules, *Am. J. Physiol.,* 249, F20, 1985.
41. **Macleod, R. J. and Hamilton, J. R.,** Separate K and Cl transport pathways are activated for regulatory volume decrease in jejunal villus cells, *Am. J. Physiol.,* 260, G405, 1991.
42. **Welling, P. A. and Linshaw, M. A.,** Importance of anion in hypotonic volume regulation of rabbit proximal straight tubule, *Am. J. Physiol.,* 255, F853, 1988.
43. **Welling, P. A. and O'Neil, R. G.,** Cell swelling activates basolateral Cl and K conductances in rabbit proximal tubule, *Am. J. Physiol.,* 258, F951, 1990.
44. **Sackin, H.,** Stretch-activated potassium channels in renal proximal tubule, *Am. J. Physiol.,* 253, F1253, 1987.
45. **Filipovic, D. and Sackin, H.,** Stretch and volume activated channels in isolated proximal tubule cells, *Am. J. Physiol.,* 262, F857, 1992.
46. **Kawahara, K.,** A stretch-activated K channel in the basolateral membrane of *Xenopus* kidney proximal tubule cells, *Pfluegers Arch.,* 415, 624, 1990.
47. **Sackin, H.,** A stretch-activated potassium channel sensitive to cell volume, *Proc. Natl. Acad. Sci. U.S.A.,* 86, 1731, 1989.
48. **Pacha, J., Frindt, G., Sackin, H., and Palmer, L.,** Apical maxi K channels in intercalated cells of CCT, *Am. J. Physiol.,* 261, F696, 1991.

49. **Stoner, L. and Morley, G.,** Effect of basolateral hypotonicity on high conductance potassium channels in the apical membrane of everted rat cortical collecting tubule, *Am. J. Physiol.,* submitted.
50. **Kirber, M. T., Ordway, R. W., Clapp, L. H., Jr., J. V. W. and Singer, J. J.,** Both membrane stretch and fatty acids directly activate large conductance Ca-activated K channels in vascular smooth muscle cells, *FEBS Lett.,* 297, 24, 1991.
51. **Martinac, B., Buechner, M., Delcour, A., Adler, J., and Kung, C.,** Pressure-sensitive ion channel in *Escherichia coli, Proc. Natl. Acad. Sci. U.S.A.,* 84, 2297, 1987.
52. **Martinac, B., Adler, J., and Kung, C.,** Mechanosensitive ion channels of *E. coli* activated by amphipaths, *Nature,* 348, 261, 1990.
53. **Falke, L. C., Edwards, K., Pickard, B., and Misler, S.,** A stretch-activated anion channel in tobacco protoplasts, *FEBS Lett.,* 237, 141, 1988.
54. **Stanton, B. A., Dietl, P., and Schwiebert, E.,** Cell volume regulation in the cortical collecting duct: stretch activated Cl channels, *J. Am. Soc. Nephrol.,* 1 (Abstr.), 692, 1990.
55. **Stanton, B. A., Mills, J. A., and Schwiebert, E. M.,** Role of the cytoskeleton in regulatory volume decrease in cortical collecting duct cells in culture, *J. Am. Soc. Nephrol.,* 2 (Abstr.), 751, 1991.
56. **Schwiebert, E. M., Karlson, K., Friedman, P. A., Dietl, P., Spielman, W. S., and Stanton, B.,** Adenosine regulates a chloride channel via protein kinase C and a G protein in a rabbit cortical collecting duct cell line, *J. Clin. Invest.,* 89, 834, 1992.
57. **Richter, E. A., Cleland, P. J., Rattigan, S., and Clark, M. G.,** Contraction-associated translocation of protein kinase C in rat skeletal muscle, *FEBS Lett.,* 217, 232, 1987.
58. **Hudson, R. L. and Schultz, S. G.,** Sodium-coupled glycine uptake by Ehrlich ascites tumor cells results in an increase in cell volume and plasma membrane channel activities, *Proc. Natl. Acad. Sci. U.S.A.,* 85, 279, 1988.
59. **Watson, P. A.,** Accumulation of cAMP and calcium in S49 mouse lymphoma cells following hyposmotic swelling, *J. Biol. Chem.,* 264, 14735, 1989.
60. **Krupinski, J., Coussen, F., Bakalyar, H., Tang, W., Feinstein, P. G., Orth, K., Slaughter, C., Reed, R., and Gilman, A.,** Adenylyl cyclase amino acid sequence: possible channel- or transport-like structure, *Science,* 244, 1558, 1989.
61. **Gustin, M. C., Zhou, X., Martinac, B., and Kung, C.,** A mechanosensative ion channel in the yeast plasma membrane, *Science,* 242, 762, 1988.
62. **Morris, C. E. and Sigurdson, W. J.,** Stretch-inactivated ion channels coexist with stretch-activated ion channels, *Science,* 243, 807, 1989.
63. **Lewis, S. A. and Clausen, C.,** Urinary proteases degrade epithelial sodium channels, *J. Membr. Biol.,* 122, 77, 1991.
64. **Colquhoun, D. and Hawkes, A. G.,** On the stochastic properties of single ion channels, *Proc. R. Soc. London Ser. B,* 211, 205, 1981.
65. **Erxleben, C.,** Stretch-activated current through single ion channels in the abdominal stretch receptor organ of the crayfish, *J. Gen. Physiol.,* 94, 1071, 1989.
66. **Sachs, F. and Lecar, H.,** Stochastic models for mechanical transduction, *Biophys. J.,* 59, 1143, 1991.
67. **Howard, J., Roberts, W. M., and Hudspeth, A. J.,** Mechanoelectrical transduction by hair cells, *Annu. Rev. Biophys. Chem.,* 17, 99, 1988.
68. **Sachs, F.,** Stretch-sensitive ion channels: an update, in *Sensory Transduction: Soc. of Gen. Physiol. 45th Annu. Symp.,* Corey, D. and Roper, S., Eds., Rockefeller University Press, New York, 1992.
69. **Kwok, R. and Evans, E.,** Thermoelasticity of large lecithin bilayer vesicles, *Biophys. J.,* 35, 637, 1981.
70. **Evans, E. A., Waugh, R., and Melnik, L.,** Elastic area compressibility modulus of red cell membrane, *Biophys. J.,* 16, 585, 1976.
71. **Sokabe, M., Sachs, F., and Jing, Z.,** Quantitative video microscopy of patch clamped membranes: stress, strain, capacitance, and stretch activation, *Biophys. J.,* 59, 722, 1991.
72. **Ruknudin, A., Song, M. J., and Sachs, F.,** The ultrastructure of patch-clamped membranes: a study using high voltage electron microscopy, *Biophys. J.,* 112, 125, 1989.
73. **Sokabe, M. and Sachs, F.,** The structure and dynamics of patch-clamped membranes: a study using differential interference contrast light microscopy, *J. Cell Biol.,* 111, 599, 1990.
74. **Lambert, I. H., Hoffmann, E. K., and Christensen, P.,** Role of prostoglandins and leukotrienes in volume regulation by Ehrlich ascites tumor cells, *J. Membr. Biol.,* 98, 247, 1987.
75. **Ordway, R. W., Walsh, J. V., and Singer, J. J.,** Arachidonic acid and other fatty acids directly activate potassium channels in smooth muscle cells, *Science,* 244, 1176, 1989.

76. **Ordway, R. W., Petrou, S., Kirber, M. T., Jr., J. V. W. and Singer, J. J.,** Two distinct mechanisms of ion channel activation by membrane stretch: evidence that endogenous fatty acids mediate stretch activation of K channels, *Biophys. J.,* 61, A391, 1992.
77. **Ordway, R. W., Singer, J. J., and Jr., J. V. W.** Direct regulation of ion channels by fatty acids, *TINS,* 14, 96, 1991.
78. **Seifert, R., Schächtele, C., Roenthal, W., and Schultz, G.,** Activation of protein kinase C by cis- and trans-fatty acids and its potentiation by diacylglycerol, *Biochem. Biophys. Res. Commun.,* 154, 20, 1988.
79. **McPhail, L. C. and Snyderman, R.,** A potential second messenger role for unsaturated fatty acids: activation of Ca-dependent protein kinase, *Science,* 224, 622, 1984.
80. **Morimoto, Y. M.,** Activation of protein kinase C by fatty acids and its dependency on Ca and phospholipid, *Cell Struct. Funct.,* 13, 45, 1988.
81. **Sheetz, M. P. and Singer, S. J.,** Biological membranes as bilayer couples. A molecular mechanism of drug-erythrocyte interactions, *Proc. Natl. Acad. Sci. U.S.A.,* 71, 4457, 1974.
82. **Branton, D.,** Membrane cytoskeletal interactions, *Cold Spring Harbor Symp. Quant. Biol.,* 46, 1, 1981.
83. **Bennett, V.,** The membrane skeleton of human erythrocytes and its implications for more complex cells, *Annu. Rev. Biochem.,* 54, 273, 1985.
84. **Cantiello, H. F., Stow, J. L., Prat, A. G., and Ausiello, D. A.,** Actin filaments regulate Na channel activity, *Am. J. Physiol.,* 261, C882, 1991.
85. **Kubalski, A., Martinac, B., Adler, J., and Kung, C.,** Altered properties of the mechanosensitive ion channel in a lipoprotein mutant of E. coli, *Biophys. J.,* 59, 455, 1991.
86. **Foskett, J. K. and Spring, K. R.,** Involvement of calcium and cytoskeleton in gallbladder epithelial cell volume regulation, *Am. J. Physiol.,* 248, C27, 1985.
87. **Cornet, M., Delpire, E., and Gilles, R.,** Study of microfilaments network during volume regulation process of cultured PC 12 cells, *Pfluegers Arch.,* 410, 223, 1987.
88. **Cornet, M., Lambert, I. H., and Hoffmann, E. K.,** Relation between cytoskeleton, hypoosmotic treatment and volume regulation in Ehrlich ascites tumor cells, *J. Membr. Biol.,* in press.
89. **Morris, C. E. and Horn, R.,** Failure to elicit neuronal macroscopic mechanosensitive currents anticipated by single-channel studies, *Science,* 251, 1246, 1991.
90. **Gustin, M. C., Sachs, F., Sigurdson, W., Ruknudin, A., Bowman, C., and Morris, C.,** Single-channel mechanosensitive currents, *Science,* 253, 800, 1991.
91. **Davis, M. J., Donovitz, J. A., and Hood, J. D.,** Stretch-activated single-channel and whole-cell currents in isolated vascular smooth muscle cells, *Biophys. J.,* 59, 236, 1991.
92. **Bear, C. E.,** A nonselective cation channel in rat liver cells is activated by membrane stretch, *Am. J. Physiol.,* 258, C421, 1990.
93. **Okada, Y., Hazma, A., and Yuan, W. L.,** Stretch-induced activation of Ca permeable ion channels is involved in the volume regulation of hypotonically swollen epithelial cells, *Neurosci. Res.,* 12, S5, 1990.
94. **Sachs, F.,** Mechanical transduction by membrane ion channels: a mini review, *Mol. Cell. Biochem.,* 104, 57, 1991.
95. **Horn, D.,** Muscarinic activation of ionic currents measured by a new whole-cell recording method, *J. Gen. Physiol.,* 92, 145, 1988.
96. **Ubl, J., Murer, H., and Kolb, H.-A.,** Simultaneous recording of cell volume, membrane current and membrane potential: effect of hypotonic shock, *Pfluegers Arch.,* 415, 381, 1989.
97. **Cemerikic, D. and Sackin, H.,** Substrate activation of mechanosensitive, whole-cell currents in renal proximal tubule, *Am. J. Physiol.,* 264, F697, 1993.
98. **Hamill, O. P. and McBride, J. D. W.,** Rapid adaptation of single mechanosensitive channels in *Xenopus* oocytes, *Proc. Natl. Acad. Sci. U.S.A.,* 89, 7462, 1992.

Chapter 14

The Cytoskeleton and Cell Volume Regulation

John W. Mills, Erik M. Schwiebert and Bruce A. Stanton

TABLE OF CONTENTS

I. INTRODUCTION

Most studies on cell volume regulation have focused on examining the recovery of cell volume after an osmotic or chemical perturbation. The majority of these investigations support the "pump-leak" hypothesis of volume regulation, which involves a balance between the entry of sodium (Na^+) down its electrochemical gradient and the energy-dependent extrusion of Na^+ by the Na^+,K^+-ATPase pump.[1,2] However, several studies have shown that other mechanisms operate as supplements to, or in parallel with, this main "pump-leak" control mechanism. These alternative mechanisms include tensile or contractile forces that act as resistive elements to volume changes,[3] bulk transport of fluid out of the cell via vesicle extrusion mechanisms,[4] gel-like properties of the cytoplasm that resist hydration and, therefore, swelling,[5] and activation of, and/or delivery of, ion transport proteins to the cell membrane during cell swelling or cell shrinkage.[6] For each of these possibilities, hypotheses have been presented to explain the role of these mechanisms in cell volume regulation. A unifying theme in each case is that elements of the cytoskeleton play an integral role in volume-regulatory processes within the cell. It is the intent of this chapter to present these models and discuss the data that support them, especially with regard to the cytoskeleton.

0-8493-4448-4/94/$0.00+$.50

II. THE CYTOSKELETON

The cytoskeleton is composed of three distinct filamentous or fibrous components: actin filaments, microtubules, and intermediate filaments. Extensive reviews of each component are listed here for reference.[7–15] However, there are certain aspects of the cytoskeleton that are pertinent to this discussion and will be reviewed here.

A. ACTIN FILAMENTS

Actin filaments (F-actin) are 5 to 7-nm-wide microfilaments that are a ubiquitous, structural component of eukaryotic cells.[9–11,14,15] The filaments are double-stranded, α-helical polymers composed of subunits of globular actin (G-actin). In non-muscle cells, G-actin monomers are in a constant state of flux, i.e., G-actin assembles onto and disassembles from the filament. The polymerization reaction is a two-stage process involving a nucleation phase in which G-actin monomers combine followed by an elongation phase in which monomers are added rapidly to the filament. ATP is required during the nucleation phase and a divalent cation is required for polymerization. Shortly after addition of G-actin to the filament, ATP is hydrolyzed.

In intact cells, the concentration of free ions and nucleotides is high enough to favor polymerization of actin. However, all actin is not polymerized due to the interaction of G-actin with profilin.[14] The sequestering of G-actin in a complex with profilin, which is sensitive to changes in phosphatidylinositol bisphosphate levels,[14,16–19] provides a rapidly available pool of G-actin monomers allowing actin filaments to grow or shrink in response to appropriate stimuli.

Although actin filaments are found throughout the cytoplasm, the majority of F-actin is concentrated in distinct domains within the cell. Most of the actin is in the cortical cytoplasm in close association with the cell membrane. Through an interaction with actin-binding proteins which interact with integral or peripheral membrane proteins, the actin filaments (and associated proteins) form a structural unit usually termed the "membrane-cytoskeleton".[20,21] The cytoskeleton has many levels of complexity depending on the cell type and on the cellular domain (i.e., apical and basolateral membranes in epithelial cells). In the red blood cell, membrane-cytoskeleton interactions have been investigated extensively.[20,21] Actin filaments are linked to the integral membrane proteins "Band 3" and glycophorin C via the actin-binding proteins spectrin, ankyrin, and protein 4.1.[20,21] This cortical meshwork is thought to contribute to the unique, biconcave shape of the erythrocyte and to its deformability.[20,21]

In epithelial cells, actin filaments have been shown to form a complex with integral membrane proteins involved in ion transport.[22,23] Moreover, the breakdown of F-actin leads to a loss of polarity in epithelial cells, especially with regard to the apical and basolateral expression of ion transporters.[24–27] The involvement of actin in epithelial cell polarity and in cell volume regulation will be discussed in detail later in this chapter.

B. MICROTUBULES

Microtubules are 25-nm-wide hollow tubes composed of 13 strands of protofilaments formed from tubulin subunits.[8,15] Microtubules can exist singly as bundles of tubules or as linked structures, such as microtubules that form a cilium or flagellum. Like actin, the assembly of tubulin involves the binding of tubulin dimers to the end of the tubule and the subsequent hydrolysis of a nucleotide. In this case, microtubule assembly requires GTP. Divalent cations (Ca^{2+} or Mg^{2+}) are required for disassembly. In many cultured animal cells, microtubules are associated with the centrosome, a microtubule-organizing center close to the nucleus. From this region, microtubules radiate out into the cytoplasm with the "plus" or polymerizing ends facing away from the centrosome. However, in MDCK cells once the fully polarized state is established, centrioles do not appear to play a role in microtubule organization. Instead, the microtubules are arranged as longitudinal bundles with the "minus" end near the apical membrane and the "plus" end toward the basal surface.[28] Microtubules do not appear to distribute preferentially into the cortical cytoplasm and there is a very low density of microtubules in the area close to the plasma membrane, the region where F-actin is most dense.

Like actin filaments, microtubules bind specifically to other proteins in the cell. These microtubule-associated proteins (MAPs), stimulating nucleation and polymerization, link microtubules to other cell constituents (organelles, centrosome), and attach microtubules to vesicles that are transported along the cytoskeleton. Moreover, the oriented movement of exo- and endocytic vesicles within the cell is, in part, mediated by specific MAPs, kinesin and dynein, which are present in vesicle membranes and provide the motors by which vesicles move along the microtubule array.[29] Although there is no evidence to support the hypothesis that microtubules could interact with the cell membrane directly, microtubule arrays may play a role in the delivery and/or retrieval of vesicles that contain ion transport proteins or regulators of ion transport.[30] In addition, it is possible that microtubules may form a functional unit with microfilaments, since MAP-2 binds to microtubules and microfilaments.[31]

C. INTERMEDIATE FILAMENTS

The term "intermediate filaments" originates from the fact that their diameter is 10 nm, "intermediate" between microfilaments (5 to 7 nm) and microtubules (25 nm).[9,12,13] Intermediate filaments are composed of five different classes of proteins that are tissue-specific. For example, desmin is found in muscle cells, whereas keratins are expressed in epithelial cells.[12] The filaments are formed from fibrous proteins that associate in an overlapping, side-to-side fashion.

Although little is known about the function of intermediate filaments, they may play a role in maintaining membrane stability. Because intermediate filaments also bind ankyrin,[32] the same protein that links spectrin and actin to many integral membrane proteins, and attaches to desmosomal plaques, it has been suggested that intermediate filaments participate in membrane-cytoskeleton, cell-to-cell, and cell-to-substrate binding.[13] In addition, intermediate filaments span the cytoplasmic space from the plasma membrane to the nucleus,[32] providing an intracellular scaffold linking all plasma membrane domains. Finally, intermediate filaments are substrates for protein kinases and phosphatases; thus, phosphorylation and dephosphorylation of these cytoskeletal elements may alter the viscoelastic properties of the cytoskeletal network.[12]

III. RESISTIVE AND CONTRACTILE FORCES: MECHANISMS OF VOLUME REGULATION

A key component in volume regulation may include "resistive" elements associated with the cell membrane. The cortical cytoskeleton, linked to the cell membrane via integral proteins, may play a role in providing resistance and, thus, regulate the *degree* of swelling under pathophysiological conditions where normal operation of pump and leak pathways are inhibited.[3,6]

Kleinzeller and colleagues have investigated the relationship between cell volume, ion fluxes, and the actin cytoskeleton in epithelial cells of the shark rectal gland.[3,33–36] Several important observations with regard to a possible role of the cytoskeleton in the prevention of excessive cell swelling have emerged from these studies. Under two different conditions, either exposure to high concentrations of extracellular K^+ or prolonged exposure to the organic mercurial agent, *p*-chloromercuribenzene sulfonate (pCMBS), cells undergo massive cell swelling. This swelling was much larger than that observed after inhibition of the Na^+,K^+-ATPase pump by ouabain. Morphological analysis revealed that the cellular architecture was greatly disrupted when massive cell swelling occurred, whereas little distortion of the cytoplasmic domain occurred during moderate swelling.

Since the cytoskeleton plays a crucial role in the maintenance of cellular architecture, the structural changes associated with the massive cell swelling could be due to an alteration in one or more components of the cytoskeleton. During massive cell swelling, there was a loss of F-actin bundles from the basolateral membrane and, in some cases, a complete loss of actin organization within the cell.[34,35] With pCMBS, the changes in F-actin organization directly correlated with severity of swelling. The high K^+-induced changes in actin were reversible, whereas the mercurial-induced changes in actin and in ion concentrations and cell volume were irreversible. Therefore, it is possible that alteration in F-actin by pCMBS was due primarily to a toxic and lethal effect on the cell unrelated to a direct interaction with actin and a volume control mechanism. Alternatively, since it is well established that mercurials bind to actin and may change the amount of F-actin,[37,38] the toxic effect could be a direct result of the mercurial-actin interaction that leads to massive cell swelling.

Another example of the correlation between changes in F-actin and a large increase in cell volume comes from studies of volume regulation in rectal gland cells exposed to anisosmotic conditions.[36] When rectal gland slices were placed in a dilute NaCl medium, the cells displayed a rapid, osmotic swelling followed by a regulatory volume decrease (RVD) mediated by the loss of organic osmolytes.[36] After 10 min in the hypotonic medium, there was a reduction in F-actin along the basolateral membrane.[36] The increase in cell volume (165% of control) corresponded to the increase in cell volume observed in high K^+ or with pCMBS and with hypotonic cell swelling in other epithelial cell types.[39,40] No change in the organization of actin or in the efflux of osmolytes was observed when the medium was made hypotonic by dilution of urea or substitution of choline for Na^+.[36] With the onset of RVD, F-actin content near the basolateral membrane gradually returned. These results suggest that the change in the actin cytoskeleton was involved in the regulation of osmolyte efflux across the basolateral membrane that occurred during RVD. Further evidence is necessary to define fully the role of F-actin in volume-induced efflux of osmolytes.

Recent studies by Linshaw and co-workers on isolated, renal proximal tubules also provide support for the role of the cytoskeleton in regulating the degree of swelling that occurs when the normal operation of the "pump-leak" mechanism of maintaining cell volume is compromised.[41,42] In the proximal tubule, ouabain-induced cell swelling can be increased by drugs that disrupt actin filaments (i.e., cytochalasin) or microtubules (i.e., vincristine).[41,42] Treatment with either cytoskeletal-disrupting agent alone or in the presence of ouabain did not change cell volume when the tubules were bathed in isotonic medium, even though disruption of microfilaments and microtubules was documented.[41] However, the additional cell swelling that occurred in ouabain-treated tubules exposed to the cytoskeletal-disrupting agents was not detected if the tubules were bathed in a medium containing protein.[41] Thus, it appears that the proximal tubule has a volume-regulatory mechanism that requires an intact microfilament-microtubule network that can partially resist swelling pressures that arise from unbalanced oncotic pressures.

On the other hand, there are studies that do not appear to agree with the hypothesis that the cytoskeleton plays a role in resisting the degree of cell swelling. For example, exposure of cell lines (PC-12 and T2) to a hypotonic medium results in a rapid swelling to 56 and 38% above control values and a disruption of actin filaments.[43,44] The RVD returns cell volume to values 16 and 12% greater than isotonic values.[43,44] If cells are exposed to cytochalasin B, a drug that disrupts F-actin, and then exposed to the hypotonic medium, the cells swell normally; however, RVD is attenuated.[43,44] Thus, disruption of actin filaments was not correlated with increased cell swelling but reduced the ability of the cells to elicit a full RVD. These experiments indicate that actin plays a role in RVD, unrelated to resisting the degree of cell swelling. No changes in microtubules were seen in dilute medium, and disruption of microtubules with colchicine did not alter cell swelling or the RVD.

Further evidence against the hypothesis that the cytoskeleton plays a role in resisting the degree of cell swelling comes from studies on HL-60 cells.[45] Hypotonic medium results in an initial swelling phase and a resultant RVD.[45] This swelling correlated with a decrease in the amount of F-actin.[45] However, cytochalasin did not change the peak swelling response nor did it alter the magnitude of the RVD, even though the drug reduced F-actin by 50%.[45] Thus, the changes in F-actin did not appear to play a key role in either the amount of swelling that occurred or in the mechanisms that control the volume recovery. It is interesting that even with a large dose of cytochalasin, the tonicity-induced changes in F-actin content in HL-60 cells still occurred. Since changes in F-actin content measured biochemically do not necessarily indicate that actin filaments throughout the cell are uniformly altered,[46–48] it would be important to determine if the changes in F-actin that occur are specific to the cortical cytoplasm.

In summary, the actin-associated cytoskeleton may, in some conditions, play a role in limiting the degree of cell swelling that can occur under pathophysiological conditions.[34,36,39–45] However, the actin-based cytoskeleton is *not* a primary mechanism for volume regulation, since disruption of actin filament organization or content does not lead to an increase in volume in cells bathed in an isotonic medium.[34,41] In fact, fragmentation of actin with cytochalasin leads to a decrease in volume in many cell types.[43,44,49] Instead, actin filaments in the cortical cytoskeleton, linked via spectrin, ankyrin, and other actin-binding proteins (e.g., myristylated acid-rich C-kinase substrate [MARCKS]) to integral membrane proteins, may act as a "safeguard" to prevent rapid, massive, and potentially lethal cell swelling during pathophysiological states. As described in later sections, actin filaments may play other, more subtle, physiological roles in the control of cell volume regulation as well.

IV. THE CYTOPLASMIC GEL AS A REGULATOR OF CELL SWELLING

The concept of the cytoplasm existing as a gel is well accepted.[5,50] In fact, some models for cell movement and cytoplasmic streaming are based on a gel-sol transition within the cell leading to changes in viscosity and pressure. This results in the movement of the cell membrane and the translocation of organelles. An additional role for this gel in volume regulation has been proposed by Lechene.[5] The reason for invoking a cytoplasmic mechanism was based on observed properties of isolated gels and the interaction of polyelectrolytes (intracellular proteins, etc.) with diffusable, inorganic ions, especially the major, intracellular cations. From this model, Lechene[5] and Wiggins[50] presented the concept of the cytoplasm balancing swelling pressures due to osmotic effects of gels and the special behavior of inorganic ions on charged surfaces. Since changes in cross-linking of cytoskeletal filaments and the structure of the cytoplasm around filamentous components are thought to play a key role in the gel-sol state, this hypothesis has some interesting aspects with regard to a key role for the cytoskeleton in cell volume regulation. However, as pointed out by Wiggins, examination of this question has significant experimental difficulties.[50] Thus, at present, there is little direct data to provide support for these hypotheses.

V. ROLE OF THE F-ACTIN SYSTEM IN EXOCYTOSIS

Van Rossum and co-workers have proposed an alternative model for the actin cytoskeleton in regulating cell volume.[4,51,52] Rather than acting as a resistive element, these investigators propose that actin filaments regulate the fusion of intracellular vesicles with the plasma membrane. This fusion of vesicles results in secretion of their contents (water, electrolytes, and secretory proteins) from the cell, thereby regulating cell volume in concert with other mechanisms. The key observations supporting this hypothesis are that (1) liver slices swollen at 1°C and then placed at 25°C still recover approximately 50% of normal volume in the presence of ouabain; (2) the ouabain-resistant volume recovery is associated with the disappearance of vesicles from the cytoplasm; (3) exposure of ouabain-treated liver slices to cytochalasins abolishes ouabain-resistant volume recovery and promotes the massive accumulation of cytoplasmic vesicles;[52] and (4) kidney slices swollen at 1°C accumulate vesicles in the apical cytoplasm which, at 25°C during volume recovery, appear to fuse with the basolateral membrane.[51] Thus, at least some of the water and electrolytes entering the cell during swelling may accumulate within endocytic vesicles. Expulsion of the vesicles via exocytosis eventually contributes to the RVD.[4] It was proposed that this exocytic mechanism involved actin filaments, since cytochalasin treatment led to an increase in the number of vesicles present in the cell.[4] An alternative explanation, discussed below, is that the exocytic vesicles contain transport proteins and perhaps water channels that, when inserted into the plasma membrane, mediate the RVD.

A study by Linshaw et al.[41] also supports a role for cytoskeleton-dependent vesicle fusion in cell volume regulation. Intravesicular volume increased in proximal tubules swollen by ouabain, cytochalasins, and vincristine. These data are compatible with the hypothesis that water, entering the cells during cell swelling, is contained in a vesicular pool. Further experiments are critical to determine if the relationship between volume and vesicle number, after cytochalasin or vincristine treatment, is altered and whether membrane fusion events between vesicles and the basolateral membrane occurs during RVD.

In contrast, Ouahbi et al. could not demonstrate a role for colchicine or cytochalasin-sensitive components in RVD in rabbit or rat kidney.[53] Slices of kidney cortex swollen by incubation in cold saline did not recover volume when incubated at room temperature in the presence of ouabain.[53] In addition, there was no effect on the RVD even after long-term incubation with cytoskeletal-disrupting agents in the presence or absence of ouabain.[53] These results suggest that the cytoskeleton does not play a role in volume control under isosmotic conditions. The role of fluid-filled vesicles was not addressed in this study.

In summary, there is significant evidence, albeit indirect, to support the possibility that at least a portion of the RVD in some epithelial cells is mediated by cytoskeletal-dependent exocytosis of fluid-filled vesicles. This mechanism is not essential for volume control, at least in some cell types, since cells exposed to high concentrations of cytochalasins can recover completely from swelling as long as the Na^+ pump is functional.[52] However, fusion of exocytic vesicles with the plasma membrane could function as an auxillary mechanism of RVD, playing a role during periods of rapid ion and water flux. Alternatively, as discussed below, some cells may insert vesicles into the plasma membrane in response to cell swelling as a means of placing key transport proteins into the plasma membrane during the RVD. Further investigation of the hypothesis that vesicles fuse with the plasma membrane and contribute to RVD would benefit from a combined approach of immunocytochemistry, biochemistry, and physiology in which changes in volume, solute and water transport, vesicle content and/or trafficking, and the content and distribution of cytoskeletal elements could be closely correlated.

VI. THE CYTOSKELETON AS A MODULATOR OF ION TRANSPORT AND CELL VOLUME REGULATION

A. EVIDENCE FOR CYTOSKELETAL REGULATION OF ION AND WATER TRANSPORT

Actin filaments and microtubules regulate the function and/or number of transport proteins involved in a variety of specialized cell functions. The acrosomal reaction in the sperm of invertebrates is a prime example. As the echinoderm sperm comes in contact with the egg, G-actin is mobilized within the head of the sperm, causing extensive polymerization of F-actin and formation of the acrosomal process.[54–57] The actin-induced extension of the sperm head into the egg is accompanied by an influx of Ca^{2+}, Na^+, and Cl^-, and an increase in water influx and sperm head volume.[54–56] H^+ efflux via Na^+/H^+ exchange and an alkalization of the cytoplasm also occurs during the acrosomal reaction and causes the release of G-actin monomers from profilin.[54] This well-studied paradigm was the first system in which a correlation between changes in the cytoskeleton, cell volume, and ion transport were demonstrated.

Similar correlations between changes in cell volume or ion transport and alterations in microtubule networks were found in other cell models.[58–60] Colchicine, by inhibiting the polymerization of microtubules, led to a 20% reduction in cell volume and a change in the shape of J774.2 macrophages.[58] This colchicine-induced reduction in cell volume was sensitive to 4-acetamido-4′-isothiocyanatostilbene-2,2′-disulfonic acid (SITS), an inhibitor of Cl^- transport, implicating a Cl^- channel or a Cl^-/HCO_3^- exchanger in cell volume control.[58] Additional evidence that microtubules may regulate ion transport events is provided by studies of Na^+ currents in squid giant axon.[59,60] When axons were perfused with colchicine, Na^+ currents decreased.[59,60] In contrast, taxol, a microtubule-stabilizing agent, increased Na^+ currents.[59] Similar results were obtained by addition of tubulin, cAMP, and a 260-kDa microtubule cross-linking protein to the cytoplasm of dialyzed axons.[59] Unfortunately, the issue of microtubule regulation of the function or number of Na^+ channels in the axolemma was not addressed in these studies.

In the collecting duct of the kidney and in the urinary bladder, the role of the cytoskeleton in the modulation of ion and water permeability of the apical membrane has received much attention. Apical membrane permeability to Na^+, H^+, H_2O, and HCO_3^- is thought to be determined by the shuttling of transport proteins between the plasma membrane and cytoplasmic vesicles.[46–48,61–70] The cytoskeleton is thought to regulate vesicle shuttling; however, the question of which cytoskeletal proteins are involved and how they are involved has not been elucidated.[71]

Disruption of microtubules with colchicine or actin filaments with cytochalasins partially inhibits vasopressin-induced water transport in amphibian urinary bladder.[30,69,72] The incidence of apical membrane intramembranous particle aggregates (thought to represent water channels or a protein closely associated with water channels) in vasopressin-treated cells is decreased with colchicine or cytochalasins.[30,69,72] Addition of colchicine after vasopressin treatment has no effect on water flow, suggesting that microtubules are involved in the delivery of the vesicles that contain water channels to the apical plasma membrane.[30] In contrast to the effects of colchicine, cytochalasins inhibit water flow before and after treatment with vasopressin.[30] Accordingly, it is thought, but not proven, that microfilaments are involved in the organization or activity of water channels in the apical membrane.[30] Alternatively, microfilaments may regulate the distribution of water channels between the apical membrane and a subapical pool of vesicles. More recent evidence from Hays and co-workers shows that vasopressin depolymerizes F-actin immediately below the surface of the apical membrane of amphibian urinary bladder granular cells and decreases F-actin content.[46–48,69,70] These results suggest that depolymerization of membrane-associated F-actin may be required for fusion of vesicles containing water channels with the apical membrane.[48] Finally, there is evidence that microtubules are involved in the transport of vesicles containing H^+-ATPases in toad urinary bladder, from the cytoplasm to the plasma membrane, since colchicine inhibits acidification by reducing the number of vesicle fusion events with the apical membrane.[72]

Recent studies have identified a role for F-actin in the regulation of epithelial Na^+ and Cl^- channels. Shapiro et al.[73] showed that stabilization of F-actin with phalloidin prevents cAMP-activated Cl^- secretion in T84 colon carcinoma cells.[73] cAMP depolymerizes and reorganizes F-actin in the basolateral membrane domain of the cell, suggesting that depolymerization of F-actin is required for cAMP-induced Cl^- secretion. It has been suggested that F-actin directly regulates cAMP-activated Cl^- channels or the delivery of Cl^- channels to the apical membrane. A recent study by Matthews and co-workers shows that F-actin modulates cAMP activation of the $Na^+/K^+/2Cl^-$ cotransporter.[74]

Recent patch clamp studies have addressed the role of cytoskeletal proteins in regulating ion channels in epithelial cells. Cantiello and co-workers have shown that cytochalasin D increased the open probability as well as the number of amiloride-sensitive, 8-pS Na^+ channels in A6 kidney cells.[75] Immunocytochemical and biochemical data indicate that amiloride-sensitive Na^+ channels colocalize with ankyrin, spectrin, and actin in the microvilli of A6 cells (see below).[76] In a preliminary report by Suzuki et al., depolarization-activated Cl^- channels in the apical membrane of rabbit proximal tubule were *inhibited* by cytochalasins.[77] However, whole cell Cl^- currents were not altered by cytochalasins.[77] Despite these findings, the physiological role of F-actin in the regulation of epithelial Na^+ and Cl^- channels is not known.

B. CYTOSKELETAL LINKAGE WITH ION TRANSPORT PROTEINS

Numerous studies have shown that ion channels and transport proteins are directly associated with the cytoskeleton.[21] Ion channels that are linked to the actin cytoskeleton by ankyrin and/or spectrin include the amiloride-sensitive Na^+ channel in A6 cells,[76] the cGMP-gated cation channel in rod photoreceptor cells,[78] the nicotinic acetylcholine receptor channel at the neuromuscular junction,[15,79] and the voltage-dependent Na^+ channel in neurons.[80] The Cl^-/HCO_3^- exchanger in erythrocytes and intercalated cells of the renal collecting duct as well as the Na^+,K^+-ATPase in renal epithelia are also linked to actin by ankyrin and spectrin.[20,22,24,81–84] This association of ion channels and transport proteins with the cytoskeleton is thought to anchor these integral membrane proteins in specific membrane domains; however, as illustrated in recent studies noted above, these cytoskeletal associations may also regulate the opening and closing of ion channels. In addition, dystrophin may regulate cation channels in skeletal muscle cells.[85] Muscle cells from *mdx* mice that lack dystrophin, a spectrin-like, membrane-associated cytoskeletal protein, manifest a decreased osmotic stability, due to an increase in Ca^{2+} conductance, possibly related to a defect in stretch inactivation of Ca^{2+}-permeable cation channels.[85,86] However, direct regulation of transport proteins and Na^+,K^+-ATPase by cytoskeletal proteins has not been demonstrated.

C. CYTOSKELETAL INVOLVEMENT IN CELL VOLUME REGULATION

Several studies have assessed the contribution of the cytoskeleton to volume regulation in a variety of cell types. In *Necturus* gallbladder cells, RVD in response to hypotonic cell swelling was inhibited by cytochalasin B but not by colchicine.[87] Similar inhibition of RVD was observed with cytochalasin B in the green crab, *Carcinus maenas*.[88] In contrast, the regulatory volume increase (RVI) in *Necturus* gallbladder was independent of cytoskeletal integrity.[87]

Mills[6] and Mills and Lubin[49] showed that dibutyryl cAMP and 3-isobutyl-1-methyl-xanthine (IBMX) caused a reduction in the volume of MDCK cells that was associated with a net loss of KCl and a rearrangement of F-actin. In contrast with the *Necturus* studies cited above, they showed that cytochalasin B reduced cell volume even when cells were bathed in an isotonic medium.[6,49] Colchicine was without effect on cell volume.[49] In HL-60 promyelocytic leukemia cells, the RVI induced by a hypertonic medium was associated with an increase in F-actin content which correlated with cell shrinkage.[45] In contrast, the RVD was associated with a decrease in F-actin content which correlated with cell swelling.[36] As discussed in detail above, hypotonic cell swelling resulted in the following series of events in shark rectal gland: (1) rearrangement and depolymerization of basolateral membrane-associated F-actin; (2) efflux of taurine during RVD; (3) efflux of Na^+ and Cl^- during initial swelling; and (4) repolymerization of F-actin after completion of RVD.[34–36] Taken together, these studies suggest that the actin filaments, but not microtubules, play a key role in cell volume regulation. The fact that cytochalasins, by disrupting F-actin, can either prevent or potentiate RVD suggests that the role of actin in cell volume regulation may be complex. An alternative explanation of the disparate effects of cytochalasins on RVD may lie in the concentration of the drug used in the studies cited above. In the studies in which cytochalasins inhibited RVD, the concentrations used were in excess of 40 μM. Hays and Franki[70] have shown that concentrations of cytochalasins greater than 1 μM can actually cause polymerization of F-actin and an increase in F-actin content due to increased fragmentation of existing actin filaments. Therefore, assessment of the dose-dependent effects of cytochalasins on volume-induced responses is important.

D. F-ACTIN INVOLVEMENT IN SWELLING AND STRETCH ACTIVATION OF ION CHANNELS

Studies by the authors of this chapter indicate that a Cl^- channel participates in the hypotonic-induced RVD of renal cortical collecting duct cells.[39,40,89–91] RVD in RCCT-28A cells, a continuous cell line derived from rabbit cortical collecting duct with phenotypic properties similar to acid-secreting intercalated cells, was inhibited by the Cl^- channel blocker, 4,4′-diisocyanatostilbene-2,2′-disulfonic acid (DIDS), and the K^+ channel blocker, barium. Single-channel, patch clamp recordings revealed that a large-conductance (305 pS) Cl^- channel was activated by mild hypotonic cell swelling (10% dilution of the bath osmolality) in cell-attached patches of the apical membrane. The time course of channel activation correlated closely with the onset and progression of RVD. In inside-out patches of the apical membrane, 305-pS Cl^- channels sensi-

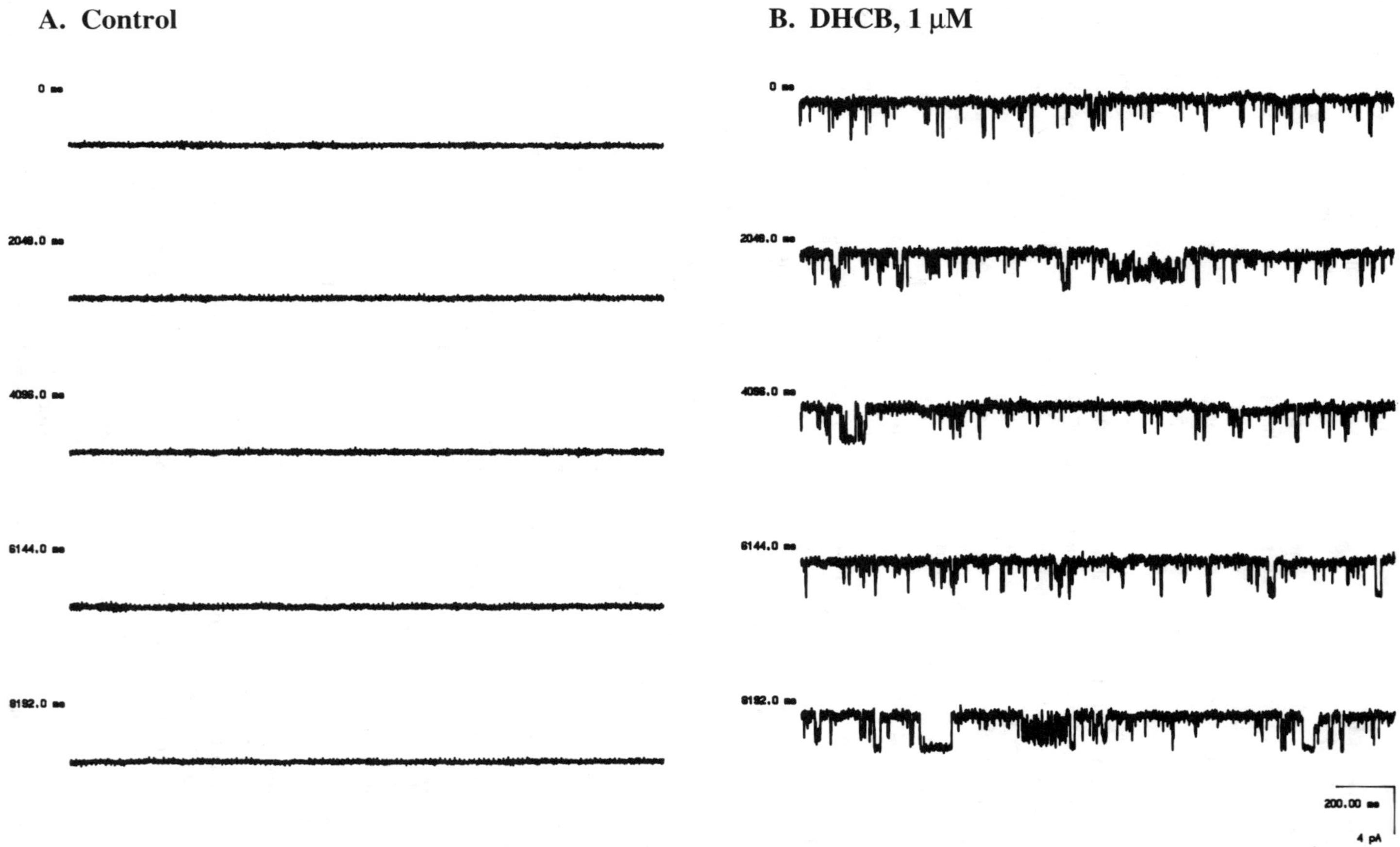

Figure 14–1 Representative channel current records illustrating activation of a 305 pS Cl^- channel in RCCT-28A cells with dihydrocytochalasin B (DHCB, 1 μ*M*). (A) Continuous 10-s recording immediately prior to addition of DHCB. (B) Continuous 10-s recording 4 min after addition of DHCB. The voltage across the inside-out, apical membrane patch was +10 mV; current and time scale bars are shown. Cl^- channel openings are indicated by upward deflections in the current records.

tive to cell swelling were also activated by negative pressure or suction applied to the patch pipette.

We investigated the mechanism by which cell swelling and membrane stretch activate the Cl^- channel.[90] Addition of dihydrocytochalasin B (DHCB; 100 n*M* to 1 μ*M*) to the cytoplasmic face of inside-out membrane patches activated the channel (Figure 14–1). DHCB increased the open probability of quiescent channels, channels activated by excision, and channels activated by negative pressure. Similar activation was obtained with cytochalasins B and D (1 μ*M* each). In contrast, phalloidin (200 p*M*), by stabilizing actin in its polymerized form, prevented suction-induced activation of the Cl^- channel. Whole cell recordings also showed swelling-induced activation of a Cl^- conductance with a linear I-V relationship. A similar Cl^- conductance was activated by DHCB (1 μ*M*); however, swelling in the presence of DHCB did not elicit an additional increase in the conductance. In contrast, phalloidin (200 p*M*) prevented swelling-induced activation of the whole cell Cl^- conductance. Taken together, these results indicate that cell swelling and membrane stretch-activate a 305-pS Cl^- channel in RCCT-28A cells by a mechanism that involves the fragmentation and depolymerization of F-actin.

In support of our patch clamp data, immunocytochemical visualization of F-actin with rhodamine phalloidin showed that hypotonic cell swelling was associated with depolymerization of F-actin (~15% reduction in F-actin content)[148] and reorganization of the actin cytoskeleton (Figure 14–2). Similar results are observed with the addition of 1 μ*M* DHCB (Figure 14–2). Microtubules were unaffected by hypotonic cell swelling of RCCT-28A cells. Preliminary results also suggest that protein kinase C (PKC) may be involved in the swelling-induced depolymerization of F-actin. Treatment with the specific PKC inhibitor, calphostin C, prevented swelling-induced effects on F-actin, whereas OAG, a permeable analogue of diacylglycerol, also altered F-actin organization similar to that seen with hypotonic cell swelling (Figure 14–3). Interestingly, PKC also activates the 305-pS Cl^- channel.[92] Whole cell recordings, like those described above, also indicate that calphostin C and pertussis toxin (PTX), an inhibitor of heterotrimeric G proteins, prevent swelling-induced activation of the Cl^- conductance. Previous results have shown that a PTX-sensitive G protein, $G_{\alpha i\text{-}3}$, activates the 305-pS Cl^- channel in RCCT-28A cells.[93] Together, these studies suggest that RVD and Cl^- channel activation in RCCT-28A cells is regulated by a complex signaling pathway involving F-actin, PKC and a PTX-sensitive G protein. Studies in progress are exploring the possibility that PKC and F-actin regulation of the swelling- and stretch-sensitive Cl^- channel may be interrelated.

Two studies support the findings described above. First, a similar, large-conductance Cl^- channel is activated by cell swelling in neuroblastoma cells.[94] Second, DHCB increases a swelling-sensitive Cl^- current in canine ventricular myocytes.[95,96] DHCB and swelling activation were not additive, suggesting that F-actin is involved in the mechanism of swelling activation.[95,96]

VII. MECHANISMS OF CYTOSKELETAL INVOLVEMENT IN CELL VOLUME REGULATION

A. MECHANICAL ALTERATION OF ION TRANSPORT PROTEIN CONFORMATION

Disruption of actin filament networks could trigger a change in protein conformation and thereby activate transport proteins.[97] Alternatively, or in addition, since the cytoskeleton maintains the structural integrity of the plasma membrane, F-actin and actin-binding proteins may regulate transporter activity by modulating membrane tension.[97–99] Guharay and Sachs[98] have suggested that actin is involved in stretch activation of K^+ channels in chick embryonic skeletal muscle.[98] Spectrin may be the cytoskeletal element involved in stretch activation of ion channels.[97] Elegant studies using high-voltage electron microscopy as well as differential interference contract microscopy indicate that cytoskeletal elements, vesicles, and organelles (i.e., a cytoplasmic gel) are associated with skeletal muscle membranes studied in the inside-out patch clamp configuration.[100–102] Challenges for future work lie in identifying the specific cytoskeletal elements involved in stretch activation of ion channels, as well as in elucidating the complex, cell type-specific mechanisms of channel activation by cytoskeletal proteins.

B. INTERACTION BETWEEN THE CYTOSKELETON AND SECOND MESSENGERS

Another mechanism by which the cytoskeleton may be involved in the regulation of ion transport, especially during RVD or RVI, lies in the interaction between the cytoskeleton and signal transduction pathways.[103] For example, phosphoinositide turnover and PKC are involved in cell volume regulation.[104] Mechanical perturbations, such as contraction, stretch, or strain of tissues, cause the release of inositol phosphates and diacylglycerol, as well as the activation of PKC.[105–107] Inhibition of PKC

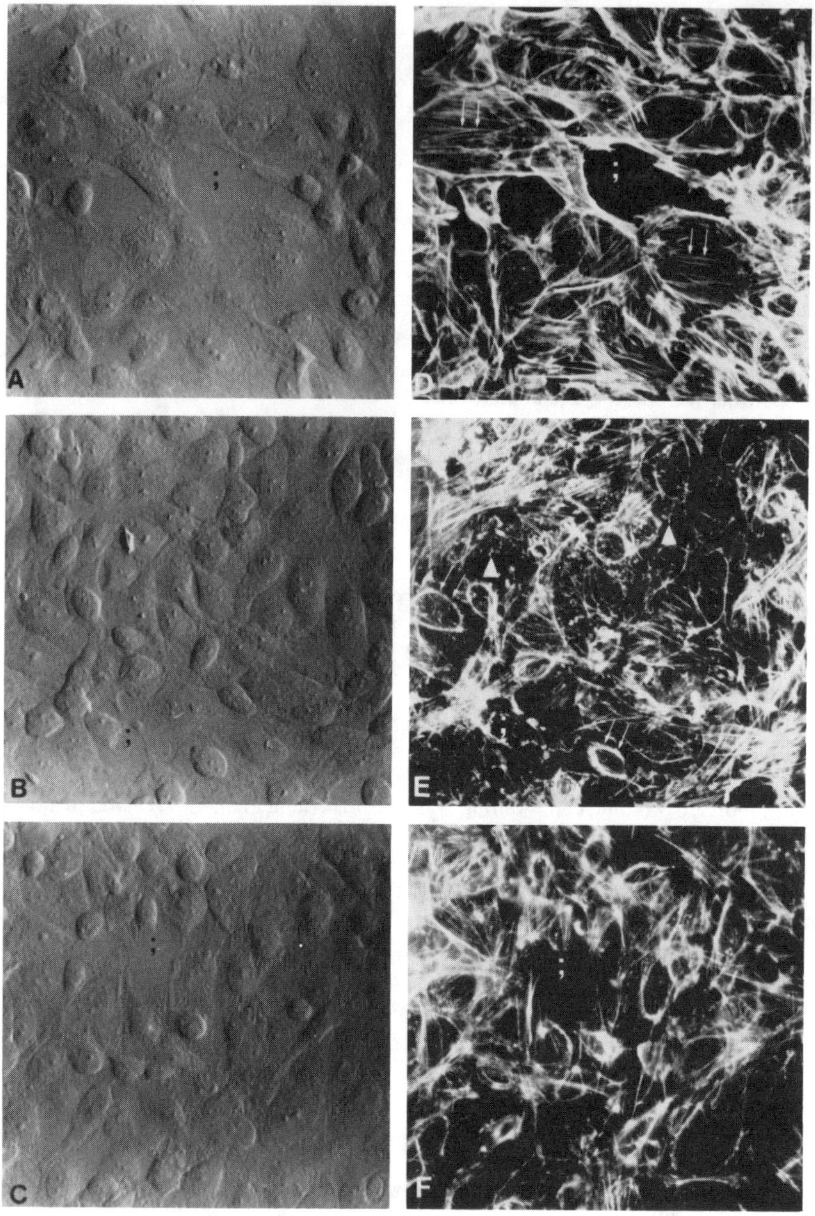

Figure 14–2 Paired photographs of RCCT-28A cells: (A–C) viewed with differential interference contrast optics; (D–F) viewed with the fluorescence optics. Cells were treated with rhodamine-phalloidin to localize F-actin ("semicolon" marks the same spot in each field). In control (isotonic) conditions (A and D), actin filaments are clearly seen throughout each cell (arrows). After 5 min in a hypotonic medium (50% reduction in osmolality; B and E), alteration of F-actin is clearly seen. Small globular aggregates of F-actin are indicated by the arrow heads, and circumferential bands of actin are seen in close proximity to nuclei (arrows). The morphology of the cells does not appear dramatically altered even though cell volume is increased by approximately 45%. When the cells are treated with DHCB (1 μM) for 5 min (C and F), there is also a clear alteration in the organization of the actin filament system. 332.5 $\times$ magnification.

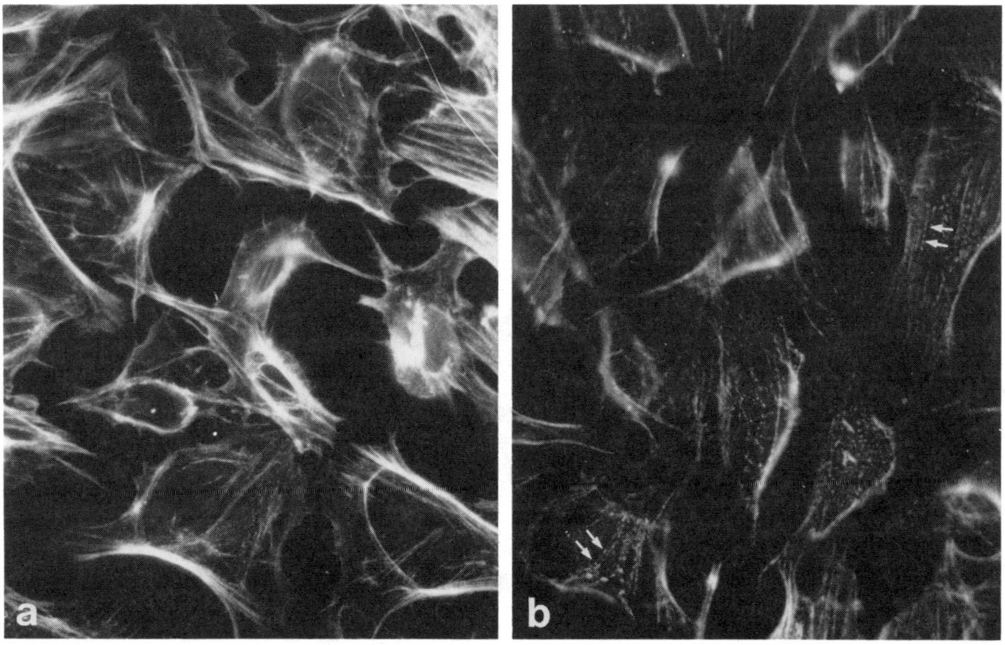

Figure 14–3 Localization of F-actin in RCCT-28A cells in the absence and presence of 1-oleoyl-2-acetyl glycerol (OAG, 100 μ*M*). (a) Control, (b) 6-min exposure of RCCT-28A cells to OAG, a permeable analogue of diacylglycerol. Note that the actin filaments are disrupted after OAG treatment, and the actin has a beaded appearance (arrows) in several cells. 450 × magnification.

prevents RVD in enterocytes swollen by increased L-alanine uptake.[108] Inhibition of PKC and calmodulin prevents RVD in astrocytes.[109] Finally, diacylglycerol promotes the reorganization and nucleation of actin at sites on the plasma membrane of *Dictyostelium*.[110]

Calcium is also a key factor in RVD.[111–115] In many cells there is an absolute requirement for an increase in intracellular calcium during cell volume regulation. The increase in intracellular calcium may be derived from IP_3-sensitive intracellular stores, since IP_3 is released during cell swelling or cyclic stretch or strain.[105] Alternatively or in addition, the rise in cell calcium may result from activation of stretch-activated cation or calcium-permeable channels in the plasma membrane, allowing calcium to enter the cell and directly or indirectly activate K^+ and Cl^- channels in the plasma membrane.[116,117] As shown by McCarty and O'Neil, the increase in calcium is only necessary during the initial minutes of RVD in renal proximal tubule.[112] Further studies are necessary to define the role of intracellular calcium in RVD; however, it is intriguing to speculate that this role may lie in the activation of channels directly or indirectly through stimulation of PKC, calmodulin, or the Ca^{2+}-calmodulin protein kinases.

Cyclic AMP generation is also affected by cell swelling and mechanical influences. Watson showed that hyposmotic swelling of S49 lymphoma cells directly stimulated adenylate cyclase and increased the concentration of intracellular cyclic AMP.[118] Both cyclic AMP and negative pressure applied to patch pipettes elicited an inhibition of K^+ channels and an activation of Na^+ channels in *Helix* neurons, suggesting that cAMP and mechanical perturbation regulate ion channels in these neurons through similar regulatory mechanisms.[119] The cytoskeleton also appears to regulate adenylate cyclase in S49 mouse lymphoma cells. Disruption of actin filaments with cytochalasin B or inhibition of microtubule assembly with colchicine or vinblastine enhances agonist- and cholera toxin-induced generation of cyclic AMP.[120,121] These results suggest that actin filaments and microtubules exert a tonic inhibitory effect on second messenger production.

Despite the wealth of knowledge that second messengers are involved in the regulation of volume control mechanisms and that the cytoskeleton may modulate second messenger production, it is not clear how second messengers and the cytoskeleton interact. Several studies have addressed this question indirectly. First, numerous studies have shown that agonist-induced increases in a variety of second messengers are associated with reorganization and/or depolymerization of the actin cytoskeleton. Agonist-induced effects on the actin cytoskeleton occur in epithelial cells, leukocytes, platelets, astrocytes, osteoblasts, gastric parietal cells, and *Dictyostelium*.[47,48,122–128] Second, serine/threonine protein kinases, specifically cyclic AMP-dependent protein kinase and PKC, phosphorylate many actin-binding proteins (MARCKS, spectrin, ankyrin, profilin) as well as G-actin and tubulin.[129–134] In every case, phosphorylation of cytoskeletal elements causes a decrease in the affinity of one cytoskeletal protein for another, a disruption of actin filament-based or microtubule-based networks, and an inability of phosphorylated G-actin to link to actin filaments. In addition, recent evidence indicates that PKC colocalizes with actin-binding proteins and actin filaments in fibroblasts and baby hamster kidney cells.[135,136] Therefore, not only do cytoskeletal proteins regulate second messenger production but cytoskeletal proteins are substrates for protein kinases. It is interesting to speculate that the cytoskeleton, specifically actin and actin-binding proteins, may be a key element in the signaling cascades regulating cell volume.

Finally, recent evidence suggests that G proteins interact directly with actin filaments and microtubules. Heterotrimeric G proteins ($G_{\alpha s}$ and $G_{\alpha i\text{-}1}$) interact with tubulin in neurons from rat cerebral cortex.[137] $G_{\alpha i\text{-}1}$, and to a lesser extent, $G_{\alpha o}$, were found to inhibit microtubule polymerization by binding to tubulin dimers.[138] Moreover, it was found that heterotrimeric G proteins and tubulin share and exchange GTP, since both tubulin and G proteins hydrolyze GTP.[139] These results suggest that G protein interaction with microtubules may serve as a component of signal transduction.[137] In neutrophils, a close association between ligand-receptor complexes, G proteins, and the actin cytoskeleton has been documented to be involved in the downregulation of receptor-mediated signaling.[140] Finally, Ca^{2+} influx required for RVD in rabbit proximal tubule cells is activated by a PTX-sensitive G protein.[141] Thus, it is likely that G proteins play an important role in cell volume regulation. Further investigation is necessary to determine whether G proteins regulate transport proteins directly or indirectly via activation of second messenger pathways.

C. CYTOSKELETAL REGULATION OF THE DELIVERY OF CHANNELS TO THE PLASMA MEMBRANE

Although the translocation and insertion of vesicles containing ion or water transport proteins from a cytoplasmic pool into the plasma membrane has been characterized in several systems, especially with regard to agonist-induced stimulation of ion transport, virtually nothing is known about the role of exocytosis in cell volume regulation. However, it has been well documented that disruption of the cortical, actin-based cytoskeleton allows the fusion of catecholamine-containing vesicles in adrenal chromaffin cells, histamine-containing vesicles in mast cells, vesicles containing intramembranous particle aggregates, and/or water channels in urinary bladder granular cells and renal collecting duct cells, and vesicles containing H^+-ATPase pumps in intercalated cells of the renal collecting duct.[30,72,142–144] Moreover, recent evidence suggests that protein phosphorylation by kinases is critical for the stimulation of exocytic transport of vesicles through the Golgi apparatus.[145,146]

VIII. CONCLUDING REMARKS

It is clear that the cytoskeleton plays a role in cell volume regulation in the majority of model systems described in this review. Two hypotheses for the involvement of actin filaments in the control of volume-regulatory processes are proposed. First, actin may play a role as a resistive element to cell swelling during pathophysiological conditions when cell volume control mechanisms are compromised. This resistance by actin is not a primary volume-regulatory mechanism and may not be subject to regulation. Rather, it may be an inherent property of the viscoelastic characteristics of the membrane-cytoskeleton complex. Second, actin may be involved directly in more subtle, physiological mechanisms by which cell swelling and associated increases in membrane tension and cytoskeletal protein interactions activate ion channels or transporters involved in RVD. Since cell swelling appears to disrupt the actin cytoskeleton in many cell types (especially membrane-associated, cortical actin filaments), depolymerization of actin filaments in specific membrane domains may trigger the activation of ion transport pathways that contribute to solute efflux during RVD. The seemingly inconsistent data appearing in the literature may be related to the diversity of mechanisms that cells utilize to regulate cell volume.[147] Combined immunocytochemical, biochemical, and physiological techniques in a single cell model should provide novel information and assess possible correlations between changes in cell volume, alterations in the content and conformation of cytoskeletal proteins, activation and/or inhibition of ion transport, and the generation and involvement of second messengers critical to cell volume regulation.

ACKNOWLEDGMENTS

Studies in the authors' laboratory were supported by the National Institutes of Health (DK-34533). We thank Jamie Russell for assistance in the preparation of the manuscript.

REFERENCES

1. **Leaf, A.,** On the mechanism of fluid exchange of tissues in vitro, *Biochem. J.*, 62, 241, 1956.
2. **Wilson, T. H.,** Ionic permeability and osmotic permeability of cells, *Science,* 120, 104, 1954.
3. **Kleinzeller, A., Ziyadeh, F. N., and Mills, J. W.,** Cell volume regulation in epithelia — with emphasis on the role of osmolytes and the cytoskeleton, in *Cell Volume Regulation,* Kinne, R. K. H., Kinne-Saffran, E., and Beyenbach, K. W., Eds., S. Karger, Basel, Switzerland, 1990, 59.
4. **Van Rossum, G. D. V., Russo, M. A., and Schisselbauer, J. C.,** Role of cytoplasmic vesicles in volume maintenance, in *Cell Volume Control: Fundamental and Comparative Aspects in Animal Cells,* Gilles, R., Kleinzeller, A., and Bolis, L., Eds., Academic Press, San Diego, 1992, 45.
5. **Lechene, C.,** Cellular volume and cytoplasmic gel, *Biol. Cell,* 55, 177, 1985.
6. **Mills, J. W.,** The cell cytoskeleton: possible role in volume control, *Curr. Top. Membr. Transport,* 30, 75, 1987.
7. **Bloom, G. S.,** Motor proteins for cytoplasmic microtubules, *Curr. Opin. Cell Biol.,* 4, 66, 1992.
8. **Caplow, M.,** Microtubule dynamics, *Curr. Opin. Cell Biol.,* 4, 58, 1992.
9. **Burridge, K., Fath, K., Kelly, T., Nuckols, G., and Turner, C.,** Focal adhesion: transmembrane junctions between the extracellular matrix and the cytoskeleton, *Annu. Rev. Cell Biol.,* 4, 487, 1988.
10. **Bretscher, A.,** Microfilament structure and function in the cortical cytoskeleton, *Annu. Rev. Cell Biol.,* 7, 337, 1991.
11. **Cooper, J. A.,** The role of actin polymerization in cell motility, *Annu. Rev. Physiol.,* 53, 585, 1991.
12. **Eriksson, J. E., Opal, P., and Goldman, R. D.,** Intermediate filament dynamics, *Curr. Opin. Cell Biol.,* 4, 99, 1992.
13. **Skalli, O. and Goldman, R. D.,** Recent insights into the assembly, dynamics, and function of intermediate filament networks, *Cell Motil. Cytoskeleton,* 19, 67, 1992.
14. **Pollard, T. D. and Cooper, J. A.,** Actin and actin-binding proteins. A critical evaluation of mechanisms and functions, *Annu. Rev. Biochem.,* 55, 987, 1986.
15. **Vale, R. D., Banker, G., and Hall, Z. W.,** The neuronal cytoskeleton, in *Molecular Neurobiology,* Hall, Z. W., Ed., Sinauer Associates, Sunderland, 1992, 247.
16. **Lassing, I. and Lindberg, U.,** Specific interaction between phosphatidylinositol-4,5-bisphosphate and profilactin, *Nature,* 318, 472, 1992.
17. **Goldschmidt-Clermont, P. J., Machesky, L. M., Baldassare, J. J., and Pollard, T. D.,** The actin-binding protein profilin binds to PIP_2 and inhibits its hydrolysis by phospholipase C, *Science,* 247, 1575, 1990.
18. **Goldschmidt-Clermont, P. J., Kim, J. W., Machesky, L. M., Rhee, S. G., and Pollard, T. D.,** Regulation of phospholipase C-gamma by profilin and tyrosine phosphorylation, *Science,* 251, 1231, 1991.
19. **Pollard, T. D., Machesky, L., and Goldschmidt-Clermont, P.,** Interaction of profilins with membrane lipids, *Curr. Top. Membr.,* 38, 217, 1991.
20. **Bennett, V.,** The membrane skeleton of human erythrocytes and its implications for more complex cells, *Annu. Rev. Biochem.,* 54, 273, 1985.
21. **Luna, E. J. and Hitt, A. L.,** Cytoskeleton-plasma membrane interactions, *Science,* 258, 955, 1992.
22. **Drenckhahn, D. and Schluter, K.,** Colocalization of band 3 with ankyrin and spectrin at the basal membrane of intercalated cells in the rat kidney, *Science,* 230, 1287, 1985.
23. **Nelson, W. J.,** Topogenesis of plasma membrane domains in polarized epithelial cells, *Curr. Opin. Cell Biol.,* 1, 660, 1989.
24. **Nelson, W. J. and Veshnock, P. J.,** Ankyrin binding to (Na+K)ATPase and implications for the organization of membrane domains in polarized cells, *Nature,* 328, 533, 1987.
25. **Molitoris, B. A., Geerdes, A., and McIntosh, J. R.,** Dissociation and redistribution of Na^+,K^+-ATPase from its surface membrane actin cytoskeletal complex during cellular ATP depletion, *J. Clin. Invest.,* 88, 462, 1991.
26. **Molitoris, B. A.,** Ischemia-induced loss of epithelial polarity: potential role of the actin cytoskeleton, *Am. J. Physiol.,* 260, F769, 1991.
27. **Morrow, J. S., Cianci, C. D., Kennedy, S. P., and Warren, S. L.,** Polarized assembly of spectrin and ankyrin in epithelial cells, *Curr. Top. Membr.,* 38, 227, 1991.

28. **Bacalloa, R., Antony, C., Dotti, C., Karsenti, E., Stelzer, E. H. K., and Simons, K.,** The subcellular organization of Madin-Darby canine kidney cells during the formation of a polarized epithelium, *J. Cell Biol.,* 109, 2817, 1989.
29. **Kuznetsov, S. A., Langford, G. M., and Weiss, D. G.,** Actin-dependent organelle movement in squid axoplasm, *Nature,* 356, 722, 1992.
30. **Pearl, M. and Taylor, A.,** Role of the cytoskeleton in the control of transcellular water flow by vasopressin in amphibian urinary bladder, *Biol. Cell,* 55, 163, 1985.
31. **Olmstead, J. B.,** Microtubule-associated proteins, *Annu. Rev. Cell Biol.,* 2, 421, 1986.
32. **Georgatos, S. D. and Blobel, G.,** Two distinct attachment sites for vimentin along the plasma membrane and the nuclear envelope in avian erythrocytes: a basis for vectorial assembly of intermediate filaments, *J. Cell Biol.,* 105, 117, 1987.
33. **Feldman, G. M., Mills, J. W., Ziyadeh, F. N., Booz, G. W., and Kleinzeller, A.,** Proprionate induces cell swelling and K^+ accumulation in shark rectal gland, *Am. J. Physiol. (Cell Physiol.),* 257, C377, 1989.
34. **Kleinzeller, A. and Mills, J. W.,** K^+-induced swelling of the dogfish shark *(Squalus acanthias)* rectal gland cells is associated with changes of the cytoskeleton, *Biochim. Biophys. Acta,* 1014, 40, 1989.
35. **Kleinzeller, A., Booz, G. W., Mills, J. W., and Ziyadeh, F. N.,** pCMBS-induced swelling of the dogfish *(Squalus acanthias)* rectal gland cells: role of the Na^+,K^+-ATPase and the cytoskeleton, *Biochim. Biophys. Acta,* 1025, 21, 1990.
36. **Ziyadeh, F. N., Mills, J. W., and Kleinzeller, A.,** Hypotonicity and cell volume regulation in shark rectal gland: role of organic osmolytes and F-actin, *Am. J. Physiol. Renal, Fluid Electrolyte Physiol.,* 262, F468, 1992.
37. **Gordon, S. G. B. R.,** Solubilization and denaturation of monomeric actin from erythrocyte membranes by p-mercuribenzenesulfonate, *Biochim. Biophys. Acta,* 1025, 43, 1990.
38. **Hays, R. M., Natochin, Y., Parnova, R., and Kleinzeller, A.,** The response of the actin cytoskeleton of dogfish *(Squalus acanthias)* rectal gland cells to hypotonicity, high potassium and mercurials, *Bull. Mt. Desert Isl. Biol. Lab.,* 29, 68, 1992.
39. **Mills, J. W., Schwiebert, E. M., and Stanton, B. A.,** Evidence for the role of actin filaments in regulating cell swelling, *J. Exp. Biol.,* in press.
40. **Schwiebert, E. M., Mills, J. W., and Stanton, B. A.,** The actin-based cytoskeleton regulates a chloride channel and cell volume in a renal cortical collecting duct cell line, *J. Biol. Chem.,* submitted.
41. **Linshaw, M. A., Macalaster, T. J., Welling, L. W., Bauman, C. A., Hebert, G., Downey, G. P., Koo, E. W. Y., and Gotlieb, A. I.,** Role of the cytoskeleton in isotonic cell volume control of rabbit proximal tubules, *Am. J. Physiol. (Renal Fluid Electrolyte Physiol.),* 261, F60, 1991.
42. **Linshaw, M. A., Fogel, C. A., Downey, G. P., Koo, E. W. Y., and Gotleib, A. I.,** Role of cytoskeleton in volume regulation of rabbit proximal tubule in dilute medium, *Am. J. Physiol. (Renal Fluid Electrolyte Physiol.),* 262, F144, 1992.
43. **Cornet, M., Delpire, E., and Gilles, R.,** Study of microfilaments network during volume regulation process of cultured PC-12 cells, *Pfluegers Arch.,* 410, 223, 1987.
44. **Cornet, M., Delpire, E., and Gilles, R.,** Relations between cell volume control, microfilaments and microtubules networks in T2 and PC12 cultured cells, *J. Physiol. (Paris),* 83, 43, 1988.
45. **Hallows, K. R., Packman, C. H., and Knauf, P. A.,** Acute cell volume changes in anisotonic media affect F-actin content of HL-60 cells, *Am. J. Physiol.,* 261, C1154, 1991.
46. **Gao, Y., Franki, N., Macaluso, F., and Hays, R. M.,** Vasopressin decreases immunogold labeling of apical actin in the toad bladder granular cell, *Am. J. Physiol.,* 263, C908, 1992.
47. **Holmgren, K., Magnusson, K.-E., Franki, N., and Hays, R. M.,** ADH-induced depolymerization of F-actin in the toad bladder granular cell: a confocal microscope study, *Am. J. Physiol. Cell Physiol.,* 262, C672, 1992.
48. **Ding, G., Franki, N., Condeelis, J., and Hays, R. M.,** Vasopressin depolymerizes F-actin in toad bladder epithelial cells, *Am. J. Physiol. Cell Physiol.,* 260, C9, 1991.
49. **Mills, J. W. and Lubin, M.,** Effect of adenosine 3′,5′-cyclic monophosphate on volume and cytoskeleton of MDCK cells, *Am. J. Physiol.,* 250, C319, 1986.
50. **Wiggins, P. M.,** Role of water in some biological processes, *Microbiol. Rev.,* 54, 432, 1992.
51. **Russo, M. A., Ernst, S. A., Kapoor, S. C., and Van Rossum, G. D. V.,** Morphological and physiological studies of rat kidney cortex slices undergoing isosmotic swelling and its reversal: a possible mechanism for ouabain-

resistant control of cell volume, *J. Membr. Biol.*, 85, 1, 1992.

52. **Van Rossum, G. D. V. and Russo, M. A.,** Ouabain-resistant mechanism of volume control and the ultrastructural organization of live slices recovering from swelling *in vitro*, *J. Membr. Biol.*, 59, 191, 1981.
53. **Ouahbi, A., Duchene, C., and Gilles, R.,** Comparative studies of volume restoration following cold-stress induced swelling in renal tissues. I. Effects of ouabain, K^+ free medium, colchicine and cytochalasin B on rat and rabbit kidney cortex slices, *Comp. Biochem. Physiol.*, 97A, 265, 1992.
54. **Tilney, L. G., Kiehart, D. P., Sardet, C., and Tilney, M.,** Polymerization of actin. IV. Role of Ca^{2+} and H^+ in the assembly of actin and in membrane fusion in the acrosomal reaction of echinoderm sperm, *J. Cell Biol.*, 77, 536, 1978.
55. **Tilney, L. G. and Inoue, S.,** Acrosomal reaction of the *Thyone* sperm. III. The relationship between actin assembly and water influx during the extension of the acrosomal process, *J. Cell Biol.*, 100, 1273, 1985.
56. **Inoue, S. and Tilney, L. G.,** Acrosomal reaction of the *Thyone* sperm. I. Changes in the sperm head visualized by high resolution microscopy, *J. Cell Biol.*, 93, 812, 1982.
57. **Schackmann, R. W., Eddy, E. M., and Shapiro, B. M.,** The acrosome reaction of *Strongylocentrotus purpuratus* sperm, *Dev. Biol.*, 65, 483, 1978.
58. **Melmed, R. N., Karanian, P. J., and Berlin, R. D.,** Control of cell volume in the J774 macrophage by microtubule disassembly and cyclic AMP, *J. Cell Biol.*, 90, 761, 1981.
59. **Matsumoto, G., Ichikawa, M., Tasaki, A., Murofushi, H., and Sakai, H.,** Axonal microtubules necessary for the generation of sodium current in squid giant axons. I. Pharmacological study on sodium current and restoration of sodium current by microtubule proteins and 260K protein, *J. Membr. Biol.*, 77, 77, 1984.
60. **Matsumoto, G., Ichikawa, M., and Tasaki, A.,** Axonal microtubules necessary for the generation of sodium current in squid giant axons. II. Effect of colchicine upon asymmetrical displacement current, *J. Membr. Biol.*, 77, 93, 1984.
61. **Rich, A., Dixon, T. E., and Clausen, C.,** Changes in membrane conductances and areas associated with bicarbonate secretion in turtle bladder, *J. Membr. Biol.*, 113, 211, 1990.
62. **Dixon, T. E., Clausen, C., Coachman, D., and Lane, B.,** Proton transport and membrane shuttling in turtle bladder epithelium, *J. Membr. Biol.*, 94, 233, 1986.
63. **Lewis, S. A. and deMoura, J. L. C.,** Incorporation of cytoplasmic vesicles into the apical membrane of mammalian urinary bladder epithelium, *Nature*, 297, 685, 1982.
64. **Loo, D. D. F., Lewis, S. A., Ifshin, M. S., and Diamond, J. M.,** Turnover, membrane insertion, and degradation of sodium channels in rabbit urinary bladder, *Science*, 221, 1288, 1983.
65. **Al-Awqati, Q.,** Regulation of membrane transport by endocytotic removal and exocytotic insertion of transporters, *Meth. Enzymol.*, 172, 49, 1989.
66. **Stetson, D. L., Lewis, S. A., Alles, W., and Wade, J. B.,** Evaluation by capacitance measurements of antidiuretic hormone induced membrane area changes in toad bladder, *Biochim. Biophys. Acta*, 689, 267, 1982.
67. **Ding, G., Franki, N., Bourguet, J., and Hays, R. M.,** Role of vesicular transport in ADH-simulated aggregate delivery, *Am. J. Physiol.*, 255, C641, 1988.
68. **Hays, R. M. and Lindberg, U.,** Actin depolymerization in the cyclic AMP-stimulated toad bladder epithelial cell, determined by the DNAse method, *FEBS Lett.*, 280, 397, 1991.
69. **Hays, R. M.,** Alteration of luminal membrane structure by antidiuretic hormone, *Am. J. Physiol. Cell Physiol.*, 245, C289, 1983.
70. **Hays, R. M. and Franki, N.,** Cytochalasin D as a model for the cytoskeletal actin of vasopressin, *J. Am. Soc. Nephrol.*, 2, 723a, 1991.
71. **Lewis, S. A. and Donaldson, P.,** Ion channels and cell volume regulation: chaos in an organized system, *News Physiol. Sci.*, 5, 112, 1990.
72. **Brown, D.,** Membrane recycling and epithelial cell function, *Am. J. Physiol.*, 256, F1, 1989.
73. **Shapiro, M., Matthews, J., Hecht, G., Delp, C., and Madara, J. L.,** Stabilization of F-actin prevents cAMP-elicited Cl^- secretion in T84 cells, *J. Clin. Invest.*, 87, 1903, 1991.
74. **Matthews, J. B., Awtrey, C. S., and Madara, J. L.,** Microfilament-dependent activation of $Na^+/K^+/2Cl^-$ cotransport by cAMP in intestinal epithelial monolayers, *J. Clin. Invest.*, 90, 1608, 1992.
75. **Cantiello, H. F., Stow, J. L., Prat, A. G., and Ausiello, D. A.,** Actin filaments regulate epithelial Na^+ channel activity, *Am. J. Physiol. Cell Physiol.*, 261, C882, 1991.
76. **Smith, P. R., Saccomani, G., Joe, E.-H., Angelides, K. J., and Benos, D. J.,** Amiloride-sensitive sodium channel is linked to the

cytoskeleton in renal epithelial cells, *Proc. Natl. Acad. Sci. U.S.A.*, 88, 6971, 1991.

77. **Suzuki, M., Kawaguchi, Y., and Sakai, O.,** Cl channel is modified by disruption of cytoskeletal filament in proximal tubule cells, *J. Am. Soc. Nephrol.*, 2, 752a, 1991.
78. **Molday, L. L., Cook, N. J., Kaupp, U. B., and Molday, R. S.,** The cGMP-gated cation channel of bovine rod photoreceptor cells is associated with a 240-kDa protein exhibiting immunochemical cross-reactivity with spectrin, *J. Biol. Chem.*, 265, 18690, 1990.
79. **Bloch, R. J. and Morrow, J. S.,** An unusual b-spectrin associated with clustered acetylcholine receptors, *J. Cell Biol.*, 108, 481, 1989.
80. **Srinivasan, Y. ,Elmer, L., Davis, J., Bennett, V., and Angelides, K.,** Ankyrin and spectrin associate with voltage-dependent sodium channels in brain, *Nature*, 333, 177, 1988.
81. **Davis, J. Q. and Bennett, V.,** The anion exchanger and Na^+K^+-ATPase interact with distinct sites on ankyrin in *in vitro* assays, *J. Biol. Chem.*, 265, 17252, 1990.
82. **Soong, C.-J., Lu, P.-W., and Tao, M.,** Analysis of band 3 cytoplasmic domain phosphorylation and association with ankyrin, *Arch. Biochem. Biophys.*, 254 (2), 509, 1987.
83. **Koob, R., Kraemer, D., Trippe, G., Aebi, U., and Drenckhahn, D.,** Association of kidney and parotid Na^+,K^+-ATPase microsomes with actin and analogs of spectrin and ankyrin, *Eur. J. Cell Biol.*, 53, 93, 1990.
84. **Morrow, J. S., Cianci, C. D., Ardito, T., Mann, A. S., and Kashgarian, M.,** Ankyrin links fodrin to the alpha subunit of Na,K-ATPase in Madin-Darby canine kidney cells and in intact renal tubule cells, *J. Cell Biol.*, 108, 455, 1989.
85. **Franco, A. and Lansman, J. B.,** Calcium entry through stretch-activated ion channels in mdx myotubes, *Nature*, 344, 670, 1990.
86. **Koenig, M., Monaco, A. P., and Kunkel, L. M.,** The complete sequence of dystrophin predicts a rod-shaped cytoskeletal protein, *Cell*, 53, 219, 1988.
87. **Foskett, J. K. and Spring, K. R.,** Involvement of calcium and cytoskeleton in gallbladder epithelial cell volume regulation, *Am. J. Physiol.*, 248, C27, 1985.
88. **Gilles, R., Delpire, E., Duchene, C., Cornet, M., and Pequeux, A.,** The effect of cytochalasin B on the volume regulation response of isolated axons of the green crab *Carcinus maenas* submitted to hypo-osmotic media, *Comp. Biochem. Physiol.*, 85A, 523, 1992.
89. **Schwiebert, E. M., Mills, J. W., and Stanton, B. A.,** The cytoskeleton regulates a volume regulatory chloride channel in cortical collecting duct cells in culture, *J. Am. Soc. Nephrol.*, 2 (Abstr.), 750a, 1991.
90. **Stanton, B. A., Mills, J. A., and Schwiebert, E. M.,** Role of the actin based cytoskeleton in regulatory volume decrease in cortical collecting duct cells in culture, *JASN*, 2 (Abstr.), 751, 1991.
91. **Schwiebert, E. M., Mills, J. W., and Stanton, B. A.,** The cytoskeleton regulates chloride channels in the apical membrane of cortical collecting duct cells, *FASEB J.*, 5 (Abstr.), 739, 1991.
92. **Schwiebert, E. M., Karlson, K. H., Friedman, P. A., Dietl, P., Spielman, W. S., and Stanton, B. A.,** Adenosine regulates a chloride channel via protein kinase C and a G protein in a rabbit cortical collecting duct cell line, *J. Clin. Invest.*, 89, 834, 1992.
93. **Schwiebert, E. M., Light, D. B., Fejes-Toth, G., Naray-Fejes-Toth, A., and Stanton, B. A.,** A GTP-binding protein activates chloride channels in a renal epithelium, *J. Biol. Chem.*, 265, 7725, 1990.
94. **Falke, L. C. and Misler, S.,** Activity of ion channels during volume regulation by clonal N1E115 neuroblastoma cells, *Proc. Natl. Acad. Sci. U.S.A.*, 86, 3919, 1989.
95. **Tseng, G.-N.,** Cell swelling activates a membrane Cl channel in canine cardiac myocytes, *Biophys. J.*, 59 (Abstr.), 91a, 1991.
96. **Tseng, G.-N.,** Cell swelling increases membrane conductance of canine cardiac cells: evidence for a volume-sensitive Cl channel, *Am. J. Physiol. Cell Physiol.*, 262, C1056, 1992.
97. **Sachs, F.,** Mechanical transduction by membrane ion channels: a mini review, *Mol. Cell. Biochem.*, 104, 57, 1991.
98. **Guharay, F. and Sachs, F.,** Stretch-activated single ion channel currents in tissue-cultured embryonic chick skeletal muscle, *J. Physiol.*, 352, 685, 1984.
99. **Sachs, F.,** Baroreceptor mechanisms at the cellular level, *Fed. Proc.*, 46 (1), 12, 1987.
100. **Sokabe, M. and Sachs, F.,** The structure and dynamics of patch-clamped membranes: a study using differential interference contrast light microscopy, *J. Cell Biol.*, 111, 599, 1990.
101. **Ruknudin, A., Song, M. J., and Sachs, F.,** The ultrastructure of patch-clamped membranes: a study using high voltage electron microscopy, *J. Cell Biol.*, 112 No. 1, 125, 1991.

102. **Sokabe, M., Sachs, F., and Jing, Z.,** Quantitative video microscopy of patch clamped membranes stress, strain, capacitance, and stretch channel activation, *Biophys. J.*, 59, 722, 1991.
103. **Watson, P. A.,** Function follows form: generation of intracellular signals by cell deformation, *FASEB J.*, 5, 2013, 1991.
104. **Grinstein, S., Furuya, W., and Bianchini, L.,** Protein kinases, phosphatases, and the control of cell volume, *News Physiol. Sci.*, 7, 232, 1992.
105. **Rosales, O. R. and Sumpio, B. E.,** Changes in cyclic strain increase inositol trisphosphate and diacylglycerol in endothelial cells, *Am. J. Physiol. Cell Physiol.*, 262, C956, 1992.
106. **von Harsdorf, R., Lang, R., Fullerton, M., Smith, A. I., and Woodcock, E. A.,** Right atrial dilatation increases inositol-(1,4,5) trisphosphate accumulation, *FEBS Lett.*, 233, 201, 1988.
107. **Richter, E. A., Cleland, P. J. F., Rattigan, S., and Clark, M. G.,** Contraction-associated translocation of protein kinase C, *FEBS Lett.*, 217, 232, 1987.
108. **MacLeod, R. J., Lembessis, P., and Hamilton, J. R.,** Effect of protein kinase C inhibitors on Cl^- conductance required for volume regulation after L-alanine cotransport, *Am. J. Physiol. Cell Physiol.*, 262, C950, 1992.
109. **Bender, A. S., Neary, J. T., Blicharska, J., Norenberg, L.-O. B., and Norenberg, M. D.,** Role of calmodulin and protein kinase C in astrocytic cell volume regulation, *J. Neurochem.*, 58, 1874, 1992.
110. **Shariff, A. and Luna, E. J.,** Diacylglycerol-stimulated formation of actin nucleation sites at plasma membranes, *Science*, 256, 245, 1992.
111. **Montrose-Rafizadeh, C. and Guggino, W. B.,** Cell volume regulation in the nephron, *Annu. Rev. Physiol.*, 52, 761, 1990.
112. **McCarty, N. A. and O'Neil, R. G.,** Dihydropyridine-sensitive cell volume regulation in proximal tubule: the calcium window, *Am. J. Physiol. Renal, Fluid Electrolyte Physiol.*, 259, F950, 1990.
113. **Wong, S. M. E. and Chase, H. E., Jr.,** Role of intracellular calcium in cellular volume regulation, *Am. J. Physiol.*, 250, C841, 1986.
114. **Rothstein, A. and Mack, E.,** Volume-activated calcium uptake: its role in cell volume regulation of Madin-Darby canine kidney cells, *Am. J. Physiol. Cell Physiol.*, 262, C339, 1992.
115. **Pierce, S. K. and Politis, A. D.,** Ca^{2+}-activated cell volume recovery mechanisms, *Annu. Rev. Physiol.*, 52, 27, 1990.
116. **Christensen, O.,** Mediation of cell volume regulation by Ca influx through stretch-activated channels, *Nature*, 330, 66, 1987.
117. **Filipovic, D. and Sackin, H.,** Stretch- and volume-activated channels in isolated proximal tubule cells, *Am. J. Physiol. Renal, Fluid Electrolyte Physiol.*, 262, F857, 1992.
118. **Watson, P. A.,** Direct stimulation of adenylate cyclase by mechanical forces in S49 mouse lymphoma cells during hyposmotic swelling, *J. Biol. Chem.*, 265, 6569, 1990.
119. **Minina, S. V., Myakotina, O. L., Avdonin, V. B., and Liberman, E. A.,** Mechanical influence and cAMP injection evoke the same reaction of neuron ionic channels, *FEBS Lett.*, 289, 224, 1991.
120. **Kennedy, M. S. and Insel, P. A.,** Inhibitors of microtubule assembly enhance b-adrenergic and prostaglandin-E_1-stimulated cyclic AMP accumulation in S49 lymphoma cells, *Mol. Pharmacol.*, 16, 215, 1979.
121. **Insel, P. A. and Koachman, A. M.,** Cytochalasin B enhances hormone and cholera toxin-stimulated cyclic AMP accumulation in S49 lymphoma cells, *J. Biol. Chem.*, 257, 9717, 1982.
122. **Omann, G. M., Porasik, M. M., and Sklar, L. A.,** Oscillating actin polymerization/depolymerization responses in human polymorphonuclear leukocytes, *J. Biol. Chem.*, 264, 16355, 1989.
123. **Howard, T. H., Wang, D., and Berkow, R. L.,** Lipopolysaccharide modulates chemotactic peptide-induced actin polymerization in neutrophils, *J. Leukocyte Biol.*, 47, 13, 1990.
124. **Carlsson, L., Markey, F., Blikstad, I., Persson, T., and Lindberg, U.,** Reorganization of actin in platelets stimulated by thrombin as measured by the DNase I inhibition assay, *Proc. Natl. Acad. Sci. U.S.A.*, 76, 6376, 1979.
125. **Egan, J. J., Gronowicz, G., and Rodan, G. A.,** Parathyroid hormone promotes the disassembly of cytoskeletal actin and myosin in cultured osteoblastic cells: mediation by cyclic AMP, *J. Cell. Biochem.*, 45, 101, 1991.
126. **Goldman, J. E. and Abramson, B.,** Cyclic AMP-induced shape changes of astrocytes are accompanied by rapid depolymerization of actin, *Brain Res.*, 528, 189, 1990.
127. **Wessels, D., Schroder, N. A., Voss, E., Hall, A. L., Condeelis, J., and Soll, D. R.,** cAMP-mediated inhibition of intracellular particle movement and actin reorganization in *Dictyostelium*, *J. Cell Biol.*, 109, 2841, 1989.
128. **Mercier, F., Reggio, H., Develliers, G., Bataille, D., and Mangest, P.,** Membrane-

cytoskeleton dynamics in rat parietal cells: mobilization of actin and spectrin upon stimulation of gastric acid secretion, *J. Cell Biol.*, 108, 441, 1989.

129. **Lutz, H. U.,** A cyclic AMP-dependent phosphorylation of spectrin dimer, *FEBS Lett.*, 169 (2), 323, 1984.

130. **Hansson, A., Skoglund, G., Lassing, I., Lindberg, U., and Ingelman-Sundberg, M.,** Protein kinase C-dependent phosphorylation of profilin is specifically stimulated by phosphatidylinositol, *Biochem. Biophys. Res. Commun.*, 150 (2), 526, 1988.

131. **LeVine, H. and Sahyoun, N. E.,** Involvement of fodrin-binding proteins in the structure of the neuronal postsynaptic density and regulation by phosphorylation, *Biochem. Biophys. Res. Commun.*, 138, 59, 1986.

132. **Sonobe, S., Takahashi, S., Hatano, S., and Kuroda, K.,** Phosphorylation of ameoba G-actin and its effect on actin polymerization, *J. Biol. Chem.*, 261, 14837, 1986.

133. **Cianci, C. D., Giorgi, M., and Morrow, J. S.,** Phosphorylation of ankyrin down-regulates its cooperative interaction with spectrin and protein 3, *J. Cell Biol.*, 37, 301, 1988.

134. **Hartwig, J. H., Thelen, M., Rosen, A., Janmey, P. A., Nairn, A. C., and Aderem, A.,** MARCKS is an actin filament crosslinking protein regulated by protein kinase C and calcium-calmodulin, *Nature*, 356, 618, 1992.

135. **Murti, K. G., Kaur, K., and Goorha, R. M.,** Protein kinase C associates with intermediate filaments and stress fibers, *Exp. Cell Res.*, 202, 36, 1992.

136. **Jakens, S., Leach, K., and Klauch, T.,** Association of type III protein kinase C with focal contacts in rat embryo fibroblasts, *J. Cell Biol.*, 109, 697, 1989.

137. **Wang, N. and Rasenick, M. M.,** Tubulin-G protein interactions involve microtubule polymerization domains, *Biochemistry*, 30, 10957, 1991.

138. **Wang, N., Yan, K., and Rasenick, M. M.,** Tubulin binds specifically to the signal-transducing proteins, $G_{s\alpha}$ and $G_{\alpha i-1}$, *J. Biol. Chem.*, 265, 1239, 1990.

139. **Rasenick, M. M. and Wang, N.,** Exchange of guanine nucleotides between tubulin and GTP-binding proteins that regulate adenylate cyclase: cytoskeletal modification of neuronal signal transduction, *J. Neurochem.*, 51, 300, 1988.

140. **Shefcyk, J., Yassin, R., Volpi, M., Molski, T. F. P., Naccache, P. H., Munoz, J. J., Becker, E. L., Feinstein, M. B., and Shaafi, R. I.,** Pertussis but not cholera toxin inhibits the stimulated increase in actin association with the cytoskeleton in rabbit neutrophils: role of the "G proteins" in stimulus-response coupling, *Biochem. Biophys. Res. Commun.*, 126, 1174, 1985.

141. **Suzuki, M., Kawahara, K., Ogawa, A., Morita, T., Kawaguchi, Y., Kurihara, S., and Sakai, O.,** $[Ca^{2+}]_i$ rises via G protein during regulatory volume decrease in rabbit proximal tubule cells, *Am. J. Physiol. Renal, Fluid Electrolyte Physiol.*, 258, F690, 1990.

142. **Sontag, J.-M., Aunis, D., and Bader, M.-F.,** Peripheral actin filaments control calcium-mediated catecholamine release from streptolysin-*O*-permeabilized chromaffin cells, *Eur. J. Cell Biol.*, 46, 316, 1988.

143. **Burgoyne, R. D. and Cheek, T. R.,** Reorganisation of peripheral actin filaments as a prelude to exocytosis, *Biosci. Rep.*, 7, 281, 1987.

144. **Aunis, D. and Bader, M.-F.,** The cytoskeleton as a barrier to exocytosis in secretory cells, *J. Exp. Biol.*, 139, 253, 1988.

145. **Davidson, H. W., McGowan, C. H., and Balch, W. E.,** Evidence for the regulation of exocytic transport by protein phosphorylation, *J. Cell Biol.*, 116, 1343, 1992.

146. **Bradbury, N. A. and Bridges, R. J.,** Endocytosis is regulated by protein kinase A, but not protein kinase C in a secretory epithelial cell line, *Biochem. Biophys. Res. Commun.*, 184, 1173, 1992.

147. **Chamberlin, M. E. and Strange, K.,** Anisosmotic cell volume regulation: a comparative view, *Am. J. Physiol.*, 257, C159, 1989.

148. **Schwiebert, E. M., Mills, J. W., and Stanton, B. A.,** unpublished observations.

Chapter 15

The Role of Calcium in the Control of Volume-Regulatory Transport Pathways

J. Kevin Foskett

TABLE OF CONTENTS

I. INTRODUCTION

Among the numerous biochemical and biophysical processes and mediators that have been proposed to serve as transduction mechanisms to couple alterations in cell volume with activation of volume-regulatory solute transport processes, perhaps none has been invoked more frequently than calcium (Ca^{2+}). This does not seem surprising. A change in intracellular Ca^{2+} concentration ($[Ca^{2+}]_i$) is a rapid response to a number of cellular stimuli, including growth factors, neurotransmitters, hormones, peptides, and toxins. $Ca^{2+}{}_i$ has been implicated as having either a direct or participatory role in numerous cell physiological processes. It would be surprising, therefore, to discover that $Ca^{2+}{}_i$ was not involved in some aspects of cellular volume regulation.

The aim of this chapter is to review the evidence for the involvement of $Ca^{2+}{}_i$ in cell volume control. This chapter will attempt to critically analyze published results with the goal of identifying whether Ca^{2+} plays a role inside the cell and, if it does, to determine whether $Ca^{2+}{}_i$ is simply permissive, or whether it plays an active role. To establish the active participation of $Ca^{2+}{}_i$, the following criteria will be considered:

1. Is the stimulus that alters cell volume, e.g., a change in the osmolarity of the medium, or a change in cell solute content under isosomotic conditions, associated with a change in $[Ca^{2+}]_i$ in the cell compartment where the volume regulatory machinery is found? If it is, are the temporal relationships among this change, the stimulus, and the response appropriate for a signaling role of $[Ca^{2+}]_i$?
2. If a change in $[Ca^{2+}]_i$ does occur, what is the basis of that change? Is Ca^{2+} influx across the plasma membrane required for or a contributor to the rise, and to what extent is Ca^{2+} mobilized from intracellular storage sites? If intracellular release of Ca^{2+} is a component of the response, what is the nature of the store that is affected?
3. What is the site of action of $Ca^{2+}{}_i$? How does a change in $[Ca^{2+}]_i$ contribute to cell volume

0-8493-4448-4/94/$0.00+$.50

control? Is a change in $[Ca^{2+}]_i$ necessary? Is it sufficient? Are there specific effector systems that can be shown to possess the required $[Ca^{2+}]_i$ sensitivities?

The most widely used approach for studying cellular volume regulation has been the acute imposition of an osmotic stress on a tissue or cells. This protocol generally involves the rapid addition or deletion of osmotically active solutes from the medium bathing the cells. As outlined in several chapters in this volume, dilution of the medium causes rapid cell swelling, whereas concentration of it causes cell shrinkage, due to flow of water into or out of the cell, respectively. The high water permeability of most cell membranes ensures that these osmotic responses are generally quite rapid. In "anisosmotic volume regulation" studies, the subsequent responses of the cell following the acute gain or loss of water are studied. Another protocol for examining cell volume regulation, which has been much less studied, involves alterations of the intracellular solute content under isosmotic conditions, by manipulations of the activities of solute transport pathways, or by replacement in the medium of impermeant solutes with permeant ones. These "isosmotic volume regulation" studies will be discussed later.

II. ANISOSMOTIC VOLUME REGULATION

A. DEPENDENCE OF VOLUME REGULATION ON Ca^{2+}

As far as I am aware, there are no reports of a Ca^{2+} dependency of activation of solute transport pathways following acute osmotic shrinkage of cells. In contrast, Ca^{2+} has been implicated in the mechanisms that underlie cell volume recovery following an acute osmotic swelling in many cell types. Table 1 summarizes the studies in which a dependence on Ca^{2+} of regulatory volume decrease (RVD) following acute osmotic swelling has been observed. It also indicates which studies have recorded elevations of $[Ca^{2+}]_i$ associated with osmotic swelling. By inspection of Table 1, it is clear that these two types of observations overlap in some studies or cell types, but in many others they do not. That is, osmotic swelling has been observed to increase $[Ca^{2+}]_i$ in some cells in which RVD was not explored, whereas many studies have implicated a Ca^{2+} dependency for RVD but have not determined $[Ca^{2+}]_i$.

Inhibition of RVD when a cell is swollen in the absence of extracellular Ca^{2+} represents the basic observation in most studies which purport to demonstrate a Ca^{2+} dependency of RVD. Dependence of a cellular response on the presence of extracellular Ca^{2+} could reflect a requirement for Ca^{2+} influx into the cell, resulting in a rise of $[Ca^{2+}]_i$, thereby triggering biochemical responses. Alternatively, extracellular Ca^{2+} might play a permissive role, for example as a cofactor for specific ion transport proteins, or in maintenance of cell integrity. In some studies, inhibition of RVD in the absence of extracellular Ca^{2+} could be mimicked in the presence of extracellular Ca^{2+} by loading the cells with a Ca^{2+} chelator,[1,30] suggesting that the extracellular Ca^{2+} dependency reflected a requirement for Ca^{2+} influx into the cell. Nevertheless, determinations of $[Ca^{2+}]_i$ during the osmotic perturbation and subsequent volume-regulatory response are necessary to conclusively establish that extracellular Ca^{2+} removal and intracellular Ca^{2+} chelation represent equivalent experimental maneuvers.

B. $[Ca^{2+}]_i$ DURING OSMOTIC PERTURBATIONS AND VOLUME REGULATION

As indicated above, evidence that Ca^{2+} plays an active role in cell volume regulation should include the demonstration that an alteration of cell volume by osmotic perturbation elicits a rise of $[Ca^{2+}]_i$ in the appropriate cellular compartment. In general, this compartment is the cytoplasm. If the volume-regulatory machinery is localized at the plasma membrane, the rise of $[Ca^{2+}]_i$ must occur, more specifically, in the cytoplasm immediately adjacent to the plasma membrane. Thus, it might be possible that the rise of $[Ca^{2+}]_i$ is so localized that it cannot be detected by methods that report $[Ca^{2+}]_i$ from the entire cytoplasm. For example, neurotransmitter activation of single salivary gland acinar cells elicits a rise of $[Ca^{2+}]_i$ and activation of Ca^{2+}-activated K^+ channels in the basolateral membrane (see Reference 50). By simultaneous electrophysiological recordings of the membrane potential, which is dominated by the activity of that conductance, and photometric measurements of $[Ca^{2+}]_i$, it was determined that the K^+ channels could detect a rise of $[Ca^{2+}]_i$ before that rise could be detected by

Table 1 **Dependence of RVD on Ca^{2+}**

Cell Type	Ca^{2+} Dependence[a]	$\Delta[Ca^{2+}]_i$[b]	Ca^{2+} Influx Pathway	Ref.
Rabbit proximal tubule	Yes — o	o, i Fura 2, 20 cells	Inhibitors — verapamil (10 μM); nifedipine (2.5, 10 μM); nitrendipine (2.5 μM); La^{3+}(500 μM)	1–3
Rabbit proximal tubule	None	i Fura 2, 40 cells		4,5
Cultured rabbit proximal tubule cells	nd	o Fura 2, single cells	Not activated by depolarization	6
Cultured rabbit proximal tubule cells	nd	o Extracellular Ca^{2+}-activated P_{K+}	nd	7,8
Inner medullary duct cells	Yes — o	nd	nd	9
Cultured inner medullary duct cells	Independent of Ca^{2+}_o	o Fura 2, cell suspensions; extracellular Ca^{2+}-activated P_{K+}	Inhibitors — verapamil (10–100 μM); nifedipine (100 μM)	10,11
Opossum kidney cells	Independent of Ca^{2+}_o		Nonselective cation channel, Ba^{2+} permeable	12,13
Goldfish proximal tubule cells	Independent of $Ca^{2+}_{o,i}$			14
MDCK cells	Independent of $Ca^{2+}_{o,i}$			15
MDCK cells	Yes — o	o Indo-1, cell suspensions	Inhibitors — La^{3+} (K_i = 1 mM); insensitive to verapamil (50 μM), nifedipine (20 μM), La^{3+}(20 μM); independent of Na^+_o	16–18
Nonpigmented ciliary cell line	±			19
Guinea pig jejunal villus enterocyte	Yes — o	nd	Inhibitors — nifedipine (5 μM), diltiazem (100 μM), Ni^{2+}, Co^{2+} (1 mM)	20
Rat hepatocytes	Independent of Ca^{2+}_o	o Indo-1, cell suspensions	Stretch-activated, Na^+, K^+, Ca^{2+} permeable, 18-pS channel; insensitive to verapamil (60 μM), nitendipine (0.1 μM), La^{3+} (1 mM)	21,22

Table 1 (continued) **Dependence of RVD on Ca^{2+}**

Cell Type	Ca^{2+} Dependence[a]	$\Delta[Ca^{2+}]_i$[b]	Ca^{2+} Influx Pathway	Ref.
Intestine 407 cells	Yes — o, i	o, i Fura 2, single cells; Ca^{2+}-activated P_{K+}	Ba^{2+}, Ca^{2+} permeable; IP_3-insensitive; inhibitors — Gd^{3+} (10–50 μM)	23–26
Rabbit gastric parietal cells	nd	i Fura 2, single cells		27
Rat salivary gland acinar cells	Independent of $Ca^{2+}_{o,i}$	i Fura 2, single cells		28
Necturus choroid plexus epithelium	nd	Extracellular Ca^{2+}-activated P_{K+}	Stretch activated (no demonstration of osmotic activation) 35–40 pS channel permeable to Ca^{2+}, Na^+, K^+; voltage, Ca^{2+}_i independent	29
Toad urinary bladder cells	Yes — o	o Quin-2, cell suspensions		30
Cultured toad urinary bladder cells	nd	o Fura 2, 40 cells	Inhibitors — verapamil (10 μM), nitrendipine (1 μM), La^{3+} (100 μM)	31
Frog urinary bladder	Yes — o	nd		32
Ehrlich ascites tumor cells	Independent of Ca^{2+}_o, slight dependence on Ca^{2+}_i	nd	Nonselective cation, Ba^{2+}-permeable channel; insensitive to Gd^{3+}, although RVD is.	33,34
Rat osteosarcoma cell line	Yes — o	o Fura 2, cell suspensions	Conductive; inhibitors — nicardipine (K_i = 0.6 μM), La^{3+} (K_i = 7 μM); diltiazem (K_i = 42 μM); verapamil (K_i = 200 μM)	35
GH_4C_1 rat pituitary cells	nd	o Fura 2, 35–40 cells	Inhibitors — verapamil (7.5 μM), nifedipine (2 μM)	36,37
Bovine chromaffin cells	nd	nd Fura 2, single cells		38
N1E115 neuroblastoma cells	Independent of Ca^{2+}_o	nd	Nonselective cation, swelling- and stretch-activated, voltage-independent 22-pS channel	39

Table 1 (continued) **Dependence of RVD on Ca^{2+}**

Cell Type	Ca^{2+} Dependence[a]	$\Delta[Ca^{2+}]_i$[b]	Ca^{2+} Influx Pathway	Ref.
Amphiuma red blood cell	Yes — o	nd Arsenazo III		40,41
Noetia red blood cell	Yes — o	o $^{45}Ca^{2+}$		42–44
Lymphocytes	Independent of $Ca^{2+}_{o,i}$	None quin-2, fura 2, $^{45}Ca^{2+}$		45–47
S49 mouse lymphoma cells	nd	nd Fura 2, cell suspensions		48
Platelets	Independent of Ca^{2+}_o	nd		49

[a] o, Dependence of RVD on extracellular Ca^{2+}; i, dependence of RVD on intracellular Ca^{2+}.
[b] o, i refers to source of Ca^{2+} that causes $[Ca^{2+}]_i$ rise, extracellular or intracellular release, respectively; second line indicates $[Ca^{2+}]_i$ measurement technique.
nd, not determined.

the $[Ca^{2+}]_i$ indicator dye in the cytoplasm.[50] Subsequently, the $[Ca^{2+}]_i$ signal spread into the bulk cytoplasm. Such localized $[Ca^{2+}]_i$ signals can sometimes remain localized and completely escape detection by cytoplasmic indicators, as evidenced by combined photometry and patch clamp electrophysiology of pancreatic acinar cells.[51] In those studies, $[Ca^{2+}]_i$ oscillations, as reflected by oscillations of a Ca^{2+}-activated Cl^- current, could be measured in the absence of any detectable indicator dye signals. The implication from such studies is that $[Ca^{2+}]_i$ might rise, but it could be undetected by some techniques, including the most common ones, which employ fluorescent $[Ca^{2+}]$ indicator dyes. Importantly, however, the failure to measure a $[Ca^{2+}]_i$ signal does not necessarily allow one to conclude that the signal exists but is too localized to measure. It is incumbent upon the investigator to determine that such localized increases do occur. Along these lines, it should be possible to detect a rise of $[Ca^{2+}]_i$ immediately under the plasma membrane by recording the activity of an ion channel whose activity is sensitive to $[Ca^{2+}]$ there, as in the studies just cited. Thus, it might not be necessary to measure $[Ca^{2+}]_i$ directly.

Several studies have detected a rise of $[Ca^{2+}]_i$ associated with osmotic swelling of cells (Table 1). Most of these studies employed fluorescence indicators, usually fura-2, to measure cytoplasmic $[Ca^{2+}]$ directly, and generally they measured $[Ca^{2+}]_i$ in cell suspensions, groups of 20 to 50 cells on coverslips, or in intact kidney tubules. $[Ca^{2+}]_i$ has been determined in single cells in only a limited number of studies. In other studies, activation of Ca^{2+}-activated K^+ channel activity or membrane potential hyperpolarization have been used as indicators.

The temporal relationships between the osmotic perturbation, the change in cell volume, and the subsequent regulatory response can help to establish whether elevated $[Ca^{2+}]_i$, triggered by cell swelling, plays a role in volume regulation. In GH_4C_1 pituitary cells, reduction of the osmotic pressure of the medium by 27% resulted in cell swelling during the subsequent 1 to 2 min.[36] The cells then regulated their volumes nearly completely back to control levels during the subsequent 3 to 4 min. Cell swelling elicited a rise in $[Ca^{2+}]_i$, from resting levels of ~150 n*M*, to nearly 500 n*M*. $[Ca^{2+}]_i$ began to rise after the cells had been exposed to the hypotonic medium for ~10 s, and then continued to increase during the subsequent 60 to 80 s. The extent and rate of rise of $[Ca^{2+}]_i$ were proportional to the degree of medium dilution, and presumably cell volume. The rise of $[Ca^{2+}]_i$ was absolutely dependent upon the presence of extracellular Ca^{2+} (Table 1), suggesting that the extent of cell swelling controls Ca^{2+} permeability of the plasma membrane. The correlation between the increase of cell size, the initiation of the rise of $[Ca^{2+}]_i$ and subsequent $[Ca^{2+}]_i$ changes, and onset of RVD suggests that a rise of $[Ca^{2+}]_i$ might mediate RVD. However, this has not been examined in these cells. To establish an active role of $[Ca^{2+}]_i$, RVD should be examined under conditions where the rise of $[Ca^{2+}]_i$ is blunted, for example by increasing intracellular Ca^{2+} buffering or by the use of plasma membrane Ca^{2+} channel blockers. Failure to observe RVD under such conditions, together with the observations detailed above, would implicate Ca^{2+} in the signal transduction pathways underlying RVD.

Swelling of cultured toad bladder epithelial cells also caused a rise of $[Ca^{2+}]_i$ which was completely dependent upon the presence of extracellular Ca^{2+}, again suggesting that cell swelling activated plasma membrane Ca^{2+} permeability.[31] Whether these cells volume-regulate or not was not determined. However, similar results were obtained in freshly isolated toad bladder cells, which do undergo RVD.[30] RVD required extracellular Ca^{2+}, and was completed in approximately 2 min. The $[Ca^{2+}]_i$ response to dilution of the medium was sluggish, rising only 50 n*M* over a 5-min period. However, the $[Ca^{2+}]$ indicator dye used, quin-2, has low fluorescence, requiring that high concentrations of the dye be loaded into the cell, thereby increasing the intracellular Ca^{2+} buffering. Enhanced Ca^{2+} buffering as a consequence of the $[Ca^{2+}]_i$ measurement may have significantly distorted the normal $[Ca^{2+}]_i$ response. Indeed, the quin-2-loaded cells displayed a significantly inhibited RVD. A similar dye-dependent inhibition of volume regulation was also reported for fura 2-loaded collapsed proximal kidney tubules.[1] Although disruption by the dyes of the normal $[Ca^{2+}]_i$ signal precluded meaningful correlations of the normal volume-regulatory and $[Ca^{2+}]_i$ responses, it did demonstrate the requirement for a rise of $[Ca^{2+}]_i$ in mediating RVD, since buffering the $[Ca^{2+}]_i$ rise inhibited it.

In addition to these cell types, osmotic swelling was associated with increases of $[Ca^{2+}]_i$ in MDCK cells,[18] cultured proximal tubule cells,[6] and the osteosarcoma cell line, UMR-106-01.[35] In all these cell types, the presence of extracellular Ca^{2+} was required for normal RVD, and $[Ca^{2+}]_i$ rose at a rate and extent that was proportional to the degree of cell swelling — 20 to 30% dilution of the medium caused $[Ca^{2+}]_i$ to rise by 150 to 300 n*M*, to 400 to 500 n*M*. In MDCK and UMR-106-01 cells, the rise of $[Ca^{2+}]_i$ began immediately upon medium dilution and reached a peak rapidly (≤1 min). In cultured proximal tubule cells, the rise commenced after a 15-s delay and reached peak values approximately 2 min later. In these cells, RVD commenced at 15 s and was completed within the subsequent 30 to 45 s. Thus, the commencements of the rise of $[Ca^{2+}]_i$ and RVD were coincident. Similarly, RVD by MDCK cells commenced following the $[Ca^{2+}]_i$ peak, which subsequently remained elevated during the entire period of RVD.

RVD by isolated collapsed rabbit renal proximal tubules, as studied by McCarty and O'Neil,[1–3] was similarly dependent upon the presence of Ca^{2+} in the extracellular medium. Swelling elicited a rapid rise in $[Ca^{2+}]_i$ immediately upon dilution of the medium, from ~150 to 300–400 n*M* at peak. Subsequently, $[Ca^{2+}]_i$ relaxed from the peak during the most rapid phase of RVD, and then remained elevated at a plateau in the continued presence of the diluted medium, while RVD progressed. In spite of the extracellular Ca^{2+} dependency of RVD, the rise in $[Ca^{2+}]_i$ was not abolished when cells were swollen in Ca^{2+}-depleted medium, albeit it was transient, suggesting that osmotic swelling resulted in the mobilization of Ca^{2+} from intracellular storage sites. This transient increase in $[Ca^{2+}]_i$, the peak of which was generally smaller than that attained in the presence of extracellular Ca^{2+}, was nevertheless insufficient to even initiate RVD. Thus, sustained Ca^{2+} entry appears to be required to support RVD. If a rise of $[Ca^{2+}]_i$ is necessary to elicit RVD, then raising $[Ca^{2+}]_i$ by other means should also be able to trigger RVD. This issue was explored by treating the cells with the Ca^{2+} ionophore, ionomycin, in the presence of extracellular Ca^{2+} in isosmotic media. Although $[Ca^{2+}]_i$ rose considerably, cell volume was nevertheless unaffected, suggesting that a rise of $[Ca^{2+}]_i$, while perhaps necessary for RVD, is not sufficient. Inhibition of RVD by swelling the cells in Ca^{2+}-free medium could be overcome by reintroducing Ca^{2+} into the hypotonic medium, but only during the first minute of exposure to hypotonicity. Interestingly, the transient rise of $[Ca^{2+}]_i$ observed when the cells were swollen in the absence of extracellular Ca^{2+} lasted at least 1 min, but was unable to support RVD, whereas the sustained $[Ca^{2+}]_i$ increase during the subsequent minutes did support it. Thus, the critical element for effective RVD would appear to be Ca^{2+} influx that occurs during the initial minute of cell swelling.

The Ca^{2+} dependency of RVD in Intestine 407 cells also appears to be more complex than a simple requirement for Ca^{2+} influx and associated rise in $[Ca^{2+}]_i$.[23–26] In these cells, removal of extracellular Ca^{2+}, or buffering of intracellular Ca^{2+} with quin-2, both abolished RVD. Hypotonicity elicited a biphasic rise in $[Ca^{2+}]_i$, measured either directly with fura-2 or indirectly by recording a K^+-dependent membrane hyperpolarization. An early $[Ca^{2+}]_i$ rise was initiated shortly after dilution of the medium, which peaked at 160 n*M* approximately 30 s later, and then subsided to close to resting levels during the next 30 s. At that time, a second, similar $[Ca^{2+}]_i$ transient occurred. Both $[Ca^{2+}]_i$ rises were dependent upon the presence of extracellular Ca^{2+}, suggesting that Ca^{2+} influx contributed to them. As discussed in more detail below, the second rise in $[Ca^{2+}]_i$ appeared to be due to Ca^{2+} release from an intracellular store, triggered by the initial influx-dependent rise of $[Ca^{2+}]_i$.

In cultured rabbit medullary thick ascending limb (MTAL) cells, hypotonicity similarly caused an extracellular Ca^{2+}-dependent rise in $[Ca^{2+}]_i$, the initial rate of change and extent of which were dependent upon the degree of medium dilution,[11] as for the cells cited above.[6,18,30,31,35,36] $[Ca^{2+}]_i$ began to rise immediately upon exposure of the cell to the diluted medium, a temporal response consistent with an intracellular mediator role of Ca^{2+}. Nevertheless, the extent and rate of RVD were unaffected by removal of extracellular Ca^{2+}. Reduction of $[Ca^{2+}]_i$ to <50 n*M* in the MTAL cells, by prolonged exposure to Ca^{2+}-free media, did slow RVD, but since those $[Ca^{2+}]_i$ were outside of the physiological range, they likely indicate at most that $Ca^{2+}{}_i$ plays a permissive role. Similarly, $[Ca^{2+}]_i$ rose in rat hepatocytes, due to enhanced Ca^{2+} influx when the medium was made hyposmotic, but RVD was independent of extracellular Ca^{2+}.[21] These data suggest the following. First, elevated $[Ca^{2+}]_i$ may mediate volume regulation, but the important source is intracellular stores, rather than extracellular influx, and furthermore the elevated $[Ca^{2+}]_i$ was localized and could not be detected by fluorescence dyes. Osmotic swelling of the MTAL cells was associated with enhanced activity of Ca^{2+}-activated K^+ channels, which was absolutely dependent upon the presence of extracellular Ca^{2+}.[8] The lack of K^+ channel activity during osmotic swelling in Ca^{2+}-free medium suggests therefore that any "invisible" $Ca^{2+}{}_i$ released from intracellular stores did not colocalize with these channels (apparently the apical membrane). Possibly, therefore, "invisible" $Ca^{2+}{}_i$ at the basolateral membrane activated undefined channels. These considerations raise an important point. If channel activity is to be used as an indicator of the $[Ca^{2+}]$ immediately under the plasma membrane, then the right channels need to be measured. That is, unless it is established that a particular channel activity is involved in the volume-regulatory response, then measurement of its activity to indicate $[Ca^{2+}]_i$ may not measure $[Ca^{2+}]_i$ at the relevant site for volume regulation. This might be especially important in epithelial cells, which have distinct apical and basolateral domains, containing specific channel types. Alternately, the ability of the cells to volume regulate in the absence of extracellular Ca^{2+} may indicate that the rise of $[Ca^{2+}]_i$ observed during osmotic swelling in the presence of extracellular Ca^{2+} may be an epiphenomenon in the RVD process. To resolve this, $[Ca^{2+}]_i$ should be clamped at physiological levels, by use of permeable Ca^{2+} chelators with the appropriate Ca^{2+} affinities, and then challenged with an osmotic shock, as described above in toad bladder cells. If the cells do not display RVD under these conditions, the hypothesis of "invisible" $Ca^{2+}{}_i$ released by an intracellular Ca^{2+} store seems likely. It would then be necessary to define the type of intracellular stores. This issue will be addressed later. If, on the other hand, the chelator-loaded cells continue to display RVD when swollen in the absence of extracellular Ca^{2+}, then the normally observed rise of $[Ca^{2+}]_i$ seems unlikely to be a mediator of RVD.

In summary, in most cells in which Ca^{2+} has been implicated in the volume-regulatory response following acute osmotic swelling, Ca^{2+} influx from the extracellular medium has been implicated as a key event. The Ca^{2+} requirement of RVD appears to reflect the need for a rise in $[Ca^{2+}]_i$. Nevertheless, many cells appear to exhibit no Ca^{2+} dependence of RVD, including some cells in which a Ca^{2+} influx and/or $[Ca^{2+}]_i$ rise is activated by cell swelling. Review of the published studies reveals that the comprehensive analyses necessary to establish an active role of $[Ca^{2+}]_i$ is lacking in many of the cells in which Ca^{2+} has been implicated.

C. CHARACTERISTICS OF VOLUME-ACTIVATED Ca^{2+} INFLUX

As discussed above, acute cell swelling appears to activate Ca^{2+} influx in several cell types. The characteristics of swelling-induced Ca^{2+} influx pathways are summarized in Table 1. Inspection of Table 1 reveals that the properties of the Ca^{2+} influx pathways that are implicated in volume regulation have been explored in only a few cell types. Furthermore, detailed characteristics of specific Ca^{2+}-permeable channels which appear to be activated by hypotonicity are known primarily in cells which (1) are not dependent upon Ca^{2+} influx for RVD or (2) have not been demonstrated to undergo Ca^{2+}-dependent volume regulation. The physiological relevance for RVD of swelling-activated Ca^{2+}-permeable channels in cells that do not depend upon Ca^{2+} influx for RVD is unknown. Likewise, the physiological significance for RVD of swelling-activated Ca^{2+}-permeable channels in cells that have not been demonstrated to undergo RVD remains to be established.

Christensen's study[29] of salamander choroid plexus epithelium established a paradigm for the role in RVD of swelling-activated Ca^{2+}-permeable channels, which has been adopted by many subsequent investigators. In that study, the patch clamp technique was employed in cell-attached and inside-out patch configurations. In both configurations, application of suction to the pipette activated a 34-pS cation-selective channel which discriminated poorly among cations, allowing permeation by monovalent as well as divalent ions, including Ca^{2+}. Internal Ca^{2+} was not necessary to activate the channel, and it displayed little voltage dependence. Sometimes the cell-attached patches also contained large-conductance K^+ channels. These channels could not be activated by suction unless the nonselective cation channels were also present in the same patch and Ca^{2+} was present in the pipette. The implication from these observations was that stretch activation of the nonselective channel allowed Ca^{2+} to permeate through the channels and then activate neighboring K^+ channels. Hypotonicity also activated the large-conductance K^+ channels in the presence but not the absence of extracellular Ca^{2+}. Taken together, these observations suggested that hypotonicity activated the same Ca^{2+}-permeable nonselective cation channels as observed to be activated by suction, and the resultant Ca^{2+} influx then activated the K^+ channels, which provided a pathway for K^+ efflux to effect volume regulation. This tidy model has been widely accepted as a paradigm for the involvement of Ca^{2+} in cellular volume regulation. Nevertheless, the model lacks key experimental support. Most importantly, no evidence was presented to demonstrate a role for the nonselective cation channels when the cells were swollen osmotically. First, there was no demonstration that the nonselective cation channels were activated by hypotonicity. Second, no attempt was made to specifically block the nonselective channel, for example with Gd^{3+}, and show that the swelling-induced increase in K^+ channel activity was likewise blocked. Finally, the demonstrations of stretch activation of the K^+ channels were always in the presence of 15 to 75 m*M* Ca^{2+} in the pipette. There was no demonstration that the nonselective cation channels have sufficient Ca^{2+} permeability to allow enough Ca^{2+} entry to activate the K^+ channels when extracellular Ca^{2+} was at physiological levels. These issues address the physiological relevance of such observations. They are important considerations since, as demonstrated in Table 1, Ca^{2+}-permeable nonselective cation channels have been observed to be activated by membrane stretch or hypotonicity in many cell types in which there is no demonstrable role for plasma membrane Ca^{2+} influx in RVD.

When those cell types that demonstrate a requirement of Ca^{2+} influx for RVD are considered, the swelling-induced Ca^{2+} influx pathways do not obviously share consistent properties. Many workers in this field have employed organic Ca^{2+} channel blockers, even in nonexcitable cells, such as epithelial cells, where voltage-dependent Ca^{2+} channels are believed not to be present. This undoubtedly reflects, in part, the absence of suitable Ca^{2+} channel blockers for these cell types. Nevertheless, Ca^{2+} channels that are activated by agonist stimulation of epithelial cells have sensitivities to a number of inorganic blockers, such as Ni^{2+}, Co^{2+}, Zn^{2+}, Mn^{2+}, Cd^{2+}, La^{3+}, and Gd^{3+}. Surprisingly, these blockers have not generally been employed in volume regulation studies. La^{3+} is generally considered to be an effective blocker of Ca^{2+} channels at micromolar concentrations. Nevertheless, relatively high concentrations were necessary to block Ca^{2+} influx activated by hypotonicity in MDCK cells[18] and rabbit proximal tubules.[1] Gd^{3+} (50 μ*M*) blocked Ca^{2+} influx in Intestine 407 cells,[26] although this concentration is generally somewhat higher than those required to block nonselective cation channels, for example. Ni^{2+} and Co^{2+}, both at 1 m*M*, blocked volume regulation of jejunal enterocytes, presumably by blocking Ca^{2+} influx.[20] The Ca^{2+} permeability in MDCK cells[17] and hepatocytes[21] appears to be insensitive to organic channel blockers. Proximal tubules, however, display a relatively high sensitivity to verapamil, and moderate sensitivities to dihydropyridines.[1–3] Toad urinary bladder epithelial cells also display similar sensitivities to these organic channel blockers,[31] and volume regulation by jejunal enterocytes was also moderately sensitive to a dihydropyridine.[20] By comparison, the Ca^{2+} permeation pathway in osteosarcoma cells, which possess voltage-activated, dihydropyridine-sensitive Ca^{2+} channels, is very sensitive to a dihydropyridine, but relatively insensitive to verapamil, or diltiazem.[35] It is important to note that sensitivities to organic Ca^{2+} channel blockers may not necessarily reflect a direct interaction of these agents with the Ca^{2+} permeation pathway. For example, verapamil inhibition of K^+ channels[52] would result in depolarization of the plasma membrane, thereby reducing the driving force for Ca^{2+} influx. Voltage-sensitive Ca^{2+} channels are completely inhibited by dihydropyridines at concentration <1 μ*M*, whereas higher concentrations have been used in volume regulation studies. At such concentrations, dihydropyridines, which are lipophilic, may have nonspecific effects.[53] Unfortunately, with the exception of Intestine 407 cells,[26] no direct electrophysiological measurements of the Ca^{2+} currents or channels have been performed in cells in which a role for Ca^{2+} influx in RVD has been established.

Regulation of the hypotonicity-induced Ca^{2+} influx pathways is largely unknown. In GH_4C_1 cells, swelling-induced Ca^{2+} influx was inhibited by a protein kinase C inhibitor or downregulation of protein kinase C,[37] suggesting that protein kinase C may be involved in the transduction pathway that couples swelling to activation of the Ca^{2+} channel.

D. ROLE OF INTRACELLULAR Ca^{2+} STORES

In most cell types studied to date, the rise of $[Ca^{2+}]_i$ associated with exposure to hypotonic solutions is a consequence of swelling-activated enhancement of Ca^{2+} influx from the external medium, as discussed above, and summarized in Table 1. Nevertheless, internal stores of Ca^{2+} appear to contribute to a $[Ca^{2+}]_i$ rise in some cells. As mentioned above, rabbit proximal tubules display a transient rise of $[Ca^{2+}]_i$ when they are swollen in the absence of extracellular Ca^{2+}.[2,5] Similarly, gastric parietal cells also display a transient rise of $[Ca^{2+}]_i$ when cells are exposed to a Ca^{2+}-free solution diluted by 45%.[27] Likewise, a rapid, transient rise of $[Ca^{2+}]_i$ is observed in single rat parotid salivary gland acinar cells exposed to a Ca^{2+}-free solution diluted by 25% (Figure 15–1). The rise of $[Ca^{2+}]_i$ was associated with the osmotic swelling phase of the response, as determined simultaneously in the same single cell. Relaxation of $[Ca^{2+}]_i$ generally preceded the initiation of RVD, and resting $[Ca^{2+}]_i$ was achieved before RVD was complete.[28]

Modulation of $[Ca^{2+}]_i$ is a ubiquitous signaling system elicited by external signals in a wide variety of cell types. In nonexcitable cells, a generally accepted model for Ca^{2+} signaling has been developed to account for ligand/receptor elevation of $[Ca^{2+}]_i$. In this scheme, receptor binding activates hydrolysis of phosphatidylinositol 4,5-bisphosphate (PIP_2) by phospholipase C, with release of the soluble second messenger, inositol (1,4,5) trisphosphate (IP_3).[54] IP_3 mobilizes Ca^{2+} from intracellular, nonmitochondrial storage sites, probably endoplasmic reticulum, by its specific interaction with a receptor, which has been isolated, cloned, and demonstrated to function as the Ca^{2+}-release channel.[55]

In salivary acinar cells and gastric parietal cells, the osmotically induced elevation of $[Ca^{2+}]_i$ appears to be due to Ca^{2+} release from IP_3-sensitive Ca^{2+} stores,[27,28] since depletion of the IP_3-sensitive store abolished the swelling-induced rise of

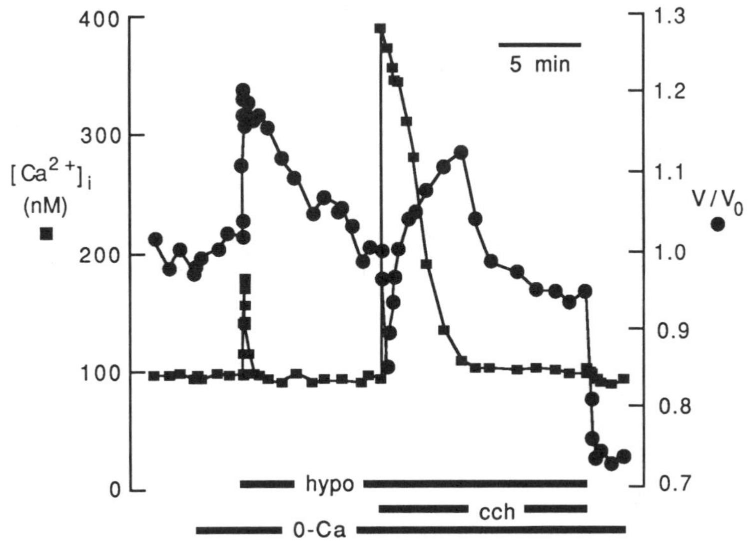

Figure 15–1 $[Ca^{2+}]_i$ and regulatory volume decrease (RVD) following osmotic swelling of a single rat salivary gland (parotid) acinar cell in the absence of extracellular Ca^{2+}. Single cell volume (V, normalized to initial volume V_0; solid circles) and $[Ca^{2+}]_i$ (solid squares) were determined simultaneously, as described in Reference 89. The cell was exposed to a Ca^{2+}-free medium (0-Ca) diluted by 25% (hypo) by removal of NaCl. Subsequent to complete RVD, exposure to 10 μM carbachol (cch) in continued presence of hypotonic medium and absence of extracellular Ca^{2+} elicited normal $[Ca^{2+}]_i$ and cell volume responses. (From Foskett, J. K. et al., *J. Exp. Zool.*, in press.)

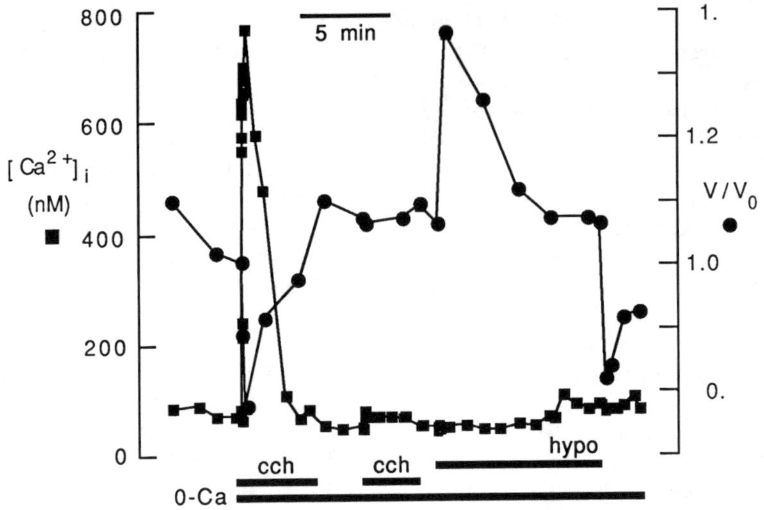

Figure 15–2. A rise of $[Ca^{2+}]_i$ is not required for RVD. Depletion of the IP_3-sensitive intracellular Ca^{2+} store by exposure to 10 μ*M* carbachol (cch) in the absence of extracellular Ca^{2+} (0-Ca) was without effect on RVD by a single salivary gland acinar cell following 25% dilution of the bath (hypo). Single cell volume (V, normalized to initial volume V_0; solid circles) and $[Ca^{2+}]_i$ (solid squares) were determined simultaneously.

$[Ca^{2+}]_i$ in Ca^{2+}-free medium. As shown in Figure 15–2, the IP_3-sensitive store in a single salivary acinar cell can be depleted by a single exposure to 10 μ*M* carbachol, a muscarinic agonist, since a subsequent exposure fails to release more Ca^{2+}. In the continued absence of Ca^{2+} in the medium, rapid dilution of the medium by 25% was without effect on $[Ca^{2+}]_i$, in contrast to the situation when the IP_3-sensitive stores had not been previously depleted. Similar results were obtained in gastric parietal cells.[27] In these two cell types, therefore, cell swelling caused a mobilization of Ca^{2+} from IP_3-sensitive stores. In neither cell type was the IP_3-sensitive store significantly depleted by swelling, however, since a subsequent exposure to an agonist elicited normal Ca^{2+} release (Figure 15–3).[27,28] Significantly, RVD in the salivary acinar cells proceeded normally in the absence of extracellular Ca^{2+} in spite of prior depletion of the IP_3-sensitive pool (Figure 15–2). Thus, in this cell type, RVD is not dependent on a rise of $[Ca^{2+}]_i$. Thus, the normal swelling-induced rise of $[Ca^{2+}]_i$ would appear to be an epiphenomenon of the RVD process. Similarly, collapsed rabbit proximal tubules studied by Breton et al.[5] responded to osmotic dilution of the medium with a rapid transient rise of $[Ca^{2+}]_i$ which was unaffected by the removal of extracellular Ca^{2+}, implying that it arose as a consequence of release from internal stores. Release from these stores could be blocked by TMB-8, but this agent was without effect on RVD. In this cell type as well, therefore, cell swelling caused a release of Ca^{2+} from intracellular stores, but the rise in $[Ca^{2+}]_i$ was not required for RVD.

The mechanisms that underly osmotic mobilization of Ca^{2+} from the IP_3-sensitive Ca^{2+} store is unknown in the cells in which it has been observed. The most obvious possibility is that hypotonicity stimulates production of IP_3. Swelling causes a transient rise of IP_3 in hepatocytes[56,57] and proximal tubule cells,[6] supporting such a suggestion. The mechanisms that might underlie cell swelling-induced enhancement of IP_3 production are unknown. One intriguing possibility has its basis in the recent observations of the coupling between cytoskeletal proteins and PIP_2. Profilin, the most abundant actin-binding protein in cells, binds PIP_2 and competes with PLC for interaction with PIP_2.[58] Tyrosine phosphorylation of PLC-γ by tyrosine-kinase growth factors overcomes the inhibitory effect of profilin, resulting in activation of PLC-γ.[59] By analogy, osmotic cell swelling might activate PLC by overcoming the inhibitory effects of profilin, perhaps by activating a tyrosine kinase, by a mechanical effect mediated by a cytoskeletal response to swelling, or some other means. PIP_2 also interacts with other actin-binding proteins, including α-actinin and filamin.[60] Osmotic swelling-induced modulation of these interactions might promote PIP_2 hydrolysis. Alternatively, cell swelling may not raise IP_3 levels in the cells in which it has been observed to mobilize Ca^{2+} from the IP_3-sensitive

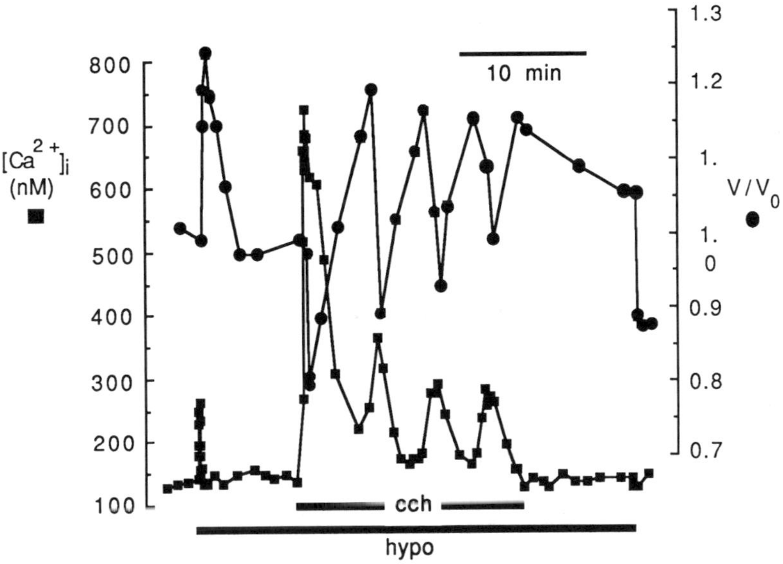

Figure 15–3. $[Ca^{2+}]_i$ and regulatory volume decrease (RVD) following osmotic swelling of a single rat salivary gland (parotid) acinar cell. Single cell volume (V, normalized to initial volume V_0; solid circles) and $[Ca^{2+}]_i$ (solid squares) were determined simultaneously. The cell was exposed to a medium diluted by 25% (hypo) by removal of NaCl. Subsequent to complete RVD, exposure to 10 μM carbachol (cch) in continued presence of hypotonic medium elicited normal $[Ca^{2+}]_i$ and cell volume oscillations. (From Foskett, J. K. et al., *J. Exp. Zool.,* in press.)

store, implying that other mechanisms are important. For example, swelling of the Ca^{2+} store may activate the release pathway, although this has not been demonstrated.

Another type of Ca^{2+} store may also participate in the $[Ca^{2+}]_i$ response to hypotonicity. As discussed earlier, osmotic swelling of Intestine 401 cells caused a biphasic rise of $[Ca^{2+}]_i$.[24,25] Both rises were dependent upon the presence of extracellular Ca^{2+}. However, the second rise could be modulated by pharmacological agents that are known to interact with the Ca^{2+} release process in muscle sarcoplasmic reticulum.[24] Another release channel, the ryanodine receptor, which is different from, albeit related to, the IP_3 receptor, is sensitive to $[Ca^{2+}]_i$, which induces its opening by a process of Ca^{2+}-induced Ca^{2+} release (CICR), causing Ca^{2+} release from the sarcoplasmic reticulum.[61] Caffeine enhances the Ca^{2+} sensitivity of the release process, whereas ryanodine inhibits the channel by binding to it and closing it or locking it into a permanently open subconducting state. The second rise of $[Ca^{2+}]_i$ following swelling of Intestine 409 cells occurred earlier in the presence of caffeine, and it was impaired in the presence of ryanodine, adenine, and procaine.[24] These data suggested that the second rise might be due to Ca^{2+} release from a Ca^{2+} store that has pharmacological sensitivities similar to sarcoplasmic reticulum. Interestingly, RVD was impaired by the agents that disrupted the second phase of the $[Ca^{2+}]_i$ response.[24] The authors suggested that swelling-activated Ca^{2+} influx triggered a release of Ca^{2+} by a CICR process, and that it was the resultant rise of $[Ca^{2+}]_i$ which was necessary for volume regulation. At the present time, this represents the only demonstration of a role or requirement for Ca^{2+} release from an intracellular store for volume regulation.

E. MECHANISMS BY WHICH $Ca^{2+}{}_i$ MODULATES VOLUME REGULATION

As discussed in detail in other chapters in this volume, RVD is mediated by the activities of various transport pathways, including K^+/H^+ exchange, KCl cotransport, separate K^+ and Cl^- channels, and organic solute permeases (in addition, see References 13 and 62 to 65). Although $Ca^{2+}{}_i$ plays a role in activating these pathways in different cell types, the issue here is which pathways are activated by $Ca^{2+}{}_i$ in cells in which it is believed to play a role in volume regulation. $Ca^{2+}{}_i$ could activate solute transport pathways by activation of quiescent channels or transporters, by upregulation of active channels or transporters, or by recruitment of new transport pathways. These effects could be mediated directly by $Ca^{2+}{}_i$, or via Ca^{2+}-dependent intermediates, such as calmodulin and protein kinases and phosphatases.

1. Ca^{2+}-Activated K^+ Channels

In most cell types examined, RVD appears to be mediated by separate K^+ and Cl^- channels.[13,62–65] Because Ca^{2+} was demonstrated many years ago to activate K^+ channels in red blood cells, the Gardos effect,[66] early workers in the field of volume regulation focused on $Ca^{2+}{}_i$ as a likely mediator of swelling-induced K^+ channel activation.[67] A critical role for Ca^{2+}-activated K^+ channel activity in RVD has been demonstrated most convincingly in Intestine 401 cells.[23,25] RVD was dependent upon a rise of $[Ca^{2+}]_i$, and this rise correlated temporally with a membrane hyperpolarization due to activation of a K^+ conductance, which was dependent upon the $[Ca^{2+}]_i$ rise. K^+ conductance could also be activated by intracellular Ca^{2+} injections, and the sensitivity of this conductance (blocked by Ba^{2+} and quinine) was similar to the block of the swelling-induced hyperpolarization and RVD. In MDCK cells,[17] cell swelling was associated with a rise of $[Ca^{2+}]_i$ and enhanced Ca^{2+} influx, and RVD was dependent upon the presence of extracellular Ca^{2+}. Swelling activated a K^+ conductance, which was rate-limiting for RVD, as evidenced by the observation that the cation ionophore gramicidin could accelerate it. Furthermore, swelling elicited an early hyperpolarization, consistent with activation of a K^+ conductance.

Activation of Ca^{2+}-activated K^+ channels by hypotonicity has been observed in cultured proximal tubule cells,[7,8] Ehrlich ascites tumor cells,[34] neuroblastoma cells,[39] cultured thick ascending limb cells,[10] choroid plexus epithelium,[29] and *Xenopus* oocytes injected with proximal tubule mRNA.[68] Nevertheless, the role of these conductances in mediating RVD was not established in those studies. K^+ conductances are activated by swelling or stretch in several cell types, where the response appears to be independent of Ca^{2+}.[4,47,69–72]

Many workers in this field have relied heavily on inhibitor sensitivities to define the types of conductances that underlie RVD. Sarkadi and Parker[63] have reviewed the parallel pharmacological sensitivities of Ca^{2+}-activated K^+ channels and those K^+ channels that are activated by cell swelling. Nevertheless, since there is no specific pharmacology for specific classes of K^+ channels, such overlapping pharmacological sensitivities must be interpreted cautiously. A review of the literature indicates that block of RVD by quinine is often cited as evidence for a role of Ca^{2+}-activated K^+ channels in the RVD response. However, quinine is a most nonspecific drug, inhibiting Cl^- channels, nonselective cation channels, K^+ channels, and affecting Na^+ and Ca^{2+} currents (see Reference 73). Thus, any effect of this class of drugs must be interpreted with extreme caution. Furthermore, other K^+ channel blockers, including TEA, Ba^{2+}, and charybdotoxin, are also not specific for Ca^{2+}-activated K^+ channels (see Reference 74 for example).

2. Ca^{2+}-Activated Cl^- Channels

Cell swelling activates Cl^- conductances in a wide variety of cell types. Nevertheless, there is no good evidence that Ca^{2+}-activated Cl^- channels are activated by cell swelling or that they play a role in cell volume regulation. Several studies have demonstrated that swelling-activated Cl^- conductances are insensitive to $[Ca^{2+}]_i$ or can be activated at $[Ca^{2+}]_i$ close to 0 (Intestine 401 cells,[23,75] ciliary epithelial cell line,[72] ventricular myocytes,[76] T84 colon carcinoma cell line,[77–79] airway epithelial cells,[77,79] MDCK cells,[17,80,81] lymphocytes,[67] Ehrlich ascites cells,[34] bovine chromaffin cells[38]) or have properties that differ from those that are activated by $[Ca^{2+}]_i$ in the same cells (lymphocytes,[82] airway epithelial cells[83,84]).

In Ehrlich ascites tumor cells, a low-conductance Cl^- channel that was activated by cell swelling was insensitive to $[Ca^{2+}]_i$ in excised patches, yet it could be activated in the cell by a Ca^{2+} ionophore. In this cell type, a volume-activated Cl^- channel might be activated by Ca^{2+}- induced intracellular mediators.[34]

3. Role of Ca^{2+}-Calmodulin

Numerous studies, excellently reviewed by Pierce and Politis,[85] have implicated Ca^{2+}/calmodulin as a mediator of RVD. In almost all cases, this conclusion has been based on the sensitivity of this process to calmodulin inhibitors. As pointed out by these authors, however, the lack of specificity of the drugs, primarily phenothiazines, requires numerous control experiments, and these have rarely been carried out. Thus, many of the studies that have purported to demonstrate a role for Ca^{2+}-calmodulin in cell volume regulation have been insufficiently controlled to be entirely confident that it indeed played a role. Perhaps the most thoroughly controlled study was that of Pierce et al.,[44] which examined the dose dependency of phenothiazines and their sulfoxide derivatives on both RVD and calmodulin inhibition in molluscan red blood cells. Furthermore, they examined the role of protein kinase C in order to rule out phenothiazine inhibition of this enzyme, and demonstrated the presence of calmodulin in the cells. The lack of specific inhibitors has certainly impeded this area of investigation. Nevertheless, more specific approaches may now be available. Specific peptide inhibitors of calmodulin and Ca^{2+}-calmodulin-dependent protein kinase (CaMK) have been developed.[86] These peptides are synthetic mimics of the autoinhibitory domains in these enzymes and are therefore quite specific. In whole cell patch clamp recordings of airway epithelial cells, Chan et al.[83] observed separate swelling-activated and Ca^{2+}-activated Cl^- currents in airway epithelial cells. By inclusion of the specific peptide inhibitor of CaMK in the pipette solution, they demonstrated that Ca^{2+}-activated Cl^- currents were dependent upon CaMK for activation by $Ca^{2+}{}_i$, whereas the swelling-activated Cl^- currents were unaffected. Unfortunately, the calmodulin peptide-inhibitor was not employed in this study; thus, although an involvement of CaMK was ruled out, a role for calmodulin was not. Because the peptide was introduced into the cells by dialysis during whole cell patch clamping, a mode which precludes volume regulation, RVD was also not examined. However, it should be possible to microinject these peptides into intact single cells and then examine their volume-regulatory behavior. Such studies, which have not yet been carried out, may provide a more specific approach to elucidate the possible involvement of calmodulin in cell volume regulation.

4. Other Ca^{2+}-Mediated Mechanisms

The cytoskeleton and arachidonic acid metabolites have been implicated in volume-regulatory responses.[13,63,64] Since $[Ca^{2+}]_i$ is known to modulate both, $[Ca^{2+}]_i$ might mediate effects on volume regulation via these pathways. The reader is referred to the chapters on cytoskeleton[87] and eicosanoids[88] in this volume.

III. ISOSMOTIC VOLUME REGULATION

A. CA^{2+}-DEPENDENT CELL VOLUME CHANGES

The activities of various plasma membrane solute influx and efflux pathways can be modulated by a wide variety of physiological mechanisms under isosmotic conditions. It is conceivable, even likely, that modulation of these activities might result in changes in total cell solute content. As is the situation in anisosmotic media during volume regulation, changes in cell solute content under isosmotic conditions will result in parallel changes in cell volume. Since many growth factors, neurotransmitters, and hormones cause a rise of $[Ca^{2+}]_i$, and $[Ca^{2+}]_i$ modulates the activities of various solute transport pathways, it is possible that such external stimuli could alter cell volume under isosmotic conditions. Surprisingly, however, regulation of cell volume under isosmotic conditions has not been studied nearly as extensively as under anisosmotic conditions.

To explore the possibility that agonist modulation of $[Ca^{2+}]_i$ and consequent $[Ca^{2+}]_i$ regulation of ion transport processes could result in changes in cell volume under isosmotic conditions, the relationship between $[Ca^{2+}]_i$ and cell volume was explored in single salivary gland acinar cells during activation of fluid secretion.[89] The parotid salivary gland from the rat normally secretes fluid at very low basal rates, but it can be rapidly stimulated to secrete large quantities of fluid, by reflex parasympathetic innervation. Elevation of $[Ca^{2+}]_i$ is believed to be the primary regulator of salivary fluid secretion. Differential interference contrast (DIC) microscopy was used to image cell volume, while the Ca^{2+} indicator dye, fura-2, simultaneously imaged using a DIC-fluorescence microscope[90] in single salivary gland acinar cells, was used to examine the effects of agonist stimulation on cell volume and $[Ca^{2+}]_i$. Agonist stimulation of fluid secretion was initially associated with a rapid tenfold increase in $[Ca^{2+}]_i$ in single acinar cells, which was associated with a rapid, substantial cell shrinkage (for examples see Figures 1 to 3), which amounted to a 20 to 25% loss of cell water within the initial 10 to 30 s of stimulation. Close examination of the relationship between the rise of $[Ca^{2+}]_i$ and the cell shrinkage indicated that the $[Ca^{2+}]_i$ rise preceded the volume loss, suggesting that elevated $[Ca^{2+}]_i$ signaled the shrinkage.

Since cell volume changes in the absence of perturbations of the osmolarity of the external medium can be interpreted to indicate that cell solute content has changed, the Ca^{2+}-induced shrinkage would be consistent with a Ca^{2+}-induced loss of cell solute, and osmotically obliged water. To account for a 20% volume loss, ~60 mOsmol/l cell water must be lost. Since these cells, as for most cell types, maintain intracellular K^+ concentration at quite high levels, and intracellular Cl^- is generally the only permeant anion in significant quantities, the most likely explanation for the shrinkage is that the cells lost ~30 mmol/l cell water KCl. Such a conclusion would be consistent with a general model of epithelial fluid secretion, in which Ca^{2+}-activated Cl^- and K^+ channels exist on the apical and basolateral membranes, respectively. Thus, the agonist-induced shrinkage likely reflects activation of these two types of ion channels.

Although cell volume changes under isosmotic conditions likely reflect changes in cell solute content, they might relect cytoskeletal contraction, vesicular water transport, change in the gelsol state of the cytoplasm, or alteration in the total quantity of, or net charge on, impermeant intracellular macromolecules.[91,92] Nevertheless, the proposition that changes in cell volume induced by stimulation in fact reflect changes in cell solute content was confirmed by simultaneous measurements of intracellular Cl^- activity ($[Cl^-]_i$) and cell volume in a single acinar cell during agonist stimulation, using a Cl-indicator dye and the DIC-fluorescence microscope.[93]

As in many cell types, the initial rapid rise of $[Ca^{2+}]_i$ following agonist treatment is primarily due to Ca^{2+} release from intracellular Ca^{2+} stores.[89,94] Subsequently, enhanced Ca^{2+} influx across the plasma membrane from the extracellular medium sustains the Ca^{2+} signal. The magnitude of the sustained phase of $[Ca^{2+}]_i$ elevation varied considerably among individual acinar cells. Generally, $[Ca^{2+}]_i$ remained considerably higher than resting levels for as long as the agonist was present. Removal of the agonist caused $[Ca^{2+}]_i$ to return rapidly to resting levels. In some cells, however, $[Ca^{2+}]_i$ returned close to resting levels in the continued presence of agonist. Despite this variability among cells in the level and kinetics of the sustained phase of the $[Ca^{2+}]_i$ response, cell volume was correlated with $[Ca^{2+}]_i$ in all cases.[89] That is, elevations of $[Ca^{2+}]_i$ that were sustained were associated with sustained shrinkage; $[Ca^{2+}]_i$ relaxation to intermediate levels was associated with the recovery of volume to intermediate levels, and transiently elevated $[Ca^{2+}]_i$ was associated with shrinkage followed by volume recovery. Volume recovery implies that cell solute content increased. Thus, activation of these cells is initially associated with solute loss and cell shrinkage, and subsequently with solute uptake and cell swelling. Other experiments demonstrated that elevated $[Ca^{2+}]_i$ was necessary (by buffering the $[Ca^{2+}]_i$) and sufficient (mimicked by a Ca^{2+} ionophore) to induce these changes in cell volume. Perhaps the most convincing demonstration of the tight coupling between $[Ca^{2+}]_i$ and cell solute content was provided by oscillating cells. $[Ca^{2+}]_i$ oscillations in acinar cells are generally sinusoidal-like with a 5- to 6-min period. These $[Ca^{2+}]_i$ oscillations were associated with coincident cell volume oscillations (Figure 15–3). Within a single cell, therefore, except for the earliest several seconds of stimulation (since the initial rise of $[Ca^{2+}]_i$ preceded any change in volume), cell solute content was directly and inversely related to $[Ca^{2+}]_i$ during the entire oscillatory period, over a wide range of $[Ca^{2+}]_i$.[89]

Given the ubiquitous use by cells of elevations of $[Ca^{2+}]_i$ to signal many cell physiological processes, the widespread occurrence among cells of Ca^{2+}-activated Cl^- and K^+ channels suggests that cell shrinkage might be a common feature associated with cell stimulation. Although this has not been examined in many cell types, some recent data suggest that this may be true. Other fluid secretory cells, including submandibular gland and sweat gland secretory coil cells, shrink upon activation by $[Ca^{2+}]_i$ elevating agonists.[95,96] Corticostatic peptides cause a nifedipine-sensitive volume reduction in jejunal villus enterocytes.[97] Furthermore, stimulation of inositol phosphate-mediated rise of $[Ca^{2+}]_i$ in Ehrlich ascites tumor cells by thrombin, histamine, or bradykinin causes a transient cell shrinkage.[64]

IV. COMPARISON OF VOLUME-REGULATORY MECHANISMS IN ISOSMOTIC AND ANISOSMOTIC MEDIA

It has been generally acknowledged by many workers in the field of volume regulation that most cells *in vivo* are unlikely to encounter rapid, large changes in osmolarity. The justification used in studies that subject cells to such perturbations *in vitro* has generally been that cells are likely to experience alterations in their volumes under isosmotic conditions as a consequence of alterations of solute transport rates or metabolism. The assumption is that osmotic manipulation of cell volume is equivalent to changes in cell volume induced by changes in solute content. Nevertheless, *a priori,* it is clear that some aspects are different in the two protocols. Since cell volume is altered as a consequence of water flow in osmotic experiments, cytoplasmic ionic strength as well as cell volume are affected. In contrast, isosmotic loss of solute with water is not associated with changes in cytoplasmic ionic strength. Furthermore, osmotic water flow in anisosmotic media will cause the concentrations of all cytoplasmic constituents to change by an equal extent, whereas that will not necessarily be the case during solute-driven water flows. There exist few experimental data regarding the differential responses of cells to cell volume changes that have been induced by osmotic means compared to those induced by changes in cell solute content under isosmotic conditions. An elegant study by Motais et al.[98] demonstrated that changes in ionic strength play a critical role in the responses of fish red blood cells to osmotic cell swelling or hormonal stimulation, suggesting that responses observed under anisosmotic conditions may not be equivalent to those employed under isosmotic conditions.

As discussed above, many cells respond to acute swelling induced by osmotic dilution of the bathing medium, by RVD mediated by loss of KCl through parallel conductances for K^+ and Cl^-. Since salivary acinar cells use an apparently similar mechanism under isosmotic conditions, which is regulated by $[Ca^{2+}]_i$, this cell type would appear to possess the necessary molecular machinery to accomplish RVD by Ca^{2+} activation of K^+ and Cl^- conductances. Does this cell type employ the same mechanisms under isosmotic and anisosmotic conditions to effect cell shrinkage? To address this issue, we examined whether acinar cells displayed RVD following osmotic swelling, and whether it was mediated by KCl loss, and furthermore whether a rise in $[Ca^{2+}]_i$ was the signal that transduced cell swelling into activation of these pathways, as in isosmotic conditions. Single salivary acinar cells displayed a complete RVD within 5 to 10 min following dilution of the medium by 25% by NaCl removal (Figures 1 to 3; and Reference 28). RVD was sensitive to K^+ and Cl^- gradients, suggesting that it is mediated by KCl efflux.[28] Nevertheless, as shown in Figure 15–2 and discussed earlier, a rise of $[Ca^{2+}]_i$ was not necessary to activate these pathways. These results suggest that the signal transduction pathways that result in activation of KCl efflux following osmotic swelling are quite unlike those employed under isosmotic conditions to also effect KCl efflux. Thus, elucidation of the signal transduction mechanisms that are activated by cell swelling would appear to have little relation to those that, similarly, cause cell shrinkage under isosmotic (physiological) conditions. A similar conclusion was reached by MacLeod et al.[20] in their studies of isolated guinea pig intestinal villus cells. These cells responded to osmotic cell swelling or to swelling induced by Na^+/nutrient cotransport in isosmotic media, by RVD mediated by K^+ and Cl^- conductances.[20,99] Nevertheless, the Ca^{2+} dependencies of the two types of RVD were different. Hypotonic RVD depended upon the presence of extracellular Ca^{2+}, and was blocked by a dihydropyridine and calmodulin inhibitors, whereas isotonic RVD was insensitive to these maneuvers.[20] Whether the same efflux pathways were activated under the two conditions is not yet known, although they are likely to be different. Nevertheless, as for salivary acinar cells, it is clear that different signal transduction pathways are employed in isosomotic vs. anisosmotic conditions to achieve the same response, loss of KCl and water. Agonist-induced elevation of $[Ca^{2+}]_i$ during RVD by single salivary gland acinar cells significantly accelerated it (our unpublished data), but this result could be explained by additive activation of the same channels, or activation of a parallel set of channels. This latter point has previously been overlooked in many studies, which have interpreted Ca^{2+} ionophore enhancement of RVD as implicating elevated $[Ca^{2+}]_i$ in the normal RVD response. For example, extensive early studies on mechanisms of cell volume regulation by lymphocytes (reviewed in Reference 65), which were important in establishing the protocols and conceptual bases for numerous subsequent studies in a wide variety of cell types, concluded that RVD was mediated by $[Ca^{2+}]_i$ activation of conductive K^+ and Cl^- losses.[65,67] This conclusion was based, first, on pharmacological sensitivities of RVD to a variety of agents consistent with the participation of $Ca^{2+}{}_i$, calmodulin, and Ca^{2+}-activated K^+ channels, and second, on the ability of elevated $[Ca^{2+}]_i$ to also induce cell shrinkage under isosmotic conditions with similar pharmacological sensitivities. Nevertheless, it is now accepted that elevated $[Ca^{2+}]_i$ is not involved in RVD in these cells.[47,65] The problem with the original studies was the use of pharmacological agents that were not specific, as discussed earlier, and the assumption that elevated $[Ca^{2+}]_i$ activation of cell shrinkage under isosmotic conditions reflected the situation under anisosmotic conditions. In addition to the volume-regulatory machinery that is activated by cell swelling in dilute media, the cells also possess Ca^{2+}-activated permeabilities which result in cell shrinkage in isosmotic media. These latter signal transduction and ion permeation pathways are unrelated to those employed during RVD.[47,65] Thus, whereas one criterion for establishment of a role for $Ca^{2+}{}_i$ in volume regulation is the demonstration that raised $[Ca^{2+}]_i$ can induce the appropriate changes in cell volume, it is insufficient to establish $[Ca^{2+}]_i$ as a mediator of the volume-activated regulatory transport pathways.

V. SUMMARY

Throughout this chapter, critical reviews of published data and conclusions have been attempted. Although a role for Ca^{2+} in volume regulation in anisosmotic media has been invoked for many cell types, the clear participation of Ca^{2+} in this process exists in only a few. A significant problem is that for most cells examined, the limited number of features of the process that has been examined precludes a clear elucidation of the role, or lack thereof, of Ca^{2+}. Recent data suggesting that $Ca^{2+}{}_i$ mobilizing extracellular signals (hormones, neurotransmitters, growth factors) may affect cell volume under isosmotic conditions raises the issue of the role of such volume changes in subsequent cell physiological responses. Thus, an important goal is to establish the cell signaling role of Ca^{2+}-induced cell volume perturbations. While anisosmotic manipulations may be able to address this important issue, preliminary evidence in a couple of cell types suggests that volume changes under anisosmotic and isosmotic conditions may involve different signal transduction mechanisms and solute translocation pathways. An important goal will be to determine the suitability of anisomotic manipulations to understand these processes.

REFERENCES

1. **McCarty, N. A. and O'Neil, R. G.,** Dihydropyridine-sensitive cell volume regulation in proximal tubule: the calcium window, *Am. J. Physiol.,* 259, F950, 1990.
2. **McCarty, N. A. and O'Neil, R. G.,** Calcium-dependent control of volume regulation in renal proximal tubule cells. I. Swelling-activated Ca^{2+} entry and release, *J. Membr. Biol.,* 123, 149, 1991.
3. **McCarty, N. A. and O'Neil, R. G.,** Calcium-dependent control of volume regulation in renal proximal tubule cells. II. Roles of dihydropyridine-sensitive and -insensitive Ca^{2+} entry pathways, *J. Membr. Biol.,* 123, 161, 1991.
4. **Beck, J. S., Breton, S., Laprade, R., and Giebisch, G.,** Volume regulation and intracellular calcium in the rabbit proximal convoluted tubule, *Am. J. Physiol.,* 260, F861, 1991.
5. **Breton, S., Beck, J. S., Cardinal, J., Giebisch, G., and Laprade, R.,** Involvement and source of calcium in volume regulatory decrease of collapsed proximal convoluted tubule, *Am. J. Physiol.,* F656, 1992.
6. **Susuki, M., Kawahara, K., Ogawa, A., Morita, T., Kawaguchi, Y., Kurihara, S., and Sakai, O.,** $[Ca^{2+}]_i$ rises via G protein during regulatory volume decrease in rabbit proximal tubule cells, *Am. J. Physiol.,* 258, F690, 1990.
7. **Dube, L., Parent, L., and Sauve, R.,** Hypotonic shock activates a maxi K^+ channel in primary cultured proximal tubule cells, *Am. J. Physiol.,* 259, F348, 1990.
8. **Kawahara, K., Ogawa, A., and Susuki, M.,** Hyposmotic activation of Ca-activated K channels in cultured rabbit kidney proximal tubule cells, *Am. J. Physiol.,* 260, F27, 1991.
9. **Bevan, C., Theiss, C., and Kinne, R. K. H.,** Role of Ca^{2+} in sorbitol release from rat inner medullary collecting duct (IMCD) cells under hypoosmotic stress, *Biochem. Biophys. Res. Commun.,* 170, 563, 1990.
10. **Taniguchi, J. and Guggino, W. B.,** Membrane stretch: a physiological stimulator of Ca^{2+}-activated K^+ channels in thick ascending limb, *Am. J. Physiol.,* 257, F347, 1989.
11. **Montrose-Rafizadeh, C. and Guggino, W. B.,** Role of intracellular calcium in volume regulation by rabbit medullary thick ascending limb cells, *Am. J. Physiol.,* 260, F402, 1991.
12. **Ubl, J., Murer, H., and Kolb, H.-A.,** Simultaneous recording of cell volume, membrane current and membrane potential: effect of hypotonic shock, *Pfluegers Arch.,* 415, 381, 1989.
13. **Hoffman, E. K. and Kolb, H.-A.,** Mechanisms of activation of regulatory volume responses after cell swelling, in *Advances in Comparative and Environmental Physiology,* Vol. 9, Gilles, R., Bolis, L., and Hoffman, E., Eds., Springer, Heidelberg, 1991, 3430.
14. **Terreros, D., Kanli, H., and Coombs, J.,** Possible role of basolateral membrane in renal tubule osmoregulation, *Am. J. Physiol.,* 258, R1022, 1990.
15. **Roy, G. and Sauve, R.,** Effect of anisosmotic media on volume, ion and amino-acid content and membrane potential of kidney cells (MDCK) in culture, *J. Membr. Biol.,* 100, 83, 1987.
16. **Rothstein, A. and Bear, C.,** Cell volume changes and the activity of the chloride conductance path, *Ann. N.Y. Acad. Sci.,* 574, 294, 1989.
17. **Rothstein, A. and Mack, E.,** Volume-activated K^+ and Cl^- pathways of dissociated epithelial cells (MDCK): role of Ca^{2+}, *Am. J. Physiol.,* 258, C827, 1990.
18. **Rothstein, A. and Mack, E.,** Volume-activated calcium uptake: its role in cell volume regulation of Madin-Darby canine kidney cells, *Am. J. Physiol.,* 262, C339, 1992.
19. **Civan, M. M., Peterson-Yantorno, K., Coca-Prados, M., and Yantorno, R. E.,** Regulatory volume decrease by cultured non-pigmented ciliary epithelial cells, *Exp. Eye Res.,* 54, 181, 1992.
20. **MacLeod, R. J., Lembessis, P., and Hamilton, J. R.,** Differences in Ca^{2+}-mediation of hypotonic and Na^+-nutrient regulatory volume decrease in suspensions of jejunal enterocytes, *J. Membr. Biol.,* 130, 23, 1992.
21. **Bear, C. E.,** A nonselective cation channel in rat liver cells is activated by membrane stretch, *Am. J. Physiol.,* 258, C421, 1990.
22. **Corrasanti, J. G., Gleeson, D., and Boyer, J. L.,** Effects of osmotic stresses on isolated rat hepatocytes. I. Ionic mechanisms of cell volume regulation, *Am. J. Physiol.,* 258, G290, 1990.
23. **Hazama, A. and Okada, Y.,** Ca^{2+} sensitivity of volume-regulatory K^+ and Cl^- channels in

cultured human epithelial cells, *J. Physiol.*, 402, 687, 1988.

24. **Hazama, A. and Okada, Y.**, Involvement of Ca^{2+}-induced Ca^{2+} release in the volume regulation of human epithelial cells exposed to a hypotonic medium, *Biochem. Biophys. Res. Commun.*, 167, 287, 1990.
25. **Hazama, A. and Okada, Y.**, Biphasic rises of cytosolic free Ca^{2+} in association with activation of K^+ and Cl^- conductance during the regulatory volume decrease in cultured human epithelial cells, *Pfluegers Arch.*, 416, 710, 1990.
26. **Okada, Y., Hazama, A., and Yuan, W.-L.**, Stretch-induced activation of Ca^{2+}-permeable ion channels is involved in the volume regulation of hypotonically swollen epithelial cells, *Neurosci. Res.*, 12 (Suppl.), S5, 1990.
27. **Negulescu, P. A., Munck, B., and Machen, T. E.**, Volume-sensitive Ca influx and release from intracellular pools in gastric parietal cells, *Am. J. Physiol.*, 263, C584, 1992.
28. **Foskett, J. K., Wong, M. M. M., Sue-A-Quan, G., and Robertson, M. A.**, Isosmotic modulation of cell volume and intracellular ion activities during stimulation of single exocrine cells, *J. Exp. Zool.*, in press.
29. **Christensen, O.**, Mediation of cell volume regulation by Ca^{2+} influx through stretch-activated channels, *Nature*, 330, 66, 1987.
30. **Wong, S. M. E. and Chase, H. S., Jr.**, Role of intracellular calcium in cellular volume regulation, *Am. J. Physiol.*, 250, C841, 1986.
31. **Wong, S. M. E., DeBell, M. C., and Chase, H. S., Jr.**, Cell swelling increases intracellular free [Ca] in cultured toad bladder cells, *Am. J. Physiol.*, 258, F292, 1990.
32. **Davis, C. W. and Finn, A. L.**, Interactions of sodium transport, cell volume, and calcium in frog urinary bladder, *J. Gen. Physiol.*, 89, 687, 1987.
33. **Hoffman, E. K., Simonsen, L. O., and Lambert, I. H.**, Volume-induced increase of K^+ and Cl^- permeabilities in Ehrlich ascites tumor cells. Role of internal Ca^{2+}, *J. Membr. Biol.*, 78, 211, 1984.
34. **Christensen, O. and Hoffman, E. K.**, Cell swelling activates K^+ and Cl^- channels as well as nonselective, stretch-activated cation channels in Ehrlich ascites tumor cells, *J. Membr. Biol.*, 129, 13, 1992.
35. **Yamaguchi, D. T., Green, J., Kleeman, C. R., and Muallem, S.**, Characterization of volume-sensitive, calcium-permeating pathway in the osteosarcoma cell line UMR-106-01, *J. Biol. Chem.*, 264, 4383, 1989.
36. **Sato, N., Wang, X., Greer, M. A., Greer, S. E., McAdams, S., and Oshima, T.**, Medium hyposmolarity stimulates prolactin secretion in GH_4C_1 cells by inducing an increase in cytosolic free calcium, *Endocrinology*, 127, 957, 1990.
37. **Sato, N., Wang, X., and Greer, M. A.**, Protein kinase C modulates cell swelling-induced Ca^{2+} influx and prolactin secretion in GH_4C_1 cells, *Mol. Cell. Endocrinol.*, 86, 137, 1992.
38. **Doroshenko, P. and Neher, E.**, Volume-sensitive chloride conductance in bovine chromaffin cell membrane, *J. Physiol.*, 449, 197, 1992.
39. **Falke, L. C. and Misler, S.**, Activity of ion channels during volume regulation by clonal N1E115 neuroblastoma cells, *Proc. Natl. Acad. Sci. U.S.A.*, 86, 3919, 1989.
40. **Cala, P. M.**, Cell volume regulation by *Amphiuma* red blood cells. The role of Ca^{2+} as a modulator of alkali metal/H^+ exchange, *J. Gen. Physiol.*, 82, 761, 1983.
41. **Cala, P. M., Mandel, L. J., and Murphy, E.**, Volume regulation by *Amphiuma* red blood cells: cytosolic free Ca and alkali metal/H exchange, *Am. J. Physiol.*, 250, C423, 1986.
42. **Smith, L. H., Jr. and Pierce, S. K.**, Cell volume regulation by molluscan erythrocytes during hypoosmotic stress: Ca^{2+} effects on ionic and organic osmolyte effluxes, *Biol. Bull.*, 173, 407, 1987.
43. **Pierce, S. K., Politis, A. D., Smith, L. H., Jr., and Rowland, L. M.**, A Ca^{2+} influx in response to hypo-osmotic stress may alter osmolyte permeability by a phenothiazine-sensitive mechanism, *Cell Calcium*, 9, 129, 1988.
44. **Pierce, S. K., Politis, A. D., Cronkite, D. H., Rowland, L. M., and Smith, L. H., Jr.**, Evidence of calmodulin involvement in cell volume recovery following hypo-osmotic stress, *Cell Calcium*, 10, 159, 1989.
45. **Bui, A. H. and Wiley, J. S.**, Cation fluxes and volume regulation by human lymphocytes, *J. Cell. Physiol.*, 108, 47, 1981.
46. **Rink, T. J., Sanchez, A., Grinstein, S., and Rothstein, A.**, Volume restoration in osmotically swollen lymphocytes does not involve changes in free Ca^{2+} concentration, *Biochim. Biophys. Acta*, 762, 593, 1983.
47. **Grinstein, S., Smith, J. D., and Miller, C.**, Calcium-independent cell volume regulation in human lymphocytes. Inhibition by charybdotoxin, *J. Gen. Physiol.*, 95, 97, 1990.
48. **Watson, P. A.**, Accumulation of cAMP and calcium in S49 mouse lymphoma cells following hyposmotic swelling, *J. Biol. Chem.*, 264, 14735, 1989.

49. **Szirmai, M., Sarkadi, B., Szasz, I., and Gardos, G.,** Volume regulatory mechanisms of human platelets, *Haematologia,* 21, 33, 1988.
50. **Foskett, J. K., Gunter-Smith, P. J., Melvin, J. E., and Turner, R. J.,** Physiological localization of an agonist-sensitive pool of Ca^{2+} in parotid acinar cells, *Proc. Natl. Acad. Sci. U.S.A.,* 86, 167, 1989.
51. **Osipchuk, Y. V., Wakui, M., Yule, D. I., Gallacher, D. V., and Petersen, O. H.,** Cytoplasmic Ca^{2+} oscillations evoked by receptor stimulation, G-protein activation, internal application of inositol trisphosphate or Ca^{2+}: simultaneous microfluorimetry and Ca^{2+} dependent Cl^- current recording in single pancreatic acinar cells, *EMBO J.,* 9, 697, 1990.
52. **Hillyard, S. D. and Van Driessche, W.,** Verapamil blocks basolateral K^+ channels in the larval frog skin, *Am. J. Physiol.,* 262, C1161, 1992.
53. **Grinstein, S. and Klip, A.,** Calcium homeostasis and the activation of calcium channels in cells of the immune system, *Bull. N.Y. Acad. Med.,* 65, 69, 1989.
54. **Berridge, M. J. and Irvine, R. F.,** Inositol phosphates and cell signalling, *Nature,* 341, 197, 1989.
55. **Taylor, C. W. and Richardson, A.,** Structure and function of inositol trisphosphate receptors, *Pharm. Ther.,* 51, 97, 1991.
56. **Baquet, A., Meijer, A. J., and Hue, L.,** Hepatocyte swelling increases inositol 1,4,5-trisphosphate, calcium and cyclic AMP concentration but antagonizes phosphorylase activation by Ca^{2+}-dependent hormones, *FEBS Lett.,* 278, 103, 1991.
57. **Dahl, S. V., Hallbrucker, C., Lang, F., and Haussinger, D.,** Role of eicosanoids, inositol phosphates and extracellular Ca^{2+} in cell-volume regulation in rat liver, *Eur. J. Biochem.,* 198, 73, 1991.
58. **Goldschmidt-Clermont, P. J., Machesky, L. M., Baldassare, J. J., and Pollard, T. D.,** The actin-binding protein profilin binds to PIP_2 and inhibits its hydrolysis by phospholipase C, *Science,* 247, 1575, 1990.
59. **Goldschmidt-Clermont, P. J., Kim, J. W., Machesky, L. M., Rhee, S. G., and Pollard, T. D.,** Regulation of phospholipase C-γ1 by profilin and tyrosine phosphorylation, *Science,* 251, 1231, 1991.
60. **Fukami, K., Furuhashi, K., Inagaki, M., Endo, T., Hatano, S., and Takenawa, T.,** Requirement of phosphatidylinositol 4,5-bisphosphate for α-actinin function, *Nature,* 359, 150, 1992.
61. **Fleisher, S. and Inui, M.,** Biochemistry and biophysics of excitation-contraction coupling, *Annu. Rev. Biophys. Biophys. Chem.,* 18, 333, 1989.
62. **Chamberlin, M. E. and Strange, K.,** Anisosmotic cell volume regulation: a comparative view, *Am. J. Physiol.,* 257, C159, 1989.
63. **Sarkadi, B. and Parker, J. C.,** Activation of ion transport pathways by changes in cell volume, *Biochim. Biophys. Acta,* 1071, 407, 1991.
64. **Hoffman, E. K., Simonsen, L. O., and Lambert, I. H.,** Cell volume regulation: intracellular transmission, in *Interaction of Cell Volume with Cell Function,* ACEP Series, Lang, F. and Haussinger, D., Eds., Springer, Heidelberg, in press.
65. **Grinstein, S. and Foskett, J. K.,** Ionic mechanisms of cell volume regulation in leukocytes, *Annu. Rev. Physiol.,* 52, 399, 1990.
66. **Gardos, G.,** The function of calcium in the potassium permeability of human erythrocytes, *Biochim. Biophys. Acta,* 30, 653, 1958.
67. **Grinstein, S., Dupre, A., and Rothstein, A.,** Volume regulation by human lymphocytes. Role of calcium, *J. Gen. Physiol.,* 79, 849, 1982.
68. **Busch, A. E., Varnum, M., Adelman, J. P., and North, R. A.,** Hypotonic solution increases the slowly activating potassium current I_{SK} expressed in *Xenopus* oocytes, *Biochem. Biophys. Res. Commun.,* 184, 804, 1992.
69. **Filipovic, D. and Sackin, H.,** A calcium-permeable stretch-activated cation channel in renal proximal tubule, *Am. J. Physiol.,* 260, F119, 1991.
70. **Bear, C. E.,** A K^+-selective channel in the colonic carcinoma cell line: CaCo-2 is activated with membrane stretch, *Biochim. Biophys. Acta,* 1069, 267, 1991.
71. **Pacha, J., Frindt, G., Sackin, H., and Palmer, L. G.,** Apical maxi K channels in intercalated cells of CCT, *Am. J. Physiol.,* 261, F696, 1991.
72. **Yantorno, R. E., Carre, D. A., Coca-Prados, M., Krupin, T., and Civan, M. M.,** Whole cell patch clamping of ciliary epithelial cells during anisosmotic swelling, *Am. J. Physiol.,* 262, C501, 1992.
73. **Gogelein, H. and Capek, K.,** Quinine inhibits chloride and nonselective cation channels in isolated rat distal colon cells, *Biochim. Biophys. Acta,* 1027, 191, 1990.
74. **Jacobs, E. R. and DeCoursey, T. E.,** Mechanisms of potassium channel block in rat alveolar epithelial cells, *J. Pharm. Exp. Ther.,* 255, 459, 1990.

75. **Kubo, M. and Okada, Y.,** Volume-regulatory Cl^- channel currents in cultured human epithelial cells, *J. Physiol.,* 456, 351, 1992.
76. **Tseng, G.-N.,** Cell swelling increases membrane conductance of canine cardiac cells: evidence for a volume-sensitive Cl channel, *Am. J. Physiol.,* 262, C1056, 1992.
77. **Solc, C. K. and Wine, J. J.,** Swelling-induced and depolarization-induced Cl^- channels in normal and cystic fibrosis epithelial cells, *Am. J. Physiol.,* 261, C5658, 1991.
78. **Worrell, R. T., Butt, A. G., Cliff, W. H., and Frizzell, R. A.,** A volume-sensitive chloride conductance in human colonic cell line T84, *Am. J. Physiol.,* 256, C1111, 1989.
79. **Galietta, L. J. V., Barone, V., Gruenert, D. C., and Romeo, G.,** A chloride conductance evoked by hypotonic shock in epithelial cells, in *The Identification of the CF (Cystic Fibrosis) Gene,* Tsui, L. C. et al., Eds., Plenum Press, New York, 1991, 307.
80. **Weiss, H. and Lang, F.,** Ion channels activated by swelling of Madin Darby canine kidney (MDCK) cells, *J. Membr. Biol.,* 126, 109, 1992.
81. **Banderali, U. and Roy, G.,** Activation of K^+ and Cl^- channels in MDCK cells during volume regulation in hypotonic media, *J. Membr. Biol.,* 126, 219, 1992.
82. **McDonald, T. V., Nghiem, P. T., Gardner, P., and Martens, C. L.,** Human lymphocytes transcribe the cystic fibrosis transmembrane conductance regulator gene and exhibit CF-defective cAMP-regulated chloride current, *J. Biol. Chem.,* 267, 3242, 1992.
83. **Chan, H.-C., Goldstein, J., and Nelson, D. J.,** Alternate pathways for chloride conductance activation in normal and cystic fibrosis airway epithelial cells, *Am. J. Physiol.,* 262, C1273, 1992.
84. **Wagner, J. A., Cozens, A. L., Schulman, H., Gruenert, D. C., Stryer, L., and Gardner, P.,** Activation of chloride channels in normal and cystic fibrosis airway epithelial cells by multifunctional calcium/calmodulin-dependent protein kinase, *Nature,* 349, 793, 1991.
85. **Pierce, S. K. and Politis, A. D.,** Ca^{2+}-activated cell volume recovery mechanisms, *Annu. Rev. Physiol.,* 52, 27, 1990.
86. **Malinow, R., Schulman, H., and Tsien, R. W.,** Inhibition of postsynaptic PKC or CaMKII blocks induction but not expression of LTP, *Science,* 245, 862, 1989.
87. **Mills, J. W., Schwiebert, E. M., and Stanton, B. A.,** The cytoskeleton and cell volume regulation, Chapter 14, this volume.
88. **Lambert, I. H.,** Eicosanoids and cell volume regulation, Chapter 16, this volume.
89. **Foskett, J. K. and Melvin, J. E.,** Activation of salivary secretion: coupling of $[Ca^{2+}]_i$ and cell volume, *Science,* 244, 1582, 1989.
90. **Foskett, J. K.,** Simultaneous Normarski and fluorescence imaging during video microscopy of cells, *Am. J. Physiol.,* 255, C566, 1988.
91. **Macknight, A. D. C. and Leaf, A.,** Regulation of cellular volume, *Physiol. Rev.,* 57, 510, 1977.
92. **Hoffman, E. K.,** Control of cell volume, in *Transport of Ions and Water in Animals,* Oschman, J. L. and Wall, B. J., Eds., Academic Press, London, 1977, 285.
93. **Foskett, J. K.,** $[Ca^{2+}]_i$ modulation of Cl^- content controls cell volume in single salivary acinar cells during fluid secretion, *Am. J. Physiol.,* 259, C998, 1990.
94. **Foskett, J. K., Roifman, C. M., and Wong, D.,** Activation of calcium oscillations by thapsigargin in parotid acinar cells, *J. Biol. Chem.,* 266, 2778, 1991.
95. **Nakahari, T., Murakami, M., Yoshida, H., Miyamoto, M., Sohma, Y., and Imai, Y.,** Decrease in rat submandibular acinar cell volume during Ach stimulation, *Am. J. Physiol.,* 258, G878, 1990.
96. **Takemura, T., Sato, F., Saga, K., Susuki, Y., and Sato, K.,** Intracellular ion concentrations and cell volume during cholinergic stimulation of eccrine secretory coil cells, *J. Membr. Biol.,* 119, 211, 1991.
97. **MacLeod, R. J., Hamilton, J. R., Bateman, A., Belcourt, D., Hu, J., Bennett, H. P. J., and Solomon, S.,** Corticostatic peptides cause nifedipine-sensitive volume reduction in jejunal villus enterocytes, *Proc. Natl. Acad. Sci. U.S.A.,* 88, 552, 1991.
98. **Motais, R., Guizouarn, H., and Garcia-Romeu, F.,** Red cell volume regulation: the pivotal role of ionic strength in controlling swelling-dependent transport systems, *Biochim. Biophys. Acta,* 1075, 169, 1991.
99. **MacLeod, R. J. and Hamilton, J. R.,** Volume regulation initiated by Na^+-nutrient cotransport in isolated mammalian villus enterocytes, *Am. J. Physiol.,* 260, G26, 1991.

Chapter 16

Eicosanoids and Cell Volume Regulation

Ian Henry Lambert

TABLE OF CONTENTS

I. EICOSANOIDS AND THEIR BIOLOGICAL EFFECTS

Eicosanoids are oxygenated metabolites of 20-carbon polyunsaturated fatty acids, e.g., dihomo-γ-linolenic acid, arachidonic acid, and the well-known "fish oil" fatty acid, eicosapentaenoic acid (EPA). Prostanoids (prostaglandins, prostacyclin, thromboxanes), leukotrienes, lipoxins, and other hydroxy fatty acids belong to this extensive family. Most eicosanoids are biologically highly potent and are formed in a cell-specific manner, i.e., a given eicosanoid-forming cell type tends to form only one of these compounds. Thus, the main metabolites of arachidonic acid is prostaglandin E_2 (PGE_2) in collecting kidney tubule cells, prostaglandin D_2 (PGD_2) in mast cells, thromboxane-A_2 (TXA_2) in blood platelets, prostacyclin (PGI_2) in vascular endothelial cells, whereas leukotrienes are primarily synthesized by neutrophils, eosinophils, monocytes, and keratinocytes.[1,2] Eicosanoids are produced in response to physiologic and nonphysiologic stimuli; they are not stored in the tissue but act as intracellular messengers or as messengers for neighboring cells. Biochemical, pharmacological, and clinical findings demonstrate an important role for eicosanoids in the physiology and pathophysiology of diverse processes, such as thrombosis and hemostasis, inflammation and immune responses, reproduction, renal function, and many other processes.[3–11]

Over the last few years attention has focused on biochemical cooperation involving unstable intermediates in eicosanoid biosynthesis. Leukocytes are able to synthesize leukotriene-A_4 (LTA_4), which in turn is converted by endothelial cells or by platelets into leukotriene-C_4 (LTC_4), leukotriene-D_4

0-8493-4448-4/94/$0.00+$.50

(LTD_4) and leukotriene-B_4 (LTB_4).[12–14] The fact that these leukotrienes are potent mediators of inflammation makes it important to analyze their immediate effect on basic properties of noninflammatory cells.

The ability to control the cell volume under "resting" conditions and to regulate the cell volume when exposed to agonists or anisosmotic media are among the most fundamental properties of animal cells. Since eicosanoids are capable of altering net movement of ions[15–27] this review will focus on the role of eicosanoids in cell volume regulation using the Ehrlich ascites tumor cell as an experimental model system.

II. ARACHIDONIC ACID MOBILIZATION AND METABOLISM

A. PHOSPHOLIPASES

Arachidonic acid is liberated from the membrane after stimulation by hormones, autocoids, neurotransmitters, growth factors, antigens, or other stimuli.[2,28–30] Several pathways exist for receptor-mediated generation of arachidonic acid. The release of arachidonic acid follows quickly after stimulation and is typically accompanied by turnover of inositol-containing phospholipids.[2] In fact, some of the arachidonic acid used for prostanoid formation is derived from the sequential hydrolysis of phosphoinositides by an inositol lipid-specific phospholipase phosphoinositidase C (PIC), which generates diacylglycerol (DAG), followed by the action of di- or monoglyceride lipase.[28,31] Alternatively, arachidonic acid can be released by diglyceride lipase from phosphatidic acid (PA) produced from hydrolysis of phospholipids by phospholipase-D (PLD). However, the major sources of arachidonic acid are the abundant glycerophospholipids, phosphatidylcholine (PC) and phosphatidylethandamine (PE), and the key enzyme is phospholipase-A_2 (PLA_2).[32] The fact that intracellular levels of arachidonic acid are extremely low[34] suggests that the enzymatic liberation of arachidonic acid from the phospholipid pools represents a rate-limiting step in the whole process of lipid mediator synthesis.

Ca^{2+} and calmodulin play a significant role in the enzymatic reactions that generate free arachidonic acid and in the oxidation of arachidonic acid into leukotrienes.[29,32,35,36] In many studies, receptor-mediated release of arachidonic acid is found to be Ca^{2+} dependent and Ca^{2+} may react with the enzyme, the substrate, or both.[29] However, optimal activity of PLA_2 in membrane preparations is in some cases only attained with millimolar concentrations of Ca^{2+}, while PLC, PLD, and diglyceride lipase appear to be less sensitive to Ca^{2+} than PLA_2.[37,38] It is, therefore, generally agreed that Ca^{2+} is necessary for the activity of various cellular phospholipases and lipases and that activators of the arachidonic acid-releasing enzymes also increase the Ca^{2+} sensitivity of the enzymes. In sheep seminal vesicular glands where phosphatidylserine (PS) and diglycerides stimulate PLC-mediated hydrolysis of phosphatidylinositol (PI), it is proposed that PI breakdown may be triggered in cells by the agonist-induced release of Ca^{2+} and that the rise in cytoplasmic Ca^{2+} may cause a redistribution of lipids, thereby transforming PI to a state where it is accessible for hydrolysis.[39] Recent studies have, however, clearly demonstrated that PLA_2 in rat platelet membrane is effectively activated by Ca^{2+} in the micromolar range, which corresponds to cytoplasmic Ca^{2+} concentrations observed in stimulated platelets.[40]

Released or exogenous arachidonic acid is either acylated into the cell membrane or converted to (1) prostanoids via the cyclooxygenase pathway; (2) leukotrienes and certain mono-, di-, and trihydroxy acids via the lipoxygenase pathways; or (3) epoxyeicosatrienoic acids via the NADPH-dependent cytochrome P-450 "epoxygenase" pathway (see Figure 16–1).

B. CYCLOOXYGENASE PATHWAY[7,45]

Nonesterified arachidonic acid is transformed into prostaglandins by the heme-requiring, membrane-associated prostaglandin endoperoxidase synthase (PES) complex. The cyclooxygenase (CO) activity of the PES complex inserts two oxygens into arachidonic acid resulting in prostaglandin-G_2 (PGG_2), while the peroxidase activity of the PES complex reduces PGG_2 to prostaglandin-H_2 (PGH_2). The CO is a self-deactivating enzyme[46] and is in itself rate-limiting to prostanoid synthesis. It is the CO activity and not the peroxidase activity that is inhibited by aspirin and indomethacin. Arachidonic acid is the preferred substrate for the CO but dihomo-γ-linolenic acid and, to a lesser extent, EPA are converted to prostaglandins.[47] The endoperoxides, which are unstable at physiological pH and temperatures, are pivotal in the formation of several other products — they can be converted enzymatically to PGI_2 and TXA_2, thromboxane-B_2 (TXB_2), or to PGE_2, prostaglandin-$F_{2\alpha}$ ($PGF_{2\alpha}$), and PGD_2 (see Figure 16–1).

Ca^{2+} has not been shown to directly regulate the activity of the PES complex or the enzymes involved in the synthesis of thromboxanes. The CO appears to be always active and an elevated level of

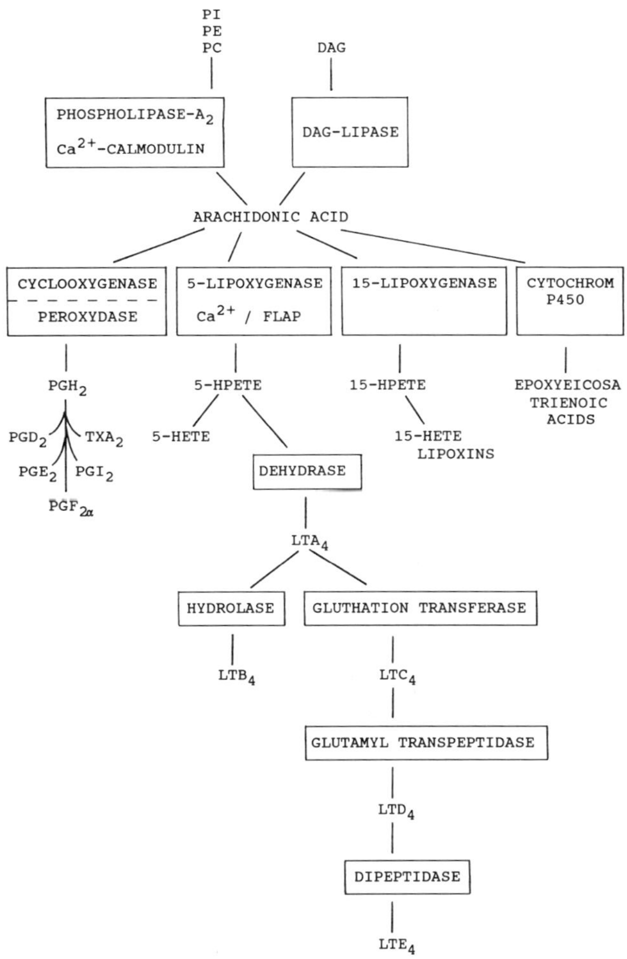

Figure 16–1 Main pathways of eicosanoid production from arachidonic acid. The Ca^{2+}-dependent steps are the Ca^{2+}/calmodulin-regulated PLA_2 and Ca^{2+}-dependent binding of 5-LIP to the 5-LIP binding protein (FLAP). PLA_2 is inhibited directly by RO 31-4493, RO 31-4639[41] and indirectly by pimozide via its inhibition of the Ca^{2+}/calmodulin system. CO is inhibited by acetylsalicylic acid, indomethacin, and naproxen. 5-LIP is inhibited directly by nordihydroguaiaretic acid (NDGA)[42] and ETH 615-139[43] and indirectly by MK 886, which prevents the binding of 5-LIP to FLAP.[44]

nonesterified arachidonic acid seems to be sufficient for an increased formation of CO products.

C. LIPOXYGENASE PATHWAY[3,7,45,48]

Leukotrienes and related mono-, di-, and trihydroxy fatty acids are formed via the lipoxygenase pathways (see Figure 16–1). There are three different mammalian lipoxygenases, 5-lipoxygenase (5-LIP), 12-LIP, 15-LIP, that catalyze the insertion of oxygen into arachidonic acid at positions 5, 12, or 15, respectively.[45] The lipoxygenases are also characterized by self-catalyzed inactivation.[46] The initial product of each of the lipoxygenase reactions 5-hydroperoxy eicosatetraeonoic acid (5-HPETE) (12-HPETE, 15-HPETE) is reduced to 5-hydroxy eicosatetraenoic acid (5-HETE) (12-HETE, 15-HETE). Alternatively, 5-HPETE is converted by removal of water to an unstable epoxide intermediate LTA_4, which is converted by a cytosolic LTA_4 hydrolase to LTB_4 or conjugated with glutathione (γ-glutamyl-cysteinyl-glycine) by a glutathion *S*-transferase to form LTC_4. In the leukotriene biosynthesis the rate-limiting steps occur at the LTA-hydrolase, which is a "suicide-type" enzyme,[5] and at the glutathion transferase.[1] Sequential loss of a glutamic acid residue (glutamyl-transpeptidase) and a glycine residue (cysteinyl-glycidase) from LTC_4 leads to the formation of LTD_4 and leukotriene-E_4 (LTE_4). The enzymes glutathion *S*-transferase, glutamyl-transpeptidase, and cysteinyl-glycidase are particulate enzymes located in the plasma membrane.[1] It is, therefore, assumed that LTD_4 is released directly to the extracellular space.

The substrate specificity for the 5-LIP pathway is quite stringent. The 5-LIP requires double bonds at carbons 5 and 8 for catalysis, whereas the dehydrase responsible for LTA synthesis and the hydrolase that mediates LTB synthesis both require additional double bonds at carbons 11 and 14.[8] On the other hand, no specific double bond configuration seems to be necessary for the gluthation transferase.[1] EPA, which has its double bonds at positions 5, 8, 11, 14, and 17, is substrate both for the lipoxygenase and for the dehydrase and, consequently, pentaene leukotrienes and lipoxins are produced by mammalian cells.[49–52]

The 5-LIP in a variety of cells, including rat basophilic leukemia cells (RBL-1), polymorphonuclear leukocytes, and macrophages, is specifically stimulated by Ca^{2+} and ATP.[1,8,45,53] According to a current model,[48,54,55] efficient leukotriene biosynthesis requires translocation of the 5-LIP from the cytosol to the cellular membrane where it binds to an integral 5-LIP-activating protein (FLAP). The membrane association of the 5-LIP requires Ca^{2+} and is reversible upon Ca^{2+} removal. The 5-LIP is assumed to undergo suicide turnover-associated inactivation, resulting in a dead enzyme that remains bound to FLAP.

D. CYTOCHROME P450, "EPOXYGENASE" PATHWAY

The epoxygenase pathway converts arachidonic acid by an NADPH-dependent mechanism to a variety of oxygenated products such as HETEs, epoxides (EETs), and their hydrolysis products. These products appear to regulate renal and ocular transport functions that rely on Na^+, K^+-ATPase and to affect platelet agregation.[10,56–58]

III. ROLE OF EICOSANOIDS IN VOLUME REGULATION IN EHRLICH ASCITES TUMOR CELLS

A. REGULATORY VOLUME DECREASE (RVD)

Vertebrate cells suspended in hypotonic media initially swell by osmotic water equilibration, but subsequently regulate their volume by net loss of KCl and concomitant loss of cell water (regulatory volume decrease, RVD). Hypotonically swollen Ehrlich cells recover their cell volume (see Figures 16–4, 16–5, and 16–7) by net loss of K^+ and Cl^- via conductive transport pathways and by net loss of taurine via an apparent leak pathway.[59–64]

In the swollen Ehrlich cells the Cl^- permeability exceeds the K^+ permeability so that the rate-limiting step in RVD is the K^+ permeability.[60,62] The effect of addition of valinomycin or gramicidin to osmotically swollen cells is, therefore, an acceleration of RVD (compare control curves in Figure 16–4, middle and right panel). After addition of sufficient gramicidin the cation permeability exceeds that for Cl^-, so that the Cl^- permeability now becomes the rate-limiting step in RVD. It is thus assumed that any drug inhibiting volume changes, after initial swelling in the presence of gramicidin, acts by blocking the swelling-induced anion pathway. On the other hand, any drug that inhibits volume changes in the absence, but not in the presence, of gramicidin can be classified as a specific inhibitor of the volume-induced K^+ pathway.

The mechanisms by which cells sense osmotically induced changes in turgor pressure, volume, and/or intracellular composition, and the mechanism by which they transduce those signals into a regulatory response is only partly understood. However, as recently reviewed by Hoffman et al.,[68] several factors, e.g., cell membrane potential, Ca^{2+}, calmodulin, eicosanoids, phospholipases, protein kinases, and the microfilament network have been assigned a regulatory function in the Ehrlich ascites tumor cells.

B. EICOSANOID SYNTHESIS

Unperturbated Ehrlich cells oxidize arachidonic acid into PGE_2, PGD_2, $PGF_{2\alpha}$, LTC_4, LTB_4, 5-HETE, and 15-HETE (Figure 16–2; Lambert et al.[21]). Hypotonic cell swelling results in stimulation of the leukotriene synthesis and a concommitant reduction in the prostaglandin synthesis (Figure 16–3, left panel). However, if the Ehrlich cells have access to sufficient arachidonic acid there is no decrease in the synthesis of prostaglandins (Figure 16–3, right panel). Thus, in the absence of exogenous arachidonic acid the metabolism of arachidonic acid metabolism via the 5-LIP pathway is favored during RVD at the expense of metabolism via the CO pathway.

C. ROLE OF PGE_2 AND LTD_4 IN REGULATION OF ION TRANSPORT

PGE_2 inhibits RVD when Na^+ is present in the hypotonic medium, but not when it is absent (Figure 16–4) and from estimation of the cell membrane potential it has been shown that PGE_2 depolarizes the plasma membrane when Na^+ is present in the medium but not when Na^+ is absent.[21] This is taken to indicate that PGE_2 stimulates the Na^+ conductance in Ehrlich cells. PGE_2 does not inhibit RVD in Na^+-free medium in the presence or absence of gramicidin (Figure 16–4, middle and right panels), i.e., PGE_2 does not affect the swelling-induced Cl^- conductance or the K^+ conductance. It, therefore, appears that the reduced synthesis of PGE_2 in Ehrlich cells seen after cell swelling (see Figure 16–3) could account for the concomitant reduction in the passive conductance for Na^+,[70] but it does not explain the concomitant stimulation of the K^+, Cl^- transport pathways. It is noted that PGE_2 stimulates Ca^{2+}-sensitive K^+ channels in the Madin-Darby canine kidney (MDCK) cells but this activation is due to PGE_2-induced release of intracellular calcium.[24]

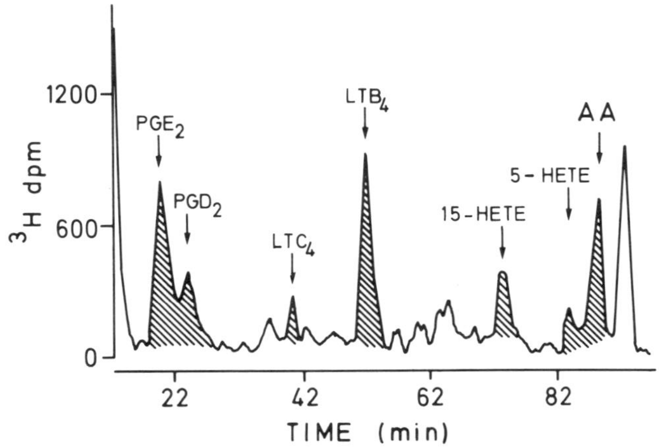

Figure 16–2 Eicosanoids synthesized in Ehrlich ascites tumor cells under isotonic conditions. Ehrlich cells were incubated for 15 min with ^{3}H-labeled arachidonic acid in isotonic NaCl medium. The metabolism of arachidonic acid was stopped by dilution of the suspension to a final concentration of 70% ethanol. Particles were removed by centrifugation and the samples were evaporated and resuspended to 30% methanol. Separation was achieved by reversed phase HPLC[69] on a Supelco column (C18, 5 μm particle size). Elution was performed with methanol: 48% (35 min), 56% (35 min), 65% (15 min), and 100% (15 min). A radiomatic radioactive flow cell detector was used to determine radioactivity in the eluant in 30-s fractions. ^{3}H-labeled standards were run in parallel for identification. The retention times for the leukotriene standards were 39 min (LTC_4), 53 min (LTD_4), and 55 min (LTB_4). (Data from Lambert and Henke.[25])

Addition of LTD_4 produces KCl and water loss from Ehrlich ascites tumor cell. In hypotonic NaCl medium this is seen as a dramatic acceleration of RVD (Figure 16–5, left panel); in isotonic NaCl medium it is seen as a significant cell shrinkage, where the cells attain a minimum cell volume about 0.9 min after addition of LTD_4 (Figure 16–5, right panel). Taking into account that K^+ movement is rate-limiting for KCl and water loss during RVD[62] and that LTD_4 accelerates RVD and stimulates K^+ loss from hypotonically swollen cells,[27] it is concluded that LTD_4 stimulates the K^+ permeability in Ehrlich cells. The LTD_4-induced cell shrinkage under isotonic conditions (Figure 16–5, right panel) is also a result of a definite net loss of KCl[20] and since the Cl^- conductance in Ehrlich cells suspended in isotonic medium is low[60,71] it is, furthermore, concluded that LTD_4 also increases the Cl^- permeability in Ehrlich ascites tumor cells.

We have demonstrated previously that Ehrlich cells lose KCl and osmotically obliged water after (1) hypotonic cell swelling;[72] (2) a rise in the free cytoplasmic Ca^{2+} concentration, accomplished by the divalent cation-ionophore A23187[61,62] or the receptor agonists bradykinin thrombin;[68,73] and (3) treatment with the sulfydryl directed agent *N*-ethylmaleimide (NEM).[74] It has been suggested that in Ehrlich cells,[61,62] as well as in lymphocytes,[75] the volume- and the Ca^{2+}-induced phenomena share a common mechanism. For the Ehrlich cells this suggestion was based on the fact that swelling-induced and A23187-induced cell shrinkage were (1) inhibited by quinine,[61,62] which blocks the Ca^{2+}-activated K^+ channels;[76] (2) unaffected by substitution of NO_3 for Cl^-;[61,62] and (3) inhibited by the anticalmodulin drug, pimozide.[62] In the case of NEM-induced KCl loss and cell shrinkage it was shown that NEM in Ehrlich cells activated an electroneutral K,Cl cotransport and that no net loss of osmotic substances and cell water could be seen after substitution of NO_3 for Cl^-.[74] LTD_4 fails to stimulate RVD in Ehrlich cells in the presence of quinine,[60] and LTD_4 still produces cell shrinkage in cells where all cellular and extracellular Cl has been substituted by NO_3.[60] Since LTD_4 also produces a transient increase in cellular Ca^{2+} when added to Ehrlich cells in isotonic medium (Figure 16–6), it is concluded that LTD_4, just like the Ca^{2+}-ionophore A23187,[61] the receptor agonists, bradykinin and thrombine,[68,73] and osmotic cell swelling, stimulates separate, Ca^{2+}-dependent K^+ and Cl^- transport pathways in Ehrlich ascites tumor cells. LTC_4, i.e., the precursor for LTD_4, and LTE_4, i.e., the metabolite of LTD_4, (see Figure 16–1) do not accelerate RVD in hypotonic medium and have no effect on the steady-state volume.[20,21]

The increased synthesis of leukotrienes seen after osmotic cell swelling (Figure 16–3) seems to be essential for the concomitant activation of the

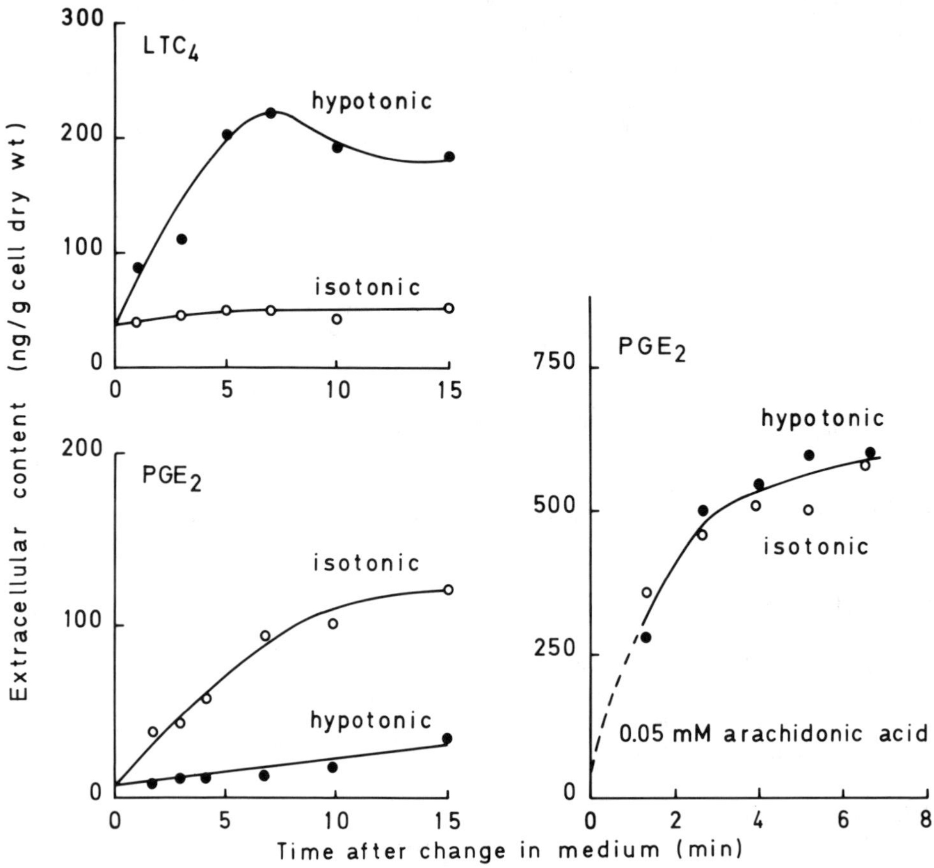

Figure 16–3 LTC_4 and PGE_2 synthesis in Ehrlich cells suspended in isotonic and hypotonic media. Ehrlich cells preincubated in standard incubation medium were spun down and resuspended in either isotonic NaCl medium (300 mOsmol) or in hypotonic NaCl medium (150 mOsmol). The release of LTC_4 and PGE_2 from the cells was followed with time by serially isolating cell-free medium by centrifugation and measuring the LTC_4 and PGE_2 concentrations in the supernatant by radioimmunoassay. The extracellular eicosanoid release (ng/g cell dry weight) was calculated from the measured concentration in the medium (ng/ml) and the cell density of the cell suspension (g dry weight/ml). *Left panel:* no exogenous arachidonic acid. In this case the medium concentration of LTC_4 10 min after transfer to the hypotonic medium approached 1 n*M*. *Right panel:* 0.05 m*M* arachidonic acid was added to the cell suspension within seconds following transfer to the experimental solution. (Data from Lambert et al.[21])

K^+ and Cl^- transport pathways. This is deduced from the data in Figure 16–7, in which it is shown that inhibition of the PLA_2, either indirectly by addition of the anticalmodulin drug diphenyl piperidine (pimozide) (Figure 16–7, upper, left panel) or more directly by addition of RO 31-4493 (Figure 16–7, upper, right panel), leads to an inhibition of the RVD response. LTD_4 in both cases is able to lift the inhibition of the PLA_2 inhibitors (Figure 16–7, upper panels). Direct inhibition of the 5-LIP by addition of ETH 615-139 (Figure 16–7, middle, left panel) or prevention of the binding of 5-LIP to FLAP by addition of MK 886 (Figure 16–7, middle, right panel) also blocks the RVD response in hypotonic medium. However, LTD_4 is not able to lift the inhibition induced by the 5-LIP inhibitors ETH 615-139 (Figure 16–7, middle, left panel), nordihydroguaiaretic acid (NDGA)[60] and the FLAP-inhibitor MK 886 (Figure 16–7, middle panels). This could be explained either by the suggestion that ETH 615-139, NDGA and MK 886, just like the LTD_4-receptor antagonist L-649923 (see Figure 16–7, lower, left panel) in the Ehrlich cells act as antagonists to the LTD_4 receptor or that LTD_4 exerts its effect via another lipoxygenase product. In the latter case, we can exclude LTB_4, LTC_4,

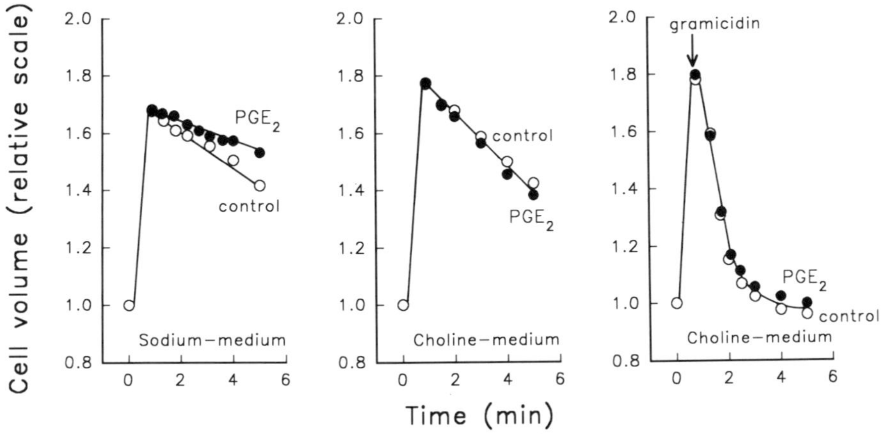

Figure 16–4 Effect of PGE_2 on RVD in hypotonic NaCl medium and in hypotonic Na-free, choline medium (choline substituted for Na) in the absence and in the presence of gramicidin. Gramicidin was added in order to ensure a high cation permeability. Ehrlich cells were preincubated in standard incubation medium and at time zero diluted 500-fold in either hypotonic standard medium (left panel) or in hypotonic choline medium (middle and right panels). The final cytocrit was 0.008%. 5 μ*M* PGE_2 was added to the cells within the first seconds after the reduction in osmolarity, and 0.5 μ*M* gramicidin (right panel) was added at the time of maximal cell swelling as indicated by the arrow. The cell volume was followed with time and given relative to the initial cell volume. (Data from Lambert et al.[21])

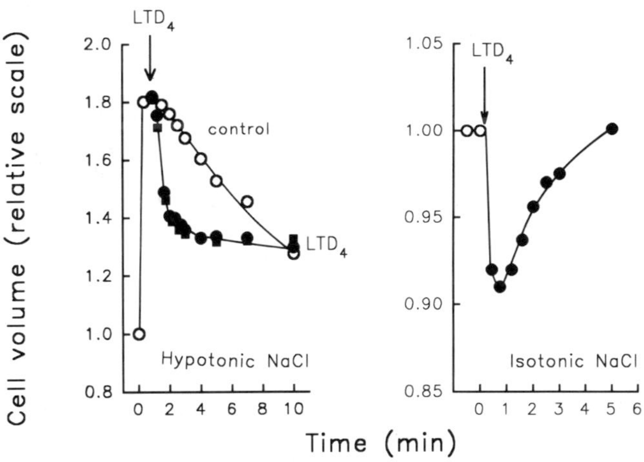

Figure 16–5 LTD_4-induced cell shrinkage in Ehrlich ascites tumor cells under hypotonic conditions (left panel) and under isotonic conditions (right panel). Cells preincubated for 35 min or more in isotonic standard NaCl medium were diluted 500-fold at time zero with hypotonic NaCl standard medium (150 mOsmol, left panel) or isotonic NaCl standard medium (300 mOsmol, right panel). The cell volume was followed with time in a Coulter counter cell sizing system and given relative to the initial cell volume in unperturbed cells. LTD_4 (20 n*M:* closed squares; 60 n*M:* closed circles) was added at the time of maximal cell swelling, i.e., 0.9 min after hypotonic challenge (left panel) or at time zero (right panel). (Data from Lambert et al.[21])

LTE_4, and 15-HETE, since they either have no effect on, or even inhibit, RVD.[20,21,25] An effect of LTD_4 via an increased PLA_2 and PES activity is also excluded because LTD_4 accelerates RVD in the presence of the PLA_2 inhibitor RO 31-4493 (Figure 16–7, upper, right panel) and in the presence of the CO inhibitor indomethacin (Figure 16–7, lower, right panel).

It is thus proposed that the increased synthesis of leukotrienes seen after osmotic cell swelling is involved in the activation of the conductive K^+ and Cl^- transport pathways and the subsequent regula-

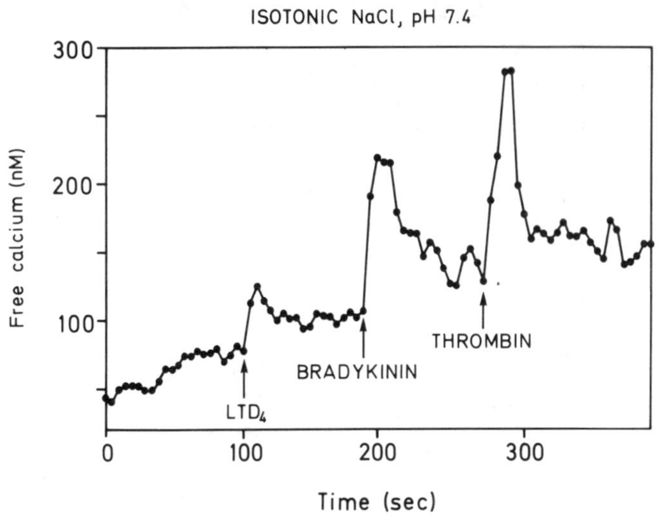

Figure 16–6 Transient increase in cytosolic Ca^{2+} in Ehrlich ascites tumor cells induced by LTD_4, bradykinin, and thrombin. LTD_4 (60 n*M*), bradykinin (0.5 m*M*), and thrombin (1 U/ml) were added at the time indicated by the arrows to a suspension of cells loaded with the fluorescent Ca^{2+} probe fura-2. The free cytosolic Ca^{2+} was measured by dual wavelength excitation at 340 and 380 nm, emission at 510 nm, and the "ratio method".[77]

tory volume decrease and that LTD_4 plays an important role as second messenger.

D. LTD_4 AS A SECOND MESSENGER INVOLVED IN ACTIVATION OF THE TAURINE LEAK PATHWAY

Organic osmolytes are implicated in cell volume regulation of animal, plant, bacterial, fungal, and protist cells exposed to anisosmotic environments.[79,80] Although the accumulation or loss of any organic solute would serve to restore the cell volume, only a restricted number of organic osmolytes are used in biological systems. The dominating part of the net loss of osmotically active nonelectrolytes following cell swelling in hypotonic medium is in Ehrlich ascites tumor cells accounted for by taurine (2-aminoethanesulfonic acid), glycine, alanine, and other small, nonessential amino acids. This decrease in amino acids accounts for approximately 30% of the total decrease in osmotically active substances.[81] It has been demonstrated that an increased degradation of amino acids accounts for one third of the cellular alanine loss,[66] while the loss of taurine, aspartic acid, glutamic acid, and glycine is the result of an increased leak of these compounds to the environment.[63]

The initiating signal for the increase in the taurine permeability in Ehrlich cells seems to be the reduction in osmolarity and the stretch of the plasma membrane (cell swelling), and not a result of the reduction in the extracellular ion concentration.[63] Taurine efflux from Ehrlich cells suspended in isotonic medium is unaffected by addition of PGE_2 (Figure 16–8), but strongly stimulated by addition of LTD_4 (Figure 16–8), and by cell swelling in hypotonic medium (Figure 16–8). The swelling-induced activation of the taurine leak pathway in hypotonic medium is strongly reduced when PLA_2 is inhibited either indirectly by addition of the anticalmodulin drug pimozide or more directly by addition of RO 31-4639 (Figure 16–8). The inhibition is lifted in both cases by addition of LTD_4.[64] Inhibition of the 5-LIP by addition of ETH 615-139 (Figure 16–8) or NDGA[64] also blocks the taurine efflux activated by cell swelling. However, the inhibition is not lifted by addition of LTD_4,[60,64] most probably because NDGA and ETH 615-139, as stated above, in Ehrlich cells act as antagonists towards the LTD_4 receptor. LTD_4 is, therefore, also regarded as a second messenger for the activation of the taurine leak pathway following cell swelling. We are presently investigating whether or not taurine and Cl^- use the same channel during RVD.

E. ROLE OF IP_3/Ca^{2+} AND DAG IN RVD

The RVD response in Ehrlich cells is associated with a rapid decrease in membrane phosphatidyl inositol 4,5-bisphosphate (PIP_2) and a substantial increase in cytoplasmic inositol 1,4,5-trisphosphate (IP_3),[65] indicating that IP_3 could play a role in Ca^{2+} mobilization during the RVD response in Ehrlich cells. Hypoosmolar cell swelling of rat liver cells is accompanied by a stimulation of inositol phosphate turnover, which could participate in triggering the subsequent K^+ efflux and volume-regulatory decrease.[26] A significant, sustained elevation of IP_3 in hepatocytes has also been reported after cell swelling induced by Na^+-dependent amino acid accumulation.[82] However, even though the time course of the swelling-induced K^+ efflux resembles the time course of the LTC_4-induced K^+ efflux under isotonic conditions, the rat liver seems not to require the

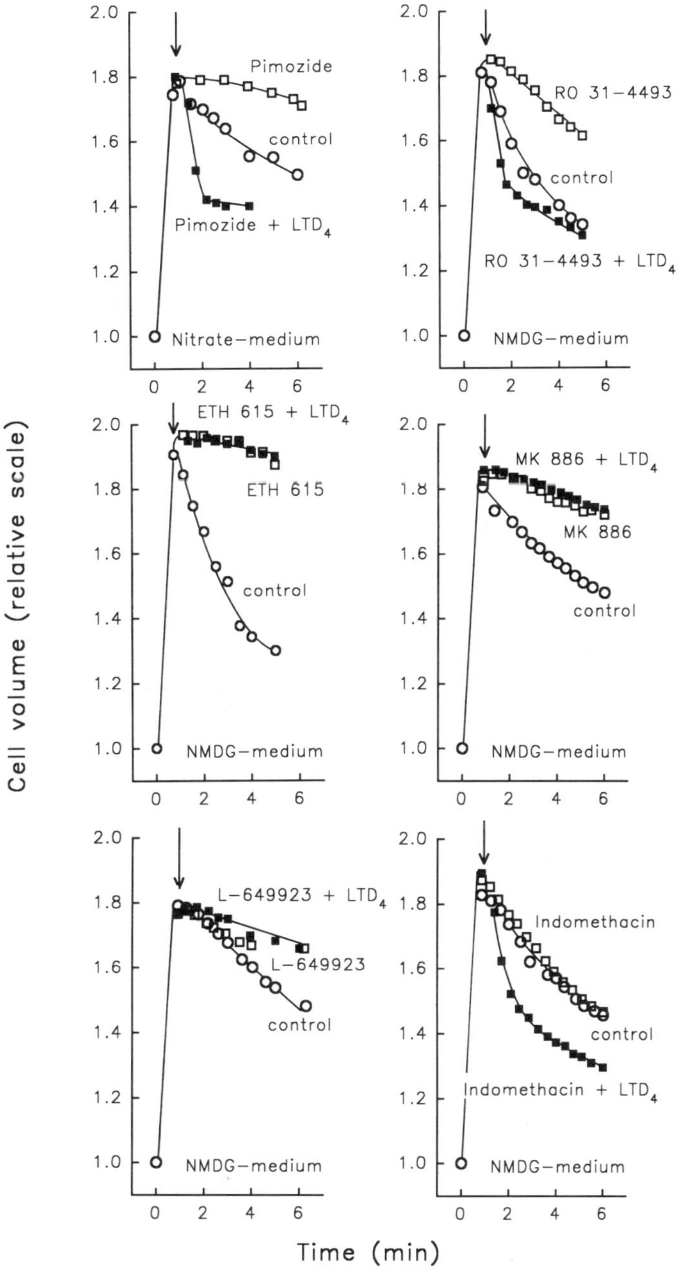

Figure 16–7 Effect of LTD_4 on cell volume in Ehrlich cells swollen in hypotonic medium in the presence of the anticalmodulin drug, pimozide, the PLA_2 inhibitor, RO 31-4493, the 5-LIP inhibitor, ETH 615-139, the FLAP-protein blocker, MK 886, the LTD_4 receptor antagonist, L-649923, and the CO inhibitor, indomethacin. Cells, preincubated in isotonic standard NaCl medium or in isotonic $NaNO_3$ medium for 40 min or more, were diluted 500-fold at time zero with hypotonic (150 mOsmol) Na-free, NMDGCl medium (*N*-methyl-D-glucammonium substituted for Na) or Cl^--free $NaNO_3$ medium (NO_3 substituted for Cl^-), respectively. In the case of NO_3 cells an additional wash of cells was performed after 20 min preincubation in order to remove excess Cl^-. The cell volume was followed with time in a Coulter counter cell sizing system and given relative to the initial cell volume in isotonic media. Pimozide (10 μ*M*, upper, left frame); RO 31-4493 (0.5 μ*M*, upper, right frame); ETH 615-139 (4 μ*M*, middle, left frame); MK 886 (1 μ*M*, middle, right frame); L-649923 (20 μ*M*, lower left frame); and indomethacin (10 μ*M*, lower, right frame), all marked with squares, were added within the first seconds after reduction in osmolarity. LTD_4 (100 n*M;* closed symbols) was added at the time of maximal cell swelling as indicated by the arrows. Control cells (open circles) contained neither inhibitor nor LTD_4. (Data from Lambert et al.[60] and Lambert and Hoffmann.[78])

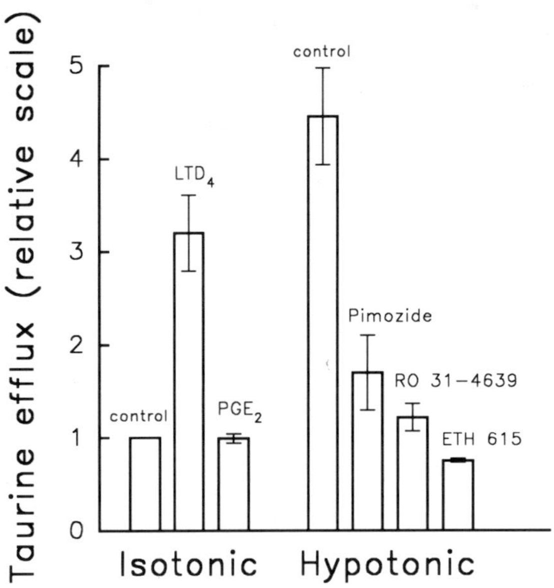

Figure 16–8 The effect of LTD_4 and PGE_2 on taurine release in Ehrlich ascites tumor cells under isotonic conditions and of PLA_2 and 5-LIP inhibitors on the swelling-induced taurine release. Ehrlich cells, equilibrated with ^{14}C-taurine in standard NaCl medium, were transferred to isotonic (300 mOsmol) or hypotonic (150 mOsmol) solutions. The taurine efflux was followed with time. LTD_4 (4 μ*M*); PGE_2 (50 μ*M*) (isotonic conditions); pimozide (12 μ*M*); RO 31-4639 (2.4 μ*M*); and ETH 615-139 (10 μ*M*) were added at time zero. The taurine efflux is given relative to the taurine efflux in isotonic standard medium. (Data from Lambert and Hoffmann.[64])

formation of eicosanoids or entry of Ca^{2+} from the extracellular space in order to perform RVD.[26]

In several cell types extracellular Ca^{2+} and Ca^{2+} entry across the cell membrane is a prerequisite for the swelling-induced RVD response. In other cell types the volume response is unaffected by removal of external Ca^{2+}, and in these cells swelling-induced release of Ca^{2+} from internal stores has been proposed to be involved in the RVD response. For a recent review of the role of Ca^{2+} in cell volume regulation, see reviews by Pierce and Pollitis[83] and Hoffmann et al.[68] In the case of Ehrlich ascites tumor cells, release of Ca^{2+} from internal stores or a modulation of the Ca^{2+} sensitivity of the conductive ion transport pathways has been suggested to play a vital role during RVD.[61] However, prolonged preincubation in Ca^{2+}-free media containing EGTA reduces the RVD response in Ehrlich cells only slightly[61] because the Ehrlich cells are able to compensate for a reduced activation of K^+ and Cl^- channels by activation of a KCl cotransport system[74] and because the depolarization of the plasma membrane, which follows Ca^{2+}-depletion, in itself stimulates taurine efflux.[64]

It is unknown whether the DAG, which is also released from PIP_2, plays an additional role in the RVD response, e.g., via a protein kinase C (PKC)-mediated modulation of the Ca^{2+} sensitivity of the transport pathways. An increase in PKC activity after cell swelling has been demonstrated recently in Ehrlich cells.[131] A role for PKC in desensitization towards LTD_4 has been proposed previously.[84] In red cells of the little skate hypotonic cell swelling produces an increase in DAG,[85] predominantly by stimulation of PC hydrolysis, and an increase in inositol 1-phosphate (IP), whereas the levels of inositol 1,4-bisphosphate (IP_2) and IP_3 were unaffected.[86] In villus epithelial cells it has been demonstrated that inhibition of PKC inhibited RVD after cell swelling induced by L-alanine uptake, whereas RVD after hypotonic cell swelling was unaffected.[87]

IV. EICOSANOID SIGNAL TRANSMISSION

A. EICOSANOID RECEPTORS

It is well known that neurotransmitters, hormones, and other ligands transmit their specific messages by binding to receptors on the cell surface. Binding of ligand brings about changes in the concentration of second messengers (cAMP, IP_3, inositol 1,3,4,5-tetrakisphosphate [IP_4], DAG, eicosanoids, or Ca^{2+}) or intracellular ions (K^+, Cl^-, Na^+) by affecting effector systems such as adenylate cyclase, cyclic nucleotide phosphodiesterase, and phospholipases, or by opening or closing Ca^{2+}, K^+, Cl^-, and Na^+ channels.

The binding sites for eicosanoids, in general, are specific and distinguish between different prostaglandins and leukotrienes; moreover, evidence has been provided for multiple receptors and receptor subclasses even within the same cell.[2,10,88–90] The binding sites are located primarily on the plasma membrane[91] and have a high affinity for the appropriate eicosanoid as identified by a dissociation constant in the nanomolar range.[89] Various leukocyte populations have high-affinity, G protein-coupled LTB_4 receptors on the surface,[92] whereas Ehrlich ascites tumor cells seem to have a LTD_4 receptor, coupled to a pertussis toxin-sensitive G protein,[27] which demonstrates strict demands to agonists, i.e., neither LTB_4, LTC_4, LTE_4, or LTD_4-methylester appear to bind to the receptor.[20,21,27] A G protein coupled to a specific LTD_4 receptor has also been demonstrated in glomerular epithelial cells.[93]

The question then arises whether all eicosanoids bind to G protein-linked receptors in order to elicit their cellular effects. In this case the multitude of biological effects produced by eicosanoids[3–11] is best understood if there are several types of receptors specific for each eicosanoid and where each receptor is coupled to a different G protein mediating a different physiological event.

B. SECONDARY MESSENGERS FOR LTD_4 SIGNAL TRANSMISSION

Several secondary messengers have been proposed to be involved in the processes that follow binding of LTD_4 to the LTD_4 receptor.[16,84,91,93–105] According to a current model,[90,106] it is suggested that the LTD_4 receptor is coupled to an inhibitory G protein. Binding of LTD_4 to the receptor activates a G protein-regulated PLC which hydrolyzes PIP_2 into the two divergent intracellular messengers, IP_3 and arachidonic acid-rich DAG. IP_3 is released into the cytosol, where it binds to specific receptors located on the surface of intracellular Ca^{2+} storage organelles, thereby triggering Ca^{2+} release.[107] DAG either promotes, in the presence of PS and Ca^{2+}, protein phosphorylation via activation of PKC[108] or it releases arachidonic acid, which is then converted to various cyclooxygenase and lipoxygenase metabolites. In some cells and tissues, subsequent to the activation of PI-PLC, the gene for a PLA_2-activating protein is induced, resulting in an increased release of arachidonic acid.[109] The effect of LTD_4 appears to be both CO-dependent and CO-independent. Samhoun and Piper[102] have demonstrated that leukotriene-induced contraction in isolated guinea pig lung and parenchyma strips were inhibited by the CO inhibitor, indomethacin, and by inhibitors of PLA_2 and thromboxane synthase, suggesting that leukotriene action is mediated via generation of thromboxane A_2. Studies with guinea pig lung have demonstrated that LTD_4 induces both CO-independent pharmacological effects (e.g., guinea pig tracheal smooth muscle contraction) and CO-dependent pharmacological effects (e.g., TXB_2 and 6-keto-$PGF_{1\alpha}$ synthesis in guinea pig lung) and that both of these effects in the guinea pig lung are probably secondary to the formation of the PLC-induced second messengers (DAG, IP_3).[99,109]

LTD_4-induced release of DAG and Ca^{2+} mobilization suggests that LTD_4 may also activate PKC or modulate existing PKC to a state of higher catalytic activity and that PKC may be involved in the subsequent intracellular events. PKC is a widespread, multifunctional enzyme and seems to play a vital role in several intracellular control mechanisms, including regulation of the apical Cl^- channel involved in transepithelial Cl^- transport in airway epithelium.[110] In the latter case phosphorylation of Cl^- channels in excised patches of membranes from airway epithelia or phosphorylation of an associated regulatory protein by cAMP-dependent protein kinase or by PKC (at low internal Ca^{2+}) results in activation of Cl^- channels while phosphorylation at high internal Ca^{2+} inactivates the Cl^- channel.[110]

V. MODEL FOR INTRACELLULAR SIGNAL TRANSMISSION INITIATED BY OSMOTIC CELL SWELLING

We have proposed a model for activation of the RVD response in Ehrlich cells (Figure 16–9): cell swelling activates nonselective stretch-activated cation channels,[59] producing Ca^{2+} entry and an increase in free cytosolic Ca^{2+}. The activation of nonselective cation channels will in itself tend to depolarize the cell membrane potential. Moreover, cell swelling activates (by an unknown mechanism) PLC, which results in hydrolysis of cell membrane PIP_2 and release of IP_3[65] and DAG. DAG stimulates PKC. The increase in cytoplasmic IP_3 stimulates the release of Ca^{2+} from intracellular stores and the formation of the Ca^{2+}-calmodulin complex. The rise in free cytosolic Ca^{2+} activates the Ca^{2+}-gated K^+ channel[59] and, moreover, activates PLA_2 and 5-LIP, thereby favoring conversion of arachidonic acid into leukotrienes, most likely LTD_4. The DAG released from PIP_2 hydrolysis also produces arachidonic acid. LTD_4 then activates both Cl^- and K^+ conductances as well as the taurine leak pathway. In Ehrlich cells the Cl^- conductance increases more than the K^+ conductance, resulting in a depolarization of the cell membrane,[60] which in itself may play a role in the activation of voltage-gated K^+ channels and the taurine efflux system. In the case of the taurine transport, taurine uptake is also reduced in hypotonic medium because of the reduced availability of Na^+, Cl^- and the depolarization of the cell membrane.[64,111,112]

That LTD_4 and TXA_2 induce Ca^{2+} transients (Figure 16–6) might constitute an important positive loop for amplification of a signal, i.e., relatively small amounts of leukotrienes bind to a cell, trigger the release of Ca^{2+}, the translocation/activation of 5-LIP and, thereby, formation of excessive

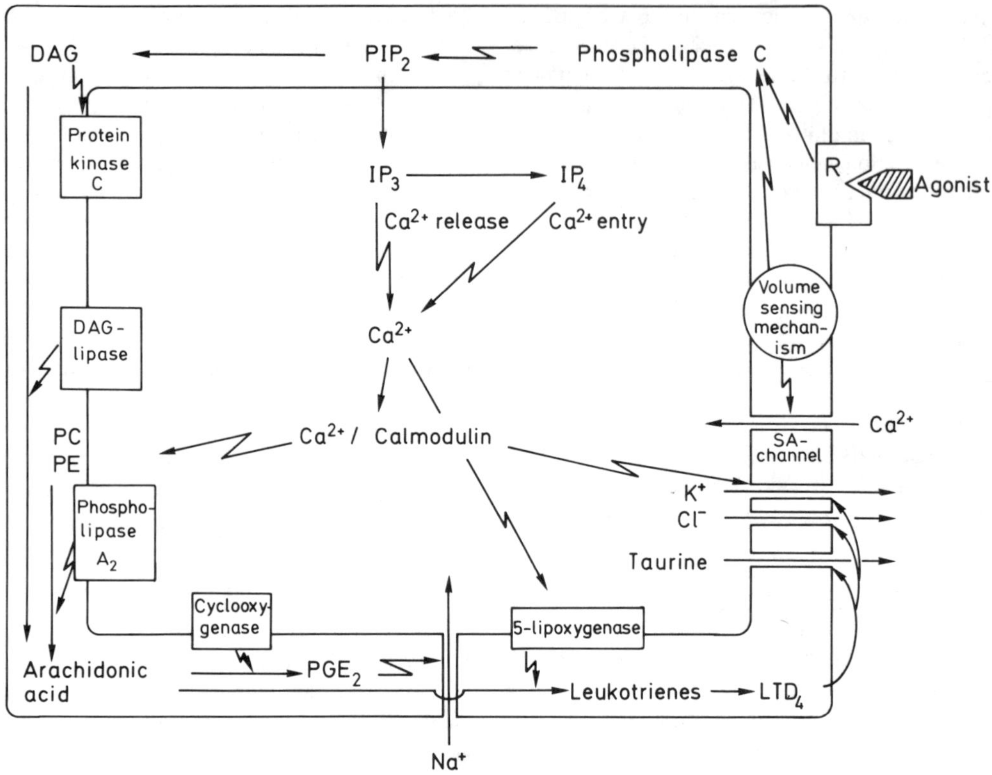

Figure 16–9 Model for the intracellular signal transmission in Ehrlich ascites tumor cells initiated by osmotic cell swelling and by agonist stimulation. Details are given in the text. (From Hoffmann, E. K. et al., in *Interaction, Cell Volume, Cell Function,* Gilles, R., Ed., ACEP Ser., Springer, Heidelberg, 1993, chap. 7.)

quantities of leukotrienes. In this case a stretch-activation of the PLA_2 could be the initial step for the intracellular signal transmission (see Figure 16–9). Other eicosanoids, such as PGI_2 and PGD_2, might, on the contrary, via an increase in cAMP, blunt an initial signal for cell activation by decreasing intracellular Ca^{2+} release.

The result of the sequence of events during the RVD response is a net loss of KCl and taurine, and also an inhibition of the Na^+ uptake mechanism(s), i.e., the Ehrlich ascites cells have provided optimal conditions for net loss of cell osmolytes, by the concomitant stimulation of osmolyte loss and inhibition of osmolyte gain.

Addition of the receptor agonist bradykinin to Swiss 3T3 fibroblasts induces an increase in inositol phosphate formation, release of arachidonic acid, and subsequent prostaglandin synthesis,[114] and addition of bradykinin to canine tracheal epithelial cells[116] and to Ehrlich ascites tumor cells produces an increase in cellular levels of calcium.[68] The increase in cellular Ca^{2+} activates the Cl^- channels in the apical membranes of canine tracheal epithelial cells[115] and stimulates the Ca^{2+}-dependent K^+ as well as Cl^- transport pathways in Ehrlich ascites tumor cells, resulting in cell shrinkage.[73] It is assumed (see Figure 16–9) that the signal cascade activated in Ehrlich cells by osmotic cell swelling is the same as the signal cascade activated by the receptor agonists bradykinin and thrombin.[73]

An important requirement of receptor-mediated signal transduction is its termination. In those systems in which the activity is linked to G proteins, termination may occur by reassociation of the inactive heterotrimer following activation of the GTPase associated with the α-subunit of the G protein. Eicosanoid receptors, like hormone receptors on cell plasma membrane, are able to regulate themselves in response to alterations in the availability of endogenous eicosanoids.[89] This indicates that eicosanoids are not only able to stimulate or inhibit cellular activity by interaction with their receptor, but they can also render the cell less responsive to individual eicosanoids over variable lengths of time by "downregulating" their own receptors. Long time exposure, i.e., longer than 5 min, of RBL-1 cells to LTD_4 results in desensitization of LTD_4 receptors followed by downregulation of LTD_4 receptor-mediated second messenger systems such as Ca^{2+} mobilization.[84] Homologous desensitization has also been demonstrated with receptors for PGE_2, PGD_2, and TXA_2.[10] Shutting off the signal can also be provided by acylation of the arachidonic acid and/or arachidonic acid metabolites (HETEs) into the phospholipids,[1] by feedback inhibition of PLA_2 activity by arachidonic acid or arachidonic acid metabolites, by reduction of cytosolic Ca^{2+} levels, and by activation of PKC. In the latter case, PKC activation has been shown to be involved in desensitization of LTD_4 receptors.[84,111]

VI. DIRECT EFFECT OF EICOSANOID PRECURSORS ON CONDUCTIVE ION TRANSPORT

In order to analyze the effect of eicosanoids on amino acid or ion transporting systems it is possible to change the arachidonic acid cascade and eicosanoid-dependent cellular functions by exogenous addition of fatty acids to isolated cells in suspension and to isolated cell patches or by nutritional modification of the eicosanoid precursor fatty acid pool.

Exogenous addition of arachidonic acid induces K^+ loss and Na^+ uptake in slices of rat brain cortex[117] and in Ehrlich ascites tumor cells.[21] Arachidonic acid also increases the Cl^- permeability in Ehrlich cells suspended in isotonic medium, leading to a significant Cl^- loss (Figure 16–10, left panel). The Na^+ and K^+ transport, provoked by arachidonic acid in Ehrlich ascites tumor cells, is via conductive ion channels and not via any Na^+/H^+ or K^+/H^+ exchange systems, nor any Cl^--dependent cotransport systems.[118] The effect of arachidonic acid on the Na^+ conductance is a nonspecific effect of an unsaturated fatty acid combined with a more specific effect mediated by cyclooxygenase metabolites of arachidonic acid, whereas the effect of arachidonic acid on K^+ transport seems to be predominantly a nonspecific detergent effect of an unsaturated fatty acid.[118] Direct activation of K^+-selective channels by nonesterified, unsaturated fatty acids has been demonstrated previously, using tight-seal, whole cell, and isolated patch recording techniques, in rat cardiac cells[119] and gastric smooth muscle cells from toads.[120]

According to the RVD-model (Figure 16–9), one would expect that addition of arachidonic acid to the Ehrlich cells swollen in hypotonic, Na-free medium should certainly accelerate the RVD response. This is not the case. Actually, exogenous addition of arachidonic acid to Ehrlich cells inhibits the swelling-induced Cl^- channels directly and instantaneously (Figure 16–10, right panel; see Lambert and co-workers[21,118]) and the subsequent RVD.[20] This effect of arachidonic acid appears to be an effect of an unsaturated fatty acid, because unsaturated fatty acids that are not turned into eicosanoids, e.g., oleic acid, are able to mimic the effect of arachidonic acid on the RVD.[20] Receptor-mediated release of arachidonic acid regulates epithelial Cl^- secretion via production of prostaglandins, which increase cellular levels of cAMP.[121] However, studies using cell-free membrane patches suggest that arachidonic acid at a free concentration of 5 μM also inactivates Cl^- channels in airway epithelia directly and reversibly.[122] It thus appears that arachidonic acid is a precursor for metabolites that activate conductive Cl^- transport and in itself inhibits the same system. The physiological significance could be a question of protection against overactivation of, e.g., ion transporting systems when the precursor is abundant.

Cheung and Ballou[123] found that human platelet activity is inhibited by unsaturated fatty acids and that the level of unsaturated fatty acids present in platelet extract was sufficient to inhibit most of the PLA_2 activity. If PLA_2 is inhibited by fatty acids, one might speculate how the inhibition is overcome when platelets are stimulated by various agonists and whether this constitutes a mechanism for feedback control.

The swelling-induced activation of the Cl^- transport pathway in Ehrlich cells is transient,[62] with inactivation within 10 min.[62] This inactivation can be explained by (1) a direct effect of the precursor on the Cl^- channel, (2) a reduction in the amount of the second messenger responsible for the Cl^- channel activation (LTD_4), (3) a reduced sensitivity of the LTD_4 receptor towards LTD_4, or by (4) the appearance of an inactivator. 15-HETE could well be considered as an inactivator for anion channels — it is synthesized in the Ehrlich cells (see Figure 16–2), it inhibits the swelling-induced Cl^- channel,[25] and it is, in the presence of exogenous arachidonic acid, most probably synthesized in surplus. Since the rate-limiting steps in leukotriene biosynthesis occur at the LTA_4-hydrolase and the glutathione transferase (see Figure 16–1), one would expect that high concentrations of arachidonic acid could lead to excess formation of LTA_4 which, combined with a relatively small conversion of LTA_4 to LTB_4 and LTC_4, could result in nonenzymatic breakdown products of LTA_4. Immunologic challenge of bone marrow-derived mouse mast cells has been shown to produce 6-*trans*-LTB_4,[124] for which the biological activity is unknown. Whether an increased synthesis

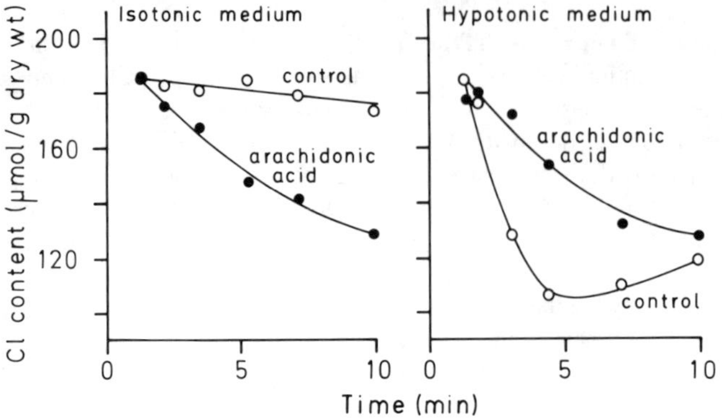

G_{Cl} ($\mu S/cm^2$)	ISOTONIC	0.8
	HYPOTONIC	43
ARACHIDONIC ACID	ISOTONIC	3.8
	HYPOTONIC	2.3

Figure 16–10 Effect of arachidonic acid on the Cl^- content in Ehrlich ascites tumor cells suspended in isotonic choline Cl^- medium and in hypotonic choline Cl^- medium. Ehrlich cells were incubated in isotonic standard NaCl medium for 40 min and washed at time zero to isotonic choline-Cl medium (left panel) or to hypotonic choline-Cl medium with half osmolarity (right panel). The final cytocrit was 7%. Arachidonic acid (0.4 m*M;* closed symbols) was added 0.2 min after the change in the medium. Control cells (open symbols) contained no arachidonic acid. The Cl^- content was measured in cell samples separated from the incubation medium by centrifugation, lysed in distilled water before deproteinization with perchloric acid. The Cl^- conductance (G_{Cl}) was estimated from the initial net loss of Cl^- and estimated values for the membrane potential. (Data from Lambert.[118])

of 15-HETE, 6-*trans*-LTB_4, or other lipoxygenase products in Ehrlich cells is involved in the inactivation of the swelling-induced Cl^- transport pathway or after addition of arachidonic acid is under investigation.

VII. EFFECT OF SUBSTITUTION OF EICOSANOID PRECURSORS IN THE PHOSPHOLIPIDS ON VOLUME REGULATION

Modification of the amount of eicosanoid precursors in the phospholipid pool can also affect the cellular eicosanoid-dependent functions. Ehrlich cells grown in mice fed on a highly saturated coconut oil diet contain considerably more saturated and monoenoic fatty acids and less polyenoic fatty acids in their plasma membrane phospholipids than cells grown in mice fed on a highly unsaturated sunflower oil diet.[125] It is also possible to increase the relative content of EPA and reduce the relative content of arachidonic acid in cellular PC and PE in Ehrlich ascites tumor cells just by adding MaxEPA to the chow fed to the host mice.[126] Analysis of the phospholipid composition of mucosal cell membranes from goldfish intestine, furthermore, has demonstrated an increased degree of unsaturation of the membrane when the environmental temperature of the fish was reduced from 30 to 6°C.[127] In this case the fatty acyl composition of ethanolamine phosphoglyceride showed a relative increase in the amount of arachidonic acid and EPA with compensatory falls in the amount of saturated palmitic and stearic acids.[127]

Increasing the saturation of the membrane lipids in Ehrlich cells increases the activation energy for sodium-dependent uptake of α-amino-isobutyric acid (AIB) and decreases the affinity of the transporter for AIB,[125] whereas enriching of the cellular phospholipids with EPA improves the RVD response in hypotonic media due to an increase in the swelling-induced K^+ and Cl^- conductances.[126] EPA is metabolized to eicosanoids and in many cases eicosanoids derived from EPA are equally or less potent than the corresponding eicosanoids synthesized from arachidonic acid.[128] PGI_3 is as potent in inhibiting platelet aggregation as PGI_2; TXA_3 is a weaker agonist for platelet aggregation and vasoconstriction when compared to TXA_2.[129] LTC_5 is equipotent to LTC_4 while LTB_5 is far less active than LTB_4 in inducing polymorphonuclear leukocyte (PMN) aggregation, degranulation, chemotaxis, and chemokinesis.[130] As demonstrated in Figure 16–7, LTD_4 is able to lift the inhibition of the RVD response provided by the anticalmodulin drug pimozide. LTD_5 appears to be equipotent to LTD_4 in lifting the inhibition of RVD by pimozide,[126] indicating that the stimulatory effect of EPA on RVD is not the result of the synthesis of a more potent messenger. Feeding the host mice with a diet supplemented with safflower oil (rich in n-6 fatty acids) instead of EPA (rich in n-3 fatty acids) also resulted in an enhanced RVD response.[126] It, therefore, seems reasonable to suggest that the effect of EPA on RVD is caused by an enhanced volume-induced leukotriene synthesis as a consequence of an increased precursor availability in the phospholipid pool.[126] It should be noted that EPA is a poor substrate for the cyclooxygenase pathway and a competitive inhibitor of arachidonic acid metabolism via this pathway.[1] This could mean that the volume-induced shift from prostaglandin to leukotriene synthesis seen during RVD (see Figure 16–3, left panel) is more pronounced in the Ehrlich cells grown in mice fed an EPA-enriched diet. Since prostaglandins inhibit the RVD response, an augmented shift from prostaglandin to leukotriene synthesis could well explain parts of the accelerated volume recovery seen in EPA-enriched cells.

In conclusion, eicosanoids play a significant role in the physiology and pathophysiology of thrombosis, inflammation, and reproduction. It is claimed that consumption of a "fish oil" diet is associated with a reduced synthesis of prostanoids and a reduced incidence of cardiovascular diseases, allergic reactions, and premature babies. However, eicosanoids also interfere with ion and amino acid transporting systems located in the plasma membrane and play, in this way, an important role in the regulation of the water content in the individual cell and in the whole organism. The ability of mammalian cells, e.g., Ehrlich ascites tumor cells, to regulate their cell volume after cell swelling in hypotonic medium depends on the availability of eicosanoid precursors and the synthesis of leukotrienes. This capability of the Ehrlich cells to volume regulate in hypotonic medium is significantly improved when the availability of eicosanoid precursors in the phospholipid pool is increased by modification of the diet fed to the host mice, but impaired when the cells are overloaded by direct exogenous addition of eicosanoid precursors. It appears that an essential fatty acid like arachidonic acid is the precursor of leukotrienes that activate Cl^--transporting channels and at the same time a potent inhibitor of the same channel.

REFERENCES

1. **Needleman, P., Turk, J., Jakschik, B. A., Morrison, A. R., and Lefkowith, J. B.,** Arachidonic acid metabolism, *Annu. Rev. Biochem.,* 55, 69, 1986.
2. **Smith, W. L.,** The eicosanoids and their biochemical mechanisms of action, *Biochem. J.,* 259, 315, 1989.
3. **Stjernschanz, J.,** The leukotrienes, *Med. Biol.,* 62, 215, 1984.
4. **Stahl, G. L., Tsao, P., Lefer, A. M., Ramphal, J. Y., and Nicolaou, K. C.,** Pharmacologic profile of lipoxins A_5 and B_5: new biologically active eicosanoids, *Eur. J. Pharmacol.,* 163, 55, 1989.
5. **Shimizu, T.,** Enzymes functional in the syntheses of leukotrienes and related compounds, *Int. J. Biochem.,* 20 (7), 661, 1988.
6. **Goetz, E. J., Brindly, L. L., and Goldman, D. W.,** Enhancement of human neutrophil adherence by synthetic leukotriene constituents of the slow-reacting substances of anaphylaxis, *Immunology,* 50, 35, 1983.
7. **Samuelsson, B.,** Leukotrienes: mediators of immediate hypersensitivity reactions and inflammation, *Science,* 220, 568, 1983.
8. **Spector, A. A., Gordon, J. A., and Moore, S. A.,** Hydroxyeicosatetraenoic acids (HETEs), *Prog. Lipid Res.,* 27, 271, 1988.
9. **Holzman, M. J.,** Arachidonic acid metabolism in airway epithelial cells, *Annu. Rev. Physiol.,* 54, 303, 1992.
10. **Nicosia, S. and Patrono, C.,** Eicosanoid biosynthesis and action: novel opportunities for pharmacological intervention, *FASEB J.,* 3, 1941, 1989.
11. **Serhan, C. N.,** Lipoxins: eicosanoids carrying intra and intercellular messages, *J. Bioenerg. Biomembr.,* 23 (1), 105, 1991.
12. **Claesson, H. E. and Haeggström, J.,** Human endothelial cells stimulate leukotriene synthesis and convert granulocyte released leukotriene A_4 into leukotriene B_4, C_4, D_4 and E_4, *Eur. J. Biochem.,* 173, 93, 1988.
13. **MacLouf, J. A. and Murphy, R. C.,** Transcellular metabolism of neutrophil-derived leukotriene A_4 by human platelets. A potential cellular source of leukotriene C_4, *J. Biol. Chem.,* 263, 174, 1988.
14. **Feinmark, S. J. and Cannon, P. J.,** Endothelial cell leukotriene C_4 synthesis results from intracellular transfer of leukotriene A_4 synthesized by polymorphonuclear leukocytes, *J. Biol. Chem.,* 261, 16466, 1988.

15. **Schaeffer, B. E. and Zadunaisky, J. A.,** Mechanism for leukotriene C_4 stimulation of chloride transport in cornea, *J. Membr. Biol.*, 93, 229, 1986.
16. **Leikauf, G. D., Ueki, I. F., Widdicombe, J. H., and Nadel, J. A.,** Alteration of chloride secretion across canine tracheal epithelium by lipoxygenase products of arachidonic acid, *Am. J. Physiol.*, 250, F47, 1986.
17. **Al-Bazzaz, F., Yadava, V. P., and Westenfelder, C.,** Modification of Na and Cl transport in canine mucosa by prostaglandins, *Am. J. Physiol.*, 240, F101, 1981.
18. **Smith, P. L., Welsh, M. J., Stoff, J. W., and Frizzell, R. A.,** Chloride secretion by canine tracheal epithelium. I. Role of intracellular cAMP levels, *J. Membr. Biol.*, 70, 217, 1982.
19. **Liedke, C. M.,** Interaction of epinephrine with isolated rabbit tracheal epithelial cells, *Am. J. Physiol.*, 251, C209, 1986.
20. **Lambert, I. H.,** Effect of arachidonic acid, fatty acids, prostaglandins, and leukotrienes on volume regulation in Ehrlich ascites tumor cells, *J. Membr. Biol.*, 98, 207, 1987.
21. **Lambert, I. H., Hoffmann, E. K., and Christensen, P.,** Role of prostaglandins and leukotrienes in volume regulation in Ehrlich ascites tumor cells, *J. Membr. Biol.*, 98, 7, 1987.
22. **Nielsen, R.,** Prostaglandin E_2 enhances the sodium conductance of exocrine glands in isolated frog skin (Rana esculenta), *J. Membr. Biol.*, 113, 31, 1990.
23. **Bjerregaard, H. F. and Nielsen, R.,** Prostaglandin E_2-stimulated glandular ion and water secretion in isolated frog skin *(Rana esculenta)*, *J. Membr. Biol.*, 97, 9, 1987.
24. **Steidl, M., Ritter, M., and Lang, F.,** Regulation of potassium conductance by prostaglandins in cultured renal epitheloid (Madin-Darby canine kidney) cells, *Pfluegers Arch.*, 418, 431, 1991.
25. **Lambert, I. H. and Henke, D. C.,** The role of hydroxyeicosa-tetraenoic acid in volume regulation in Ehrlich ascites tumor cells, *Acta Physiol. Scand.*, 143, 43A, 1991.
26. **Dahl, S. V., Hallbrucker, C., Lang, F., and Häussinger, D.,** Role of eicosanoids, inositol phosphates and extracellular Ca^{2+} in cell-volume regulation of rat liver, *Eur. J. Biochem.*, 198, 73, 1991.
27. **Lambert, I. H.,** Leukotriene-D_4 induced cell shrinkage in Ehrlich ascites tumor cells, *J. Membr. Biol.*, 108, 165, 1989.
28. **Irvine, R. F.,** How is the level of free arachidonic acid controlled in mammals?, *Biochem. J.*, 204, 3, 1982.
29. **Feinstein, M. B. and Sha'afi, R. I.,** Role of calcium in arachidonic acid metabolism and in the actions of arachidonic acid-derived metabolites, Calcium Cell Function, IV, 337, 1983.
30. **Berridge, M. J.,** Inositol trisphosphate and diacylglycerol: two interacting second messengers, *Annu. Rev. Biochem.*, 56, 159, 1987.
31. **Neufeld, E. J. and Majerus, P. W.,** Arachidonate release and phosphatidic acid turnover in stimulated human platelets, *J. Biol. Chem.*, 258, 2461, 1983.
32. **Van den Bosch, H.,** Intracellular phospholipase A, *Biochim. Biophys. Acta*, 604, 191, 1980.
34. **Kunze, H. and Vogt, W.,** Significance of phospholipase A for prostaglandin formation, *Ann. N.Y. Acad. Sci.*, 180, 123, 1971.
35. **Wong, P. Y. K. and Cheung, W. Y.,** Calmodulin stimulates human platelet phospholipase A_2, *Biochem. Biophys. Res. Commun.*, 90, 473, 1979.
36. **Chang, J., Musser, J. H., and McGregor, H.,** Phospholipase A_2: function and pharmacological regulation, *Biochem. Pharmacol.*, 36 (15), 2429, 1987.
37. **Taki, T. and Kanfer, J. N.,** Partial purification and properties of a rat brain phospholipase D, *J. Biol. Chem.*, 254, 9761, 1979.
38. **Wolf, R. A. and Gross, R. W.,** Gross. Identification of neutral active phospholipase C which hydrolyzes choline glycerophospholipids and plasmalogen selective phospholipase A_2 in canine myocardium, *J. Biol. Chem.*, 260, 7295, 1985.
39. **Hofmann, S. L. and Majerus, P. W.,** Modulation of phosphatidylinositol-specific phospholipase C activity by phospholipid interactions, diglycerides and calcium ions, *J. Biol. Chem.*, 257 (23), 14359, 1982.
40. **Nakashima, S., Suganuma, A., Matsui, A. A., Hattori, H., Satto, M., Takenawa, A., and Nozawa, Y.,** Primary role of calcium ions in arachidonic acid release from rat platelet membranes, *Biochem. J.*, 259, 139, 1989.
41. **Henderson, L. M., Chappell, J. B., and Jones, T. G.,** Superoxide generation is inhibited by phospholipase A_2 inhibitors. Role for phospholipase A_2 in the activation of the NADPH oxidase, *Biochem. J.*, 264, 249, 1989.
42. **Cashman, J. R.,** Leukotriene biosynthesis inhibitor, *Pharm. Res.*, 6, 253, 1985.
43. **Kirstein, D., Thomsen, M. K., and Ahnfelt-Rønne, I.,** Inhibition of leukotriene biosynthesis and polymorphonuclear leukocyte functions by orally active quinolylmethoxyphenylamines, *Pharmacol. Toxicol.*, 68, 125, 1991.

44. **Rouzer, C. A., Ford-Hutchinson, A. W., Morton, H. E., and Gillard, J. W.,** MK886, a potent and specific leukotriene biosynthesis inhibitor blocks and reverses the membrane association of 5-lipoxygenase in ionophore-challenged leukocytes, *J. Biol. Chem.*, 265 (3), 1436, 1990.
45. **Yamamoto, S.,** Mammalian lipoxygenases: molecular and catalytic properties. Prostaglandins, leukotrienes and essential fatty acids, 35, 219, 1989.
46. **Yamamoto, S.,** Enzymes in the arachidonic acid cascade, in *Prostaglandins and Related Substances,* Pace-Asciak, C. and Granstöm, E., Eds., Elsevier, Amsterdam, 1983, 171.
47. **Van Dorp, D. A.,** Aspects of the biosynthesis of prostaglandins, *Prog. Biochem. Pharmacol.*, 3, 71, 1967.
48. **Sigal, E.,** The molecular biology of mammalian arachidonic acid metabolism, *Am. J. Physiol.*, 260, L13, 1991.
49. **Prescot, S. M.,** The effect of eicosapentaenoic acid on leukotriene B production by human neutrophils, *J. Biol. Chem.*, 259, 7615, 1984.
50. **Leich, A. G., Lee, T. H., Ringel, E. W., Prickett, J. D., B. Robinson, D. R., Tyne, S. G., Corey, E. J., Drazen, J. M., and Austen, Lewis, R. A.,** Immunologically induced generation of tetraene and pentaene leukotrienes in the peritoneal cavities of menhaden-fed rats, *J. Immunol.*, 132, 2559, 1984.
51. **Budowski, P.,** ω3-fatty acids in health and diseases, *World Rev. Nutr. Diet.*, 57, 214, 1988.
52. **Wong, P. Y. K., Hughes, R., and Lam, B.,** Lipoxins: a novel group of trihydroxy pentaenes of eicosapentaenoic acid derived from porcine leukocytes, *Biochem. Biophys. Res. Commun.*, 126, 763, 1985.
53. **Rouzer, C. A. and Samuelson, B.,** On the nature of the 5-lipoxygenase reaction in human leukocytes: enzyme purification and requirement for multiple stimulatory factors, *Proc. Natl. Acad. Sci. U.S.A.*, 82, 6040, 1985.
54. **Ford-Hutchinson, A. W.,** Inhibition of leukotrine biosynthesis, *Ann. N.Y. Acad. Sci.*, 629, 133, 1991.
55. **Samuelsson, B., Haeggström, J. Z., and Wetterholm, A.,** Leukotriene biosynthesis, *Ann. N.Y. Acad. Sci.*, 629, 89, 1991.
56. **Schwartzman, M. L., Balazy, M., Masferrer, J., Abraham, N. C., McGiff, J. C., and Murphy, R. C.,** 12 (HETE): a cytochrome P450-dependent arachidonate metabolite that inhibits Na,K-ATPase in the cornea, *Proc. Natl. Acad. Sci. U.S.A.*, 84, 8125, 1987.
57. **Montellano, P. R. O.,** Cytochrome P-450 catalysis: radical intermediates and dehydrogenation reactions, *TIPS,* 10, 354, 1989.
58. **Capdevila, J. H., Mosset, P., Yadagiri, P., Lumin, S., and Falck, J. R.,** NADPH-dependent microsomal metabolism of 14,15-epoxy eicosatrienoic acid to diepoxides and epoxyalcohols, *Arch. Biochem. Biophys.*, 261, 122, 1988.
59. **Christensen, O. and Hoffmann, E. K.,** Cell swelling activates K and Cl channels as well as nonselective, stretch-activated cation channels in Ehrlich ascites tumor cells, *J. Membr. Biol.*, 129, 13, 1992.
60. **Lambert, I. H., Hoffmann, E. K., and Jørgensen, F.,** Membrane potential, anion and cation conductances in Ehrlich ascites tumor cells, *J. Membr. Biol.*, 111, 113, 1989.
61. **Hoffmann, E. K., Simonsen, L. O., and Lambert, I. H.,** Volume-induced increase of K^+ and Cl^- permeabilities in Ehrlich ascites tumor cells. Role of internal Ca^{2+}, *J. Membr. Biol.*, 78, 211, 1984.
62. **Hoffmann, E. K., Lambert, I. H., and Simonsen, L. O.,** Separate, Ca^{2+} activated K^+ and Cl^- transport pathways in Ehrlich ascites tumor cells, *J. Membr. Biol.*, 91, 227, 1986.
63. **Hoffmann, E. K. and Lambert, I. H.,** Amino acid transport and cell volume regulation in Ehrlich ascites tumour cells, *J. Physiol. (London),* 338, 613, 1983.
64. **Lambert, I. H. and Hoffmann, E. K.,** Regulation of taurine transport in Ehrlich ascites tumor cells, *J. Membr. Biol.*, 131, 67, 1993.
65. **Christensen, S., Hoffmann, E. K., Saermark, T., and Simonsen, L. O.,** Inositol trisphosphate may be a second messenger in regulatory volume decrease in Ehrlich mouse ascites-tumour cells, *J. Physiol. (London),* 403, 109P, 1988.
66. **Lambert, I. H. and Hoffmann, E. K.,** Amino acid metabolism and protein turnover under different osmotic conditions in Ehrlich ascites tumor cells, *Mol. Physiol.*, 2, 273, 1982.
67. **Livne, A. and Hoffmann, E. K.,** Cytoplasmatic acidification and activation of Na^+/H^+ exchange during regulatory volume decrease in Ehrlich ascites tumor cells, *J. Membr. Biol.*, 114, 153, 1990.
68. **Hoffmann, E. K., Simonsen, L. O., and Lambert, I. H.,** Cell volume regulation: intracellular transmission, in *Interaction, Cell Volume, Cell Function,* Gilles, R., Ed., ACEP Series, Springer, Heidelberg, 1993, Chap. 7.
69. **Henke, D. C., Kouzan, S., and Eling, T. E.,** Analysis of leukotrienes, prostaglandins, and other oxygenated metabolites of arachidonic

acid by high-performance liquid chromatography, *Anal. Biochem.*, 140, 87, 1984.

70. **Hoffmann, E. K.,** Regulation of cell volume by selective changes in the leak permeabilities of Ehrlich ascites tumor cells, in *Alfred Benzon Symposium. XI. Osmotic and Volume Regulation,* Jørgensen, C. B. and Skadhauge, E., Eds., Munksgaard, Copenhagen, 1978, 397.
71. **Hoffmann, E. K., Simonsen, L. O., and Sjøholm, C.,** Membrane potential, chloride exchange and chloride conductance in Ehrlich mouse ascites tumor cells, *J. Physiol. (London),* 296, 61, 1979.
72. **Hendil, K. B. and Hoffmann, E. K.,** Cell volume regulation in Ehrlich ascites tumor cells, *J. Cell Physiol.,* 84, 115, 1974.
73. **Simonsen, L. O., Brown, A. M., Christensen, S., Harbak, H., Svane, P. C., and Hoffmann, E. K.,** Thrombin and bradykinin mimic the volume response induced by cell swelling in Ehrlich mouse ascites tumour cells, *Renal Physiol. Biochem.,* 13, 176, 1990.
74. **Kramhøft, B., Lambert, I. H., Hoffmann, E. K., and Jørgensen, F.,** Activation of Cl-dependent K transport in Ehrlich ascites tumor cells, *Am. J. Physiol.,* 251, C369, 1986.
75. **Grinstein, S., Cohen, S., Sarkadi, B., and Rothstein, A.,** Induction of ^{86}Rb fluxes by Ca^{2+} and volume changes in thymocytes and their isolated membranes, *J. Cell Physiol.,* 116, 352, 1983.
76. **Valdeolmillos, M., Garcia-Sancho, J., and Herreros, B.,** Ca^{2+}-dependent K transport in the Ehrlich ascites tumor cell, *Biochim. Biophys. Acta,* 685, 273, 1982.
77. **Grynkiewicz, G., Poenie, M., and Tsien, R. Y.,** A new generation of Ca^{2+} indicators with greatly improved fluorescence properties, *J. Biol. Chem.,* 260(6), 3440, 1985.
78. **Lambert, I. H. and Hoffmann, E. K.,** The role of phospholipase-A_2 and 5-lipoxygenase in the activation of K and Cl channels and the taurine leak pathway in Ehrlich ascites tumor cells, *Acta Physiol. Scand.,* 143 (1), 33A, 1991.
79. **Chamberlin, M. E. and Strange, K.,** Anisosmotic cell volume regulation: a comparative view, *Am. J. Physiol.,* 257, C159, 1989.
80. **Law, R. O.,** Amino acids as volume-regulatory osmolytes in mammalian cells, *Comp. Biochem. Physiol.,* 99A, 263, 1991.
81. **Hoffmann, E. K. and Hendil, K. B.,** The role of amino acids and taurine in isosmotic intracellular regulation in Ehrlich ascites mouse tumour cells, *J. Comp. Physiol.,* 108, 279, 1976.
82. **Baquet, A., Meijer, A. J., and Hue, L.,** Hepatocyte swelling increases inositol 1,4,5-trisphosphate, calcium and cyclic AMP concentration but antagonizes phosphorylase activation by Ca^{2+}-dependent hormones, *FEBS Lett.,* 278 (1), 103, 1991.
83. **Pierce, S. K. and Politis, A. D.,** Ca^{2+}-activated cell volume recovery mechanisms, *Annu. Rev. Physiol.,* 52, 27, 1990.
84. **Vegesna, R. V. K., Mong, S., and Crooke, S. T.,** Leukotriene-D_4-induced activation of protein kinase C in rat basophilic leukemia cells, *Eur. J. Pharmacol.,* 147, 387, 1988.
85. **Musch, M. W. and Goldstein, L.,** Hypotonicity stimulates phosphatidyl choline hydrolysis and generates diacylglycerol in erythrocytes, *J. Biol. Chem.,* 265 (22), 13055, 1990.
86. **McConnel, F. M. and Goldstein, L.,** Intracellular signals and volume regulatory response in skate erythrocytes, *Am. J. Physiol.,* 255, R982, 1988.
87. **MacLeod, R. J. and Hamilton, J. R.,** Separate K and Cl transport pathways are activated for regulatory volume decrease in jejunal villus cells, *Am. J. Physiol.,* 260, G405, 1991.
88. **Feurstein, G.,** Autonomic pharmacology of leukotrienes, *J. Auton. Pharmac.,* 5, 149, 1985.
89. **Robertson, R. P.,** Characterization and regulation of prostaglandin and leukotriene receptors: an overview, *Prostaglandins,* 31, 395, 1986.
90. **Cristol, J. P., Provencal, B., and Sirois, P.,** Leukotriene receptors, *J. Receptor Res.,* 9 (4&5), 341, 1989.
91. **Sarau, H. S., Mong, S., Foley, J. J., Wu, H.-L., and Crooke, S. T.,** Identification and characterization of leukotriene D_4 receptors and signal transduction processes in rat basophilic leukemia cells, *J. Biol. Chem.,* 262 (9), 4034, 1987.
92. **Ford-Hutchinson, A. W.,** Regulation of the production and action of leukotrienes by MK-571 and MK-886, *Adv. Prostaglandin, Thromboxane, Leukotriene Res.,* 21, 9, 1990.
93. **Baud, L., Perez, J., Cherqui, G., Cragoe, E. J., and Ardaillou, R.,** Leukotriene D_4-induced proliferation of glomerular epithelial cells: PKC- and Na^+-H^+ exchanger-mediated response, *Am. J. Physiol.,* 257, C232, 1989.
94. **Badr, K. F., Mong, S., Hoover, R. L., Schwartzberg, M., Ebert, J., Jacobson, H. R., and Harris, R. C.,** Leukotriene D_4 binding and signal transduction in rat glomerular mesanglial cells, *Am. J. Physiol.,* 257, F280, 1989.
95. **Yousufzai, S. Y. K., Tachado, S. D., and Abdel-Latif, A. A.,** Species differences in the

effect of leukotriene D_4 on inositol trisphosphate accumulation, cyclic AMP formation and contraction in the iris sphincter of the mammalian eye, *Prostaglandins,* 39 (3), 227, 1990.

96. **Mong, S., Miller, J., Wu, H.-L., and Crooke, S. T.,** Leukotriene D_4 receptor mediated hydrolysis of phosphoinositide and mobilization of calcium in sheep tracheal smooth muscle cells, *J. Pharmacol. Exp. Ther.,* 244, 508, 1988.

97. **Mong, S., Hoffmann, K., Wu, H.-L., and Crooke, S. T.,** Leukotriene-induced hydrolysis of inositol lipids in guinea pig lung: mechanism of signal transduction for leukotriene-D_4 receptors, *Mol. Pharmacol.,* 31, 35, 1987.

98. **Sjölander, A., Grönroos, E., Hammerström, S., and Anderson, T.,** Leukotriene D_4 and E_4 induce transmembrane signaling in human epithelial cells, *J. Biol. Chem.,* 265, 20976, 1990.

99. **Mong, S., Wu, H.-L., Clark, M. A., Gleason, J. G., and Crooke, S. T.,** Leukotriene-D_4 receptor-mediated synthesis and release of arachidonic acid metabolites in guinea pig lung: induction of thromboxane and prostacycline biosynthesis by leukotriene D_4, *J. Pharmacol. Exp. Ther.,* 239 (1), 63, 1986.

100. **Clark, M. A., Littlejohn, D., Conway, T. M., Mong, S., Steiner, S., and Crooke, S. T.,** Leukotriene D_4 treatment of bovine aortic endothelial cells and murine smooth muscle cells in culture results in an increase in phospholipase A_2 acitivity, *J. Biol. Chem.,* 261 (23), 10713, 1986.

101. **Piper, P. J. and Samhoun, M. N.,** The mechanism of action of leukotriene C_4 and D_4 in guinea-pig isolated perfused lung and parenchymal strips of guinea-pig, rabbit and rat, *Prostaglandins,* 21, 793, 1981.

102. **Samhoun, M. N. and Piper, P. J.,** Actions and interactions of lipoxygenase and cyclooxygenase products in respiratory and vascular tissues, *Prostaglandins Leukotrienes Med.,* 13, 79, 1984.

103. **Dahlen, S. E.,** The significance of liberated cyclo-oxygenase products for the pulmonary and cardiovascular actions of leukotriene C_4 in the guinea-pig depends upon the route of administration, *Acta Physiol. Scand.,* 118, 415, 1983.

104. **Farrukh, I. S., Sciutto, A. M., Spannhake, E. W., Gurtner, G. H., and Michael, J. R.,** Leukotriene D_4 increases pulmonary vascular permeability and pressure by different mechanisms in the rabbit, *Am. Rev. Respir. Dis.,* 134, 229, 1986.

105. **Noonan, T. C. and Malik, A. B.,** Pulmonary vascular response to leukotriene D_4 in unanethetized sheep: role of thromboxane, *J. Appl. Physiol.,* 60 (3), 765, 1986.

106. **Crooke, S. T., Mattern, M., Sarau, H. M., Winkler, J. D., Balcarek, J., Wong, A., and Bennett, C. D.,** The signal transduction system of the leukotriene D_4 receptor, *TIPS,* 10, 103, 1989.

107. **Berridge, M. J. and Irvine, R. F.,** Inositol phosphates and cell signalling, *Nature,* 341, 197, 1989.

108. **Nishizuka, Y.,** Studies and perspectives of protein kinase C, *Science,* 233, 305, 1986.

109. **Clark, M. A., Cook, M., Mong, S., and Crooke, S. T.,** The binding of leukotriene C_4 and leukotriene D_4 to membranes of a smooth muscle cell line (BC_3H_1) and evidence that leukotriene induced contraction in these cells is mediated by thromboxane, protein and RNA synthesis, *Eur. J. Pharmacol.,* 116, 207, 1985.

110. **Welsh, M. J.,** Abnormal regulation of ion channel in cystic fibrosis epithelia, *FASEB J.,* 4, 2718, 1990.

111. **Winkler, J. D., Mong, S., and Crooke, S. T.,** Leukotriene D_4-induced homologous desensitization of calcium mobilization in rat basophilic leukemia cells, *J. Pharmacol. Exp. Ther.,* 244, 449, 1987.

112. **Lambert, I. H.,** Na^+-dependent taurine uptake in Ehrlich ascites tumour cells, *Mol. Physiol.,* 6, 233, 1984.

113. **Lambert, I. H.,** Taurine transport in Ehrlich ascites tumour cells. Specificity and chloride dependence, *Mol. Physiol.,* 7, 323, 1985.

114. **Regoli, D. and Barabe, J.,** Pharmacology of bradykinin and related kinins, *J. Pharmacol. Rev.,* 32, 1, 1980.

115. **Clancy, J. P., McCann, J. D., Li, M., and Welsh, M. J.,** Calcium-dependent regulation of airway epithelial chloride channels, *Am. J. Physiol.,* 258, L25, 1990.

116. **McCann, J. D., Bhalla, R. C., and Welsh, M. J.,** Release of intracellular calcium by two different second messengers in airway epithelium, *Am. J. Physiol.,* 257, L116, 1989.

117. **Chan, P. H. and Fishman, R. A.,** Brain edema: induction in cortical slices by polyunsaturated fatty acids, *Science,* 201, 358, 1978.

118. **Lambert, I. H.,** Effect of arachidonic acid on conductive Na, K and anion transport pathways in Ehrlich ascites tumor cells under isotonic and hypotonic conditions, *Cell. Physiol. Biochem.,* 1, 177, 1991.

119. **Kim, D. and Clapham, D. E.,** Potassium channels in cardiac cells activated by arachidonic

acid and phospholipids, *Science,* 244, 1174, 1989.

120. **Ordway, R. W., Walsh, S. V., and Singer, J. J.,** Arachidonic acid and other fatty acids directly activate potassium channels in smooth muscle cells, *Science,* 244, 1176, 1989.

121. **Welsh, M. J.,** Electrolyte transport by airway epithelia, *Physiol. Rev.,* 67, 1143, 1987.

122. **Anderson, M. P. and Welsh, M. J.,** Fatty acids inhibit apical membrane chloride channels in airway epithelia, *Proc. Natl. Acad. Sci. U.S.A.,* 87, 7334, 1990.

123. **Cheung, W. Y. and Ballou, L. R.,** Substrate-specific forms of phospholipase A_2 in human platelets, *Adv. Prostaglandin, Thromboxane Leukotriene Res.,* 19, 584, 1989.

124. **Wei, Y. and Jakschik, B. A.,** Interaction of macrophages and mast cells, *Fed. Proc.,* 44, 986, 1985.

125. **Kaduce, T. L., Awad, A. B., Fontenelle, L. J., and Spector, A. A.,** Effect of fatty acid saturation on α-aminoisobutyric acid transport in Ehrlich ascites cells, *J. Biol. Chem.,* 252, 6624, 1977.

126. **Lauritzen, L., Hoffmann, E. K., Hansen, H. S., and Jensen, B.,** Dietary (n-3) and (n-6) fatty acids are equipotent in stimulating volume regulation in Ehrlich ascites tumor cells, *Am. J. Physiol.,* 264, C109, 1993.

127. **Smith, M. W.,** Membrane transport in fish intestine, *Biochem. Physiol.,* 75A (3), 325, 1983.

128. **Kinsella, J. E., Lokesh, B., and Stone, R. A.,** Dietary n-3 polyunsaturated fatty acids and amelioration of cardiovascular disease: possible mechanisms, *Am. J. Physiol. Nutr.,* 52, 1, 1990.

129. **Lagarde, M., Gualde, N., and Rigaud, M.,** Metabolic interactions between eicosanoids in blood and vascular cells, *Biochem. J.,* 257, 313, 1989.

130. **Terano, T., Hirai, A., Tamura, Y., Yoshida, S., Salmon, J. A., and Moncada, S.,** Effect of eicosapentaenoic acid on eicosanoids formation by stimulated human polymorphonuclear leukocytes, *Prog. Lipids Res.,* 25, 129, 1986.

131. **Larsen, A. K., Jensen, B., and Hoffman, E. K.,** Activation of protein Kinase C during volume regulation in Ehrlich mouse ascites tumor cells, *Biochim. Biophys. Acta,* submitted.

Chapter 17

Involvement of a G Protein in the Shrinkage-Induced Activation of Na-H Exchange in Barnacle Muscle Fibers

Walter F. Boron, Emilia M. Hogan and Bruce A. Davis

TABLE OF CONTENTS

I. INTRODUCTION

Our laboratory's experimental interest in cell volume regulation evolved quite by accident. We were intrigued by Parker's observation that the Na-H exchanger of dog erythrocytes is activated not only by low intracellular pH (pH_i) and shrinkage, but also by intracellular Li^+ at concentrations between ~10 and ~100 m*M*.[1] One of the possibilities raised for this Li^+ effect is that the Li^+ might substitute for H^+ at the hypothesized intracellular modifier site of the Na-H exchanger.[2] We, at the time, were studying the Na^+-dependent $Cl\text{-}HCO_3$ exchanger in muscle fibers of the giant barnacle, *Balanus nubilus*. The Na^+-dependent $Cl\text{-}HCO_3$ exchanger, like the Na-H exchanger, is an acid-extruding transporter (i.e., it moves acid-base equivalents out of the cell). Previous work had shown that this transporter was responsible for almost all acid extrusion in the barnacle muscle fiber.[3,4] Moreover, it was with the Na^+-dependent $Cl\text{-}HCO_3$ exchanger that the concept first evolved (see Reference 5) that an acid extruder might be activated in all of its modes of operation (e.g., acid extrusion, Cl-Cl exchange, Na-Na exchange) by low pH_i per se. Therefore, it was natural to ask whether intracellular Li^+ might also activate the barnacle muscle Na^+-dependent Cl-HCO_3 exchanger at normal pH_i. As expected, we found that internal Li^+ did indeed stimulate acid

0-8493-4448-4/94/$0.00+$.50

extrusion in the barnacle muscle fiber. However, we were surprised to find that the stimulated acid-extrusion mechanism was not the Na^+-dependent Cl-HCO_3 exchanger, but a latent Na-H exchanger. We later showed that this Na-H exchanger is activated by low pH_i and shrinkage.

Our experimental approach with barnacle muscle fibers would be ideally suited to exploring some aspects of the signal transduction mechanism by which shrinkage activates Na-H exchange. As outlined in Figure 17–1, we work with internally dialyzed muscle fibers. A single fiber is oriented horizontally in a chamber, and cannulated at both ends. A length of dialysis tubing (outer diameter, ~200 µm; molecular weight cut-off, ~6000 Da) is inserted in one cannula, down the length of the fiber, and out the opposite cannula. By pumping dialysis fluid (DF) through this tubing, we are able to exert a fair degree of control over the internal environment.[6] Parallel to this dialysis tubing, we insert a glass pH-sensitive microelectrode[7] through one cannula, and an open-tipped microelectrode filled with 3 *M* KCl through the other. Thus, it is possible to dialyze the cell to a predetermined composition, halt dialysis, and monitor changes in pH_i and membrane voltage. Molecules too large to diffuse through the dialysis tube can be microinjected into a cannulated cell.

In the remainder of this chapter, we review our work on the control of Na-H exchange in the muscle fiber of the giant barnacle, with special emphasis on experiments that point to the involvement of a G protein in the signal transduction process leading to the activation of Na-H exchange by shrinkage. We also suggest experiments that could more definitively implicate a G protein in the signal transduction process. Finally, we speculate on the relative roles of Na-H exchange in cell pH and cell volume regulation.

II. EFFECT OF INTRACELLULAR Li^+ ON Na-H EXCHANGE

A. EFFECT OF Li^+ ON THE RECOVERY OF pH_i FROM AN ACID LOAD IN THE PRESENCE OF CO_2/HCO_3^-

In our first experiments, we examined the effects of Li^+ on the recovery of pH_i from an acid load in barnacle muscle fibers (BMFs), a recovery we expected to be mediated almost exclusively by the Na^+-dependent Cl-HCO_3 exchanger.[3,4,8] We acid loaded the BMFs by dialyzing them for at least 90 min with a DF having a pH of 6.5. Both the DF and the artificial seawater (ASW), which bathed the exterior of the fibers, were Na^+-free. Figure 17–2A illustrates an experiment in which a BMF was dialyzed with a control (i.e., Li^+-free) DF. Only the final minutes of dialysis are shown in the figure (segment *ab*). After dialysis was halted, pH_i slowly drifted upward *(bc)*. When we simultaneously introduced 440 m*M* Na^+ and 10 m*M* HCO_3^- (~0.4% CO_2) into the ASW, pH_i recovered very rapidly and stabilized at a value of ~7.3 *(cd)*. In nine such experiments, the final pH_i for control cells averaged 7.32 ± 0.02 (SEM). When we added 0.5 m*M* SITS, a blocker of Na^+-dependent Cl-HCO_3 exchange, in the continued presence of Na^+ and HCO_3^-, pH_i declined slowly *(de)*. These results are similar to those previously reported for BMFs.[3,4] In particular, earlier work had shown that application of SITS after recovery of pH_i from an acid load unmasks a background acidification.[3]

Figure 17–2B shows the results of an experiment similar to that of Figure 17–2A, except that the DF contained 100 m*M* Li^+(Li^+ replacing K^+). The general features of the pH_i recovery are similar to those in Figure 17–2A. However, there are three differences. First, the Na^+ and HCO_3^--dependent component of the acid extrusion rate (J), computed from the initial rate of pH_i increase at point *a,* is nearly 30% higher in the Li^+-dialyzed cells. Although this difference did not reach statistical significance with the limited number of cells in our study, data presented below make it likely that the difference is in fact real, and due to stimulation of Na-H exchange. Second, the maximal pH_i achieved after the pH_i recovery averaged 7.49 ± 0.05 in eight Li^+-dialyzed fibers, significantly higher than in the controls. Finally, the addition of SITS did not unmask an acidification *(de),* as would have been expected if the Li^+ had stimulated Na^+-dependent Cl-HCO_3 exchange.

B. EFFECT OF Li^+ ON Na-H EXCHANGE NEAR THE RESTING pH_i

Figures 17–3A and B show a pair of experiments in which we examined the effect of Li^+ in BMFs in which Na^+-dependent Cl-HCO_3 exchange had been minimized in three ways. First, the BMFs were pretreated during the dialysis period with 0.5 m*M* SITS. Second, the cells were dialyzed to a pH_i of ~7.35, where Na^+-dependent Cl-HCO_3 exchange is minimal.[3] Finally, the experiments were conducted in the nominal absence of CO_2/HCO_3^-. The cell in Figure 17–3A, dialyzed with a Li^+-free DF at pH 7.35, exhibited minimal evidence of Na-H exchange. Thus, after dialysis was halted (point *b*) and pH_i stabilized *(bc),* the addition of 50 m*M* Na^+ elicited only a slow pH_i increase *(cd)* that was blocked by 1 m*M* amiloride *(de)*. In contrast, the cell in Figure 17–3B, dialyzed with 100 m*M* Li^+, had considerable Na-H exchange activity. When dialysis was halted, pH_i drifted downward rather rapidly *(bc),* presumably reflecting the

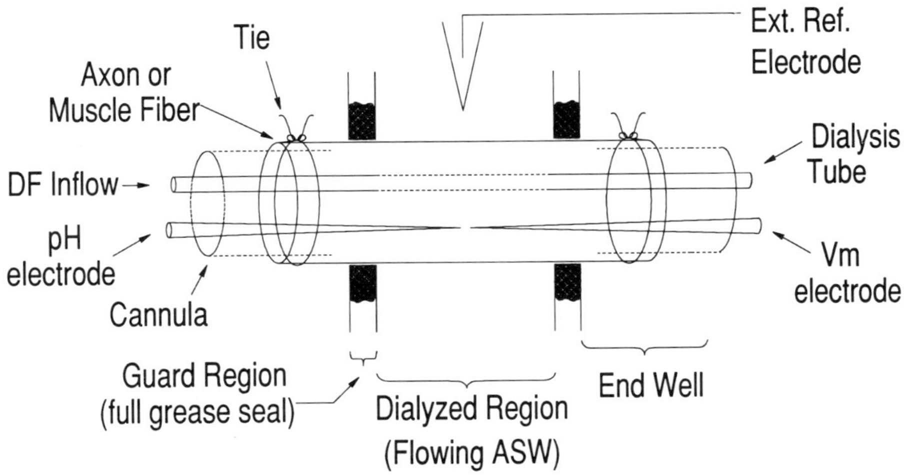

Figure 17–1 Schematic diagram of an internally dialyzed giant barnacle muscle fiber.

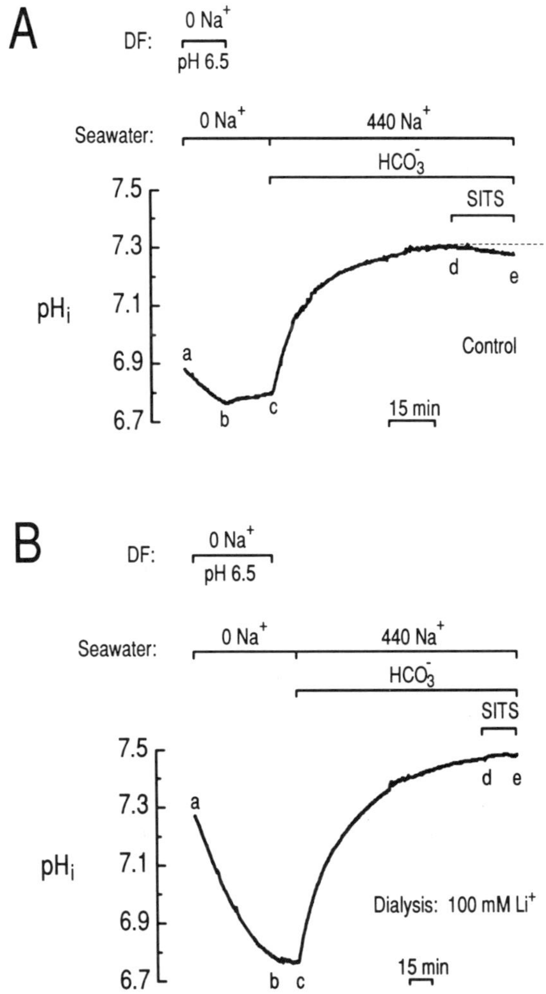

Figure 17–2 The effect of Li^+ on the pH_i recovery from an acid load in the presence of CO_2/HCO_3^-. (A) Fiber dialyzed with a Li^+-free solution. The cell was dialyzed for 90 min with a Na^+-free DF buffered to pH 6.5 with 100 m*M* MES; only the last few minutes of this dialysis period is shown (segment *ab*). The ASW was Na^+- and HCO_3^--free until point *c*. Introducing 440 m*M* Na^+ and 10 m*M* HCO_3^- caused pH_i to recover rapidly, and level off at ~7.33. SITS (0.5 m*M*) was added to the ASW at *d*, causing a slow pH_i decrease. (B) Fiber dialyzed with a Li^+-containing solution. This experiment was similar to that shown in (A) except that the DF contained 100 m*M* Li^+. The maximal pH_i achieved (point *d*) was greater than in the absence of Li^+. Also, Li^+ failed to elicit a pH_i decrease. (From Davis, B. A. et al., *Am. J. Physiol.*, 263, C246, 1991. With permission of the American Physiological Society.)

exchange of intracellular Li^+ for extracellular H^+. However, adding 50 m*M* Na^+ to the ASW converted this decrease to a rapid pH_i increase *(cd)* that was blocked reversibly by amiloride *(de)*. Thus, dialysis with 100 m*M* Li^+ stimulates an Na-H exchanger that is otherwise minimally active near the resting pH_i. These experiments indicate that it was Na-H exchange that was responsible for the high final pH_i

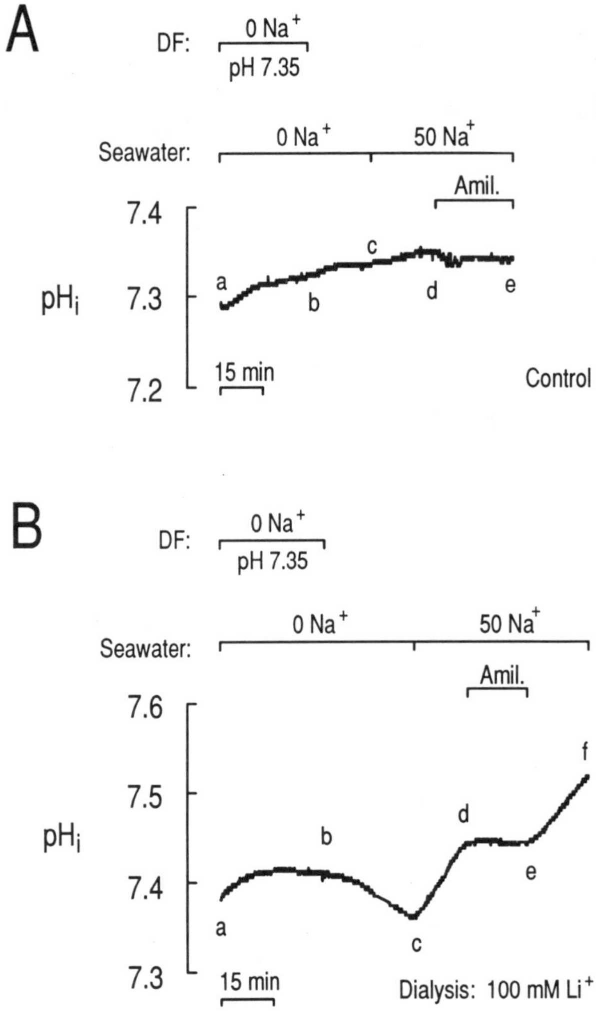

Figure 17–3 The effect of Li^+ on the pH_i of fibers dialyzed with a fluid at pH 7.35 and assayed in the nominal absence of CO_2/HCO_3^-. (A) Fiber dialyzed with a Li^+-free fluid. The fiber was dialyzed for 90 min with a Na^+-free DF buffered to pH 7.35 with 100 m*M* HEPES; only the last few minutes of this dialysis period is shown (segment *ab*). At point *c,* switching the ASW to one containing 50 m*M* Na^+ caused a slow pH_i recovery *(cd).* During segment *de,* 1 m*M* amiloride was added to the ASW. (B) Fiber dialyzed with a Li^+-containing fluid. The protocol was similar to that in (A) except that the DF contained 100 m*M* Li^+ The pH_i decrease after the halt of dialysis (segment *bc*) probably reflects the exchange of internal Li^+ for external H^+. Addition of 50 m*M* Na^+ reversed this pH_i decline, causing a rapid pH_i increase *(cd)* that was blocked by amiloride *(de).* Removing amiloride partially reversed this inhibition *(ef).* (From Davis, B. A. et al., *Am. J. Physiol.,* 263, C246, 1991. With permission of the American Physiological Society.)

shown in Figure 17–2B, in which a Li^+-dialyzed fiber was exposed to CO_2/HCO^-_3.

C. EFFECT OF Li^+ ON THE pH_i DEPENDENCE OF Na-H EXCHANGE

Figure 17–4 summarizes experiments such as those shown in Figures 17–3A and B, as well as others that were identical except for the low pH_i to which they were dialyzed. The results show that, in control cells (open circles), the flux due to Na-H exchange ($J_{Na\text{-}H}$) is minimal near the resting pH_i, and rises to ~60 μ*M*/min at pH_i 6.8. In Li^+-dialyzed cells, however, $J_{Na\text{-}H}$ is in the vicinity of 170 μ*M*/min over the entire pH_i range. It is perhaps worthwhile to note that, in control cells with an uninhibited Na^+-dependent Cl-HCO_3 exchanger (see Figure 17–2A), the total acid extrusion rate at pH_i 6.8 was 1069 μ*M*/min. Thus, at pH_i 6.8, Na-H exchange accounts for only ~6% of the total acid-extruding capacity of BMFs. We conclude that high levels of intracellular Li^+ activate a Na-H exchanger that otherwise makes a minimal contribution to pH_i regulation.

III. EFFECT OF SHRINKAGE ON Na-H EXCHANGE

A. ACTIVATION OF Na-H EXCHANGE BY SHRINKAGE OVER A BROAD pH_i RANGE

The latent Na-H exchanger in BMFs, activated by high levels of intracellular Li^+, probably fills a physiological role. Considering Parker's[1] results on dog erythrocytes, it is likely that this role is to mediate a volume-regulatory increase (VRI) in response to cell shrinkage. We therefore examined the effect on $J_{Na\text{-}H}$ of shrinking BMFs in hypertonic ASW (975 → 1600 mOsmol/kg). Our experimental protocol was similar to that used in the experiments that generated the pH_i dependence of $J_{Na\text{-}H}$ in Figure 17–4. We found that shrinkage stimulates Na-H exchange even

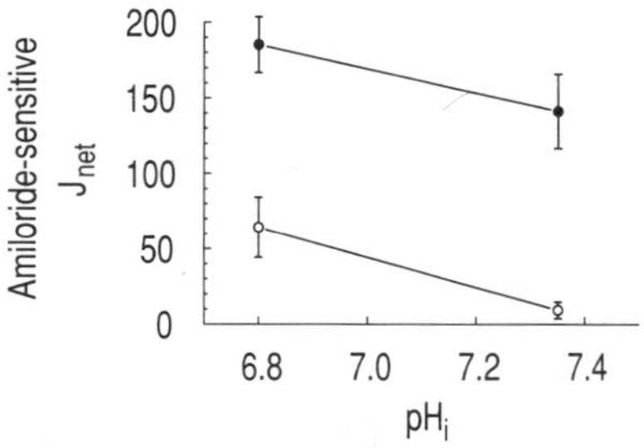

Figure 17–4 The dependence of the amiloride-sensitive net acid-extrusion rate ($J_{Na\text{-}H}$) on pH_i. The data were taken from experiments such as those shown in Figure 17–3, in which cells were dialyzed to pH_i values of either 7.35 or 6.8. The DFs contained either no Li^+ (open circles) or 100 m*M* Li^+ (filled circles). (From Davis, B. A. et al., *Am. J. Physiol.,* 263, C246, 1991. With permission of the American Physiological Society.)

more than high levels of intracellular Li^+. In unshrunken cells, lowering pH_i from 7.2 to 6.8 caused $J_{Na\text{-}H}$ to increase from ~0 to 25 μ*M*/min. On the other hand, in shrunken BMFs, a similar decrease in pH_i caused $J_{Na\text{-}H}$ to increase from ~172 to ~304 μ*M*/min. Thus, over a broad pH_i range, shrinkage caused a substantial stimulation of Na-H exchange activity.[9]

B. EFFECT OF INTRACELLULAR Cl^- DEPLETION ON THE ACTIVATION OF Na-H EXCHANGE BY LOW pH_i AND BY SHRINKAGE

Because Parker had found that intra- and/or extracellular Cl^- is required for the shrinkage-induced activation of Na-H exchange in dog erythrocytes,[10,11] we decided to examine whether Cl^- is similarly required in BMFs. In one set of experiments, we dialyzed cells to a pH_i of ~6.8 using a Cl^--free DF (glutamate replacing Cl^-) and a Cl^--free ASW (gluconate replacing Cl^-). As anticipated from the erythrocyte data, we found that Na-H exchange in shrunken cells was reduced ~88% by the bilateral absence of Cl^-. However, we were surprised to find that, even under normotonic conditions (i.e., in the absence of shrinkage), Na-H exchange was reduced ~50% in the bilateral absence of Cl^-. Removing Cl^- from the ASW, but leaving the Cl^- in the DF, had little effect on $J_{Na\text{-}H}$ under either hyper- or normotonic conditions. Thus, these results extend the previous work on erythrocytes by showing that it is specifically Cl^- on the inside of the cell that is crucial for the shrinkage response.

IV. INVOLVEMENT OF A G PROTEIN IN THE SHRINKAGE-INDUCED ACTIVATION OF Na-H EXCHANGE

A. EFFECT OF DIALYZING WITH GDPβS.

Because it can be internally dialyzed, and can also be injected with nondialyzable molecules, the BMF is potentially a powerful preparation in which to study the signal transduction process by which shrinkage leads to activation of the Na-H exchanger. To explore the possible role of a G protein in the shrinkage-induced activation of Na-H exchange, we examined the effect of predialyzing BMFs with GDPβS. This drug antagonizes G protein activation by competing with GTP for binding to G proteins,[12] and possibly by decreasing GTP synthesis.[13] Figure 17–5A illustrates a control experiment. The cell was dialyzed for 90 min with a Na^+-free DF at pH 7.2, the terminal portion of which is shown in the figure *(ab)*. During dialysis, as well as the postdialysis stabilization period *(bc)*, the ASW contained SITS and lacked Na^+. Adding 50 m*M* Na^+ to the ASW after the stabilization period had virtually no effect on pH_i in this unshrunken cell *(cd)*, indicating a lack of Na-H exchange activity. We then simultaneously removed the Na^+ and raised the ASW osmolality from 975 to 1600 mOsmol/kg. This elicited a small, self-limited increase in pH_i *(de)*. In other experiments, in which we first removed the Na^+ and then increased osmolality in the absence of Na^+, we found that the increase in ASW osmolality was accompanied by a pH_i increase virtually identical to that shown in Figure 17–5A. In the experiment shown in Figure 17–5A, adding 50 m*M* Na^+ under hypertonic conditions caused a rapid increase in pH_i *(ef)* that was largely blocked by 1 m*M* amiloride. Thus, as noted above, Na-H exchange near the resting pH_i is negligible unless the BMF is shrunken.

The effects of dialyzing with GDPβS are shown in Figure 17–5B. The protocol of this experiment was identical to that employed in Figure 17–5A, except that the DF contained 1 m*M* GDPβS. As in the control cell, the addition of 50 m*M* Na^+ under normotonic conditions had little effect on pH_i *(cd)*. Also as in Figure 17–5A, the simultaneous removal of Na^+ and exposure to a hypertonic ASW elicited a small pH_i increase. However, a distinct difference with this experiment, in which the cell was dialyzed with GDPβS, was that the addition of Na^+ under hypertonic conditions had a negligible effect on pH_i, and the subsequent addition of amiloride was similarly ineffective. In seven cells, shrinkage increased the Na^+-dependent flux (comparing *ef* and *cd*) by an average of 72 ± 34 μ*M*/min (not significantly differ-

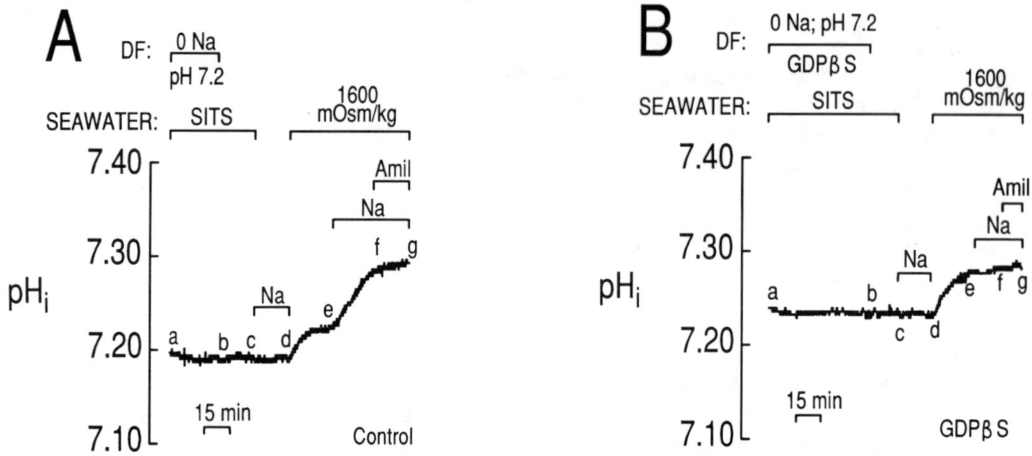

Figure 17–5 The effect of GDPβS on the shrinkage-induced activation of Na-H exchange. (A) Stimulation of Na-H exchange by hypertonicity. The addition of 50 m*M* Na^+ had little effect on pH_i under normotonic conditions (975 mOsmol/kg; segment *cd*), but under hypertonic conditions (1600 mOsmol/kg) caused a rapid pH_i increase *(ef)* that was blocked by 1 m*M* amiloride *(fg)*. (B) Effect of GDPβS. The protocol was the same as in (A), except that the fiber was dialyzed with a fluid containing 1 m*M* GDPβS *(ab)*. Addition of 50 m*M* Na^+ under hypertonic conditions had little effect on pH_i *(ef)*. (From Davis, B. A. et al., *Am. J. Physiol.,* 263, C246, 1991. With permission of the American Physiological Society.)

ent from zero). In eight matched controls not dialyzed with GDPβS, shrinkage increased the Na^+-dependent flux by an average of 261 ± 78 μ*M*/min. This control value is significantly greater than the one obtained in GDPβS-dialyzed cells ($P < .05$). Comparing the Na^+-dependent fluxes in control and GDPβS-dialyzed cells, we see that GDPβS reduced the shrinkage-induced stimulation of Na-H exchange by 72%. The most straightforward explanation for these data is that GDPβS interfered with a G protein necessary for linking shrinkage to the activation of Na-H exchange.

B. EFFECT OF DIALYZING WITH GTPγS IN UNSHRUNKEN CELLS

If a G protein is involved in the signal transduction system linking shrinkage to the activation of the Na-H exchanger, then we ought to be able to bypass shrinkage and activate the G protein directly with GTPγS. GTPγS is a GTP analogue that binds to the α-subunit of G proteins, but is hydrolyzed only slowly. GTPγS therefore causes prolonged activation of these α-subunits.[14,15] Figure 17–6C shows an experiment in which the BMF was dialyzed with a DF of pH 7.2 containing 1 m*M* GTPγS. Our protocol was similar to that used in the first portion of Figures 5A and 5B. The terminal portion of the dialysis period is shown in Figure 17–6A *(ab)*. After the postdialysis stabilization period *(bc)*, the addition of 50 m*M* Na^+ under normotonic conditions led to a rapid pH_i increase *(cd)* that was largely blocked by amiloride *(de)*. In five such experiments, the Na^+-dependent flux (comparing *cd* and *bc*) averaged 209 ± 53 μ*M*/min, which is significantly higher ($P < .005$) than the value in five matched controls not dialyzed with GTPγS, 13 ± 11 μ*M*/min. The amiloride-sensitive flux (comparing *de* and *cd*) averaged 196 ± 58 μ*M*/min in GTPγS-dialyzed cells, which is significantly greater than the flux of 3 ± 3 μ*M*/min in matched controls ($P < .005$). Thus, in unshrunken cells, GTPγS causes a substantial activation of Na-H exchange even near the resting pH_i, where the exchanger normally exhibits little activity.

C. EFFECT, IN UNSHRUNKEN CELLS, OF MICROINJECTING CHOLERA TOXIN

Cholera toxin (CTX) ADP-ribosylates and thereby causes prolonged activation of the α-subunit of some types of G proteins.[15–18] To examine the possibility that the BMF Na-H exchanger can be activated by a CTX-activatable G protein, we microinjected cells with activated CTX to a final estimated concentration of 0.3 μ*M*. After incubating the cells 16 to 24 h at 4°C to allow for diffusion of the toxin, we assayed Na-H exchange using a protocol identical to that employed in Figure 17–6A, except that the DF did not contain GTPγS. A representative experiment is shown in Figure 17–6B. The first

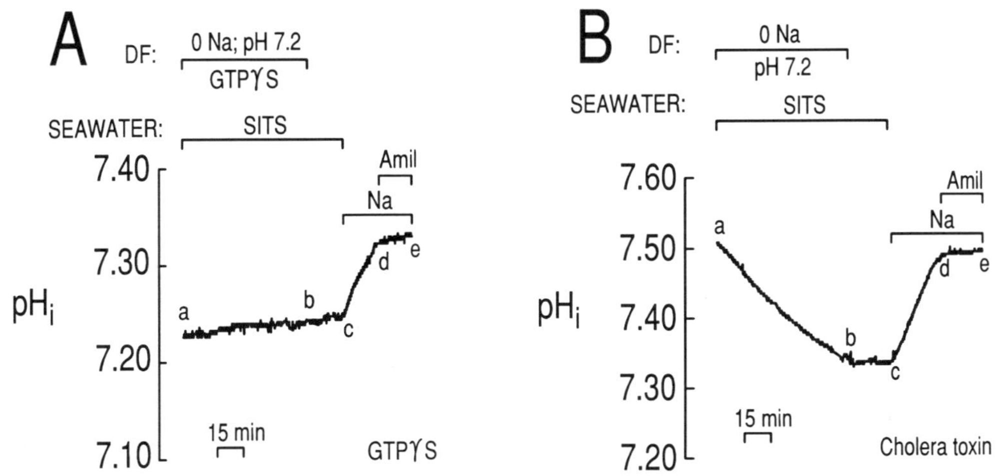

Figure 17–6 The effects of GTPγS and cholera toxin on Na-H exchange activity. (A) Stimulation of Na-H exchange by GTPγS. The cell was dialyzed for 90 min (*ab* is the terminal portion) with a DF containing 1 m*M* GTPγS. Adding 50 m*M* Na^+ under normotonic conditions caused a rapid pH$_i$ increase *(cd)* that was blocked by 1 m*M* amiloride (Amil; *de*). In matched control cells (not shown), both the Na^+-dependent and amiloride-sensitive pH$_i$ increases were negligible. (B) Stimulation of Na-H exchange by cholera toxin (CTX). CTX-injected cells had high initial values of pH$_i$. Dialysis for 90 min with a DF of pH 7.2 returned pH$_i$ towards its normal range *(ab)*. The addition of 50 m*M* Na^+ under normotonic conditions caused a rapid pH$_i$ increase *(cd)* that was blocked by 1 m*M* amiloride *(de)*. In sham-injected control cells (not shown), both the Na^+-dependent and amiloride-sensitive pH$_i$ increases were negligible. (From Davis, B. A. et al., *Am. J. Physiol.*, 263, C246, 1991. With permission of the American Physiological Society.)

indication of the effect of CTX on Na-H exchange was that the initial pH$_i$ (~7.6) was substantially higher than in sham-injected control cells (~7.3). As shown in Figure 17–6B, the period of dialysis in DF with a pH of 7.2 only reduced pH$_i$ to ~7.35. After the postdialysis stabilization period *(bc)*, the addition of 50 m*M* Na^+ to the ASW elicited an extremely rapid pH$_i$ increase *(cd)* that was blocked by amiloride *(de)*. In five CTX-injected cells, the Na^+-dependent and amiloride-sensitive fluxes were 362 ± 53 μ*M*/min and 320 ± 37 μ*M*/min, respectively. These values are the highest we have observed for Na-H exchange in BMFs at any pH$_i$, and substantially higher than the comparable values observed in sham-injected controls, 93 ± 21 μ*M*/min and 59 ± 26 μ*M*/min, respectively. Thus, like hypertonicity and GTPγS, CTX produces a substantial stimulation of Na-H exchange at pH$_i$s near the resting value. In fact, the degree of stimulation is even higher than that produced by shrinkage or GTPγS. However, we caution that our observations do not prove that the G protein responsible for the CTX-induced activation of Na-H exchange is the same one responsible for shrinkage-induced activation of Na-H exchange.

In other experiments (not shown), we examined the effects of pertussis toxin (PTX), which ADP-ribosylates the α-subunits of the G proteins G_i[19,20] and G_t.[21,22] The effect is to reduce the G protein/receptor interaction, and thereby inhibit the signal transduction pathway.[15] We found that PTX failed to stimulate Na-H exchange under normotonic conditions, and failed to inhibit Na-H exchange under hypertonic conditions. Thus, it seems that a PTX-sensitive G protein is not involved in the transduction of the shrinkage signal to the Na-H exchanger in BMFs.

D. EFFECT OF RAISING [cAMP]$_i$

Our observation that the shrinkage-induced stimulation of Na-H exchange is mimicked by CTX immediately suggested that the G protein involved in the signal transduction pathway might include a CTX-sensitive α_s-subunit. Because adenylyl cyclase is the classic downstream effector for such CTX-sensitive α_s-subunits, we explored the possibility that the formation of 3′,5′ cyclic adenosine monophosphate (cAMP) might be a key second messenger for the shrinkage signal transduction pathway. In one set of experiments, we dialyzed BMFs with 20 μ*M* cAMP and 0.5 m*M* 3-isobutyl-1-methylxanthine (IBMX). In four experiments on these unshrunken cells, we found that the amiloride-sensitive J_{net} was indistinguishable from zero. In a second series of experiments, we instead dialyzed with 50 μ*M* dibutyryl cAMP plus IBMX. Here we found that the mean amiloride-sensitive J_{net} averaged 28 ± 7 μ*M*/min for five cells studied under normotonic conditions. This modest flux is far smaller than the fluxes observed at comparable pH_i values in cells dialyzed with Li^+ (~140 μ*M*/min), shrunken cells (170 μ*M*/min in shrinkage study, 261 μ*M*/min in GDPβS study), cells dialyzed with GTPγS (196 μ*M*/min), or cells injected with CTX (320 μ*M*/min). Earlier work on BMFs had shown that cAMP stimulates both Cl^- and acid-base transport,[23] indicating the presence of cAMP-dependent signal transduction pathways in this preparation. Thus, it does not appear that a cAMP-dependent pathway can account for the activation of Na-H exchange produced by Li^+, shrinkage, GTPγS, or CTX.

E. EFFECT OF STIMULATING PROTEIN KINASE C

It has been appreciated for some time that hormones and growth factors acting through protein kinase C (PKC) can activate the Na-H exchanger.[24–34] On the other hand, evidence from Grinstein's laboratory indicates that PKC is not involved in shrinkage-induced activation of Na-H exchange.[35] Because stimulation of PKC involves the G protein-mediated activation of phospholipase C, a trivial explanation for the stimulation of Na-H exchange by GTPγS in BMFs would be that the GTPγS activates a PKC-dependent pathway. To test this hypothesis, we exposed BMFs to an ASW containing 100 n*M* phorbol 12-myristate 13-acetate (PMA). The cells had been dialyzed previously to a pH_i of ~7.2 with our standard Na^+-free DF. We found that the amiloride-sensitive J_{net} averaged only 16 ± 5 μ*M*/min in five experiments. This value is far below those obtained in Li^+-dialyzed cells, shrunken cells, GTPγS-dialyzed cells, or CTX-injected cells. Thus, we cannot invoke a PKC-dependent pathway to account for the activation of Na-H exchange by any of the aforementioned treatments.

V. DISCUSSION

A. WHAT IS THE PHYSIOLOGICAL ROLE OF THE Na-H EXCHANGER?

There is little doubt that, even in the presence of CO_2/HCO^-_3, the Na-H exchanger in some cells is the major acid-extrusion mechanism. Thus, even if the Na-H exchanger in these cells is activated by shrinkage, one could make the case that the major role of the exchanger is pH_i regulation. However, for some cells that have been studied in our laboratory, it is not at all clear that pH_i regulation is the *raison d'etre* of the Na-H exchanger. As discussed above, the acid-extruding capacity of the Na-H exchanger in the unshrunken BMF is trivial compared to that of the Na^+-dependent $Cl\text{-}HCO_3$ exchanger. Renal mesangial cells have a Na-H exchanger that appears to be stimulated substantially by shrinkage.[36a] However, in unshrunken mesangial cells, which have a resting pH_i in CO_2/HCO^-_3 of ~7.25, the activity of the Na-H exchanger at pH_i 6.6 is only about half that of the cell's major acid-extruder, the Na^+-dependent $Cl\text{-}HCO_3$ exchanger.[36] Preliminary work from our laboratory on *Xenopus* oocytes indicates that Na-H exchange activity is negligible unless the cell is shrunken in a hypertonic medium.[37] In these three cells, it would appear that the major benefit conferred by the Na-H exchanger is cell volume, rather than cell pH, regulation.

Why should a Na-H exchanger ever be expressed at the same time as a second acid extruder, such as a Na^+-dependent $Cl\text{-}HCO_3$ exchanger or an H^+ pump? One possibility is that the Na-H exchanger was heavily expressed in undifferentiated progenitor cells, and that its presence in the fully differentiated cell is merely a vestige of its past history. Another possibility is that it is beneficial to the cell to be able to call upon acid extruders with different effects on cell volume. For example, Na-H exchange activity increases both pH_i and volume. At the other extreme is the uncoupled H^+ pump, which increases pH_i without any effect on cell volume. This occurs because most of the H^+ extruded from the cell is derived from buffers, making the extrusion of H^+ osmotically silent. Between these extremes is the Na^+-dependent $Cl\text{-}HCO_3$ exchanger, which mediates the influx of Na^+ and HCO^-_3, as well as the efflux of H^+ and Cl^- (or an equivalent process). From an osmotic standpoint, the fluxes of Na^+ and Cl^- cancel one another. As noted above, the efflux of H^+ is osmotically silent. This leaves the influx of HCO^-_3, which might be regarded as an osmotic "whimp". Although some of the incoming HCO^-_3 remains HCO^-_3, and thus is osmotically active, the rest of the incoming HCO^-_3 is osmotically silent. This silent HCO^-_3 neutralizes H^+ (derived from intracellular buffers, and thus osmotically silent) to form H_2CO_3, which in turn dissociates into CO_2 and H_2O, both of which exit the cell.

B. WHAT IS THE RELATIONSHIP BETWEEN THE ACTIVATION OF Na-H EXCHANGE BY LOW pH_i vs. SHRINKAGE?

Although the BMF Na-H exchange rate is increased from 0 to 25 μ*M*/min by decreasing pH_i from 7.2 to 6.8 in unshrunken cells, it is increased from 172 to 304 μ*M*/min by decreasing pH_i to a similar extent in shrunken cells. Thus, at least for our chosen degrees of acidification (0.4 pH units, or a 2.5-fold increase in $[H^+]$) and shrinkage (975 → 1600 mOsmol/kg, or a 1.6-fold increase in osmolality), shrinkage is a far more potent stimulus of Na-H exchange than intracellular acidity. Moreover, as can be computed from the figures provided above, the pH_i dependence of the exchanger is far higher in shrunken (330 μ*M*/s per pH unit) than in unshrunken cells (62 μ*M*/s per pH unit). These observations raise the question of whether the Na-H exchanger could be completely inactivated by a sufficient degree of cell *swelling*. If so, such a result would suggest that the prime trigger for Na-H exchange is shrinkage, with intracellular acidity merely promoting the activity of a transporter previously activated by shrinkage. According to this scenario, baseline activity of the exchanger in ostensibly unshrunken cells could be explained if the cell sensed this "resting" volume as being slightly below the threshold for activating the shrinkage signal transduction cascade.

C. HOW IS SHRINKAGE SENSED?

The mechanism by which a cell senses a change in cell volume is not known for any volume-regulatory process. As discussed elsewhere in this book (see Chapter 13), stretch-activated channels may mediate volume-regulatory decrease (VRD), although this remains to be firmly established (see References 39). The observation that Cl^-,[10] specifically intracellular Cl^-,[40] is required for shrinkage activation of the Na-H exchanger raises the possibility that a change in $[Cl^-]_i$ could be the parameter sensed during cell shrinkage. On the other hand, as discussed in Chapters 10 and 18, work on dog red blood cells (RBCs) suggests that a change in total protein concentration may be involved in the sensing of cell shrinkage.[41]

D. HOW IS THE SHRINKAGE SIGNAL TRANSDUCED TO THE Na-H EXCHANGER?

The work summarized in this chapter is consistent with the hypothesis that a heterotrimeric G protein is part of the signal transduction system linking shrinkage to activation of the Na-H exchanger. The left-hand portion of Figure 17–7 summarizes how shrinkage could accelerate the turnover of a G protein, and lead to the activation of a downstream effector. If a G protein is indeed involved, it might interact with a classic seven-membrane-spanning receptor, as is the case for the adrenergic receptors and rhodopsin.[42] Such a hypothetical receptor could sense a soluble intracellular ligand. Alternatively, the hypothetical receptor could, like rhodopsin, have a permanently bound ligand. In this case, the interaction between the receptor and G protein could be triggered by a change in the receptor's environment, such as a change in the concentration of an intracellular solute or a change in a cytoskeletal interaction.

As far as the distal end of the signal transduction process is concerned, work on the K/Cl cotransporter (a VRD mechanism) suggests that swelling stimulates K/Cl cotransport by causing a dephosphorylation,[43,44] as discussed in Chapters 12 and 18. For example, okadaic acid (OKA), which inhibits certain phosphatases,[45,46] blocks the swelling-induced activation of K/Cl cotransport. Kinetic data indicate that the putative dephosphorylation is slow, whereas the putative phosphorylation that would reverse swelling-induced activation of the K/Cl cotransporter is fast. Conversely, the shrinkage-induced activation of the Na-H exchanger in dog RBCs is fast, whereas the inactivation is slow. For this and another VRI mechanism (Na/K/Cl cotransporter), inhibition of phosphatases either activates the transporter in the absence of shrinkage,[47,48] or prolongs activation after shrunken cells are swollen.[48] These VRI and VRD data have suggested the model summarized in the right-hand portion of Figure 17–7. Here, shrinkage elicits phosphorylation that stimulates VRI systems, but inhibits VRD systems. Swelling would have the opposite set of reciprocal effects on VRI/VRD systems. Preliminary data from our laboratory suggest that OKA may accentuate the shrinkage-induced stimulation of Na-H exchange in barnacle muscle. If this is indeed true, then these data would link shrinkage-induced G protein activation and a phosphorylation event in a single cell type. Although the model in Figure 17–7 is drawn as if the only volume-related signal is shrinkage (or lack thereof), one could imagine a system in which shrinkage and swelling are sensed separately, and have reciprocal effects on a system of kinases and phosphatases. A third alternative is a system in which swelling would be the only volume-related signal sensed by the cell, with shrinkage being detected as a lack of swelling.

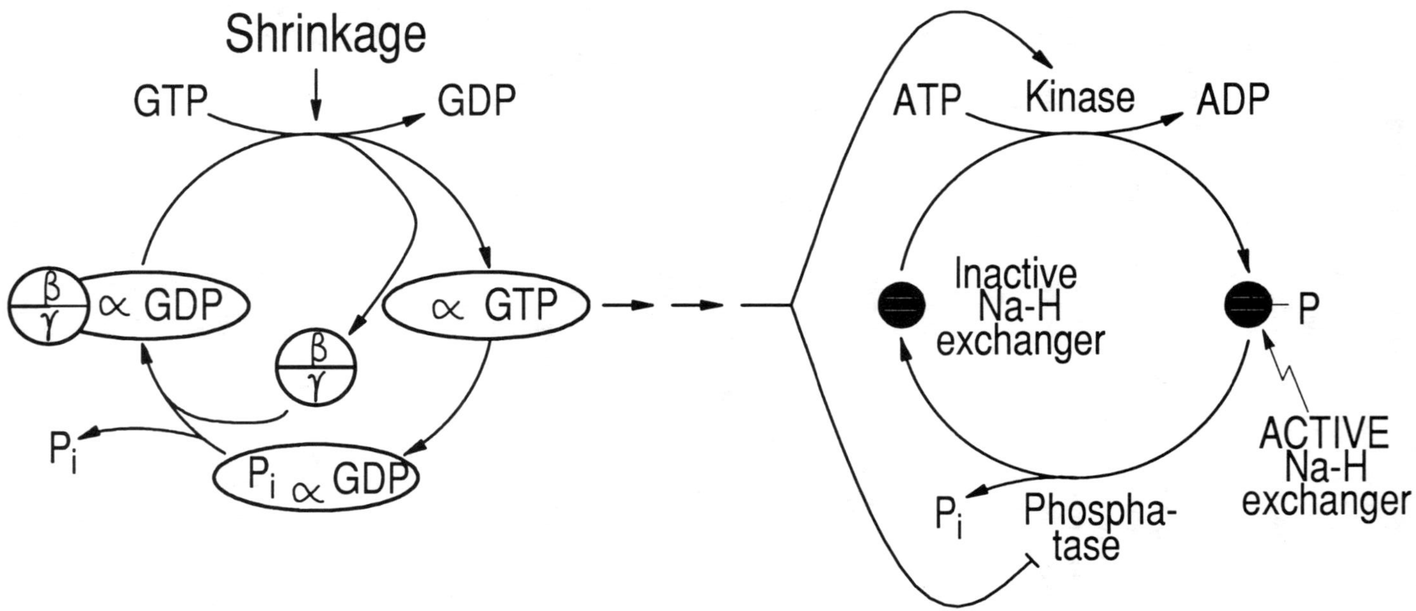

Figure 17–7 A model of how shrinkage might accelerate G protein turnover, and thereby lead to the activation of the Na-H exchanger via phosphorylation. In the absence of shrinkage, the Na-H exchanger would be dephosphorylated and return to its baseline state.

REFERENCES

1. **Parker, J. C.,** Interactions of lithium and protons with the sodium-proton exchanger of dog red blood cells, *J. Gen. Physiol.*, 87, 189, 1986.
2. **Aronson, P. S., Nee, J., and Suhm, M. A.,** Modifier role of internal H^+ in activating the Na^+-H^+ exchanger in renal microvillus membrane vesicles, *Nature*, 299, 161, 1982.
3. **Boron, W. F., McCormick, W. C., and Roos, A.,** pH regulation in barnacle muscle fibers: dependence on intracellular and extracellular pH, *Am. J. Physiol.*, 237, C185, 1979.
4. **Boron, W. F., McCormick, W. C., and Roos, A.,** pH regulation in barnacle muscle fibers: dependence on extracellular sodium and bicarbonate, *Am. J. Physiol.*, 240, C80, 1981.
5. **Roos, A. and Boron, W. F.,** Intracellular pH, *Physiol. Rev.*, 61, 296, 1981.
6. **Brinley, F. J., Jr., and Mullins, L. J.,** Sodium extrusion by internally dialyzed squid axons, *J. Gen. Physiol.*, 50, 2303, 1967.
7. **Hinke, J. A. M.,** Cation-selective microelectrodes for intracellular use, in *Glass Electrodes for Hydrogen and Other Cations. Principle and Practice,* Eisenman, G., Ed., Marcel Dekker, New York, 1967, 464.
8. **Boron, W. F.,** Intracellular pH transients in giant barnacle muscle fibers, *Am. J. Physiol.*, 233, C61, 1977.
9. **Davis, B. A., Hogan, E. M., and Boron, W. F.,** Activation of Na-H exchange by shrinkage in barnacle muscle fibers, *Am. J. Physiol.*, submitted.
10. **Parker, J. C.,** Volume-responsive sodium movements in dog red blood cells, *Am. J. Physiol.*, 244, C324, 1983.
11. **Parker, J. C.,** Glutaraldehyde fixation of sodium transport in dog red blood cells, *J. Gen. Physiol.*, 84, 789, 1984.
12. **Eckstein, F.,** Guanosine 5′-O-(2-thiodiphosphate), *J. Biol. Chem.*, 254, 9829, 1979.
13. **Breitwieser, G. E. and Szabo, G.,** Mechanism of muscarinic receptor-induced K^+ channel activation as revealed by hydrolysis-resistant GTP analogues, *J. Gen. Physiol.*, 91, 469, 1988.
14. **Northup, J. K., Smigel, M. D., Sternweis, P. C., and Gilman, A. G.,** The subunits of the stimulatory regulatory component of adenylate cyclase, *J. Biol. Chem.*, 258, 11369, 1983.
15. **Gilman, A. G.,** G Proteins: transducers of receptor-generated signals, *Annu. Rev. Biochem.*, 56, 615, 1987.
16. **Abood, M. E., Hurley, J. B., Pappone, M.-C., Bourne, H. R., and Stryer, L.,** Functional homology between signal-coupling proteins, *J. Biol. Chem.*, 257, 10540, 1982.
17. **Cassel, D. and Pfeuffer, T.,** Mechanism of cholera toxin action: covalent modification of the guanyl nucleotide-binding protein of the adenylate cyclase system, *Proc. Natl. Acad. Sci. U.S.A.*, 75, 2669, 1978.
18. **Kahn, R. A. and Gilman, A. G.,** ADP-ribosylation of Gs promotes the dissociation of its alpha and beta subunits, *J. Biol. Chem.*, 259, 6235, 1984.
19. **Katada, T. and Ui, M.,** Direct modification of the membrane adenylate cyclase system by islet-activating protein due to ADP-ribosylation of a membrane protein, *Proc. Natl. Acad. Sci U.S.A.*, 79, 3129, 1982.
20. **Katada, T. and Ui, M.,** ADP ribosylation of the specific membrane protein of C6 cells by islet-activating protein associated with modification of adenylate cyclase activity, *J. Biol. Chem.*, 257, 7210, 1982.
21. **Bitensky, M. W., Aloni, B., Vetury, S., and Matuo, Y.,** Light- and GTP-activated photoreceptor phosphodiesterase: regulation by a light-activated GTPase and identification of rhodopsin as the phosphodiesterase binding site, *Adv. Cyclic Nucleotide Res.*, 9, 553, 1978.
22. **Yee, R. and Liebman, P. A.,** Light-activated phosphodiesterase of the rod outer segment, *J. Biol. Chem.*, 253, 8902, 1978.
23. **Boron, W. F., Russell, J. M., Brodwick, M. S., Keifer, D. W., and Roos, A.,** Influence of cyclic AMP on intracellular pH regulation and chloride fluxes in barnacle muscle fibers, *Nature*, 276, 511, 1978.
24. **Grinstein, S., Cohen, S., Goetz, J. D., Rothstein, A., Mellors, A., and Gelfand, E. W.,** Activation of the Na^+-H^+ antiport by changes in cell volume and by phorbol esters; possible role of protein kinase, in *Na^+-H^+ Exchange, Intracellular pH and Cell Function,* Peter, S., Ed., Academic Press, New York, 1986, 115.
25. **Lopez-Rivas, A. and Rozengurt, E.,** Serum rapidly mobilizes calcium from an intracellular pool in quiescent fibroblastic cells, *Biochem. Biophys. Res. Commun.*, 114, 240, 1983.
26. **Moolenaar, W. H., Tertoolen, L. G. J., and de Laat, S. W.,** Phorbol ester and diacylglycerol mimic growth factors in raising cytoplasmic pH, *Nature*, 312, 371, 1984.
27. **Moolenaar, W. H.,** Effects of growth factors on intracellular pH regulation, *Annu. Rev. Physiol.*, 48, 363, 1986.

28. **Rosoff, P. M. and Cantley, L. C.,** Stimulation of the T3-T cell receptor-associated Ca^{2+} influx enhances the activity of the Na^+/H^+ exchanger in a leukemic human T cell line, *J. Biol. Chem.,* 260, 14053, 1985.
29. **Rozengurt, E.,** Early signals in the mitogenic response, *Science,* 234, 161, 1986.
30. **Villereal, M. L., Mix-Muldoon, L. L., Vincentini, L. M., Jamieson, G. A., and Owen, N. E.,** Mechanisms of growth factor stimulation of Na^+-H^+ exchange in cultured fibroblasts, in *Na^+-H^+ Exchange, Intracellular pH, and Cell Function,* Aronson, P. S. and Boron, W. F., Eds., Academic Press, New York, 1986, 175.
31. **Vincentini, L. M. and Villereal, M. L.,** Inositol phosphates turnover, cytosolic Ca^+ and pH: putative signals for the control of cell growth, *Life Sci.,* 38, 2269, 1986.
32. **Muldoon, L., Jamieson, A., Jr., Kao, A., Palfrey, H., and Villereal, M.,** Mitogen stimulation of Na^+-H^+ exchange: differential involvement of protein kinase C, *Am. J. Physiol.,* 253, C219, 1987.
33. **Slotki, I. N., Schwartz, J. H., and Alexander, E. A.,** Na^+-H^+ exchange is stimulated by protein kinase C activation in inner medullary collecting duct cells, *Am. J. Physiol.,* 259, F666, 1990.
34. **Besterman, J. M., May, W. S., Jr., LeVine, I. I. I., Cragoe, E. J., Jr., and Cuatrecasas, P.,** Amiloride inhibits phorbol ester-simulated Na^+/H^+ exchange and protein kinase C, *J. Biol. Chem.,* 260, 1155, 1985.
35. **Grinstein, S., Mack, E., and Mills, G. B.,** Osmotic activation of the Na^+/H^+ antiport in protein kinase C-depleted lymphocytes, *Biochem. Biophys. Res. Commun.,* 134, 8, 1986.
36. **Ganz, M. B., Boyarsky, G., Sterzel, R. B., and Boron, W. F.,** Arginine vasopressin enhances pH_i regulation in the presence of HCO_3^- by stimulating three acid-base transport systems, *Nature,* 337, 648, 1989.

36a. **Boyarsky, G., Schlue, W. R., and Boron, W. F.,** unpublished data.

37. **Hogan, E. M., Romer, M. F., Davis, B. A., and Boron, W. F.,** Shrinkage-induced activation of Na-H exchange in *Xenopus* oocytes, *FASEB J.,* in press.
38. **Hoffmann, E. K. and Simonsen, L. O.,** Membrane mechanisms in volume and pH regulation in vertebrate cells, *Physiol. Rev.,* 69, 315, 1989.
39. **Montrose-Rafizadeh, C. and Guggino, W. B.,** Cell volume regulation in the nephron, *Annu. Rev. Physiol.,* 52, 761, 1990.
40. **Davis, B. A., Hogan, E. M., and Boron, W. F.,** Intracellular Cl^- dependence of hypertonically activated Na-H exchange activity in barnacle muscle, *Biophys. J.,* 57, 594a, 1990.
41. **Colclasure, G. and Parker, J. C.,** Cytosolic protein concentration is the primary volume signal in dog red cells, *J. Gen. Physiol.,* 98, 881, 1991.
42. **Birnbaumer, L., Abramowitz, J., and Brown, A. M.,** Receptor-effector coupling by G proteins, *Biochim. Biophys. Acta,* [???], 163, 1990.
43. **Jennings, M. L. and Al-Rohil, N.,** Kinetics of activation and inactivation of swelling-stimulated K^+/Cl^- transport. The volume-sensitive parameter is the rate constant for inactivation, *J. Gen. Physiol.,* 95, 1021, 1990.
44. **Jennings, M. L. and Schulz, R. K.,** Okadaic acid inhibition of KCl cotransport. Evidence that protein dephosphorylation is necessary for activation of transport by either cell swelling or N-ethylmaleimide, *J. Gen. Physiol.,* 97, 799, 1991.
45. **Cohen, P., Holmes, C. F. B., and Tsukitani, Y.,** Okadaic acid: a new probe for the study of cellular regulation, *TIBS,* 15, 98, 1990.
46. **Ishihara, H., Martin, B. L., Brautigan, D. L., Karaki, H., Ozaki, H., Kato, Y., Fusetani, N., Watabe, S., Hashimoto, K., Uemura, D., and Hartshorne, D. J.,** Calyculin A and okadaic acid: inhibitors of protein phosphatase activity, *Biochem. Biophys. Res. Commun.,* 159, 871, 1989.
47. **Pewitt, E. B., Hegde, R. S., Haas, M., and Palfrey, H. C.,** The regulation of Na/K/2Cl cotransport and bumetanide binding in avian erythrocytes by protein phosphorylation and dephosphorylation, *J. Biol. Chem.,* 265, 20747, 1991.
48. **Parker, J. C., Colclasure, C., and McManus, T. J.,** Coordinated regulation of shrinkage-induced Na/H exchange and swelling-induced K-Cl cotransport in dog red blood cells, *J. Gen. Physiol.,* 98, 869, 1991.
49. **Davis, B. A., Hogan, E. M., and Boron, W. F.,** Activation of Na-H exchange by intracellular lithium in barnacle muscle fibers, *Am. J. Physiol.,* 263, C246, 1991.

Chapter 18

Coordinated Regulation of Volume-Activated Transport Pathways

John C. Parker

TABLE OF CONTENTS

I. INTRODUCTION

In this chapter the evidence is considered that in nonepithelial cells — specifically, red cells — the responses of various transporters to volume perturbation are in some way coordinated. The concept of coordination among transporters may have originated with epithelial physiologists. If there were no cross-talk between the basolateral and apical membranes, epithelia might respond to secretory or absorptive stimuli by bursting or shrinking irreversibly as a result of unequal flow of solute and water across the two cell surfaces.[1,2]

The "null hypothesis" for the present discussion is represented in Figure 18–1A. It is well known that cell swelling activates some transporters whereas cell shrinkage activates others. One possibility (**A**) is that the two types of transporters interact only via their mutual dependence on cell volume. The hypothesis reviewed here is represented schematically by Figure 18–1B. According to this view there is a "regulator" in the cells consisting of a volume-sensing mechanism plus a signaling apparatus that communicates with both swelling- and shrinkage-induced transporters. Much of the evidence concerns the experimental manipulation of set points, or set point volumes (we use the terms interchangeably). For a particular volume-activated membrane transporter, the set point can be defined as that volume at which the transporter is activated or deactivated. The set point is usually demonstrated by experiments of the type shown in Figures 2 and 3, in which the water content of cells is controlled by varying the tonicity of the suspending medium or by manipulating cell solute content. Transport rates are plotted as some function of cell volume.

Figure 18–2 shows sodium influx in dog red cells as influenced by changes in their water content. Swelling cells above 68% water had no effect on sodium flux, but as the cells were progressively dehydrated the sodium flux increased. The set point in this experiment was about 66 to 67% cell water,

0-8493-4448-4/94/$0.00+$.50

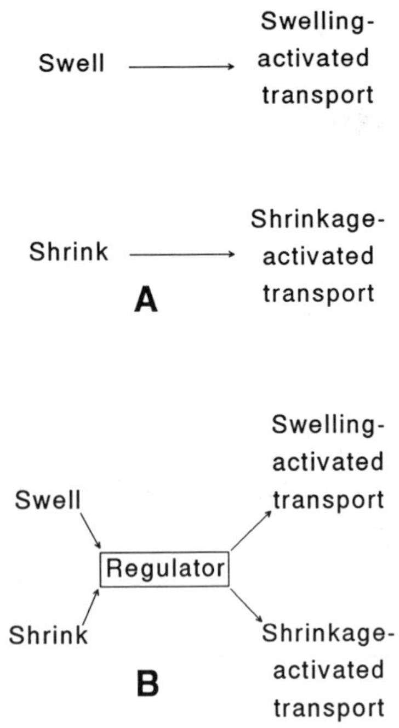

Figure 18–1 Schematic representation of the linkage, or lack thereof, between swelling-induced and shrinkage-induced transporters. (A) Shows the two transporters responding to cell swelling or shrinkage independently; (B) indicates that the responses are in some way coordinated.

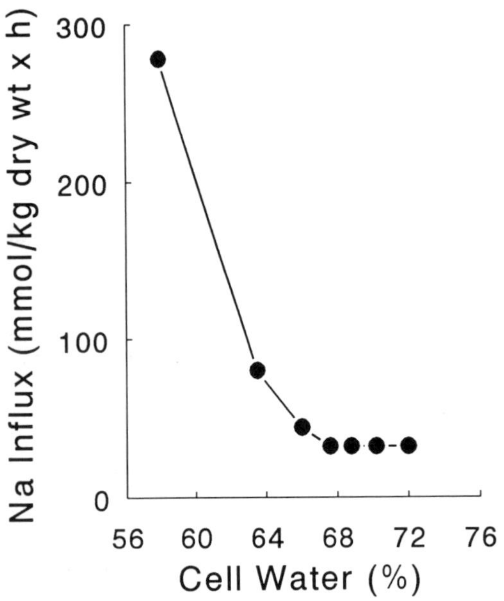

Figure 18–2 Na influx in dog red cells as a function of cell water content. The set point in this experiment is about 66% cell water.

which is the approximate value found in fresh red cells suspended in plasma.

The set point for a swelling-induced transporter in dog red cells is shown in Figure 18–3. Potassium influx was not affected by shrinking cells below a volume of about 66% water, but swelling them above this level increased the flux progressively.

One way to decide between the alternative model schemes in Figure 18–1 is to determine whether a perturbation that influences the set point for a shrinkage-induced ion flux will have any effect on the set

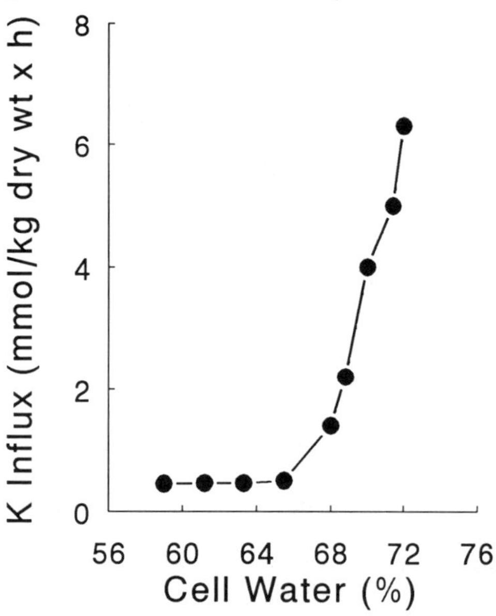

Figure 18–3 K influx in dog red cells as a function of cell water content. The set point in this experiment is about 66% cell water.

point of a swelling-induced flux. If the set points vary independently, then there is no reason to think the two volume-responsive transport mechanisms are linked. However, if the set points vary concurrently, then it is possible the transporters are under a common regulatory influence.

In the following sections it is argued that in the red cells of some species there is mutual interaction of swelling-induced and shrinkage-induced transporters. Although the existence of a coordinating system can be demonstrated, it cannot be characterized in any detail.

II. TRANSPORTERS ACTIVATED BY SWELLING OR SHRINKING CELLS

Most of the evidence for coordination between volume-activated transporters involves duck and dog red cells (Table 1). Characterization of the transport pathways has resulted from the work of many laboratories.[3–5] We have chosen the term "volume-activated" rather than "volume regulatory" because the physiological role of these responses is not clear in all cases. When duck red cells are shrunken in media containing physiological concentrations of sodium and potassium, Na/K/2Cl cotransport is activated, but the driving forces are such that little or no net ion or water movement occurs. Dog red cells have a low potassium concentration. When they are swollen in media containing 4 to 5 m*M* potassium, the K/Cl cotransporter is activated, but cell ion and water content are hardly affected, because the transmembrane potassium gradient is so small. Cytoplasmic potassium is high in most cell types, however, and in this more usual circumstance swelling-activated K/Cl cotransport would have a volume-regulatory function.

Table 1 **Volume-Activated Transporters in Duck and Dog Red Cells**

Species	Shrinkage-Activated	Swelling-Activated
Duck	Na/K/2Cl cotransport	K/Cl cotransport
Dog	Na/H exchange	K/Cl cotransport

III. EVIDENCE FOR COORDINATION BETWEEN SHRINKAGE- AND SWELLING-INDUCED TRANSPORTERS

A. MANIPULATION OF THE SET POINT VOLUME

The transport pathways listed in Table 1 can be influenced by a variety of experimental conditions. Amiloride and bumetanide, for example, inhibit Na/H exchange and Na/K/2Cl cotransport, respectively, by acting directly on the transport proteins. In this section a heterogeneous group of perturbations is described, many of which were once thought to have direct stimulatory or inhibitory effects. On closer study, however, these interventions have been found to act indirectly, by changing the set point at which the transporters are activated. Furthermore, the direction of change is the same for both swelling- and shrinkage-induced transporters. These characteristics point to a common regulatory mechanism of the type schematized in Figure 18–1B.

1. Cytosolic Lithium Concentration

Lithium-loaded dog red cells showed an increase in the set point volume for Na/H exchange.[6] Lithium loading also raised the set point for K/Cl cotransport.[7] As shown in Figure 18–4, lithium-loaded dog red cells behave as though they were shrunken: Na/H exchange was activated even at a normal water content of 66%, and one had to swell the cells to abnormally high volumes to turn off Na/H exchange and turn on K/Cl cotransport.

2. Anion Effects

Replacement of chloride by anions of the lyotropic series, e.g., nitrate and thiocyanate, caused a downward shift of the set point volume in dog red cells. In 100% thiocyanate media the set point for the Na/H exchanger is so low that the transporter is virtually unresponsive to cell shrinkage.[8] The effect of anion substitutions on K/Cl cotransport can only be studied in the presence of a halide ion. Nevertheless, partial replacement of chloride by thiocyanate caused stimulation of this pathway and a change in its set point to values of cell water content considerably below 60% (Figure 18–5).[9] Similar effects of thiocyanate on K/Cl cotransport in sheep red cells have been reported.[10]

3. Cytosolic Magnesium Concentration

Lauf[11] reported that red cells (of sheep), when depleted of magnesium by incubation in ethylene diamine tetraacetate (EDTA)-containing media in the presence of A23187, showed a large stimulation of K/Cl cotransport. Cells loaded with magnesium showed inhibition of this pathway. In contrast to the findings with K/Cl cotransport, Na/K/2Cl cotransport of (ferret) red cells was inhibited by cytosolic magnesium depletion and stimulated by

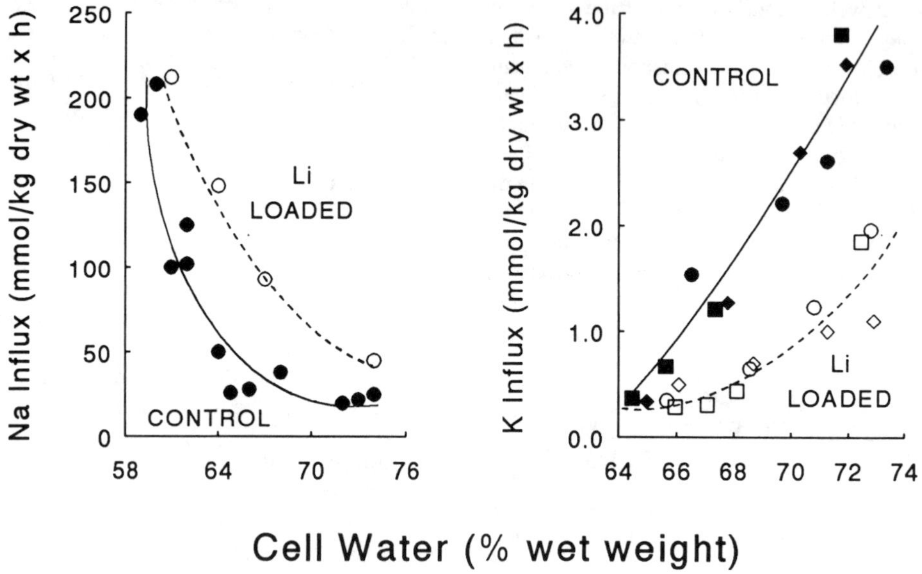

Figure 18–4 Na influx (left panel) and K influx (right panel) in dog red cells as functions of cell water content, in the presence (open symbols, dashed lines) and absence (closed symbols, solid lines) of lithium loading. Note that the set point for both volume-activated transporters is shifted to a higher volume. (From Parker, J. C. et al., *J. Gen. Physiol.,* 96, 1141, 1990. With permission.)

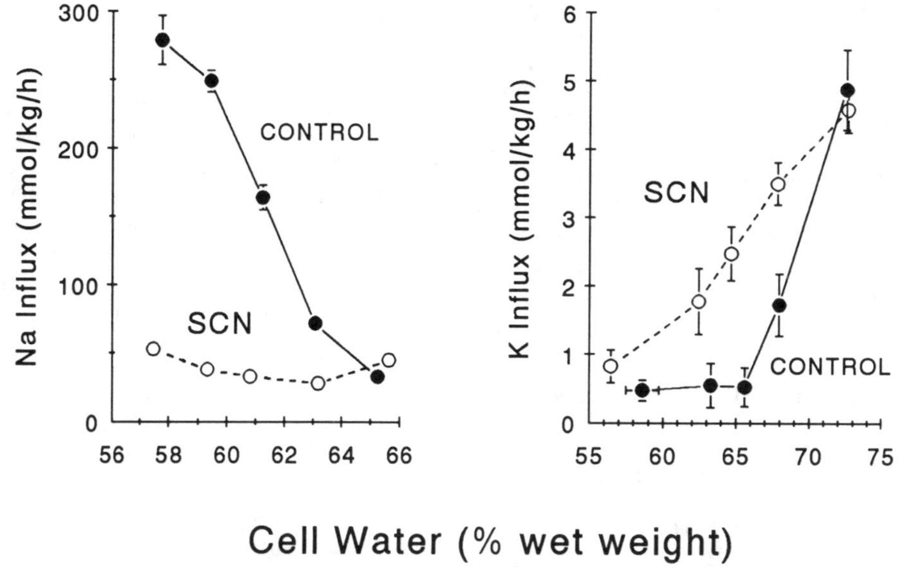

Figure 18–5 Na influx (left panel) and K influx (right panel) in dog red cells as functions of cell water content, in the presence (open symbols, dashed lines) and absence (closed symbols, solid lines) of thiocyanate. Note that the set point for both volume-activated transporters is shifted to a lower volume. (From Parker, J. C. and Colclasure, G. C., *Am. J. Physiol.,* 262, C418, 1992. With permission.)

magnesium loading.[12] Starke and McManus[13] showed that varying magnesium in duck red cells had the same qualitative influences on K/Cl cotransport and Na/K/2Cl cotransport as those described in sheep and ferret cells. Moreover, these authors pointed out that the magnesium effects were exerted on the set point volumes of the two transporters. Thus, magnesium-depleted duck red cells showed a decrease in set point volume. At a normal water content they behaved as though they were swollen: the K/Cl cotransporter was activated. Na/K/2Cl cotransport did not turn on until cells were shrunken well below the set point volume seen in magnesium-replete cells. Conversely, magnesium-loaded duck red cells had an increase in set point volume. At a normal water content these cells showed activation of Na/K/2Cl cotransport as if they were responding to shrinkage. The cells had to be swollen to supraphysiologic volumes before Na/K/2Cl cotransport was turned off and K/Cl cotransport was turned on.[13,14]

Dog red cells show a response to manipulation of cytosolic magnesium similar to that of duck red cells in that the set point for both shrinkage- and swelling-induced transport is affected in the same way.[7,15] The influence of magnesium depletion on dog red cells (Figure 18–6) is like that of thiocyanate substitution (Figure 18–5), in that the set points for both Na/H exchange and K/Cl cotransport are reduced. Magnesium loading (Figure 18–6) mimics the effect of lithium loading (Figure 18–4): the set points are increased. In the duck experiments the alteration in set points for shrinkage- and swelling-induced ion fluxes could plausibly be ascribed to a change in specificity of a single membrane cotransporter from Na/K/2Cl to K/Cl, but in dog red cells this mechanism for coordinated response of volume-activated transporters seems unlikely, since Na/H exchange and K/Cl cotransport carry no ions in common.

4. Hypoxia

A stimulatory effect of hypoxia on monovalent cation permeability of avian red cells was described by Tosteson and Robertson in 1956.[16] McManus[17] reported that, whereas hypoxia induced by aerating duck red cell suspensions with nitrogen increased cation flux, replacement of oxygen by carbon monoxide had no effect on permeability. Moreover, cells preequilibrated with carbon monoxide failed to increase their cation transport rate when subsequently aerated with nitrogen.[18] These results suggested that the effect of deoxygenation on transport might depend on the change in the conformation of hemoglobin to its deoxy form, since cells treated with carbon monoxide are locked into the oxy conformation. Subsequent work in mammalian red cells[13,14,19–22] has led to the concept that the effects of deoxygenation on cation transport are mediated via changes in the binding of organic phosphates to hemoglobin, which in turn influence the cytosolic free magnesium concentration, as schematized in Figure 18–7.

The cytosol of red cells contains millimolar concentrations of ATP, plus 2,3-diphosphoglycerate

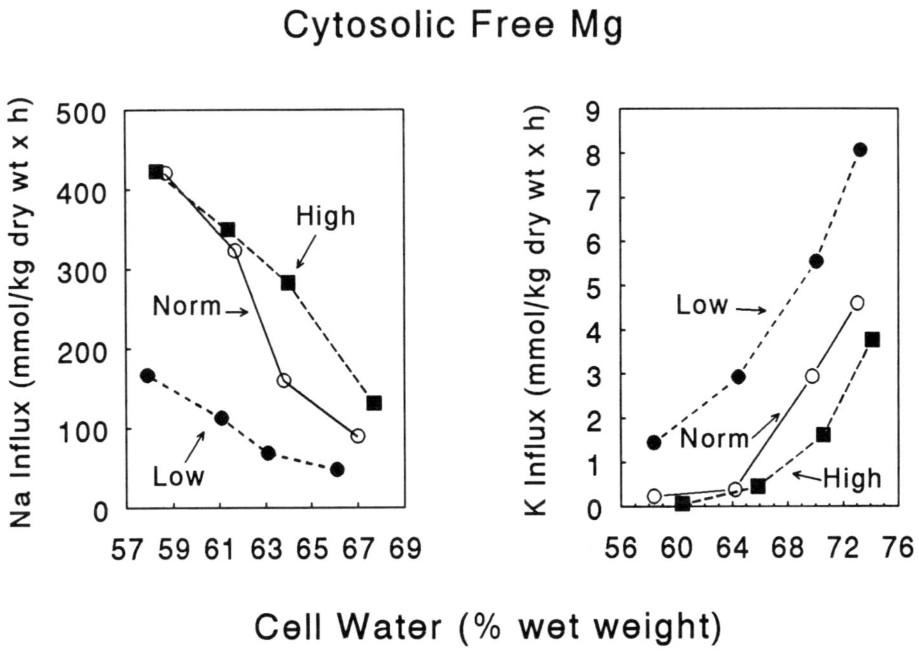

Figure 18–6 Na influx (left panel) and K influx (right panel) in dog red cells as functions of cell water content, in the presence of high, normal, and low cytosolic free magnesium concentrations. For quantitative information regarding these experiments, see Reference 7.

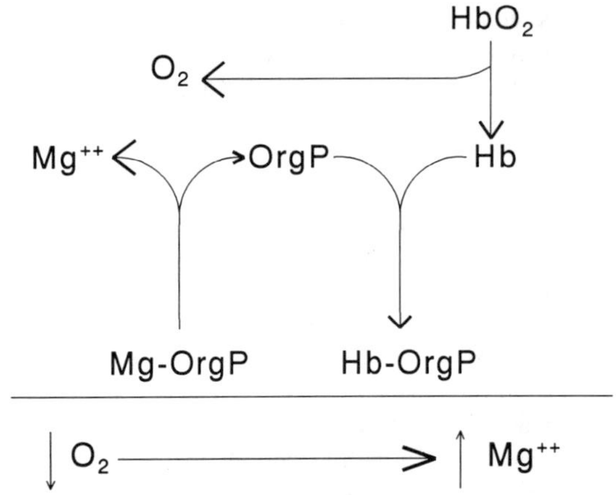

Figure 18–7 Schema indicating the relationships between oxygenated hemoglobin, deoxygenated hemoglobin, organic phosphates, and free magnesium in the red cell cytosol. See text.

(mammals) or inositol pentaphosphate (adult birds). These organic phosphates (OrgP in Figure 18–7) are bound by deoxy-but not by oxyhemoglobin. In well-oxygenated red cells the organic phosphates are free to chelate magnesium, and the cytosolic level of free Mg is low. Upon deoxygenation, however, the organic phosphates bind to hemoglobin and lose their affinity for magnesium. The cytosolic level of free magnesium rises. As a consequence, the volume set point for Na/K/2Cl cotransport increases. Thus, in ferret red cells at physiological volume, deoxygenation increases the bumetanide-sensitive fluxes of Na and K.[23] The rise in free magnesium also raises the set point for K/Cl cotransport, and movements of ions through this pathway are inhibited, as demonstrated in deoxygenated human sickle red cells.[22] In duck red cells, which exhibit high activities of both Na/K/2Cl and K/Cl cotransport (Table 1), the reciprocal effects of hypoxia are particularly well shown.[14]

5. Permeant Nonelectrolytes

Incubation of both dog and human red cells with solutions that are isotonic with respect to salt concentration but containing 0.1 to 0.6 *M* urea has no effect on cell water content, because urea equilibrates rapidly across the membrane.[24] Cells bathed in physiologic saline solutions containing these concentrations of urea or formamide behave as if swollen.[25] Within 2 min after exposure to the urea-containing solution the cells had the same water content as controls suspended in the same solution minus urea. However, K/Cl cotransport was demonstrable even at normal volumes in the urea-incubated cells, and their Na/H exchanger did not function unless they were shrunken far below the volume at which this pathway is normally activated. Thus, urea has an effect on the set point for both pathways (Figure 18–8). Human red cells incubated in physiologic medium supplemented with 0.5 *M* urea behaved similarly. Being high-potassium cells they sustained a net loss of potassium and water, provided chloride was present, suggesting that their K/Cl cotransporter had been activated.[26] The meaning of these observations for normal red cell physiology is uncertain. At the tip of the renal medulla in a human under conditions of maximum water conservation, urea concentrations as high as 600 m*M* can occur, but the impact of this on the red cells circulating through that region has not been studied.

6. Catecholamines

Ørskov[27] reported a stimulatory effect of catecholamines on salt transport in red cells of pigeons. Subsequent work, summarized by Kregenow[3] and Haas and McManus,[28] has led to the following conclusions. In the presence of catecholamines the volume set point for duck red cells is greatly altered: at normal volume the shrinkage-induced Na/K/2Cl cotransporter is turned on; one has to swell the cells to a higher volume than normal in order to deactivate Na/K/2Cl cotransport and stimulate K/Cl cotransport. The effects of catecholamines are duplicated by cyclic AMP and opposed by the β-adrenergic antagonist, propranolol. These observations may explain the finding that the steady-state water content of red cells in fresh whole blood from ducks exceeds that of washed red cells in saline media of the same tonicity. Presumably, catecholamines released into the animal's circulation at the time of venipuncture influence the set point of the transporters in such a way that the cell seeks a higher volume. The same effect can be seen in duck red cells incubated in saline solutions containing norepinephrine.

7. Protein Phosphatase Inhibitors

The field of transport regulation was greatly stimulated by the discovery of okadaic acid and its

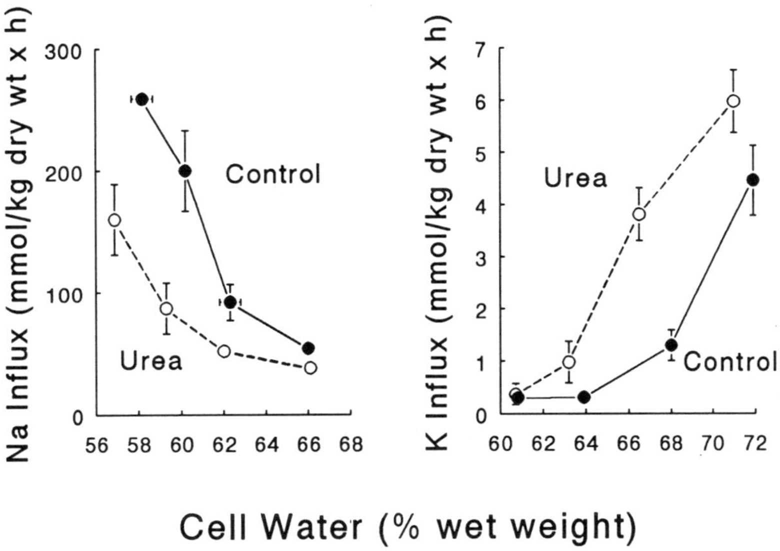

Figure 18–8 Na influx (left panel) and K influx (right panel) in dog red cells as functions of cell water content, in the presence (open symbols, dashed lines) and absence (closed symbols, solid lines) of 300 m*M* urea. Note that the set point for both volume-activated transporters is shifted to a lower volume. (From Parker, J. C., *Am. J. Physiol.,* in press. With permission.)

congeners, compounds that freely enter intact cells and inhibit certain serine and threonine protein phosphatases.[29] Pewitt et al.[30] showed that okadaic acid stimulated the shrinkage-induced Na/K/2Cl cotransporter in duck red cells. Jennings and Schulz[31] and Kaji and Tsukitani[32] showed that this agent inhibited swelling-induced K/Cl cotransport in mammalian red cells. In dog red cells okadaic acid increased the set point volume for both shrinkage-activated Na/H exchange and swelling-activated K/Cl cotransport.[33] These actions of okadaic acid (Figure 18–9) constitute further evidence that swelling- and shrinkage-induced pathways are coordinated.

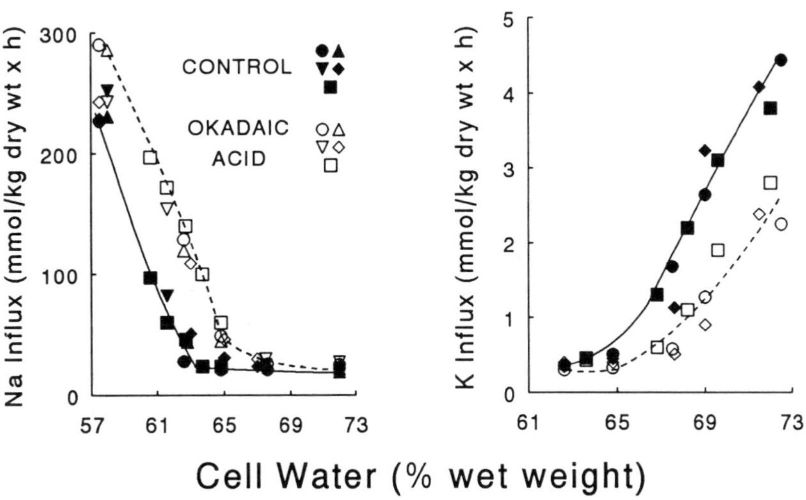

Figure 18–9 Na influx (left panel) and K influx (right panel) in dog red cells as functions of cell water content, in the presence (open symbols, dashed lines) and absence (closed symbols, solid lines) of okadaic acid. Note that the set point for both volume-activated transporters is shifted to a higher volume. (From Parker, J. C. et al., *J. Gen. Physiol.,* 98, 869, 1991. With permission.)

B. FIXATION OF TRANSPORTERS

The Na/H exchanger of dog red cells could be irreversibly activated if, while the cells were shrunken, they were exposed to a low concentration of glutaraldehyde. When the agent was applied to swollen cells, the Na/H exchanger could not be activated by shrinking.[34] The non-cross-linking agent, *N*-phenylmaleimide, had a similar action[35] and, furthermore, was shown to influence both swelling- and shrinking-activated transporters. Thus, *N*-phenylmaleimide applied to shrunken dog red cells fixed Na/H exchange on and K/Cl cotransport off, whereas just the reverse effects were seen when the agent was applied to swollen cells.[9] Both pathways were fixed at about the same concentration of *N*-phenylmaleimide. Solubilized membranes of dog red cells exposed to *N*-phenylmaleimide showed differences in the electrophoretic pattern of glyceraldehyde-3-phosphate dehydrogenase (G3PDH) (band 6) depending on the volume of the cells at the time of fixation.[9] The meaning of this finding is not clear at present. One might speculate as follows: G3PDH is bound to band 3. Band 3 has a binding site for 4,4′-diiothio-cyanatostilbene-2,2′-disulfonic acid (DIDS). DIDS inhibits several apparently independent swelling-induced transport pathways in trout red cells.[36] Perhaps the interaction between band 3 and G3PDH is involved in the signaling pathway between volume perturbation and the activation of transporters. Whatever the explanation, the ability of *N*-phenylmaleimide to fix both swelling-induced K/Cl cotransport and shrinkage-induced Na/H exchange constitutes further evidence that the two transporters are in some way coordinated or linked, as schematized in Figure 18–1B.

C. ON-OFF KINETICS

Jennings and Al-Rohil[37] showed in rabbit red cells that following a sudden reduction in tonicity of the bathing medium, there was a 5- to 10-min transient before the K/Cl cotransporter became fully activated. Sudden shrinkage of swollen cells resulted in prompt deactivation of the K/Cl cotransporter, with no measurable lag period. From this observation, and from the finding that activation of swelling-induced potassium transport was delayed further by vanadate and fluoride, it was concluded that regulation of K/Cl cotransport must be controlled by an equilibrium between the phosphorylated and dephosphorylated forms of some protein and that the effect of swelling cells must be to inhibit the putative kinase rather than to stimulate the phosphatase.[37]

This elegant experiment left open the question whether the reaction responsible for the slow activation of swelling-induced K/Cl cotransport involved the transporter itself or some regulator. Evidence in favor of the latter possibility was obtained in a study with dog red cells. It was found that the swelling-induced events (turning on of K/Cl cotransport and turning off of Na/H exchange) had about the same 10-min transient as described for activation of K/Cl cotransport in rabbit red cells. The triggering of shrinkage-induced responses (turning off of K/Cl cotransport and turning on of Na/H exchange) occurred too rapidly for the transient to be resolved by the methods used.[33] The similar pattern of activation/deactivation kinetics for both swelling- and shrinkage-induced pathways constitutes further evidence for some sort of mutual regulation, as suggested in Figure 18–1B.

D. THE ROLE OF CYTOSOLIC PROTEIN CONCENTRATION

Experiments in dog red cells and their ghosts indicate that the signal for activation of shrinkage-induced Na/H exchange and swelling-induced K/Cl cotransport originates with the attainment of a critical cytosolic protein concentration, rather than a certain cell volume.[38,39] The intracellular protein concentration in the resealed ghosts that defines the set point for K/Cl cotransport is the same as that for Na/H exchange, suggesting that the two transporters are under the control of a common regulator.[39]

IV. SENSITIVITY OF VOLUME-ACTIVATED TRANSPORTERS TO OTHER STIMULI

All the manipulations discussed in Section III.A (magnesium, lithium, catecholamines, hypoxia, etc.) appear to exert their effects by changing the set point volume for the transporters. Are there agents (other than pharmacologic inhibitors) that can act on the transporters independently of the putative volume-regulatory apparatus posited in Figure 18–1B? The only manipulation we know of that may have such an independent effect is alteration of cytosolic pH. The evidence for this comes from experiments with dog red cells rendered incapable of responding to cell shrinkage, either by magnesium depletion or by preincubation with *N*-phenylmaleimide (See Section III.B). In this circumstance the Na/H exchanger could still be turned on by acidification of the cytosol, a well-known allosteric activator of this transporter.[15,35]

V. FEATURES OF A VOLUME SENSOR/TRANSDUCER REGULATOR MECHANISM

The discussion above outlines evidence that there may be a volume-coordinating system in cells that communicates with both swelling- and shrinkage-induced transporters. Our minimal hypothesis, outlined in Figure 18–10, is that there is a regulator protein (R in Figure 18–10), the phosphorylated and dephosphorylated forms of which are in equilibrium. When the protein, R, is phosphorylated, shrinkage-induced transporters are stimulated and swelling-induced transporters inhibited. The state of the transporters is reversed when the protein is dephosphorylated. We postulate that all the manipulations listed in Figure 18–10 (volume change, magnesium, lithium, anions, urea) exert their actions by altering the equilibrium between R and R-P.

There are many problems and clues in regard to this system. The molecular events during the activation of transporters are not known. An intriguing observation is that the shrinkage-induced stimulation of Na/K/2Cl cotransport in duck red cells involves an increase in the number of bumetanide binding sites, suggesting either that latent transporters are activated or that binding of the inhibitor is stimulated when the transporter becomes active.[40] The effects of lithium are possibly related to the role of this cation as an inhibitor of the phosphoinositide signaling system,[41] or even an interaction with G proteins.[42,43] The effects of catecholamines and cyclic nucleotides in duck red cells suggest an interaction with the A-kinase system.

The explanations for other phenomena, such as the actions of lyotropic anions and urea in shifting the set point, remain obscure. An interesting possibility is that some of them act as "protein perturbants".

Long incubation of proteins with molar concentrations of urea or chaotropic ions like thiocyanate can cause irreversible protein denaturation. With lower concentration of these agents, however, prompt and reversible effects may occur.[44–46] The first kind of perturbation involves major alterations of the protein's tertiary structure and disruption of intramolecular bonds. The second is thought to arise from milder interactions of the solutes with protein side chains and/or changes in water structure.[44,46] Some of these effects might alter the volume of individual protein molecules[46,47] and so influence their ability to crowd or be crowded in solutions. At present it is not possible to tell whether the urea- or thiocyanate-induced changes in volume set point are due to effects on hemoglobin. It is clear, however, that small alterations in hemoglobin structure can affect cell volume perception: witness the profound influence of the substitution of lysine for glutamate in the β-chain of hemoglobin C. Red cells containing this variant live in a dehydrated condition; when brought to a normal water content, K/Cl cotransport becomes active, and the cells rapidly lose solute and water until they reach their usual steady state, in which their cytoplasm is abnormally concentrated.[48] The effect of hemoglobin C is not due solely to its altered electrical charge, because red cells containing other mutants with the same amino acid substitution but at different loci on the globin molecule do not show this phenomenon of altered set point volume.[49]

The sites of action of the agents listed in Figure 18–10 are thus not clear. However, the observation

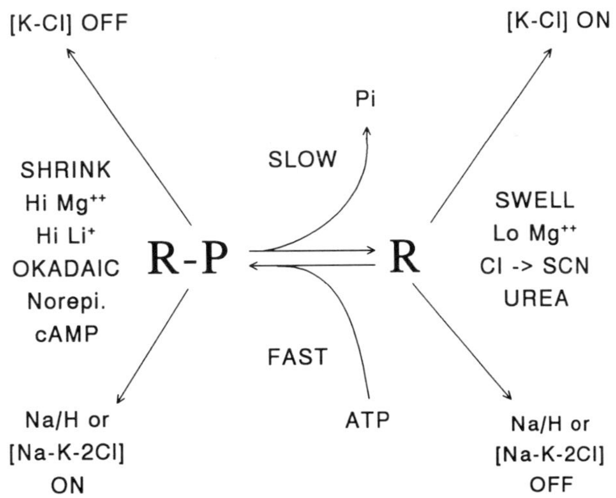

Figure 18–10 Schema showing the responses of a putative volume-regulatory transducer to swelling, shrinking, and other influences discussed in this chapter.

that the set point volumes for different ion flux pathways can be influenced in a coordinated way by so many diverse maneuvers suggests that changes in cell water content impinge on a regulatory mechanism that comprises not only membrane transporters, but also a volume sensor and a signaling system that orchestrates the cell's responses.

REFERENCES

1. **Schultz, S. G.,** Homocellular regulatory mechanisms in sodium-transporting epithelia, avoidance of extinction by "flush-through", *Am. J. Physiol.,* 241, F579, 1981.
2. **Schultz, S. G.,** Membrane cross-talk in sodium-absorbing epithelial cells, in *The Kidney: Physiology and Pathophysiology,* 2nd ed., Seldin, D. W. and Giebisch, G., Eds., Raven Press, New York, 1992, 287.
3. **Kregenow, F. M.,** Osmoregulatory salt transporting mechanisms. Control of cell volume in anisotonic media, *Annu. Rev. Physiol.,* 43, 493, 1981.
4. **McManus, T. J., Haas, M., Starke, L. C., and Lytle, C. Y.,** The duck red cell model of volume-sensitive, chloride-dependent cation transport, *Ann. N.Y. Acad. Sci.,* 456, 186, 1985.
5. **Sarkadi, B. and Parker, J. C.,** Activation of ion transport pathways by changes in cell volume, *Biochim. Biophys. Acta (Rev. Biomembr.),* 1071, 407, 1991.
6. **Parker, J. C.,** Interactions of lithium and protons with the sodium-proton exchanger of dog red blood cells, *J. Gen. Physiol.,* 87, 189, 1986.
7. **Parker, J. C., McManus, T. J., Starke, L. C., and Gitelman, H. J.,** Coordinated regulation of Na/H exchange and [K-Cl] cotransport in dog red cells, *J. Gen. Physiol.,* 96, 1141, 1990.
8. **Parker, J. C.,** Volume-responsive sodium movements in dog red blood cells, *Am. J. Physiol.,* 244, C324, 1983.
9. **Parker, J. C. and Colclasure, G. C.,** Actions of thiocyanate and N-phenylmaleimide on volume-responsive Na and K transport in dog red cells, *Am. J. Physiol.,* 262, C418, 1992.
10. **Lauf, P. K.,** Foreign anions modulate volume set point of sheep erythrocyte K-Cl cotransport, *Am. J. Physiol.,* 260, C503, 1991.
11. **Lauf, P. K.,** Passive KCl fluxes in low-K sheep erythrocytes: modulation by A23187 and bivalent cations, *Am. J. Physiol.,* 249, C271, 1985.
12. **Flatman, P. W.,** The effects of magnesium and potassium transport in ferret red cells, *J. Physiol. (London),* 397, 471, 1988.
13. **Starke, L. C. and McManus, T. J.,** Intracellular free magnesium determines the volume regulatory set point in duck red cells, *FASEB J.,* 4, A818, 1990.
14. **Starke, L. C.,** Regulation of [Na-K-2Cl] and [K-Cl] Cotransport in Duck Red Cells, Ph.D. thesis, Duke University, Durham, NC, 1989.
15. **Parker, J. C., Gitelman, H. J., and McManus, T. J.,** Role of magnesium in the activation of Na/H exchange in dog red cells, *Am. J. Physiol.,* 257, C1038, 1989.
16. **Tosteson, D. C. and Robertson, J. S.,** Potassium transport in duck red cells, *J. Cell. Comp. Physiol.,* 47, 147, 1956.
17. **McManus, T. J.,** Comparative biology of red cells, *Fed. Proc.,* 26, 1821, 1967.
18. **McManus, T. J. and Allen, D. W.,** Some factors which alter passive cation movements in duck red cells, in *Metabolism and Membrane Permeability of Erythrocytes and Thrombocytes,* Deutsch, E., Gerlach, E., and Moser, K., Eds., George Thieme Verlag, Stuttgart, 1968, 428.
19. **Bunn, H. F., Ransill, B. J., and Chao, A.,** The interaction between erythrocyte organic phosphates, magnesium ion, and hemoglobin, *J. Biol. Chem.,* 246, 5273, 1971.
20. **Gupta, R. K., Benkovic, J. L., and Rose, Z. B.,** The determination of the free magnesium level in the human red blood cell by ^{31}P NMR, *J. Biol. Chem.,* 253, 6172, 1978.
21. **Flatman, P. W.,** The effect of buffer composition and deoxygenation on the concentration of ionized magnesium inside human red blood cells, *J. Physiol.,* 300, 19, 1980.
22. **Canessa, M., Fabry, M. E., and Nagel, R. L.,** Deoxygenation inhibits the volume-stimulated, Cl-dependent K efflux in SS and young AA cells: a cytosolic Mg modulation, *Blood,* 70, 1861, 1987.
23. **Flatman, P. W.,** Regulation of potassium transport by magnesium, in *Regulation of Potassium Transport Across Biological Membranes,* Reuss, L., Russell, J. M., and Szabo, G., Eds., University of Texas Press, Austin, 1990, 191.
24. **Macey, R. I.,** Transport of water and urea in red blood cells, *Am. J. Physiol.,* 246, C195, 1984.
25. **Parker, J. C.,** Urea alters the set point volume for [K-Cl] cotransport, Na/H exchange, and Ca/Na exchange in dog red cells, *Am. J. Physiol.,* in press.
26. **Orringer, E. P., Brockenbrough, J. S., and Parker, J. C.,** Effects of permeant amides on volume-activated [K-Cl] cotransport and Na/H exchange in mammalian red cells, *Blood,* 80, 379a, 1992.

27. **Ørskov, S. L.,** Experiments on the influence of adrenaline and noradrenaline on the potassium absorption of red blood cells from pigeons and frogs, *Acta Physiol. Scand.,* 37, 229, 1956.
28. **Haas, M. and McManus, T. J.,** Effect of norepinephrine on swelling-induced potassium transport in duck red cells, *J. Gen. Physiol.,* 85, 649, 1985.
29. **Cohen, P., Holmes, C. F. B., and Tsukitani, Y.,** Okadaic acid: a new probe for the study of cellular regulation, *Trends Biochem. Sci.,* 15, 98, 1990.
30. **Pewett, E. B., Hegde, R. S., Haas, M., and Palfrey, H. C.,** The regulation of Na/K/2Cl cotransport and bumetanide binding in avian erythrocytes by protein phosphorylation and dephosphorylation, *J. Biol. Chem.,* 265, 20747, 1990.
31. **Jennings, M. L. and Schulz, R. K.,** Okadaic acid inhibition of KCl cotransport. Evidence that protein dephosphorylation is necessary for activation of transport by either cell swelling or N-ethylmaleimide, *J. Gen. Physiol.,* 97, 799, 1991.
32. **Kaji, D. and Tsukitani, Y.,** Role of protein phosphatase in activation of [K-Cl] cotransport in human erythrocytes, *Am. J. Physiol.,* 260, C178, 1991.
33. **Parker, J. C., Colclasure, G. C., and McManus, T. J.,** Coordinated regulation of shrinkage-induced Na/H exchange and swelling-induced [K-Cl] cotransport in dog red cells: further evidence from activation kinetics and phosphatase inhibitors, *J. Gen. Physiol.,* 98, 869, 1991.
34. **Parker, J. C.,** Glutaraldehyde fixation of sodium transport in dog red blood cells, *J. Gen. Physiol.,* 84, 789, 1984.
35. **Parker, J. C. and Glosson, P. S.,** Interactions of the sodium-proton exchange mechanism in dog red blood cells with N-phenylmaleimide, *Am. J. Physiol.,* 253, C60, 1987.
36. **Garcia-Romeu, F., Cossins, A. R., and Motais, R.,** Cell volume regulation by trout erythrocytes: characteristics of the transport systems activated by hypotonic swelling, *J. Physiol. (London),* 440, 547, 1991.
37. **Jennings, M. L. and Al-Rohil, N.,** Kinetics of activation and inactivation of swelling-stimulated K^+/Cl^- transport. The volume-sensitive parameter is the rate constant for inactivation, *J. Gen. Physiol.,* 95, 1021, 1990.
38. **Colclasure, G. C. and Parker, J. C.,** Cytosolic protein concentration is the primary volume signal in dog red cells, *J. Gen. Physiol.,* 98, 881, 1991.
39. **Colclasure, G. C. and Parker, J. C.,** Cytosolic protein concentration is the primary volume signal for swelling-induced [K-Cl] cotransport in dog red cells, *J. Gen. Physiol.,* 100, 1, 1992.
40. **Haas, M.,** Properties and diversity of (Na-K-Cl) cotransporters, *Annu. Rev. Physiol.,* 51, 443, 1989.
41. **Nahorski, S. R., Jenkinson, S., and Challis, R. A. J.,** Disruption of phosphoinositide signalling by lithium, *Biochem. Soc. Trans.,* 20, 430, 1992.
42. **Avissar, S., Schreiber, G., Danon, A., and Belmaker, R. H.,** Lithium inhibits adrenergic and cholinergic increases in GTP binding in rat cortex, *Nature,* 331, 440, 1988.
43. **Davis, B. A., Hogan, E. M., and Boron, W. F.,** Role of G proteins in stimulation of Na/H exchange by cell shrinkage, *Am. J. Physiol.,* 262, C533, 1992.
44. **Yancey, P. H., Clark, M. E., Hand, S. C., Bowlus, R. D., and Somero, G. N.,** Living with water stress: evolution of osmolyte systems, *Science,* 217, 1214, 1982.
45. **Rajagopalan, K. V., Fridovich, I., and Handler, P.,** Competitive inhibition of enzyme activity by urea, *J. Biol. Chem.,* 236, 1059, 1961.
46. **Low, P. S.,** Molecular basis of the biological compatibility of nature's osmolytes, in *Transport Processes, Iono- and Osmoregulation,* Gilles, R. and Gilles-Baillien, M., Eds., Springer-Verlag, Berlin, 1985, 469.
47. **Brandts, J. F.,** Conformations of proteins in water and in aqueous mixtures, in *Structure and Stability of Biological Macromolecules,* Timasheff, S. N. and Fasman, G. D., Eds., Marcel Dekker, New York, 1969, 213.
48. **Brugnara, C., Kopin, A. S., Bunn, H. F., and Tosteson, D. C.,** Regulation of cation content and cell volume in erythrocytes from patients with homozygous hemoglobin C disease, *J. Clin. Invest.,* 75, 1608, 1985.
49. **Olivieri, O., Vitoux, D., Galacteros, F., Bachir, D., Blouquit, Y., Beuzard, Y., and Brugnara, C.,** Hemoglobin variants and activity of the [K-Cl] cotransport system in human erythrocytes, *Blood,* 79, 793, 1992.

Part IV
Role of Gene Expression in Cell Volume Control

Chapter 19

Regulation of Eukaryotic Gene Expression

Arlyn García-Pérez

TABLE OF CONTENTS

I. INTRODUCTION

The purpose of this chapter is to provide a general overview of the major processes involved in gene expression and its regulation. The intention is also to introduce the reader unfamiliar with molecular biology to the special nomenclature of this discipline, in order to facilitate the study of the chapters included in this book section. Important terminology is printed **bold** throughout the chapter.

The phrase "regulation of gene expression" generally encompasses transcriptional, posttranscriptional, and translational mechanisms of regulation. It does not include posttranslational modifications.[1,2] Transcriptional regulation is emphasized in this chapter because it is the most studied. The mechanisms discussed are the ones generalized to higher eukaryotes, and more specifically to higher animals, not necessarily to plants. Although many of these mechanisms were originally found in prokaryotes (e.g., eubacteria) and most have been best defined in lower eukaryotes (e.g., yeast and *Candida elegans*), the aspects unique to these organisms will not be addressed. Chapters 20 and 21 discuss bacteria and plants, respectively. In addition, because most of the references for this chapter are reviews of each of the topics included here, the reader is encouraged to consult these for in-depth treatments.

0-8493-4448-4/94/$0.00+$.50

II. TRANSCRIPTIONAL REGULATION OF GENE EXPRESSION

A. EUKARYOTIC GENOME STRUCTURE

The vast majority of our present knowledge of regulation of gene expression is based on studies of cloned single genes, which are naked (without proteins). However, it is important to remember that naturally occurring genes are packaged in **chromosomes.**

In the late 1930s, the two main chromosomal components were found: **deoxyribonucleic acid (DNA)** and **histones,** positively charged, small proteins that, being basic, were presumed to neutralize the acidity of the DNA.[1] It is now known that a chromosome consists of a single molecule of duplex (double-stranded) DNA to which are attached an approximately equal mass of proteins that are involved in maintaining chromosome structure and regulating gene expression.[1,2] It is estimated that the single *Escherichia coli* chromosome contains a DNA molecule of approximately 4 million **base pairs (bp),** while the DNA molecule of each of the larger chromosomes of higher plants and animals is estimated to be 80 million bp long (i.e., 80,000 kb).[1] Although the chromosome is the discrete unit of the genome and many genes are present and active in each chromosome, it is the single gene, structurally defined as **the transcription unit,** that will be discussed here.[2]

In 1953 what is now known as "the central dogma of molecular biology — that information flows from DNA to RNA to protein" was established.[1] The relationship was conceived as follows: single DNA chains served as **templates** for either complementary DNA molecules (during **replication**) or complementary RNA molecules (during **transcription**). The RNA molecules then served as templates for **translation,** so called because the process involved conversion of the nucleotide language of RNA into the amino acid language of proteins.[1] That same year Watson and Crick proposed the double-helical structure for DNA.[3–5] The discoveries of the semiconservative mechanism of DNA replication (1958), the triplet nature of the genetic code (1961), and the existence of **messenger RNA (mRNA)** (1961) followed.[1] Then, the work of Jacob and Monod on the genes involved in lactose metabolism in *E. coli* provided our fundamental knowledge of the basic molecular principles involved in the control of gene expression.[6,7]

1. Exons and Introns

In 1967, eubacterial genes and their proteins **(gene products)** were found to be **colinear,** that is, the continuous sequence of nucleotides in the gene corresponded exactly with the continuous sequence of amino acids in the protein.[2] In 1977, this simple view of the gene was modified with the discovery of interruptions in eukaryotic genes.[1,2,8,9] When comparing a nuclear eukaryotic gene with its mRNA, the gene included additional sequences within the coding region that interrupted the sequence that represented the protein. The portions of DNA that are transcribed in an interrupted gene fall into two categories (Figure 19–1): the intervening sequences, which are missing in the mRNA, are **introns,** while the regions that remain represented in the mRNA are **exons,** so called because the processed mRNA without the introns "exits" from the nucleus into the cytoplasm for translation.[1,2] Thus, in eukaryotes the process of gene expression requires a new step that does not occur in eubacteria. As depicted in Figure 19–1, eukaryotic DNA is first transcribed colinearly to produce a precursor RNA molecule **(primary RNA transcript)** from which the intron sequences must be excised to obtain the mRNA, which consists only of a series of exons to be translated into protein. The excision process is termed **RNA splicing.** As shall be discussed, RNA splicing is an additional control point for regulation of gene expression. In reference to cloned genes, **complementary DNA (cDNA)** is that synthesized *in vitro* from spliced mRNA, while **genomic DNA** is usually naked chromosomal DNA (exons plus introns).

In addition to most of the genes of higher eukaryotes, many of the genes of animal viruses have introns. Interestingly, the process of splicing was discovered in adenovirus.[10,11] Among interrupted genes, both the number and size of introns vary widely. But in general, the exons comprise a smaller portion of the total length of a gene than the introns. Therefore, introns fill a greater proportion of the higher eukaryotic genome than exons.[2] It is hypothesized that the long length of many introns helps to ensure that the much shorter coding sequences have a lower probability of undergoing genetic crossing over during evolution, so they remain intact.[1] Imprecise recombination would be tolerated within introns but would disrupt the coding sequence in exons, resulting in aberrated gene expression. For a given gene, introns usually occur at homologous positions (relative to the coding region) across species, although the length of the introns often varies among the species. In these genes, the sequences of exons (coding sequences mainly, much less so in untranslated regions) are conserved, while those of introns vary greatly. In addition, genes that are evolutionarily related have similar organizations,

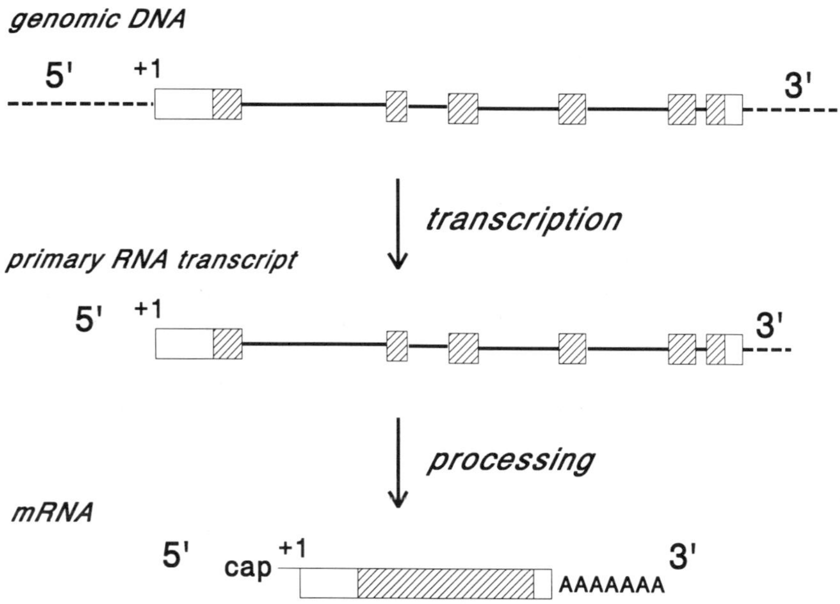

Figure 19–1 Schematic representation of mRNA synthesis. Broken lines represent 5′ and 3′ flanking regions of the gene shown. The boxes represent the exons; hatched portions are coding regions, empty portions are 5′ and 3′ UTRs. The transcription startsite is indicated by +1. The string of As represent the poly(A) tail.

with conservation of the positions of at least some of the introns. A few genes in higher eukaryotes and most of the genes in *Saccharomyces cerevisiae* (yeast) are intronless.[1]

2. 5′/3′ Untranslated Regions and 5′/3′ Flanking Regions

DNA and RNA chains have a direction defined by the orientation of their sugar-phosphate backbone.[1] The nucleotides in the chain are linked together by regular phosphodiester bonds formed between carbon atom number 3 in one sugar moiety and carbon atom number 5 in the next. The end of the chain lacking a phosphodiester at a number 5 carbon is called the **5′ end,** while the end lacking it at a number 3 carbon is called the **3′ end.** All naturally synthesized DNA (by **DNA polymerases**) and RNA (by **RNA polymerases**) chains grow in the **5′-to-3′ direction.** Translation on the mRNA template occurs in the same direction, and the resulting protein grows from **amino terminus** to **carboxy terminus;** that is, the amino terminus of a protein corresponds to the 5′-most **codon** (nucleotide triplet that "codes" for a specific amino acid) of the mRNA.

Noncoding regions of the processed mRNA, which are not translated, are called **untranslated regions (UTRs)** (Figure 19–1). The **transcription startsite (+1)** is the first nucleotide at the **5′UTR.** To the contrary, transcription does not end at the last nucleotide of the **3′UTR,** but rather extends beyond, and the 3′ end of the transcript is generated during processing by cleavage with endonucleases. Sequences beyond the UTRs in both directions are the **flanking regions (5′ or 3′)** of a gene. These are not transcribed (except for the extension beyond the 3′UTR just described). Most ***cis*** **control elements** have been found in flanking regions, primarily in the 5′ flanking region. The term **upstream** is commonly used to mean towards the 5′ end of the coding strand, while **downstream** signifies towards the 3′ end.

B. TRANSCRIPTION INITIATION

The principal point of regulation of transcription is during initiation. Formation of an initiation complex is an absolute requirement for transcription. The complex consists of many components that must not only interact with the gene to be transcribed, but in many cases also with one another. It is inferred that regulation can be accomplished by modulation of each of the individual components.[2,7,12–16] Because the bulk of the studies to date have concentrated in defining what the processes required for initiation are and not much is known about generalized mechanisms of their regulation, only the processes will be addressed here.

1. RNA Polymerase II

There are three RNA polymerases found in eukaryotic cells.[2] **RNA polymerase II (RNA pol II)** is the enzyme that is responsible for synthesizing **heterogeneous nuclear RNA (hnRNA),** the pool of primary RNA transcripts that are the precursors of mRNA (Figure 19–1). RNA pol II is located in the nucleoplasm, the part of the nucleus that excludes the nucleolus. All eukaryotic RNA polymerases, including RNA pol II, are large protein aggregates of 500 kDa or more. Their subunit compositions are complex. Each enzyme has two large subunits of approximately 200 and 140 kDa. In addition, they have <10 smaller subunits ranging in size from 10 to 90 kDa. Because it has not been possible to reconstitute an active RNA polymerase from purified subunits, there is no evidence as to which subunits represent catalytic activities and which have regulatory functions. Further, because the eukaryotic RNA polymerases do not display DNA sequence-specific binding capabilities by themselves, it is believed that they are part of a basic transcription apparatus that functions as directed by transcription factors.

2. *Cis* Control Elements

An mRNA encoding a protein contains the information needed for translation by ribosomes, but the flanking regions of the gene (mostly 5′) have been identified to contain specific nucleotides that control gene expression. Because these sequences affect the activity only of a gene on its own molecule of DNA they are termed ***cis* control elements** or ***cis* loci.**[1,2] It appears that the major point of control of eukaryotic gene expression is at the level of initiation of transcription. There are two types of *cis* control elements that affect transcription initiation: **promoters** and **enhancers.** The sequencing and characterization of many of these elements have led to the identification of certain DNA sequence motifs or **consensus sequences** which help to locate them and study their effects on newly cloned genes. Signal-responsive elements (**response elements**) are presented separately here because some are best categorized as promoter elements, whereas others fit the definition of enhancers.

a. Promoters

Promoters contain the sites at which RNA polymerase binds to DNA and thus fixes the start of transcription.[1,2,7,12] Therefore, promoters are always found in specific positions (upstream) and orientations with respect to the gene they govern. Three short discrete **modules** with distinct consensus sequences have been found within promoters in many different tissues: the **TATA box,** the **CAAT box,** and the **GC box.** None of these modules appears to be uniquely essential for promoter function. Some promoters lack a TATA box, while others lack a CAAT box and/or have no GC boxes.

The **TATA box,** located about 30 bp upstream (-30) of the transcription startsite (+1), is the most common promoter module. It is the only promoter module in eukaryotes with a relatively fixed location with respect to the starsite. It has been found in mammals, birds, amphibians, insects, and plants. The consensus sequence of the TATA box consists of six to eight contiguous A-T base pairs. The TATA box is usually surrounded by GC-rich sequences, a configuration that could be a factor in its function. Except for its position further upstream from the startsite, the TATA box is almost identical to the -10 element of bacterial promoters.

The **CAAT box** is named directly after its concensus sequence. It is often located close to -80, but it can function at distances from the startsite that vary considerably. It may also function in either orientation. In genes where it is present, the CAAT box tends to play the strongest role in determining the efficiency of the promoter.[2]

The **GC box** contains the sequence GGGCGG and is often found 50 to 100 bp upstream of the startsite. Promoters often contain multiple GC boxes, which can occur in either orientation.[2,7,12]

In general, the promoter modules can be ascribed two different types of function with regard to their effect on RNA polymerase.[2,7,12] The modules farther upstream (CAAT and GC boxes) strongly influence the frequency of transcription initiation and often associate with proteins that act as **positive regulators.** On the other hand, the definition of the startsite depends on the TATA box. It has been shown that its deletion causes the site of initiation to become erratic, although there is only a small overall reduction in transcription. Promoters without TATA boxes usually lack unique startsites; initiation occurs instead at any one of a cluster of startsites.

One of the puzzles of promoter organization is that, although the promoter conveys directional information (transcription proceeds only downstream from it) and, therefore, has an orientation requirement with respect to the startsite, the CAAT and GC boxes are able to function in either orientation. Additional levels of complexity are found in genes that have multiple promoters, each of which can be responsive to different forms of regulation.[7]

b. Enhancers

Enhancers, so called because the activity of promoters associated with them is highly increased, are also DNA sequences that regulate transcription of genes.[2,7] Unlike promoters, neither the position nor the orientation of enhancers is fixed in relation to the gene they regulate. That is, enhancers can exert their effect at long distances (thousands of bp) away and upstream or downstream from the gene. In addition, because enhancers have been shown to stimulate any promoter placed in their vicinity, their activity is promoter-dependent (a promoter must be present for transcription to take place) but not promoter-specific. Consensus sequences of some commonly found enhancers are discussed below with the description of the transcription factors that bind to them.

c. Response elements

Response elements are modules that uniquely identify particular groups of genes.[2] These include the **heat shock response element (HRE),** the **glucocorticoid response element (GRE),** and the **metal response element (MRE).** Response elements have similar characteristics to other promoter or enhancer modules. They contain short consensus sequences to be recognized by specific factors. They are not present at fixed distances from the startsite, but are usually in the region <200 bp upstream of it (GREs are found further upstream). The presence of a single module is usually sufficient to confer the regulatory response, but multiple copies may be found close to each other.

3. *Trans*-Activating Factors

Trans-activating factors are DNA sequence-specific polypeptides that regulate gene transcription by binding *cis* elements. They are also called **transcription factors.** The transcription factors necessary for formation of the transcriptional initiation complex are constitutive in all eukaryotic cells. Additional transcription factors can be ubiquitous, tissue-specific, or signal-specific.

The best-defined transcription factors are positive regulators. These transcription factors must be able not only to recognize and bind their target DNA sequences located in promoters, enhancers, or other regulatory elements, but also, having bound to the DNA, bind RNA polymerase or other transcription factors to exert their effect. Some structural motifs responsible for these activities have been found in the transcription factors, but almost nothing is known about how the molecular interactions result in increased transcription. The motifs are usually quite short and comprise only a small part of the protein structure. These motifs include **zinc fingers** and **leucine zippers,** which will be described for some of the transcription factors that they have been found in.

Although little is known about negative regulators in eukaryotes, it is presumed that, unlike positive regulators, these need only bind the DNA itself because their presence alone may serve to prevent gene expression.[2,15,16]

a. Constitutive components of the transcriptional complex

Initiation of transcription requires the constitutive transcription factors to build a complex on the promoter that can be joined by RNA polymerase.[2,7] The factors bind in a specific order. The first step in the complex formation is binding of **TFIID** to the TATA box present in most promoters. **TFIIA** then binds upstream of, but in contact with, TFIID. **TFIIB** then joins the complex, but is only loosely associated with the DNA. RNA pol II is able to join the complex at this point by binding to TFIID and the DNA. An additional transcription factor, **TFIIE,** then associates to the complex. Both TFIIB and TFIIE have been shown to interact directly with RNA pol II. Since the binding of TFIID to the TATA box is required for transcriptional complex formation, including the binding of RNA pol II, what fulfills the function of TFIID in the few genes lacking a TATA box is unknown.[2,7] In addition, because these factors are still incompletely purified, it is difficult to equate each factor with a defined protein.[2]

b. Ubiquitous transcription factors

Ubiquitous transcription factors are those that bind modules that occur in many promoters. When such modules are identified in the promoter of a newly cloned gene, the ubiquitous factors are assumed to be available to bind the module and thus regulate that promoter's function.[2] These modules include the CAAT and GC boxes and, in some cases, what is known as **the octamer** (ATTTGCAT). Binding of ubiquitous factors to their modules may be necessary for activation of the promoter in which the module occurs. Thus, these factors are regarded as **positive regulator proteins.** The ubiquitous transcription factors are the best characterized because most of them have been purified. Binding of multiple ubiquitous factors to the promoter may build a structure that, together with the constitutive factors and RNA pol II, constitutes the **initiation complex.** The variety of modules in a functional promoter, and the variation of their locations relative to the startsite, suggests that these factors can interact with one another by protein-protein interactions in many ways.

GC boxes are exclusively recognized by the factor **SP1.**[2] Although the sequence of GC boxes is specific (GGGCGG), the context in which they are found may be different in every promoter. For example, GC boxes in the thymidine kinase promoter are adjacent to a CAAT box and a TATA box, while in the SV40 promoter there is a tandem series of GC boxes upstream of a TATA box. This suggests that SP1 must be able to interact with factors bound to these different elements. SP1 is a monomer of 105 kDa that makes contact with one strand of DNA over a binding site of approximately 20 bp that includes at least one GC box. The nucleotides in the SP1 binding site that flank the required GC box recognition sequence probably influence binding, because GC boxes have been found to bind SP1 with different affinities.

SP1 is a **finger protein,** that is, its DNA-binding domain contains **zinc fingers.**[2,7,12,13] Each zinc finger has a characteristic pattern of cysteine and histidine residues that constitute the zinc-binding site. This structural motif is named after the loop of amino acids that protrudes from the zinc-binding site and is also called a **Cys_2/His_2 finger.** Zinc fingers are usually found in series. For example, SP1 has three zinc fingers in tandem. The finger structure itself comprises approximately 23 amino acids, and the linker between fingers is usually 7 or 8 amino acids. The finger conformation is required for SP1 to bind DNA.

Unlike the GC box, which is only recognized by SP1, the CAAT box may be recognized by multiple factors.[2,7,12,13] Among these are CTF/NF1, a family of proteins found first in HeLa cells and adenovirus. These proteins bind the CAAT box sequence GCCAAT present in many genes. The DNA-binding domain of these approximately 55-kDa proteins lies toward the amino terminus and comprises a basic region that is probably α-helical in conformation. Another transcription factor, **CP1** also binds CAAT boxes with high affinity, but favors different permutations of the sequence. Other proteins also bind CAAT boxes, including two from rat liver. Although little is known about the exact details of recognition, the fact that a variety of factors recognize CAAT boxes points out that "the identification of a conserved module does not inevitably imply that it will always be recognized by the same protein in all promoters."[2]

The octamer, a commonly found module of sequence ATTTGCAT, is also recognized by more than one transcription factor. However, only one, **Oct-1,** is ubiquitous. Other factors that recognize the octamer have been found only in lymphoid cells.[2,12,13] Oct-1 is a 90- to 100-kDa protein.

c. *Tissue- or signal-specific transcription factors*

Tissue-specific factors are only expressed in particular tissues. One of the best characterized is C/EBP, which is expressed in liver.[2,7] C/EBP is a dimer that binds both a CAAT box and an enhancer. It was the first transcription factor found to contain the leucine zipper structural motif. The leucine zipper domain is divided into two regions, both apparently helical. One part is a region containing a leucine residue every seventh amino acid, which results in alignment of all the leucines along one face of the helix. This face is a hydrophobic surface that can then interact with an equivalent structure, forming a dimer. Amino-terminal to the leucines is a region rich in positively charged amino acids (basic). This region is believed to be responsible for DNA binding and specificity.[1,2,7]

Signal-specific transcription factors are those that are responsive to extracellular signals.[1,2,7,12,13] The signals for which a corresponding responsive element has been found include steroid hormone receptors. The glucocorticoid receptor is the best understood. Glucocorticoid (and other steroids) can enter the cell through the cell membrane by simple diffusion. Within the cell, the glucocorticoid binds its receptor. The localization of free receptors is not entirely clear, but it is suggested that they are in equilibrium between the nucleus and cytoplasm. However, the hormone-receptor complex is always found in the nucleus. The receptor part of the complex can bind the GRE and enhance the transcription of the gene the GRE is associated with. The term enhance is not used loosely, because the GRE fits the definition of enhancer. It may be several kilobases upstream or downstream of the promoter which it affects. Other signal-specific transcription factors and their corresponding responsive elements fit the definition of a promoter module rather than an enhancer (e.g., the heatshock response element). The DNA-binding domain of the receptor that defines specifity for the GRE is a stretch of amino acids that form two contiguous Cys^2/Cys^2 fingers. This structural motif is similar to the zinc fingers described above, except that cysteines are found in place of the histidines. However, the two types of fingers are not interchangeable for functional capacity.[2]

C. TRANSCRIPTION ELONGATION

Control of the elongation phase of transcription is a recognized regulatory mechanism in prokaryotes.[1,2] To the contrary, no generalized mechanisms of elongation regulation have yet been discovered in higher eukaryotes. However, examples of apparent conditional transcription elongation blockage have been reported in some eukaryotic genes.[17] In several *Drosophila* promoters, transcription by RNA pol II is initiated and paused prior to induction of gene expression, suggesting that release of an arrested polymerase, as opposed to polymerase recruitment to a disengaged promoter, may be a key mechanism of transcriptional regulation.[17] In addition, RNA pol II apparently can be modified (e.g., phosphorylated) at stages subsequent to promoter recruitment and establishment of the pre-initiation complex. It has been suggested that such modifications may alter the enzyme's transcription elongation activity. However, it must be noted that mechanisms equivalent to bacterial **attenuation** and **antitermination** have not been clearly demonstrated in eukaryotic cells.[2]

D. TRANSCRIPTION TERMINATION

The signals for termination or the process involved are poorly understood.[1,2] The major difficulty in analyzing transcripts is the uncertainty about the actual site of termination. Since the 3′ end used for polyadenylation is generated by cleavage, the 3′ region of the primary transcript remains uncharacterized. It is possible that termination by RNA polymerase is not a specific process.[2] In some transcription units, termination occurs more than 1 kb downstream of the 3′ end of the mature mRNA. It has been suggested that instead of using specific terminator sequences, RNA pol II may stop synthesizing RNA within multiple sites located in rather long "terminator regions". However, the nature of these sites is unknown.[2]

III. POSTTRANSCRIPTIONAL REGULATION OF GENE EXPRESSION

A. RNA PROCESSING

The primary RNA transcript resulting from gene transcription (hnRNA) must be processed into mature mRNA before it can be exported to the cytoplasm for translation. Processing involves several steps. **Polyadenylation** and **capping** may occur before **RNA splicing,** since some hnRNAs with these modifications at the 3′ and 5′ ends have been found.[2]

Capping occurs very rapidly after the start of transcription, so that hnRNAs that are not yet terminated may be capped.[1,2] **Polyadenylation** involves two steps.[1,2] First, an endonuclease recognizes the six-base sequence AAUAAA, called the **polyadenylation signal,** which is found at the 3′ region of the RNA and then clips the RNA strand at a site 10 to 30 bases downstream. Second, a poly(A) polymerase adds the **poly(A) tail,** strings of 100 to 300 adenine nucleotides.

During **RNA splicing** the introns are precisely removed and the exons are joined into a contiguous mRNA molecule.[2] The introns of RNA pol II transcription products conform to the **GT-AG rule.** This term describes the very extreme bases of the introns, which appear to be the only region of high sequence conservation. The introns start with the dinucleotide GT and end with the dinucleotide AG. Note that the junctions have different sequences, so they define the intron directionally. They are named, proceeding from left (or 5′) to right (or 3′) along the intron, the **left** or **donor** splicing site or the **right** or **acceptor** site. Only these two short sequences and the sequence CTGAC, located approximately 30 bases upstream of the acceptor site, appear to be necessary for nuclear splicing. Interaction of these three sequences result in a **lariat** structure required for proper splicing. The process takes place in **spliceosomes,** large ribonucleoprotein particles that form on the intact precursor RNA, akin to ribosomal assembly on mRNA. **Alternative splicing,** the mechanism by which a given primary transcript can give rise to several different mRNAs, is the principal mechanism of regulation during RNA processing. *Trans*-acting factors that regulate alternative splicing have been described recently.[18]

There is virtually no information about the exit of mRNA to the cytoplasm.[2] It is possible that emigration is linked spatially to processing. The nuclear pores through which proteins enter the nucleus may be the route for mRNA export, but this is speculative.

B. mRNA TURNOVER

The half-life of a mRNA represents the balance between its transcription rate and its degradation rate or turnover rate. About 50% of the mRNA in a mammalian cell in tissue culture has a half-life of approximately 6 h.[2] The other 50% has an even longer half-life of 12 to 24 h. Of course, some individual mRNAs may have short half-lives, but, in general, the pool of mRNA in the eukaryotic cytoplasm is reasonably stable.

The turnover rate of eukaryotic mRNAs is determined by complex interactions that depend on the primary and secondary structure of the mRNA, its rate of translation, its location in the cytoplasm, and the levels and types of ribonucleases in the cell at a given moment. Although many examples of processes that affect mRNA turnover have been accumulating, the regulatory mechanisms involved are ill-defined. Two recent reviews are recommended.[19,20]

IV. TRANSLATIONAL REGULATION OF GENE EXPRESSION

The template that provides the specific sequence for the synthesis of a protein is a molecule of processed (or mature) mRNA. Protein synthesis occurs in the cytoplasm of eukaryotic cells. Each nucleotide triplet in the mRNA represents a **codon** which defines a specific amino acid. Translation is mediated by an "adaptor" molecule. This adaptor is **transfer RNA (tRNA),** a small RNA molecule (75 to 85 bases long) which has an intricate secondary structure called the **cloverleaf** form.[2]

Transfer RNA has two properties that are crucial for its function as the adaptor: (1) it is covalently linked to the one amino acid it represents, and (2) it contains the **anticodon,** a trinucleotide sequence that is complementary to the codon that defines its amino acid. The anticodon enables the tRNA to recognize the appropriate codon in mRNA by complementary base pairing.

When a tRNA is first **charged** with the amino acid corresponding to its anticodon, it becomes **aminoacyl-tRNA.** The amino acid is linked by an ester bond from its carboxyl group to a hydroxyl group on the ribose of the last base of the tRNA (which is always adenine). This reaction is catalyzed by **aminoacyl-tRNA synthetase.** There are at least 20 of these enzymes; each can recognize only a single amino acid and all the tRNAs with the corresponding anticodon. Once a tRNA has been charged, the amino acid has no further influence on its specificity, which is determined exclusively by the anticodon.[2]

The individual amino acids specified by codons-anticodons in the mRNA-tRNA are assembled into protein in the **ribosome,** a ribonucleoprotein composed of two subunits, one being approximately twice the size of the other. The subunits are named according to their approximate rate of sedimentation. Ribosomes of the eukaryotic cytoplasm are composed of **40S** and **60S** subunits. Each subunit consists of several proteins associated with a long RNA molecule, termed **ribosomal RNA (rRNA).**

To accomplish the sequential synthesis of a protein, the ribosome moves along the mRNA template. A ribosome attaches to mRNA near its 5′ end and moves along towards the 3′ end as each codon is translated into an amino acid. Unlike bacteria, in which the Shine-Dalgarno sequence provides a specific ribosome-attachment site, eukaryotic cytoplasmic ribosomes do not bind directly to a concensus sequence. Instead, the first feature they recognize is the methylated **cap** at the 5′ end. As the ribosome proceeds from the 5′ end to the 3′ end, the appropriate aminoacyl-tRNAs associate with it, donating their amino acids to the **nascent protein.**

After a ribosome starts moving downstream on the mRNA, another ribosome may initiate protein synthesis on that same mRNA molecule. The assembly in which several ribosomes are attached at different sites along the length of a mRNA at one time is called the **polyribosome** or **polysome.** Because virtually all eukaryotic mRNAs are **monocistronic** (i.e., they encode a single protein), a polysome is synthesizing the same protein throughout, but the protein is at different stages of completion. The average eukaryotic mRNA has fewer than 10 ribosomes attached at one time.

Like transcription, translation is divided into three distinct stages: initiation, elongation, and termination. As in transcriptional control, the best-documented examples of translational regulation are at the initiation level.[21–23] However, there is some evidence for translation regulation at the elongation stage.[24] The mechanisms involved in both types of regulation have been extensively reviewed recently, and only one example will be discussed here.[22–24] However, it is important to describe some of the details of the basic mechanisms necessary for translation because they may suggest the putative targets for its regulation.

A. TRANSLATION INITIATION

Initiation involves the reactions that precede formation of the peptide bond between the first two amino acids of the nascent protein.[2] It requires binding of the ribosome to the mRNA to form an initiation complex that includes the first aminoacyl-tRNA. Initiation is the slowest stage of translation.

First the 5′ end of the mRNA must be recognized and bound to the 40S subunit of the ribosome. This process is understood in outline. As in bacteria, there are multiple specific **initiation factors,** called **eIFs** (the "e" stands for eukaryotic) that assist throughout initiation. The number of eIFs is much greater than the number of IFs in bacteria.[2,21–23] The first step in the process is formation of the **ternary complex** containing the **initiator tRNA (Met-tRNA$_i$),** GTP, and **eIF2.** This complex then associates directly with a free 40S subunit. This reaction is independent of mRNA. In the meantime, the factor **eIF4A,** a multimer that includes a **cap binding protein (CPB),** recognizes the cap at the 5′ end of the mRNA, and unwinds any secondary structure that may exist in the first 15 bases of the mRNA. Shortly after, and with the help of other eIFs, the 40S ternary complex binds to the mRNA. Release of some of the eIFs, including eIF2, is required before the 60S subunit can attach and initiation is completed. Of all the eIFs, most is known about eIF2, which has been strongly implicated in translational regulation.[21,23]

An outstanding example of translational regulation is the inhibition of ferritin translation by an iron-responsive mRNA-binding protein (IRE-BP) in mammalian cells.[21,25] Ferritin is an iron-sequestering protein that is regulated by iron availability.[25] The regulation is accomplished by the binding of IRE-BP near the 5′ end of ferritin mRNA, which results in translation inhibition. The IRE-BP target site on the ferritin mRNA, called the iron-response element (IRE), is characterized by a conserved stem-and-loop secondary structure. The IRE-BP has been found to have aconitase activity. The fact that aconitase contains a labile iron-sulfhydryl complex suggests a likely mechanism by which IRE-BP may sense cellular iron levels. The reversible binding of IRE-BP to ferritin mRNA is thought to relate to the reversible oxidation of a critical sulfhydryl group. When iron is scarce (and thus ferritin is not needed to sequester it), the reduced form of IRE-BP binds the IRE and inhibits ferritin translation. On the other hand, when iron is abundant the oxidized IRE-BP dissociates from the IRE and ferritin translation is allowed to proceed.[21]

B. TRANSLATION ELONGATION

Elongation includes all the reactions from synthesis of the first peptide bond to addition of the last amino acid.[2] Amino acids are added to the nascent protein one at a time. This is the fastest stage of translation.

Only two tRNAs are involved in peptide bond formation at any moment. Each tRNA sits in a distinct site of the ribosome complex. The **A site** or **entry site** is the only site that can be entered by an incoming aminoacyl-tRNA. Prior to the entry of the tRNA, the codon representing the next amino acid to be added to the peptide is exposed at the A site. The codon representing the last amino acid to have been added lies in the **P site** or **donor site.** This site is occupied by **peptidyl-tRNA,** a tRNA carrying the entire nascent peptide synthesized up to this point.

When both sites are occupied, peptide bond formation occurs by a reaction in which the entire nascent peptide carried by the peptidyl-tRNA in the P site is transferred to the single amino acid on the aminoacyl-tRNA in the A site. This transfer results in a ribosome with a **deacylated-tRNA** (devoid of any amino acid) in the P site, and a new peptidyl-tRNA in the A site. This peptidyl-tRNA carries a nascent polypeptide that is longer by one amino acid than the peptidyl-tRNA that had been in the P site. The ribosome then moves one triplet downstream on the mRNA. Its movement transfers the deacylated-tRNA out of the P site and moves the peptidyl-tRNA into it, exposing the next codon to be translated at a newly vacant A site. A new aminoacyl-tRNA can then enter and the cycle is repeated.

As with initiation, several protein factors are necessary for all the steps described to occur. These are **eukaryotic elongation factors (eEFs).** A number of covalent modifications of these eEFs have been suggested as principal mechanisms for regulation of translation at the elongation stage.[24]

C. TRANSLATION TERMINATION

Termination includes release of the polypeptide chain and dissociation of the ribosome from the mRNA. Not much is known about the mechanisms involved except that a single **release factor (eRF)** is involved in eukaryotes.[2] Termination begins when the ribosome exposes a **termination codon** at the A site. Termination codons are not represented by tRNAs. Thus, the A site remains vacant. This state may be a signal for eRF to assist in cleavage of the polypeptide from the peptidyl-tRNA in the P site. However, this is highly speculative. In addition, it is not known yet how the ribosome dissociates from the mRNA. Since so little is known about translation termination, its role in regulation of gene expression is undefined.

REFERENCES

1. **Watson, J. D., Gilman, M., Witkowski, J., and Zoller, M.,** *Recombinant DNA,* 2nd ed., W. H. Freeman, New York, 1992, chaps. 2–4, 8, 9.
2. **Lewin, B.,** *Genes IV,* 4th ed., Cell Press, Cambridge, MA and Oxford University Press, New York, 1990, chap. 6–10, 25, 29–31, and glossary.
3. **Watson, J. D. and Crick, F. H. C.,** Molecular structure of nucleic acids: a structure for deoxyribose nucleic acid, *Nature,* 171, 737, 1953.
4. **Watson, J. D. and Crick, F. H. C.,** Genetical implications of the structure of deoxyribonucleic acid, *Nature,* 171, 964, 1953.
5. **Hall, S. S.,** Old school ties: Watson, Crick, and 40 years of DNA, *Science,* 259, 1532, 1993.
6. **Jacob, F. and Monod, J.,** Genetic regulatory mechanism in the synthesis of proteins, *J. Mol. Biol.,* 3, 318, 1961.
7. **Evan, G.,** Regulation of gene expression, *Br. Med. Bull.,* 47, 116, 1991.
8. **Breathnach, R., Mandell, J. L., and Chambon, P.,** Ovalbumin gene is split in chicken DNA, *Nature,* 270, 314, 1977.
9. **Jeffreys, A. J. and Flavell, R. A.,** The rabbit beta-globin gene contains a large insert in the coding sequence, *Cell,* 12, 1097, 1977.

10. **Berget, S. M., Moore, C., and Sharp, P.,** Spliced segments of the 5′ termini of adenovirus-2 mRNA, *Proc. Natl. Acad. Sci. U.S.A.,* 74, 3171, 1977.
11. **Chow, L. T., Gelinas, R. E., Broker, T. R., and Roberts, R. J.,** An amazing sequence arrangement at the 5′ ends of adenovirus-2 messenger RNA, *Cell,* 12, 1, 1977.
12. **Nussinov, R.,** Sequence signals in eukaryotic upstream regions, *Crit. Rev. Biochem. Mol. Biol.,* 25, 185, 1990.
13. **Latchman, D. S.,** Eukaryotic transcription factors, *Biochem. J.,* 270, 281, 1990.
14. **Roeder, R. G.,** The complexities of eukaryotic transcription initiation: regulation of preinitiation complex assembly, *Trends Biochem. Sci.,* 16, 402, 1991.
15. **Jackson, M. E.,** Negative regulation of eukaryotic transcription, *J. Cell Sci.,* 100, 1, 1991.
16. **Goodbourn, S.,** Negative regulation of transcriptional initiation in eukaryotes, *Biochim. Biophys. Acta,* 1032, 53, 1990.
17. **Spencer, C. A. and Groudine, M.,** Transcription elongation and eukaryotic gene regulation, *Oncogene,* 5, 777, 1990.
18. **Mattox, W., Ryner, L., and Baker, B. S.,** Autoregulation and multifunctionality among trans-acting factors that regulate alternative pre-mRNA processing, *J. Biol. Chem.,* 267, 19023, 1992.
19. **Peltz, S. W., Brewer, G., Bernstein, P., Hart, P. A., and Ross, J.,** Regulation of mRNA turnover in eukaryotic cells, *Crit. Rev. Eukaryotic Gene Expression,* 1, 99, 1991.
20. **Saini, K. S., Summerhayes, I. C., and Thomas, P.,** Molecular events regulating messenger RNA stability in eukaryotes, *Mol. Cell. Biochem.,* 96, 15, 1990.
21. **Kozak, M.,** Regulation of translation in eukaryotic systems, *Annu. Rev. Cell Biol.,* 8, 197, 1992.
22. **Nygard, O. and Nilsson, L.,** Translational dynamics, *Eur. J. Biochem.,* 191, 1, 1990.
23. **Rhoads, R. E.,** Regulation of eukaryotic protein synthesis by initiation factors, *J. Biol. Chem.,* 268, 3017, 1993.
24. **Ryazanov, A. G., Rudkin, B. B., and Spirin, A. S.,** Regulation of protein synthesis at the elongation stage, *FEBS Lett.,* 285, 170, 1991.
25. **Klausner, R. D. and Harford, J. B.,** *Cis-trans* models for post-transcriptional gene regulation, *Science,* 246, 870, 1989.

Chapter 20

Gene Expression and Osmoregulation in Bacteria

Harry H. Yim and Merna R. Villarejo

Table of Contents

I. UNIVERSAL STRATEGIES FOR OSMOTIC ADAPTATION

All organisms, from bacteria to plants and animals, respond to hyperosmotic stress by accumulating intracellular solutes. These solutes increase the osmotic strength of the cytoplasm to counterbalance the high extracellular osmolarity. In bacteria and plants, the cytoplasmic osmolytes provide positive turgor pressure; the outward force generated by a higher osmotic pressure inside the cell than outside. Turgor pressure is required by these cells to drive cell growth and division, but does not osmotically lyse the cells, because they have rigid cell walls. In animal cells, which lack a cell wall, osmolytes are accumulated to maintain cell volume. Thus, an important function of the osmotic stress response is to increase the osmotic strength of the cytoplasmic milieu so that cellular pressure or volume remains constant.

Because osmolytes may have to be accumulated to high concentrations, they must be compatible with biological processes. Potassium and/or sodium are the initial solutes accumulated by hyperosmotically stressed cells.[1] However, high concentrations of these inorganic cations usually perturb cellular functions, so they are used only as short-term osmolytes. The exceptions are the extreme halophilic archaebacteria, which require a high internal concentration of potassium.[2] Mammalian cells initially accumulate Na^+ and K^+, but the internal cation concentration falls with increasing sorbitol levels, suggesting that sorbitol replaces the cations.[3] This pattern of osmolyte accumulation is similar to that found in eubacteria, in which inorganic cations are gradually replaced by more compatible organic solutes.[4]

Throughout the phylogenetic kingdoms, only a few types of osmolytes are utilized as compatible solutes.[5–7] The three main classes of organic solutes are quaternary amines, amino acids, and polyols (in this paper we will refer to sugars also as polyols).[5] The quaternary amine, *N,N,N,*-trimethylglycine (glycine betaine) is the most effective osmoprotective compound for a variety of organisms,[8,9] and is distributed widely throughout nature. It is synthesized in the liver of animals and in the chloroplasts of plants.[10,11] The renal medullas of rats and rabbits take up betaine by a sodium-dependent transport system and accumulate it as a compatible solute to aid in volume regulation.[12] Glycine betaine is

0-8493-4448-4/94/$0.00+$.50

also present in human urine, and may contribute to bladder infections by enhancing *Escherichia coli*'s ability to grow in hypertonic urine.[13–15] The chemical properties of glycine betaine make it an ideal universal osmolyte. It is highly soluble, zwitterionic, and nonperturbing to enzymatic function; it even stabilizes enzymatic activity against the effects of high salt and urea.[5,16]

Amino acids constitute a major class of osmolytes, but only certain amino acids, such as taurine, ectoine, glutamate, and proline, are involved in cell volume regulation (see references cited in Reference 1). Others, such as lysine and arginine, are not accumulated in hyperosmotically stressed cells, perhaps because they are detrimental to enzyme function.[16] Proline appears to be an especially important osmotic stress amino acid, as it is accumulated by salt-stressed plants[5,17,18] and bacteria.[8] Like glycine betaine, it is relatively nonperturbing to enzyme activity.[6]

Increased polyol concentration correlates strongly with the osmotic tolerance of a diverse range of organisms, including bacteria, fungi, algae, and higher plants.[5,6] In mammalian renal cells, antidiuresis induces sorbitol accumulation mediated by the combined actions of increased synthesis from glucose and decreased sorbitol permeability (reviewed in Reference 3). Trehalose, a nonreducing disaccharide, is also synthesized from glucose and can be accumulated by *E. coli*,[19,20] as well as dehydrated yeast, bacterial spores, and nematodes (see references cited in Reference 21).

Because the hyperosmotic stress response is well conserved throughout nature, simple organisms are likely to be useful paradigms for more complex systems. Bacteria have the advantages of short generation time and relatively simple biochemistry and genetics. The enteric bacteria *E. coli* and *Salmonella typhimurium* are particularly appropriate to use as model organisms because they occur naturally in high-osmolarity environments, such as the colon, urine, and seawater. They also accumulate many of the same osmolytes as higher organisms (see above) and may sense the same signal to regulate osmotically controlled genes. In this review, we will focus on the regulatory mechanisms of osmotically responsive bacterial genes.

II. REGULATION OF TRANSCRIPTION IN BACTERIA

Bacteria, like all organisms, are capable of sensing and responding to a variety of chemical and physical changes in their environment. They generally adjust to these changes by selectively expressing genes that enable them to adapt to the new conditions. In the case of increased osmotic pressure, bacteria preadapted to high osmolarity become significantly more resistant to an extreme osmotic shock (2.5 *M* NaCl)[22] and survive better in high-osmolarity medium (our unpublished results). Chloramphenicol added during prestress sensitizes the cells to the extreme salt shock, indicating that the adaptation is dependent on *de novo* protein synthesis.[22] This observation suggests that genes induced by increased osmolarity are necessary for survival at high osmolarity. How these genes are regulated is central to understanding the overall mechanism of osmotic adaptation.

Bacterial gene expression is regulated primarily by altering transcription. To illustrate the steps involved in the stimulus-gene expression pathway, a generalized picture is shown in Figure 20–1. A stimulus (heat, nutrients, osmotic pressure change) is perceived by a sensor, which then transduces a signal to an effector protein. The signal modifies, covalently or noncovalently, the effector so that its DNA binding specificity is altered. In so doing, the effector alters the ability of RNA polymerase to express a target gene and transcription increases or decreases. Some regulatory systems do not contain all the parts of this model and other systems have two or more functions combined into a single protein. Various mechanisms used to regulate bacterial genes are described below.

A. GENERALIZED TRANSCRIPTION

In *E. coli* and *S. typhimurium*, transcription is mediated by RNA polymerase, a multisubunit complex composed of two functional components, the core polymerase and the sigma factor. The core polymerase is the catalytic component of the polymerase, while the sigma factor is responsible for promoter recognition. There are a number of different sigma factors, each recognizing a particular promoter sequence. *E. coli* promoters recognized by the major, housekeeping form of RNA polymerase, $E\sigma^{70}$, consist of two hexameric sequences located approximately 10 and 35 nucleotides upstream of the transcriptional start site (designated as +1).[23] The canonical $E\sigma^{70}$ promoter has a -35 sequence of TTGACA, a 17-nucleotide spacer and a -10 sequence of TATAAT (Figure 20–2). Generally, σ^{70} promoters that have less homology with the consensus sequence are transcriptionally less active than promoters that are closer to consensus. A positive activator protein can enhance transcription from a weak promoter or a repressor can inhibit transcription from a strong promoter.

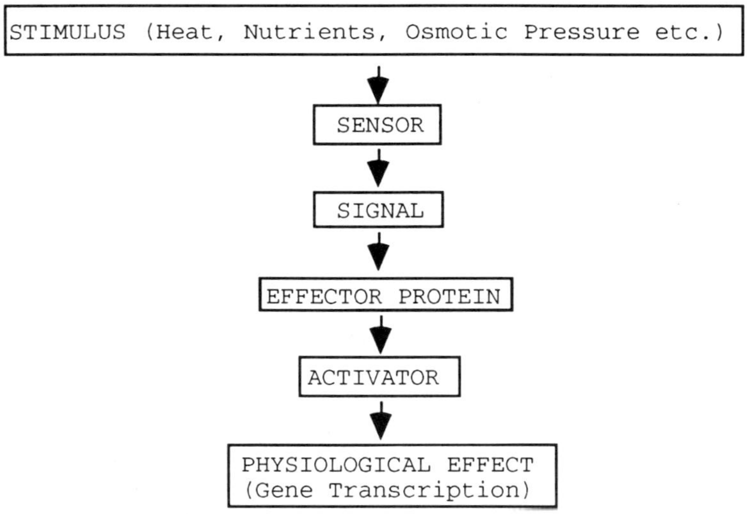

Figure 20–1 Stimulus-response pathway for gene expression.

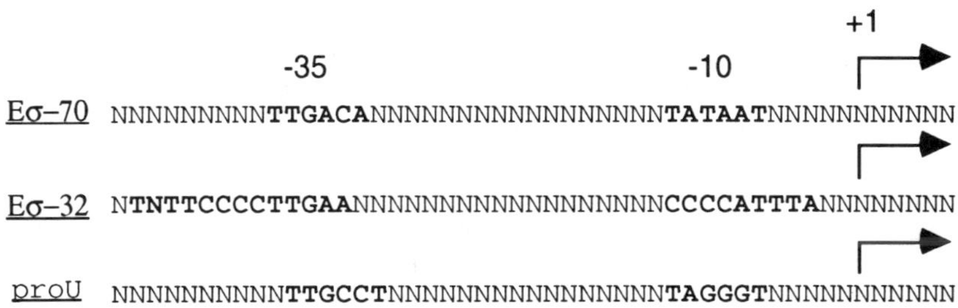

Figure 20–2 Sequences of the consensus *E. coli* Eσ^{70} and Eσ^{32} promoters, and the *proU* promoter.

B. POSITIVE AND NEGATIVE REGULATION

Positive regulation is usually mediated by an activator protein binding to the DNA just upstream of the -35 region.[24] It is thought that the activator facilitates transcription by making physical contact with RNA polymerase, stabilizing its interaction with the promoter region. Mechanistically, negative regulation is very similar to positive regulation, except that binding of the effector protein (repressor) to DNA interferes with RNA polymerase function. The classic example of repression is the lactose operon. LacI protein is the repressor that keeps expression of the lactose utilization genes low when lactose is absent. It does this by binding to a portion of the lac promoter, preventing binding of RNA polymerase. When lactose is available, LacI binds the inducer, changing conformation so that the repressor is released from the DNA and transcription of the *lac* operon can occur.

C. TWO-COMPONENT REGULATORY SYSTEMS

A number of genes are transcriptionally regulated by so-called "two-component regulatory networks" (for a review see Reference 25). These networks are generally composed of two proteins; a sensor protein and a regulator effector protein. The sensors and regulators share a high degree of amino acid sequence homology with other sensors and regulators, suggesting these systems constitute a conserved family of proteins. The C-terminal end of the sensor protein is conserved and encodes a histidine kinase, while the N-terminal is somewhat variable and is thought to be the sensory portion of the protein.[25] The stimulus-regulatory response pathway begins by the sensor protein detecting a stimulus, autophosphorylating itself, and transferring the phosphate group to the regulator. The regulator, now activated, can bind to specific DNA sequences and activate transcription.

D. GLOBAL REGULATION

Genes expressed in response to a common environmental stimulus constitute a stimulon. If the mechanism of gene activation involves a common regulatory protein, the network is termed a regulon. The heat shock regulon is a good example of this type of global regulation.

Virtually all organisms exposed to a heat treatment induce a set of proteins that enable the organism to adapt and survive the temperature increase. Amino acid sequence homology of the heat shock proteins has revealed extensive conservation throughout the phylogenetic kingdoms (see References in 26). Many of the biological responses involved with heat stress were first identified in *E. coli,* illustrating the usefulness of this organism as a model system.

In *E. coli,* expression of the heat shock genes is dependent upon the regulatory protein RpoH, the so-called heat shock sigma factor (also called sigma-32) (reviewed in Reference 27). Normally, cellular levels of RpoH are low and the transcription of heat shock genes is low. Heat stress increases the concentration of RpoH, which complexes with RNA polymerase and transcribes genes with heat shock promoter sequences,[28,29] which have a very different consensus sequence than $E\sigma^{70}$ promoters[30] (Figure 20–2). The heat shock genes are dispersed throughout the *E. coli* genome and their products are involved in diverse functions, such as RNA transcription, protein degradation, DNA replication, protein synthesis, and protein folding.[31]

It is apparent that the key to regulating the heat shock response is to control RpoH levels. But what controls RpoH levels and what is the signal? An interesting hypothesis has been put forth by Van Bogelen and Neidhardt that suggests the ribosome is the sensor for temperature and the state of the ribosome may itself be the signal.[32] The hypothesis is based upon a number of correlations between translational capacity of the cell and the expression of heat shock proteins. For example, a number of chemical agents induce some heat shock proteins and also affect translation.[32] Definitive proof of the proposal has not been presented, but the idea of a large, multisubunit enzyme regulating gene expression is intriguing and may be similar to a proposed model for regulating osmotically controlled genes in *E. coli* (see below).

III. REGULATION OF THE OSMOTIC STIMULON

An increase in environmental osmotic pressure triggers a number of physiological and biochemical changes in *E. coli* (for a review see Reference 4). Initially, water diffuses out of the cell, resulting in plasmolysis and loss of turgor pressure. To restore turgor, the cell increases the concentration of cytoplasmic solutes, primarily by accumulating potassium ions[20,33] and synthesizing glutamate to act as the charge-neutralizing counterion.[20,34,35] However, as stated above, K^+ is deleterious to many cellular processes and is replaced with compatible solutes. These organic osmolytes include glycine betaine, proline, and trehalose (reviewed in Reference 4). Many of the genes preferentially expressed during hyperosmotic stress encode proteins that enable the cell to accumulate these solutes.

Two-dimensional gel electrophoresis shows that synthesis of more than 22 proteins is induced by hyperosmotic stress.[22,36] In addition, 11 or 12 genetic loci have been identified as being regulated by elevated osmolarity (see references cited in References 4, 37, and 38). How is the expression of these loci regulated? Recent evidence suggests at least three different mechanisms are used to control osmotically responsive genes.

A. EFFECTOR/SENSOR REGULATION

Historically, the first osmotically regulated genes studied in *E. coli* were OmpC and OmpF, the two major porin proteins in the outer membrane. They are coordinately regulated by a two-component regulatory system made up of the sensor protein EnvZ and regulatory protein OmpR (for a review see Reference 39). At low osmolarity, OmpF is preferentially expressed and OmpC is repressed; at high osmolarity, OmpC is induced and OmpF repressed. EnvZ is a transmembrane protein which somehow senses the osmotic signal and modulates OmpR activity, probably through phosphorylation.[40] It is thought that phosphorylated OmpR induces the expression of OmpC while repressing the expression of OmpF. What EnvZ senses and how it modulates its kinase activity is unclear.

Another operon regulated by a two-component regulatory system is *kdpABC,* which encodes a high-affinity potassium transport system. *kdp* expression is activated under conditions requiring additional K^+, such as reduced turgor pressure and potassium starvation.[41] The regulatory proteins for the *kdp* operon are KdpD and KdpE,[42] whose predicted amino acid sequences are homologous to the two-component family of proteins.[43] The KdpD protein is believed to be the membrane-bound sensor protein and KdpE is thought to be the effector protein that stimulates transcription at the *kdpABC* promoter. KdpD has autophosphorylation activity and can phosphorylate KdpE,[44] while KdpE binds to a site upstream of the *kdpABC* promoter required for proper *kdp* transcription.[45] The mechanism by which KdpD senses the reduced turgor pressure is unknown. However, Epstein has proposed a model by which KdpD senses changes in the distance between the inner and outer membranes.[41]

B. DUAL-SIGNAL GENES

When bacterial cells run out of nutrients, cell division stops and their metabolism adapts to the starvation conditions. These so-called stationary-phase cells are resistant to a variety of environmental stresses, including high osmolarity.[22,46,47] Essential for resistance, and the expression of proteins specific to stationary phase, is the putative sigma factor *rpoS (katF)*.[46] Of the 32 proteins induced during stationary phase, 5 are also induced by osmotic stress during active growth.[22] This type of cross-regulation between osmotic stress and growth phase signals has also been demonstrated at the level of gene expression; specifically with *osmB*, the *otsBA* operon, *osmA*, and *osmY*.[38,48,49] Recently, it has been reported that cells with mutations in *rpoS* are unable to induce *osmY* or the *otsBA* operon,[49–50a] and express *osmA* and *osmB* at significantly reduced levels during stationary-phase growth.[49] How *rpoS* regulates these genes is unknown, but the fact that *rpoS*-dependent promoters do not share sequence homologies argues against direct interaction of RpoS and promoter DNA.

C. *proU* REGULATION

Among the genes induced by high osmotic conditions, probably the best-characterized locus is the *proU* operon. Expression is independent of *ompR/envZ*[51,52] and *rpoS*.[50] Mutations that affect *proU* expression are either *cis*-acting mutations in the promoter region,[53–57] or nonspecific mutations with pleiotrophic effects.[53–55] Therefore, either no regulatory protein exists or its loss is lethal. *proU* expression is controlled at the transcriptional level[4] and *in vitro*, requires the housekeeping sigma factor, σ^{70}.[58] As mentioned above, the canonical σ^{70} promoter consists of a -35 sequence of TTGACA, a 17-nucleotide spacer, and a -10 sequence of TATAAT.[23] However, the *proU* promoter shares only marginal homology with the *consensus* σ^{70} promoter sequence. The *proU* promoter in *E. coli* has a -35 sequence of TTGCCT, a 16-nucleotide spacer, and a -10 sequence of TAGGGT[55,56,59–61] (Figure 20–2). Because of the poor homology between the *proU* promoter and the consensus σ^{70} promoter, one would predict it to be poorly transcribed. At normal osmolarity, transcription is indeed very low. However, when the osmolarity is increased, *proU* is induced more than 100-fold and is expressed at quite high levels.[62,63] What occurs during the hyperosmotic stress to make the *proU* promoter so transcriptionally active? To date, three models have been proposed to explain *proU* induction.

1. Supercoiling Model

Higgins et al. first claimed that there was a correlation between the superhelicity of a reporter plasmid and *proU* expression.[54] Increasing the osmolarity of the medium increases both *proU* expression and the superhelical density of reporter plasmids. The superhelical density of DNA is controlled by two enzymes, topoisomerase and gyrase. Mutations in topoisomerase I *(topA)*, will affect plasmid supercoiling and *proU* expression.[54] Another locus, named *osmZ*, also alters *proU* expression and supercoiling of reporter plasmids when mutated.[54] *osmZ* encodes a “histone-like” DNA binding protein also known as H-NS(H1a).[64,65] Because of its association with DNA, H-NS has been implicated in modulating DNA conformation, perhaps by constraining supercoils. These correlations were formulated into a model that suggested changes in DNA supercoiling regulates *proU* transcription.

If increased supercoiling is the primary signal for *proU* induction, there should be a cause and effect relationship between superhelical density and *proU* expression. Many other environmental factors, such as anaerobiosis, heat stress, and extreme pH, which increase supercoiling,[66] should also induce *proU*. However, they do not. Anaerobiosis does not induce *proU*, even when cells were grown in absolutely anaerobic conditions.[66,67] Plasmids isolated from cells grown at high osmolarity do have higher superhelical density than plasmids isolated from cells grown in standard medium.[67] The supercoiling theory would predict that the more supercoiled plasmids would be better templates for expression. However, these plasmids were no more adept at *proU*-directed transcription and coupled transcription/translation in vitro.[67,68] Finally, *proU* expression was not affected by a gyrase mutation *in vivo* or by gyrase inhibitors *in vitro*.[67] Collectively, these results cast serious doubt on the idea that supercoiling directly regulates *proU* expression. Rather, the supercoiled state of DNA may influence general transcriptional ability.[69]

2. Downstream Silencer Model

In addition to the *proU* promoter sequence, DNA sequences downstream of the promoter appear to modulate the expression of *proU* in both *E. coli* and *S. typhimurium*.[56,60,70] Deletions of the region extending approximately 200 to 300 base pairs into the coding region of the first structural gene increase the basal level of *proU* expression without significantly altering the induced level.[56,70] How this region represses *proU* expression at standard osmolarity is unknown. One hypothesis implicates this region in supercoiling changes in response to osmolarity.[56] In contrast, Overdier et al. hypothesize that this region is a binding site for a regulatory protein that sequesters the *proU* promoter under normal osmotic conditions.[70,71] These models have yet to be tested.

3. Activation by Intracellular Potassium

The final hypothesis regarding *proU* regulation was originally proposed by Higgins et al.[72] and Epstein.[41] They postulated that intracellular potassium concentration acts as a signal for *proU* activation. As mentioned, cells rapidly accumulate potassium,[20] and synthesize glutamate[19,20,35] in response to hyperosmotic stress. If the K^+ uptake is prevented by a mutation in the uptake systems or by growth in K^+-limiting medium, a *proU-lacZ* fusion cannot be induced by increased osmolarity.[73] Another *in vivo* correlation between K^+ and *proU* expression is observed when glycine betaine is added to hyperosmotically stressed cells. Glycine betaine replaces the potassium within the cells,[74] and also reduces *proU* expression.[63,73]

To determine if there is a cause and effect relationship between high K^+ concentration and *proU* expression, our laboratory investigated *proU* expression *in vitro.* At first, we utilized an S-30 coupled transcription/translation system.[75] The S-30 extract contains all the necessary machinery (enzymes, cofactors, etc.) needed for transcription and translation, but lacks a cell envelope and DNA. When an exogenous source of DNA is added, genes encoded on that template can be expressed by the S-30 extract. Gene expression can be visualized by adding radioactive ^{35}S-methionine to the mixture and examining the protein products on a SDS-polyacrylamide gel. The S-30 extract, in standard buffers, was programmed with a plasmid carrying the ampicillin resistance gene (β-lactamase; *bla*) and a *proU-lacZ* fusion gene. The net result was the same as observed *in vivo* under isotonic conditions; β-lactamase was expressed strongly, and the *proU-lacZ* fusion expressed only slightly.[75] When potassium glutamate was added, mimicking the high osmolarity state, expression of *bla* decreased and *proU-lacZ* expression increased.[75] *proU* induction was specific for potassium glutamate,[75] the ion pair normally accumulated by hyperosmotically stressed *E. coli.* Confirmation of our results was independently reported by other investigators.[58] These findings reinforce the idea that *proU* regulation is independent of turgor pressure or transmembrane signaling, since the S-30 extract is a cell-free system lacking a membrane, and indicate that *proU* expression is responsive to intracellular ion concentrations.

How does potassium glutamate induce *proU* expression? One hypothesis is that there is a positive regulatory protein in the S-30 extract that is activated by elevated potassium concentrations. To test this model, we used a defined *in vitro* transcription system consisting of buffer, RNA polymerase, ribonucleotides, RNase inhibitor, and a circular plasmid containing a *proU* promoter-reporter gene fusion. Presumably the putative K^+-activated factor from the S-30 extract would have to be added back to observe regulated transcription from *proU.* In standard buffer, *bla* transcripts were easily detected but *proU* transcripts were barely detectable[68] (Figure 20–3). When potassium glutamate was added, *bla* expression was repressed, while *proU* transcription was strongly induced[68] (Figure 20–3). Recall that the system was not supplemented with any S-30 extract and is nonetheless capable of specifically inducing *proU* transcription by elevating K^+. Like the S-30 system, *proU* induction was specific for potassium, and in particular, the ion pair potassium glutamate.[68] This simple, *in vitro* transcription assay successfully reproduced the results of the S-30 experiments. These results strongly suggest a novel and simple mechanism to regulate gene expression: physiological changes in the intracellular ionic composition can specifically alter the transcription of a gene.

Having established that potassium glutamate stimulates transcription from the *proU* promoter, we are utilizing a number of footprinting reagents to detect changes in RNA polymerase-DNA interactions in the presence of K^+. Preliminary results with $KMnO_4$ footprinting reveals that open complex formation at the *proU* promoter is stimulated by potassium glutamate. Isomerization to an open promoter complex is characterized by separation of the DNA strands to a "transcription bubble": 15 to 20 base pairs of DNA near the startsite of transcription become single stranded. The reagent potassium permanganate ($KMnO_4$), reacts and modifies unpaired thymidine nucleotides which are detected by primer extension. The extent of open complex formation is directly related to the potassium glutamate concentration, contributing to enhanced transcription.

The downstream negative regulatory site of the *proU* promoter could also be affected by potassium *in vivo.* A putative DNA binding protein could be stripped off by high potassium, freeing the regulatory region and allowing transcription to occur. A possible candidate for such a regulatory protein is

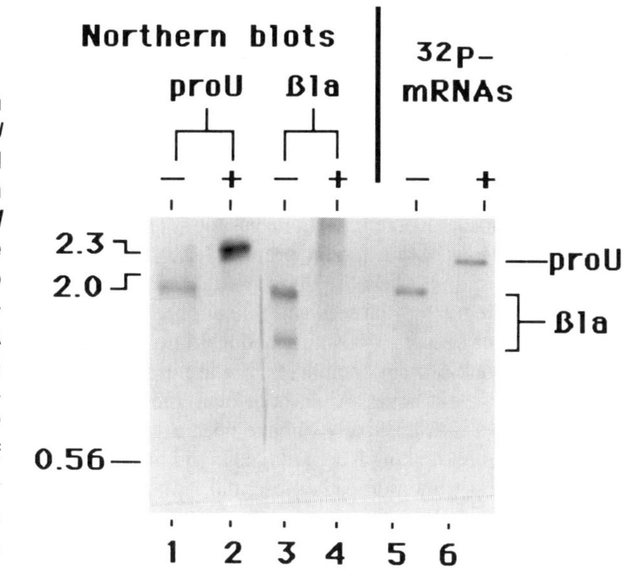

Figure 20–3 Effect of potassium glutamate on *in vitro* transcription of *proU* and *bla*. A plasmid containing *proU* and *bla* was added to a purified transcription system with (+) and without (-) 0.3 *M* potassium glutamate. Transcripts were electrophoresced and transferred to nylon membrane. To detect the transcripts, the RNA were hybridized to DNA probes specific to *proU* (lanes 1 and 2) and *bla* (lanes 3 and 4), or the transcripts were labeled directly with ^{32}P (lanes 5 and 6). The migration of ^{32}P-labeled DNA molecular weight markers is shown on the left. (From Prince, W. S. and Villarejo, M. R., *J. Biol. Chem.*, 265, 17673, 1990. With permission.)

OsmZ, the "histone-like" DNA binding protein. Mutations in *osmZ* increase *proU* expression in normal-osmolarity medium approximately 10- to 20-fold.[54] The same expression occurs when the downstream silencer region is deleted,[56,70] raising the possibility that *osmZ* and the silencer sequences are interrelated.

4. Model for Potassium Activation of *proU* Transcription

Based on the correlative and *in vitro* data, we propose the following scenario for K^+-activated transcription of *proU*. Upon hyperosmotic shock, water diffuses out of the cell, causing the cell to plasmolyze. Adaptation begins with the accumulation of potassium and the synthesis of glutamate, elevating the internal osmolarity of the cell. The increased concentration of potassium glutamate stimulates the transcription of *proU* by enhancing open complex formation and possibly relieving inhibition at the downstream site. *proU* transcripts are translated and the glycine betaine transport system is expressed. If glycine betaine is present in the medium, it is accumulated and replaces potassium glutamate. As the concentration of the signal compound, potassium glutamate, drops, the expression of *proU* will be reduced, providing the feedback response. If no glycine betaine is present, *proU* will be expressed at high steady-state levels. The regulation of *proU* is likely to involve both the promoter and the downstream regulatory element, each contributing approximately 10-fold to the induction and thereby conferring 100-fold total inducibility to the gene.

D. A POTASSIUM REGULON?

Is the *proU* regulatory mechanism unique or is K^+ stimulation of transcription a general mechanism for sensing elevated osmolarity? The *osmC* gene is osmotically regulated and encodes a 14,468-Da protein.[76] We have found that *osmC* expression is stimulated by K^+ glutamate in the S-30 system[67] and in the transcription system (our unpublished results). Therefore, *in vitro,* expression of another osmotically regulated gene is influenced by K^+ glutamate.

The *bet* and *proP* loci are also possible candidates for this type of regulation. The function of the *bet* operon is the same as that of the *proU* operon, accumulation of glycine betaine during hyperosmotic stress. *proP* transports proline, and some glycine betaine, under stress conditions. It is tempting to speculate that all these loci would use the same signal. Of course, the regulation of the *bet* genes is probably more complex than *proU,* as it involves a repressor gene, *betI*[77] and responds to additional stimuli.[78] Like *proU, proP* is quickly induced by hyperosmotic stress, although its induction ratio is only tenfold.[79] The gene encoding *proP* has been cloned and sequenced, revealing a putative promoter with many features that are strikingly similar to the *proU* promoter sequence.[84] A potentially significant difference between the *proP* and *proU* regulatory regions is that the *proP* regulatory region apparently lacks the downstream silencer site. The tenfold induction from the *proP* promoter is consistent with the approximately tenfold induction at the *proU* promoter.

It is apparent that the genes comprising the osmotic stimulon are regulated by a variety of factors, ranging from two component systems, negative regulatory proteins, an alternate sigma factor, repressor DNA sequences, and ionic strength. However, common to all these models is RNA polymerase. It is possible that RNA polymerase itself may be the osmotic regulatory protein and the sensor of osmotic stress through a conformational change dependent on potassium concentration. This idea is analogous to the model proposed for the heat shock stimulon, where the ribosome acts in the dual roles of regulatory protein and sensor of temperature.[31] Both sensors (ribosomes and RNA polymerase) are multiprotein complexes that are associated with nucleic acids. Also, both heat shock and osmotic stress were likely to have been a part of primordial life, making it advantageous to sense these stimuli by proteins already present in the cell.

Another way potassium concentration could regulate osmotically responsive genes is by altering the specificity or activity of their regulatory proteins. For example, the EnvZ sensor protein has both a kinase and phosphatase activity that modifies the activity of OmpR by phosporylation. However, how EnvZ senses changes in osmolarity is not known. Many have proposed that the membrane-spanning region and a domain external to the cell are important for sensing changes in osmolarity. Alternatively, EnvZ could be sensitive to the intracellular potassium concentration, which could affect its kinase or phosphatase activity. Consistent with this model is the effect of glycine betaine on OmpF expression. Normally, OmpF is repressed at high osmolarity. However, this repression can be overcome by the addition of glycine betaine,[63] which replaces the potassium in the cell. The activity of glutamate dehydrogenase, which synthesizes glutamate, and trehalose synthase, responsible for trehalose synthesis, are stimulated at higher K^+ concentrations.[34,80] It is worth noting that these enzymes synthesize solutes required by hyperosmotically stressed cells. Thus, K^+ concentration is known to affect the activity of proteins important for adaptation to elevated osmotic pressure in the environment.

Could K^+ be involved in osmotic regulation of *rpoS*-dependent genes? The growth-phase dependent activation is probably due to increased levels of RpoS, since during carbon starvation an *rpoS::lacZ* fusion is induced approximately fivefold.[81] However, the basal concentration of RpoS is not altered by medium osmolarity.[50a] Therefore, the osmotic stimulation of genes within the *rpoS*-dependent regulon may be a result of RpoS modification to an activated state under conditions of increased potassium concentration. Alternatively, it has been proposed that growth rate is the signal. Severe hyperosmotic stress reduces growth rate, which may mimic the stationary phase. However, not all *rpoS*-dependent genes are induced by hyperosmotic stress,[50a] suggesting there are subgroups of *rpoS*-dependent genes. The protein products of the osmotic subgroup of genes contribute to adaptation to osmotic and growth rate stresses. For example, the *otsBA* operon, which encodes trehalose synthase, is induced by osmolarity and growth phase,[50] trehalose is both a compatible solute for *E. coli*[19] and involved in stationary-phase thermotolerance.[49]

Ionic strength may also contribute to gene regulation in mammalian cells. In rabbit renal inner medulla cells, sorbitol is accumulated during NaCl stress. The enzyme responsible for increasing sorbitol is aldose reductase, which converts glucose into sorbitol. When NaCl increases in the growth medium, aldose reductase transcription increases, resulting in increased aldose reductase synthesis.[82] Like *proU* transcription, aldose reductase expression correlates with cellular ionic strength, particularly with the sum of cellular sodium plus potassium concentrations.[83] Thus, ionic strength may play an important role in activating transcription of osmotic stress genes not only in bacteria but across the phylogenetic kingdoms.

REFERENCES

1. **Chamberlin, M. E. and Strange, K.,** Anisosmotic cell volume regulation: a comparative view, *Am. J. Physiol.,* 257, C159, 1989.
2. **Russell, N. J.,** Adaptive modifications in membranes of halotolerant and halophilic microorganisms, *J. Bioenerg. Biomembr.,* 21, 93, 1989.
3. **Burg, M.,** Molecular basis for accumulation of compatible osmolytes in mammalian cells, in *Water and Life,* Somero, G. N., Osmond, C. B., and Bolis, C. L., Eds., Springer-Verlag, Berlin, 1992, 33.
4. **Csonka, L. N.,** Physiological and genetic responses of bacteria to osmotic stress, *Microbiol. Rev.,* 53, 121, 1989.
5. **Yancey, P. H., Clark, M. E., Hand, S. C., Bowlus, R. D., and Somero, G. N.,** Living with water stress: evolution of osmolyte systems, *Science,* 217, 1214, 1982.
6. **Borowitzka, L. J.,** Glycerol and other carbohydrate osmotic effectors, in *Transport Processes, Iono- and Osmoregulation,* Gilles-Baillien, R. G. A. M., New York, 1985, 437.
7. **Somero, G. N.,** Adapting to water stress: convergence on common solutions, in *Water and*

Life, Somero, G. N., Osmond, C. B., and Bolis, C. L., Eds., Spring-Verlag, Berlin, 1992, 33.
8. **Le Rudulier, D., Strøm, A. R., Dandekar, A. M., Smith, L. T., and Valentine, R. C.,** Molecular biology of osmoregulation, *Science*, 224, 1064, 1984.
9. **Petronini, P. G., De Angelis, E. M., Borghetti, P., Borghetti, A. F., and Wheeler, K. P.,** Modulation by betaine of cellular responses to osmotic stress, *Biochem. J.*, 282(1), 69, 1992.
10. **Hanson, A. D. H., May, M. A., Grumet, R., Bode, J., Jamieson, G. J., and Rhoads, D.,** Betaine synthesis in chenopods: localization in chloroplasts, *Proc. Natl. Acad. Sci. U.S.A.*, 82, 3678, 1985.
11. **Weigel, P., Weretilnik, E. A., and Hanson, A. D.,** Betaine aldehyde oxidation by spinach chloroplasts, *Plant Physiol.*, 82, 753, 1986.
12. **Bagnasco, S., Balaban, R., Fales, H., Yang, Y.-M., and Burg, M. B.,** Predominant osmotically active organic solutes in rat and rabbit renal medullas, *J. Biol. Chem.*, 261, 5872, 1986.
13. **Chambers, S. T. and Kunin, C. M.,** Isolation of glycine betaine and proline betaine from human urine. Assessment of their role as osmoprotective agents for bacteria and kidney, *J. Clin. Invest.*, 79, 731, 1987a.
14. **Chambers, S. T. and Kunin, C. M.,** Osmoprotective activity for *E. coli* in mammalian renal inner medulla and urine: correlation of glycine and proline betaines and sorbitol with response to osmotic stress, *J. Clin. Invest.*, 80, 1255, 1987b.
15. **Kunin, C. M., Hua, T. H., White, L. V., and Villarejo, M.,** Growth of *Escherichia coli* in human urine. Role of salt tolerance and accumulation of glycine betaine, *J. Infect. Dis.*, 166, 1311, 1992.
16. **Bowlus, D. R. and Somero, G. N.,** Solute compatibility with enzyme function and structure: rationales for the selection of osmotic agents and end-products of anaerobic metabolism in marine invertebrates, *J. Exp. Zool.*, 208, 137, 1979.
17. **Fourege, F., Le Rudulier, D., and Streeter, J. G.,** Effects of salt stress on amino acid, organic acid, and carbohydrate composition of roots, bacteroids, and cytosol of alfalfa (*Medicago, sativa* L.), *Plant Physiol.*, 96, 1228, 1991.
18. **Stewart, G. R. and Lee, J. A.,** The role of proline accumulation in halophytes, *Planta*, 120, 1974.
19. **Larsen, P. I., Sydnes, L. K., Landfald, B., and Strom, A. R.,** Osmoregulation in *Escherichia coli* by accumulation of organic osmolytes: betaines, glutamic acid, and trehalose, *Arch. Microbiol.*, 147, 1, 1987.
20. **Dinnbier, U., Limpinsel, E., Schmid, R., and Bakker, E. P.,** Transient accumulation of potassium glutamate and its replacement by trehalose during adaptation of growing cells of *Escherichia coli* K-12 to elevated sodium chloride concentrations, *Arch. Microbiol.*, 150, 348, 1988.
21. **Van Laere, A.,** Trehalose, reserve and/or stress metabolite?, *FEMS Microbiol. Rev.*, 63, 201, 1989.
22. **Jenkins, D. E., Chaison, S. A., and Matin, A.,** Starvation-induced cross protection against osmotic challenge in *Escherichia coli*, *J. Bacteriol.*, 172, 2779, 1990.
23. **Hawley, D. K. and McClure, W. R.,** Compilation and analysis of *Escherichia coli* promoter DNA sequences, *Nucleic Acids Res.*, 11, 6745, 1983.
24. **Gralla, J. D.,** Transcriptional control — lessons from an *E. coli* promoter data base, *Cell*, 66, 415, 1991.
25. **Stock, J. B., Ninfa, A., and Stock, A. M.,** Protein phosphorylation and regulation of adaptive responses in bacteria, *Microbiol. Rev.*, 53, 450, 1989.
26. **Watson, K.,** Microbial stress proteins, *Adv. Microbial Physiol.*, 31, 183, 1990.
27. **Neidhardt, F. C., VanBogelen, R. A., and Vaughn, V.,** The genetics and regulation of the heat shock response, *Annu. Rev. Gen.*, 18, 295, 1985.
28. **Grossman, A. D., Erikson, J. W., and Gross, C. A.,** The *htpR* gene product of *E. coli* is a sigma factor for heat-shock promoters, *Cell*, 38, 383, 1984.
29. **Bloom, M., Skelly, S., VanBogelen, R., Neidhardt, F. C., Brot, N., and Weissbach, H.,** In vitro effects of the *Escherichia coli* heat shock regulatory protein on expression of heat shock genes, *J. Bacteriol.*, 166, 380, 1986.
30. **Cowing, D. W., Bardwell, J. C. A., Craig, E. A., Woolford, C., Hendrix, R. W., and Gross, C. A.,** Consensus sequence for *Escherichia coli* heat shock promoters, *Proc. Natl. Acad. Sci. U.S.A.*, 82, 2679, 1985.
31. **Neidhardt, F. C. and VanBogelen, R. A.,** Heat shock response, in *Escherichia coli and Salmonella typhimurium: Cellular and Molecular Biology*, Vol. 2, Neidhart, F. C., Ingraham, J. L., Low, K. B., Magasanik, B., Schaechter, M., and Umbarger, H. E., Eds., Washington, D. C., 1987, 1334.

32. **Van Bogelen, R. A. and Neidhardt, F. C.,** Ribosomes as sensors of heat and cold shock in *Escherichia coli, Proc. Natl. Acad. Sci. U.S.A.,* 87, 5589, 1990.
33. **Epstein, W. and Schultz, S. G.,** Cation transport in *Escherichia coli.* V. Regulation of cation content, *J. Gen. Physiol.,* 49, 221, 1965.
34. **Measures, J. C.,** Role of amino acids in osmoregulation of non-halophilic bacteria, *Nature,* 257, 398, 1975.
35. **Richey, B., Cayley, D. S., Mossing, M. C., Kolka, C., Anderson, C. F., Farrar, T. C., and Record, M. T., Jr.,** Variability of the intracellular ionic environment of *Escherichia coli.* Differences between *in vitro* and *in vivo* effects of ion concentrations on protein-DNA interactions and gene expression, *J. Biol. Chem.,* 262, 7157, 1987.
36. **Botsford, J. L.,** Analysis of protein expression in response to osmotic stress in *Escherichia coli, FEMS Microbiol. Lett.,* 72, 355, 1990.
37. **Gutierrez, C., Barondess, J., Manoil, C., and Beckwith, J.,** The use of transposon Tn*phoA* to detect genes for cell envelope proteins subject to a common regulatory stimulus. Analysis of osmotically regulated genes in *Escherichia coli, J. Mol. Biol.,* 195, 289, 1987.
38. **Yim, H. H. and Villarejo, M.,** *osmY,* a new hyperosmotically inducible gene, encodes a periplasmic protein in *Escherichia coli, J. Bacteriol.,* 174, 3637, 1992.
39. **Mizuno, T. and Mizushima, S.,** Signal transduction and gene regulation through the phosphorylation of two regulatory components: the molecular basis for the osmotic regulation of the porin genes, *Mol. Microbiol.,* 4, 1077, 1990.
40. **Igo, M. M. and Silhavy, T. J.,** EnvZ, a transmembrane environmental sensor of *Escherichia coli* K-12, is phosphorylated in vitro, *J. Bacteriol.,* 170, 5971, 1988.
41. **Epstein, W.,** Osmoregulation by potassium transport in *Escherichia coli, FEMS Microbiol. Rev.,* 39, 73, 1986.
42. **Polarek, J. W., Williams, G., and Epstein, W.,** The products of the *kdpDE* operon are required for expression of the Kdp ATPase of *Escherichia coli, J. Bacteriol.,* 174, 2145, 1992.
43. **Walderhaug, M. O., Polarek, J. W., Voelkner, P., Daniel, J. M., Hesse, J. E., Altendorf, K., and Epstein W.,** KdpD and KdpE, proteins that control expression of the *kdpABC* operon, are members of the two-component sensor-effector class of regulators, *J. Bacteriol.,* 174, 2152, 1992.
44. **Nakashima, K., Sugiura, A., Momoi, H., and Mizuno, T.,** Phosphotransfer signal transduction between two regulatory factors involved in the osmoregulated *kdp* operon in *Escherichia coli, Mol. Microbiol.,* 6, 1777, 1992.
45. **Sugiura, A., Nakashima, K., Tanaka, K., and Mizuno, T.,** Clarification of the structural and functional features of the osmoregulated *kdp* operon of *Escherichia coli, Mol. Microbiol.,* 6, 1769, 1992.
46. **McCann, M. P., Kidwell, J. P., and Matin, A.,** The putative σ factor KatF has a central role in development of starvation-mediated resistance in *Escherichia coli, J. Bacteriol.,* 173, 4188, 1991.
47. **Matin, A.,** The molecular basis of carbon-starvation-induced general resistance in *Escherichia coli, Mol. Microbiol.,* 5, 3, 1991.
48. **Jung, J. U., Gutierrez, C., Martin, F., Ardourel, M., and Villarejo, M.,** Transcription of *osmB,* a gene encoding an *Escherichia coli* lipoprotein, is regulated by dual signals. Osmotic stress and stationary phase, *J. Biol. Chem.,* 265, 10574, 1990.
49. **Hengge-Aronis, R., Klein, W., Lange, R., Rimmele, M., and Boos, W.,** Trehalose synthesis genes are controlled by the putative sigma factor encoded by *rpoS* and are involved in stationary-phase thermotolerance in *Escherichia coli, J. Bacteriol.,* 173, 7918, 1991.
50. **Kaasen, I., Falkenberg, P., Stryvold, O. B., and Strom, A. R.,** Molecular cloning and physical mapping of the *otsBA* genes, which encode the osmoregulatory trehalose pathway of *Escherichia coli:* evidence that transcription is activated by KatF (AppR), *J. Bacteriol.,* 174, 889, 1992.

50a. **Hengge-Aronis, R.,** personal communication.

51. **May, G., Faatz, E., Villarejo, M., and Bremer, E.,** Binding protein dependent transport of glycine betaine and its osmotic regulation in *Escherichia coli* K12, *Mol. Gen. Genet.,* 205, 225, 1986.
52. **Cairney, J., Booth, I. R., and Higgens, C. F.,** Osmoregulation of gene expression in *Salmonella typhimurium: proU* encodes an osmotically induced betaine transport system, *J. Bacteriol.,* 164, 1224, 1985b.
53. **Druger-Liotta, J., Prange, V. J., Overdier, D. G., and Csonka, L. N.,** Selection of mutations that alter the osmotic control of transcription of the *Salmonella typhimurium proU* operon, *J. Bacteriol.,* 169, 2449, 1987.
54. **Higgins, C. F., Dorman, C. J., Stirling, D. A., Waddell, L., Booth, I. R., May, G., and Bremer, E.,** A physiological role for DNA

supercoiling in the osmotic regulation of gene expression in *S. typhimurium and E. coli*, *Cell*, 52, 569, 1988.

55. **May, G., Faatz, E., Lucht, J. M., Haardt, M., Bolliger, M., and Bremer, E.,** Characterization of the osmoregulated *Escherichia coli proU* promoter and identification of ProV as a membrane-associated protein, *Mol. Microbiol.*, 3, 1521, 1989.
56. **Dattananda, C. S., Rajkumari, K., and Gowrishankar, J.,** Multiple mechanisms contribute to osmotic inducibility of proU operon expression in Escherichia coli: demonstration of two osmoresponsive promoters and of a negative regulatory element within the first structural gene, *J. Bacteriol.*, 173, 7481, 1991.
57. **Lucht, J. M. and Bremer, E.,** Characterization of mutations affecting the osmoregulated *proU* promoter of *Escherichia coli* and identification of 5′ sequences required for high-level expression, *J. Bacteriol.*, 173, 801, 1991.
58. **Jovanovich, S. B., Record, M. T., Jr., and Burgess, R. R.,** In an *Escherichia coli* coupled transcription-translation system, expression of the osmoregulated gene *proU* is stimulated at elevated potassium concentrations and by an extract from cells grown at high osmolality, *J. Biol. Chem.*, 264, 7821, 1989.
59. **Gowrishankar, J.,** Nucleotide sequence of the osmoregulatory *proU* operon of *Escherichia coli*, *J. Bacteriol.*, 171, 1923, 1989.
60. **Overdier, D. G., Olsen, E. R., Erickson, B. D., Ederer, M. M., and Csonka, L. N.,** Nucleotide sequence of the transcriptional control region of the osmotically regulated *proU* operon of *Salmonella typhimurium* and identification of the 5′ endpoint of the *proU* mRNA, *J. Bacteriol.*, 171, 4694, 1989.
61. **Stirling, D. A., Hulton, C. S. J., Wadell, L., Park, S. F., Stewart, G. S. A. B., Booth, I. R., and Higgins, C. F.,** Molecular characterization of the *proU* loci of *Salmonella typhimurium* and *Escherichia coli* encoding osmoregulated glycine betaine transport systems, *Mol. Microbiol.*, 3, 1025, 1989.
62. **Bremer, E., Silhavy, T. J., Weisemann, J. M., and Weinstock, G. M.,** *plac*Mu: a transposable derivative of bacteriophage lambda for creating *lacZ* protein fusions in a single step, *J. Bacteriol.*, 158, 1084, 1984.
63. **Barron, A., May, G., Bremer, E., and Villarejo, M.,** Regulation of envelope protein composition during adaptation to osmotic stress in *Escherichia coli*, *J. Bacteriol.*, 167, 433, 1986.
64. **Hulton, C. S. J., Seirafi, A., Hinton, J. C. D., Sidebotham, J. M., Waddel, L., Pavitt, G. D., Owen-Hughes, T., Spassky, A., Buc, H., and Higgins, C. F.,** Histone-like protein H1 (H-NS), DNA supercoiling and gene expression in bacteria, *Cell*, 63, 631, 1990.
65. **May, G., Dersch, P., Haardt, M., Middendorf, A., and Bremer, E.,** The *osmZ (bglY)* gene encodes the DNA-binding protein H-NS (H1a), a component of the *Escherichia coli* K-12 nucleoid, *Mol. Gen. Genet.*, 224, 81, 1990.
66. **Ni Bhriain, N., Dorman, C. J., and Higgins, C. F.,** An overlap between osmotic and anaerobic stress responses: a potential role for DNA supercoiling in the coordinate regulation of gene expression, *Mol. Microbiol.*, 3, 933, 1989.
67. **Ramirez, R. M. and Villarejo, M.,** Osmotic signal transduction to *proU* is independent of DNA supercoiling in *Escherichia coli*, *J. Bacteriol.*, 173, 879, 1991.
68. **Prince, W. S. and Villarejo, M. R.,** Osmotic control of *proU* transcription is mediated through direct action of potassium glutamate on the transcription complex, *J. Biol. Chem.*, 265, 17673, 1990.
69. **Pruss, G. J. and Drlica, K.,** DNA supercoiling and prokaryotic transcription, *Cell*, 56, 521, 1989.
70. **Overdier, D. G. and Csonka, L. N.,** A transcriptional silencer downstream of the osmotically controlled *proU* operon of *Salmonella typhimurium*, *Proc. Natl. Acad. Sci. U.S.A.*, 89, 3140, 1992.
71. **Overdier, D. G., Fletcher, S., and Csonka, L.,** Osmotic control of transcription of the *proU* operon of *Salmonella typhimurium*, *Water and Life*, Somero, G. N., Osmond, C. B., and Bolis, C. L., Eds., Springer-Verlag, Berlin, 1990, 61.
72. **Higgins, C. F., Cairney, J., Stirling, D. A., Sutherland, L., and Booth, I. R.,** Osmotic regulation of gene expression: ionic strength as an intracellular signal?, *Trends Biochem. Sci.*, 12, 339, 1987.
73. **Sutherland, L., Cairney, J., Elmore, M., Booth, I. R., and Higgins, C. F.,** Osmotic regulation of transcription: induction of the *proU* betaine transport gene is dependent on accumulation of intracellular potassium, *J. Bacteriol.*, 168, 805, 1986.
74. **Cayley, S., Lewis, B. A., and Record, M. T.,** Origins of the osmoprotective properties of betaine and proline in *Escherichia coli* K-12, *J. Bacteriol.*, 174, 1586, 1992.
75. **Ramirez, R. M., Prince, W. S., Bremer, E., and Villarejo, M.,** In vitro reconstitution of osmoregulated expression of *proU* of *Escherichia coli*, *Proc. Natl. Acad. Sci. U.S.A.*, 86, 1153, 1989.

76. **Gutierrez, C. and Devedjian, J. C.,** Osmotic induction of gene *osmC* expression in *Escherichia coli* K12, *J. Mol. Biol.*, 220, 959, 1991.
77. **Lamark, T., Kaasen, I., Eshoo, M. W., Falkenberg, P., McDougall, J., and Strøm, A. R.,** DNA sequence and analysis of the *bet* genes encoding the osmoregulatory choline-glycine betaine pathway of *Escherichia coli, Mol. Microbiol.*, 5, 1049, 1991.
78. **Eshoo, M.,** *lac* fusion analysis of the *bet* genes of *Escherichia coli:* regulation by osmolarity, temperature, oxygen, choline, and glycine betaine, *J. Bacteriol.*, 170, 5208, 1988.
79. **Jovanovich, S. B., Martinell, M., Record, M. T., Jr., and Burgess, R. R.,** Rapid response to osmotic upshift by osmoregulated genes in *Escherichia coli* and *Salmonella typhimurium, J. Bacteriol.*, 170, 534, 1988.
80. **Giaever, H. M., Styrvold, O. B., Kaasen, I., and Strom, A. R.,** Biochemical and genetic characterization of osmoregulatory trehalose synthesis in *Escherichia coli, J. Bacteriol.*, 170, 2841, 1988.
81. **Lange, R. and Hengge-Aronis, R.,** Identification of a central regulator of stationary-phase gene expression in *Escherichia coli, Mol. Microbiol.*, 5, 49, 1991.
82. **Smardo, J. F. L., Burg, M. B., and Garcia-Perez, A. R.,** Kidney aldose reductase gene transcription is osmotically regulated, *Am. J. Physiol.*, 262, C776, 1992.
83. **Uchida, S., Garcia-Perez, A., Murphy, H., and Burg, M.,** Signal for induction of aldose reductase in renal medullary cells by high external NaCl, *Am. J. Physiol.*, 256, C614, 1989.
84. **Wood, J.,** personal communication.

Chapter 21

Genetic Control of Osmoregulation in Plants

David Rhodes and Yiannis Samaras

TABLE OF CONTENTS

I. INTRODUCTION: DEFINITION(S) OF OSMOREGULATION

Osmoregulation in higher plants is defined by Blum[1] as follows:

> As water is being removed from the cell, osmotic potential is reduced due to the simple effect of solute concentration. However, if during the course of cellular water loss solutes are actively accumulated, osmotic potential would be reduced beyond the rate dictated by the mere effect of concentration. Such accumulation of solutes during the development of water deficit is termed "osmotic adjustment" or "osmoregulation".

This definition is similar to that used by Morgan.[2] The terms "osmotic adjustment" and "osmoregulation" are essentially used synonymously.[1,2,3] However, Munns,[4] states that the term "osmotic adjustment" is *erroneously* called "osmoregulation" in higher plants. Munns[4] cautions that the term "osmoregulation" should be used carefully, if at all. According to Munns, "osmotic adjustment" is usually defined as an increase in osmotic pressure of cell sap resulting from more solutes per cell, rather than from a lower cell volume.[4] However, the term "osmoregulation" in the strict sense means the *regulation of osmotic pressure,* which, in algae, is distinct from *turgor* or *volume* regulation.[4,5] Munns[4] argues that evidence for turgor regulation in higher plants is sparse, and until it is proven that turgor is regulated by a feedback loop controlling turgor via changes in osmotic pressure, it seems inadvisable to use terms that imply regulation (such as "osmoregulation") to describe the simple observation of osmotic adjustment. Munns further argues that many observations of osmotic adjustment in higher plants may be merely the consequence of growth inhibition (e.g., decreased utilization of solutes in growth processes such as cell wall and protein synthesis, and/or reduced pool dilution caused by a reduction in cell expansion),[4] implying that such adjustment is not regulated. This contrasts with the view of Zimmermann:[6]

> . . . research into turgor- (and osmo-) regulation has revealed that the basic mechanisms underlying the adjustment of internal osmotic pressure in response to osmotic stress or growth demand may be comparable and under the direct control of turgor pressure, both in higher plants and algal cells, despite their anatomical,

0-8493-4448-4/94/$0.00+$.50

ultrastructural, and physiological differences.

In prokaryotes the term *osmoregulation* refers to active processes (including solute accumulation) carried out during adaptation to the osmotic strength of the environment.[7] These processes are inextricably linked to general ionic and metabolic regulation, to turgor and volume regulation, and to regulation of growth rate.[7] In cyanobacteria, the term osmoregulation refers to the process of adjusting intracellular solute and solvent levels, and encompasses several possible mechanisms, including turgor regulation and volume regulation.[8] Blomberg and Adler[9] also adhere to the commonly adopted term osmoregulation to describe the ability of fungal cells to adjust the total number of intracellular solute molecules, without making any assumptions concerning the mechanisms (turgor and/or volume regulation) by which the process is controlled. Against the recommendation of Munns,[4] we use the term *osmoregulation* in higher plants as it has been commonly used in microorganisms, to descrcibe adjustments in intracellular solute and solvent levels,[7–9] and has been commonly adopted in other reviews of this process in higher plants.[1,2,6,10,11]

In eukaryotic marine algae[12] and other eukaryotic organisms discussed elsewhere in this volume, hyperosmotic shock first leads to changes in ionic contents via changes in the activity of ion transporters and channels. This is followed by an increase in levels of organic compatible solutes, mediated either by increased uptake of the solutes (if present in the external medium), activation of enzymes, or increased *de novo* synthesis of enzymes involved in compatible solute synthesis.[12] Later, as steady state is approached, the metabolically adverse ion concentrations are further reduced in favor of the accumulated compatible solutes.[12] In higher plants similar mechanisms may operate. Initial adjustments to the osmotic strength of the environment may involve modulation of ligand-, voltage-, stretch-, and/or osmotic pressure-gated ion channels.[13] Longer-term adjustments may involve changes of gene expression affecting the steady-state levels of mRNA for proteins involved in solute transport and/or compartmentation (e.g., H^+-[14] and Ca^{2+}-ATPases[15]) and proteins involved in the synthesis of compatible osmolytes (e.g., glycine betaine[16,17] or pinitol[18]).

In prokaryotes considerable progress in elucidating the mechanisms involved in osmoregulation has been made by the characterization of mutants with altered solute synthesis and/or transport properties, clearly illustrating that the process of osmoregulation is under genetic control.[7] In fungi, specific genes affecting solute transport and/or synthesis have also been identified which have a profound effect on osmoregulation and osmotolerance.[9] The recent observation that alternative alleles of a single locus determine osmoregulation in wheat[19,20] clearly indicate that genetic approaches can assist in dissecting the underlying osmoregulatory mechanisms of higher plant cells and tissues, as they have in prokaryotes[7] and fungi.[9] This chapter will consider primarily recent progress in the characterization of genes which determine *active* solute accumulation by higher plant cells or tissues in response to water deficits and/or salinity stress, including genes implicated in osmoregulation that have been characterized by conventional Mendelian genetic approaches,[19,21] as well as genes of putative importance in osmoregulation that have recently been isolated by nontraditional molecular cloning approaches.[16–18,22] Because of the vast body of literature in this rapidly advancing field, we rely heavily on many recent reviews of this diverse subject matter.

We believe that any discussion of osmoregulation in higher plants would be incomplete without first considering the specific osmoregulatory mechanisms of guard cells, which determine stomatal aperture and hence leaf gas exchange. Although guard cell physiology is not always included in discussions of plant osmoregulation/osmotic adjustment (see, for example, References 2 and 4), it is notable that the highly specialized osmoregulatory mechanisms of these cells have been the subject of intensive investigations with respect to the signal transduction pathways involved in stomatal movements.[23] These pathways deserve special emphasis, since they share many features in common with animal cells,[23] and may be pertinent to other plant cells that have been less intensively studied. Hellebust[10] notes:

> The massive literature on stomatal action indicates the complexity of osmoregulation in guard cells with its dependence on ion concentrations and fluxes, metabolic energy, pH, CO_2, compartmentation of ions and metabolites, hormones, and interactions of these factors. Osmoregulation in other plant cells may prove to be almost equally complex.

II. STOMATAL CONTROL

Stomatal regulation involves rapid changes in solute concentrations of guard cells which affect guard cell osmotic pressure, turgor and volume, modulating leaf transpirational water loss, CO_2 uptake, and, hence, photosynthesis.[24–28] By affecting transpirational water loss and CO_2 fixation, stomatal movements have a marked effect on the carbon economy and water relations of the leaf (and plant) as a whole. Stomatal movements are responsive to environmental cues and phytohormones, and are modifiable by mutation.[29]

Stomatal opening occurs by light-activated increases in the turgor pressure of guard cells, which causes them to change shape so that the stomatal pore between them widens.[23] These increases in turgor pressure oppose increases in cellular osmotic pressure that result from uptake of K^+.[23] K^+ uptake occurs by a chemiosmotic mechanism in response to light-activated extrusion of H^+ outward across the plasma membrane of the guard cell.[23] The initial changes in cellular membrane potential lead to the opening of inward-rectifying K^+ channels, after which K^+ is taken up along its electrochemical gradient.[23] Changes in membrane potential resulting from K^+ may be balanced by accumulation of Cl^- ions by guard cells, and/or by the synthesis of malic acid via the reaction catalyzed by phospho*enol* pyruvate carboxylase (PEPCase).[23] The carbon skeletons for malic acid synthesis are derived primarily from osmotically inactive starch. The malic acid acts to buffer increases in cytosolic pH caused by H^+ extrusion.[23] Malic acid accumulation is stimulated by alkaline pH, whereas malate decarboxylation is stimulated by acid pH.[30]

Possible signal transduction processes in stomatal regulation have been discussed recently.[23] Three second messengers (in addition to Ca^{2+}) have been considered by Tallman: 3′,5′-cAMP, inositol 1,4,5-trisphosphate, and 1,2-diacylglycerols.[23] Inositol 1,4,5-trisphosphate causes release of Ca^{2+} from internal stores via activation of vacuolar Ca^{2+} channels,[31] elevating cytosolic Ca^{2+} concentration. Increased cytosolic Ca^{2+} would result in activation of a voltage-dependent depolarizing conductance to anions (which in turn depolarizes the membrane to potentials that trigger opening of an outward-rectifying K^+ channel[23]) and reversible inactivation of K^+-influx channels,[23,31] leading to stomatal closure.

Light-induced changes in concentrations of 1,2-diacylglycerols would activate protein kinase C, which in turn would catalyze phosphorylation of the plasma membrane H^+-translocating ATPase.[23] The resulting extrusion of H^+ outward across the guard cell plasma membrane would result in hyperpolarization of the membrane, causing inward-rectifying, voltage-gated channels to open, leading to stomatal opening.[23] The opening of the latter channels would be dependent on Ca^{2+}, which, in turn, would be regulated by the activity of G protein,[23] which activates phosphatidylinositol 4,5-bisphosphate (PIP_2)-specific phospholipase C, responsible for cleaving PIP_2 to inositol 1,4,5-trisphosphate and 1,2-diacylglycerol.[31–33] As in mammalian cells, an intimate interaction between polyphosphoinositides and components of the cytoskeleton has been demonstrated in plant cells.[32]

According to Tallman, evidence for the involvement of cAMP in the signal transduction pathway of stomatal regulation is "equivocal".[23] However, it may be significant that a potassium channel gene has been cloned recently from *Arabidopsis thaliana*[22] and a region of this K^+ channel gene has been shown to share homology with mammalian ion channels and protein kinases that are regulated by cGMP and cAMP, and with the catabolite gene activator protein of *Escherichia coli,* which includes a cAMP binding site.[22]

It is notable that Ca^{2+} is considered to act as a secondary messenger in signal transduction in plant cells other than guard cells.[33] Ca^{2+}, by its interaction with calmodulin, regulates the activity of protein kinases and phosphatases.[33,34] These in turn regulate the activity of a large number of enzymes.[33,35,36] Ca^{2+} concentration can affect transcription directly or affect both transcription and translation through the activity of protein kinases.[33] Little is yet known about the extent to which Ca^{2+}-binding proteins, such as calmodulin, or the activity of Ca^{2+}-ATPases affect cytosolic Ca^{2+} concentrations in guard cells,[23] and whether or not changes in intracellular Ca^{2+} may modulate the transcription of genes encoding key enzymes of carbon metabolism, or the activity of these enzymes via reversible phosphorylation in guard cells.[23] Reversible phosphorylation of PEPCase (affecting the regulatory properties of the enzyme) has been reported for Crassulacean Acid Metabolism (CAM) plants,[35,36] and this has prompted investigations of similar control mechanisms in guard cells. Evidence for two different forms of PEPCase in guard cells has been reported,[37] but further work is needed to clarify this point,[23] and to determine if reversible phosphorylation and Ca^{2+} play a role in this control. It is noteworthy that whereas cyclic nucleotides (cAMP, cGMP), Ca^{2+}, and calmodulin are the major effectors of protein phosphorylation in animals, of these only Ca^{2+} and calmodulin have been shown to affect protein phosphorylation in plants.[35] Nitrate, or a product of its metabolism, has been implicated in the control of cytosolic protein kinases and the reversible phosphorylation of enzymes of carbon metabolism (including PEPCase) in plants.[38]

Several types of channels have been characterized in plant cells other than guard cells. These include ligand-gated channels (e.g. a Ca^{2+}-selective channel in the tonoplast of red beet which is opened by micromolar concentrations of inositol 1,4,5-trisphosphate[39]), voltage-gated channels,[40] and stretch-activated channels.[41] The latter may be involved in osmoregulation.[41] A hydrostatic and osmotic pressure activated channel of the tonoplast of red beet has also been described.[13] The latter channel could be gated *in vivo* by a change of osmolality without a change of turgor pressure in the cell, and represents a possible example of an osmoreceptor in plant cells.[13] It is possible that such hydrostatic and osmotic pressure activated channels could play an important role in coupling ion and water fluxes between the cytosolic and vacuolar compartments of guard cells during stomatal opening and closing, or in other plant cells in response to alterations of external water potential.[13]

The plant hormone abscisic acid (ABA), which increases in level in response to water stress, promotes stomatal closure.[28] ABA binds to proteins in the plasmalemma and causes a rapid inhibition of proton excretion, and rapid excretion of K^+.[28] ABA triggers a rapid release of osmotica from the vacuole through tonoplast, cytoplasm, and plasmalemma.[28] Starch that disappears during stomatal opening reappears.[28] The simplest explanation of this phenomenon is that ABA inhibits the proton-translocating H^+-ATPase at the plasmalemma,[28] but other modes of action cannot yet be eliminated.[28] ABA causes a rapid increase in cytosolic Ca^{2+} of guard cells,[31,42,43] which could elicit stomatal closure via the pathway described above.[23,31,42] ABA has been shown to open a channel in the plasma membrane that is permeable to both K^+ and Ca^{2+}.[44] However, ABA treatment also causes an increase of cytosolic pH,[42] which may in turn modulate ion channels.[45] It is not yet clear whether ABA-induced alkalinization is by modification of one or more of the H^+-translocating transport processes, or by activation of a biochemical process that consumes H^+, for example decarboxylation of malate, which occurs rapidly in response to ABA treatment.[28,42]

ABA is synthesized in both shoots and roots in response to water stress.[46] Transport of ABA via the xylem from the roots to the shoots provides a mechanism whereby water deficits sensed in roots can regulate stomatal aperture and hence leaf transpiration rate.[46] However, it is now clear that changes of ABA compartmentation in the leaf, mediated by water stress-induced changes in pH gradients between cytoplasm and apoplast, can lead to stomatal closure before any net increase in total ABA level can be measured in the leaf as a whole, and before loss of leaf turgor occurs.[46] ABA regulates the expression of a large number of genes in higher plants.[47] Many of these ABA-regulated genes have now been cloned and sequenced, but the precise functions of many of the gene products remain to be determined.[47] The osmotic stress- and ABA-inducible protein, osmotin, in tobacco has antifungal activity.[48] It is suggested that the translational and posttranslational control of this protein may be adapted to respond to the general symptoms of pathogen invasion, such as turgor loss.[48] Several wound-inducible genes are also induced by ABA.[49] Moreover, methyljasmonate also induces accumulation of these transcripts, suggesting a role for jasmonic acid as an intermediate in the signaling pathway that leads from ABA accumulation in response to wounding to the transcriptional activation of the genes.[49] DNA sequences flanking the coding regions that confer ABA responsiveness (ABA- responsive elements) to ABA-inducible genes have been identified, and are providing access to the sequence-specific DNA binding proteins that promote transcription.[47] Such elements may prove to be of key importance in genetic engineering of plant osmoregulation, whereby transgenes can be placed under the specific control of ABA.

Several ABA-deficient mutants of higher plants have been identified[29,50] and have been invaluable in studies of the role of ABA in regulation of plant gene expression.[47,49] These include mutants blocked at a number of different steps in the synthesis of ABA from carotenoids.[50] These mutants are typically extremely prone to wilting due to inadequate control over stomatal aperture and, hence, transpiration.[50] The best-characterized mutants are those with lesions in the last step of the pathway: ABA-aldehyde oxidation to ABA. The enzyme responsible for ABA-aldehyde oxidation is thought to be a molybdoenzyme, since the *nar-2* mutation in barley, which results in impaired molybdenum cofactor formation (affecting the activity of a number of molybdoenzymes, including nitrate reductase, xanthine dehydrogenase, and various aldehyde oxidases), causes a reduced capacity to accumulate ABA.[50] Heterologous transposon tagging is currently being explored as a method for cloning of ABA biosynthetic genes.[50] The *vp1* gene of maize, which causes ABA insensitivity, has been cloned by transposon tagging.[51] The *vp1* gene may encode a novel transcriptional activator.[52]

III. VARIABILITY FOR OSMOREGULATION BETWEEN PLANT SPECIES

There are marked differences between plant species and cultivars with respect to osmotic adjustment and with respect to the major solutes contributing to osmotic potential and turgor maintenance. Osmotic adjustment is a time-dependent process but has finite limits.[1] Species and cultivars differ with respect to the rate of solute accumulation, the types of solutes accumulated, and the limit of solute accumulation.[1,2,10,53] This inter- and intraspecies variability for solute accumulation offers fertile ground for dissecting the molecular genetic determinants of osmoregulation in the plant kingdom, although our understanding of this variability remains far from complete.

Species differences in osmoregulation have been reviewed[2,10,53] and are particularly apparent in the comparison of higher plant halophytes and glycophytes. Halophytes are adapted to environments of very low water potentials and high salt concentrations. Survival of angiospermous halophytes at high salinity is often accompanied by a high ion content.[53] In general, halophytes are distinguished from glycophytes by their ability to accumulate (and tolerate) high ion concentrations, particularly in leaf cells.[53] However, it is clear that among halophytes there are marked differences between species with respect to inorganic ion accumulation and their Na^+:K^+ ratios.[54,55] Ion concentrations in the shoots of halophytes appear to be regulated and maintained within fine limits, perhaps as a consequence of the limitation of shoot growth by the supply of ions from the roots (determined primarily by the radial passage of ions across the root cortex).[56] Glycophytes have no specialized mechanisms for dealing with high external ion concentrations, and can regulate their leaf ion content only to a moderate extent by ion exclusion.[10,53] The growth sensitivity of nonhalophytes to salinity stress is usually associated with either ion excess in the expanded leaves (perhaps in part a function of inadequate cellular compartmentation of ions) or by water deficits in the expanding tissue.[57]

The degree to which different plant species (and cultivars) can tolerate high concentrations of salt is clearly under genetic control.[58–60] The feasibility of transferring genes determining salt tolerance from wild species to commercial species has been demonstrated by traditional genetic approaches.[58,59] Specific chromosome locations of genes determining salt tolerance and ion uptake have been identified,[58,61,62] although the precise functions of these genes have yet to be determined.

In recent years there has been an explosion of research on ion carriers and channels in higher plants, as discussed above in relation to stomatal control. The reader is referred to the following comprehensive reviews[63–65] and recent publications[13–15,22,66–70] in this rapidly developing field of central importance to ion homeostasis and osmoregulation in plants. Of particular significance are recent advances in the cloning of H^+-ATPase[14,69] and Ca^{2+}-ATPase[15] genes and K^+ channel genes[22] from higher plants, permitting the evaluation of the regulation of these genes in relation to the osmotic strength of the environment and in response to phytohormones,[14,15] and expression of these genes in heterologous systems.[22,66] Regenerated plants of carrot containing an antisense construct of the vacuolar H^+-ATPase A subunit, exhibit markedly altered leaf morphologies and reduced cell expansion, clearly indicating that the vacuolar H^+-ATPase plays a central role in vacuolar solute accumulation, osmotic water uptake, and cell expansion.[69] Continued application of these transformation and antisense technologies[69] can be expected to yield important new information concerning the functions of specific carriers and channels, and the fundamental differences in properties and regulation of these proteins between glycophytes and halophytes. A novel yeast gene, *HAL1*, has been isolated recently, which, upon overexpression, improves growth of yeast under salt stress.[71] This gene is implicated in Na^+/K^+ discrimination, although the protein encoded by *HAL1* does not appear to be a membrane protein or a nuclear transcription factor.[71] Antibodies against the HAL1 protein cross react with similar proteins in plants. These proteins are inducible by NaCl stress and ABA.[71]

There is increasing evidence to suggest that the inorganic ions accumulated by angiospermous halophytes must be contained largely within the vacuoles, leaving cytoplasmic ion concentrations a half to one third of that of the vacuolar values.[53,56] This compartmentation of ions is associated with the accumulation of relatively large amounts of organic solutes in the cytoplasm to balance cytoplasmic and vacuolar water potentials.[10,53,72–77] It is now well established that certain solutes are more “compatible” to macromolecular structure and function than others.[11,73,78–80] Inorganic ions readily enter the hydration sphere of proteins, favoring protein unfolding and denaturation.[78] Certain organic solutes, however, are excluded from the hydration sphere of proteins and are more compatible to the maintenance of folded protein structures than equivalent concentrations of inorganic ions.[78,79] Effective osmoregulation in higher plants requires not only net accumulation of solutes, but also the appropriate compartmentation of solutes between organelles; specifically, the sequestering of “noncompatible” solutes in the vacuole, and the accumulation of “compatible” solutes in cytoplasmic compartments.[11,72,73–77,81,82] Genes regulating the accumulation of compatible organic solutes in the cytoplasm are of considerable interest in terms of elucidating the fundamental genetic differences between halophytes and glycophytes with respect to osmoregulation.

The capacity for accumulating organic solutes is a general feature of halophytic higher plants, but species differ markedly with respect to the types of organic osmolytes accumulated.[54] Briens and Larher[54] have recognized three main physiotypes: (1) plants accumulating predominantly carbohydrates (sucrose and/or polyols); (2) plants accumulating both carbohydrates and nitrogenous compounds (including proline, glycine betaine, or β-alanine betaine); and (3) plants predominantly accumulating nitrogenous compounds.[54] Certain halophytes also accumulate sulfur-containing compounds (β- dimethylsulfoniopropionate [DMSP] or choline-*O*-sulfate) to high levels.[83–87] It is also clear that different species within the same genus can differ markedly for the specific organic solutes they accumulate. For example, among *Spartina* species, DMSP is accumulated by *S. alterniflora*[83] and *S. anglica,*[84,85] but not by *S. patens* or *S. cynosuroides*.[83] *Limonium* species accumulate choline-*O*-sulfate and either glycine betaine or β-alanine betaine,[85] but not both betaines. The enzymological and genetic bases for these species differences in solute accumulation merit further investigation.[82]

The compatible solute, glycine betaine, is accumulated to high levels in the chloroplasts of chenopods in response to salinity stress,[74,76,77] and this may play an important role in the maintenance of photosynthesis by protecting the photosynthetic apparatus against the inhibitory effects of salt.[88–90] The pathway of synthesis of glycine betaine involves a two-step oxidation of choline via the intermediate betaine aldehyde.[91–93] The first enzyme, choline monooxygenase, has been partially purified from spinach, and has been shown to be a ferredoxin-dependent enzyme utilizing molecular oxygen as substrate.[94] The enzyme is stimulated by Mg^{2+}, has a native molecular weight of approximately 98,000 and a pH optimum close to 8.[82,94] The second enzyme, betaine aldehyde dehydrogenase, has a strong preference for NAD^+, and is a dimeric protein with subunits of molecular weight 60,000.[82] Both enzymes are chloroplast localized in chenopods,[91,92] and both are induced by salinity stress.[16,17,94] Betaine aldehyde dehydrogenase clones containing the entire protein coding sequence have recently been isolated from spinach[16] and sugar beet.[17]

It has been suggested that the capacity to synthesize glycine betaine is an archetypal angiosperm characteristic, expressed strongly by some living plants, weakly by others.[82,95] Introduction of heterologous genes encoding glycine betaine biosynthetic enzymes into non-glycine betaine-accumulating species, such as rice, represents an attractive possibility for genetically manipulating salt and/or drought tolerance of such species.[73,96] Similarly, it may soon be possible to genetically engineer the capacity to accumulate the compatible solute, proline, in higher plants. A gene encoding Δ^1-pyrolline-5-carboxylate reductase has recently been cloned from soybean and overexpressed in tobacco.[97] Moreover, a gene encoding a putative bifunctional γ-glutamyl kinase and γ-glutamyl phosphate reductase has recently been cloned from tomato,[98] opening up the possibility of overexpression of the entire proline biosynthetic pathway in targeted plant cells and/or organelles in order to determine the effects of such overexpression on osmoregulation and salt or drought tolerance. A bacterial gene encoding a key enzyme of mannitol biosynthesis has been introduced recently into tobacco; a species that does not normally accumulate this compound.[99] The transgenic plants have been shown to accumulate mannitol, demonstrating the feasibility of engineering plant osmolyte accumulation.[99]

The widespread occurrence of orders of flowering plants with halophytic species is "indicative of a polyphyletic origin perhaps involving relatively few changes in the genome".[53] In this context it may be significant that extremely halotolerant cell cultures have been derived from glycophytes such as tobacco.[100] After adaptation to NaCl, the growth response of the tobacco cells resembles that of halophytic growth behavior.[100] Adaptation to NaCl involves a marked reduction in cell volume, a marked reduction in cell expansion rate despite turgor maintenance, and marked changes of gene expression.[14,15,48,100,101] Osmotic adjustment in the salt-adapted tobacco cells is modulated by ABA,[102] and involves processes analogous to those typical of halophytic higher plants, including accumulation of Na^+ and Cl^- primarily in the vacuole,[103] and accumulation of organic solutes (mostly sucrose and amino acids, particularly proline),[104] presumably in the cytoplasm. Proline accumulation in salt-adapted tobacco cells involves at least a 10-fold increase in the rate of synthesis of proline from glutamic acid.[105] The precise number (and functions) of stable genetic alterations responsible for facilitating adaptation to salinity stress in tobacco cells remains to be determined. Such genetic analyses may be hindered by the fact that salt adaptation induces polyploidy in tobacco cells.[106]

Certain higher plant species exhibit Crassulacean Acid Metabolism (CAM) which allows night-time carbon fixation. CAM is characterized by significant

CO_2 assimilation catalyzed by PEPCase (with stomates open) in the dark, leading to the synthesis of free malic acid.[107] The acid, stored in the cell vacuole, is decarboxylated in the subsequent light period, and the CO_2 released (with stomates closed) is then refixed by photosynthesis via the catalytic action of ribulose-bisphosphate carboxylase-oxygenase.[107] CAM plants tend to exhibit low stomatal frequency and extraordinarily high cuticular resistance, and it may be these morphological features, rather than physiological features, that determine the ability of CAM plants to restrict water loss in extremely arid environments.[107] The osmoregulatory mechanisms of CAM plants must be specialized to accommodate large diurnal changes in malic acid levels (from 20 mmol kg^{-1} fresh weight at the end of the light period to 220 mmol kg^{-1} fresh weight at the end of the dark period in certain species).[108] Reversible phosphorylation of PEPCase may be an important component of this control.[35,36] The carbon sources for the accumulated malic acid may include starch, glucans (low-molecular-weight polymers of glucose), and soluble sugars.[109] The accumulation of malic acid in the vacuole in the dark at the expense of insoluble carbohydrates in the cytoplasm has a significant impact on cell water relations, and may account for an increase of several bars in osmotic potential of the vacuole.[107] This requires that the levels of some cytoplasmic solutes must oscillate in parallel with the malic acid levels in the vacuoles, but because of the very small relative volume of cytoplasm in CAM plant leaf cells, the absolute amounts involved in such changes may be small.[110]

The facultative halophyte *Mesembryanthemum crystallinum* undergoes a switch from C_3 photosynthesis to CAM in response to salinity stress, and is proving to be an excellent model system for studies of the molecular biology of CAM metabolism, salt tolerance, and gene regulation by environmental stimuli and phytohormones.[18,111–116] In addition to the marked changes in expression of genes encoding CAM pathway enzymes, salinity stress also induces proline accumulation,[116] and *myo*-inositol *O*-methyl transferase, which catalyzes the first step in the synthesis of the cyclic sugar alcohol, pinitol, a compatible solute that may play a significant role in leaf cytoplasmic osmoregulation in *M. crystallinum*.[18,111,113,117] Cytokinins and NaCl, when sensed in the roots, elicit the transport of a signal to aerial leaves where the signal initiates the stress responses of proline accumulation and the accumulation of a specific isoform of PEPCase.[116] The identity of the second messenger(s) influenced by cytokinin and NaCl is presently unknown.[116]

IV. VARIABILITY FOR OSMOREGULATION WITHIN PLANT SPECIES

Among crop plants substantial differences between cultivars or landraces in terms of osmoregulation have been reported. For barley,[118] wheat,[19,20,119–123] sorghum,[1,124–127] millet,[125] and chickpea,[128] these cultivar and landrace differences in osmotic adjustment or osmoregulation appear to be significantly correlated with drought resistance. However, not all studies of this kind confirm that osmotic adjustment is always associated with higher productivity in osmotic stress environments.[129–133]

In wheat, cultivar differences in osmoregulation have been attributed to alternative alleles of a single locus (probably located on chromosome 7A), with high osmoregulation being recessive.[19] Tentative evidence for a primary role of potassium in the single-gene-controlled difference in osmoregulation between wheat genotypes has been presented.[20] This gene appears to be distinct from the gene(s) determining solute potential differences between barley isopopulations.[132,134,135] In barley, the gene(s) involved appear to be highly pleiotropic in their action, simultaneously influencing growth, total solute potential, and glycine betaine level.[132,135] In sorghum, at least three genes have been identified that confer high osmotic adjustment; one behaves as a recessive gene, a second shows additive dominance, and a third has not yet been characterized.[148] The precise functions of these genes remain unknown. Conceivably, each locus determining osmotic adjustment could have different metabolic costs or yield penalties, and this may explain some of the conflicting reports concerning the adaptive significance of osmotic adjustment in crop plants based on comparisons between cultivars. Ideally, near-isogenic lines or isopopulations differing solely for the gene of interest are required to assess the specific contributions of each locus to osmotic adjustment and stress tolerance, and to assess possible trade-offs with productivity under optimum conditions.

A number of glycine betaine-deficient maize genotypes have been identified.[21,136–138] Recessive alleles of a single locus determine glycine betaine deficiency in maize.[138] This deficiency is associated with a metabolic block at the choline → betaine aldehyde step in the glycine betaine biosynthetic pathway.[138] Near-isogenic maize inbreds differing for glycine betaine content are currently being developed,[137,138] and these should prove useful in evaluating the effects of this trait on water relations, osmotic adjustment, and dry matter and grain yield in dry or saline environments.

V. POSSIBLE FUNCTIONS OF OSMOREGULATION IN CROP PLANTS

As reviewed recently, leaf osmotic adjustment or osmoregulation during water stress can permit maintenance of a higher turgor potential at a given leaf water potential.[1,2] Thus, zero turgor will occur at a lower water potential in an osmotically adjusted leaf tissue.[1,2] This follows from the relationship between leaf water potential (Ψ_l), pressure (turgor) potential (Ψ_p), osmotic potential (Ψ_s), and matric potential (Ψ_m):[1]

$$\Psi_l = \Psi_p + \Psi_s + \Psi_m \quad (1)$$

It can be seen from Equation 1 that in order to maintain a positive turgor ($\Psi_p > 0$), $\Psi_s + \Psi_m$ must be $< \Psi_l$. Neglecting the contribution of Ψ_m, solute accumulation (generating a more negative Ψ_s) clearly represents a mechanism for maintaining a positive turgor pressure at low Ψ_l. Controversy surrounding the role(s) of osmotic adjustment in higher plants stems not from the fact that osmotic adjustment can contribute to maintenance of leaf turgor, but rather from conflicting reports that turgor is required for maintenance of photosynthesis and limits cell expansion growth at low leaf water potentials.[1,2,4] One interpretation is that osmotic adjustment in leaf tissues is an adaptive response to osmotic stress favoring continued growth, photosynthetic productivity, and dry matter or grain yield at low leaf water potentials.[1,2] However, Munns[4] argues that turgor probably does not limit leaf expansion growth or stomatal conductance, and that when water is limited there is little advantage in terms of carbon gain of keeping stomates open and using the soil water more quickly than of closing them and using it more slowly. Munns[4] further argues that osmotic adjustment is not essential for maintaining leaf turgor, since a reduction in stomatal aperture can accomplish this. It is clear, however, that even after turgor loss, a leaf or other organ with higher osmotic adjustment will have a higher water content than a leaf with lower osmotic adjustment.[2] Thus, osmotic adjustment may ensure *survival* of expanding tissues or inflorescences, and/or more rapid recovery upon alleviation of the osmotic stress.[2] On this point, Munns[4] agrees: osmotic adjustment might be an adaptation for surviving stress rather than for growing during it. Osmotic adjustment may protect membranes against functional or structural perturbations by stress,[1] and may enable the maintenance of a positive carbon balance during water stress; in sorghum, the carbon stored during stress is immediately available for growth upon rehydration and recovery.[139] The metabolic cost of storing photosynthate and using it for osmotic adjustment is less than the cost of converting it to new biomass.[140]

As leaf turgor approaches zero, the synthesis of ABA is markedly induced.[2] Increased translocation of ABA from leaves to inflorescences can lead to floret sterility and/or senescence, even though turgor may be fully maintained in the inflorescence.[2] Reduction in inflorescence senescence associated with osmotic adjustment in leaf tissue may in part account for differences in grain yield between genotypes of wheat and sorghum with contrasting levels of osmoregulation.[2,119–121]

During water stress, osmotic adjustment in root apical meristems may be particularly important in maintenance of root growth required for extraction of soil moisture,[2] and therefore the ability to assimilate more carbon and acquire nutrients, which could in turn influence yield.[2,4] It is conceivable that leaf osmotic adjustment[2,119,121] is an indicator of root osmotic adjustment,[4] and hence the efficiency of extraction of soil moisture. In this context it is notable that increased endogenous ABA maintains primary root growth but inhibits shoot growth of maize seedlings at low water potentials,[141] as determined from growth analyses of wild type and a vivaparous *(vp 5)* mutant of maize defective in ABA synthesis, and the use of fluoridone, an inhibitor of ABA synthesis.[141] Thus, the same hormonal signal affecting floret sterility and/or senescence, stomatal aperture, and gene expression, may be involved in the differential control of shoot and root growth at low water potentials, with shoot growth being more severely inhibited by ABA.[141] This phenomenon may in part account for the often-observed increase in root-to-shoot ratio during prolonged water stress.[3]

Osmotic adjustment in maize roots is intrinsically linked to growth and solute deposition rates, with potassium, hexose, and proline being the principal solutes accumulated.[142–144] Proline accumulation may be particularly important in osmotic adjustment in the apical 2 mm of the maize root tip, accounting for up to half the decrease in osmotic potential, and equal to the magnitude of the value of turgor.[144,145] It is likely that the large increase in proline concentration (10- to 40-fold) in the root apical region at low water potentials (resulting from increased rates of net proline deposition, rather than decreased volume expansion) is essential for the maintenance of cell expansion in this region.[144,145] Proline deposition rate in maize roots may in turn be regulated by ABA.[145a] The work of Sharp et al.[141–145] clearly illustrates that root osmotic adjustment is finely spatially regulated and is not simply the consequence of growth inhibition.[4]

Ludlow and Muchow[124] have recently summarized the ways in which osmotic adjustment in shoots, panicles, and roots may influence plant processes and grain yield in sorghum. Leaf osmotic adjustment maintains leaf turgor, which may in turn maintain leaf growth, photosynthesis, and transpiration, providing more solutes for growth and dry matter production and delaying the production of ABA and hence floral abortion. Root osmotic adjustment maintains root turgor and hence root growth, contributing to maintenance of water and nutrient uptake required for maintenance of leaf turgor and transpiration. Panicle osmotic adjustment maintains panicle turgor, thereby assisting in head exertion, reducing floral abortion and facilitating retranslocation of preanthesis dry matter.[124]

VI. CONCLUDING DISCUSSION

Large gaps still remain in our knowledge of the mechanisms by which osmotic stress is perceived and transduced into changes of metabolic rates in plant cells. However, emerging are striking similarities between the animal and guard cell signal transduction pathways with respect to the involvement of Ca^{2+} and inositol 1,4,5-trisphosphate as second messengers in osmoregulation.[23,32,33] Although ion pumps, carriers, and channels of higher plants show extensive sequence homology with other eukaryote equivalents,[14,15,22] it is likely that the coordinated regulation of these transport proteins in higher plant cells may be specialized to accommodate: (1) light as an environmental stimulus;[23] (2) the large vacuole of higher plant cells (i.e., the need to compartmentalize compatible solutes in the cytoplasm and noncompatible ions in the vacuole);[11] and (3) the unique hormonal cues employed by plants to sense environmental perturbations.[43] The central importance of ABA in the osmoregulatory mechanisms of higher plants contrasts with other organisms that have been reviewed in this volume. ABA may modulate plant metabolism and gene expression in higher plants either directly[47] or indirectly via altering cytoplasmic Ca^{2+} concentrations,[43] influencing protein phosphorylation.[33–36] Cytokinins[116] and jasmonic acids[49] have also been implicated in the regulation of plant gene expression in response to the osmotic strength of the environment and/or in the ABA-mediated transcriptional activation of certain genes. In addition, the gaseous plant hormone, ethylene, can markedly alter gene expression in a calcium-dependent pathway.[147] The multiple hormonal and environmental signals affecting gene expression in plants (see e.g., the expression of osmotin[48]) pose interesting questions concerning the interconnections between the signal transduction pathways that control osmoregulation, and those that control resistance to pathogens. These pathways may share common second messengers and hormonal signals.

Despite the unique hormonal cues employed by higher plants to respond to the osmotic strength of the environment, there are clearly many similarities in the osmoregulatory mechanisms of higher plants and other biological systems (including bacteria, cyanobacteria, eukaryotic marine algae, fungi, and animals) with respect to the array of compatible osmotic solutes employed for cytoplasmic osmotic adjustment.[7–9,11,12,72,79,82,146] Rapid progress can be expected within the next decade in the characterization and molecular cloning of genes encoding transport proteins[14,15,22,66,69] and genes determining compatible osmolyte accumulation (including genes conferring proline,[97,98] glycine betaine,[16,17] and sugar alcohol accumulation[18]) in higher plants. Under- and overexpression of these genes in heterologous plant systems is likely to provide valuable new information concerning the mechanisms of osmoregulation in plants and the elastic limits of plant osmotolerance.

ACKNOWLEDGMENTS

This chapter represents Purdue University Agricultural Experiment Station Journal Article # 13581. We thank the U.S. Department of Agriculture (contract # 91–37100–5873), Pioneer Hi-Bred International Inc., and the McKnight Foundation for financial support.

REFERENCES

1. **Blum, A.,** *Plant Breeding for Stress Environments,* CRC Press, Boca Raton, FL, 1988, chap. 3.
2. **Morgan, J. M.,** Osmoregulation and water stress in higher plants, *Annu. Rev. Plant Physiol.,* 35, 299, 1984.
3. **Hsiao, T. C.,** Plant responses to water stress, *Annu. Rev. Plant Physiol.,* 24, 519, 1973.
4. **Munns, R.,** Why measure osmotic adjustment?, *Aust. J. Plant Physiol.,* 15, 717, 1988.
5. **Reed, R. H.,** Use and abuse of osmo-terminology, *Plant Cell Environ.,* 7, 165, 1984.
6. **Zimmermann, U.,** Physics of turgor- and osmoregulation, *Annu. Rev. Plant Physiol.,* 29, 121, 1978.
7. **Csonka, L. N. and Hanson, A. D.,** Prokaryotic osmoregulation: genetics and physiology, *Annu. Rev. Microbiol.,* 45, 569, 1991.

8. **Borowitzka, L. A.,** Osmoregulation in blue-green algae, *Prog. Phycol. Res.*, 4, 243, 1986.
9. **Blomberg, A. and Adler, L.,** Physiology of osmotolerance in fungi, *Adv. Microbial Physiol.*, 33, 145, 1992.
10. **Hellebust, J. A.,** Osmoregulation, *Annu. Rev. Plant Physiol.*, 27, 485, 1976.
11. **Borowitzka, L. J.,** Solute accumulation and regulation of cell water activity, in *The Physiology and Biochemistry of Drought Resistance in Plants,* Paleg, L. G. and Aspinall, D., Eds., Academic Press, Sydney, 1981, 97.
12. **Kirst, G. O.,** Salinity tolerance of eukaryotic marine algae, *Annu. Rev. Plant Physiol. Plant Mol. Biol.*, 40, 21, 1989.
13. **Alexandre, J. and Lassalles, J.-P.,** Hydrostatic and osmotic pressure activated channel in plant vacuole, *Biophys. J.*, 60, 1326, 1991.
14. **Narasimhan, M. L., Binzel, M. L., Perez-Prat, E., Chen, Z., Nelson, D. E., Singh, N. K., Bressan, R. A., and Hasegawa, P. M.,** NaCl regulation of tonoplast ATPase 70-kilodalton subunit mRNA in tobacco cells, *Plant Physiol.*, 97, 562, 1991.
15. **Perez-Prat, E., Narasimhan, M. L., Binzel, M. L., Botella, M. A., Chen, Z., Valpuesta, V., Bressan, R. A., and Hasegawa, P. M.,** Induction of a putative Ca^{2+}-ATPase mRNA in NaCl adapted cells, *Plant Physiol.,* 100, 1471, 1992.
16. **Weretilnyk, E. A. and Hanson, A. D.,** Molecular cloning of a plant betaine-aldehyde dehydrogenase, an enzyme implicated in adaptation to salinity and drought, *Proc. Natl. Acad. Sci. U.S.A.*, 87, 2745, 1990.
17. **McCue, K. F. and Hanson, A. D.,** Salt-inducible betaine aldehyde dehydrogenase from sugar beet: cDNA cloning and expression, *Plant Mol. Biol.*, 18, 1, 1992.
18. **Vernon, D. M. and Bohnert, H. J.,** A novel methyl transferase induced by osmotic stress in the facultative halophyte *Mesembryanthemum crystallinum, EMBO J.*, 11, 2077, 1992.
19. **Morgan, J. M.,** A gene controlling differences in osmoregulation in wheat, *Aust. J. Plant Physiol.*, 18, 249, 1991.
20. **Morgan, J. M.,** Osmotic components and properties associated with genotypic differences in osmoregulation in wheat, *Aust. J. Plant Physiol.*, 19, 67, 1992.
21. **Rhodes, D. and Rich, P. J.,** Preliminary genetic studies of the phenotype of betaine deficiency in *Zea mays* L., *Plant Physiol.*, 88, 102, 1988.
22. **Sentenac, H., Bonneaud, N., Minet, M., Lacroute, F., Salmon, J.-M., Gaymard, F., and Grignon, C.,** Cloning and expression in yeast of a plant potassium ion transport system, *Science,* 256, 663, 1992.
23. **Tallman, G.,** The chemiosmotic model of stomatal opening revisited, *Crit. Rev. Plant Sci.*, 11, 35, 1992.
24. **Raschke, K.,** Stomatal action, *Annu. Rev. Plant Physiol.*, 26, 309, 1975.
25. **Sharpe, P. J. H., Wu, H., and Spence, R. D.,** Stomatal mechanics, in *Stomatal Function,* Zeiger, E., Farquhar, G. D., and Cowan, I. R., Eds., Stanford University Press, Stanford, CA, 1987, 91.
26. **Outlaw, W. H.,** An introduction to carbon metabolism in guard cells, in *Stomatal Function,* Zeiger, E., Farquhar, G. D., and Cowan, I. R., Eds., Stanford University Press, Stanford, CA, 1987, 115.
27. **MacRobbie, E. A. C.,** Ionic relations of guard cells, in *Stomatal Function,* Zeiger, E., Farquhar, G. D., and Cowan, I. R., Eds., Stanford University Press, Stanford, CA, 1987, 125.
28. **Raschke, K.,** Action of abscisic acid on guard cells, in *Stomatal Function,* Zeiger, E., Farquhar, G. D., and Cowan, I. R., Eds., Stanford University Press, Stanford, CA, 1987, 253.
29. **Jones, H. G.,** Breeding for stomatal characters, in *Stomatal Function,* Zeiger, E., Farquhar, G. D., and Cowan, I. R., Eds., Stanford University Press, Stanford, CA, 1987, 431.
30. **Smith, F. A. and Raven, J. A.,** Intracellular pH and its regulation, *Annu. Rev. Plant Physiol.*, 30, 289, 1979.
31. **McAinsh, M. R., Brownlee, C., Sarsağ, M., Webb, A. A. R., and Hetherington, A. M.,** Involvement of second messengers in the action of ABA, in *Abscisic Acid: Physiology and Biochemistry,* Davies, W. J. and Jones, H. G., Eds., BIOS Scientific, Oxford, 1991, 137.
32. **Xu, P., Lloyd, C. W., Staiger, C. J., and Drøbak, B. K.,** Association of phosphatidylinositol 4-kinase with the plant cytoskeleton, *Plant Cell,* 4, 941, 1992.
33. **Poovaiah, B. W. and Reddy, A. S. N.,** Calcium messenger system in plants, *Crit. Rev. Plant Sci.*, 6, 47, 1987.
34. **Roberts, D. M. and Harmon, A. C.,** Calcium-modulated proteins: targets of intracellular calcium signals in higher plants, *Annu. Rev. Plant Physiol. Plant Mol. Biol.*, 43, 375, 1992.
35. **Budde, R. J. A. and Chollet, R.,** Regulation of enzyme activity in plants by reversible phosphorylation, *Physiol. Plant.*, 72, 435, 1988.

36. **Ranjeva, R. and Boudet, A. M.,** Phosphorylation of proteins in plants: regulatory effects and potential involvement in stimulus/response coupling, *Annu. Rev. Plant Physiol.*, 38, 73, 1987.
37. **Michalke, B. and Schnabl, H.,** Modulation of the activity of phosphoenolpyruvate carboxylase during potassium-induced swelling of guard-cell protoplasts of *Vicia faba* L. after light and dark treatments, *Planta*, 180, 188, 1990.
38. **Champigny, M.-L. and Foyer, C.,** Nitrate activation of cytosolic protein kinases diverts photosynthetic carbon from sucrose to amino acid biosynthesis: basis for a new concept, *Plant Physiol.*, 100, 7, 1992.
39. **Alexandre, J., Lassalles, J. P., and Kado, R. T.,** Opening of Ca^{2+} channels in isolated red beet root vacuole membrane by inositol 1,4,5-trisphosphate, *Nature (London)*, 343, 567, 1990.
40. **Hedrich, R. and Neher, E.,** Cytoplasmic calcium regulates voltage-dependent ion channels in plant vacuoles, *Nature (London)*, 329, 833, 1987.
41. **Falke, L. C., Edwards, K. L., Pickard, B. G., and Misler, S.,** A stretch-activated anion channel in tobacco protoplasts, *FEBS Lett.*, 237, 141, 1988.
42. **MacRobbie, E. A. C.,** Effect of ABA on ion transport and stomatal regulation, in *Abscisic Acid: Physiology and Biochemistry*, Davies, W. J. and Jones, H. G., Eds., BIOS Scientific, Oxford, 1991, 153.
43. **McAinsh, M. R., Brownlee, C., and Hetherington, A. M.,** Visualizing changes in cytosolic-free Ca^{2+} during the response of stomatal guard cells to abscisic acid, *Plant Cell*, 4, 1113, 1992.
44. **Schroeder, J. I. and Hagiwara, S.,** Repetitive increases in cytosolic Ca^{2+} of guard cells by abscisic acid activation of nonselective Ca^{2+} permeable channels, *Proc. Natl. Acad. Sci. U.S.A.*, 87, 9305, 1990.
45. **Blatt, M. R.,** K^+ channels of stomatal guard cells: characteristics of the inward rectifier and its control by pH, *J. Gen. Physiol.*, 99, 615, 1992.
46. **Hartung, W. and Davies, W. J.,** Drought-induced changes in physiology and ABA, in *Abscisic Acid: Physiology and Biochemistry*, Davies, W. J. and Jones, H. G., Eds., BIOS Scientific, Oxford, 1991, 63.
47. **Bray, E. A.,** Regulation of gene expression by endogenous ABA during drought stress, in *Abscisic Acid: Physiology and Biochemistry*, Davies, W. J. and Jones, H. G., Eds., BIOS Scientific, Oxford, 1991, 81.
48. **LaRosa, P. C., Chen, Z., Nelson, D. E., Singh, N. K., Hasegawa, P. M., and Bressan, R. A.,** Osmotin gene expression is posttranscriptionally regulated, *Plant Physiol.*, 100, 409, 1992.
49. **Hildmann, T., Ebneth, M., Peña-Cortés, H., Sánchez-Serrano, J. J., Willmitzer, L., and Prat, S.,** General roles of abscisic and jasmonic acids in gene activation as a result of mechanical wounding, *Plant Cell*, 4, 1157, 1992.
50. **Taylor, I. B.,** Genetics of ABA synthesis, in *Abscisic Acid: Physiology and Biochemistry*, Davies, W. J. and Jones, H. G., Eds., BIOS Scientific, Oxford, 1991, 23.
51. **McCarty, D. R., Carson, C. B., Stinard, P. S., and Robertson, D. S.,** Molecular analysis of *viviparous-1:* an abscisic acid-insensitive mutant of maize, *Plant Cell*, 1, 523, 1989.
52. **McCarty, D. R., Hattori, T., Carson, C. B., Vasil, V., Lazar, M., and Vasil, I. K.,** The *vivaparous-1* developmental gene of maize encodes a novel transcriptional activator, *Cell*, 66, 895, 1991.
53. **Flowers, T. J., Troke, P. F., and Yeo, A. R.,** The mechanism of salt tolerance in halophytes, *Annu. Rev. Plant Physiol.*, 28, 89, 1977.
54. **Briens, M. and Larher, F.,** Osmoregulation in halophytic higher plants: a comparative study of soluble carbohydrates, polyols, betaines and free proline, *Plant Cell Environ.*, 5, 287, 1982.
55. **Rozema, J.,** Growth, water and ion relationships of halophytic monocotyledonae and dicotyledonae; a unified concept, *Aquatic Bot.*, 39, 17, 1991.
56. **Flowers, T. J., Hajibagheri, M. A., and Clipson, N. J. W.,** Halophytes, *Q. Rev. Biol.*, 61, 313, 1986.
57. **Greenway, H. and Munns, R.,** Mechanisms of salt tolerance in nonhalophytes, *Annu. Rev. Plant Physiol.*, 31, 149, 1980.
58. **Omielan, J. A., Epstein, E., and Dvořák, J.,** Salt tolerance and ionic relations of wheat as affected by individual chromosomes of salt-tolerant *Lophopyrum elongatum*, *Genome*, 34, 961, 1991.
59. **Epstein, E., Norlyn, J. D., Rush, D. W., Kingsbury, R. W., Kelley, D. B., Cunningham, G. A., and Wrona, A. F.,** Saline culture of crops: a genetic approach, *Science*, 210, 399, 1980.
60. **Schachtman, D. P. and Munns, R.,** Sodium accumulation in leaves of *Triticum* species that differ in salt tolerance, *Aust. J. Plant Physiol.*, 19, 331, 1992.

61. **Forster, B. P., Phillips, M. S., Miller, T. E., Baird, E., and Powell, W.,** Chromosome location of genes controlling salt (NaCl) and vigour in *Hordeum vulgare* and *H. chilense, Heredity,* 65, 99, 1990.
62. **Gorham, J., Hardy, C., Wyn Jones, R. G., Joppa, L. R., and Law, C. N.,** Chromosomal location of a K/Na discrimination character in the D genome of wheat, *Theor. Appl. Genet.,* 74, 584, 1987.
63. **Lüttge, U. and Clarkson, D. T.,** Mineral nutrition: potassium, in *Progress in Botany,* Vol. 50, Behnke, H.-D., Esser, K., Kubitzki, K., Runge, M., and Ziegler, H., Eds., Springer-Verlag, Berlin, 1989, 51.
64. **Hedrich, R. and Schroeder, J. I.,** The physiology of ion channels and electrogenic pumps in higher plants, *Annu. Rev. Plant Physiol.,* 40, 539, 1989.
65. **Tyerman, S. D.,** Anion channels in plants, *Annu. Rev. Plant Physiol. Plant Mol. Biol.,* 43, 351, 1992.
66. **Cao, Y., Anderova, M., Crawford, N. M., and Schroeder, J. I.,** Expression of an outward-rectifying potassium channel from maize mRNA and complementary RNA in Xenopus oocytes, *Plant Cell,* 4, 961, 1992.
67. **Schroeder, J. I. and Fang, H. H.,** Inward-rectifying K^+ channels in guard cells provide a mechanism for low-affinity K^+ uptake, *Proc. Natl. Acad. Sci. U.S.A.,* 88, 11583, 1991.
68. **Ward, J. M. and Sze, H.,** Proton transport activity of the purified vacuolar H^+-ATPase from oats: direct stimulation by Cl^-, *Plant Physiol.,* 99, 925, 1992.
69. **Gogarten, J. P., Fichmann, J., Braun, Y., Morgan, L., Styles, P., Taiz, S. L., DeLapp, K., and Taiz, L.,** The use of antisense mRNA to inhibit the tonoplast H^+ ATPase in carrot, *Plant Cell,* 4, 851, 1992.
70. **Schroeder, J. L. and Thuleau, P.,** Ca^{2+} channels in higher plant cells, *Plant Cell,* 3, 555, 1991.
71. **Gaxiola, R., Larrinoa, I. F., Villalba, J. M., and Serrano, R.,** A novel and conserved salt-induced protein is an important determinant of salt tolerance in yeast, *EMBO J.,* 11, 3157, 1992.
72. **Wyn Jones, R. G., Storey, R., Leigh, R. A., Ahmad, N., and Pollard, A.,** A hypothesis on cytoplasmic osmoregulation, in *Regulation of Cell Membrane Activities in Higher Plants,* Marrè, E. and Ciferri, O., Eds., Elsevier/North-Holland, Amsterdam, 1977, 121.
73. **McCue, K. F. and Hanson, A. D.,** Drought and salt tolerance: towards understanding and application, *Trends Biotechnol.,* 8, 358, 1990.
74. **Schröppel-Meier, G. and Kaiser, W. M.,** Ion homeostasis in chloroplasts under salinity and mineral deficiency. I. Solute concentrations in leaves and chloroplasts from spinach plants under NaCl or $NaNO_3$ salinity, *Plant Physiol.,* 87, 822, 1988.
75. **Matoh, T., Watanabe, J., and Takahashi, E.,** Sodium, potassium, chloride, and betaine concentrations in isolated vacuoles from salt-grown *Atriplex gmelini* leaves, *Plant Physiol.,* 84, 173, 1987.
76. **Génard, H., Le Saos, J., Billard, J.-P., Trémolières, A., and Boucaud, J.,** Effect of salinity on lipid composition, glycine betaine content and photosynthetic activity in chloroplasts of *Suaeda maritima, Plant Physiol. Biochem.,* 29, 421, 1991.
77. **Robinson, S. P. and Jones, G. P.,** Accumulation of glycinebetaine in chloroplasts provides osmotic adjustment during salt stress, *Aust. J. Plant Physiol.,* 13, 659, 1986.
78. **Low, P. S.,** Molecular basis of the biological compatibility of nature's osmolytes, in *Transport Processes, Iono- and Osmoregulation,* Gilles, R. and Gilles-Baillien, M., Eds., Springer-Verlag, Berlin, 1985, 469.
79. **Yancey, P. H., Clark, M. E., Hand, S. C., Bowlus, R. D., and Somero, G. N.,** Living with water stress: evolution of osmolyte systems, *Science,* 217, 1214, 1982.
80. **Somero, G. N.,** Protons, osmolytes, and fitness of internal mileu for protein function, *Am. J. Physiol.,* 251, R197, 1986.
81. **Wyn Jones, R. G. and Storey, R.,** Betaines, in *The Physiology and Biochemistry of Drought Resistance in Plants,* Paleg, L. G. and Aspinall, D., Eds., Academic Press, Sydney, 1981, 171.
82. **Rhodes, D. and Hanson, A. D.,** Quaternary ammonium and tertiary sulfonium compounds in higher plants, *Annu. Rev. Plant Physiol. Plant Mol. Biol.,* 44, 357, 1993.
83. **Dacey, J. W. H., King, G. M., and Wakeham, S. G.,** Factors controlling emission of dimethylsulfide from salt marshes, *Nature,* 330, 643, 1987.
84. **van Diggelen, J., Rozema, J., Dickson, D. M. J., and Broekman, R.,** β-3-Dimethylsulphoniopropionate, proline and quaternary ammonium compounds in *Spartina anglica* in relation to sodium chloride, nitrogen and sulphur, *New Phytol.,* 103, 573, 1986.
85. **Stewart, G. R., Larher, F., Ahmad, I., and Lee, J. A.,** Nitrogen metabolism and salt-tolerance in higher plant halophytes, in *Ecologi-*

cal Processes in Coastal Environments, Jefferies, R. L. and Davy, A. J., Eds., Blackwell, Oxford, 1979, 211.

86. **Hanson, A. D., Rathinasabapathi, B., Chamberlin, B., and Gage, D. A.,** Comparative physiological evidence that β-alanine betaine and choline-O-sulfate act as compatible osmolytes in halophytic *Limonium* species, *Plant Physiol.,* 97, 1199, 1991.

87. **Hanson, A. D. and Gage, D. A.,** Identification and determination by fast atom bombardment mass spectrometry of the compatible solute, choline-O-sulfate in *Limonium* species and other halophytes, *Aust. J. Plant Physiol.,* 18, 317, 1991.

88. **Homann, P. H.,** Stabilization of the water oxidizing polypeptide assembly on Photosystem II membranes by osmolytes and other solutes, *Photosynthesis Res.,* 33, 29, 1992.

89. **Papageorgiou, G. C., Fujimura, Y., and Murata, N.,** Protection of the oxygen-evolving Photosystem II complex by glycinebetaine, *Biochim. Biophys. Acta,* 1057, 361, 1991.

90. **Incharoensakdi, A., Takabe, T., and Akazawa, T.,** Effect of betaine on enzyme activity and subunit interaction of ribulose-1,5-bisphosphate carboxylase/oxygenase from *Aphanothece halophytica, Plant Physiol.,* 81, 1044, 1986.

91. **Hanson, A. D., May, A. M., Grumet, R., Bode, J., Jamieson, G. C., and Rhodes, D.,** Betaine synthesis in chenopods: localization in chloroplasts, *Proc. Natl. Acad. Sci. U.S.A.,* 82, 3678, 1985.

92. **Weigel, P., Weretilnyk, E. A., and Hanson, A. D.,** Betaine aldehyde oxidation by spinach chloroplasts, *Plant Physiol.,* 82, 753, 1986.

93. **Lerma, C., Hanson, A. D., and Rhodes, D.,** Oxygen-18 and deuterium labeling studies of choline oxidation by spinach and sugar beet, *Plant Physiol.,* 88, 695, 1988.

94. **Brouquisse, R., Weigel, P., Rhodes, D., Yocum, C. F., and Hanson, A. D.,** Evidence for a ferredoxin-dependent choline monooxygenase from spinach chloroplast stroma, *Plant Physiol.,* 90, 322, 1989.

95. **Weretilnyk, E. A., Bednarek, S., McCue, K. F., Rhodes, D., and Hanson, A. D.,** Comparative biochemical and immunological studies of the glycine betaine synthesis pathway in diverse families of dicotyledons, *Planta,* 178, 342, 1989.

96. **Rathinasabapathi, B., Gage, D. A., Mackill, D. J., and Hanson, A. D.,** Cultivated and wild rices do not accumulate glycinebetaine due to deficiencies in two biosynthetic steps, *Crop Sci.,* 33, 534, 1993.

97. **Szoke, A., Miao, G.-H., Hong, Z., and Verma, D. P. S.,** Subcellular location of Δ^1-pyrroline-5-carboxylate reductase in root/nodule and leaf of soybean, *Plant Physiol.,* 99, 1642, 1992.

98. **Garcia-Rios, M. G., LaRosa, P. C., Bressan, R. A., Csonka, L. N., and Hanquier, J. M.,** Cloning by complementation of the γ-glutamyl kinase gene from a tomato expression library, *Int. Soc. Plant Mol. Biol. 3rd Int. Cong.,* 1507, 1991.

99. **Tarczynski, M. C., Jensen, R. G., and Bohnert, H. J.,** Expression of a bacterial *mtlD* gene in transgenic tobacco leads to production and accumulation of mannitol, *Proc. Natl. Acad. Sci. U.S.A.,* 89, 2600, 1992.

100. **Hasegawa, P. M., Bressan, R. A., and Handa, A. K.,** Cellular mechanisms of salinity tolerance, *Hortic. Sci.,* 21, 1317, 1986.

101. **Singh, N. K., Nelson, D. E., Kuhn, D., Hasegawa, P. M., and Bressan, R. A.,** Molecular cloning of osmotin and regulation of its expression by ABA and adaptation to low water potential, *Plant Physiol.,* 90, 1096, 1989.

102. **LaRosa, P. C., Hasegawa, P. M., Rhodes, D., Clithero, J. M., Watad, A.-E. A., and Bressan, R. A.,** Abscisic acid stimulated osmotic adjustment and its involvement in adaptation of tobacco cells to NaCl, *Plant Physiol.,* 85, 174, 1987.

103. **Binzel, M. L., Hess, F. D., Bressan, R. A., and Hasegawa, P. M.,** Intracellular compartmentation of ions in salt adapted tobacco cells, *Plant Physiol.,* 86, 607, 1988.

104. **Binzel, M. L., Hasegawa, P. M., Rhodes, D., Handa, S., Handa, A. K., and Bressan, R. A.,** Solute accumulation in tobacco cells adapted to NaCl, *Plant Physiol.,* 84, 1408, 1987.

105. **Rhodes, D. and Handa, S.,** Amino acid metabolism in relation to osmotic adjustment in plant cells, in *Environmental Stress in Plants: Biochemical and Physiological Mechanisms,* NATO ASI Series, Vol. G19, Cherry, J. H., Ed., Springer-Verlag, Berlin, 1989, 41.

106. **Kononowicz, A. K., Floryanowicz-Czekalska, K., Clithero, J., Meyers, A., Hasegawa, P. M., and Bressan, R. A.,** Chromosome number and DNA content of tobacco cells adapted to NaCl, *Plant Cell Rep.,* 8, 672, 1990.

107. **Osmond, C. B.,** Crassulacean acid metabolism: a curiosity in context, *Annu. Rev. Plant Physiol.,* 29, 379, 1978.

108. **Lüttge, U., Smith, J. A. C., and Marigo, G.,** Membrane transport, osmoregulation, and the

control of CAM, in *Crassulacean Acid Metabolism,* Ting, I. P. and Gibbs, M., Eds., Waverly Press, Baltimore, 1982, 69.

109. **Black, C. C., Carnal, N. W., and Kenyon, W. H.,** Compartmentation and the regulation of CAM, in *Crassulacean Acid Metabolism,* Ting, I. P. and Gibbs, M., Eds., Waverly Press, Baltimore, 1982, 51.
110. **Lüttge, U. and Smith, J. A. C.,** Structural, biophysical, and biochemical aspects of the role of leaves in plant adaptation to salinity and water stress, in *Salinity Tolerance in Plants: Strategies for Crop Improvement,* Staples, R. C. and Toenniessen, G. H., Eds., John Wiley & Sons, New York, 1984, 125.
111. **Paul, M. J. and Cockburn, W.,** Pinitol, a compatible solute in *Mesembryanthemum crystallinum* L.?, *J. Exp. Bot.*, 40, 1093, 1989.
112. **McElwain, E. F., Bohnert, H. J., and Thomas, J. C.,** Light moderates the induction of phospho*enol*pyruvate carboxylase by NaCl and abscisic acid in *Mesembryanthemum crystallinum, Plant Physiol.*, 99, 1261, 1992.
113. **Vernon, D. M. and Bohnert, H. J.,** Increased expression of a *myo*-inositol methyl transferase in *Mesembryanthemum crystallinum* is part of a stress response distinct from Crassulacean Acid Metabolism induction, *Plant Physiol.,* 99, 1695, 1992.
114. **Schmitt, J. M. and Piepenbrock, M.,** Regulation of phosphoenolpyruvate carboxylase and Crassulacean Acid Metabolism induction in *Mesembryanthemum crystallinum* L. by cytokinin: modulation of gene expression by roots?, *Plant Physiol.,* 99, 1664, 1992.
115. **Chu, C., Dai, Z. Y., Ku, M. S. B., and Edwards, G. E.,** Induction of Crassulacean Acid Metabolism in the facultative halophyte *Mesembryanthemum crystallinum* L. by abscisic acid, *Plant Physiol.,* 93, 1253, 1990.
116. **Thomas, J. C., McElwain, E. F., and Bohnert, H. J.,** Convergent induction of osmotic stress-responses, *Plant Physiol.,* 100, 416, 1992.
117. **Sommer, C., Thonke, B., and Popp, M.,** The compatibility of D-pinitol and 1D-1-O-methyl-muco-inositol with malate dehydrogenase activity, *Bot. Acta,* 103, 270, 1990.
118. **Blum, A.,** Osmotic adjustment and growth of barley genotypes under drought stress, *Crop Sci.,* 29, 230, 1989.
119. **Morgan, J. M. and King, R. W.,** Association between loss of leaf turgor, abscisic acid levels and seed set in two wheat cultivars, *Aust. J. Plant Physiol.,* 11, 143, 1984.
120. **Morgan, J. M. and Condon, A. G.,** Water use, grain yield, and osmoregulation in wheat, *Aust. J. Plant Physiol.,* 13, 523, 1986.
121. **Morgan, J. M., Hare, R. A., and Fletcher, R. J.,** Genetic variation in osmoregulation in bread and durum wheats and its relationship to grain yield in a range of field environments, *Aust. J. Agric. Res.,* 37, 449, 1986.
122. **Morgan, J. M.,** Osmoregulation as a selection criterion for drought tolerance in wheat, *Aust. J. Agric. Res.,* 34, 607, 1983.
123. **Blum, A., Mayer, J., and Gozlan, G.,** Associations between plant production and some physiological components of drought resistance in wheat, *Plant Cell Environ.,* 6, 219, 1983.
124. **Ludlow, M. M. and Muchow, R. C.,** A critical evaluation of traits for improving crop yields in water-limited environments, *Adv. Agron.,* 43, 107, 1990.
125. **Blum, A. and Sullivan, C. Y.,** The comparative drought resistance of landraces of sorghum and millet from dry and humid regions, *Ann. Bot.,* 57, 835, 1986.
126. **Santamaria, J. M., Ludlow, M. M., and Fukai, S.,** Contribution of osmotic adjustment to grain yield in *Sorghum bicolor* (L.) Moench under water-limited conditions. I. Water stress before anthesis, *Aust. J. Agric. Res.,* 41, 51, 1990.
127. **Ludlow, M. M., Santamaria, J. M., and Fukai, S.,** Contribution of osmotic adjustment to grain yield in *Sorghum bicolor* (L.) Moench under water-limited conditions. II. Water stress after anthesis, *Aust. J. Agric. Res.,* 41, 67, 1990.
128. **Morgan, J. M., Rodriguez-Maribona, B., and Knights, E. J.,** Adaptation to water-deficit in chickpea breeding lines by osmoregulation: relationship to grain-yields in the field, *Field Crops Res.,* 27, 61, 1991.
129. **Flower, D. J., Rani, A. U., and Peacock, J. M.,** Influence of osmotic adjustment on the growth, stomatal conductance and light interception of contrasting sorghum lines in a harsh environment, *Aust. J. Plant Physiol.,* 17, 91, 1990.
130. **Blum, A., Golan, G., Mayer, J., Sinmena, B., Shpiler, L., and Burra, J.,** The drought response of landraces of wheat from the northern Negev Desert in Israel, *Euphytica,* 43, 87, 1989.
131. **Quisenberry, J. E., Cartwright, G. B., and McMichael, B. L.,** Genetic relationships between turgor maintenance and growth in cotton germplasm, *Crop Sci.,* 24, 479, 1984.

132. **Grumet, R., Albrechtsen, R. S., and Hanson, A. D.,** Growth and yield of barley isopopulations differing in solute potential, *Crop Sci.,* 27, 991, 1987.
133. **Bolaños, J. and Edmeades, G. O.,** Value of selection for osmotic potential in tropical maize, *Agron. J.,* 83, 948, 1991.
134. **Hanson, A. D. and Grumet, R.,** Betaine accumulation: metabolic pathways and genetics, in *Cellular and Molecular Biology of Plant Stress,* Key, J. L. and Kosuge, T., Eds., Alan R. Liss, New York, 1985, 71.
135. **Grumet, R. and Hanson, A. D.,** Genetic evidence for an osmoregulatory function of glycinebetaine accumulation in barley, *Aust. J. Plant Physiol.,* 13, 353, 1986.
136. **Brunk, D. G., Rich, P. J., and Rhodes, D.,** Genotypic variation for glycinebetaine among public inbreds of maize, *Plant Physiol.,* 91, 1122, 1989.
137. **Rhodes, D., Rich, P. J., Brunk, D. G., Ju, G. C., Rhodes, J. C., Pauly, M. H., and Hansen, L. A.,** Development of two isogenic sweet corn hybrids differing for glycinebetaine content, *Plant Physiol.,* 91, 1112, 1989.
138. **Lerma, C., Rich, P. J., Ju, G. C., Yang, W.-J., Hanson, A. D., and Rhodes, D.,** Betaine deficiency in maize: complementation tests and metabolic basis, *Plant Physiol.,* 95, 1113, 1991.
139. **McCree, K. J., Kallsen, C. E., and Richardson, S. G.,** Carbon balance of sorghum plants during osmotic adjustment to water stress, *Plant Physiol.,* 76, 898, 1984.
140. **McCree, K. J.,** Whole-plant carbon balance during osmotic adjustment to drought and salinity stress, *Aust. J. Plant Physiol.,* 13, 33, 1986.
141. **Saab, I. N., Sharp, R. E., Pritchard, J., and Voetberg, G. S.,** Increased endogenous abscisic acid maintains primary root growth and inhibits shoot growth of maize seedlings at low water potentials, *Plant Physiol.,* 93, 1329, 1990.
142. **Sharp, R. E., Silk, W. K., and Hsiao, T. C.,** Growth of the maize primary root at low water potentials. I. Spatial distribution of expansive growth, *Plant Physiol.,* 87, 50, 1988.
143. **Sharp, R. E., Hsiao, T. C., and Silk, W. K.,** Growth of the maize primary root at low water potentials. II. Role of growth and deposition of hexose and potassium in osmotic adjustment, *Plant Physiol.,* 93, 1337, 1990.
144. **Voetberg, G. S. and Sharp, R. E.,** Growth of the maize primary root at low water potentials. III. Role of increased proline deposition in osmotic adjustment, *Plant Physiol.,* 96, 1125, 1991.
145. **Spollen, W. G. and Sharp, R. E.,** Spatial distribution of turgor and root growth at low water potentials, *Plant Physiol.,* 96, 438, 1991.
145a. **Sharp, R. E.,** personal communication.
146. **Anthoni, U., Christophersen, C., Hougaard, L., and Nielsen, P. H.,** Quaternary ammonium compounds in the biosphere — an example of a versatile adaptive strategy, *Comp. Biochem. Physiol.,* 99B, 1, 1991.
147. **Raz, V. and Fluhr, R.,** Calcium requirement for ethylene-dependent responses, *Plant Cell,* 4, 1123, 1992.
148. **Ludlow, M. M.,** personal communication.

Chapter 22

Transcription Factors and Stress Proteins in Renal Epithelial Cell Adaptation to Hyperosmotic Stress

David M. Cohen and Steven R. Gullans

TABLE OF CONTENTS

I. THE IMMEDIATE-EARLY GENES AND HYPEROSMOTIC STRESS

The renal medulla is the only mammalian tissue normally exposed to elevated (and fluctuating) total osmolality as a consequence of the renal concentrating mechanism. NaCl and urea contribute the bulk of this excess osmolality. After cells *in vivo* and *in vitro* are acutely subjected to hyperosmotic stress, intracellular accumulation of osmolytes typically requires hours to days, as does upregulation of the proteins responsible for osmolyte synthesis or uptake.[1] For this reason, the nature of the earliest genetic events associated with the cellular response to hyperosmotic stress is of great interest.

A. THE IMMEDIATE-EARLY GENES

The immediate-early genes (IEGs) comprise a diverse group of genes that can be rapidly and transiently upregulated at the mRNA level in response to a particular stimulus, even in the absence of new protein synthesis.[2,3] Inducers of an IEG response are diverse as well; the best-studied examples include serum, growth factors,[3–5] and phorbol esters,[4–7] although physical stimuli, such as heat shock, can also induce IEG expression.[6,8] An important subset of IEGs encode known or putative transcription factors — nuclear proteins that bind DNA in a sequence-specific fashion and alter the rate at which associated genes are transcribed into mRNA. As such, they have been likened to a "third messenger system"[9] capable of initiating a coordinated *program* of gene expression in response to a single stimulus. Examples of this class of IEG include the nuclear protooncogenes c-*fos* and c-*myc*,[10] the zinc finger protein *Egr-1*,[3,11,12] and others. Based upon our data, we hypothesize that hyperosmotic stress in renal epithelial cells increases expression of a subset of IEGs, which in turn activates expression of a series of effector genes comprising an adaptive response to hyperosmotic stress.

In our investigations we have focused principally upon c-*fos* and *Egr-1*. The product of the c-*fos* protooncogene encodes a 380-amino acid nuclear protein with widely studied transcriptional activating ability.[13] c-*Fos* is induced at the transcriptional level within minutes of mitogenic stimulation in a variety of cell types,[5] as well as in

0-8493-4448-4/94/$0.00+$.50

numerous other setting.[13] c-*Fos* (or one of its homologs) interacts with the protooncogene c-*jun* (or its homologs) through their respective leucine zipper motifs;[14–17] the resulting heterodimer constitutes the site-specific DNA binding factor and transcriptional activator known as AP-1.[18–20] The consensus sequence to which AP-1 binds (T(T/G)A(C/G)TCA) is found in the promoter region of many genes, including c-*fos* itself[20] and c-*jun*.[21,22]

Egr-1 induction parallels that of c-*fos* in many respects. *Egr-1* is rapidly induced after mitogen stimulation with a time course similar to c-*fos*, although *Egr-1* generally accumulates to a considerably higher level.[3] The *Egr-1* gene encodes a nuclear phosphoprotein containing a zinc finger DNA binding motif.[23–25] Unlike c-*fos*, *Egr-1* does not require (hetero)dimerization prior to interacting with its cognate GCG(C/G/T)GGGCG consensus sequence.[23,24,26] The ability of *Egr-1* protein to activate transcription has been confirmed *in vitro*.[24]

B. IEG INDUCTION BY HYPEROSMOTIC STRESS

We have examined the induction of IEGs by hyperosmotic stress in cell culture models using renal epithelial MDCK and LLC-PK_1 cells. Cells were grown to confluence and then growth-suppressed with relative serum-deprivation (1% SCS) for 48 h. Confluent monolayers were used for two reasons: (1) to minimize background serum- and growth-associated IEG expression;[27] and (2) because the use of quiescent high-density cultures of renal epithelial cells has been advocated as a means of mimicking the relatively low mitotic activity of renal cells *in vivo*.[28] Cells were then treated with control medium or medium supplemented with hyperosmotic solute, usually in the range of 200 mOsmol. Expression of IEG mRNA was examined by Northern analysis at various time points after stimulation and then quantitated with scanning densitometry of autoradiograms. NaCl (200 mOsmol) induced a two- to threefold upregulation of *Egr-1* and c-*fos* mRNA at 2 h of treatment but not at 30 min in MDCK cells; control treatment and treatment with the hyperosmotic but readily membrane-permeant solute, glycerol, failed to do so[27] (Figures 22–1 and 22–2).

Because urea represents the other principal renal medullary solute, its ability to upregulate *Egr-1* and c-*fos* mRNA was examined. Interestingly, 200-mOsmol urea, unlike the other readily membrane-permeant solute, glycerol, rapidly and markedly upregulated these genes, almost to the same degree as the potent inducer TPA, a phorbol ester. Expres-

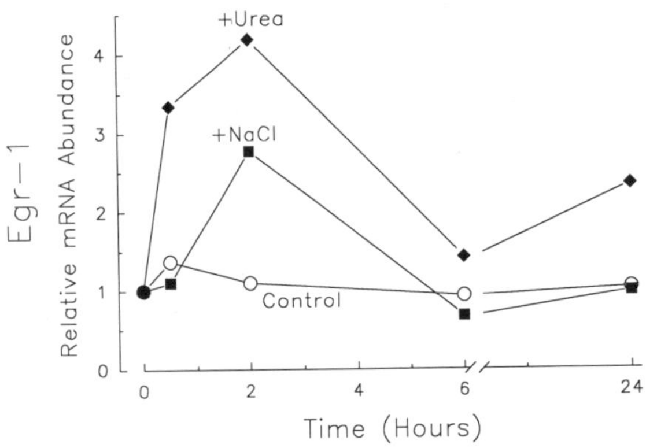

Figure 22–1 Time course of *Egr-1* mRNA abundance (measured densitometrically) in response to hyperosmotic stress induced by various solutes in serum-deprived MDCK cells.

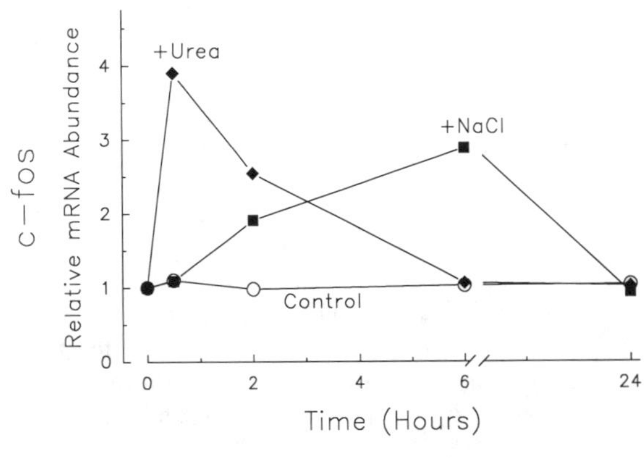

Figure 22–2 Time course of c-*fos* mRNA abundance in response to hyperosmotic stress induced by various solutes in serum-deprived MDCK cells.

sion was increased within 30 min of exposure to hyperosmotic urea and then returned essentially to baseline by 6 h of treatment[29] (Figures 22–1 and 22–2). A similar time course of IEG induction has been observed by others in response to mitogenic stimulation in numerous models.[3–5] Preliminary evidence indicates that *Egr-1* expression is also increased at the *protein* level by urea treatment of renal epithelial cells. In addition to the MDCK cell line, the porcine renal epithelial LLC-PK_1 cell line also exhibited a marked increase in *Egr-1* mRNA expression in response to urea treatment. The bovine glomerular mesangial cell, a renal but nonepithelial cell type, failed to exhibit this phenomenon, as did the human colon carcinoma cell line T84, an epithelial but nonrenal cell type. It therefore appeared that the ability to respond to hyperosmotic urea with enhanced expression of *Egr-1* was restricted to cells of renal epithelial origin.[29] Because renal epithelial cells may be uniquely exposed to elevated (and fluctuating) concentrations of urea, the teleological significance of a specialized adaptive response to hyperosmotic urea is clear. The potential mechanisms of this upregulation are addressed below.

Apart from our observations in cell culture, there are no direct observations linking hyperosmotic stress and IEG expression in a physiological system. There are several examples of renal IEG expression at the mRNA and protein levels in animal models in response to various other stressors, including folate nephropathy,[30,31] recovery from acute ischemic renal failure,[32] and the compensatory contralateral response to both unilateral ureteral obstruction[33] and uninephrectomy.[34] These models, however, are primarily associated with renal hypertrophy (and potentially the attendant mitogenesis); none has been directly correlated with changes in osmolality.

C. THE MECHANISM OF HYPEROSMOTIC STRESS-INDUCED IEG EXPRESSION

The mechanism underlying hyperosmotic stress-induced IEG expression is obscure. Because multiple adverse stimuli have been shown to induce IEG expression, the nonspecific cytotoxic effect of hyperosmotic stress was determined; hyperosmotic stimuli in the range employed were not cytotoxic as determined by LDH release into cell supernatants.[27] Because IEG expression may be induced by an inhibition of protein synthesis alone,[35,36] the ability of hyperosmotic stress to inhibit protein synthesis was examined. Hyperosmotic NaCl has been shown by us and others[27,37–39] to inhibit the rate of protein synthesis as measured by [^{3}H]leucine incorporation; therefore, it was impossible to dissociate NaCl-induced IEG expression from a potential nonspecific effect attributable to a generalized NaCl-induced inhibition of protein synthesis. Interestingly, urea failed to inhibit protein synthesis in the concentrations at which it stimulated IEG expression.[29] Therefore, at least in the case of urea, hyperosmotic stress-induced IEG expression is not a consequence of inhibited protein synthesis.

IEG expression also appears to be an invariant accompaniment of mitogenesis; it was in this context that mammalian IEG regulation was initially elucidated.[4,5] It was observed that hyperosmotic NaCl potently inhibits DNA synthesis in renal epithelial cells;[27] therefore, the gene induction observed is not a consequence of mitogenesis. With urea treatment, however, the picture is not as clear. In addition to stimulating IEG expression in a renal epithelial cell-specific fashion, urea also increases DNA synthesis, as measured by both[^{3}H]thymidine incorporation and fluorimetric total DNA quantitation.[40] Urea, however, appears not to be a mitogen for renal epithelial cells because an increase in cell number could not be demonstrated.[40] Furthermore, urea failed to alter the cell cycle distribution of treated cells, arguing against a proproliferative effect.[40] Therefore, hyperosmotic stress-inducible IEG expression is not a consequence of mitogenesis, but rather, likely represents a specific adaptive response to hyperosmotic stress.

To define at the molecular level whether urea-inducible IEG upregulation is a result of enhanced transcription or increased mRNA stability, actinomycin D studies were performed. Messenger RNA half-life was unaffected by urea treatment,[29] and was in agreement with published data.[41,42] Transcriptional upregulation would appear to be the likely mechanism of increased message abundance.

Because hyperosmotic stress-induced IEG expression appears specific and is likely transcriptionally mediated, we hypothesize that hyperosmotic urea (and perhaps other solutes as well) activates transcription of *Egr-1* and/or c-*fos* via either a known or novel *cis*-acting regulatory element in the promoters of these genes. A number of functional *cis*-acting elements have already been identified in these promoters, to which transcriptional regulatory proteins may bind and influence gene expression in a complex fashion. Both promoters contain the serum response element (SRE) or elements closely resembling it,[26,43–45] which likely confers the parallel induction kinetics of c-*fos* and *Egr-1* in several models of mitogenesis, including cells of renal origin.[3,46] The SRE mediates both *Egr-1* and c-*fos* induction by growth factors, serum, and TPA.[47–50] The SRE is bound by the serum response factor (SRF) which may require phosphorylation for transcriptional activation.[13,44,51] Both promoters also contain at least one putative AP-1 site, which binds the fos/jun heterodimer.[18–20] Although this site confers TPA-inducibility to c-*jun* (and hence is known as the TPA-response element), it is *not* involved in TPA induction of c-*fos*[13] or *Egr-1*.[50] The c-*fos* promoter also contains a cAMP response element (CRE),[52] which is bound by a protein kinase A (PKA)-phosphorylated form of the CRE binding protein (CREB).[13,53] Calcium- and cAMP-mediated c-*fos* induction occur via the CRE;[49,53] the function of two CRE-like elements in the *Egr-1* promoter is unknown.[45] Other putative regulatory sites have been identified;[45] however, functional data are lacking.

We hypothesize that hyperosmotic urea activates IEG transcription by altering the conformation, and hence DNA-binding affinity, of a *known* or *novel trans*-acting factor. Apart from our observations on IEG expression, no effects of urea upon eukaryotic gene expression *in vivo* have been described. As urea stimulation, like heat shock, represents a potentially denaturing stimulus,[54] in contrast to most biochemical stimuli (e.g., peptide growth factors), a urea-induced conformational effect upon an existing or latent nuclear DNA binding activity can readily be envisioned. A DNA binding protein, apparently unrelated to the cellular response to urea, has been identified which has remarkable affinity for its *cis*-acting element despite extraordinarily high concentrations of urea *in vitro*.[55] In addition, it has been demonstrated that elevated concentrations of urea, albeit in excess of those employed in our model (0.5 to 1.0 *M*), can *activate* binding of the heat shock transcription factor HSF (v.i.) to its cognate heat shock element.[56] Furthermore, in this same model the presence of deuterium oxide, a known stabilizer of protein structure, *inhibited* urea-induced activation of HSF.[56] While exclusively *in vitro* observations, these examples nonetheless provide important precedents for urea-resistant and urea-inducible DNA binding activity *in vivo*. A change in such binding affinity (and hence promoter occupancy) could readily alter the rate of gene transcription. Furthermore, since the methylamine osmolyte glycerophosphorylcholine (GPC) accumulates with urea treatment[57,58] and likely counteracts the protein-denaturing effect of urea,[1,54,59–61] it is conceivable that a urea-induced conformational change in a DNA binding protein may be reversed by intracellular methylamine accumulation. This could serve as a transcriptional "switch" that is activated by urea and inactivated after methylamines have accumulated to a sufficiently protective level.

II. THE HEAT SHOCK RESPONSE AND HYPEROSMOTIC STRESS

A. THE HEAT SHOCK RESPONSE

The heat shock response is an extremely highly conserved genetic program inducible by a variety of chemical and physical stimuli, including heat shock, oxidative stress, and heavy metals.[62] In mammals, the HSP70 family of stress proteins includes HSP70, HSC70, and GRP78; in the human, mouse, and rat, a fourth protein, HSP72, is also detectable.[63] Although members of the HSP70 family exhibit differences in intracellular localization and degree of stress inducibility, in general they appear to be instrumental in the folding and membrane translocation of nascent polypeptides, the stabilization and refolding of denatured or malformed proteins, and the regulation of protein turnover.[64] HSP70 is one of the principal constituents of the mammalian heat shock response[65] and can be promptly upregulated at the mRNA level even in the absence of new protein synthesis, much like an IEG.[66] For these reasons, and because hyperosmotic stress, from an evolutionary standpoint, represents an environmental stressor conceivably as ubiquitous as heat shock, we examined the ability of hyperosmotic stress to upregulate mRNA encoding HSP70 in renal epithelial cells.

B. HSP70 AND HYPEROSMOTIC STRESS

In our investigation,[27] treatment of confluent serum-deprived MDCK cells with 100 m*M* NaCl caused a sustained fivefold increase in HSP70 message abundance (Figure 22–3). Interestingly, equiosmolar hyperosmotic stress with the permeant solutes glycerol and urea (200 m*M*) failed to affect HSP70 mRNA level, as did vehicle treatment. In addition, this NaCl-inducible upregulation was blunted by the protein synthesis inhibitor cycloheximide (unpublished observation). With the exception of studies in tomatoes, there are very few reports of the effects of hyperosmolality upon HSP expression. Using BALB/c3T3 fibroblasts, Tanaka and co-workers[67] examined the time-dependent effects of hyperosmolar medium on HSP70 and HSP72 expression, as well as several GRPs and the glucose permease, GLUT1. They observed that hypertonic stress increased expression of HSP70, GRP78, and GRP95 with mRNA levels increasing 4- to 50-fold by 8 h of treatment. Emami and co-workers[68] noted that both renal ischemia and heat shock induce renal expression of immunoreactive HSP70. Interestingly, in their study the renal pa-

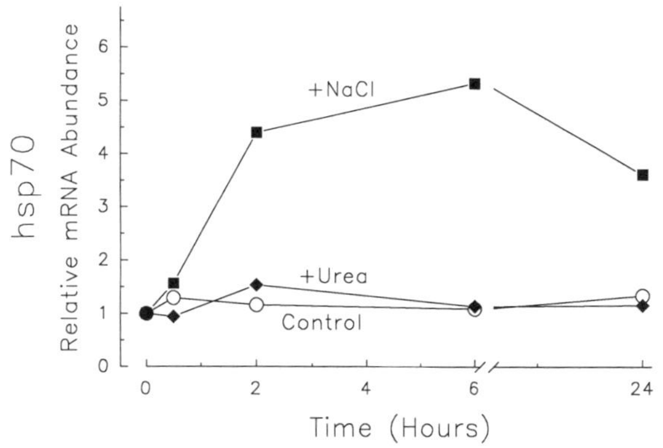

Figure 22–3 Time course of HSP70 mRNA abundance in serum-deprived MDCK cells treated with various solutes.

pilla, which is likely subjected to the most hyperosmotic environment, exhibited the most rapid and robust increase in HSP70 protein content. Recently, Cowley and co-workers[69] observed upregulation of HSP70 mRNA in the murine inner medulla in three different models of antidiuresis resulting in increased inner medullary sodium and urea concentrations. Therefore it appears that both *in vitro* and *in vivo* hyperosmotic stress increase HSP70 expression at the mRNA level.

C. IEG EXPRESSION AND THE HEAT SHOCK RESPONSE

The relationship between the IEG expression and the heat shock response is complex. Heat shock itself can increase both c-*fos*[70,71] and *Egr-1* mRNA (unpublished observation). As with IEG expression, HSP70 mRNA expression has been associated with mitogenesis[72,73] and, more specifically, with the S-phase of the cell cycle.[74] Therefore, sequential activation of IEGs and HSP70 in the setting of cell proliferation would not be unexpected. We have described sequential expression of IEG and HSP70 mRNA despite a marked inhibition of DNA synthesis in NaCl-treated MDCK cells. Several *in vivo* rat models are associated with a similar "cascade" induction of members of the IEG and HSP70 gene families, including ventricular myocardial pressure overload,[75] postischemic hepatic reperfusion,[76] and postcastration regression of the ventral prostate gland.[77] The relationship of these models to mitogenesis remains undefined.

D. TRANSCRIPTIONAL REGULATION OF THE HEAT SHOCK RESPONSE

The mechanism of activation of HSP70 is the subject of much recent investigation. Although heat shock can induce expression of the IEG c-*myc*,[78] whose protein product can, in turn, activate transcription from the HSP70 promoter,[79] stress protein expression in eukaryotes is more typically regulated by posttranslational modification of a constitutively expressed transcription factor known as the heat shock factor (HSF). We observed that hyperosmotic stress failed to upregulate expression of c-*myc* mRNA,[27] arguing in favor of the latter mechanism of HSP70 induction in response to hyperosmotic stress.

The HSF interacts with double-stranded DNA in a sequence-specific fashion at heat shock elements (HSEs) found in the promoters of all heat-shock inducible genes. Two distinct HSFs, HSF1 and HSF2, have been cloned from human and mouse tissues,[80–82] and exhibit temperature-sensitive interaction with the HSE *in vitro*.[82] Because our data implied a requirement for protein synthesis in achieving a maximal HSP70 response to hyperosmotic NaCl, we examined the ability of hyperosmotic stress to alter HSF message abundance. Supplemental NaCl (100 m*M*) failed to alter mRNA abundance of either HSF1 or HSF2 in serum-deprived MDCK cells (unpublished observation); however, its effect upon posttranslational modification, and hence activation, of constitutively expressed HSF is unknown. Interestingly, the human HSP70 gene, which is serum-responsive,[73] contains an upstream regulatory element resembling the SRE shared by *Egr-1* and c-*fos*.[83] Whereas it is difficult to extrapolate to the hyperosmotic stress-inducible member of the canine HSP70 family, it is conceivable that *Egr-1*, c-*fos*, and HSP70 — all of which are upregulated at the mRNA level within 2 h of NaCl-induced hyperosmotic stress — are inducible by a single *trans*-acting factor.

E. PHYSIOLOGICAL RELEVANCE TO HYPEROSMOTIC STRESS

A potential role for members of the HSP70 family in adaptation to hyperosmotic stress can readily be envisioned from their reputed role in governing the assumption and maintenance of protein conformation.[64] It is conceivable that HSP70 may play a role in protein stabilization in the face of the elevated and potentially denaturing intracellular ionic concentrations that accompany acute volume regulation in hyperosmotic stress.[84] In addition, the timing of HSP70 mRNA induction is such that it appears to span the "window" between IEG expression and eventual osmolyte accumulation. Although functional data are lacking, it is tempting to speculate that members of the HSP family may serve in a protective capacity until adaptation to hyperosmotic stress is completed.

III. A MODEL OF ADAPTATION TO HYPEROSMOTIC STRESS

Figure 22–4 summarizes our current model of hyperosmotic stress-induced IEG and stress protein expression. Hyperosmotic NaCl and urea directly or indirectly alter intracellular protein conformation and, thereby, influence intracellular protein-protein interactions and intranuclear DNA-protein interactions. The resultant change in affinity of constitutively expressed transcription factors for their cognate consensus sequences permits accelerated transcription of IEGs such as *Egr-1* and c-*fos*. Similarly, increased transcription and synthesis of proteins responsible for posttranslational modification (and activation) of constitutively expressed HSF may also occur, and mediate the observed increase in stress protein expression. This interpretation is supported by the partial dependence of hyperosmotic stress-inducible HSP70 expression (unlike IEG expression) upon protein synthesis. Alternatively, altered protein-protein interactions may result in direct activation of latent HSF. Augmented expression or activation of transcription factors, in turn, results in increased expression of a set of effector genes. IEGs such as c-*fos* and *Egr-1*, as well as others yet to be identified, induce the synthesis of proteins essential for *adaptation* to hyperosmotic stress, including enzymes necessary for osmolyte synthesis, as well as transport proteins mediating uptake of osmolytes or osmolyte precursors. In parallel, augmented activation of HSF increases expression of stress proteins which afford *protection* from the potentially denaturing influence of elevated intracellular ionic strength or urea concentration until osmolyte accumulation can be achieved. Assumption of intracellular volume regulation with nonperturbing or even counteracting osmolytes, as well as the presence of protein-stabilizing chaperonins together ameliorate the alteration in intracellular protein conformation and provide a negative feedback loop.

ACKNOWLEDGMENTS

S. R. Gullans is an Established Investigator of the American Heart Association. This work was supported by National Institutes of Health Grants DK-36031 (SRG) and DK-08288 (DMC).

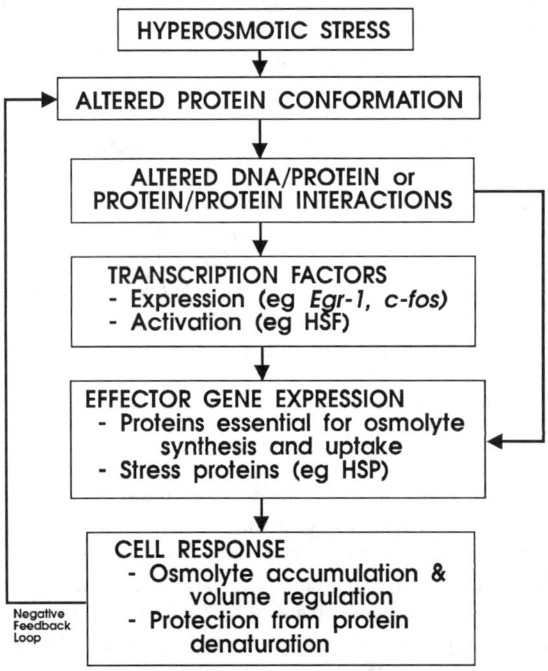

Figure 22–4 Model depicting proposed sequence of events following exposure of renal epithelial cells to hyperosmotic stress, culminating in the adaptation to and protection from altered intracellular protein conformation (see text).

REFERENCES

1. **Garcia-Perez, A. and Burg, M. B.,** Role of organic osmolytes in adaptation of renal cells to high osmolality, *J. Membr. Biol.,* 119, 1, 1991.
2. **Greenberg, M. E., Hermanowski, A. L., and Ziff, E. B.,** Effect of protein synthesis inhibitors on growth factor activation of c-fos, c-myc, and actin gene transcription, *Mol. Cell. Biol.,* 6, 1050, 1986.
3. **Sukhatme, V. P., Kartha, S., Toback, F. G., Taub, R., Hoover, R. G., and Tsai-Morris, C.,** A novel early growth response gene rapidly induced by fibroblast, epithelial cell and lymphocyte mitogens, *Oncogene Res.,* 1, 343, 1987.
4. **Greenberg, M. E. and Ziff, E. B.,** Stimulation of 3T3 cells induces transcription of the c-*fos* proto-oncogene, *Nature,* 311, 433, 1984.
5. **Kruijer, W., Cooper, J. A., Hunter, T., and Verma, I. M.,** Platelet-derived growth factor induces rapid but transient expression of the c-fos gene and protein, *Nature,* 312, 711, 1984.
6. **Bukh, A., Martinez-Valdez, H., Freedman, S. J., Freedman, M. H., and Cohen, A.,** The expression of c-*fos,* c-*jun,* and c-*myc* genes is regulated by heat shock in human lymphoid cells, *J. Immunol.,* 144, 4835, 1990.
7. **Lim, R. W., Varnum, B. C., and Herschman, H. R.,** Cloning of tetradecanoyl phorbol ester-induced "primary response" sequences and their expression in density arrested Swiss 3T3 cells and a TPA non-proliferative variant, *Oncogene,* 1, 263, 1987.
8. **Andrews, G. K., Harding, M. A., Calvet, J. P., and Adamson, E. D.,** The heat shock response in Hela cells is accompanied by elevated expression of the c-*fos* proto-oncogene, *Mol. Cell. Biol.,* 7, 3452, 1987.
9. **Rangnekar, V. M., Aplin, A. C., and Sukhatme, V. P.,** The serum and TPA responsive promoter and intron-exon structure of EGR2, a human early growth response gene encoding a zinc finger protein, *Nucleic Acids Res.,* 18, 2749, 1990.
10. **Johnson, P. F. and McKnight, S. L.,** Eukaryotic transcriptional regulatory proteins, *Annu. Rev. Biochem.,* 58, 799, 1989.
11. **Christy, B. A., Lau, L. F., and Nathans, D.,** A gene activated in mouse 3T3 cells by serum growth factors encodes a protein with "zinc-finger" sequences, *Proc. Natl. Acad. Sci. U.S.A.,* 85, 7857, 1988.
12. **Milbrandt, J.,** A nerve growth factor-induced gene encodes a possible transcriptional regulatory protein, *Science,* 238, 797, 1987.
13. **Herschman, H. R.,** Primary response genes induced by growth factors and tumor promoters, *Annu. Rev. Biochem.,* 60, 281, 1991.
14. **Chiu, R., Boyle, W. J., Meek, J., Smeal, T., Hunter, T., and Karin, M.,** The c-*fos* protein interacts with c-*jun*/AP-1 to stimulate transcription of AP-1 responsive genes, *Cell,* 54, 541, 1988.
15. **Turner, R. and Tjian, R.,** Leucine repeats and an adjacent DNA binding domain mediate the formation of functional c-*fos*-c-*jun* heterodimers, *Science,* 243, 1689, 1989.
16. **Halazonetis, T. D., Georgopoulos, G., Greenberg, M. E., and Leder, P.,** c-*Jun* dimerizes with itself and with c-*fos,* forming complexes of different DNA binding affinities, *Cell,* 55, 917, 1988.
17. **Landschulz, W. H., Johnson, P. F., and McKnight, S. L.,** The leucine zipper: a hypothetical structure common to a new class of DNA binding proteins, *Science,* 240, 1759, 1988.
18. **Bohmann, D., Bos, T. J., Admon, A., Nishimura, T., Vogt, P. K., and Tjian, R.,** Human proto-oncogene c-*jun* encodes a DNA-binding protein with structural and functional properties of transcription factor AP-1, *Science,* 238, 1386, 1987.
19. **Rauscher, F. J., III, Cohen, D. R., Curran, T., Bos, T. J., Vogt, P. K., Bohmann, D., Tjian, R., and Franza, B. R., Jr.,** Fos-associated protein p39 is the product of the c-*jun* proto-oncogene, *Science,* 240, 1010, 1988.
20. **Sassoni-Corsi, P., Lamph, W. W., Kamps, M., and Verma, I. M.,** Fos-associated cellular p39 is related to nuclear transcription factor AP-1, *Cell,* 54, 553, 1988.
21. **Hattori, K., Angel, P., LeBeau, M. M., and Karin, M.,** Structure and chromosomal location of the functional intronless human JUN proto-oncogene, *Proc. Natl. Acad. Sci. U.S.A.,* 85, 9148, 1988.
22. **Angel, P., Hattori, K., Smeal, T., and Karin, M.,** The *jun* proto-oncogene is positively autoregulated by its product, Jun/AP-1, *Cell,* 55, 875, 1988.
23. **Cao, X., Koski, R. A., Gashler, A., McKiernan, M., Morris, C. F., Gaffney, R., Hay, R. V., and Sukhatme, V. P.,** Identification and characterization of the *Egr-1* gene product, a DNA-binding zinc finger protein induced by differentiation and growth signals, *Mol. Cell. Biol.,* 10, 1931, 1990.
24. **Lemaire, P., Vesque, C., Schmitt, J., Stunnenberg, H., Frank, R., and Charnay,**

P., The serum-inducible mouse gene Krox-24 encodes a sequence-specific transcriptional activator, *Mol. Cell. Biol.,* 10, 3456, 1990.

25. **Waters, C. M., Hancock, D. C., and Evan, G. I.,** Identification and characterization of the *egr-1* gene product as an inducible, short-lived, nuclear phosphoprotein, *Oncogene,* 5, 669, 1990.
26. **Christy, B. and Nathans, D.,** DNA binding site of the growth factor-inducible protein Zif268, *Proc. Natl. Acad. Sci. U.S.A.,* 86, 8737, 1989.
27. **Cohen, D. M., Wasserman, J. C., and Gullans, S. R.,** Immediate early gene and HSP70 expression in hyperosmotic stress in MDCK cells, *Am. J. Physiol.,* 261, C594, 1991.
28. **Toback, F. G.,** Induction of growth in kidney epithelial cells in culture by Na^+, *Proc. Natl. Acad. Sci. U.S.A.,* 77, 6654, 1980.
29. **Cohen, D. M. and Gullans, S. R.,** Urea induces *Egr-1* and c-*fos* expression in renal epithelial cells, *Am. J. Physiol. (Renal, Fluid, and Electrolyte Physiol.),* in press.
30. **Asselin, C. and Marcu, K. B.,** Mode of c-*myc* gene regulation in folic acid-induced kidney regeneration, *Oncogene Res.,* 5, 67, 1989.
31. **Cowley, B. D., Chadwick, J. J., Grantham, J. J., and Calvet, J. P.,** Sequential protooncogene expression in regenerating kidney following acute renal injury, *J. Biol. Chem.,* 264, 8389, 1989.
32. **Ouellette, A. J., Malt, R. A., Sukhatme, V. P., and Bonventre, J. W.,** Expression of two "immediate early" genes, *Egr1* and c-*fos,* in response to renal ischemia and during compensatory renal hypertrophy in mice, *J. Clin. Invest.,* 85, 766, 1990.
33. **Sawczuk, I. S., Hoke, G., Olsson, C. A., Connor, J., and Buttyan, R.,** Gene expression in response to acute unilateral ureteral obstruction, *Kidney Int.,* 35, 1315, 1989.
34. **Nakamura, T., Ebihara, I., Tomino, Y., Koide, H., Kikuchi, K., and Koiso, K.,** Gene expression of growth-related proteins and ECM constituents in response to unilateral nephrectomy, *Am. J. Physiol.,* 262, F389, 1992.
35. **Greenberg, M. E., Hermanowski, A. L., and Ziff, E. B.,** Effect of protein synthesis inhibitors on growth factor activation of c-fos, c-myc, and actin gene transcription, *Mol. Cell. Biol.,* 6, 1050, 1986.
36. **Sukhatme, V. P., Kartha, S., Toback, F. G., Taub, R., Hoover, R. G., and Tsai-Morris, C.,** A novel early growth response gene rapidly induced by fibroblast, epithelial cell and lymphocyte mitogens, *Oncogene Res.,* 1, 343, 1987.
37. **Stubblefield, E. and Mueller, G. C.,** Effects of sodium chloride concentration on growth, biochemical composition, and metabolism of HeLa cells, *Cancer Res.,* 20, 1646, 1960.
38. **Robbins, E., Pederson, T., and Klein, P.,** Comparison of mitotic phenomena and effects induced by hypertonic solutions in HeLa cells, *J. Cell Biol.,* 44, 400, 1970.
39. **Wheatley, D. N.,** Cell growth and division in hypertonic medium, *Exp. Cell Res.,* 87, 219, 1974.
40. **Cohen, D. M. and Gullans, S. R.,** Urea selectively increases DNA synthesis in renal epithelial cells, *Am. J. Physiol. (Renal, Fluid, and Electrolyte Physiol.),* in press.
41. **Liu, J., Lacy, J., Sukhatme, V. P., and Coleman, D. L.,** Granulocyte-macrophage colony-stimulating factor induces transcriptional activation of *Egr-1* in murine peritoneal macrophages, *J. Biol. Chem.,* 266, 5929, 1991.
42. **Sukhatme, V. P.,** Early transcriptional events in cell growth: the Egr family, *J. Am. Soc. Nephrol.,* 1, 859, 1990.
43. **Treisman, R.,** Transient accumulation of c-*fos* RNA following serum stimulation requires a conserved 5′ element and c-fos 3′ sequences, *Cell,* 42, 889, 1985.
44. **Treisman, R.,** Identification of a protein-binding site that mediates transcriptional response of the c-*fos* gene to serum factors, *Cell,* 46, 567, 1986.
45. **Tsai-Morris, C., Cao, X., Sukhatme, V. P.,** 5′ flanking sequence and genomic structure of *Egr-1,* a murine mitogen inducible zinc finger encoding gene, *Nucleic Acids Res.,* 16, 8835, 1988.
46. **Sellmayer, A., Uedelhoven, W. M., Weber, P. C., and Bonventre, J. V.,** Endogenous noncyclooxygenase metabolites of arachidonic acid modulate growth and mRNA levels of immediate early response genes in rat mesangial cells, *J. Biol. Chem.,* 266, 3800, 1991.
47. **Greenberg, M. E., Siegfried, Z., and Ziff, E. B.,** Mutation of the c-*fos* gene dyad symmetry element inhibits serum inducibility of transcription in vivo and the nuclear regulatory factor binding in vitro, *Mol. Cell. Biol.,* 7, 1217, 1987.
48. **Gilman, M. Z.,** The c-*fos* serum response element responds to protein kinase C-dependent and -independent signals but not to cyclic AMP, *Genes Dev.,* 2, 394, 1988.
49. **Sheng, M., Dougan, S. T., McFadden, G., and Greenberg, M. E.,** Calcium and growth factor pathways of c-*fos* transcriptional activa-

tion require distinct upstream regulatory elements, *Mol. Cell. Biol.*, 8, 2787, 1988.

50. **Christy, B. and Nathans, D.,** Functional serum response elements upstream of the growth factor-inducible gene *zif*268, *Mol. Cell. Biol.*, 9, 4889, 1989.
51. **Treisman, R.,** Identification and purification of a polypeptide that binds to the c-*fos* serum response element, *EMBO J.*, 6, 2711, 1987.
52. **Montminy, M. R., Sevarino, K. A., Wagner, J. A., Mandel, G., and Goodman, R. H.,** Identification of a cyclic-AMP-responsive element within the rat somatostatin gene, *Proc. Natl. Acad. Sci. U.S.A.*, 83, 6682, 1986.
53. **Sassoni-Corsi, P., Visvader, J., Ferland, L., Mellon, P. L., and Verma, I. M.,** Induction of proto-oncogene *fos* transcription through the adenylate cyclase pathway: characterization of a cAMP-responsive element, *Genes Dev.*, 2, 1529, 1988.
54. **Yancey, P. H., Clark, M. E., Hand, S. C., Bowlus, R. D., and Somero, G. N.,** Living with water stress: evolution of osmolyte systems, *Science*, 217, 1214, 1982.
55. **Jack, R. S.,** An unusually stable DNA binding protein can locate its specific binding site in the presence of high concentrations of urea, *Biochem. Biophys. Res. Commun.*, 169, 840, 1990.
56. **Mosser, D. D., Kotzbauer, P. T., Sarge, K. D., and Morimoto, R. I.,** In vitro activation of heat shock transcription factor DNA-binding by calcium and biochemical conditions that affect protein conformation, *Proc. Natl. Acad. Sci. U.S.A.*, 87, 3748, 1990.
57. **Moriyama, T., Garcia-Perez, A., and Burg, M. B.,** Factors affecting the ratio of different organic osmolytes in renal medullary cells, *Am. J. Physiol.*, 259, F847, 1990.
58. **Nakanishi, T. and Burg, M. B.,** Osmoregulation of glycerophosphorylcholine content of mammalian renal cells, *Am. J. Physiol.*, 257, C795, 1989.
59. **Yancey, P. H. and Burg, M. B.,** Counteracting effects of urea and betaine in mammalian cells in culture, *Am. J. Physiol.*, 258, R198, 1990.
60. **Yancey, P. H. and Somero, G. N.,** Methylamine osmoregulatory solutes of elasmobranch fishes counteract urea inhibition of enzymes, *J. Exp. Zool.*, 212, 205, 1980.
61. **Arakawa, T. and Timasheff, S. N.,** The stabilization of proteins by osmolytes, *Biophys. J.*, 47, 411, 1985.
62. **Lindquist, S.,** The heat-shock response, *Annu. Rev. Biochem.*, 55, 1151, 1986.
63. **Nover, L.,** *Heat Shock Response*, CRC Press, Boca Raton, FL, 1991.
64. **Gething, M.-J. and Sambrook, J.,** Protein folding in the cell, *Nature*, 355, 33, 1992.
65. **Lindquist, S. and Craig, E. A.,** The heat-shock proteins, *Annu. Rev. Genet.*, 22, 631, 1988.
66. **Widelitz, R. B., Duffy, J. J., and Gerner, E. W.,** Accumulation of heat shock protein 70 RNA and its relationship to protein synthesis after heat shock in mammalian cells, *Exp. Cell Res.*, 168, 539, 1987.
67. **Tanaka, K., Jay, G., and Isselbacher, K.,** Expression of heat-shock and glucose-regulated genes: differential effects of glucose starvation and hypertonicity, *Biochim. Biophys. Acta*, 950, 138, 1988.
68. **Emami, A., Schwartz, J. H., and Borkin, S. C.,** Transient ischemia or heat stress induces a cytoprotectant protein in rat kidney, *Am. J. Physiol.*, 260, F479, 1991.
69. **Cowley, B. D., Douglass, D., and Muessel, M. J.,** Osmotic regulation of HSP70 RNA and prostaglandin synthase RNA *in vitro* and *in vivo*, *J. Am. Soc. Nephrol.*, 3 (Abstr.), 826, 1992.
70. **Andrews, G. K., Harding, M. A., Calvet, J. P., and Adamson, E. D.,** The heat shock response in HeLa cells is accompanied by elevated expression of the c-*fos* proto-oncogene, *Mol. Cell. Biol.*, 7, 3452, 1987.
71. **Bukh, A., H. Martinez-Valdez, S. J., Freedman, M. H., Freedman, S. J., and Cohen, A.,** The expression of c-*fos*, c-*jun*, and c-*myc* genes is regulated by heat shock in human lymphoid cells, *J. Immunol.*, 144, 4835, 1990.
72. **Ferris, D. K., Harel-Bellan, A., Morimoto, R. I., Welch, W. J., and Farrar, W. L.,** Mitogen and lymphokine stimulation of heat shock proteins in T-lymphocytes, *Proc. Natl. Acad. Sci. U.S.A.*, 85, 3850, 1988.
73. **Wu, B. J. and Morimoto, R. I.,** Transcription of the human *hsp70* gene is induced by serum stimulation, *Proc. Natl. Acad. Sci. U.S.A.*, 82, 6070, 1985.
74. **Milarski, K. L. and Morimoto, R. I.,** Expression of human *hsp70* during the synthetic phase of the cell cycle, *Proc. Natl. Acad. Sci. U.S.A.*, 83, 9517, 1986.
75. **Izumo, S., Nadal-Ginard, B., and Mahdavi, V.,** Protooncogene induction and reprogramming of cardiac gene expression produced by pressure overload, *Proc. Natl. Acad. Sci. U.S.A.*, 85, 339, 1988.
76. **Schiaffonati, L., Rappocciolo, E., Tacchini, L., Cairo, G., and Bernelli-Zazzera, A.,**

Reprogramming of gene expression in postischemic rat liver: induction of proto-oncogenes and hsp 70 gene family, *J. Cell. Physiol.*, 143, 79, 1990.

77. **Buttyan, R., Zakeri, Z., Lockshin, R., and Wolgemuth, D.,** Cascade induction of c-*fos,* c-*myc,* and heat shock 70K transcripts during regression of the rat ventral prostate gland, *Mol. Endocrinol.*, 2, 650, 1988.
78. **Cajone, F., Salina, M., and Bernelli-Zazzera, A.,** C-*myc* gene expression in heat-adapted and heat-shocked cells, *Cell Biol. Int. Rep.*, 12, 549, 1988.
79. **Kingston, R. E., Baldwin, A. S., Jr., and Sharp, P. A.,** Regulation of heat shock protein 70 gene expression by c-*myc, Nature,* 312, 280, 1984.
80. **Rabindran, S. K., Giorgi, G., Clos, J., and Wu, C.,** Molecular cloning and expression of a human heat shock factor, HSF1, *Proc. Natl. Acad. Sci. U.S.A.*, 99, 6906, 1991.
81. **Scheutz, T. J., Gallo, G. J., Sheldon, L., Tempst, P., and Kingston, R. E.,** Isolation of a cDNA for HSF2: evidence for two heat shock factor genes in humans, *Proc. Natl. Acad. Sci. U.S.A.*, 88, 6911, 1991.
82. **Sarge, K. D., Zimarino, V., Holm, K., Wu, C., and Morimoto, R. I.,** Cloning and characterization of two mouse heat shock factors with distinct inducible and constitutive DNA-binding ability, *Genes Dev.*, 5, 1902, 1991.
83. **Wu, B. J., Kingston, R. E., and Morimoto, R. I.,** Human *hsp70* promoter contains at least two distinct regulatory domains, *Proc. Natl. Acad. Sci. U.S.A.*, 83, 629, 1986.
84. **Uchida, S., Garcia-Perez, A., Murphy, H., and Burg, M. B.,** Signal for induction of aldose reductase in renal medullary cells by high external NaCl, *Am. J. Physiol. (Cell Physiol.),* C614, 1989.

Chapter 23

Aldose Reductase Gene Expression and Osmoregulation in Mammalian Renal Cells

Arlyn García-Pérez and Joan D. Ferraris

TABLE OF CONTENTS

I. SORBITOL

A. INTRODUCTION

Sorbitol, the sugar alcohol or polyol of D-glucose, is a compatible organic osmolyte. As such, it is accumulated preferentially over inorganic ions by cells exposed to long-term hyperosmotic stress.[1] The renal medulla is the only mammalian tissue that is routinely exposed, under physiologic conditions, to a variable and mostly high extracellular osmolality. The cells of the renal medulla cope with this *milieu* by regulating the accumulation and dissipation of five principal organic osmolytes: sorbitol, inositol, betaine, glycerophosphocholine, and taurine.[2] The role of sorbitol as a compatible, osmotically active, organic solute was proposed upon its identification in rat and rabbit inner medullas.[3] Although many aspects of the mechanisms involved in the osmoregulated accumulation and dissipation of sorbitol have been confirmed *in vivo,* they were originally defined by using a single renal medullary cell line.[2,4–14] This chapter will concentrate on these findings.

B. SORBITOL METABOLISM

Sorbitol is synthesized from glucose in a reaction catalyzed by aldose reductase (EC 1.1.1.21) (Figure 23–1). It can then be oxidized to produce fructose in a reaction catalyzed by sorbitol dehydrogenase (EC 1.1.1.14). Sorbitol catabolism by the action of sorbitol dehydrogenase is not a major determinant of osmotically regulated intracellular sorbitol content.[2] On the other hand, sorbitol synthesis through increased aldose reductase activity was the first mechanism of osmolyte accumulation in renal cells to be defined as osmotically regulated.[5]

C. SORBITOL ACCUMULATION IN RENAL CELLS UNDER HYPERTONIC STRESS

PAP-HT25, a cell line derived from rabbit inner medullary epithelium, is adapted to survive and grow in hypertonic medium (500 mOsmol/kg H_2O).[4] These cells were observed to accumulate enough organic osmolyte (mainly sorbitol) when in high NaCl medium so that cell volume and intracellular Na and K remain virtually normal.[5] Thus, by balancing the osmotic pressure of the high extracellular NaCl, sorbitol allowed the cells to maintain a normal intracellular ionic composition and volume.

PAP-HT25 cells maintained in isotonic medium (300 mOsmol/kg H_2O) accumulate little sorbitol.[5] When these cells are switched to high NaCl medium, sorbitol accumulates slowly, reaching a maximum by 4 d.[7] Increasing medium osmolality with raffinose, a trisaccharide that does not enter cells

0-8493-4448-4/94/$0.00+$.50

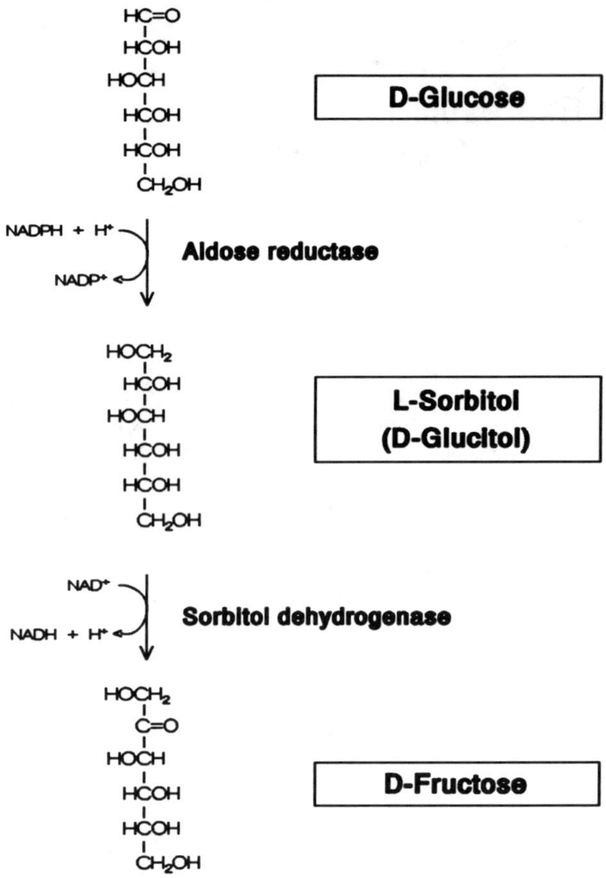

Figure 23–1 Sorbitol metabolic pathway: sorbitol synthesis and degradation. (From García-Pérez, A. and Burg, M. B., *Physiol. Rev.,* 71, 1081, 1991. With permission.)

readily, is as effective at inducing sorbitol accumulation as is NaCl. Therefore, neither extracellular sodium nor chloride need be elevated for induction of sorbitol accumulation. In contrast, raising the medium osmolality with either urea or glycerol, which enter these cells readily, fails to increase cell sorbitol. Therefore, increased extracellular osmolality, per se, is not a sufficient stimulus. Apparently, an increase in effective osmotic pressure is required.

PAP-HT25 cells accumulate sorbitol after osmotic induction because aldose reductase activity increases.[5–7] When cells grown in isotonic medium are switched to hypertonic medium, aldose reductase activity significantly increases by 6 h after the switch, half-maximal activity occurs by 2 d, and maximal activity is achieved by 3 d.[7] Sorbitol dehydrogenase activity is virtually undetectable in the cells in hypertonic medium and it does not increase with decreases in medium osmolality.[6]

The intracellular signal for induction of aldose reductase activity and subsequent sorbitol accumulation has been investigated by comparing the effects of using different solutes to raise medium osmolality.[7] As discussed above for sorbitol accumulation, solutes that do not readily penetrate the cells (NaCl and raffinose) induce aldose reductase activity, while those that do (urea and glycerol) fail to induce activity. When osmolality is increased by adding NaCl or raffinose, cell water decreases, and cell sodium and potassium concentrations increase. The changes persist for at least 24 h, until sorbitol content begins to rise and, consequently, cell volume is restored. Addition of urea, on the other hand, does not cause any significant change in cell water or electrolyte concentration, just as it fails to induce increased aldose reductase activity. Based on these correlations, the trigger for induction of aldose reductase appears to be prolonged extracellular hypertonicity.

It was postulated that the mechanism by which prolonged extracellular hypertonicity increases intracellular aldose reductase activity might involve either the reduction in cell volume or the increase in intracellular Na or K, both of which precede the rise in enzyme activity.[7] To distinguish between these possibilities, ouabain was used. Ouabain generally decreases cell potassium and increases cell sodium and volume. Medium osmolality was increased by adding NaCl either without ouabain or with different concentrations of the drug. Aldose reductase activity did not correlate significantly with cell volume or sodium. It did correlate, however, with cell potassium concentration and, even better, with the sum of cell potassium plus sodium concentration. Thus, it was postulated that the signal that ultimately induces sorbitol accumulation during prolonged extracellular hypertonicity is a rise in the total concentration of sodium plus potassium salts (or ionic strength) inside the cells. Inherent in this hypothesis is the prediction that a reduction of the intracellular signal should decrease the response. The results of recent experiments to be discussed later in this chapter are consistent with the hypothesis.

Initially, sorbitol appeared to be the only organic osmolyte accumulated by PAP-HT25 cells in hyperosmotic medium.[4–7] The definition of the mechanisms of accumulation of betaine, inositol, and glycerophosphocholine in another renal cell line, Madin-Darby canine kidney (MDCK) cells, led to the reevaluation of the initial observation. PAP-HT25 cells have been shown to accumulate all four osmolytes under hyperosmotic conditions.[15] The effect of accumulation of each of the osmolytes on the others has been examined. The most striking relationship exists between sorbitol and betaine. Betaine accumulates in PAP-HT25 cells in hyperosmotic medium only by uptake from the medium. If betaine is absent in the medium, it does not accumulate. As shown in Figure 2, under these circumstances sorbitol is the predominant osmolyte. As medium betaine is increased, intracellular betaine content rises and sorbitol content falls in a virtually reciprocal fashion. Therefore, it appears that the osmotically induced accumulation of betaine and sorbitol share the same intracellular signal. As betaine accumulation increases, the intracellular signal inducing sorbitol accumulation decreases and sorbitol content falls.

D. SORBITOL DISSIPATION FROM RENAL CELLS AS EXTRACELLULAR TONICITY DECREASES

When PAP-HT25 cells adapted to hypertonic medium are switched to isotonic medium, sorbitol exits rapidly from the cells. The PAP-HT25 cells maintained at a constant osmolality in either hypertonic or isotonic medium have a very low permeability to sorbitol.[6] However, sudden decreases in extracellular tonicity cause rapid, transient increases in sorbitol permeability. This occurs when the cells are switched either from hypertonic to isotonic medium or isotonic to hypotonic medium.[12] When PAP-HT25 cells, grown under hypertonic conditions in which sorbitol is the predominant osmolyte, are switched to isotonic medium, permeability to sorbitol increases 80-fold, and approximately one half of the cell sorbitol exits into the medium within 15 min. As sorbitol and water exit the cells, the sorbitol efflux rate decreases again to approximately twice the original value in the hypertonic medium. Thus, the fall in cell sorbitol becomes much slower after 15 min. Nevertheless, cell sorbitol is reduced to <25% of the initial value after 24 h and is close to baseline after 3 d, despite the persistence of aldose reductase and sorbitol synthesis, as shown in Figure 23–3.[6]

The results are similar when the cells are grown under conditions in which they also contain large amounts of the other organic osmolytes in addition to sorbitol. In this case, approximately 30% of the cell sorbitol effluxes within 10 min of switching the cells to isotonic medium, and the other osmolytes, especially betaine, also exit from the cells.[15] The rapid efflux of sorbitol (and the other osmolytes) ensures that the cells do not remain greatly swollen for very long after the osmolality decreases abruptly. This response is analogous to the rapid efflux of KCl that occurs in many different types of cells during volume-regulatory decrease after a fall in extracellular osmolality.

After losing sorbitol to the medium, PAP-HT25 cells can reaccumulate it relatively rapidly by synthesis, if needed, because aldose reductase remains high (Figure 23–3). Although gene transcription, mRNA abundance, and protein syn-

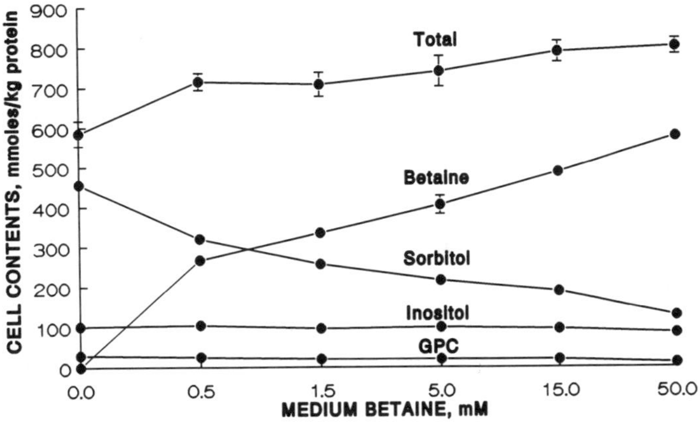

Figure 23–2 Effect of different concentrations of betaine in medium. PAP-HT25 cells were grown at 300 mOsmol/kg H_2O without betaine and then switched for 4 d to medium made hypertonic (500 mOsmol/kg H_2O) by adding NaCl and a variable concentration of betaine. (From Moriyama, T., García-Pérez, A., and Burg, M. B., *Am. J. Physiol.*, 259, F847, 1990. With permission.)

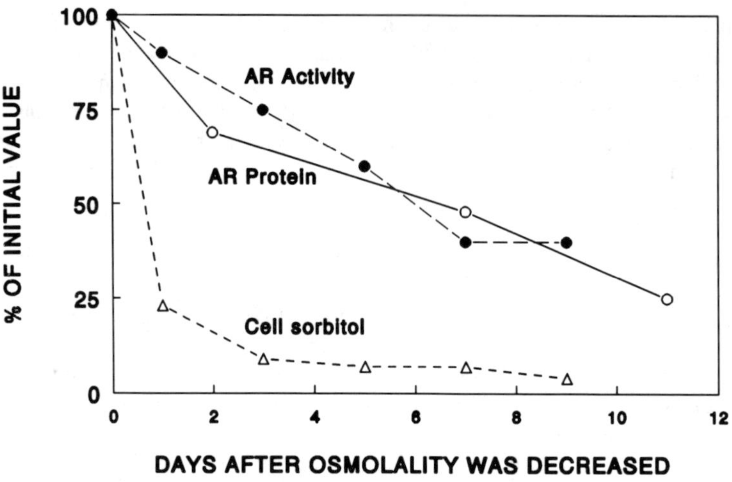

Figure 23–3 Effect of switching PAP-HT25 cells maintained in hypertonic medium to isotonic medium. (From Bagnasco, S. M., Murphy, H. R., Bedford, J. J., and Burg, M. B., *Am. J. Physiol.,* 254, C788, 1988. With permission.)

thesis rate of aldose reductase fall rapidly after a decrease in medium osmolality, the enzyme protein, which is quite long-lived, persists for days to weeks.[9,10,16]

The mechanism by which the permeability of sorbitol in PAP-HT25 cells increases after a decrease in medium tonicity has been examined further.[12–14] No saturation of the sorbitol efflux is apparent with sorbitol concentrations of up to 315 m*M*, and there is no apparent competitive inhibition by high concentrations of other polyols.[12] Therefore, if mediated transport is involved, it must have a very high K_m. The increase in sorbitol efflux from PAP-HT25 cells is prevented by lowering the temperature to 0°C.[12] In addition, a number of inhibitors of the sorbitol efflux have been identified. Eicosa-5,8,11,14-tetraynoic acid (ETYA; 100 μ*M*), a tetraacetylenic derivative of arachidonic acid that is a general inhibitor of arachidonic acid metabolism, reduces the increase in sorbitol efflux by 82%.[14] Sorbitol efflux was not affected by inhibitors of cyclooxygenase and lipoxygenase, but was reduced by 91% with ketoconazole (100 μ*M*) and by 63% with SKF-525A (100 μ*M*), suggesting that the cytochrome P-450 limb of the arachidonic acid cascade may be involved in the sorbitol efflux process.[14]

II. ALDOSE REDUCTASE

A. ALDOSE REDUCTASE ACTIVITY

The PAP-HT25 cells accumulate sorbitol in response to osmotic induction because aldose reductase activity increases.[5–7] When cells grown in isotonic medium are switched to hypertonic medium, aldose reductase activity and cell sorbitol content increase slowly (Figure 23–4).[7] There is no significant increase in aldose reductase activity before 6 h, half-maximal activity is reached after about 2 d, and maximal activity is not attained until 3 d after osmotic induction.[7]

Aldose reductase (EC 1.1.1.21) is a member of the family of aldo-keto reductases. These cytosolic (soluble) enzymes are monomeric, NADPH-dependent reductases that preferentially reduce aromatic aldehydes (e.g., *p*-nitrobenzaldehyde) over nonaromatic aldehydes and aldoses (e.g., DL-glyceraldehyde, D-glucose).[17] Most closely related to aldose reductase is aldehyde reductase (EC 1.1.1.2). These separate enzyme activities were originally distinguished by their different substrate specificity and inhibitor sensitivity.[18,19] Thus, aldehyde reductase is typically inhibited by valproic acid and barbiturates, whereas aldose reductase is only slightly so.[17,19,20]

Aldose reductase has been purified to homogeneity from many different mammalian tissues and from PAP-HT25 cells in hypertonic medium.[2,8] The enzymes from the various tissues are monomers with apparent molecular weights ranging from 35,000 to 40,000; pI is 4.7 to 5.9; K_m for DL-glyceraldehyde is in the micromolar range, whereas K_m for D-glucose is 230 to 650 m*M*.

The amino acid sequences of aldose reductases from several sources, including PAP-HT25 cells, have been determined.[2,10] Because the aldose reductase protein is blocked at the amino terminus, most of the reported amino acid sequences are deduced from cloned cDNA sequences. Clones with full coding regions define a protein 316 amino acids long, including the initiator methionine. There is a high degree of identity (>84%) among the reported protein sequences. In addition, the mRNAs encoding these proteins in different tissues and species have >80% identity in their coding regions, but identity is typically low (<35%) in the untranslated regions. On the other hand, deduced sequences for aldehyde reductase define a protein 325 amino acids long that has a 52% identity with aldose reductase of the same tissue and species (human placenta).[2,21]

B. EFFECT OF EXTRACELLULAR TONICITY ON ALDOSE REDUCTASE PROTEIN SYNTHESIS

Aldose reductase activity rises in PAP-HT25 cells after osmotic induction because the amount of enzyme protein increases.[17] The amount of enzyme rises because its synthesis rate is increased.[9] Aldose reductase synthesis and degradation were measured by pulse-chase with ^{35}S-methionine, followed by immunoprecipitation with an aldose reductase-specific antibody.[8,9] As shown in Figure 23–4, synthesis rate increases to two times the level in isotonic medium after 6 h of exposure of PAP-HT25 cells to hypertonic medium. It peaks by 24 h at approximately 15 times, then it decreases to a steady-state level that is approximately 6 times greater than that of cells in isotonic medium. The degradation of aldose reductase protein is slow (half-life is about 6 d) and does not differ between PAP-HT25 cells in isotonic and hypertonic medium.[9] Therefore, the rise in the amount of aldose reductase protein in hypertonic medium is due entirely to the increased rate of its synthesis.

C. OSMOTIC REGULATION OF ALDOSE REDUCTASE GENE EXPRESSION

To determine how medium hypertonicity induces aldose reductase protein synthesis, aldose reductase cDNA was cloned from PAP-HT25 cells, and the cloned cDNA was used as a probe to measure mRNA abundance by Northern blot hybridization.[10] When the cells are exposed to hypertonic medium, aldose reductase mRNA increases. The time course and magnitude of the increase closely resemble that of aldose reductase protein synthesis rate (Figure 23–4). Therefore, the increased translation of enzyme protein is attributable to a rise in its mRNA abundance.

The rise in aldose reductase mRNA levels is a result of increased aldose reductase gene transcription rather than increased stability of the mRNA.[16] Nuclear run-on assays were used to assess the changes in aldose reductase transcription rate after PAP-HT25 cells in isotonic medium are switched to hypertonic medium. This type of assay measures the relative increase in the number of transcripts initiated for a

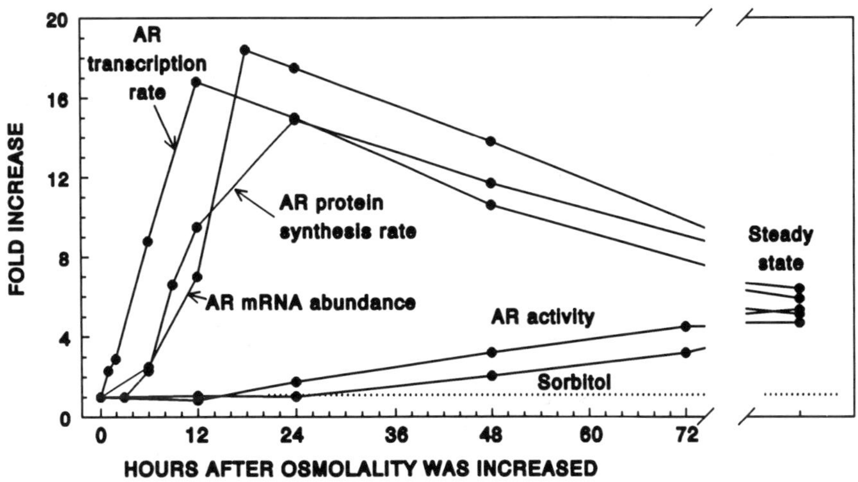

Figure 23–4 Effect on PAP-HT25 cells of increasing medium tonicity from isotonic (300 mOsmol/kg H_2O) to hypertonic (500 mOsmol/kg H_2O) by adding NaCl. The dotted line represents the level in cells maintained in isotonic medium. Hypertonic steady-state levels were measured in cells maintained and passaged in hypertonic medium. (From Smardo, F. L., Burg, M. B., and García-Pérez, A., *Am. J. Physiol.*, 262, C776, 1992. With permission.)

specific gene in a set time. Most eukaryotic regulation of gene transcription occurs by affecting this rate of transcript initiation. Procedurally, after the cells are exposed to isotonic or hypertonic medium for a given period of time, their nuclei are harvested at 0°C. At this temperature transcription has stopped. The reaction is then allowed to continue *in vitro* for a set time in the presence of ^{32}P-uridine triphosphate in order to label all gene transcripts that were initiated, and not terminated, during the cell treatment period. The transcribed mRNAs are then isolated and the gene transcript of interest (aldose reductase mRNA in this case) is selected by hybridization to its specific cDNA.

Figure 23–4 shows that aldose reductase gene transcription begins to rise within 1 h, reaches a peak increase of approximately 15-fold by 12 to 24 h, and then decreases to a steady state of fivefold within a few days. Transcription precedes the increases in aldose reductase mRNA abundance and protein synthesis by a few hours, and is large enough to entirely explain their induction. Further, the measured half-lives of aldose reductase mRNA in cells in isotonic vs. hypertonic medium are not significantly different.[16] Therefore, aldose reductase mRNA is osmotically regulated through changes in its transcription.

As discussed in Section I.C. of this chapter, it is postulated that the signal that ultimately induces sorbitol accumulation during prolonged extracellular hypertonicity is a rise in the total concentration of sodium plus potassium salts (or ionic strength) inside the cells. Inherent in this hypothesis is the prediction that a reduction of the intracellular signal should decrease the response. The results of recent experiments are consistent with the hypothesis.[9,16] As notable in Figure 23–4, when PAP-HT25 cells are switched from isotonic to hypertonic medium, aldose reductase transcription rises slowly for 12 h and then begins to fall. A similar but somewhat delayed time course is seen for aldose reductase mRNA abundance and translation. The decline in aldose reductase transcription and translation has been shown to coincide with increased sorbitol accumulation.[9,16] As shown in Figure 23–5, when sorbitol accumulation is blocked with the aldose reductase inhibitor Tolrestat (*N*-methyl-5-trifluoromethyl-6-methoxy-naphthoylglycine), aldose reductase transcription and translation continue to increase past 24 h. Presumably, sorbitol accumulation decreases the intracellular sodium plus potassium concentration through the restoration of cell volume, and this decrease in the intracellular signal then reduces aldose reductase transcription. Further support for this explanation comes from experiments in which the cells switched to hypertonic medium were allowed to accumulate large amounts of another compatible organic osmolyte, betaine, which is chemically different from sorbitol.[16] The usual increase in aldose reductase gene transcription was markedly inhibited, as expected if betaine, like sorbitol, lowers the intracellular signal (ionic strength) by restoring cell volume (Figure 23–5). Thus, these results are consistent with the hypothesis that increased intracellular ionic strength is the signal that ultimately results in increased sorbitol accumulation.

D. ALDOSE REDUCTASE GENE

To understand how the intracellular signal produced by extracellular hypertonicity results in increased transcription of the aldose reductase gene, it is necessary to isolate the gene and characterize its regulatory regions. The aldose reductase gene has been cloned from the human, rat, and mouse.[22–24] Although the size of the gene varies from 11 to 18 kb among these species, its structure is constant. For all species the gene is composed of ten exons

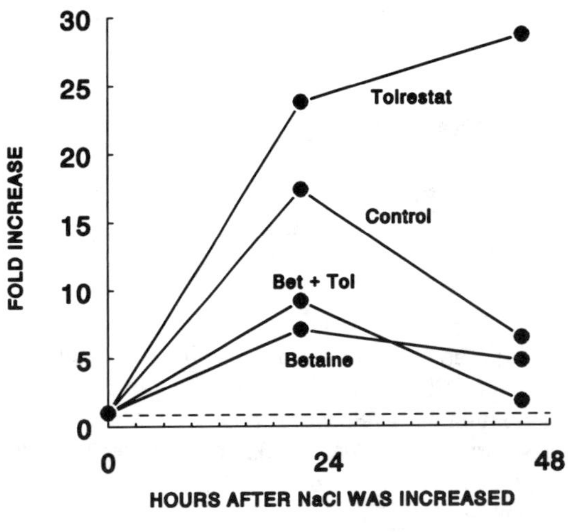

Figure 23–5 Effect on aldose reductase transcription rate in PAP-HT25 cells adapted to isotonic medium after switching them to hypertonic medium at zero hour with either Tolrestat (10 *M*), betaine (5 m*M*), or both added to the medium. (From Smardo, F. L., Burg, M. B., and García-Pérez, A., *Am. J. Physiol.*, 262, C776, 1992. With permission.)

and nine introns and the splice sites for the introns occur at the same positions in all the species cloned. A TATA box and a CAAT box upstream of the transcription initiation site is found on all cloned aldose reductase genes. However, a region involved in the osmotic response has yet to be reported.

III. A MODEL FOR CELL ADAPTATION TO OSMOTIC STRESS THROUGH SORBITOL ACCUMULATION AND DISSIPATION

Figure 23–6 shows a model for various states of renal medullary cell adaptation to altered extracellular osmolality.[25] The different states of adaptation are illustrated for sorbitol. The model is meant to apply both to cells *in vivo* and in tissue culture.

As mentioned above, once aldose reductase has been induced in tissue culture, it is very stable (half-life of about 1 week, regardless of the osmolality). This is the adapted state. Renal medullary cells *in vivo* apparently can become similarly adapted. The urine osmolality and medullary extracellular NaCl concentration remain high in animals that are chronically antidiuretic because they are not ingesting much fluid. Under those conditions, aldose reductase activity is high and does not decrease within hours to days after a diuresis starts.[2,26]

The adapted cells can rapidly control their sorbitol content, however, by varying their permeability to it. The following sequence is proposed. When a diuresis ensues and the renal medullary extracellular NaCl concentration falls, the cells swell (Figure 23–6, *step 1*) and sorbitol permeability increases rapidly *(step 2)*.[26] The resultant efflux of sorbitol and water restores cell volume without any change in aldose reductase *(step 3)*. Although its production continues, sorbitol does not accumulate in the cells, because its permeability remains high enough that the sorbitol exits the cells as fast as it is formed. If, within a few days, the diuresis ends and extracellular NaCl increases, then the permeability decreases and the sorbitol that is being synthesized is retained in the cells, increasing sorbitol content over a few hours to compensate for the higher NaCl *(step 4)*. Sorbitol accumulates rapidly because the machinery for synthesizing it has been maintained during the transient antidiuretic state.

On the other hand, when the diuresis is prolonged and extracellular NaCl remains low, aldose reductase activity falls slowly over several days as the enzyme degrades, and the cells eventually become unadapted *(step 5)*. This occurs in pathological conditions like diabetes insipidus. If the condition causing the diuresis is corrected, then the renal medullary extracellular NaCl increases, as the kidney makes more concentrated urine *(step 6)*. This triggers the transcription of aldose reductase mRNA and translation of enzyme protein *(step 7)*, returning the cells to the adapted state within several days *(step 8)*. This sequence of events occurs in Brattleboro rats, which

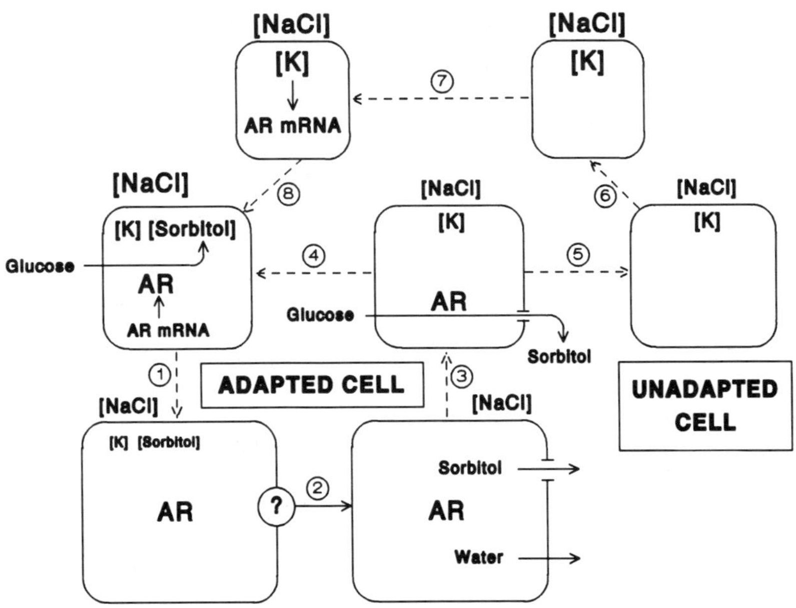

Figure 23–6 A model for cell adaptation to osmotic stress through sorbitol accumulation and dissipation. (From García-Pérez, A. and Burg, M. B., *Hypertension,* 16, 595, 1990. With permission.)

have congenital diabetes insipidus because of lack of antidiuretic hormone (ADH). When they are given ADH, their renal medullary sodium, aldose reductase mRNA, aldose reductase activity, and sorbitol (all of which are initially low) increase considerably within a few days.[26–28] This *in vivo* process is slow, however. Sorbitol rises gradually during the first 3 d and is even higher after 12 d.

The model in Figure 23–6 refers specifically to sorbitol. However, the concept of adapted and unadapted states applies to all five renal organic osmolytes. Although the mechanisms for accumulating betaine, inositol, taurine, and glycerophosphocholine differ from that for sorbitol, all are induced relatively slowly. In contrast, cellular levels of all five osmolytes decrease rapidly by efflux when extracellular osmolality falls.[2]

REFERENCES

1. **Yancey, P., Clark, M., Hand, S., Bowlus, R., and Somero, G.,** Living with water stress: evolution of osmolyte systems, *Science,* 217, 1214, 1982.
2. **García-Pérez, A. and Burg, M. B.,** Renal medullary organic osmolytes, *Physiol. Rev.,* 71, 1081, 1991.
3. **Bagnasco, S., Balaban, R., Fales, H., Yang, Y.-M., and Burg, M. B.,** Predominant osmotically active organic solutes in rat and rabbit renal medullas, *J. Biol. Chem.,* 261, 5872, 1986.
4. **Uchida, S., Green, N., Coon, H., Triche, T., Mims, S., and Burg, M. B.,** High NaCl induces stable changes in phenotype and karyotype of renal cells in culture, *Am. J. Physiol.,* 253, C230, 1987.
5. **Bagnasco, S., Uchida, S., Balaban, R., Kador, P., and Burg, M. B.,** Induction of aldose reductase and sorbitol in renal inner medullary cells by elevated extracellular NaCl, *Proc. Natl. Acad. Sci. U.S.A.,* 84, 1718, 1987.
6. **Bagnasco, S. M., Murphy, H. R., Bedford, J. J., and Burg, M. B.,** Osmoregulation by slow changes in aldose reductase and rapid changes in sorbitol flux, *Am. J. Physiol.,* 254, C788, 1988.
7. **Uchida, S., García-Pérez, A., Murphy, H. R., and Burg, M. B.,** Signal for induction of aldose reductase in renal medullary cells by high external NaCl, *Am. J. Physiol.,* 256, C614, 1989.
8. **Bedford, J. J., Bagnasco, S. M., Kador, P. F., Harris, W. H., Jr., and Burg, M. B.,** Characterization and purification of a mammalian osmoregulatory protein, aldose reductase, induced in renal medullary cells by high extracellular NaCl, *J. Biol. Chem.,* 262, 14255, 1987.
9. **Moriyama, T., García-Pérez, A., and Burg, M. B.,** Osmotic regulation of aldose reductase protein synthesis in renal medullary cells, *J. Biol. Chem.,* 264, 16810, 1989.
10. **García-Pérez, A., Martin, B., Murphy, H. R., Uchida, S., Murer, H., Cowley, B. D., Handler, J. S., and Burg, M. B.,** Molecular cloning of cDNA coding for kidney aldose reductase: regulation of specific mRNA accumulation by NaCl-mediated osmotic stress, *J. Biol. Chem.,* 264, 16815, 1989.
11. **Yancey, P. H., Burg, M. B., and Bagnasco, S. M.,** Effects of NaCl, glucose and aldose reductase inhibitors on cloning efficiency of renal medullary cells, *Am. J. Physiol.,* 258, C156, 1990.
12. **Siebens, A. and Spring, K.,** A novel sorbitol transport mechanism in cultured renal papillary epithelial cells, *Am. J. Physiol.,* 257, F937, 1989.
13. **Garty, H., Furlong, T. J., Ellis, D. E., and Spring, K. R.,** The sorbitol permease: an apical membrane transporter in cultured renal papillary epithelial cells, *Am. J. Physiol.,* 260, F650, 1991.
14. **Furlong, T. J., Moriyama, T., and Spring, K. R.,** Activation of osmolyte efflux from cultured renal papillary epithelial cells, *J. Membr. Biol.,* 123, 269, 1991.
15. **Moriyama, T., García-Pérez, A., and Burg, M. B.,** Factors affecting the ration of the different organic osmolytes in renal medullary cells, *Am. J. Physiol.,* 259, F847, 1990.
16. **Smardo, F. L., Burg, M. B., and García-Pérez, A.,** Kidney aldose reductase gene transcription is osmotically regulated, *Am. J. Physiol.,* 262, C776, 1992.
17. **Flynn, T. G.,** Aldehyde reductases: monomeric NADPH-dependent oxidoreductases with multifunctional potential, *Biochem. Pharmacol.,* 31, 2705, 1982.
18. **Kador, P., Robinson, W., and Kinoshita, J.,** The pharmacology of aldose reductase inhibitors, *Annu. Rev. Pharmacol. Toxicol.,* 25, 691, 1985.
19. **Srivastava, S. K., Das, B., Hair, G. A., Gracy, R. W., Awasthi, S., Ansari, N. H., and Petrash, J. M.,** Interrelationships among human aldoketo reductases: immunochemical, kinetic and structural properties, *Biochim. Biophys. Acta,* 840, 334, 1985.

20. **Morjana, N. A. and Flynn, T. G.,** Aldose reductase from human psoas muscle: purification, substrate specificity, immunological characterization, and effect of drugs and inhibitors, *J. Biol. Chem.,* 264, 2906, 1989.
21. **Bohren, K. M., Bullock, B., Wermuth, B., and Gabbay, K. H.,** The aldo-keto reductase superfamily: cDNAs and deduced amino acid sequences of human aldehyde and aldose reductases, *J. Biol. Chem.,* 264, 9547, 1989.
22. **Graham, A., Brown, L., Hedge, P. J., Gammack, A. J., and Markham, A. F.,** Structure of the human aldose reductase gene, *J. Biol. Chem.,* 266, 6872, 1991.
23. **Graham, C., Szpirer, C., Levan, G., and Carper, D.,** Characterization of the aldose reductase-encoding gene family in rat, *Gene,* 107, 259, 1991.
24. **Pailhoux, E., Veyssiere, G., Fabre, S., Tournaire, C., and Jean, C.,** The genomic organization and DNA sequence of the mouse vas deferens androgen-regulated protein gene, *J. Steroid Biochem. Mol. Biol.,* 42, 561, 1992.
25. **García-Pérez, A. and Burg, M. B.,** Importance of organic osmolytes for osmoregulation by renal medullary cells, *Hypertension,* 16, 595, 1990.
26. **Cowley, B. D., Ferraris, J. D., Carper, D., and Burg, M. B.,** In vivo osmoregulation of aldose reductase mRNA, protein and sorbitol in renal medulla, *Am. J. Physiol.,* 258, F154, 1990.
27. **Schmolke, M., Beck, F., and Guder, W.,** Effect of antidiuretic hormone on renal organic osmolytes in Brattleboro rats, *Am. J. Physiol.,* 257, F732, 1989.
28. **Blumenfeld, J., Hebert, S., Heilig, C., Balschi, J., Stromski, M., and Gullans, S.,** Organic osmolytes in inner medulla of the Brattleboro rat: effects of ADH and dehydration, *Am. J. Physiol.,* 256, F916, 1989.

Chapter 24

Molecular Regulation of Mammalian Osmolyte Transporters

H. Moo Kwon

TABLE OF CONTENTS

I. INTRODUCTION

As described in the previous chapters, *myo*-inositol (MI), betaine, and taurine are major organic osmolytes in the renal medulla of mammals. Madin-Darby canine kidney (MDCK) cells, an epithelial cell line originated from an unknown site of a dog kidney, have been very useful in studying the mechanisms concerning the osmoregulatory accumulation of these osmolytes. Recent advances in understanding the mechanism for the accumulation of the three osmolytes will be discussed here. Unlike the sorbitol accumulation in PAP-HT25 cells, which is mediated by synthesis from glucose, MDCK cells can amass MI, betaine, and taurine only when they are present in the culture medium.[1-3] When cultured in media lacking these substances for an extended period of time (>7 d), cells will be depleted of these osmolytes, indicating that MDCK cells possess little capacity to synthesize any of these substances. In hypertonic media intracellular concentrations of the osmolytes can be over 50 m*M*, which is >1,000-fold medium concentration. Such a remarkable ability to concentrate the osmolytes to high concentration is essential for the MDCK cells to grow in hypertonic media.[3]

The MDCK cells take up each of the osmolytes by a specific Na^+- or Na^+/Cl^--dependent transport. The membrane proteins involved are Na^+/MI cotransporter, Na^+/Cl^-/betaine cotransporter, and Na^+/Cl^-/taurine cotransporter. They are secondary active transporters capable of moving the substrates uphill against a concentration gradient and are driven by the electrochemical potential of Na^+ or Na^+ plus Cl^- across the plasma membrane. They are localized exclusively to the basolateral membrane, with the exception of the taurine transporter, which is present also in the apical membrane.[3,4] Apical taurine transport is much smaller than that of the basolateral membrane in magnitude and is not regulated by medium tonicity.

All of the transporters mentioned are regulated by medium tonicity, indicating that they are responsible for the osmoregulatory accumulation of osmolytes in the cells. Figure 24–1 shows MI content of MDCK cells when medium osmolality is raised to 500 mOsmol/kg H_2O by adding raffinose.[4] The cell MI increases after a few hours and then rapidly climbs to a new steady-state level in 24 h and remains at that level as long as the hypertonicity is imposed. Also shown is a time course of the Na^+-dependent MI uptake rate across the basolateral membrane. The uptake rate, a measure of the number of functioning transporters, exhibits a dramatic, over 20-fold increase in 24 h. Thereafter, it decreases but remains elevated compared to the isotonic control. Average transport rate measured in the presence of 10 μM MI during the period of the initial 24 h after switch is around 70 pmol/min/mg (Figure 24–1). Given that the medium used in Figure 24–1 contains 200 μM MI and K_m of the Na^+-coupled MI transport in the MDCK cells is ~30 μM, the actual rate of transport should be about sixfold that seen in Figure 24–1 or 25 nmol/h/mg protein, which is sufficient to account for the ~400 nmol/mg protein of MI accumulated in the cells during the first day. Inhibition of Na^+-

0-8493-4448-4/94/$0.00+$.50

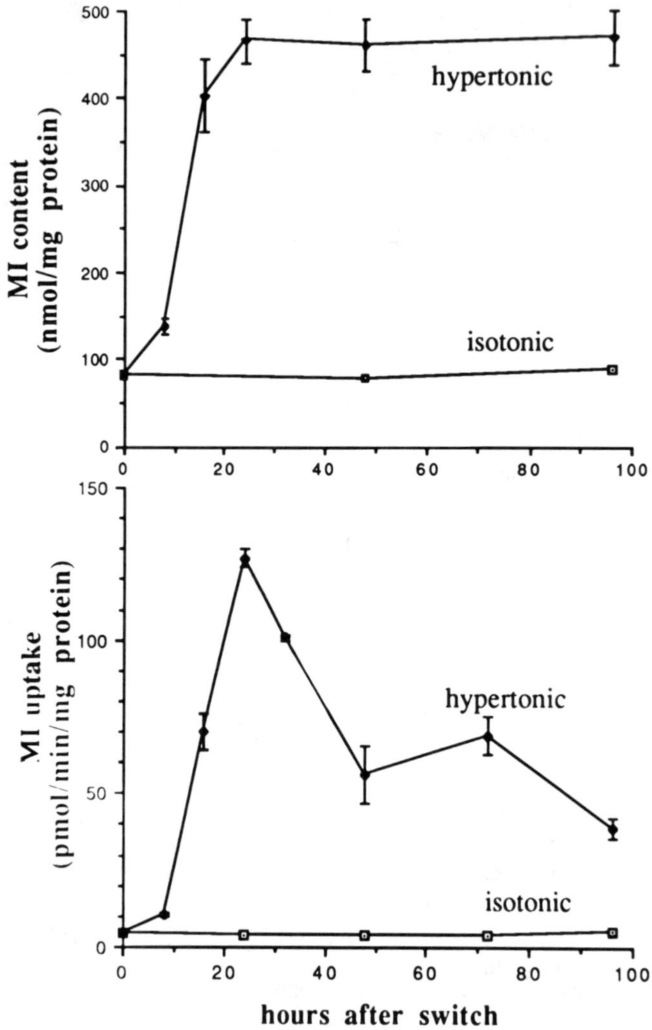

Figure 24–1 Time course of *myo*-inositol (MI) content (upper panel) and Na^+-dependent MI uptake of the MDCK cells (bottom panel). Cells were grown to confluence on filters and medium osmolality was increased to 500 mOsmol/kg H_2O by adding raffinose (hypertonic) or left unchanged (isotonic). MI content and initial rate of Na^+-dependent MI uptake on the basolateral side in the presence of 10 μM MI were measured at various points. (From Yamauchi, A. et al., *Am. J. Physiol.*, 261, F197, 1991. With permission.)

dependent MI transport using phlorizin completely blocked hypertonicity-induced MI accumulation in the MDCK cells.[5] It is evident that the upregulation of the transporter is sufficient to achieve the observed MI accumulation in hypertonic medium. Similar conclusions can be drawn for betaine and taurine accumulation in the MDCK cells.

The kidney is capable of *de novo* synthesis and catabolism of MI, betaine, and taurine.[6,7] At present, it is not clear whether any of these synthetic and catabolic processes plays any role in osmolyte accumulation in the renal medullary cells. There is, however, good evidence that transport plays a significant role. Slices made from cortex and medulla of kidney can concentrate betaine from medium in a Na^+-dependent fashion, indicating that basolaterally located Na^+-dependent betaine transport is present throughout the kidney. The transport activity in the slices made from the medulla, not the ones from the cortex, is stimulated by hypertonicity, suggesting that the betaine transporters in the medulla may be different from those of the cortex (see below) and that the medullary transporter may be at least partially responsible for betaine accumulation in medullary cells during antidiuresis.[8] Although both cortical and medullary slices can also concentrate MI in a sodium-dependent fashion,[9,10] regulation by medium tonicity has not been described in these preparations. Rat inner medullary collecting tubules in primary culture displayed Na^+-coupled MI uptake that is stimulated by medium hypertonicity with a time course very similar to that of the MDCK cells.[11] The downregulation of the transport when the medium tonicity is returned to isotonic level was also very similar to that of the MDCK cells. The specific mRNAs for the Na^+-coupled MI and betaine transporters are expressed exclusively or most abundantly in renal medulla (see below). Taken together, it would be safe to conclude that Na^+-coupled transporters are responsible for at least part of the osmoregulatory accumulation of MI and betaine by renal medullary cells. Regulation of these transporters by medium tonicity appears to be the major mechanism underlying the osmoregulatory accumulation of these solutes.

II. MOLECULAR CLONING AND STRUCTURE OF OSMOLYTE TRANSPORTERS

A. MOLECULAR CLONING

Since no information was available regarding the sequence of the transporters and antibodies to the transporters were not available, we chose to clone the cotransporters using expression in *Xenopus laevis* oocytes. Injection into the oocytes of poly(A)$^+$RNA extracted from the hypertonic MDCK cells results in expression of the Na^+ dependent transport of the three osmolytes — MI, betaine, and taurine.[12–14] When the poly(A)$^+$RNA was size-fractionated, one fraction with median RNA size of ~4 kb was most enriched for the induction of Na^+-coupled MI and taurine transport and another fraction with median size of ~2.8 kb was most enriched for the induction of Na^+-coupled betaine transport in oocytes. A directional cDNA library was constructed from each of the two size-selected poly(A)$^+$RNA fractions in a plasmid vector, pSPORT1 (BRL). Use of reverse transcriptase devoid of RNase H activity (SuperScript reverse transcriptase, BRL) for the first strand cDNA synthesis was essential to obtain good quality (long inserts) cDNA libraries.

Different screening strategies were successful for cloning the cDNAs. The first strategy made use of the capacity of oocytes to express functional transporters.[15,16] Pools of colonies (1000 colonies per pool) were grown separately and plasmid DNA was isolated. A capped transcript was made from each pool and injected into oocytes. A number of pools were screened and eventually a pool was identified whose transcript induced Na^+-dependent MI or betaine transport in oocytes. Standard sib-selection led to isolation of single clone for MI (pSMIT, *s*odium/*MI* co*t*ransporter)[15] and betaine (pBGT1, *b*etaine/*G*ABA *t*ransporter; see below for GABA).[16] The cRNA synthesized from each of the clones induced Na^+-dependent MI or betaine transport in oocytes indistinguishable from that of the basolateral membrane of the MDCK cells with respect to their affinity to substrates, ionic requirements, and sensitivity to inhibitors. Sequencing of the cDNAs revealed amino acid sequences homologous to the previous cloned transporter families (see below). It was concluded that the cDNAs code for the major (if not exclusive) transport proteins of the MDCK cells specific for MI and betaine rather than activators that stimulate dormant endogenous transporters residing in oocytes.

The "direct expression" strategy was not successful for the taurine transporter even after screening more than 50 pools (i.e., 50,000 colonies). This was unexpected because taurine transport expression following injection of MDCK poly(A)$^+$RNA was an order of magnitude higher than MI or betaine transport expression. A different, and eventually successful, strategy was devised based on the assumption that the taurine transporter is a member of the same gene family as the MDCK cell betaine/GABA transporter and the brain neurotransmitter transporters for GABA[17] and noradrenaline.[18] All of these transporters are dependent on Cl^- as well as Na^+ (see below). Degenerate PCR (polymerase chain reaction) primers based on regions of highly conserved sequence in the three cDNAs (see Figure 24–3) were used for RT-PCR (reverse transcription-PCR) on the poly(A)$^+$RNA prepared from the hypertonic MDCK cells. The RT-PCR product included a cDNA with a sequence that was different but highly similar to those of the gene family members. The new cDNA sequence recognized a 7-kb mRNA on a Northern blot of the MDCK poly(A)$^+$RNA whose abundance was higher in hypertonic cells than isotonic cells. Since all of these observations appeared appropriate for the cDNA of the taurine transporter, we screened the pools of cDNA by performing PCR using the degenerate primers. One out of ten pools yielded the putative taurine cDNA product. A standard sib-selection using the PCR assay led to isolation of a plasmid, pNCT (*N*a$^+$- and *C*l$^-$-dependent *t*aurine transporter).[19] The isolated clone contained a 6.2-kb insert. Injection into oocytes of cRNA synthesized using pNCT as a template elicited a tremendous taurine uptake but did not elicit betaine uptake. Ionic requirements and other characteristics of the induced taurine uptake were again indistinguishable from those of the MDCK cells and the deduced amino acid sequence confirmed its membership in the Na^+- and Cl^--coupled transporter gene family.

B. STRUCTURE AND TISSUE EXPRESSION

There was a long open reading frame in SMIT predicting a polypeptide of 718 amino acids. The protein exhibits highly significant sequence (Figure 24–2) and hydropathy profile similarities to the Na^+/glucose cotransporter[20] and Na^+/nucleoside cotransporter.[21] These proteins share 40 to 60% amino acid identity with each other and display weak sequence homology (20 to 30%) to the Na^+-coupled transporters of *Escherichia coli*. The five

```
SMIT   .......... .......... ...MRAVLET ADIAIVALYF ILVMCIGFFA  50
SGLT1  ....MDSSTW SPKTTAVTRP VETHELIRNA ADISIIVIYF VVVMAVGLWA
SNST1  MEEHMEAGSR LGLG...... .DHGALIDNP ADIAVIAAYF LLVIGVGLWS
                                       ***     **   *   *

SMIT   MWKSNRSTVS GYFLAGRSMT WVAIGASLFV SNIGSEHFIG LAGSGAASGF  100
SGLT1  MFSTNRGTVG GFFLAGRSMV WWPIGASLFA SNIGSGHFVG LAGTGAASGI
SNST1  MCRTNRGTVG GYFLAGRSMV WWPVGASLFA SNIGSGHFVG LAGTGAANGL
       *   ** **  * *******  *   *****  ***** ** * *** *** *

SMIT   AVGAWEFNAL LLLQLLGWVF IPIYIRSGVY TMPEYLSKRF GGHRIQVYFA  150
SGLT1  AIGGFEWNAL VLVVVLGWLF VPIYIKAGVV TMPEYLRKRF GGQRIQVYLS
SNST1  AVAGFEWNAL FVVLLLGWLF APVYLTAGVI TMPQYLRKRF GGHRIRLYLS
       *    * ***      *** *  * *   **  *** ** *** ** **  *

SMIT   ALSLILYIFT KLSVDLYSGA LFIQESLGWN LYVSVILLIG MTALLTVTGG  200
SGLT1  LLSLLLYIFT KISADIFSGA IFINLALGLN LYLAIFLLLA ITALYTITGG
SNST1  VLSLFLYIFT KISVDMFSGA VFIQQALGWN IYASVIALLG ITMVYTVTGG
        *** ***** * * *  ***  **   ** *  *    **    *   * ***

SMIT   LVAVIYTDTL QALLMIVGAL TLMIISMMEI GGFEEVKRRY MLASPNVTSI  250
SGLT1  LAAVIYTDTL QTVIMLVGSL ILTGFAFHEV GGYDAFMEKY MKAIPTI..V
SNST1  LAALMYTDTV QTFVIIAGAF ILTGYAFHEV GGYSGLFDKY MGAMTSLT.V
       * *  ****  *      *    *      *  **       * * *

SMIT   LLTYNLSNTN SCNVHPKKDA LKMLRNPTDE DVPWPGFVLG QTPASVWYWC  300
SGLT1  SDGN..TTFQ EKCYTPRADS FHIFRDPLTG DLPWPGFIFG MSILTLWYWC
SNST1  SEDPAVGNIS SSCYRPRPDS YHLLRDPVTG DLPWPALLLG LTIVSGWYWC
                       *  *      * *    * ***    *       ****

SMIT   ADQVIVQRVL AAKNIAHAKG STLMAGFLKL LPMFIIVVPG MISRILFADD  350
SGLT1  TDQVIVQRCL SAKNMSHVKG GCILCGYLKL MPMFIMVMPG MISRILYTEK
SNST1  SDQVIVQRCL AGRNLTHIKA GCILCGYLKL TMPFLMVMPG MISRILYPDE
        ******* *    *  * *       * ***    *  * ** ******
                                              +
SMIT   IACINPEHCM QVCGSRAGCS NIAYPRLVMK LVPVGLRGLM MAVMIAALMS  400
SGLT1  IACVVPSECE KYCGTKVGCT NIAYPTLVVE LMPNGLRGLM LSVMLASLMS
SNST1  VACVAPEVCK RVCGTEVGCS NIAYPRLVVK LMPNGLRGLM LAVMLAALMS
        **  *  *    **   **  ***** **  * * ******   ** * ***
                                   +    +    ++
SMIT   DLDSIFNSAS TIFTLDVYKL IRRSASSREL MIVGRIFVAF MVVISIAWVP  450
SGLT1  SLTSIFNSAS TLFTMDIYAK VRKRASEKEL MIAGRLFILV LIGISIAWVP
SNST1  SLASIFNSSS TLFTMDIYT. LRPRAGEGEL LLVGRLWVVF IVAVSVAWLP
        * ***** * * ** * *   *  *   **    **          * ** *

SMIT   IIVEMQGGQM YLYIQEVADY LTPPVAALFL LAIFWKRCNE QGAFYGGMAG  500
SGLT1  IVQSAQSGQL FDYIQSITSY LGPPIAAVFL LAIFWKRVNE PGAFWGLILG
SNST1  VVQAAQGGQL FDYIQSVSSY LAPPVSAVFV VALFVPRVNE KGAFWGLIGG
            * **    ***    * * **  * *  * *  * **  *** *   *

SMIT   FVLGAVRLTL AFAYRAPECD QPDNRPGFIK DIHYMYVATA LFWVTGLITV  550
SGLT1  LLIGISRMIT EFAYGTGSCM EPSNCPTIIC GVHYLYFAII LFAISFITIV
SNST1  LLMGLARLIP EFSFGTGSCV RPSACPAFLC RVHYLYFAIV LFFCSGLLII
          *  *     *      *   *   *       ** * *   **

SMIT   IVSLLTPPPT KEQIRTTTFW SKKSLVVKES CSPKDEPYKM QEKSILRCSE  600
SGLT1  VISLLTKPIP DVHLYRLCWS LRNS...... .........K EERIDLDAEE
SNST1  IVSLCTAPIP RKHLHRLVFS LRHS...... .........K EEREDLDADE
         ** * *                 *                  *   *   *

SMIT   NSEATNHIIP NGKSEDSIKG LQPEDVNLLV TCREEGNPVA SLGHSEAETP  650
SGLT1  EN......IQ EGPKETIEIE TQVPE..... .......... ..........
SNST1  LEAPASPPVQ NGRPEHAVEM EEPQA..... .......... ..........
                   *  *

SMIT   VDAYSNGQAA LMGEKERKKE TEDGGRYWKF IDWFCGFKSK SLSKRSLRDL  700
SGLT1  .......... .......... .KKKGIFRRA YDLFCGLEQ. ..HGAPKMTE
SNST1  .......... .......... .PGPGLFRQC LLWFCGMNRG RAGGPAPPTQ
                                  *         ***

SMIT   MEEEAVCLQM LE..EPPQVK LILNIGLFAV CSLGIFMFVY FSL
SGLT1  EEEKAMKMKM TDTSEKPLWR TVLNVNGIIL VTVAVFCHAY FA.
SNST1  EEEAAAARRL EDINEDPRWS RVVNLNALLM MAVAMFFWGF YA.
        ** *          * *       *          *
```

Figure 24–2 Amino acid sequence of the canine Na+/MI cotransporter (SMIT)[15] compared with the human Na+/glucose cotransporter (SGLT1)[20] and the rabbit Na+/nucleoside cotransporter (SNST1).[21] Putative transmembrane domains of SMIT are underlined. An aspartate residue in SGLT1 that is mutated in some patients with glucose/galactose malabsorption is double-underlined.[23] Five conserved residues among *E. coli* and mammalian sodium-coupled transporters are noted by plus signs.[22] Asterisks are placed below amino acids that are identical in all of the transporter sequences.

amino acids conserved among mammalian and bacterial transporters[22] are conserved in SMIT (Figure 24–2). The Asp[23] residue in the Na+/glucose cotransporter that is mutated in the genetic disease glucose/galactose malabsorption[23] is conserved in mammalian transporters but not in the bacterial transporters. It appears that SMIT is a member of the Na+-coupled transporter gene family of mammals. Several phosphorylation consensus sites for cAMP-dependent protein kinase and protein kinase C are found in the putative cytoplasmic domains. The significance of these phosphorylation sites is not known.

Northern analysis of poly (A)+RNA from various dog tissues using the SMIT probe revealed hybridizations, in order of relative abundance, in kidney medulla, kidney cortex, and the brain.[15] The absence of hybridization in intestinal mucosa suggests that SMIT represents an isoform of the Na/MI cotransporter in the basolateral membrane of renal medullary epithelia and the plasma membrane of nonepithelial tissues, such as brain and mesangial cells in the kidney cortex, rather than an isotype present in the apical membrane of proximal tubule and intestinal mucosa. The presence of the SMIT mRNA in kidney medulla and brain is consistent with its proposed role in osmolyte regulation in those tissues.[11,24]

The most surprising finding obtained from the cloning of the osmolyte transporters was that the betaine transporter is a member of the neurotransmitter (GABA,[17] noradrenaline,[18] and others[19]) transporter family that requires Cl^- as well as Na^+ for the transport function. This observation led to the recognition that the betaine transporter in the MDCK cells is Cl^-- (and Na^+-) dependent and that it also transports GABA (hence the name, betaine/GABA transporter).[16] Chloride dependence of the taurine transporter was recognized before the cloning of the protein and this information was critical for successful cloning (see above). As shown in Figure 24–3, BGT1 and NCT exhibit highly significant sequence similarity to the GABA and noradrenaline transporters. The sequence similarity extends to other recently cloned members of the family (not shown). While there is 56% amino acid identity between the BGT1 and NCT, they share 40 to 50% amino acid identity with the neurotransmitter transporters. Like SMIT, several phosphorylation consensus sites are noted in BGT1 and NCT but their significance is not clear at this point.

The mRNA of BGT1 is clearly expressed in kidney medulla, but is not detected in kidney cortex and several other tissues, including mucosa of the small intestine, epididymis, and brain.[16] The order of the NCT mRNA abundance among tissues is kidney medulla ≅ kidney cortex > ileal mucosa > brain > other tissues.[19] The wide tissue distribu-

```
BGT1  MDRKVAVPED GPPVVSWLPE .........E GEKL...DQE G........E   50
NCT   ....MATKEK LQCLKDFHKD ILKPSPGKSP GTRP...EDE A........E
GAT1  ....MATDNS KVADGQISTE VSEAPVASDK PKILVVKVQK K........A
NAT   MLLARMNPQV QPENNGADTG PEQPLRARKT AELLVVKERN GVQCLLAPRD

BGT1  DQVKDRGQWT NKMEFVLSVA GEIIGLGNVW RFPYLCYKNG GGAFFIPYFI  100
NCT   GKPPQREKWS SKIDFVLSVA GGFVGLGNVW RFPYLCYKNG GGAFLIPYFI
GAT1  GDLPDRDTWK GRFDFLMSCV GYAIGLGNVW RFPYLCGKNG GGAFLIPYFL
NAT   GDAQPRETWG KKIDFLLSVV GFAVDLANVW RFPYLCYKNG GGAFLIPYTL
           *  *      *  *   *    * *** ****** *** ********

BGT1  FFFTCGIPVF FLEVALGQYT SQGSVTAWRK ICPLLQGIGL ASVVIESYLN  150
NCT   FLFGGGLPVF FLEVIIGQYT SEGGITCWEK ICPLFSGIGY ASIVIVSLLN
GAT1  TLIFAGVPLF LLECSLGQYT SIGGLGVW.K LAPMFKGVGL AAAVLSFWLN
NAT   FLIIAGMPLF YMELALGQYN REGAATVW.K ICPFFKGVGY AVILIALYVG
           * * *   *   ***    *    * *   *   * *  *

BGT1  IYYIIILAWA LFYLFSSFTS ELPWTTCTNT WNTEHCMD.. FLNHS..GAR  200
NCT   IYYVIILAWA TYYLFQSFQS ELPWAHCNHS WNTPQCMEDT MRKNK..SLW
GAT1  IYYIVIISWA IYYLYNSFTT TLPWKQCDNP WNTDRCF... .......SNY
NAT   FYYNVIIAWS LYYLFSSFTL NLPWTDCGHT WNSPNCTDPK LLNGSVLGNH
       **  *  *    **  **    ***       **   *

BGT1  TATSSENFTS PVMEFWERRV LGI..TSGIH DLGALRWELA LCLLLAWLIC  250
NCT   ITLSTKNFTS PVTEFWERNV LSL..SSGID DPGSLKWDLA LCLLLVWLVC
GAT1  SLVNTTNMTS AVVEFWERNM HQM..TDGLD KPGQIRWPLA ITLAIAWVLV
NAT   TKYSKYKFT. PAAEFYERGV LHLHESSGIH DIGLPQWQLL LCLMVVVIVL
              *     ** **          *     *   * *    *

BGT1  YFCIWKGVKT TGKVVYFTAT FPYLMLVILL IRGITLPGAY QGVIYYLKPD  300
NCT   FFCIWKGVKS TGKVVYFTAT FPFAMLLVLL VRGLTLPGAG AGIKFYLYPD
GAT1  YFCIWKGVGW TGKVVYFSAT YPYIMLIILF FRGVTLPGAK EGILFYITPN
NAT   YFSLWKGVKT SGKVVWITAT LPYFVLFVLL VHGVTLPGAS NGINAYLHID
       *  ****    ****   **  *   *  *    * *****       *

BGT1  LLRLKDPQVW MDAGTQIFFS FAICQGCLTA LGSYNKYHNN CYRDSIALCF  350
NCT   ISRLEDPQVW IDAGTQIFFS YAICLGAMTS LGSYNKYKYN SYRDCMLLGC
GAT1  FRKLSDSEVW LDAATQIFFS YGLGLGSLIA LGSYNSFHNN VYRDSIIVCC
NAT   FYRLKEATVW IDAATQIFFS LGAGFGVLIA FASYNKFDNN CYRDALLTSS
         *    **  ** ******              ***    *  ***

BGT1  LNSATSFAAG FVVFSILGFM AQEQGLPISE VAESGPGLAF IAFPKAVTMM  400
NCT   LNSGTSFVSG FAIFSILGFM AQEQGVDIAD VAESGPGLAF IAYPKAVTMM
GAT1  INSCTSMFAG FVIFSIVGFM AHVTKRSIAD VAASGPGLAF LAYPEAVTQL
NAT   INCITSFVSG FAIFSILGYM AHEHKVNIED VATEGAGLVF ILYPEAISTL
       *  **   * *  *** * * *      *   **  * ** *    * *

BGT1  PLSQLWSCLF FIMLIFLGLD SQFVCVECLV TASMDMFPSQ LRKSGRRELL  450
NCT   PLPTFWSILF FIMLLLLGLD SQFVEVEGQV TSLVDLYPSF LRKGFRREIF
GAT1  PISPLWAILF FSMLLMLGID SQFCTVEGFI TALVDEYPRL LRN..RRELF
NAT   SGSTFWAVVF FVMLLALGLD SSMGGMEAVI TGLADDFQVL KR...HRKLF
           *   * * **  ** * *     *    *   *       *    *

BGT1  ILAIAVFCYL AGLFLVTEGG MYIFQLFDYY ASSGICLLFL AMFEVICISW  500
NCT   IAFMCSISYL LGLSMVTEGG MYVFQLFDYY AASGVCLLWV AFFECFVIAW
GAT1  IAAVCIVSYL IGLSNITQGG IYVFKLFDYY SASGMSLLFL VFFECVSISW
NAT   TFGVTFSTFL LALFCITKGG IYVLTLLDTF AA.GTSILFA VLMEAIGVSW
               *   *   * **  *   * *      *  **      *     *

BGT1  VYGADRFYDN IEDMIGYRPW PLVKISWLFL TPGLCLATFL FSLSQYTPLK  550
NCT   IYGSDNLYDG IEDMIGYRPG PWMKYSWAVV TPVLCVGCFI FSLVKYVPLT
GAT1  FYGVNRFYDN IQEMVGSRPC IWWKLCWSFF TPIIVAGVFL FSAVQMTPLT
NAT   FYGVDRFSND IQQMMGFRPG LYWRLCWKFV SPAFLLFVVV VSIINFKPLT
       **        *  * * **        *     *          *     **

BGT1  YNNIYVYPPW GYSIGWFLAL SSMICVPLFV IITLLKTRGS FKKRLRQLTT  600
NCT   YNKVYVYPTW AIGLGWSLAL SSMMCVPLVM VIRLCQTEGP FLVRLKYLLT
GAT1  MGS.YVFPKW GQGVGWLMAL SSMVLIPGYM AYMFLTLKGS LKQRLQVMI.
NAT   YDD.YIFPPW ANWVGWGIAL SSMVLVPIYV IYKFLSTQGS LWERLAYGIT
          *  * *     **  ** ***   *            *     **

BGT1  PDPS...... .......LPQ PKQHLYLDGG T..SQDCGPS PTKEG.....  650
NCT   PREPNRWAVE REGATLQLPP GRERGSHETD PHHSRDHDVS SWADGPLPCC
GAT1  .......... .........Q PSEDIVRPEN GPEQPQAGSS ASKEAYI...
NAT   PENEHHLVAQ RDIRQFQLQH WLAI...... .......... ..........

BGT1  .LIVGEKETH L......... ....
NCT   LLTLDSHRTR FTELSICTRI VFFF
GAT1  .......... .......... ....
NAT   .......... .......... ....
```

Figure 24–3 Comparison of amino acid sequences of the canine betaine/GABA (BGT1)[16] and taurine transporters (NCT),[19] rat brain GABA (GAT1),[17] and human noradrenaline transporters (NAT).[18] Proposed transmembrane domains are underlined in BGT1 and asterisks are placed below amino acids that are identical in all of the sequences. Two lines under the asterisks represent highly conserved regions used for generation of PCR primers (see text).

tion of NCT mRNA is not unexpected in light of the wide distribution of the taurine transporter.[7] The predominant expression of the mRNA for the BGT1 and NCT in the kidney medulla is consistent with their role in the osmoregulatory accumulation of betaine and taurine in medullary cells.

III. REGULATION OF OSMOLYTE TRANSPORTER mRNA BY HYPERTONICITY

Initial Northern analysis of poly(A)$^+$RNA prepared from isotonic and hypertonic MDCK cells revealed that the mRNAs of the osmolyte transporters are markedly elevated by hypertonicity,[15,16,19] consistent with the earlier observations that RNA extracted from hypertonic MDCK cells induced higher rates of osmolyte transport in oocytes than RNA extracted from isotonic cells.[12–14] It has been postulated that the elevation of the transporter mRNA by the hypertonicity leads to the increased osmolyte transport activity. In order to test this hypothesis more directly in the MDCK cell model, cells grown on plastic dishes were used to measure rates of osmolyte transport and mRNA abundance.[25] Both dot blot hybridization and ribonuclease protection assays were used to measure specific mRNA yielding the same results. The MDCK cells on plastic dishes, when compared to the cells grown on filters (Figure 24–1), displayed less dramatic stimulation but the same time course of the osmolyte transport by medium hypertonicity (Figure 24–4). Although lower transport rates were obtained, the cells on plastic tissue culture dishes were chosen in this series of studies because those culture conditions were used for analysis of mRNA and transcription rate.

Figure 24–5 shows the time course of the change in abundance of SMIT mRNA after switching medium osmolality.[25] mRNA abundance increases very rapidly after increasing osmolality and reaches a peak at 16 h after the switch. After the peak, the mRNA abundance falls somewhat and then remains at that level. When cells cultured in hypertonic medium for 4 d are shifted back to isotonic medium, the SMIT mRNA abundance falls rapidly and returns to the isotonic level in 12 h. The same results are obtained with the betaine uptake rates and the abundance of the BGT1 mRNA.[26] These results demonstrate that the changes in the MI and betaine transporter mRNA abundance precede the corresponding changes in the transport rates (Figure 24–4), supporting the view that changes in mRNA abundance lead to changes in the synthesis and amount of the osmolyte transporters functioning in the plasma membrane.[11–14]

The magnitude of the increase in the MI transport and cell MI content depends on the solute used to increase the osmolality of the medium.[1,2,27] To determine whether the different effectiveness of

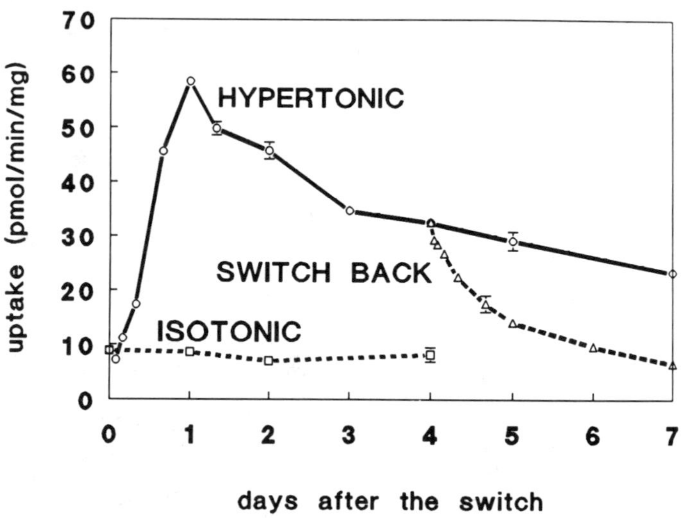

Figure 24–4 Time course of the rate of MI uptake in MDCK cells following a shift in medium osmolality. On day 0, confluent cells cultured in isotonic medium on a plastic surface were switched to the same medium made hypertonic (500 mOsmol/kg H_2O) by addition of raffinose. Isotonic cells were maintained in isotonic medium. On day 4, some dishes of hypertonic cells were returned to isotonic medium ("switch back"). (From Yamauchi, A. et al., *Am. J. Physiol.,* in press. With permission.)

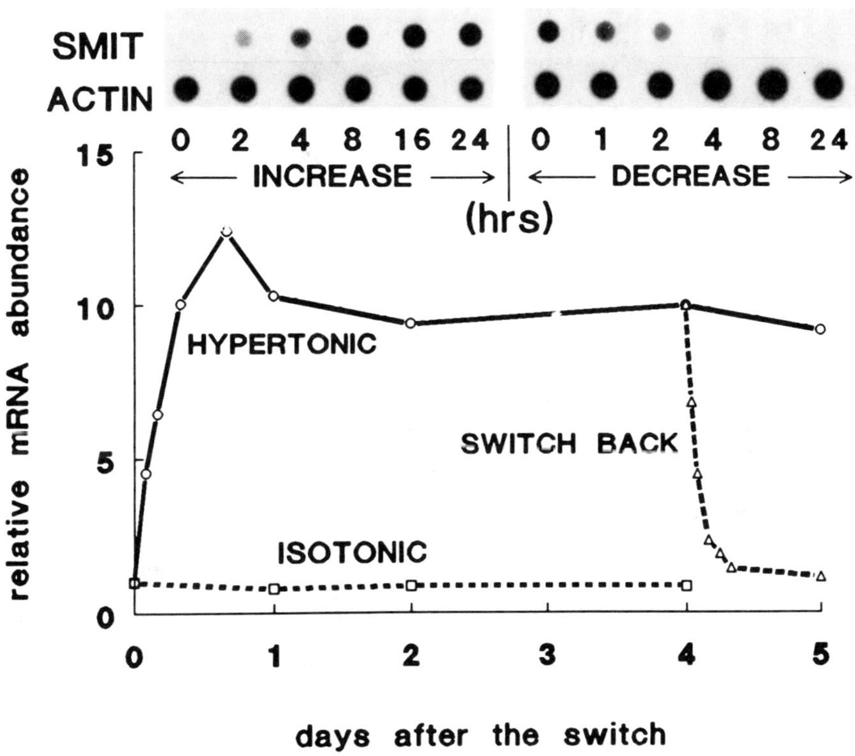

Figure 24–5 Time course of abundance of mRNA for the Na+/MI cotransporter. The cells were treated as in Figure 24–6, RNA was extracted at various points, and mRNA of the transporter was measured. Autoradiography of dot blot analysis for the Na+/MI cotransporter (SMIT) and β-actin mRNA (ACTIN) is shown on the top of the figure. Numbers are hours after increasing (INCREASE) or decreasing (DECREASE) osmolality. Others are the same as in Figure 24–4. (Data from Reference 25.)

different hypertonic solutes is due to regulation of the abundance of cotransporter mRNA, the SMIT mRNA in cells shifted for 24 h to media made hyperosmolar by the addition of 200 mOsmol/kg H_2O of different solutes was measured (Figure 24–6). Raffinose, NaCl, and mannitol increase transport rate and mRNA abundance. Urea, a permeant solute, does not change the mRNA abundance nor the transport rate when added alone. These results indicate that hypertonicity, not hyperosmolality, is the stimulus for the osmolyte accumulation. While the relative increase in transport rate is about half the relative increase in mRNA abundance, changes in the two parameters are highly correlated. The same results are obtained with betaine transport and abundance of BGT1 mRNA.[26]

Accumulation of MI by PAP-HT25 cells, a continuous cell line derived from rabbit kidney medulla, increases with increasing hypertonicity up to about 400 mOsmol/kg H_2O, and then levels off despite further increase in tonicity.[27] Since uptake is the major determinant of MI content in hypertonic PAP-HT25 and MDCK cells, uptake of MI and the abundance of mRNA for the cotransporter in MDCK cells exposed to different degrees of hypertonicity for 24 h were measured (Figure 24–7). The rate of MI transport increases as medium osmolality increases up to 450 mOsmol/kg H_2O, but does not rise further at higher osmolality, similar to the pattern of MI accumulation in PAP-HT25 cells in media of increasing osmolality. In contrast, the abundance of SMIT mRNA continued to increase throughout the range of tonicity examined. The results indicate that at the tonicity of 450 mOsmol/kg H_2O and greater, expression of Na/MI cotransporter activity is regulated to a common maximal value at steps after accumulation of the cotransporter mRNA.

When MDCK cells are switched to hypertonic medium that does not contain MI, the MI uptake rate is higher than in cells switched to the same

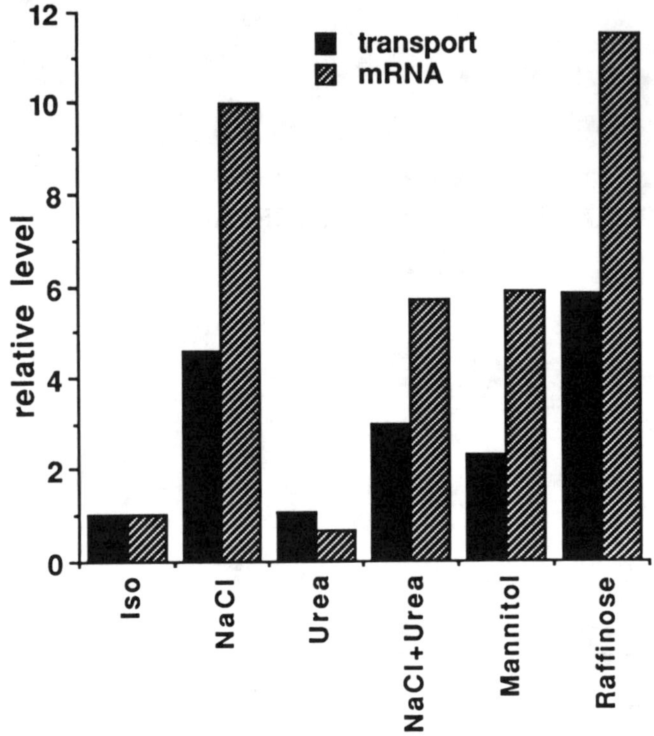

Figure 24–6 Effects of different solutes on the MI transport rate and abundance of SMIT mRNA.[28] MDCK cells cultured in isotonic medium on a plastic surface were switched to medium made hypertonic (500 mOsmol/kg H_2O) by addition of solutes indicated. Medium osmolality was 700 mOsmol/kg H_2O in "NaCl+Urea" by adding 200 mOsmol/kg H_2O of NaCl and 200 mOsmol/kg H_2O of urea. Initial uptake rate and abundance of mRNA were measured 24 h after the switch. Values are relative to levels in isotonic cells.

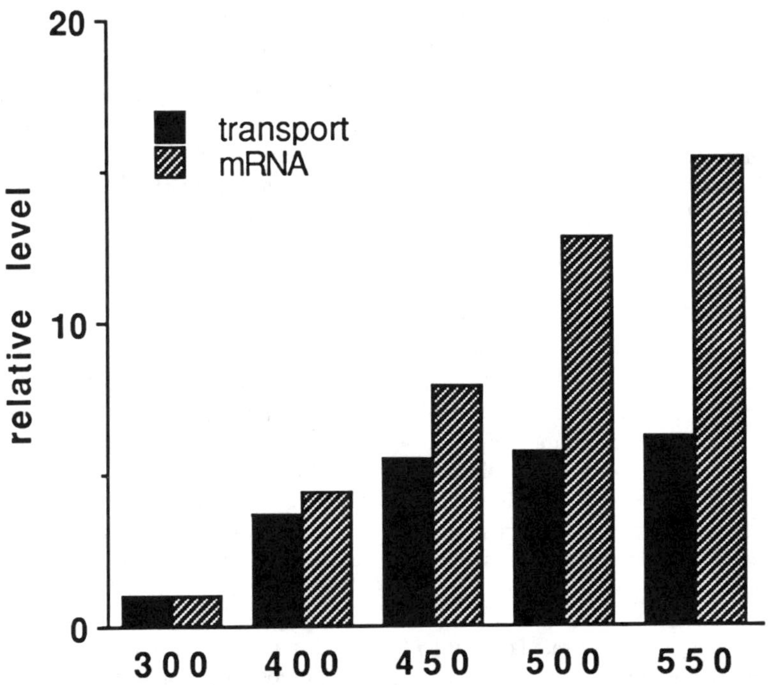

Figure 24–7 Effect of medium osmolality on the relationship between MI transport rate and abundance of SMIT mRNA.[28] MDCK cells cultured in isotonic medium (300 mOsmol/kg H_2O) were switched to a range of hypertonic media (400 to 550 mOsmol/kg H_2O by adding raffinose as indicated under the bars). Transport rate and mRNA were measured 24 h after the switch. Others are the same as those in Figure 24–6.

medium containing MI.[1] To evaluate the role of mRNA abundance in the increased expression of the MI transport observed under those conditions, some of the cells were switched to hypertonic medium that did not contain MI and others to the usual hypertonic medium containing MI. We then compared the abundance of SMIT mRNA and the rate of MI transport (Figure 24–8). Omission of MI from the medium, thereby preventing MI accumulation, not only increases the SMIT mRNA abundance and the MI transport rate further but also makes the relative increase in MI transport a greater fraction of the relative increase in abundance of the SMIT mRNA. It appears that accumulation of MI in cells regulates MI transport at steps after mRNA abundance, i.e., translation/processing of the transport protein or directly at the protein level, as well as the abundance of SMIT mRNA.

In summary, mRNA abundance of the transporters plays a major role in the regulation of activity of the osmolyte transporters by hypertonicity. Presumably, the increase in mRNA abundance leads to an increase in the translation/processing of the transporters as integral membrane proteins. Additional regulation at translational or posttranslational sites is also likely. Antibodies to the transporter proteins are needed to evaluate the posttranscriptional events directly.

IV. REGULATION OF TRANSCRIPTION OF THE OSMOLYTE TRANSPORTER GENES BY HYPERTONICITY

To delineate the role of transcription in the increase of SMIT mRNA abundance observed in hypertonic MDCK cells, the transcription rate of the gene was measured in isolated nuclei collected from cells switched to hypertonic medium. The time course of the increase in transcription rate of the Na/MI cotransporter gene shows virtually the same pattern as that of mRNA abundance (Figure 24–9). The peak, reached 16 h after the switch, is 20 times the level in isotonic cells. When the hypertonic cells were switched back to isotonic medium, the transcription rate fell to isotonic levels in 8 h. Exactly

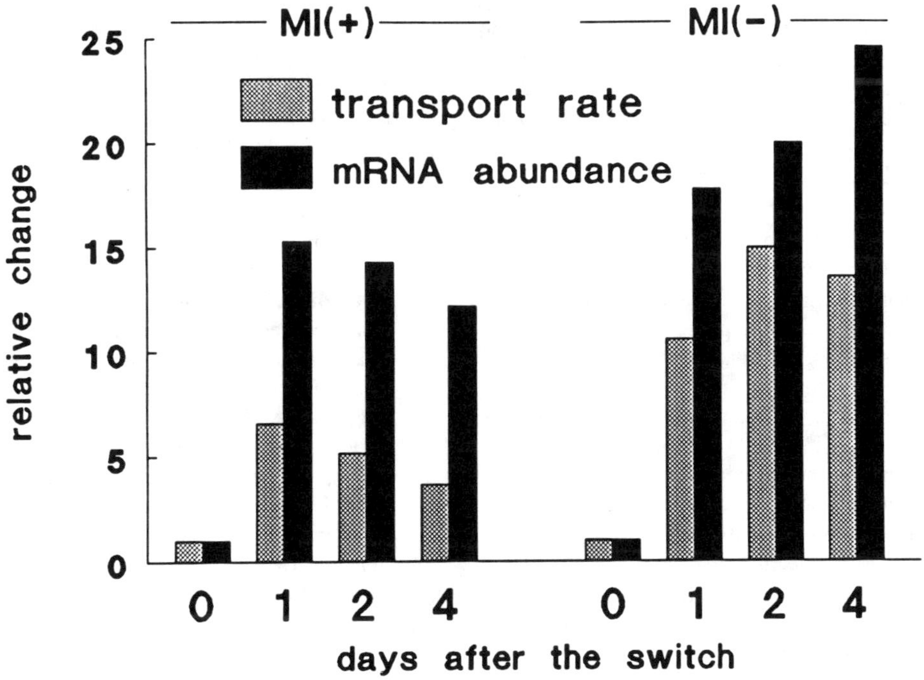

Figure 24–8 Effect of removal of MI from the medium on the rate of transport of MI and the abundance of SMIT mRNA.[29] MDCK cells cultured in isotonic medium containing MI were switched to medium with [MI(+)] or without MI [MI(-)] that had been made hypertonic (500 mOsmol/kg H_2O) by addition of raffinose. The rates of MI transport and mRNA abundance were measured 1, 2, and 4 d after the switch. Values are expressed as relative to the level in cells before the switch.

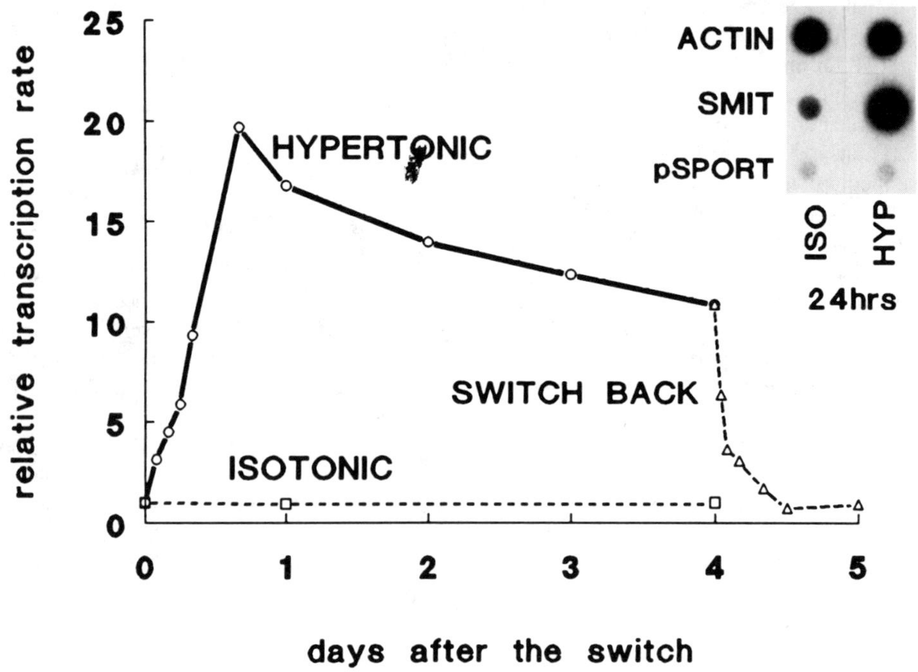

Figure 24–9 Time course of transcription of the Na^+/MI cotransporter gene. The cells were treated as in Figure 24–4, nuclei were isolated at various time points, and "run-off" transcripts were measured. Autoradiography of nuclear run-on assay for the genes of β-actin (ACTIN) and Na^+/MI cotransporter (SMIT) is inset on the top right of the figure. pSPORT is the vector for both cDNA inserts and is used as a blank. Results from isotonic (ISO) and hypertonic (HYP) cells on day 1 (24 h) are shown. (Data from Reference 25.)

the same results were obtained with the transcription of the Na/Cl/betaine cotransporter gene.[26] In view of the similar time course of the rate of transcription of the genes for the cotransporters and their mRNA abundance, we suggested that transcription is primarily responsible for the change of the mRNA abundance of the osmolyte transporters (Figure 24–10).

Transfection into CHO cells of the SMIT cDNA placed under a heterologous promoter results in expression of the Na/MI cotransporter.[30] The activity of the expressed transporter is not regulated by hypertonicity, indicating that the intact gene with its own regulatory elements (promotor, enhancer, etc.) is required for the stimulation of transcription to occur. Cloning of the transporter genes will be necessary to delineate the mechanisms underlying the genetic response to hypertonicity.

REFERENCES

1. **Nakanishi, T., Turner, R. J., and Burg, M. B.,** Osmoregulatory changes in myo-inositol transport by renal cells, *Proc. Natl. Acad. Sci. U.S.A.*, 86, 6002, 1989.
2. **Nakanishi, T., Turner, R. J., and Burg, M. B.,** Osmoregulation of betaine transport in mammalian renal medullary cells, *Am. J. Physiol.*, 258, C1061, 1990.
3. **Uchida, S., Nakanishi, T., Kwon, H. M., Preston, A. S., and Handler, J. S.,** Taurine behaves as an osmolyte in Madin-Darby Canine kidney cells: protection by polarized, regulated transport of taurine, *J. Clin. Invest.*, 88, 656, 1991.
4. **Yamauchi, A., Kwon, H. M., Uchida, S., Preston, A. S., and Handler, J. S.,** Myo-inositol and betaine transporters regulated by tonicity are basolateral in MDCK cells, *Am. J. Physiol.*, 261, F197, 1991.
5. **Heilig, C. W., Brenner, R. M., Yu, A. S., Kone, B. C., and Gullans, S. R.,** Modulation of osmolytes in MDCK cells by solutes, inhibitors, and vasopressin, *Am. J. Physiol.*, 259, F653, 1990.

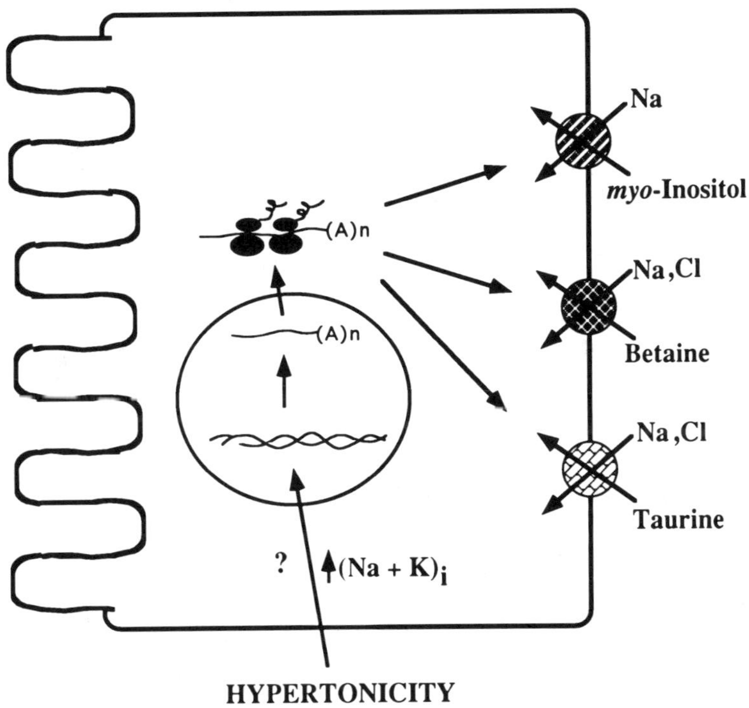

Figure 24–10 Mechanisms of osmolyte accumulation in the MDCK cells. Hypertonicity increases transcription of the genes for the osmolyte transporters. Enhanced transcription will lead to increase in mRNA and subsequent increase in the functioning transporters in the plasma membrane.

6. **Garcia-Perez, A. and Burg, M. B.,** Renal medullary organic osmolytes, *Physiol. Rev.,* 71, 1081, 1991.
7. **Chesney, R. W.,** Taurine: its biological role and clinical implications, *Adv. Pediatr.,* 32, 1, 1985.
8. **Lohr, J. W., Pochal, M. A., and Acara, M.,** Osmoregulatory betaine uptake by rat renal medullary slices, *J. Am. Soc. Nephrol.,* 2, 879, 1991.
9. **Hauser, G.,** Myo-inositol transport in slices of rat kidney cortex. II. Effects of the ionic composition of the medium, *Biochim. Biophys. Acta,* 173, 267, 1969.
10. **Goldstein, L.,** Organic solute profiles and transport in the rat renal medulla, *Am. J. Kidney Dis.,* 14, 310, 1989.
11. **Veis, J. H., Molitoris, B. A., Teitelbaum, I., Mansour, J. A., and Berl, T.,** Myo-inositol uptake by rat cultured inner medullary collecting tubule cells: effect of osmolality, *Am. J. Physiol.,* 260, F619, 1991.
12. **Kwon, H. M., Yamauchi, A., Uchida, S., Robey, R. B., Garcia-Perez, A., Burg, M. B., and Handler, J. S.,** Renal Na^+/myo-inositol cotransporter mRNA expression in Xenopus oocyte: regulation by hypertonicity, *Am. J. Physiol.,* 260, F258, 1991.
13. **Uchida, S., Kwon, H. M., Preston, A. S., and Handler, J. S.,** Expression of Madin-Darby kidney cell Na^+- and Cl^--dependent taurine transporter in Xenopus laevis oocytes, *J. Biol. Chem.,* 266, 9605, 1991.
14. **Robey, R. B., Kwon, H. M., Handler, J. S., Garcia-Perez, A., and Burg, M. B.,** Induction of glycinebetaine uptake in Xenopus oocytes by injection of poly(A) RNA from renal cells exposed to high extracellular NaCl, *J. Biol. Chem.,* 266, 10400, 1991.
15. **Kwon, H. M., Yamauchi, A., Uchida, S., Preston, A. S., Garcia-Perez, A., Burg, M. B., and Handler, J. S.,** Cloning of the cDNA for a Na^+/myo-inositol cotransporter, a hypertonicity stress protein, *J. Biol. Chem.,* 267, 6297, 1992.
16. **Yamauchi, A., Uchida, S., Kwon, H. M., Preston, A. S., Robey, R. B., Garcia-Perez,**

A., Burg, M. B., and Handler, J. S., Cloning of a Na^+- and Cl^--dependent betaine transporter that is regulated by hypertonicity, *J. Biol. Chem.,* 267, 649, 1992.

17. **Guastella, J., Nelson, N., Nelson, H., Czyzyk, L., Keynan, S., Miedel, M. C., Davidson, N., Lester, H. A., and Kanner, B. I.,** Cloning and expression of a rat brain GABA transporter, *Science,* 249, 1303, 1990.
18. **Pacholczyk, T., Balakely, R. D., and Amara, S. G.,** Expression cloning of a cocaine- and antidepressant-sensitive human noradrenaline transporter, *Nature,* 350, 350, 1991.
19. **Uchida, S., Kwon, H. M., Yamauchi, A., Preston, A. S., Marumo, F., and Handler, J. S.,** Molecular cloning of the cDNA for an MDCK cell Na^+- and Cl^--dependent taurine transporter that is regulated by hypertonicity, *Proc. Natl. Acad. Sci. U.S.A.,* 89, 8230, 1992.
20. **Hediger, M. A., Turk, E., and Wright, E. M.,** Homology of the human intestinal Na^+/glucose and *Escherichia coli* Na/proline cotransporters, *Proc. Natl. Acad. Sci. U.S.A.,* 86, 5748, 1989.
21. **Pajor, A. M. and Wright, E. W.,** Cloning and functional expression of a mammalian Na^+/nucleoside cotransporter, *J. Biol. Chem.,* 267, 3557, 1992.
22. **Deguchi, Y., Yamato, I., and Anraku, Y.,** Nucleotide sequence of *gltS,* the Na^+/glutamate symport carrier gene of *Escherichia coli* B, *J. Biol. Chem.,* 265, 21704, 1990.
23. **Turk, E., Zabel, B., Mundlos, S., Dyer, J., and Wright, E. M.,** Glucose/galactose malabsorption caused by a defect in the Na/glucose cotransporter, *Nature,* 350, 354, 1991.
24. **Strange, K., Morrison, R., Heilig, C. W., DiPietro, S., and Gullans, S. R.,** Up-regulation of inositol transport mediates inositol accumulation in hyper-osmolar brain cells, *Am. J. Physiol.,* 260, C784, 1991.
25. **Yamauchi, A., Uchida, S., Preston, A. S., Kwon, H. M., and Handler, J. S.,** Hypertonicity stimulates transcription of the gene for the Na^+/myo-inositol cotransporter in MDCK cells, *Am. J. Physiol.,* 264, F20, 1993.
26. **Uchida, S., Yamauchi, A., Preston, A. S., Kwon, H. M., and Handler, J. S.,** Medium tonicity regulates expression of the Na^+- and Cl^--dependent betaine transporter in Madin-Darby canine kidney cells by increasing transcription of the transporter gene, *J. Clin. Invest.,* 91, 1604, 1993.
27. **Moriyama, T., Garcia-Perez, A., and Burg, M. B.,** Factors affecting the ratio of different organic osmolytes in renal medullary cells, *Am. J. Physiol.,* 259, F847, 1990.
28. **Yamauchi, A., Kwon, H. M., and Handler, J. S.,** unpublished observations, 1992.
29. **Yamauchi, A., Uchida, S., Kwon, H. M., and Handler, J. S.,** unpublished observations, 1992.
30. **Preston, A. S., Kwon, H. M., and Handler, J. S.,** unpublished observations, 1992.

Index

INDEX

P

R

S

T

U

V

W